ADVANCED SIGNAL INTEGRITY FOR HIGH-SPEED DIGITAL DESIGNS

ADVANCED SIGNAL INTEGRITY FOR HIGH-SPEED DIGITAL DESIGNS

STEPHEN H. HALL

HOWARD L. HECK

IEEE

Celebrating 125 Years
of Engineering the Future

WILEY

A JOHN WILEY & SONS, INC., PUBLICATION

Published by John Wiley & Sons, Inc., Hoboken, New Jersey.
Published simultaneously in Canada.

For general information on our other products and services or for technical support, please contact our Customer Care Department within the United States at (800) 762-2974, outside the United States at (317) 572-3993 or fax (317) 572-4002.

Wiley also publishes its books in a variety of electronic formats. Some content that appears in print may not be available in electronic formats. For more information about Wiley products, visit our web site at www.wiley.com.

Library of Congress Cataloging-in-Publication Data:

Hall, Stephen H.
Advanced signal integrity for high-speed digital designs / Stephen H. Hall, Howard L. Heck.
p. cm.
Includes bibliographical references and index.
ISBN 978-0-470-19235-1 (cloth)
1. Digital electronics. 2. Logic designs. I. Heck, Howard L. II. Title.
TK7868.D5H298 2009
621′.381—dc22

2008027977

CONTENTS

PREFACE

Technology has progressed to a point where digital design has entered a new realm that requires design techniques and concepts that baffle even some of the most seasoned digital system designers. The fact is that state-of-the-art digital systems such as personal computers cannot be designed without a thorough understanding of advanced signal integrity. As computer technology evolves, high-speed interconnect phenomena that designers historically have ignored begin to dominate performance, and unforeseen problems arise that dramatically increase the complexity of design. Consequently, every new generation of computer design requires an understanding of new signal integrity issues that were previously not critical and new design techniques that were previously not necessary. In modern and future systems, an incomplete understanding of high-speed interconnect phenomena will literally result in the progress of digital computing coming to a standstill.

In this book we leverage theory and techniques from fields such as applied physics, communications, and microwave engineering and apply them to the field of high-speed digital design, creating an optimal combination between theory and practical applications. Although some basic material is covered, we assume that readers are well acquainted with basic electromagnetic theory, vector calculus, differential equations, statistics, and transmission-line analysis. In this book we build on the traditional knowledge base and discuss advanced topics ranging from electromagnetic theory for signal integrity to equalization methods that compensate for signal integrity problems with circuitry as required to design modern and future digital systems. Detailed theory is presented in the context of real-life design examples so that it can be applied immediately by practicing engineers, yet provides more than enough technical content to facilitate complete understanding of the concepts.

Features

1. Visual description of theoretical concepts wherever possible, so each chapter includes numerous figures to help reinforce the concepts discussed.
2. Rigorous coverage of theory and use of practical examples to demonstrate how to use the theory in practical, real-world applications.
3. Summary of the electromagnetic theory concepts required to comprehend signal integrity.
4. Rigorous development of transmission-line and crosstalk theory to build a fundamental understanding and then apply the theory to real-world problems.
5. Development of physically consistent dielectric and conductor models to account for frequency-dependent properties, surface roughness, and physical anomalies due to manufacturing and environmental effects.
6. Description of differential signaling at a practical and a theoretical level.
7. Explanation of the mathematical limits of models such as causality, passivity, stability, and reality that must be obeyed to ensure that simulations remain consistent with nature.
8. Full description of network theory, including S-parameters and frequency-domain analysis.
9. Coverage of topics such as nonideal current return paths, tabular modeling, and via resonance.
10. Covers the basics of I/O design and channel equalization.
11. Methods for modeling and budgeting of timing jitter and noise.
12. System analysis techniques for handling large numbers of variables using response surface modeling.

Contemporary signaling systems continue to offer new problems to solve. Engineers who can solve these problems will define the future. This book will equip readers with the necessary practical understanding to contend with contemporary problems in high-speed digital design and with enough theory to see beyond the book to solve problems that the authors have not yet encountered.

ACKNOWLEDGMENTS

Many people have contributed directly or indirectly to the writing of this book. We have been fortunate to keep the company of excellent engineers, scientists, and peers, whose technical knowledge, wisdom, and mentorship have truly inspired us throughout our careers. Without these contributions, this book would not have been possible. Our gratitude is immeasurable.

Among the direct contributors are: *Guy Barnes* of Ansoft Corporation, who reviewed our mathematics, provided simulation data for Chapter 10, and provided the graphics for the front cover. *Stephen Hall, Sr.*, Stephen's father, for educating us on how molecules interact with externally applied electric fields and for writing that section in Chapter 6. *Paul Huray* of the University of South Carolina, who educated us on complicated aspects of electromagnetic theory and acted as a role model for academic excellence. *Yun Ling* of Intel Corporation, who patiently reviewed our mathematics and politely corrected our mistakes. *Kevin Slattery* of Intel Corporation, for reviewing Chapter 2 and strongly encouraging us to write the book. *Chaitanya Sreerama Burt* of Intel Corporation, who both reviewed chapters and double-checked all of our calculations in Chapter 2. *Steven Pytel* of Ansoft Corporation, who graciously reviewed Chapter 5, correcting errors and making excellent recommendations for improvements. *Daniel Hua*, for checking results and helping to solve the difficult differential equations. *Luis Armenta* of Intel Corporation, for reviewing the dielectric models in Chapter 6. *Paul Hamilton* of Tyco, for reviewing Chapter 6. *Gerardo Romo* of Intel Corporation, who reviewed Chapters 4 and 12. *Michael Mirmak* of Intel Corporation and current chair of the IBIS open forum, who reviewed Chapter 11. *Richard Allred* of Intel Corporation, who reviewed Chapters 13 and 14. *Pelle Fornberg* and *Adam Norman*, who reviewed the jitter and peak distortion material in Chapter 13.

Stephen would also like to thank: *Dorothy Hall*, Stephen's mother, for a lifetime of encouragement and for instilling the enthusiasm, discipline, and drive needed not only to become an engineer, but to write this book. *Garrett Hall*, not only for reviewing chapters and double-checking results, but for providing much of the material and motivation for the introduction and for being the most encouraging, devoted, trustworthy, and principled friend that a big brother could ever have.

Howard would like thank *Eric Dibble* of Lockheed-Martin Corporation and *Martin Rausch* of Intel Corporation for supporting his efforts to break into the signal integrity field and for guiding his professional and personal growth. He also thanks *Dr. Ricardo Suarez-Gartner* of Intel Corporation for encouraging him to write the book.

Finally, Stephen dedicates his contributions to this book to his lifelong companion, best friend, wife, and jewel of his life, *Jodi*, with all his love and gratitude, as well as his little girls, *Emily* and *Julia*, who bring him more joy than he could have thought possible.

Howard dedicates his contributions to his beautiful wife, *Lisa*, and sons, *Tyler* and *Nick*, whose support and love provide him the day-to-day happiness that makes each day a gift.

Hillsboro, OR
July 2008

STEPHEN H. HALL
HOWARD L. HECK

1

INTRODUCTION: THE IMPORTANCE OF SIGNAL INTEGRITY

1.1 COMPUTING POWER: PAST AND FUTURE

It is estimated that sometime between the years 2025 and 2050, commonplace personal computers will exceed the calculation power of a human brain. Further extrapolation based on historical trends indicates that a single commonplace computer could exceed the computational power of the human race sometime between 2060 and 2100. Are such vast increases in computational power possible in less than 100 years? We cannot say for certain because it is impossible to predict the future. However, hindsight is always 20/20, and if we subscribe to the notion that history tends to repeat itself, we can look at the progress of computational capabilities over the last century to see if historical data support a rate sufficient to achieve such performance. Hans Moravec, a researcher from the Robotics Institute at Carnegie Melon University, estimated that a computer would require 100 million megainstructions per second (MIPS) to mimic sufficiently closely the behavior of a human brain [Moravec, 1998]. Based on the

Advanced Signal Integrity for High-Speed Digital Design, By Stephen H. Hall and Howard L. Heck

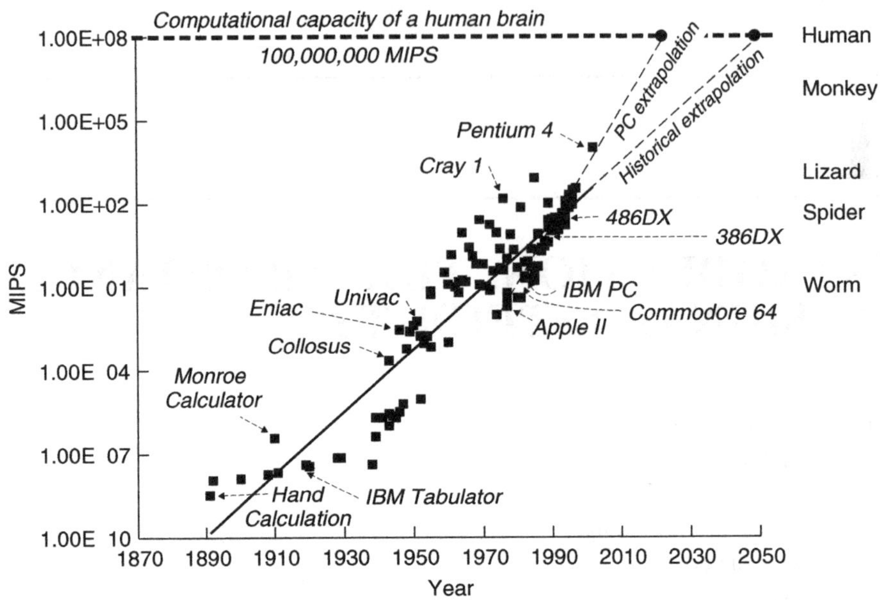

Figure 1-1 Historical computational power and extrapolation into the future. (Adapted from Moravec [1998].)

number of neurons, he was also able to compare the current state of computer technology to the estimated computational power of animal brains. These data points outline a particularly interesting way to examine the history of computational power while level-setting computer performance against the brainpower of common animals.

Figure 1-1 plots the computational power for mechanical and electrical computers used over the last 100 years. Some of the more interesting data points are labeled on the plot, ranging from hand calculation (ca. 1/100,000,000 MIPS) to the Pentium 4 processor of 2002 (10,000 MIPS), which is only two orders of magnitude away from the estimated brain power of a monkey (1,000,000 MIPS). As the plot indicates, computers comparable to the human brain could appear as early as the 2020s based on the extrapolation of personal computer (PC) performance over the last three decades. If the historical data for the entire twentieth century are used, the time frame is extended to 2050. The predictions get even more outrageous if we extend the extrapolation to the estimated computational power of all humans presently on Earth (ca. 6 billion), which would require 6×10^{17} MIPS. Such a computer could exist by 2060, as shown in Figure 1-2. The question is: Can the historical pace of development be sustained? Observation of the data indicates that the historical trend shows no sign of slowing. In fact, the rate appears to be increasing.

However, one often encounters articles by knowledgeable people in the computer industry who believe that the trend cannot be sustained and that the decades of exponential growth must stop. In 1998 it was estimated in such an article that

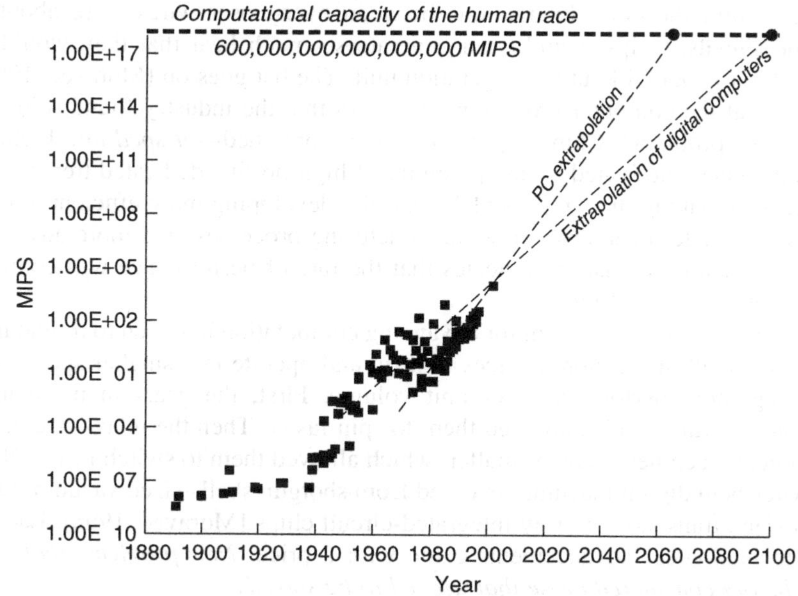

Figure 1-2 Historical computational power and extrapolation into the future.

commonplace printed circuit boards (PCBs) built on an FR4 dielectric could not support bus speeds faster than 300 MHz [Porter, 1998]. Current designs using FR4 substrates exceed that bus speed in commonplace personal computers by almost 10-fold (PCI Express Gen 2 buses run at 5 giga-transfers per second, which has a fundamental frequency of 2.5 GHz). History is filled with "experts" who mispredicted the future:

> Heavier-than-air flying machines are impossible.
> —Lord Kelvin, British mathematician and physicist, president of the British Royal Society, 1895

> Fooling around with alternating current is just a waste of time. Nobody will use it, ever.
> —Thomas Edison, American inventor, 1889

> Rail travel at high speed is not possible because passengers, unable to breathe, would die of asphyxia.
> —Dr Dionysys Larder (1793–1859), professor of natural philosophy and astronomy, University College London

In the mid-1970s, integrated circuits held approximately 10,000 components, which was enough to construct an entire computer with devices as small as 3 μm. Experienced engineers worried that semiconductor technology had reached its pinnacle. Three micrometers was barely larger than the wavelength of the light used to sculpt the chip. Interactions between ever-closer wires were about to ruin the signals. Chips would soon generate so much heat that they would be impossible to cool without a refrigeration unit. The list goes on [Moravec, 1998].

A look at the computer growth graph shows that the industry found solutions to all those problems. Chip progress not only continued—*it sped up*. Technology companies, motivated by the potential of high profits, dedicated tremendous resources to making the "impossible" possible: developing more efficient transistor designs, better heat sinks, new manufacturing processes, and more advanced analysis techniques. History indicates that the rate of performance will continue to grow at exponential rates.

Historically, the mechanism for advancing computation has been to miniaturize components, allowing more devices to fit in and operate in a smaller space, thus producing more performance per unit volume. First, the gears in mechanical calculators shrunk, which allowed them to spin faster. Then the relays in electromechanical machines became smaller, which allowed them to switch faster. Next, the switches in digital machines evolved from shotgun shell–sized vacuum tubes, to pea-sized transistors, to tiny integrated-circuit chips [Moravec, 1998]. Each of these technological advancements came with a price: *New problems that were never before considered arose that needed to be solved*.

How does this relate to signal integrity? The field of signal integrity arose directly from the exponential growth of computing power. A computer system is comprised of many integral components in addition to the processor, such as the memory, cache, and chip set. The interconnections between these parts within a computer system are known collectively as system *buses*. Essentially, a bus is an integrated set of interconnections used to transfer data between different parts of a digital system. Accordingly, to capitalize on the benefits of increased processor power, system buses must also operate at higher data transfer rates. For example, if the memory bus fails to transmit data at a sufficiently fast rate, the processor simply sits idle until data are available. This bottleneck would negate much of the performance gained from a more powerful processor. Subsequently, it is vital that the bus performance scale correspondingly with processor performance.

1.2 THE PROBLEM

The two mechanisms used historically to scale bus design to feed the growing performance of computer processors have been speed and width. *Speed* facilitates higher information transfer rates by sending more bits in a given amount of time. *Width* facilitates more information transfer by sending more bits in parallel. From now on, the rate of information transfer on a bus will be referred to as the *bus bandwidth*.

Increasing the bus speed to overcome bandwidth limitations becomes problematic for many reasons. As bus frequencies increase in speed, the pathways that comprise the bus, called *interconnects*, begin to exhibit high-frequency behavior, which thoroughly puzzles many conventional digital designers. What is required is complete comprehension of the relevant analog techniques and theory commonly used in microwave system and radio designs applied carefully to the digital realm. As the operating frequencies of digital systems increase, these analog effects become more prevalent and severely impede the overall performance if not resolved properly. Furthermore, increased bus speed usually requires more power, which is a precious commodity, especially in mobile designs such as laptops, which rely on battery power.

Increasing the bus width to overcome bandwidth roadblocks is self-limiting. Practical mechanical limitations arise quickly due to increased pin counts on packages, sockets, connectors, and the shortcomings of PCB technology. Furthermore, interactions between closely spaced interconnects lower the signal-to-noise level, making clean data transmission more difficult. Since Moore's law results in computer performance doubling every 18 months, and the bus bandwidth must scale in proportion, doubling the number of signals in the bus to facilitate the required bandwidth provides a solution to the problem that lasts less than two years. Increasing the width of the bus is simply a short-term "band-aid." At some point, faster bus speeds will be required.

The problem is that as bus designs become both wider and faster and form factors shrink to provide more computational power per unit volume, the assumptions used for past designs become outdated and new techniques must be developed. As a result, the field of signal integrity is evolving continuously to encompass new effects that were not relevant to earlier designs. Modern bus designs have become so fast that the designer must calculate the voltage and timing numbers to a resolution as small as a few picoseconds and a few millivolts. This degree of resolution was unheard of in computer designs just a few years ago. Just to put this problem in perspective, the light that is reflected off your nose takes a little over 85 ps to travel to the surface of your eye, which is well over 10 times the required timing resolution of some modern bus designs. This dramatic decrease in bus timing requirements leads to several problems. First, the number of effects that must be accounted for in the design stage increases. This is because effects that were either second order, or ignored completely in previous designs, begin to dominate the performance. Consequently, the total number of variables that must be accounted for increases, which makes the problem more difficult. Furthermore, the new variables are often very difficult or impossible to model using conventional methods. So we have not only many more variables to worry about, but most of the new variables are very difficult to model correctly. Finally, to top it off, the laboratory equipment currently available is often insufficient to resolve these small timing variations, making it difficult or impossible to verify the completed design or to correlate the models to reality.

1.3 THE BASICS

As the reader undoubtedly knows, the basic idea in digital design is to communicate information with signals representing 1's or 0's. Typically, this involves sending and receiving series of trapezoidal voltage signals in which a high voltage is a 1 and a low voltage is a 0. The conductive paths carrying the digital signals are known as *interconnects*. The interconnect includes the entire electrical pathway from the chip sending a signal to the chip receiving the signal. This includes the chip packages, connectors, sockets, transmission lines, and vias. A group of interconnects is referred to as a *bus*. The region of voltage where a digital receiver distinguishes between a high and a low voltage is known as the *threshold region*. Within this region, the receiver will either switch high or switch low. On the silicon, the actual switching voltages vary with temperature, supply voltage, silicon process, and other variables. From the system designer's point of view, there are usually high- and low-voltage thresholds, known as V_{ih} and V_{il}, associated with the receiving silicon, above and below which a high or low value is guaranteed to be received under all conditions. Thus, the designer must guarantee that the system can, under all conditions, deliver high voltages that do not, even briefly, fall below V_{ih}, and low voltages that remain below V_{il}, to ensure the integrity of the data.

To maximize the speed of operation of a digital system, the timing uncertainty of a transition through the threshold region must be minimized. This means that the rise or fall time of the digital signal must be as fast as possible. Ideally, an infinitely fast edge rate would be used, although there are many practical problems that prevent this. Realistically, edge rates as fast as 35 ps are encountered in real systems. The reader can use Fourier analysis to verify that the quicker the edge rate, the higher the frequencies that are found in the spectrum of the signal. Herein lies a clue to the difficulty. Every conductor has a frequency-dependent capacitance, inductance, conductance, and resistance. At a high-enough frequency, none of these things are negligible. Thus, a wire is no longer a wire but a distributed, frequency-dependent parasitic element that has delay and a transient impedance profile that can cause distortions and glitches to manifest themselves on the waveform propagating from the driving chip to the receiving chip. The wire is now an element that is coupled to everything around it, including power and ground structures, heat sinks, other traces, and even the wireless network. The signal is not contained in the conductor itself but is, instead, carried in the local electric and magnetic fields around the conductor. The signals on one interconnect will affect, and be effected by, the signals on another. The inductance, capacitance, and resistance of all the structures in the vicinity of the interconnect have vital roles in the simple task of guaranteeing proper signaling transitions with appropriate timing at the receiver.

One of the most difficult aspects of high-speed design is the fact that there are many codependent variables that affect the outcome of a digital design. Some of the variables are controllable, and others force the designer to live with the random variation. One of the difficulties in high-speed design is how to handle the many variables, whether they are controllable or uncontrollable. Often, simplifications can be made by neglecting or assuming values for variables, but this can lead to unknown failures down the road for which it will not be possible after the fact to locate the root cause. As timing becomes more constrained, the simplifications of the past are rapidly dwindling in utility to the modern designer. In this book we also show how to incorporate a large number of variables that would otherwise make the problem intractable. Without a methodology for handling the large number of variables, a design ultimately incorporates some guesswork, no matter how much the designer understands the system physically. The final step in handling all the variables is often the most difficult part and the one most readily ignored by designers. A designer crippled by the inability to handle large numbers of variables will ultimately resort to proving a few "point solutions" instead and hope that they plausibly represent all known conditions. Although such methods are sometimes unavoidable, this can be a dangerous guessing game. Of course, a certain amount of guesswork is always present in design, but the goal of the system designer should be to minimize uncertainty.

1.4 A NEW REALM OF BUS DESIGN

Technology has progressed to a point where digital design has entered a new realm, where new design techniques and concepts are required that baffle even the most seasoned digital system designers. The fact is that present and future state-of-the-art digital systems, such as personal computers, cannot be designed without a thorough understanding of the principles outlined in this book. Why hasn't this been a problem before? The answer is that digital designers didn't need to understand these things. But digital circuits are reaching speeds where design will not be possible without an understanding of this subject. Seasoned engineers face the threat of becoming a legacy if they do not adapt to the new design space. This book will help practicing engineers adapt.

From the Monroe calculator to the Pentium, from punch cards to flash memory, from vacuum tubes to integrated circuits, computer performance is increasing at an exponential rate. In this book we address the needs of the contemporary digital designer as he or she encounters the numerous new challenges with modern and future high-speed digital systems and is forced to learn material previously not needed. As the conventional digital designer transitions to faster designs, he or she will indeed experience a completely different view of logic signals at high speeds. This book will help to make sense of the ugly, distorted, and smeared waveforms produced in a high-speed digital system.

1.5 SCOPE OF THE BOOK

This book was written to be an advanced study in signal integrity. Although some basic material is covered, it is assumed that the reader is well acquainted with basic electromagnetic theory, vector calculus, differential equations, statistics, and transmission-line analysis. The book builds on the traditional knowledge base and covers topics required to design present and future digital systems.

1.6 SUMMARY

All of this leads to the present situation: There are new problems to solve. Engineers who can solve these problems will define the future. This book will equip readers with the necessary practical understanding to contend with contemporary problems of modern high-speed digital design with enough theory to see beyond this book and solve problems that the authors have not yet encountered.

ERRATA

Inevitably, some errors will slip past the layers of review. Although they will be remedied in subsequent printings, it is useful to summarize the corrections in one place. The errors, along with the corrections, will be posted at ftp://ftp.wiley.com/public/sci_tech_med/high_speed_design.

REFERENCES

Moravec, Hans, 1998, When will computer hardware match the human brain?, *Journal of Evolution and Technology*, vol. 1.

Porter, Chris, 1998, High chip speeds spell end for FR4, *Electronic Times*, Mar. 30.

2

ELECTROMAGNETIC FUNDAMENTALS FOR SIGNAL INTEGRITY

Advanced Signal Integrity for High-Speed Digital Design, By Stephen H. Hall and Howard L. Heck

Much of signal integrity is based heavily in electromagnetic theory. Various aspects of this theory are found in numerous books on a variety of topics, such as microwaves, electromagnetics, optics, and mathematics. To rely on these books to form a basis of the fundamental understanding of signal integrity would result in a confusing disarray of conflicting assumptions, notations, and conventions. Although it is assumed that readers have a basic understanding of electromagnetics, the presentation of Maxwell's equations and subsequent solutions in the form most often used in signal integrity will minimize confusion and help readers extract the relevance from the haze of mathematical calculation often encountered in generalized electromagnetic textbooks. It is also convenient to summarize, in one place, the underlying physics that forms the basis of succeeding chapters. In this section we present Maxwell's equations and the underlying electromagnetic theory needed for signal integrity. The concepts are used and expanded on in several subsequent chapters. This analysis does not constitute a complete theoretical study; however, it does present the fundamental electromagnetic concepts needed to develop the basis of signal integrity theory. As the book progresses, this material will be built on to describe more advanced concepts as they are applied to real-world examples.

Initially, the most common vector operators are reviewed briefly. This is important because Maxwell's equations will be presented in differential form and a fundamental understanding of the vector operators will allow readers to visualize the behavior of electromagnetic fields. Next, the equations that govern a plane wave propagating in free space are derived directly from Maxwell's equations. Then the concepts of wave propagation, intrinsic impedance, and the speed of light are derived. Next, the theory of electrostatics and magnetostatics is covered to explain the physical meaning of an electric and a magnetic field, the energy they contain, and how they relate to specific circuit elements, such as inductance and capacitance, used in later chapters. Finally, we discuss the power carried by electromagnetic waves and how they react when propagating into different materials, such as metal or other dielectric regions. Other aspects of electromagnetic theory are covered in later chapters, but the basis of that analysis is defined here.

2.1 MAXWELL'S EQUATIONS

Electromagnetic theory is described by Maxwell's equations, published originally in 1873. In this section we outline the fundamentals of electromagnetic theory that we will need for the remainder of the book. Since a broad study of Maxwell's equations is beyond our scope in this text, we present only the necessary information that applies directly to the specific problems addressed here. Note that it is assumed that readers have completed basic electromagnetic theory classes as a prerequisite to this material.

The differential form of Maxwell's equations in SI units is summarized in equations (2-1) through (2-4). The equivalent integral forms of Maxwell's equations are presented throughout when necessary; however, emphasis is placed on the differential forms when convenient because they lend themselves to better intuitive understanding:

$$\nabla \times \vec{E} + \frac{\partial \vec{B}}{\partial t} = 0 \qquad \text{(Faraday's law)} \qquad (2\text{-}1)$$

$$\nabla \times \vec{H} = \vec{J} + \frac{\partial \vec{D}}{\partial t} \qquad \text{(Ampère's law)} \qquad (2\text{-}2)$$

$$\nabla \cdot \vec{D} = \rho \qquad \text{(Gauss's law)} \qquad (2\text{-}3)$$

$$\nabla \cdot \vec{B} = 0 \qquad \text{(Gauss's law for magnetism)} \qquad (2\text{-}4)$$

where $\vec{E}$ = electric field intensity (V/m)
$\vec{H}$ = magnetic field intensity (A/m), where

$$\vec{H} = \frac{\vec{B}}{\mu_0} - \vec{M} \qquad (2\text{-}5)$$

$\vec{M}$ = magnetization density (A/m)
$\vec{B}$ = magnetic flux intensity (Wb/m^2)
$\vec{J}$ = current density (A/m^2)
ρ = charge density (C/m^3)
$\vec{D}$ = electric flux density (C/m^2) where

$$\vec{D} = \varepsilon_0 \vec{E} + \vec{P} \qquad (2\text{-}6)$$

$\vec{P}$ = electric polarization density (C/m^2)
ε_0 = permittivity of free space (8.85×10^{-12} F/m)
μ_0 = permeability of free space ($4\pi \times 10^{-7}$ H/m)

Electromagnetic fields are derived from the movement of charge, so $\vec{J}$ and ρ are the ultimate sources that induce the electric and magnetic fields, while the other quantities are responses. Note that the magnetization density, $\vec{M}$, does not exist in nature, as it is a mere mathematical convenience. Realistic sources of magnetic current are caused by an electric current loop, as opposed to the flow of a magnetic charge, which is discussed in detail in Section 2.5. The magnetization density is included here only for completeness, as it is typically not used for the applications covered in this book. The constants ε_0 and μ_0 dictate the electromagnetic properties of free space, such as the speed of light and the intrinsic impedance, both of which are discussed in more detail in Section 2.3.

Equations (2-1) through (2-6) are not sufficient to describe the electromagnetic properties of general materials; they must be supplemented with relations that comprehend the properties of media other than free space. Specifically, equation (2-7) comprehends the finite conductivity of metal, equation (2-8) accounts for the magnetic properties of a material, and equation (2-9) describes how the dielectric will respond to an applied electric field:

$$\vec{J} = \sigma \vec{E} \tag{2-7}$$

$$\vec{B} = \mu_r \mu_0 \vec{H} = \mu \vec{H} \tag{2-8}$$

$$\vec{P} = \varepsilon_0(\varepsilon_r - 1)\vec{E} \tag{2-9}$$

where $\vec{J}$ is the current density (A/m^2), σ the conductivity of a medium (e.g., a metal) (S/m), μ_r the relative permeability, and ε_r the relative permittivity (also known as the *relative dielectric constant*). Note that both μ_r and ε_r are unitless quantities. The convention used in this book is to represent the equivalent relative permittivity and permeability as

$$\mu = \mu_r \mu_0 \tag{2-10a}$$

$$\varepsilon = \varepsilon_r \varepsilon_0 \tag{2-10b}$$

For use in high-speed digital design, materials included typically have descriptive coefficients σ, μ_r, and ε_r that are *linear*, meaning that they do not change as a function of the applied field. However, for many realistic dielectric materials used to construct circuit boards in digital designs, the descriptive coefficients are not *homogeneous* (independent of position) or *isotropic* (independent of direction), meaning that extreme care must be taken to ensure that the material properties are accounted for properly, which is explored in detail in Chapter 6. Furthermore, the descriptive coefficients often exhibit strong frequency dependence, the nature of which is explored throughout the book.

Although this brief review of Maxwell's equations may initially seem intimidating, throughout the book the theory will be simplified and applied directly to the solution of practical real-world problems, allowing readers to extract the important concepts from the haze of mathematical calculations so that an intuitive understanding can be conveyed.

2.2 COMMON VECTOR OPERATORS

Maxwell's equations are presented in differential form using many vector operators, which simplifies their representation. The vector operators describe the behavior of the fields and how they interact, so a full understanding of the meaning behind these operators will allow readers to visualize the problem, which will lead to an intuitive understanding. In engineering, the most valuable tool is a comprehensive understanding of the concept. In electromagnetics and signal integrity, the concept is often best understood through visualization of the fields. The vector operators used in this book are reviewed here, with an emphasis on how they affect the fields visually.

2.2.1 Vector

In physics and in vector calculus, a *vector* is a concept characterized by a magnitude and a direction. A component of a vector is the influence of that vector in a given direction. A vector is often described by a fixed number of components that sum uniquely to the total vector. When used in this role, the choice of directions is dependent on the particular coordinate system being used: Cartesian coordinates, spherical coordinates, or polar coordinates. A common example of a vector is force, because it has a magnitude and a direction. Whenever possible, the problems and analysis are presented in rectangular (Cartesian) coordinates. When the geometry of the problem dictates coordinate transformation into a spherical or cylindrical coordinate system, the relevant transformations are given in Appendix A.

Recall that if the vector $\vec{A}$ was located at the point in space $P(x,y,z)$, with the components $\vec{a}_x A_1$, $\vec{a}_y A_2$, and $\vec{a}_z A_3$, where $\vec{a}_m$ is a unit vector along the axis m, the expression for $\vec{A}$ could be written

$$\vec{A} = \vec{a}_x A_x + \vec{a}_y A_y + \vec{a}_z A_z \tag{2-11a}$$

where the magnitude of the vector is shown as

$$A = \sqrt{A_x^2 + A_y^2 + A_z^2} \tag{2-11b}$$

This vector is shown graphically in Figure 2-1.

2.2.2 Dot Product

The *dot product* of two vectors $\vec{A}$ and $\vec{B}$ is a metric of how much parallelism exists between vectors:

$$\vec{A} \cdot \vec{B} \equiv AB\cos\phi \tag{2-12}$$

where ϕ is the angle between $\vec{A}$ and $\vec{B}$. Note that if ϕ is 90°, then $\vec{A} \cdot \vec{B} = 0$, and if ϕ is 0° (the vectors are parallel), then $\vec{A} \cdot \vec{B} = AB$, which is simply the product of their magnitudes.

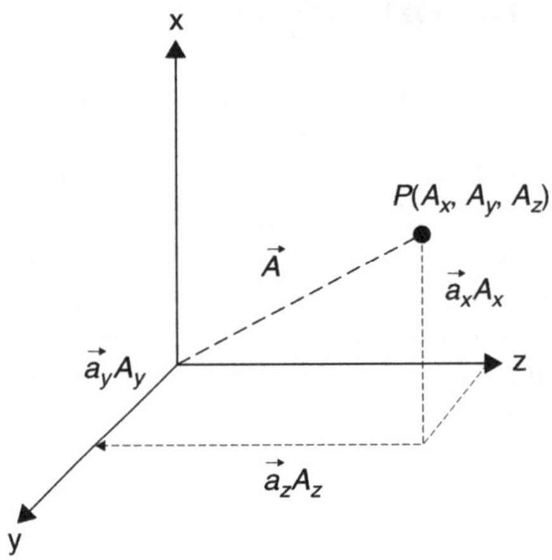

Figure 2-1 Graphical representation of a vector in rectangular coordinates.

If the vectors $\vec{A}$ and $\vec{B}$ are expressed in terms of their generalized orthogonal components as in (2-11) and the expression is expanded, the dot product can be calculated as

$$\vec{A} \cdot \vec{B} = A_x B_x + A_y B_y + A_z B_z \tag{2-13}$$

2.2.3 Cross Product

Similarly, the *cross product* of $\vec{A}$ and $\vec{B}$ is a measure of how orthogonal the vectors are, as shown by

$$\vec{A} \times \vec{B} \equiv (AB \sin \phi)\vec{a}_n \tag{2-14}$$

where $\vec{a}_n$ is a unit vector normal to the plane containing $\vec{A}$ and $\vec{B}$. If ϕ is 0°, then $\vec{A} \times \vec{B} = 0$, and if ϕ is 90° (the vectors are at a right angle to each other), then $\vec{A} \times \vec{B} = AB\vec{a}_n$ with a direction perpendicular to both $\vec{A}$ and $\vec{B}$, with the ambiguity of direction resolved by means of the right-hand rule, as shown in Figure 2-2. If the angle between $\vec{A}$ and $\vec{B}$ is something other than 0° or 90°, the following determinant form of the cross product applies:

$$\vec{A} \times \vec{B} = \begin{vmatrix} \vec{a}_x & \vec{a}_y & \vec{a}_z \\ A_x & A_y & A_z \\ B_x & B_y & B_z \end{vmatrix} \tag{2-15}$$

which simplifies to

$$A \times B = \vec{a}_x(A_y B_z - A_z B_y) + \vec{a}_y(A_z B_x - A_x B_z) + \vec{a}_z(A_x B_y - A_y B_x) \tag{2-16}$$

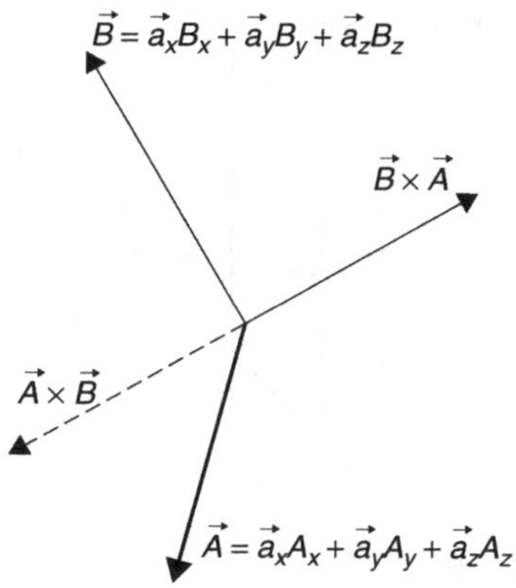

Figure 2-2 Graphical representation of the cross product.

2.2.4 Vector and Scalar Fields

In electromagnetic theory, a *field* is defined as a mathematical function of space and time. Fields can be classified as either scalar or vector. A *scalar field* has a specific value (magnitude) at every point in a region of space at each instance in time. Figure 2-3 shows two examples of a scalar field, temperature in a block of material and the voltage across a resistive strip. Note that each point $P(x,y,z)$, there exists a corresponding temperature $T(x,y,z)$ or voltage $v(x)$ at any instant in time. Other examples of scalar fields are pressure and density. A *vector field* has a variable magnitude and direction at any point in time, as illustrated with Figure 2-4. Note that the velocity and direction of the fluid inside the pipe changes in the vicinity of the neck-down region, so the magnitude and direction (phase) of the vectors that describe the motion of the fluid at a given instant in time are a function of the position in space. Other examples of vector fields are acceleration and electric and magnetic fields.

2.2.5 Flux

A vector field, $\vec{F}(x,y,z,t)$, can be represented graphically by depicting a large number of individual vectors with a specific magnitude and phase (direction); however, this is cumbersome. A more useful method for representation of a vector field is to use the concept of flux. *Flux* is a measure of how many field vectors pass though a surface in space, as depicted in Figure 2-5a. The net flux of vector field $\vec{F}$ through surface S is shown as

$$\psi = \int_S \vec{F} \cdot d\vec{s} \tag{2-17}$$

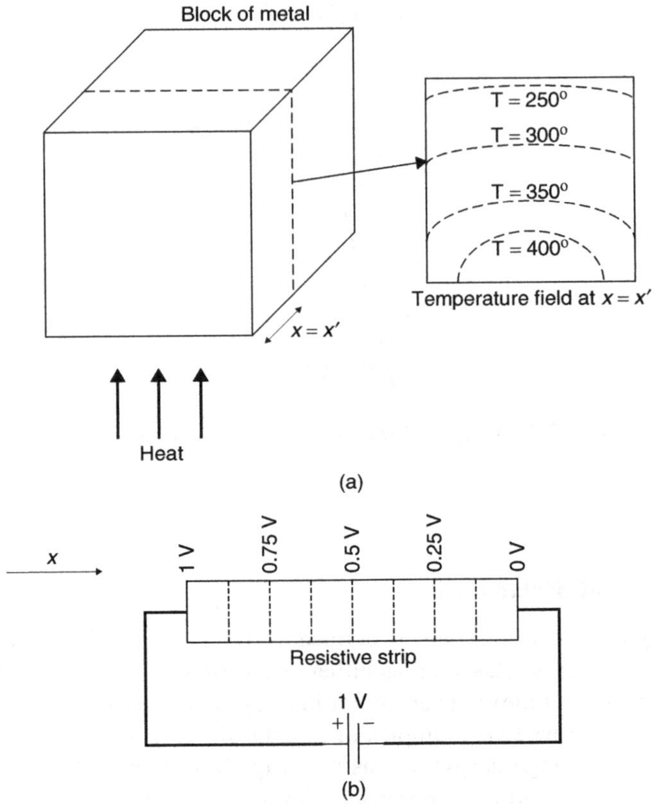

Figure 2-3 Examples of scalar fields: (a) temperature; (b) voltage.

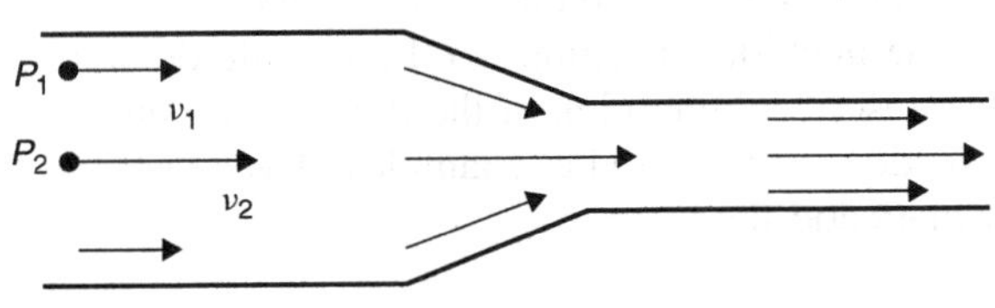

Figure 2-4 Example of a vector field: fluid velocity.

A flux plot can replace vectors with a system of flux lines that are created in accordance with the following rules:

1. The transverse density of the flux lines agrees with the magnitudes of the vectors. So, in Figure 2-5b, the velocity of the fluid is slower near the pipe wall, necessitating that the flux lines be drawn farther apart than in the middle, where the fluid flow is faster.
2. The direction of the flux lines must agree with the direction of the vectors.

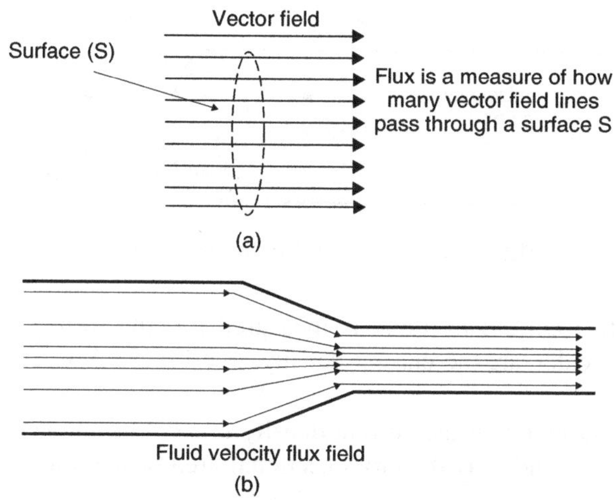

Figure 2-5 (a) Definition of flux; (b) example of a flux field.

Flux is, however, useful for more than just simplifying a vector field. If a surface S is drawn in a region of space that includes flux lines, the number of flux lines passing through that surface is a measure of several physical quantities, such as current or power flow. Note that if (2-17) is integrated over a closed surface, the net flux will always be zero, assuming that no sources exist within the volume of the closed surface. This is because the same number of flux lines enter the volume as exit it.

To illustrate the utility of the flux concept with an example, consider current flow in a wire. Suppose that a wire contains electric charges of density $\rho(\text{C/m}^3)$ in a region and the charges have a velocity $v(\text{m/s})$. The current density in the region is calculated as

$$\vec{J} = \rho\vec{v} \quad \text{A/m}^2 \tag{2-18}$$

the instantaneous rate of charge flow per unit cross-sectional area at point P in space. For n points in space with charge densities ρ_i and velocities v_i, the current density becomes

$$\vec{J} = \sum_{i=1}^{n} \rho_i \vec{v}_i \quad \text{A/m}^2 \tag{2-19}$$

Therefore, the total current flowing through a surface (e.g., the cross section of the wire) is the sum of all the current density functions within the area of the surface times the surface area. This calculates the total number of vectors ($\vec{J}$) passing though the cross-sectional surface S of the wire, which is flux. Therefore, the flux of the current density function is the current flowing through area S and is calculated as

$$\psi_i = i = \int_S \vec{J} \cdot d\vec{s} \qquad A \tag{2-20}$$

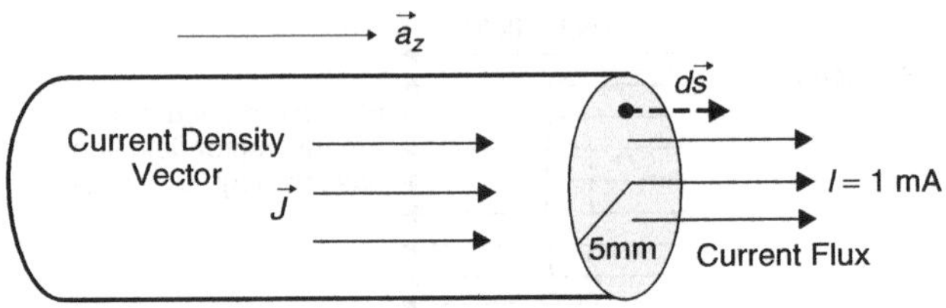

Figure 2-6 Current flux through a wire.

Example 2-1 If a current of 1 mA is measured flowing through a wire with a radius of 5 mm, calculate the current density. See Figure 2-6.

SOLUTION Assume that the current density is constant in the cross section so that $\vec{J} = \vec{a}_z J = J_z$ and A is the cross-sectional area of the wire.

$$i = \int_S J_z \cdot \vec{a}_z \, ds = J_z \int_S ds = J_z A = J_z(\pi r^2)$$

Therefore,

$$J_z = \frac{i}{\pi r^2} \approx 12.7 \text{ A/m}^2$$

2.2.6 Gradient

The vector operator ∇, pronounced *del*, is shorthand for the *gradient* of a scalar field. In simple terms, *the gradient is the space rate of change of a scalar field*. In rectangular coordinates, the gradient of a function f is

$$\nabla f = \vec{a}_x \frac{\partial f}{\partial x} + \vec{a}_y \frac{\partial f}{\partial y} + \vec{a}_z \frac{\partial f}{\partial z} \tag{2-21}$$

Subsequently, the gradient constructs a vector field from a scalar field.

2.2.7 Divergence

The *divergence* of a vector field $\vec{F}$ is a measure of the outward flux per unit volume. For example, if $\vec{F}$ is represented by a continuous system of unbroken flux lines in a volume region, the region is said to be source-free and divergenceless. However, if $\vec{F}$ is discontinuous through the volume region or contains broken flux lines, the region contains sources of flux fields and has a nonzero divergence. The divergence of $\vec{F}(x,y,z,t)$ is

$$\nabla \cdot \vec{F}(x, y, z, t) = \frac{\partial F_x}{\partial x} + \frac{\partial F_y}{\partial y} + \frac{\partial F_z}{\partial z} \tag{2-22}$$

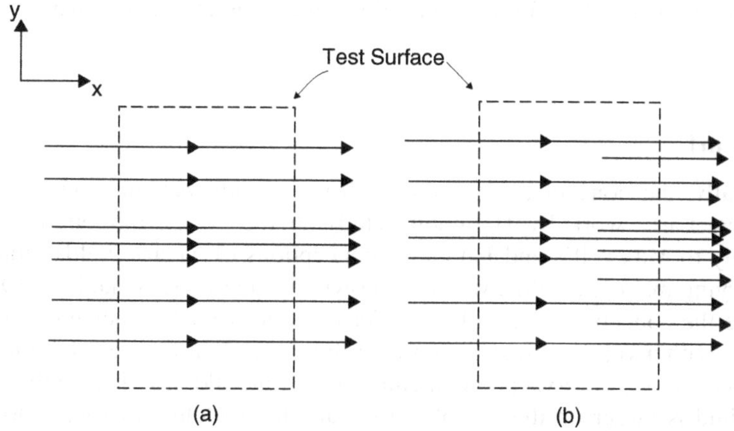

Figure 2-7 (a) Example of a divergence-free flux plot (incoming flux = outgoing flux); (b) flux plot with nonzero divergence (outgoing flux is greater than incoming flux, indicating that there are sources in the test region).

Figure 2-7a is an example of the flux plot for a divergence-free field. Inspection reveals why the divergence is zero because the vectors do not seem to converge or emerge from any source points. Additionally, a test closed surface (the box) placed in the region will have zero net flux emanating from it, because the flux lines going into the test region equal the flux lines leaving it. Figure 2-7b has nonzero divergence. Note that the emergence or convergence of the field vectors from source points is a common characteristic of a field with finite divergence. Simply put, if a region contains a source of a field, it will have a positive divergence. In Figure 2-7b, the nonzero divergence is evident from inspection because discontinuous flux lines are required to represent the increasing density with x, yielding a net nonzero flux. In other words, there are more flux lines emanating from the closed surface than are entering it, meaning that a source of the field must exist inside the test region.

An understanding of the meaning of divergence allows us to gain some insight into Gauss's laws:

$$\nabla \cdot \vec{D} = \rho \qquad \text{(Gauss's law)} \tag{2-3}$$

$$\nabla \cdot \vec{B} = 0 \qquad \text{(Gauss's law for magnetism)} \tag{2-4}$$

Note that the divergence of $\vec{D}$, which equals $\varepsilon\vec{E}$, is nonzero and equal to the charge density, which implies that the source of the electric field is an electrical charge. Equivalently, if the electric field terminates abruptly, the termination must be an electric charge. Conversely, the divergence of $\vec{B}$ is zero, indicating that there is no magnetic equivalent to the electric charge and that the magnetic field is always source-free. For a test surface, the number of flux lines entering the surface must equal the flux leaving it, and there are no abrupt terminations of the magnetic field. Therefore, the flux lines of a magnetic field consist of closed lines.

2.2.8 Curl

Historically, the concept of *curl* comes from a mathematical model of hydrodynamics. Early work by Helmholtz studying the vortex motion of fluid led ultimately to Maxwell's and Faraday's conceptions of electric fields induced by time-varying magnetic fields, which is shown in equation (2-1) [Johnk, 1988]. To visualize the concept of the curl, consider a paddle wheel immersed in a stream of water, with a velocity field as shown in Figure 2-8. In Figure 2-8a, the paddle is oriented along the z-axis perpendicular to the water flow, and since the velocity of the fluid is larger on the top of the paddle, the paddle will rotate clockwise, and therefore has a finite curl along the z-axis, with a direction pointing into the page as determined using the right-hand rule. Similarly, if the paddle is rotated so that it is oriented along the x-axis, as in Figure 2-8b, the paddle will not rotate, and the curl is zero.

The curl of $\vec{F}(x,y,z,t)$ in determinant form (in rectangular coordinates) is shown as

$$\nabla \times \vec{F}(x, y, z, t) = \begin{vmatrix} \vec{a}_x & \vec{a}_y & \vec{a}_z \\ \dfrac{\partial}{\partial x} & \dfrac{\partial}{\partial y} & \dfrac{\partial}{\partial z} \\ F_x & F_y & F_z \end{vmatrix} \tag{2-23}$$

which simplifies to

$$\nabla \times \vec{F} = \vec{a}_x \left(\frac{\partial F_z}{\partial y} - \frac{\partial F_y}{\partial z} \right) + \vec{a}_y \left(\frac{\partial F_x}{\partial z} - \frac{\partial F_z}{\partial x} \right) + \vec{a}_z \left(\frac{\partial F_y}{\partial x} - \frac{\partial F_x}{\partial y} \right) \tag{2-24}$$

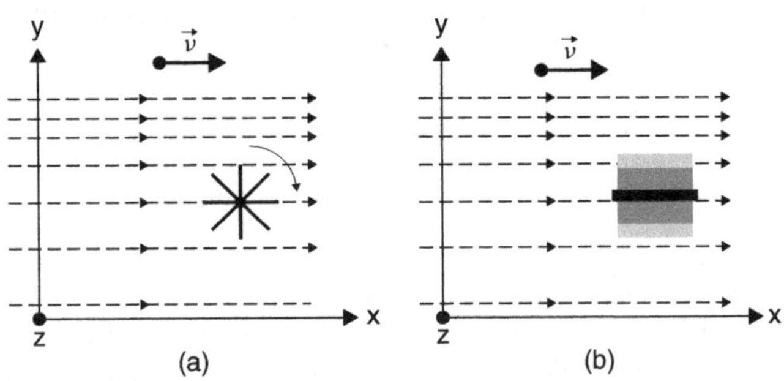

Figure 2-8 (a) The fluid velocity field causes the paddle wheel to rotate when it is oriented orthogonal to the field, giving it a nonzero curl with a direction pointing into the page ($-z$); (b) when the paddle wheel is parallel to the field it has a zero curl because the field will not make it rotate.

Simply put, if the curl is finite, a field will be induced that possesses circulation. This allows us intuitively to understand Faraday's and Ampère's laws:

$$\nabla \times \vec{E} + \frac{\partial \vec{B}}{\partial t} = 0 \qquad \text{(Faraday's law)}$$

$$\nabla \times \vec{H} = \vec{J} + \frac{\partial \vec{D}}{\partial t} \qquad \text{(Ampère's law)}$$

Faraday's law says that a time-varying magnetic field will induce an electric field that possesses circulation around $\vec{B}$. More intuitively, if we examine Ampère's law for a steady-state current, it reduces to

$$\nabla \times \vec{H} = \vec{J} \tag{2-25}$$

Equation (2-25) implies that a current flowing in a wire will induce a magnetic field that circulates around the wire, which is consistent with Gauss's law for magnetism (2-4), which implies that the flux lines of a magnetic field must consist of closed lines.

Example 2-2 Calculate the magnetic field of a current I flowing through an infinitely long wire of radius a. Show that the current flowing in the wire induces a magnetic field that circulates around the z-axis. See Figure 2-9.

SOLUTION To solve this problem it is necessary to present the integral form of Ampère's law for static fields:

$$\oint_l \frac{\vec{B}}{\mu_0} \cdot dl = \int_S \vec{J} \cdot ds = i \tag{2-26}$$

Switching to a cylindrical coordinate system, $\vec{B} = a_\phi B_\phi$ and $dl = a_\phi r d\phi$, yielding

$$\int_0^{2\pi} \frac{B_\phi}{\mu_0} r d\phi = \frac{2\pi r B_\phi}{\mu_0} = i$$

$$B_{\phi(r>a)} = \frac{i\mu_0}{2\pi r} \qquad \text{for } r > a$$

To calculate the magnetic field inside the conductor, only the amount of current passing through a percentage of the wire area must be considered. This is achieved by expressing the current in terms of an area ratio:

$$\int_0^{2\pi} \frac{B_\phi}{\mu_0} r\, d\phi = \frac{2\pi r B_\phi}{\mu_0} = i \frac{\pi r^2}{\pi a^2}$$

$$B_{\phi(r<a)} = \frac{i\mu_0 r}{2\pi a^2} \qquad \text{for } r < a$$

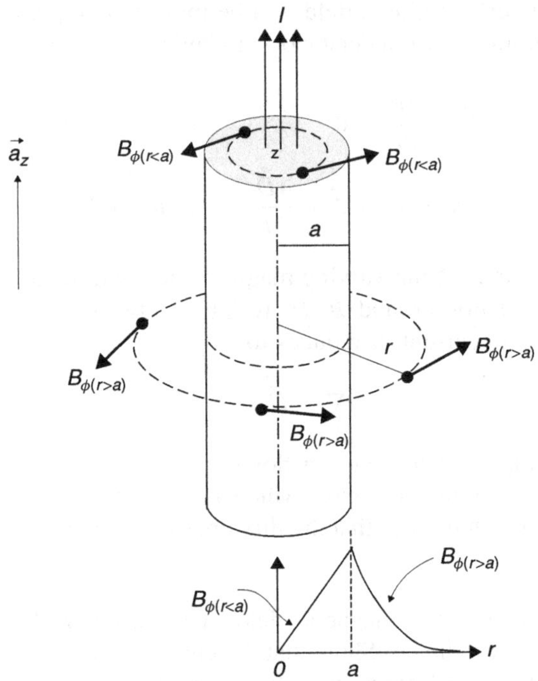

Figure 2-9 How the magnetic field will rotate around a wire carrying current.

The analysis above shows that the magnetic field has only a ϕ-component that is perpendicular to the current flow, proving that the magnetic field will wrap around the wire. Since the current flow is inducing the magnetic field, its intensity will increase until r becomes greater than the wire radius a. When examining fields outside the wire radius (where no current is flowing), the magnetic fields will decrease, as shown in Figure 2-9.

We can confirm that $\vec{B}$ circulates around the wire by calculating the curl. The curl of the magnetic field inside the wire can be calculated using the differential form of Ampère's law for the static case:

$$\nabla \times \vec{H} = \nabla \times \frac{\vec{B}}{\mu_0} = \vec{J}$$

The curl of $\vec{F}$ in cylindrical coordinate is (from Appendix A)

$$\nabla \times \vec{F} = \vec{a}_r \left[\frac{1}{r} \frac{\partial F_z}{\partial \phi} - \frac{\partial F_\phi}{\partial z} \right] + \vec{a}_\phi \left[\frac{\partial F_r}{\partial z} - \frac{\partial F_z}{\partial r} \right] + \vec{a}_z \left[\frac{1}{r} \frac{\partial (r F_\phi)}{\partial r} - \frac{1}{r} \frac{\partial F_r}{\partial \phi} \right]$$

The solution of the integral form of Ampère's law shows that the only component of the magnetic field in the ϕ-direction is a function of r. Consequently, the curl becomes

$$\nabla \times \vec{B} = \begin{cases} \vec{a}_z \left[\dfrac{1}{r} \dfrac{\partial (r B_\phi)}{\partial r} \right] \\ \vec{a}_z \left[\dfrac{1}{r} \dfrac{\partial \left(r(\mu_0 i r / 2\pi a^2) \right)}{\partial r} \right] = \dfrac{\mu_0 i}{\pi a^2} \vec{a}_z \qquad \text{for } r < a \end{cases}$$

which says that the magnetic field with a ϕ-component will be induced that circulates around a wire when current I is flowing in the z-direction.

For $r > a$, the curl becomes

$$\nabla \times \vec{B} = \vec{a}_z \left[\frac{1}{r} \frac{\partial \left(r(\mu_0 i / 2\pi r) \right)}{\partial r} \right] = 0$$

The curl of the magnetic field outside the wire is zero. This does not mean that the magnetic field does not circulate around the wire outside the conductor (it certainly does). The zero curl result is simply due to the fact that the area outside the conductor does not contain any current density ($\vec{J} = 0$), and therefore Ampère's law states that the curl of the magnetic field must be zero.

2.3 WAVE PROPAGATION

When studying Maxwell's equations, it becomes apparent that Faraday's and Ampère's laws (the two curl equations), which state, respectively, that a changing magnetic field will produce an electric field and a changing electric field will produce a magnetic field, are responsible for the propagation of an electromagnetic wave. In this section we derive equations that regulate electromagnetic wave propagation in a simple source-free medium. In the study of signal integrity, the propagation of waves in packages, on printed circuit boards, though cables, and between power and ground planes constitutes a very large portion of the discipline. In fact, communication between components in a high-speed digital design necessitates the intentional propagation of electromagnetic waves guided by transmission lines and the prevention of energy propagation across unintentional pathways (such as crosstalk) or in unwanted signal propagation modes. Without a detailed study of wave propagation, the study of signal integrity would become impossible.

2.3.1 Wave Equation

In subsequent chapters it will become necessary to analyze electromagnetic wave propagation only in terms of magnetic or electric fields because they are related directly to the voltage and current propagating on transmission lines, through vias, or across planes. The wave equation forms the basis for calculating critical electrical properties, such as crosstalk, reflections, standing waves, and different modes of propagation in multiconductor systems (e.g., a bus). We begin by manipulating Faraday's and Ampère's laws using some useful vector identities:

$$\nabla \times \vec{E} + \frac{\partial \vec{B}}{\partial t} = 0 \qquad \text{(Faraday's law)} \tag{2-1}$$

$$\nabla \times \vec{H} = \vec{J} + \frac{\partial \vec{D}}{\partial t} \qquad \text{(Ampère's law)} \tag{2-2}$$

Taking the curl of (2-1) produces

$$\nabla \times (\nabla \times \vec{E}) = -\nabla \times \frac{\partial \vec{B}}{\partial t}$$

Since $\vec{B} = \mu_r \mu_0 \vec{H}$ [from (2-8)], the equation above can be written in terms of the electric field by substituting (2-2) into the right-hand part:

$$\nabla \times (\nabla \times \vec{E}) = -\nabla \times \frac{\partial \vec{B}}{\partial t} = -\frac{\partial(\nabla \times \mu \vec{H})}{\partial t} = -\mu \frac{\partial}{\partial t}\left(\vec{J} + \frac{\partial \vec{D}}{\partial t}\right)$$

where $\mu = \mu_r \mu_0$.

If it is assumed that the region of wave propagation is source-free, the current density $\vec{J}$ is zero. Combining equations (2-6) and (2-9) yields the relation $\vec{D} = \varepsilon_r \varepsilon_0 \vec{E} = \varepsilon \vec{E}$ and allows the equation to be expressed only in terms of $\vec{E}$:

$$\nabla \times (\nabla \times \vec{E}) = -\mu\varepsilon \frac{\partial^2 \vec{E}}{\partial t^2}$$

The formula can be simplified further by using the following vector identity (see Appendix A):

$$\nabla \times (\nabla \times \vec{E}) = \nabla(\nabla \cdot \vec{E}) - \nabla^2 \vec{E}$$

Since we have assumed a source-free medium, the charge density is zero ($\rho = 0$), Gauss's law reduces to $\nabla \cdot \vec{E} = 0$, yielding equation (2-27), which is known as the *wave equation for the electric field*:

$$\nabla^2 \vec{E} - \mu\varepsilon \frac{\partial^2 \vec{E}}{\partial t^2} = 0 \tag{2-27}$$

Using the identical technique, the *wave equation for the magnetic field* can be derived:

$$\nabla^2 \vec{H} - \varepsilon\mu \frac{\partial^2 \vec{H}}{\partial t^2} = 0 \tag{2-28}$$

Note that equations (2-27) and (2-28) are similar except that the order of multiplication of $\mu\varepsilon$ is reversed. The order of multiplication for this derivation was preserved because it will become important when using matrices to calculate the solution for waves propagating on multiple transmission lines in Chapter 4.

2.3.2 Relation Between *E* and *H* and the Transverse Electromagnetic Mode

The wave equations (2-27) and (2-28) are presented in their most general form, where the fields have components in four dimensions: x, y, z, and *time*. However, for the vast majority of signal integrity analysis, the wave equations (and all of Maxwell's equations) can be simplified so that the fields have only one nonzero component that varies with one spatial coordinate. For example, the electric field $\vec{E}(x, y, z, t)$ can be reduced to $\vec{a}_x E_x(z, t)$. A good example of this is the magnetic field that was calculated in Example 2-2, where the magnetic flux intensity $\vec{B}$ had only one component in the ϕ-direction.

Although the wave equations (2-27) and (2-28) were derived separately, they are coupled and interdependent. For example, if the electric field is restricted so that $\vec{E} = \vec{a}_x E_x(z, t)$, similar restrictions on the magnetic field cannot be chosen arbitrarily. In this case, since (2-27) was derived using the entire set of Maxwell's equations, once the electric field is restricted, the magnetic field is already determined. Thus, the proper way to calculate the magnetic field is to derive it from the electric field. For example, if a wave is propagating in a source-free medium in the z-direction and its electric field only has a component in the x-direction $[\vec{E} = \vec{a}_x E_x\ (z,t)]$, the magnetic field can be calculated from(2-1) and (2-2):

$$\nabla \times \vec{E} + \frac{\partial \vec{B}}{\partial t} = 0 \tag{2-1}$$

Since we have restricted $\vec{E}$ so that it varies only with z $(\partial \vec{E}/\partial x = \partial \vec{E}/\partial y = 0)$, equation (2-24) shows that the curl of $\vec{E}$ can only produce components in the x- and y-directions.

$$\vec{a}_y \frac{\partial E_x}{\partial z} + \vec{a}_x \frac{\partial B_x}{\partial t} + \vec{a}_y \frac{\partial B_y}{\partial t} = 0$$

Since $\vec{B} = \mu \vec{H}$,

$$\vec{a}_y \frac{\partial E_x}{\partial z} + \vec{a}_x \mu \frac{\partial H_x}{\partial t} + \vec{a}_y \mu \frac{\partial H_y}{\partial t} = 0$$

Grouping into vector components yields

$$\vec{a}_y \frac{\partial E_x}{\partial z} + \vec{a}_y \mu \frac{\partial H_y}{\partial t} = 0$$

$$\vec{a}_x \mu \frac{\partial H_x}{\partial t} = 0$$

From Ampère's law (2-2),

$$\nabla \times \vec{H} = \vec{J} + \frac{\partial \vec{D}}{\partial t}$$

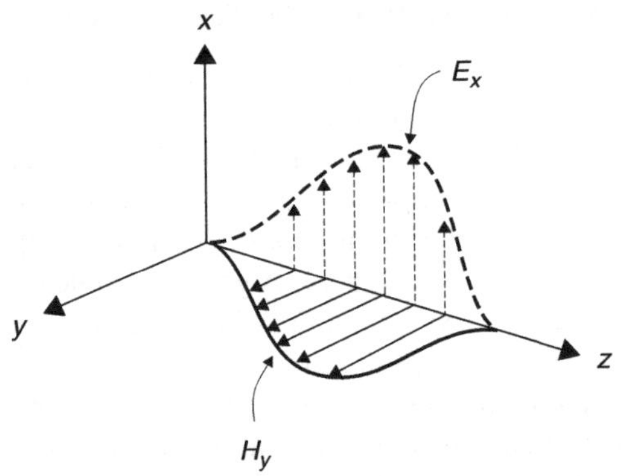

Figure 2-10 How the electric and magnetic fields are related as a TEM electromagnetic pulse propagates through space along the z-axis.

where $\vec{J} = 0$ for a source-free medium and $\vec{D} = \varepsilon\vec{E}$ (for now, we assume that $\vec{P} = 0$):

$$\vec{a}_y \frac{\partial H_x}{\partial z} - \vec{a}_x \frac{\partial H_y}{\partial z} = 0 + \vec{a}_x \varepsilon \frac{\partial E_x}{\partial t}$$

Grouping into vector components yields

$$-\vec{a}_x \frac{\partial H_y}{\partial z} = \vec{a}_x \varepsilon \frac{\partial E_x}{\partial t}$$

$$\vec{a}_y \frac{\partial H_x}{\partial z} = 0$$

The nonzero components of the equations above can be grouped to see the contributions in both the x- and y-directions.

$$\vec{a}_y \left(\frac{\partial E_x}{\partial z} = -\mu \frac{\partial H_y}{\partial t} \right) \tag{2-29}$$

$$\vec{a}_x \left(\varepsilon \frac{\partial E_x}{\partial t} = -\frac{\partial H_y}{\partial z} \right) \tag{2-30}$$

Equations (2-29) and (2-30) symbolize an important concept used throughout signal integrity analysis, which is that *the electric and magnetic fields are orthogonal and there are no components in the z-direction*. When waves propagate in this manner, it is called the *transverse electromagnetic mode* (TEM). Figure 2-10 depicts the relationship between the electric and magnetic fields as the TEM electromagnetic pulse propagates along the z-axis.

It should be noted that waves are not always restricted to propagate only in TEM mode because some structures (such as a microstrip) distort the waveform so that a small portion of the fields will have a z-component that lies outside the $x-y$ plane. However, the relationship between the wavelength and the structure sizes in practical systems allows us to assume that the waves are propagating in TEM mode until very high frequencies. Measurements have confirmed that the TEM assumption remains valid to at least 50 GHz for typical transmission-line structures used in contemporary digital designs. The validity of the TEM assumption for transmission lines is discussed further in Chapter 3.

2.3.3 Time-Harmonic Fields

A simplification of Maxwell's equations can be made if the time variation is assumed to be steady-state sinusoidal or time harmonic in nature. Although perfect sinusoidal waveforms are rarely encountered in digital design, the trapezoidal digital pulses usually employed can be constructed from a series of sinusoidal waveforms via the Fourier transform, making this general simplification particularly useful. Time-harmonic electromagnetic fields will be generated whenever their charge and current sources also have densities that have a sinusoidal variation with time. Assuming that the sinusoidal sources are steady state permits the assumption that both $\vec{E}$ and $\vec{B}$ also reach steady state and vary according to $\cos(\omega t + \theta_E)$ and $\cos(\omega t + \theta_B)$, where $\omega = 2\pi f$ and θ is the phase of either the electric or the magnetic field.

Generally, a sinusoidal waveform can be represented as

$$\cos\phi + j\sin\phi = e^{j\phi} \tag{2-31}$$

so the sinusoidal form of a time-harmonic field will vary according to the complex exponential factor $e^{j\omega t}$, which leads to a reduction of Maxwell's equations from a function space and time to simply space:

$$\vec{E}(x, y, z, t) = \vec{E}(x, y, z)e^{j\omega t} \tag{2-32a}$$

$$\vec{B}(x, y, z, t) = \vec{B}(x, y, z)e^{j\omega t} \tag{2-32b}$$

Equations (2-32) allow Maxwell's equations to be rewritten as

$$\nabla \times (\vec{E}e^{j\omega t}) + \frac{\partial \vec{B}(e^{j\omega t})}{\partial t} = 0$$

$$\nabla \times (\vec{H}e^{j\omega t}) = \vec{J}e^{j\omega t} + \frac{\partial(\vec{D}e^{j\omega t})}{\partial t}$$

$$\nabla \cdot (\vec{D}e^{j\omega t}) = \rho e^{j\omega t}$$

$$\nabla \cdot (\vec{B}e^{j\omega t}) = 0$$

The curl and the gradient affect only space-dependent functions, and the $e^{j\omega t}$ is operated on only by the partial time derivatives. Therefore, after canceling all the extraneous $e^{j\omega t}$ terms, the *time-harmonic form of Maxwell's equations*, where time has been eliminated, is shown.

$$\nabla \times \vec{E} + j\omega\vec{B} = 0 \tag{2-33}$$

$$\nabla \times \vec{H} = \vec{J} + j\omega\vec{D} \tag{2-34}$$

$$\nabla \cdot \vec{D} = \rho \tag{2-35}$$

$$\nabla \cdot \vec{B} = 0 \tag{2-36}$$

Note that the time variation of the fields can be restored by multiplying by $e^{j\omega t}$ and taking the real part:

$$\vec{F}(x, y, z, t) = Re[\vec{F}(x, y, z)e^{j\omega t}] \tag{2-37}$$

2.3.4 Propagation of Time-Harmonic Plane Waves

As will be demonstrated in subsequent chapters, the propagation of time-harmonic plane waves is of particular importance for the study of transmission-line or other guided-wave structures. This allows us to study a simplified subset of Maxwell's equations where propagation is restricted to one direction (usually along the z-axis) and time is removed as described in Section 2.3.3. A plane wave is defined so that propagation occurs in only one direction (z) and the fields do not vary with time in the x- and y-directions. If the fields were observed at an instant in time, they would be constant in the $x-y$ plane for any given point z and would change for different values of z or t. Figure 2-11 depicts a plane wave propagating in the z-direction.

To study the behavior of time-harmonic plane waves, it is necessary to re-derive the wave equation from the time-harmonic form of Maxwell's

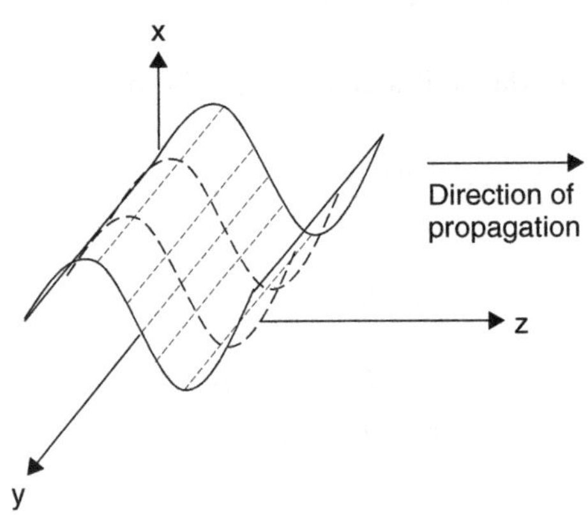

Figure 2-11 Plane wave propagating in the z-direction.

equations using the procedure employed in Section 2.3.1. Again, assume a source-free, linear, homogeneous medium:

$$\nabla \times (\nabla \times \vec{E}) = -j\omega\mu(\nabla \times H)$$

The formula can be further simplified by using the following vector identity (Appendix A):

$$\nabla \times (\nabla \times \vec{E}) = \nabla(\nabla \cdot \vec{E}) - \nabla^2 \vec{E}$$

Since we have assumed a source-free medium, the charge density is zero ($\rho = 0$) and Gauss's law reduces to $\nabla \cdot \vec{E} = 0$, yielding

$$\nabla^2 \vec{E} + j^2\omega^2\mu\varepsilon\vec{E} = \nabla^2\vec{E} - \omega^2\mu\varepsilon\vec{E} = 0 \tag{2-38}$$

Substituting $\gamma^2 = \omega^2\mu\varepsilon$ yields

$$\nabla^2\vec{E} - \gamma^2\vec{E} = 0 \tag{2-39}$$

which is the *time-harmonic plane-wave equation* for the electric field, where γ is known as the propagation constant.

If the solution is limited to plane waves propagating in the z-direction that have an electric field component only in the x-direction, the wave equation becomes (see Appendix A)

$$\begin{aligned} \nabla^2\vec{E} = \frac{\partial^2 E_x}{\partial x^2} + \frac{\partial^2 E_x}{\partial y^2} + \frac{\partial^2 E_x}{\partial z^2} &= \frac{\partial^2 E_x}{\partial z^2} - \gamma^2 E_x = 0 \\ &\frac{\partial^2 E_x}{\partial z^2} - \gamma^2 E_x = 0 \end{aligned} \tag{2-40}$$

which is a second-order ordinary differential equation with the general solution

$$E_x = C_1 e^{-\gamma z} + C_2 e^{\gamma z} \tag{2-41}$$

where C_1 and C_2 are determined by the boundary conditions of the particular problem.

As discussed in Chapter 3, equation (2-41) and its magnetic field equivalent will prove to be particularly important for signal integrity because they describe the propagation of a signal on a transmission line. The first term, $C_1 e^{-\gamma z}$, describes completely the forward-traveling part of the wave propagating in the z-direction (i.e., down the length of the transmission line), and the second term, $C_2 e^{+\gamma z}$, describes the propagation of the rearward-traveling wave in the $-z$-direction. Observing equation (2-41) allows the definition of an important term, the *propagation constant*:

$$\gamma = \alpha + j\beta \tag{2-42}$$

The terms in (2-42) have special meanings used throughout the book to describe the medium where the electromagnetic wave is propagating, whether it is in free space, is an infinite dielectric, or is a transmission line. Specifically, α is the *loss term*, which describes signal attenuation as it propagates through the medium. The loss term accounts for the fact that real-world metals are not infinitely conductive (except superconductors) and dielectrics are not perfect insulators (except free space), both of which are discussed in detail in Chapters 5 and 6. The imaginary portion of (2-42), β, called the *phase constant*, essentially dictates the speed at which the electromagnetic wave will travel in the medium. To visualize these waves propagating as described in (2-41), it is necessary first to recover the time dependency removed in Section 2.3.3. Considering only the forward-propagating component of a wave in a vacuum, replacing C_1 with the magnitude of the electric field, restoring the time dependency as in (2-37), and applying the identity of equation (2-31) yields

$$E(z,t) = \mathrm{Re}(E_x^+ e^{-\gamma z} e^{j\omega t}) = \mathrm{Re}(E_x^+ e^{-\alpha z} e^{-j\beta z} e^{j\omega t}) = e^{-\alpha z} E_x^+ \cos(\omega t - \beta z) \tag{2-43}$$

Assuming that the loss term is zero ($\alpha = 0$), Figure 2-12 depicts successive snapshots of a wave propagating though space. To determine how fast the wave is propagating, it is necessary to observe the cosine term for a small duration of time Δt. Since the wave is propagating, a small change in time will be proportional to a small change in distance Δz, which means that an observer moving with the wave will experience no phase change because she is moving at the phase velocity (v_p). Setting the term inside the cosine of (2-43) to a constant ($\omega t - \beta z = \text{constant}$) and differentiating allows the definition of the phase velocity from the cosine term in (2-43):

$$v_p = \frac{dz}{dt} = \frac{\omega}{\beta} \qquad \text{m/s} \tag{2-44}$$

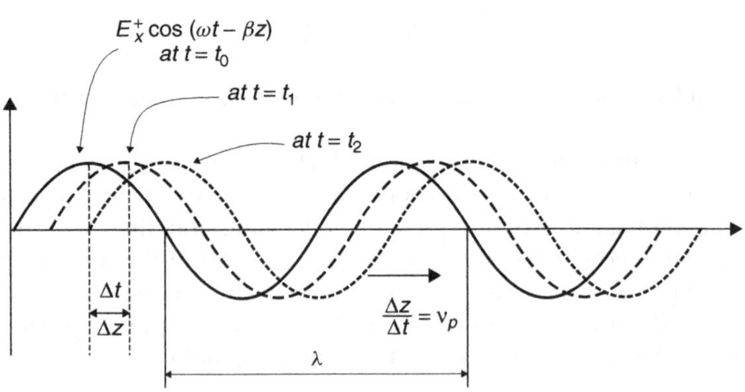

Figure 2-12 Snapshots in time of a plane wave propagating along the z-axis, showing the definition of phase velocity.

The relationship between the frequency and its wavelengths is calculated based on the speed of light, which is the phase velocity (v_p) in a vacuum:

$$f = \frac{c}{\lambda} \qquad \text{hertz} \tag{2-45}$$

Since $\omega = 2\pi f$ and c is the speed of light in a vacuum (ca. 3×10^8 m/s), equation (2-45) can be substituted into (2-44) to obtain a useful formula for β in terms of the wavelength λ:

$$c = \frac{\omega}{\beta} = \frac{2\pi c}{\beta\lambda} \rightarrow \beta = \frac{2\pi}{\lambda} \qquad \text{rad/m} \tag{2-46}$$

The speed of light in a vacuum is defined as the inverse of the square root of the product of the permeability and the permittivity of free space:

$$c \equiv \frac{1}{\sqrt{\mu_0 \varepsilon_0}} \qquad \text{m/s} \tag{2-47}$$

Calculation of λ in terms of (2-47) allows the phase constant β to be rewritten in terms of the properties of free space:

$$\beta = 2\pi f \sqrt{\mu_0 \varepsilon_0} = \omega \sqrt{\mu_0 \varepsilon_0} \qquad \text{rad/m} \tag{2-48}$$

This is expanded on later in this chapter to include propagation of a wave in a dielectric medium.

Now that the propagation constant has been defined, (2-43) can be rewritten in physical terms, assuming free space (which is lossless, so $\alpha = 0$):

$$E(z,t) = \text{Re}(E_x^+ e^{-j\omega z\sqrt{\mu_0\varepsilon_0}} e^{j\omega t}) = E_x^+ \cos(\omega t - \omega z\sqrt{\mu_0 \varepsilon_0}) \tag{2-49}$$

Since (2-49) is a solution to the wave equation, the magnetic field is found simply by using Faraday's law ($\nabla \times \vec{E} + j\omega\vec{B} = 0$):

$$\begin{aligned} \frac{\partial}{\partial z}(E_x^+ e^{-j\omega z\sqrt{\mu_0\varepsilon_0}}) e^{j\omega t} &= -j\omega\mu_0 H_y^+ \\ H_y^+ = \frac{\sqrt{\mu_0\varepsilon_0}}{\mu_0} E_x^+ e^{-j\omega z\sqrt{\mu_0\varepsilon_0}} e^{j\omega t} &= \frac{1}{\eta_0} E_x^+ e^{-j\omega z\sqrt{\mu_0\varepsilon_0}} e^{j\omega t} \\ &= \frac{1}{\eta_0} E_x^+ \cos(\omega t - \omega z\sqrt{\mu_0\varepsilon_0}) \end{aligned} \tag{2-50}$$

where η_0 is the *intrinsic impedance of free space* and has a value of 377 Ω:

$$\eta_0 \equiv \sqrt{\frac{\mu_0}{\varepsilon_0}} = 377\ \Omega \tag{2-51}$$

Equations (2-49) and (2-50) describe how a plane wave propagates in free space. The intrinsic impedance and the speed of light are constants that describe how the electromagnetic wave will propagate through the medium. The speed of light defines the phase delay of the wave, and the intrinsic impedance describes the relationship between the electric and magnetic fields. However, for wave propagation in other media, such as the dielectric of a printed circuit board (PCB), the speed of light and the intrinsic impedance are calculated using the relative permittivity ε_r and relative permeability μ_r, which simply describe the properties of the material relative to free-space values. Note that both μ_r and ε_r are unitless values that are real numbers for loss-free media but become complex for lossy media, as described in Chapters 5 and 6. The *speed of light* (referred to as the *phase velocity* for media other than free space) and the intrinsic impedance in a medium is calculated as

$$v_p = \frac{1}{\sqrt{\mu_r \mu_0 \varepsilon_r \varepsilon_0}} = \frac{c}{\sqrt{\mu_r \varepsilon_r}} \qquad \text{m/s} \tag{2-52}$$

$$\eta \equiv \sqrt{\frac{\mu_r \mu_0}{\varepsilon_r \varepsilon_0}} = \sqrt{\frac{\mu}{\varepsilon}} = \frac{E}{H} \qquad \text{ohms} \tag{2-53}$$

Note that for free space, μ_r and ε_r are both defined to be unity.

Equations (2-54) and (2-55) summarize the TEM plane waves of both the electric and magnetic fields in general form, with the time dependency removed:

$$E_x(z) = E_x^+ e^{-\gamma z} + E_x^- e^{\gamma z} \tag{2-54}$$

$$H_y(z) = \frac{1}{\eta}(E_x^+ e^{-\gamma z} - E_x^- e^{\gamma z}) \tag{2-55}$$

where $\gamma = \alpha + j\beta$ is the propagation constant; α describes how the signal is attenuated by conductor and dielectric losses, described in full detail in Chapters 5 and 6; and β is the phase constant, as defined by (2-46) when the phase velocity in (2-52) is substituted for the speed of light in a vacuum *(c)*. Note that the second term in (2-55) is negative. This is because the sign of the exponent for the reverse traveling wave does not cancel the negative sign in Faraday's law as it did for the forward-traveling wave in equation (2-50) when the derivative with respect to z was calculated. The terms E_x^+ and E_x^- describe the directions of each component of a propagating wave. For example, the total propagating wave could have a portion of the electric field propagating in the $+z$-direction and another propagating in the $-z$-direction. Figure 2-13 depicts a time-harmonic TEM plane wave propagating along the z-axis.

2.4 ELECTROSTATICS

Electrostatics is the study of stationary charge distributions. A complete understanding of electrostatics is essential for the high-speed digital designer because

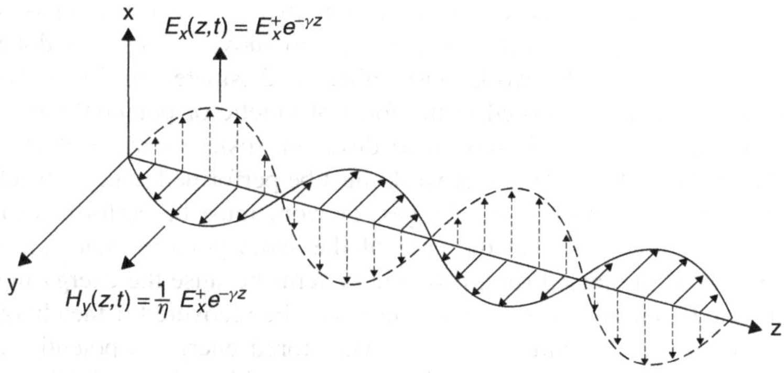

Figure 2-13 Time-harmonic TEM plane wave propagating down the z-axis.

it forms the basis of fundamental signal integrity theory and promotes an intuitive understanding of how electric fields behave. In this chapter we (1) define the electric field, (2) describe how energy is stored in the electric field, and (3) define capacitance, which is the circuit element used in circuit models to represent the energy stored in an electric field. The vast majority of signal integrity analysis performed in the industry today uses electrostatic techniques to calculate critical design variables such as transmission-line impedance, phase velocities, and effective dielectric permittivities. The concepts introduced in this section are expanded on in later chapters to describe a myriad of concepts.

When a weightlifter hoists a barbell over his head, the energy expended (or the work done against the gravitational field) is stored in the form of potential energy. The gravitational potential energy can be recovered by lowering the barbell to the ground. Similarly, when two infinitely separated charges of the same polarity are brought together, the charges will experience a repulsive force whose magnitude depends on the distance between the charges. The existence of this force is described by saying that a charge q with units of coulombs (C) produces an electric field in the region surrounding it. When the electric fields of two charges of the same polarity begin to interact, a force will be generated that will push the charges apart. Therefore, the region surrounding a charge is permeated by a force field known as the *electric field*, defined fundamentally as force per unit charge, with units of newtons per coulomb (N/C). Note that because a volt is defined as joules per coulomb (J/C), newtons per coulomb is equivalent to volts per meter (V/m), which are the units commonly used to describe an electric field.

$$
\begin{aligned}
&\mathrm{V} = \frac{\mathrm{J}}{\mathrm{C}} \rightarrow \mathrm{C} = \frac{\mathrm{J}}{\mathrm{V}} \\
&1\mathrm{J} = 1\mathrm{kg} \cdot \mathrm{m}^2/\mathrm{s}^2 \\
&1\mathrm{N} = 1\mathrm{kg} \cdot \mathrm{m}/\mathrm{s}^2 \\
&\frac{\mathrm{N}}{\mathrm{C}} = \frac{\mathrm{N} \cdot \mathrm{V}}{\mathrm{J}} = \frac{1\mathrm{kg} \cdot \mathrm{m}/\mathrm{s}^2 \cdot \mathrm{V}}{1\mathrm{kg} \cdot \mathrm{m}^2/\mathrm{s}^2} = \frac{\mathrm{V}}{\mathrm{m}}
\end{aligned}
$$

The description above makes the argument that an electric field is a field of force. If that force field acts upon a body to move it, work is done. The energy used to perform the work must either be dissipated by losses (friction in mechanical systems) or stored in the form of kinetic or potential energy. For example, if a test charge q is moved to different positions in an electric field produced by a fixed charge Q, either work must be performed to keep the charges separated (in the case of opposite charges) or work must be performed to bring the charges closer together (in the case of the same polarity charges). In this case, no energy is dissipated (in a loss-free system) because the energy in stored in the separated configuration and the energy can be recovered if the charges are allowed to return to the initial positions. The stored energy is potential energy because it depends on the position of the charges within the field. The concept of *scalar electric potential*, which will now be derived, provides a metric to describe the work or energy required to move charges from one point to another inside an electrostatic field.

The discussion above describes how an electric field is produced when two charges are brought into the vicinity of each other. If we assume that one charge (Q) is stationary and the other charge (q) is moved toward the stationary charge from point a to point b, the work can be calculated as force × distance. To calculate the work done while moving the charge along a path, the following line integral is used:

$$\vec{W}_{a\to b} = -\int_a^b \vec{F}\cdot d\vec{l} \qquad \text{joules} \tag{2-56}$$

Note that the minus sign is necessary because (2-56) represents work being done against the field. The dot product accounts for the fact that it takes zero work to move the charge perpendicular to the field because there is no opposing force in that direction.

Since a fundamental unit of an electric field is newtons per coulomb (described above), the force field in (2-56) can be rewritten in terms of the electric field as

$$\vec{E} = \left[\frac{\vec{F}}{q}\right]_{a\to b} \qquad \text{N/C or V/m} \tag{2-57}$$

$$\left[\frac{W}{q}\right]_{a\to b} = v(b) - v(a) = -\int_a^b \vec{E}\cdot d\vec{l} \qquad \text{V} \tag{2-58}$$

where q is a test charge being moved from point a to point b within the vicinity of a fixed charge Q, with units of coulombs. To calculate the electric field generated by a point charge, the integral form of Gauss's law is used:

$$\oint_S \varepsilon\vec{E}\cdot d\vec{s} = \int_V \rho\, dV = Q_{\text{enc}} \tag{2-59}$$

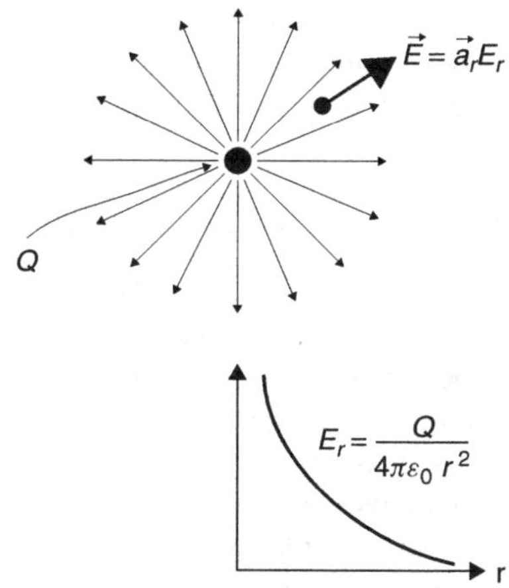

Figure 2-14 Electric field generated by a point charge Q.

where is the volume density of the charge within volume V in C/m^3 and Q_{enc} is the total charge enclosed by the surface S and contained in the volume V in units of coulombs.

As shown in Figure 2-14, a point charge has an electric field that radiates out in all directions, necessitating the use of spherical coordinates. In spherical coordinates, the ϕ and θ components of the electric field are zero, leaving only an r component directed outward from the point charge. Therefore, since the surface area of a sphere is $4\pi r^2$, the electric field around a point charge in free space is derived from (2-59):

$$\vec{E} = \vec{a}_r E_r = \frac{Q}{4\pi \varepsilon_0 r^2} \qquad \text{V/m} \tag{2-60}$$

Substituting (2-60) into (2-58) allows the calculation of the work done per unit charge when a charge is moved along r from point a to point b in an electrostatic field. Note that if the test charge was moved along ϕ or θ, there would be no work done because $\vec{E} \cdot d\vec{l} = 0$; however, along the radial component, $\vec{E} \cdot d\vec{l} = E_r dr$.

$$\begin{aligned} \left[\frac{W}{q}\right]_{a\to b} &= -\int_a^b \vec{E} \cdot d\vec{l} = -\int_a^b \frac{Q}{4\pi \varepsilon_0 r^2}\, dr \\ &= \frac{Q}{4\pi \varepsilon_0}\left(\frac{1}{r_b} - \frac{1}{r_a}\right) = \Phi(a) - \Phi(b) \qquad \text{V} \end{aligned} \tag{2-61}$$

Therefore, we can define the electrostatic potential as work done to move a charge in an electrostatic field from point a to point b:

$$\Phi_{ab} = \frac{Q}{4\pi \varepsilon_0 r_{ab}} \qquad \text{V} \tag{2-62}$$

where r_{ab} denotes the radial distance that the stationary charge moved and Φ_{ab} is known as the *electrostatic potential* or voltage between points a and b (don't confuse the symbol for potential Φ with the polar coordinate variable ϕ).

2.4.1 Electrostatic Scalar Potential in Terms of an Electric Field

For electrical engineers, it is often convenient to think of fields in terms of more familiar circuit concepts, which are usually described in terms of potential differences (i.e., voltage) between points in a circuit. Therefore, a relationship between the electrostatic potential and the electric fields that is described in terms of a scalar function will prove to be useful in the study of signal integrity. To derive this relationship, it is convenient to sidestep the complexities of spherical coordinates and think in terms of simple rectangular coordinates. From equations (2-61) and (2-62) it is obvious that the differential amount of work done in moving a charge in an electrostatic field is directly proportional to the potential difference:

$$\begin{aligned} dW &= q\Phi(x + \Delta x, y + \Delta y, z + \Delta z) - q\Phi(x, y, z) \\ &= q\left(\frac{\partial \Phi}{\partial x}\Delta x + \frac{\partial \Phi}{\partial y}\Delta y + \frac{\partial \Phi}{\partial x}\Delta z\right) \end{aligned} \tag{2-63}$$

However, from equation (2-58),

$$W = -\int_a^b q\vec{E}\cdot d\vec{l} \rightarrow dW = -q\vec{E}\cdot d\vec{l}$$

In rectangular coordinates, $d\vec{l} = \vec{a}_x\Delta x + \vec{a}_y\Delta y + \vec{a}_z\Delta z$. Thus, (2-63) can be simplified using the definition of the dot product in (2-13).

$$\begin{aligned} dW &= q\left(\frac{\partial \Phi}{\partial x}\Delta x + \frac{\partial \Phi}{\partial y}\Delta y + \frac{\partial \Phi}{\partial x}\Delta z\right) = -q\vec{E}\cdot d\vec{l} \\ &= q\left(\vec{a}_x\frac{\partial \Phi}{\partial x} + \vec{a}_y\frac{\partial \Phi}{\partial y} + \vec{a}_z\frac{\partial \Phi}{\partial z}\right)\cdot(\vec{a}_x\Delta x + \vec{a}_y\Delta y + \vec{a}_z\Delta z) \\ &= -q\vec{E}\cdot(\vec{a}_x\Delta x + \vec{a}_y\Delta y + \vec{a}_z\Delta z) \\ \vec{E} &= -\left(\vec{a}_x\frac{\partial \Phi}{\partial x} + \vec{a}_y\frac{\partial \Phi}{\partial y} + \vec{a}_z\frac{\partial \Phi}{\partial z}\right) \end{aligned} \tag{2-64}$$

Note that the form of the last term in (2-64) is identical to (2-21), which is the gradient, which gives a very useful relation between the electric field and the electrostatic potential:

$$\vec{E} = -\nabla\Phi \tag{2-65}$$

Also note that (2-65) can be derived using vector identities, which is significantly simpler but does not provide any intuition. The alternative derivation is shown here because we employ a similar technique when studying magnetostatics in the next section.

For an electrostatic field, Ampère's law is reduced to $\nabla \times \vec{E} = 0$ because the field does not vary with time. For any differentiable scalar function, the following vector identity holds true (from Appendix A):

$$\nabla \times \nabla \psi = 0$$

Therefore, since $\nabla \times \vec{E} = 0$, $\vec{E}$ must be derivable from the gradient of a scalar function. Since (2-61) shows a relationship between the electrostatic potential and the electric field, a leap of logic says that the scalar function in the vector identity must be the electrostatic potential:

$$\nabla \times \vec{E} = \nabla \times (-\nabla \Phi) = 0 \tag{2-66}$$

Equation (2-65), known as the *electrostatic scalar potential*, is used often when solving electrostatic problems such as transmission-line impedance or calculating the effective dielectric constant of a microstrip, as we demonstrate in Chapter 3.

2.4.2 Energy in an Electric Field

To calculate the energy stored in an electric field, it is necessary to begin with a stationary charge (q_1) in free space that is infinitely far way from any other charges. It takes no work to move the first charge into position ($W_1 = 0$) because there are no other nearby charges to provide electrostatic repulsion. Then we calculate how much work it takes to bring another charge (q_2) into the vicinity of the first charge using (2-61) and (2-62):

$$W_2 = q_2 \Phi_{12} = \frac{q_1 q_2}{4\pi \varepsilon_0 r_{12}} \tag{2-67}$$

If we bring another charge, q_3, into the vicinity of q_1 and q_2, the work is calculated as

$$W_3 = q_3(\Phi_{13} + \Phi_{23}) = q_3 \frac{1}{4\pi \varepsilon_0} \left(\frac{q_1}{r_{13}} + \frac{q_2}{r_{23}} \right)$$

Thus, the total work done to bring the three charges together is

$$\begin{aligned} W_{\text{tot}} &= W_1 + W_2 + W_3 = 0 + \frac{q_1 q_2}{4\pi \varepsilon_0 r_{12}} + \frac{q_3}{4\pi \varepsilon_0} \left(\frac{q_1}{r_{13}} + \frac{q_2}{r_{23}} \right) \\ &= \frac{1}{4\pi \varepsilon_0} \left(\frac{q_1 q_2}{r_{12}} + \frac{q_1 q_3}{r_{13}} + \frac{q_2 q_3}{r_{23}} \right) \end{aligned}$$

which can be generalized into the following double summation to account for n charges:

$$W_{\text{tot}} = \frac{1}{4\pi\varepsilon_0}\sum_{i=1}^{n}\sum_{\substack{j=1\\j>i}}^{n}\frac{q_i q_j}{r_{ij}} \tag{2-68}$$

The limit on the j term exists to ensure that terms are not counted twice. For example, without the limit the term Φ_{12} would be counted twice because $\Phi_{12} = \Phi_{21}$, which simply means that it takes the same amount of work to move q_1 into the vicinity of q_2 as it does to move q_2 into the vicinity of q_1. Equation (2-68) can be simplified by allowing the terms to be counted twice and dividing by 2 to compensate for the double counting of terms:

$$W_{\text{tot}} = \frac{1}{2}\left(\frac{1}{4\pi\varepsilon_0}\sum_{i=1}^{n}\sum_{\substack{j=1\\j\neq i}}^{n}\frac{q_i q_j}{r_{ij}}\right) = \frac{1}{2}\sum_{i=1}^{n}\sum_{\substack{j=1\\j\neq i}}^{n} q_i\Phi_{ij} \tag{2-69}$$

Equation (2-69) is the work it takes to assemble n charges. It is interesting to consider where the energy is stored in an accumulation of charges. It is analogous to a mechanical system of a compressed spring with a weight on each end. If the weights are forced together, the energy is stored in the stressed state of the spring. Therefore, just like the spring example, the charges, which will tend to repel each other, will have stored energy that is a function of the proximity of the charges and the properties of an electric field.

To calculate the energy stored in a continuous charge distribution, and thus an electric field, it is more convenient to express the charge in terms of the unit volume, the potential in terms of a continuous function, and to take the limit as $n \to \infty$, which allows us to write (2-69) in terms of an integral,

$$W_{\text{tot}} = W_e = \tfrac{1}{2}\int_V \rho(r)\Phi(r)\,dV \qquad \text{joules} \tag{2-70}$$

where ρ (r) is the charge density in units of C/m^3.

It is useful to express (2-70) in terms of the electric field. To do this, Gauss's law is used to express the charge density in terms of the electric field:

$$\nabla\cdot\vec{D} = \nabla\cdot\varepsilon\vec{E} = \rho \to W_e = \tfrac{1}{2}\int_V (\nabla\cdot\varepsilon\vec{E})\Phi(r)\,dV \tag{2-71}$$

This equation can be simplified using the flowing vector identity (Appendix A):

$$\nabla\cdot\psi\vec{a} = \vec{a}\cdot\nabla\psi + \psi(\nabla\cdot\vec{a}) \to \psi(\nabla\cdot\vec{a}) = \nabla\cdot\psi\vec{a} - \vec{a}\cdot\nabla\psi$$

The identity can be rewritten in terms of the electrostatic vector potential, substituting $\psi = \Phi$ and $\vec{a} = \varepsilon\vec{E}$:

$$W_e = \frac{\varepsilon}{2}\int_V (\nabla \cdot \Phi E - E \cdot \nabla\Phi)\, dV$$

Since $\vec{E} = -\nabla\Phi$, the equation can be simplified further:

$$W_e = \frac{\varepsilon}{2}\int_V (\nabla \cdot \Phi\vec{E} + \vec{E} \cdot \vec{E})\, dV \tag{2-72}$$

The divergence theorem of vector calculus states (Appendix A) that

$$\int_V (\nabla \cdot \vec{F})\, dV = \oint_S \vec{F} \cdot d\vec{s}$$

allowing further simplification of (2-72):

$$W_e = \frac{\varepsilon}{2}\oint_S \Phi\vec{E} \cdot \vec{n}\, ds + \frac{\varepsilon}{2}\int_V (\vec{E} \cdot \vec{E})\, dV \tag{2-73}$$

where $\vec{n}$ is a unit vector normal to the surface.

Equation (2-73) describes the total energy in an electric field that is induced by a volume of charges. If we expand the volume of integration to include all space, any contribution of (2-73) outside the charge distribution will contribute nothing to the work done since there are no charges in that space. Also note that if the volume of integration is chosen to be infinity, the surface integral disappears. To understand why, remember that the surface integral sums all the contributions evaluated at the surface. Since $\Phi \propto 1/r$ [equation (2-62)], $E \propto 1/r^2$ [equation (2-60)], and $ds \propto r^2$, the limit of $\Phi\vec{E} \cdot \vec{n} ds$ is proportional to $(1/r)(1/r^2)r^2$, whose limit goes to zero when r is infinity and the surface integral disappears. The volume integral includes contributions over the entire volume, not only on the surface. Subsequently, equation (2-73) can be reduced to (2-74), which is the work done by accumulating charges to create an electric field:

$$W_e = \frac{\varepsilon}{2}\int_{\text{all space}} (E^2)\, dV \qquad \text{joules} \tag{2-74}$$

This leads to the definition of the *volume energy density*, which expresses the stored energy in a charge distribution in terms of the electric field:

$$w_e = \frac{\varepsilon}{2}E^2 \qquad \text{joules/m}^3 \tag{2-75}$$

2.4.3 Capacitance

In circuit terms, the quantity associated with storing energy in an electric field is capacitance. To define capacitance, imagine two conductors, with a charge of $+Q$ on one and of $-Q$ on the other. If we assume that the voltage is constant over each conductor, the potential difference (voltage) between them is calculated as

$$v(b) - v(a) = -\int_a^b \vec{E} \cdot d\vec{l} \qquad \text{V} \tag{2-58}$$

We show that $\vec{E}$ is proportional to Q:

$$\vec{E} = \vec{a}_r E_r = \frac{Q}{4\pi \varepsilon_0 r^2} \qquad \text{V/m} \tag{2-60}$$

Since $\vec{E}$ is proportional to both Q and v, we can define a constant of proportionality that relates Q and v. The constant of proportionality is defined to be the *capacitance*:

$$C \equiv \frac{Q}{v} \qquad \text{farads} \tag{2-76}$$

where Q is the total charge in coulombs and v is the voltage potential between the conductors, given in units of farads, defined as 1 coulomb per volt. Capacitance depends purely on the geometry of the structures and the value of the dielectric permittivity. Note that v is defined as the potential of the positive conductor minus the negative conductor and that Q is the charge on the positive conductor. Therefore, *capacitance is always a positive value*.

Example 2-3 Consider the case where two conductive plates of area A are oriented parallel to each other separated by a distance d. Assume that we place a charge of $+Q$ on the top plate and $-Q$ on the bottom plate and assume that the charges will spread out evenly (a reasonable assumption, assuming a good conductor). Then the surface charge density becomes

$\rho = Q/A \quad (\text{C/m}^2)$. Calculate the capacitance.

SOLUTION Using the integral form of Gauss's law (2-59), we can calculate the electric field:

$$\oint_s \varepsilon \vec{E} \cdot d\vec{s} = \int_V \rho \, dV$$

where dV in this case refers to the volume. Since we are considering the charge distribution on a surface, $d\vec{s} = dV = \vec{n}A$ (where $\vec{n}$ is the unit normal vector to the plate), we can write the electric field in terms of the area and dielectric permittivity:

$$\varepsilon E A = \frac{Q}{A} A \rightarrow E = \frac{Q}{\varepsilon A}$$

The voltage is evaluated with equation (2-58), where $dl = x(a) - x(b) = d$, which is the distance between the plates:

$$v = \int_a^b \vec{E} \cdot d\vec{l} = \frac{Q}{\varepsilon A} d$$

Therefore, the capacitance between two parallel plates is

$$C = \frac{Q}{v} = \frac{Q}{(Q/\varepsilon A)\, d} = \frac{\varepsilon A}{d} \qquad \text{farads}$$

where A is the area of the parallel plates, v the voltage, and d the distance between them.

2.4.4 Energy Stored in a Capacitor

The process of storing energy in a capacitor involves electric charges of equal magnitude, but opposite polarity, building up on each plate. As long as the capacitor holds a charge, it is storing energy. To calculate how much energy is stored in a capacitor, consider how much energy it would take to transport a single charge from the positive plate to the negative plate. From equation (2-58) we know that voltage is the work done to move a charge from point a to b,

$$\left[\frac{W}{q}\right]_{a\to b} = v$$

and capacitance is defined as

$$C \equiv \frac{Q}{v} \qquad \text{farads} \tag{2-76}$$

Therefore, the amount of work needed to move one charge q of the total charge Q from plate a to plate b is

$$dW = v\, dq = \frac{q}{C}\, dq$$

To calculate the total work done to charge up the capacitor to a value of Q, all charges must be moved:

$$W = \int_0^Q \frac{q}{C}\, dq = \frac{1}{2}\frac{Q^2}{C}$$

However, from (2-69), $Q = Cv$. Therefore, the energy stored in a capacitor with a final potential of v is

$$W = \frac{1}{2} C v^2 \qquad \text{joules} \tag{2-77}$$

2.5 MAGNETOSTATICS

In Section 2.4 we discussed the problem of classical electrostatics, where we defined the electric field in terms of the force that a collection of charges exert on each other. In the case of electrostatics, we considered only cases where the charges are at rest. Now it is time to consider the forces that exist between charges in motion.

To begin this discussion, let's consider an experiment that most people performed in high school physics class. If you recall, direct current (dc) (from a battery) driving through a coil will produce an electromagnet. The magnetic field produced by such a configuration can be descried by Ampère's law, $\nabla \times \vec{H} = \vec{J}$. Note that the time dependence of the electric field $(\partial D/\partial t)$ has been eliminated because we are considering only a dc flow. Ampère's law tells us that a steady-state current $\vec{J}$ will induce a magnetic field $\vec{H}$ that circulates around the wire. As described in Example 2-2, the direction of the circulation can be determined using the right-hand rule. If the thumb points in the direction of the current flow, the fingers of the right hand will curl around in the direction of the magnetic field. Subsequently, it is easy to imagine the form of the magnetic field from a single loop of current in our electromagnet, as shown in Figure 2-15a.

Now consider a tiny elemental loop of current circulating around a point that will induce a small magnetic field as shown in Figure 2-15b. This small current loop will produce a small magnetic field that is analogous to an electric charge. In fact, historically, scientists initially speculated that there was a magnetic charge analogous to the electric charge described earlier. However, experimental evidence suggests overwhelmingly that magnetic charges do not exist. Magnetic fields are not generated by the forces that magnetic charges exert on each other; rather, magnetic fields are generated by current loops.

Similar to our description of the forces between charges in electrostatics, consider an isolated tiny current loop (l_0) which will induce a magnetic field and behave like a small electromagnet. Now, bring another tiny current loop (l_1) from infinity into the magnetic field of l_0. If the orientations of the current loops are similar, it will take work to push them together because magnets exert forces on one another similar to electric charges. Like poles will repel each other, and unlike poles will attract. Each "electromagnet" has its own north and south

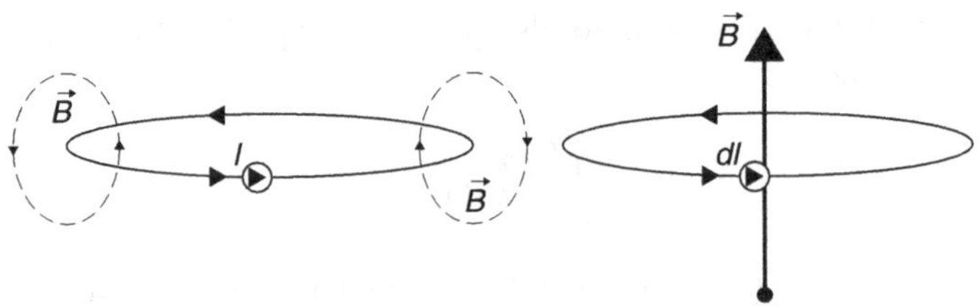

Figure 2-15 (a) Magnetic field generated by a loop of current; (b) elemental current loop analogous to an electric charge.

poles. It is interesting to note that the fundamental source of the magnetic field is the moving charge Q that constitutes the steady-state current. Subsequently, when l_1 is moved into the proximity of l_0, the force induced between the two electromagnets is caused by the charge (Q) of l_1 moving in the magnetic field of l_0 and is described by the Lorenz force law:

$$\vec{F}_m = Q(\vec{v} \times \vec{B}) \tag{2-78a}$$

A charge moving in the presence of both an electric and a magnetic field produces a force calculated as

$$\vec{F}_m = Q(\vec{E} + \vec{v} \times \vec{B}) \tag{2-78b}$$

The implications of (2-78) are that the force is perpendicular to both the velocity $\vec{v}$ of the charge q and the magnetic field $\vec{B}$. The magnitude of the force is $F = qvB\sin\theta$, where θ is the angle between the velocity vector and the magnetic field. Because $\sin(0) = 0$, this implies that the magnetic force on a stationary charge or a charge moving parallel to the magnetic field is zero.

From the force relationship in (2-78) it can be deduced that the units of magnetic field are newton · seconds/coulomb · meter or newtons/ampere · meter. This unit is named the tesla. It is a large unit, and the smaller unit gauss is used for small fields such as Earth's magnetic field. A tesla is 10,000 G. Earth's magnetic field is on the order of 0.5 G.

To make this concept more apparent, the force can be defined in terms of the current, which is the flow of 1 C of charge per second:

$$1\text{A} = 1\text{C/s} \tag{2-79}$$

If we consider the current flowing along on a differential slice of a wire (dl), we can write (2-78) in terms of the current. Since Qv has units of C (m/s), which is the same as A · m, Qv can be simplified to $I\,dl$:

$$Qv = Q\frac{\Delta l}{s} = Q\frac{dl}{s} \rightarrow I dl$$

This allows us to write the force in terms of both the current and the magnetic field:

$$\vec{F}_m = \int (\vec{I} \times \vec{B})\, dl \tag{2-80}$$

Equation (2-80) says that the force caused by the magnetic field will be perpendicular to the current flow and the magnetic field.

When deriving the energy in an electric field in Section 2.4, we calculated the amount of work done against the electric field to bring an accumulation of charges together. However, when calculating the energy in the magnetic field, a different approach must be taken because of a unique feature of equation (2-78).

If a charge Q moves an amount $d\vec{l} = \vec{v}\,dt$ in a magnetic field, the amount of work done is

$$dW_m = \vec{F}_m \cdot d\vec{l} = Q(\vec{v} \times \vec{B}) \cdot \vec{v}\,dt = 0 \qquad (2\text{-}81)$$

Note that $\vec{v} \times \vec{B}$ is perpendicular to the flow of the current, which follows the path of $d\vec{l}$. In other words, *the work done by the magnetic field is zero* because the force is perpendicular to the moving charge. Therefore, the magnetic field can change the direction of the moving particle, but it cannot speed it up or slow it down. This concept may be confusing, especially when considering the simple electromagnets that we all played with in high school, because we all know that we can use an electromagnet to pick up a paper clip. Since we are moving the mass of the paper clip against Earth's gravitational field, we know that work is being done. However, if the work is not being done by the magnetic field, what is doing the work? The answer is demonstrated in the following example.

Example 2-4 Consider a long wire carrying current I_1 in the presence of a rigid rectangular loop carrying current I_2, as shown in Figure 2-16. The long wire will generate a magnetic field as calculated in Example 2-2:

$$B_1 = \frac{I_1 \mu_0}{2\pi r}$$

Calculate the magnetic force.

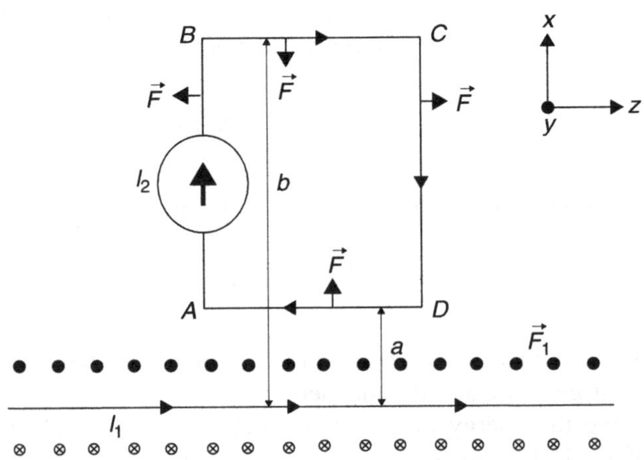

Figure 2-16 Forces generated on a wire loop in the vicinity of a magnetic field generated from a wire.

SOLUTION According to equation (2-80), the force exerted on the loop carrying current I_2 in the presence of $\vec{B}_1$ is

$$\vec{F}_{\text{loop}} = \int (\vec{I}_2 \times \vec{B}_1)\, dl = I_2 \frac{I_1 \mu_0}{2\pi} \left(\int_A^B \vec{a}_x\, dx \times \frac{\vec{a}_y}{x} + \int_B^C \vec{a}_z\, dz \times \frac{\vec{a}_y}{b} + \int_C^D -\vec{a}_x\, dx \times \frac{\vec{a}_y}{x} + \int_D^A -\vec{a}_z\, dz \times \frac{\vec{a}_y}{a} \right)$$

Note that the segments AB and CD are equal but opposite, so they cancel.

$$\vec{F}_{\text{loop}} = I_2 \frac{I_1 \mu_0}{2\pi} \left(\int_B^C \vec{a}_z\, dz \times \frac{\vec{a}_y}{b} + \int_D^A -\vec{a}_z\, dz \times \frac{\vec{a}_y}{a} \right)$$

From the right-hand rule, the cross products are as follows:

$$\vec{a}_z \times \vec{a}_y = -\vec{a}_x$$
$$-\vec{a}_z \times \vec{a}_y = \vec{a}_x$$

Therefore, the force is reduced to

$$\vec{F}_{\text{loop}} = I_2 \frac{I_1 \mu_0}{2\pi} \left[-\vec{a}_x \frac{1}{b}(B - C) + \vec{a}_x \frac{1}{a}(D - A) \right]$$

Since segments $BC = DA$, we can call this length d:

$$\vec{F}_{\text{loop}} = \vec{a}_x I_2 \frac{I_1 \mu_0\, d}{2\pi} \left(\frac{1}{a} - \frac{1}{b} \right)$$

Therefore, the loop will be pushed away from the wire in the direction of $\vec{a}_x$.

Note that the magnetic force has caused the wire loop to move. Since work is force × distance, it would be easy to conclude that the magnetic force has performed work. However, equation (2-81) explicitly states that the magnetic field can do no work. What is performing work? To answer this question, consider the force vector on a single segment of the loop as soon as it begins to move. Remember that the force is perpendicular to the direction of the current flow, and the current flow is defined by the movement of charge. When the loop moves, the direction of the current flow I_2 will be altered. To understand this, Figure 2-17 shows that the direction in which a single charge in the loop will travel when the loop is moved in the $+x$-direction. Instead of moving from right to left, it is moved up and to the left because the loop is moving in the $+x$-direction. This will cause the force vector, which must remain perpendicular to the current flow, to tilt to the right, as shown in Figure 2-17. When the force vector tilts, the component $\vec{a}_z F_z$ opposes the charge flow of the current I_2 in the loop. For I_2 to remain constant, the source of the current must overcome this force. *This leads us to the conclusion that the power source is performing the work!* The magnetic field simply alters the direction of the force vector.

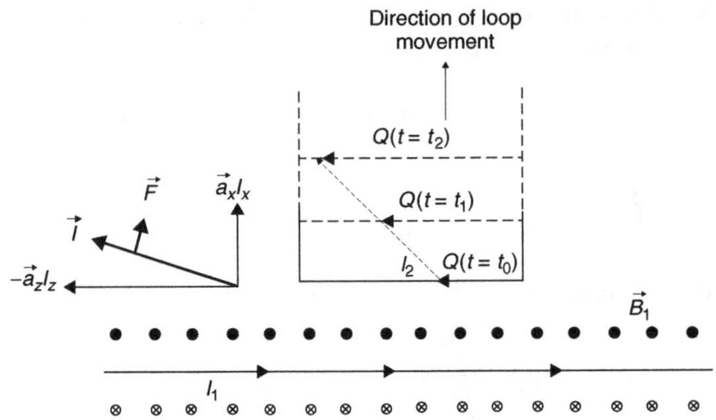

Figure 2-17 Forces generated on a wire loop in the vicinity of a magnetic field generated from a wire.

2.5.1 Magnetic Vector Potential

Because magnetic fields can do no work, we cannot calculate the energy stored in a magnetic field in the same manner as we did for the electric field (i.e., by calculating the work done to accumulate a distribution of elemental current loops), so another approach is needed. In this section we develop the concept of the *magnetic vector potential*. The magnetic vector potential is used to calculate inductance, which is in turn used to calculate the energy stored in a magnetic field.

The basic laws that rule magnetostatics are the time-invariant forms of Ampère's and Gauss's laws for magnetism:

$$\nabla \times \vec{H} = \vec{J} \tag{2-82a}$$

$$\nabla \cdot \vec{B} = 0 \tag{2-82b}$$

If the following vector identity is applied (from Appendix A),

$$\nabla \cdot (\nabla \times \vec{A}) = 0$$

it implies that $\vec{B}$ can be written in terms of the curl of a vector $\vec{A}$. This is called the *magnetic vector potential*, which sometimes helps to simplify calculations such as inductance:

$$\vec{B} = \nabla \times \vec{A} \tag{2-83}$$

To calculate the form of $\vec{A}$, it is first necessary to introduce the most basic law of magnetostatics, the *Biot–Savart law*, which describes how the $\vec{B}$ field at a given point is produced by the moving charges in the vicinity of that point. During application of this law, we consider currents that are either static or very slowly varying with time. The Biot–Savart law is given by [Inan and Inan, 1998]

$$d\vec{B}_p = \frac{\mu_0 I d\vec{l} \times \vec{R}}{4\pi R^2} \tag{2-84}$$

where I is the current, $\vec{R}$ a unit vector pointing from the location of the differential current element $Id\vec{l}$ to the point P, and $R = |\vec{r} - \vec{r}'|$ the distance between the current element and point P. Note that a primed quantity represents the position vector or the coordinates of the source points and the unprimed quantities represent the position vector or the coordinates of the point where $\vec{B}$ is being evaluated. Note that the cross product in (2-84) indicates that $d\vec{B}$ is perpendicular to both $Id\vec{l}'$ and $\vec{R}$, with the orientation being described by the right-hand rule. Also note that the field generated by $Id\vec{l}'$ will fall off with the square of the distance.

Note that the current element $Id\vec{l}'$ is a small part of a closed current loop and that an arbitrarily shaped loop can be constructed by the superposition of many such closed elemental loops of current, as shown in Figure 2-18. Subsequently, the Biot–Savart law can be written in terms of an integral:

$$\vec{B} = \frac{\mu_0}{4\pi} \oint_C \frac{Id\vec{l}' \times \vec{R}}{R^2} \tag{2-85}$$

Careful observation of (2-85) allows us to calculate the form of the magnetic vector potential $\vec{A}$. Equation (2-83) states that the magnetic field is the curl of $\vec{A}$. The trick is to manipulate (2-85) into the form of a curl so that $\vec{A}$ can be found. The right-hand term of (2-85) has a term that can be equated to the gradient of $1/R$ in spherical coordinates (see Appendix A):

$$\nabla\left(\frac{1}{R}\right) = \vec{R}\frac{\partial}{\partial R}\left(\frac{1}{R}\right) + \vec{\theta}\frac{1}{R}\frac{\partial}{\partial \theta}\left(\frac{1}{R}\right) + \vec{\phi}\frac{1}{R\sin\theta}\frac{\partial}{\partial \phi}\left(\frac{1}{R}\right) = -\vec{R}\frac{1}{R^2} \tag{2-86}$$

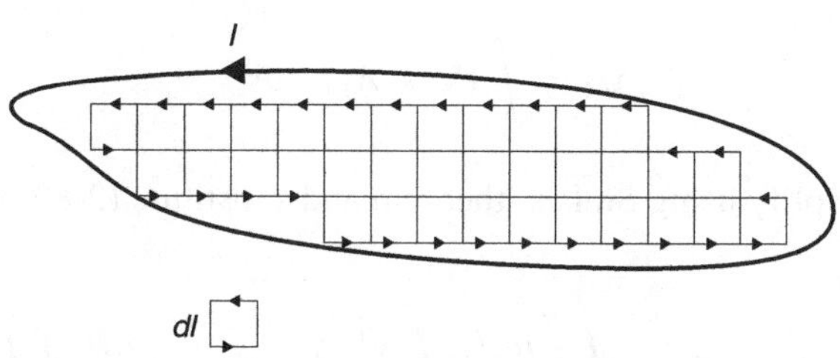

Figure 2-18 Any current loop can be constructed from many elemental current loops dI.

Substituting (2-86) into (2-85) gives

$$\vec{B} = \frac{\mu_0}{4\pi} \oint_C -I d\vec{l}' \times \nabla \left(\frac{1}{R} \right) \tag{2-87}$$

The del operator (∇) does not operate on source variables and therefore does not affect $Id\vec{l}'$. Therefore, it is acceptable to rewrite (2-87) in the form of a curl:

$$\vec{B} = \nabla \times \frac{\mu_0 I}{4\pi} \oint_C \frac{d\vec{l}'}{R} \tag{2-88}$$

Also note that the negative sign has been eliminated because it indicates current direction, which is comprehended by the vector $Id\vec{l}'$. Comparing (2-88) to the definition of the magnetic vector potential in (2-83), we can deduce the form of $\vec{A}$:

$$\vec{A} = \frac{\mu_0 I}{4\pi} \oint_C \frac{d\vec{l}}{R} \tag{2-89}$$

2.5.2 Inductance

Suppose that two current loops are in close proximity to each other as shown in Figure 2-19. If a steady-state current (I_1) is flowing in loop 1, it produces a magnetic field B_1, as predicted by (2-85). If some of the magnetic field lines pass though loop 2, the flux passing through loop 2 is calculated using the form of equation (2-17):

$$\psi_2 = \int \vec{B}_1 \cdot d\vec{s}_2 \tag{2-90}$$

Note that the flux in loop 2 is proportional to $\vec{B}_1$ and is therefore also proportional to I_1. This allows us to define a constant of proportionality, more commonly known as the *mutual inductance*:

$$L_{21} \equiv \frac{\psi_2}{I_1} \tag{2-91}$$

Substituting the (2-83) into (2-90) gives the flux passing through loop 2 in terms of the magnetic vector potential, where s_2 is the surface enclosed by loop 2:

$$\psi_2 = \int (\nabla \times A_1) \cdot d\vec{s}_2 \tag{2-92}$$

Now we can simplify using Stokes' theorem and substitute (2-89) for $\vec{A}$ [Jackson, 1999]:

$$\psi_2 = \int (\nabla \times A_1) \cdot d\vec{s}_2 = \oint \left(\frac{\mu_0 I_1}{4\pi} \oint_C \frac{d\vec{l}_1}{R} \right) \cdot d\vec{l}_2 = \frac{\mu_0 I_1}{4\pi} \oint \left(\oint \frac{d\vec{l}}{R} \right) \cdot d\vec{l}_2 \tag{2-93}$$

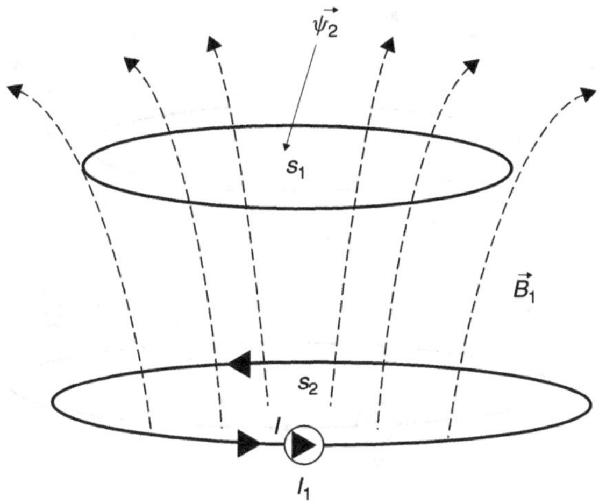

Figure 2-19 Mutual inductance caused by magnetic flux from loop 1 passing though loop 2.

Since $\psi_2 = L_{21}I_1$ [from (2-91)], the mutual inductance between loops 1 and 2 is determined by dividing (2-93) by the current in loop 1:

$$L_{21} = \frac{\mu_0}{4\pi} \oint \oint \frac{d\vec{l}_1 \cdot d\vec{l}_2}{R} \tag{2-94}$$

Equation (2-94), called the *Neumann formula*, involves integration around both loops 1 and 2. Note two very important concepts that can be derived from equation (2-94).

1. The mutual inductance is a function of the size, shape, and distance between the two loops.
2. The mutual inductance from loop 1 to loop 2 (L_{21}) is identical to the mutual inductance from loop 2 to loop 1 (L_{12}).

If we consider the implications of Faraday's law ($\nabla \times \vec{E} + \partial \vec{B}/\partial t = 0$), another very important concept used throughout signal integrity can be surmised. For simplification, let's assume that we only have a component of the electric field in the x-direction and it is propagating along z so that $\vec{E} = \vec{a}_x E_x(z, t)$. Faraday's law is then reduced to

$$\nabla \times \vec{E} + \frac{\partial \vec{B}}{\partial t} = \vec{a}_y \frac{\partial E_x}{\partial z} + \frac{\partial \vec{B}}{\partial t} = 0 \tag{2-95}$$

Equation (2-95) says that a time-varying magnetic flux $\vec{B}$ generated from loop 1 and passing through loop 2 will induce an electric field and subsequently a

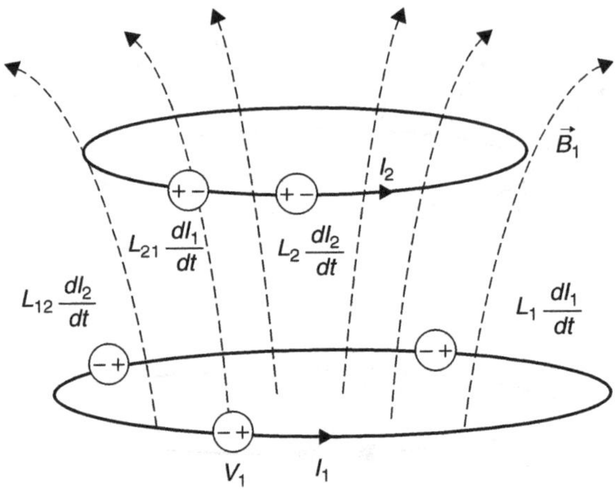

Figure 2-20 Voltage sources induced by the time-varying currents in loop 1.

voltage in loop 2. Faraday's law can be rewritten in terms of circuit parameters:

$$v_2 = -\frac{d\psi_2}{dt} = -L_{21}\frac{dI_1}{dt} \tag{2-96}$$

Therefore, every time you change the current in loop 1, an electromotive force (i.e., a voltage) is induced in loop 2, which causes current to flow. In addition to inducing a voltage on loops in the vicinity, changing current in loop 1 will change the magnetic flux flowing through itself and, consequently, induce a voltage in itself. This is called the *self-inductance:*

$$L_{11} \equiv \frac{\psi_1}{I_1} \tag{2-97}$$

Similar to the mutual inductance, if the current changes a voltage is induced in the loop:

$$v = -L_{11}\frac{dI_1}{dt} \tag{2-98}$$

By observing (2-98), the units of inductance are determined to be volt · seconds per ampere, also known as henries. Figure 2-20 shows the voltage elements superimposed on the magnetically coupled loops corresponding to the mutual and self-inductance values calculated with (2-96) and (2-98). Note that the voltage sources were chosen to be positive and the negative sign is accounted for with the direction of current flow.

Example 2-5 Equivalent Circuit of a Magnetically Coupled System The magnetically coupled circuit can be understood more easily if conventional circuit theory is applied. Figure 2-21 shows the circuit for the magnetically coupled

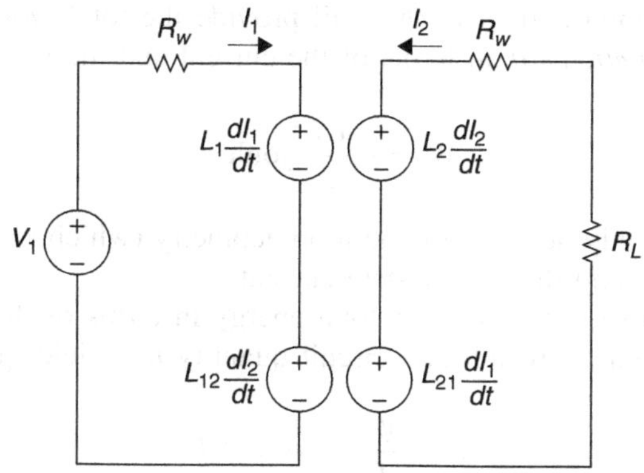

Figure 2-21 Circuit for Example 2-5, showing magnetically coupled loops.

loops, assuming that the second loop is terminated in a resistor R_L and each wire has a resistance of R_w. Kirchhoff's voltage relations can be written for each loop:

Loop 1:

$$v_1 = R_w I_1 + L_1 \frac{dI_1}{dt} + L_{12}\frac{dI_2}{dt}$$

Loop 2:

$$(R_w + R_L)I_2 + L_2\frac{dI_2}{dt} + L_{21}\frac{dI_1}{dt} = 0$$

Note that inductance is always a positive quantity that can be compared to a mass in a mechanical system. Large masses are difficult to move, making it difficult to accelerate in any direction. Similarly, the greater the inductance, the more difficult it is to change the current because of the back emf (back voltage) generated in a direction to oppose the current, which is enforced by the negative sign in (2-98). This is called *Lenz's law*.

2.5.3 Energy in a Magnetic Field

If a circuit has a finite amount of inductance, it takes energy to make a current flow because it requires work to overcome the back EMF voltage described by Lenz's law. The work done on a charge against this back EMF is $-v$, from (2-98). The negative sign dictates that it is work being done against the EMF, not work done by the EMF. Since current is charge flow per unit time, the work done per unit time is derived from (2-58) and given by

$$W = -vq \rightarrow \frac{dW}{dt} = -vI = -\left(L\frac{dI}{dt}\right)I \tag{2-99}$$

Integrating the second half of (2-99) will provide the total work done and thus *energy in the magnetic field* in terms of the current and the inductance:

$$W = -\frac{LI^2}{2} \text{ joules} \tag{2-100}$$

Note that (2-100) only depends on the loop geometry (which is where the inductance comes from) and the steady-state current.

It is often useful to represent the total energy in terms of the magnetic field. To translate inductance, which is a circuit quantity to a field quantity, equation (2-97) is used.

$$L = \frac{\psi}{I} \rightarrow \psi = LI \tag{2-101}$$

Recalling equations (2-90) and (2-92) and invoking Stoke's theorem as in (2-93), the flux can be equated as follows:

$$\psi = LI = \int \vec{B} \cdot d\vec{s} = \int (\nabla \times \vec{A}) \cdot d\vec{s} = \oint \vec{A} \cdot d\vec{l} \tag{2-102}$$

Therefore, the work calculated in (2-100) can be rewritten

$$W = \tfrac{1}{2}LI^2 = \tfrac{1}{2}I \oint \vec{A} \cdot d\vec{l} = \tfrac{1}{2} \oint (\vec{A} \cdot \vec{I})\, dl \tag{2-103}$$

Expressing (2-103) in a volume integral allows us to use Ampère's law ($\nabla \times \vec{B}/\mu_0 = \vec{J}$) in place of I:

$$W = \frac{1}{2} \oint (\vec{A} \cdot \vec{I})\, dl = \frac{1}{2} \int_V (\vec{A} \cdot \vec{J})\, dV = \frac{1}{2} \int_V \left(\vec{A} \cdot \nabla \times \frac{\vec{B}}{\mu_0}\right) dV \tag{2-104}$$

Conjuring up more mathematical trickery from Jackson [1999] allows us to rewrite (2-104) in a simplified form using the following vector identity:

$$\nabla \cdot (\vec{A} \times \vec{B}) = \vec{B} \cdot (\nabla \times \vec{A}) - \vec{A} \cdot (\nabla \times \vec{B}) \tag{2-105}$$

Rearranging the terms and utilizing the definition of the vector magnetic potential,

$$\begin{aligned} \vec{A} \cdot (\nabla \times \vec{B}) &= \vec{B} \cdot (\nabla \times \vec{A}) - \nabla \cdot (\vec{A} \times \vec{B}) \\ \nabla \times \vec{A} &= \vec{B} \\ \vec{A} \cdot (\nabla \times \vec{B}) &= \vec{B} \cdot \vec{B} - \nabla \cdot (\vec{A} \times \vec{B}) \end{aligned}$$

allows the work to be expressed in terms of the magnetic fields:

$$W_m = \frac{1}{2\mu_0} \int_V [\vec{B} \cdot \vec{B} - \nabla \cdot (\vec{A} \times \vec{B})]\, dV$$

which can be simplified:

$$W_m = \frac{1}{2\mu_0} \int_{\text{all space}} B^2 \, dV \tag{2-106}$$

This leads to the definition of the *volume energy density*, which expresses the stored energy in a magnetic field:

$$w_m = \frac{B^2}{2\mu_0} \quad \text{joules/m}^3 \tag{2-107}$$

2.6 POWER FLOW AND THE POYNTING VECTOR

As electromagnetic waves propagate through space, they carry power. The goal of this section is to derive a relationship between the electric and the magnetic field vectors and the power transferred. To begin this derivation, the energy stored in both the electric and magnetic fields must be quantified within a volume of space. Consider Figure 2-22, which depicts a cubic volume of space with an electromagnetic plane wave propagating through it. To quantify the total power, which is the transfer of energy over time, propagating through the cube, all sources of energy must be accounted for. The power balance equation can be expressed as

$$P_A = P_S - P_L - P_{\text{EM}} \tag{2-108}$$

where P_A is the power flowing through the surface of the far end of the cube of space with area A, as shown in Figure 2-22, P_S represents any sources of power within the cube, P_L represents the losses within the cube that dissipate power as heat (such as resistive losses), and P_{EM} represent the power contained within the electromagnetic waves that propagate into the cube. If we assume a source-free and loss-free medium, the power flowing into the cube must equal the power flowing out of the cube, where the minus sign convention is chosen because the power is flowing out of the surface:

$$P_A = -P_{\text{EM}} \tag{2-109}$$

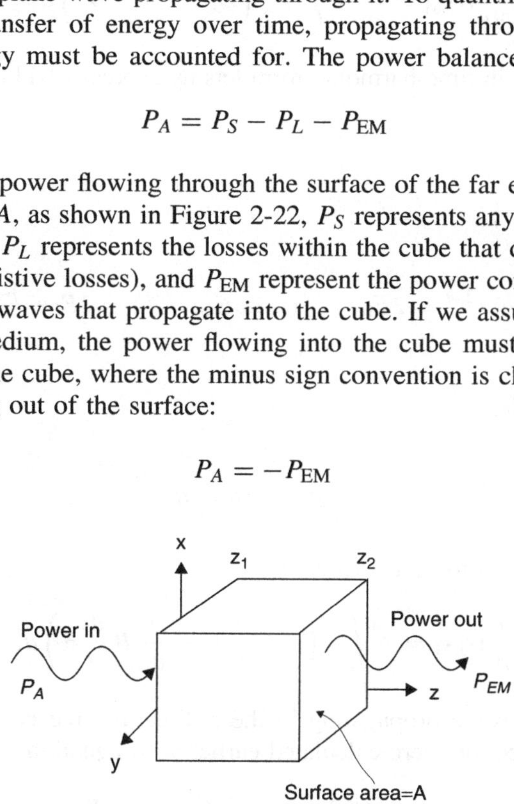

Figure 2-22 Volume of space used to calculate power flow and the Poynting vector.

To quantify (2-109) in terms of the electric and magnetic fields, we begin with the volume energy densities for an electric and a magnetic field, which were derived earlier and listed here for convenience.

$$w_e = \frac{\varepsilon}{2}E^2 \qquad \text{joules/m}^3 \tag{2-75}$$

$$w_m = \frac{B^2}{2\mu_0} \qquad \text{joules/m}^3 \tag{2-107}$$

Integrating the sum of (2-75) and (2-107) over the volume of the cube will give total energy:

$$W_{\text{EM}} = A\int_{z_1}^{z_2} \frac{1}{2}\left(\varepsilon_0 E^2 + \frac{1}{\mu_0}B^2\right) dz \tag{2-110}$$

Since power is energy transfer per unit time interval, P_{EM} is calculated by taking the time derivative:

$$\frac{\partial}{\partial t}W_{\text{EM}} = \frac{\partial}{\partial t}A\int_{z_1}^{z_2} \frac{1}{2}\left(\varepsilon_0 E^2 + \frac{1}{\mu_0}B^2\right) dz \tag{2-111}$$

Defining the fields in time-harmonic form lets us express (2-111) in terms of the partial time derivatives:

$$E = E_0 e^{-j\omega t}$$

$$\frac{\partial}{\partial t}E = -j\omega E_0 e^{-j\omega t}$$

$$\frac{\partial}{\partial t}E^2 = -j2\omega E_0^2 e^{-j2\omega t} = 2E_0 e^{-j\omega t}(-j\omega E_0 e^{-j\omega t}) = 2E_0 e^{-j\omega t}\left(\frac{\partial}{\partial t}E_0 e^{-j\omega t}\right)$$

$$\frac{\partial}{\partial t}E^2 = 2E\frac{\partial}{\partial t}E$$

Similarly,

$$\frac{\partial}{\partial t}B^2 = 2B\frac{\partial}{\partial t}B$$

which allows (2-111) to be rewritten as

$$\frac{\partial}{\partial t}W_{\text{EM}} = A\int_{z_1}^{z_2} \left(\varepsilon_0 E\frac{\partial}{\partial t}E + \frac{1}{\mu_0}B\frac{\partial}{\partial t}B\right) dz \tag{2-112}$$

Now, for a plane wave propagating in the z-direction, the contributions of the fields in the y-direction were calculated earlier with equation (2-29):

$$\vec{a}_y\left(\frac{\partial E_x}{\partial z} = -\mu\frac{\partial H_y}{\partial t}\right) = \vec{a}_y\left(\frac{\partial E_x}{\partial z} = -\frac{\partial B_y}{\partial t}\right) \tag{2-29}$$

and the contributions in the x-direction are given by

$$\vec{a}_x\left(\varepsilon\frac{\partial E_x}{\partial t} = -\frac{\partial H_y}{\partial z}\right) = \vec{a}_x\left(\mu\varepsilon\frac{\partial E_x}{\partial t} = -\frac{\partial B_y}{\partial z}\right) \tag{2-30}$$

Note that $\vec{J} = 0$ since it is assumed that there are no sources in the volume under consideration.

Substituting (2-29) and (2-30) into (2-112) allows us to simplify the equation:

$$\begin{aligned}\frac{\partial}{\partial t}W_{\mathrm{EM}} &= A\int_{z_1}^{z_2}\left[\varepsilon_0 E_x\left(-\frac{1}{\mu_0\varepsilon_0}\frac{\partial B_y}{\partial z}\right) + \frac{1}{\mu_0}B_y\left(-\frac{\partial E_x}{\partial z}\right)\right]\mathrm{d}z \\ &= -A\int_{z_1}^{z_2}\left[E_x\left(\frac{\partial H_y}{\partial z}\right) + H_y\left(\frac{\partial E_x}{\partial z}\right)\right]\mathrm{d}z\end{aligned} \tag{2-113}$$

Note that the form of the integrand in (2-113) looks like a curl multiplied by a vector. This tells us that (2-113) might be simplified if we use the vector identity introduced earlier:

$$\nabla\cdot(\vec{E}\times\vec{H}) = \vec{H}\cdot(\nabla\times\vec{E}) - \vec{E}\cdot(\nabla\times\vec{H}) \tag{2-105}$$

Note that from (2-24), $\vec{a}_x(\partial H_y/\partial z) = -\nabla\times\vec{H}$ and $\vec{a}_y(\partial E_x/\partial z) = \nabla\times\vec{E}$. Therefore,

$$\vec{E}\cdot(\nabla\times\vec{H}) = -\frac{\partial H_y}{\partial z}$$

$$\vec{H}\cdot(\nabla\times\vec{E}) = \frac{\partial E_x}{\partial z}$$

This allows us to rewrite (2-113) in terms of the cross product of the electric and magnetic fields:

$$P_{\mathrm{EM}} = -A\int_{z_1}^{z_2}(\vec{H}\cdot\nabla\times\vec{E} - \vec{E}\cdot\nabla\times\vec{H})\,dz = -A\int_{z_1}^{z_2}\nabla\cdot(\vec{E}\times\vec{H})\,dz \tag{2-114}$$

Note that the final term in (2-114) is equivalent to a volume integral, where $A\,dz = dV$. This allows us to use the divergence theorem to eliminate the del operator:

$$\int_V(\nabla\cdot\vec{F})\,dV = \oint_S\vec{F}\cdot d\vec{s} \qquad \text{(divergence theorem)}$$

$$\begin{aligned}P_{\mathrm{EM}} &= -A\int_{z_1}^{z_2}(\nabla\cdot(\vec{E}\times\vec{H}))\,dz = -\int_V\nabla\cdot(\vec{E}\times\vec{H})\,dV \\ &= -\oint_S(\vec{E}\times\vec{H})\cdot d\vec{s}\end{aligned}$$

This allows us to define the *Poynting vector*, which represents the flow of power per unit area through a surface S at an instant in time.

$$\vec{S} = \vec{E} \times \vec{H} \qquad \mathrm{W/m^2} \tag{2-115}$$

Note that the direction of the power flow is perpendicular to both $\vec{E}$ and $\vec{H}$.

2.6.1 Time-Averaged Values

When considering the electromagnetic power delivered by a sinusoidal time-varying field, practical measurement considerations tend to favor the time-averaged value of the power rather than the instantaneous value described in (2-115). This is because the time-averaged power entering a passive network, as measured with a watt-meter, is a measure of the power dissipated by heat in all the resistive circuit elements. In the laboratory, the time average of a time-harmonic function is taken over an interval of many periods. For a steady-state sinusoidal function, the average of one period will be the same as the average over many periods, since each period looks identical. The *time average* of the Poynting vector is defined as the area under the function for one period, divided by the duration of the cycle:

$$\vec{S}_{\mathrm{ave}} = \frac{A_{\mathrm{period}}}{T_{\mathrm{period}}} = \frac{1}{T}\int_0^T \vec{S}(x, y, z, t)\, dt \tag{2-116}$$

Although the math is not shown here, Jackson [1999] shows the derivation of the time average value for the Poynting vector:

$$\vec{S}_{\mathrm{ave}} = \tfrac{1}{2}\mathrm{Re}(\vec{E} \times \vec{H}^*) \qquad \mathrm{W/m^2} \tag{2-117}$$

where the star represents the complex conjugate.

It is sometimes useful to represent the magnitude of the Poynting vector in terms of the electric or magnetic fields and the intrinsic impedance defined in equation (2-53). Integrating sinusoidal functions as described in (2-116) will yield the time-averaged Poytning vector in terms of the $\vec{E}$ and $\vec{H}$ fields [Johnk, 1988]:

$$\vec{E} = \vec{a}_x E^+ \cos(\omega t - \beta z) \tag{2-118}$$

$$\vec{H} = \vec{a}_y \frac{H^+}{\eta} \cos(\omega t - \beta z) \tag{2-119}$$

$$\vec{S} = \vec{E} \times \vec{H} = [\vec{a}_x E^+ \cos(\omega t - \beta z)] \times \left[\vec{a}_y \frac{E^+}{\eta} \cos(\omega t - \beta z)\right]$$

$$= \vec{a}_z \frac{(E^+)^2}{\eta} \cos^2(\omega t - \beta z) \tag{2-120}$$

To obtain the time-averaged value of (2-120), remember that $\sin^2\theta + \cos^2\theta = 1$. Therefore, over a complete cycle, the average of $\cos^2\theta$ is equal to the average of $\sin^2\theta$ and $\sin^2\theta = \cos^2\theta = \frac{1}{2}$. Subsequently, the time-averaged value of a cosine-squared term over a complete cycle is $\frac{1}{2}$:

$$\vec{S}_{\text{ave}} = \vec{a}_z \frac{(E^+)^2}{2\eta} \tag{2-121}$$

Similarly, the Poynting vector can be expressed in terms of the magnetic field:

$$\begin{aligned} \vec{S} &= \vec{E} \times \vec{H} = [\vec{a}_x \eta E^+ \cos(\omega t - \beta z)] \times [\vec{a}_y H^+ \cos(\omega t - \beta z)] \\ &= \vec{a}_z \eta (H^+)^2 \cos^2(\omega t - \beta z) \\ \vec{S}_{\text{ave}} &= \vec{a}_z \eta \frac{(H^+)^2}{2} \end{aligned} \tag{2-122}$$

2.7 REFLECTIONS OF ELECTROMAGNETIC WAVES

So far, we have considered the propagation of electromagnetic waves in a simple, infinitely large medium. However, most practical problems involve waves propagating in multiple dielectric media. Since each medium will have different electric characteristics, it is essential to understand how a propagating electromagnetic wave will behave when it enters a region where the properties of the medium change.

Generally, when an electromagnetic plane wave propagating in medium A enters region B, where the properties of the dielectric change, two things happen: (1) a portion of the wave is reflected away from region B, and (2) a portion of the wave is transmitted into region B. When these plane waves encounter planar interfaces, both the reflected and transmitted waves are also planar, so their directions, amplitudes, and phase constants can easily be calculated. The simultaneous existence of both the transmitted and reflected waves is a direct result of the boundary conditions that must be satisfied when solving Maxwell's equations at the interface between the two regions. We begin with a plane wave incident on a perfect conductor.

2.7.1 Plane Wave Incident on a Perfect Conductor

Consider a plane wave propagating in medium A in the z-direction. Assume that the medium in region A is a simple, loss-free medium and medium B is a perfectly conducting metal plane, as shown in Figure 2-23. Assume that the electric field is oriented in the x-direction, necessitating that the magnetic field

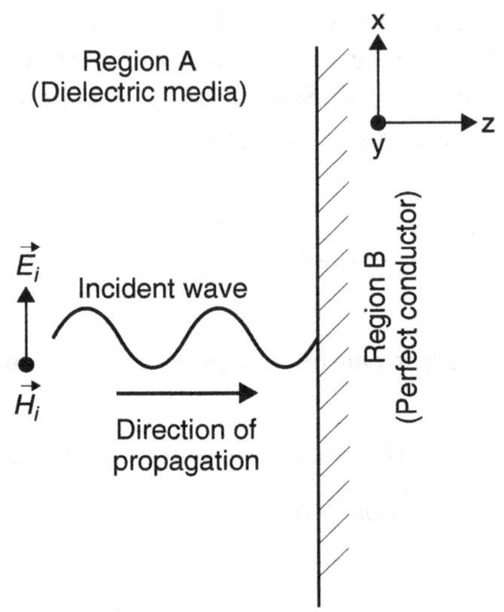

Figure 2-23 Incident electromagnetic wave propagating in dielectric region A and impinging on a perfect conductor.

be oriented in the y-direction. The incident fields propagating in medium A are given by

$$\vec{E}_i(z) = \vec{a}_x E_i e^{-j\omega z\sqrt{\mu_0 \varepsilon_0}} \tag{2-123}$$

$$\vec{H}_i(z) = \vec{a}_y \frac{1}{\eta} E_i e^{-j\omega z\sqrt{\mu_0 \varepsilon_0}} e^{j\omega t} \tag{2-124}$$

Since the boundary is normal to the incident waves, the reflected portion will be reflected back in the $-z$-direction. Subsequently, the reflected electric field will have the form

$$\vec{E}_r(z) = \vec{a}_x E_r e^{-j\omega(-z)\sqrt{\mu_0 \varepsilon_0}} = \vec{a}_x E_r e^{+j\omega z\sqrt{\mu_0 \varepsilon_0}} \tag{2-125}$$

Since the electric field of the reflected wave is propagating in the $-z$-direction, the magnetic field must also flip and point in the $-y$-direction, to maintain the proper relationship between the electric and magnetic fields:

$$\vec{H}_r(z) = -\vec{a}_y \frac{1}{\eta} E_r e^{+j\omega z\sqrt{\mu_0 \varepsilon_0}} e^{j\omega t} \tag{2-126}$$

To calculate the values of the reflected and transmitted waves we must consider what happens when an electromagnetic field impinges on a perfect conductor. Equation (2-7) implies that an electric field impinging on a conductor with conductivity σ will produce a current density $\vec{J}$:

$$\vec{J} = \sigma \vec{E} \tag{2-7}$$

For a perfect conductor, $\sigma = \infty$. However, since the real world limits currents to finite densities, equation (2-7) implies that infinite conductivity will produce infinite currents inside a perfect conductor, which is impossible; therefore, the electric field inside the perfect conductor must be zero. If the electric field is zero, the magnetic fields must also be zero. This allows us to deduce the boundary conditions for both $\vec{E}$ and $\vec{H}$ at the interface between a dielectric medium and a perfect electrical conductor (PEC). Since the fields in region A are finite and the fields in region B (the PEC) are zero, we can deduce that the wave impinging on the conductor must induce a wave equal but opposite to the incident wave at the surface, so the fields in the conductor are zero. The boundary condition at the surface ($z = 0$) of the PEC requires that the tangential electric field must vanish for all x and y to ensure that the electric field inside the conductor is zero. Applying the boundary conditions to both the incident and the reflected portions of the electric field gives

$$\vec{E}(z = 0) = \vec{E}_i(z = 0) + \vec{E}_r(z = 0) = \vec{a}_x E_i + \vec{a}_x E_r = 0$$

which produces the relationship between the incident and reflected electric fields for a electromagnetic wave impinging on a PEC:

$$\vec{a}_x E_i = -\vec{a}_x E_r \tag{2-127}$$

This means that when an electromagnetic wave is incident normal to a perfect electrical conducting plane traveling in the $+z$-direction, it will experience a 100% reflection back toward the $-z$-direction with the same magnitude as the incident wave with a negative amplitude. This is the same as saying that the magnitude of the reflected wave will remain constant but the phase will be shifted by 180°. This is shown in Figure 2-24.

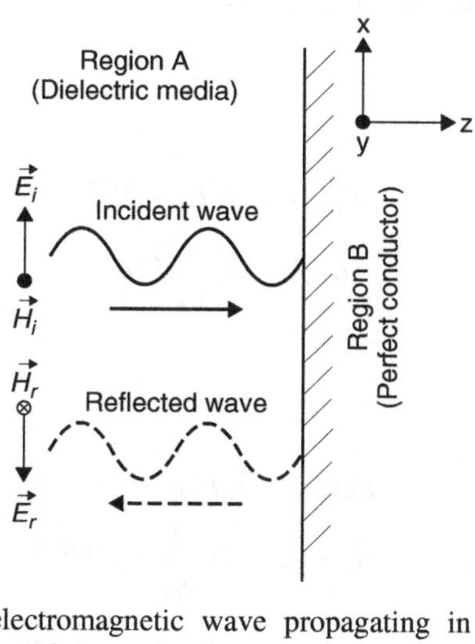

Figure 2-24 Incident electromagnetic wave propagating in dielectric region A and impinging on a perfect conductor, showing that 100% of the wave is reflected back in the $-z$-direction.

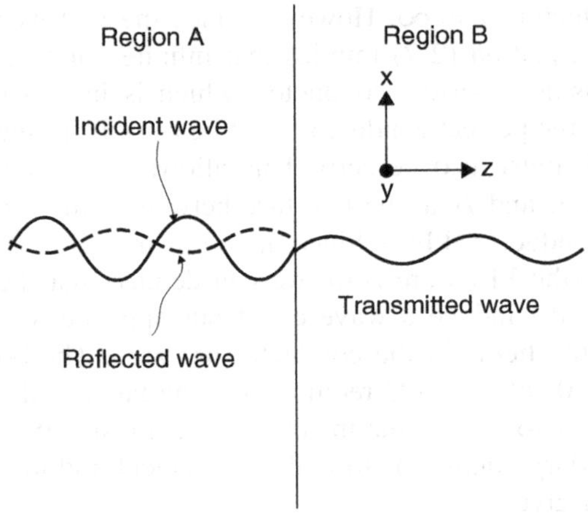

Figure 2-25 Incident electromagnetic wave propagating in dielectric region A and impinging on dielectric region B, showing that a portion of the wave is reflected back in the $-z$-direction and a portion is transmitted in the $+z$-direction.

2.7.2 Plane Wave Incident on a Lossless Dielectric

In Section 2.7.1 we considered the special case where a propagating electromagnetic plane wave was incident on a plane boundary where the second medium was a perfect conductor. Now we consider the more general case where the second medium is a lossless dielectric. When a plane wave impinges on an area with a different dielectric, a portion of the wave is reflected and a portion of the wave in transmitted into the new medium, where it continues to propagate, as shown in Figure 2-25.

If we assume that the incident fields are represented by phasors, we can separate the total plane wave into three parts:

1. Incident wave:

$$\vec{E}_i(z) = \vec{a}_x E_i e^{-j\beta_A z}$$
$$\vec{H}_i(z) = \vec{a}_y \frac{1}{\eta_A} E_i e^{-j\beta_A z}$$

2. Reflected wave:

$$\vec{E}_r(z) = \vec{a}_x E_r e^{+j\beta_A z}$$
$$\vec{H}_r(z) = -\vec{a}_y \frac{1}{\eta_A} E_r e^{+j\beta_A z}$$

3. Transmitted wave:

$$\vec{E}_t(z) = \vec{a}_x E_t e^{-j\beta_B z}$$

$$\vec{H}_t(z) = \vec{a}_y \frac{1}{\eta_B} E_t e^{-j\beta_B z}$$

where $\beta_A = \omega\sqrt{\mu_A \varepsilon_A}$, $\beta_B = \omega\sqrt{\mu_B \varepsilon_B}$, $\eta_A = \omega\sqrt{\mu_A/\varepsilon_A}$, and $\eta_B = \omega\sqrt{\mu_B/\varepsilon_B}$ (where $\mu_x = \mu_{rx}\mu_0$ and $\varepsilon_x = \varepsilon_{rx}\varepsilon_0$) are the phase constants and intrinsic impedances of regions A and B, respectively.

When a wave intersects a boundary between two lossless dielectric regions, the tangential components of both the electric and magnetic fields across the interface must remain continuous. In other words, the tangential component of the fields cannot change instantaneously. In our particular scenario, the plane wave is propagating in TEM mode in the z-direction, so both the electric and magnetic fields are oriented parallel (i.e., tangent) to the boundary of the dielectric interface. Subsequently, we can say that at the interface ($z = 0$), the sum of the incident and reflected waves must equal the transmitted wave:

$$\vec{E}_t(z = 0) = \vec{E}_i(z = 0) + \vec{E}_r(z = 0) \rightarrow E_t = E_i + E_r \tag{2-128}$$

$$\vec{H}_t(z = 0) = \vec{H}_i(z = 0) + H_r(z = 0) \rightarrow \frac{E_t}{\eta_2} = \frac{E_i}{\eta_1} - \frac{E_r}{\eta_1} \tag{2-129}$$

Since the incident waves are known, we can solve (2-128) and (2-129) simultaneously for the transmitted and reflected portions of the wave:

$$E_t = E_i \frac{2\eta_B}{\eta_B + \eta_A} \tag{2-130}$$

$$E_r = E_i \frac{\eta_B - \eta_A}{\eta_B + \eta_A} \tag{2-131}$$

Equations (2-130) and (2-131) lead to the definition of the reflection and transmission coefficients:

$$\Gamma \equiv \frac{E_r}{E_i} = \frac{\eta_B - \eta_A}{\eta_B + \eta_A} \tag{2-132}$$

$$T \equiv \frac{E_t}{E_i} = \frac{2\eta_B}{\eta_B + \eta_A} = 1 + \Gamma \tag{2-133}$$

The reflection coefficient is a measure of how much of the wave is reflected back off the intersection between the two media, and the transmission coefficient tells how much of the wave is transmitted. If the reflection coefficient is zero, it means that the intrinsic impedance in the two regions is identical. If the intrinsic impedances are not equal, the reflection coefficient will be finite. These equations are used heavily in Chapter 3 when analyzing multiple reflections on transmission-line structures.

REFERENCES

Huray, Paul G., 2009, *Foundations of Signal Integrity*, Wiley, Hoboken, NJ.

Inan, Umran S., and Aziz S. Inan, 1998, *Engineering Electromagnetics*, Addison Wesley Longman, Reading, MA.

Jackson, J. D., 1999, *Classical Electrodynamics*, 3rd ed., Wiley, New York.

Johnk, Carl T. A., 1988, *Electromagnetic Theory and Waves*, Wiley, New York.

PROBLEMS

2-1 For a TEM plane wave traveling in the z-direction where $E_x = 60$ V/m, if $\lambda = 20$ cm and $\nu = 1.5 \times 10^8$, determine the frequency of the wave, the relative dielectric permittivity (ε_r), the phase constant, and the intrinsic impedance.

2-2 For the wave in Problem 2-1, write formulas for the time-domain expressions for the E and H fields, and plot them.

2-3 For the plane wave in Problem 2-1, how much power is being transported?

2-4 If a wire with a diameter of 1 in. is carrying a dc current of 1000 A, how much power would be delivered to the 100-Ω resistive load in the loop shown in Figure 2-26?

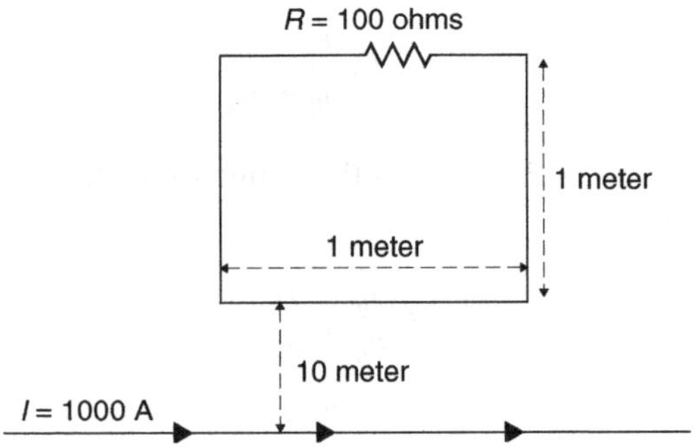

Figure 2-26

2-5 If two identical point charges are separated by 10 in. in free space and are experiencing a repulsion force of 1 N, what is the magnitude of each charge?

2-6 Assume that two infinitely long parallel wires are separated by a distance of 5 in. with currents that are equal but opposite. Find the current if the force per unit length is 0.25 N/m.

2-7 Determine the energy needed to assemble three point charges in free space from infinity to the following (x,y,z) coordinates: (0,0,0), (1 mm, 0, 0), (−1 mm, 0, 0).

2-8 For a parallel-plate capacitor in free space ($\varepsilon_r = 1$) with total change Q on one plate and—Q on the other with plate areas of A separated by the distance d, calculate the external force required to keep the plates stationary.

2-9 If a plane wave is propagating in the z-direction in free space and it impinges on a planar dielectric medium where $\varepsilon_r = 9.6$, how much of the wave is propagated? How much is reflected? If the electric field incident on the boundary has a value of 60 V/m, how much power is carried in the initial wave? How much power is reflected and transmitted? Show that power is conserved.

2-10 If the electric field in an area of space is known to be $\vec{E} = br^3\vec{a}_r$ (where b is a constant) in spherical coordinates, find the charge density ρ and the total charge within a sphere of diameter d.

3

IDEAL TRANSMISSION-LINE FUNDAMENTALS

Advanced Signal Integrity for High-Speed Digital Design, By Stephen H. Hall and Howard L. Heck

Transmission lines are simple electrical structures that consist of an insulating layer of dielectric material sandwiched between two layers of metal (usually, a nonferrous metal such as copper). In high-speed digital design, transmission lines are used for communicating between electrical components, such as a microprocessor and a memory module, or chip set. In modern designs such as computer systems, the data transfer rates are so high that the width (in time) of the digital pulses are small compared to the time it takes to propagate a signal from one end of a transmission line to another. As a result, it is quite common to have more than one bit of information propagating on the transmission line at any given instant in time, where the transmission line is essentially "storing" information before the receiving circuitry can latch the data. Consequently, to preserve the quality of the digital pulse stream to a degree where it can be captured with no errors at the receiving agent, great attention must be given to the construction and design of the transmission lines so that the electrical characteristics are controlled and predictable. Designing a successful bus for high-data-rate information transfer requires a thorough understanding of how the signals propagate on transmission lines. In this chapter we introduce basic transmission-line structures typically used in digital systems and present fundamental transmission-line theory for the ideal case.

3.1 TRANSMISSION-LINE STRUCTURES

Transmission lines come in many shapes and sizes. One of the most common transmission lines is the coaxial cable, which is used with cable television. When designing a high-speed digital system, transmission lines are usually manufactured on a printed circuit board (PCB) or a multichip module (MCM), which typically consists of conductive traces buried in or attached to a dielectric with one or more reference power and ground planes. The metal typically used is copper (although the copper is often electroplated with silver or nickel to prevent corrosion) and the dielectric is often FR4, a fiberglass–resin composite discussed in detail in Chapter 6. The two most common types of transmission lines used in digital designs are microstrips and striplines. A *microstrip* is routed on an outside layer of the PCB and has only one reference plane. There are two types of microstrips, buried and nonburied. A *buried* (sometimes called *embedded*) *microstrip* is simply a transmission line that is embedded into the dielectric but

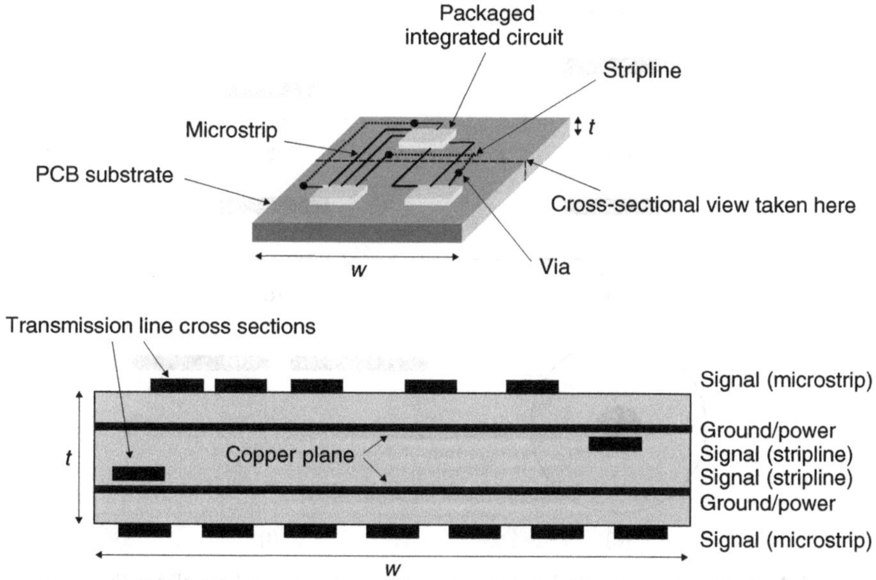

Figure 3-1 Typical application of transmission lines in a digital design.

still has only one reference plane. A *stripline* is routed on an inside layer and has two reference planes. Figure 3-1 represents a PCB with traces routed between the various components on both internal (stripline) and external (microstrip) layers. The accompanying cross section is taken at the given mark so that the position of the transmission lines relative to the ground and power planes can be seen. Transmission lines in the book are often represented in cross-section form because they are useful for calculating and visualizing the various transmission-line parameters described.

Multiple-layer PCBs such as the one depicted in Figure 3-1 can provide a variety of stripline and microstrip structures. Control of the conductor and dielectric layers (which is referred to as the *stackup*) is required to make the electrical characteristics of the transmission line predictable. These basic electrical characteristics are defined in this chapter and referred to as *transmission-line parameters*. Figure 3-2 shows several transmission-line structures discussed in the text [Hall, 2000].

3.2 WAVE PROPAGATION ON LOSS-FREE TRANSMISSION LINES

Transmission lines are designed to guide electromagnetic waves from one point to another for the purpose of information transfer. For practical applications, the electromagnetic wave can be approximated as planar because the wavelength is usually much larger than the electrical delay across the width of the transmission line. This means that there will be no variation in the electric or magnetic

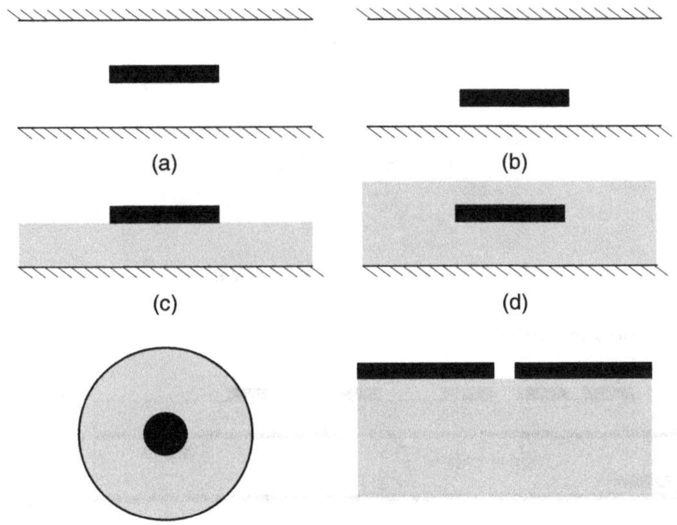

Figure 3-2 Common transmission-line structures: (a) balanced stripline; (b) asymmetrical stripline; (c) microstrip line; (d) buried microstrip line; (e) coaxial line; (f) slotline.

fields across the width of the transmission line. The approximation of a planar electromagnetic wave will become inadequate when the transmission-line width is no longer small compared to the wavelength. For a typical transmission line on a common PCB used in motherboard designs for personal computers, the bandwidth of the plane-wave approximation can be estimated using (2-45) and (2-52) by calculating the frequency where the wavelength equals the width of the conductor. For typical applications, the dielectric constant of FR4 ($\varepsilon_r = 4$) and a trace width of 5 mils (127 μm) can be used to calculate the frequency where the wavelength equals the width of the transmission-line signal conductor:

$$f_{\text{plane}} = \frac{c/\sqrt{\varepsilon_r}}{\lambda} = \frac{3.0 \times 10^8 \text{ m/s}/\sqrt{4}}{127 \times 10^{-6} \text{ m}} = 1181 \text{ GHz}$$

The fundamental frequency of the fastest digital design in a product at the time of this writing is less than 20 GHz. Consequently, the analysis of electromagnetic plane waves described in Section 2.3 applies for digital transmission lines because $f \ll f_{\text{plane}}$.

3.2.1 Electric and Magnetic Fields on a Transmission Line

On a transmission line, information is transferred from one component to the other by guiding electromagnetic energy from point A to point B. To gain an intuitive understanding of how a signal propagates on a transmission line, we must understand how the electric and magnetic field patterns are distributed. To do this, we first derive the boundary conditions of the electric field at the interface between a dielectric and a perfect conductor, which allows us to draw the electric field patterns for typical transmission line cross sections in a uniform dielectric. Next, we explore how the electric field behaves at a dielectric boundary, such as the interface between the air and the board material in a microstrip transmission line. Finally, we use the relationships for a TEM wave derived in Chapter 2 to obtain the magnetic field pattern from the electric field.

When a voltage is applied between the signal conductor and the reference plane on one end of a transmission line, an electric field is established as described in Section 2.4. The voltage is calculated by the line integral of the electric field between the signal conductor and the reference plane:

$$v = -\int_a^b \vec{E} \cdot d\vec{l} \tag{3-1}$$

As discussed in Section 2.3.2, once the electric field is established, the properties of the magnetic field can be calculated using Faraday's and Ampère's laws (2-1) and (2-2). From Section 2.3.2, the relationship between the electric and magnetic fields was shown always to remain orthogonal:

$$\vec{a}_y \left(\frac{\partial E_x}{\partial z} = -\mu \frac{\partial H_y}{\partial t} \right) \tag{2-29}$$

$$\vec{a}_x \left(\varepsilon \frac{\partial E_x}{\partial t} = -\frac{\partial H_y}{\partial z} \right) \tag{2-30}$$

Note that since we are applying a voltage source to generate the electric field, and voltage is defined in terms of the work done per unit charge, equation (2-58) implies that charges must be present on both the signal and the reference conductors:

$$\left[\frac{W}{q}\right]_{a \to b} = v(b) - v(a) = -\int_a^b \vec{E} \cdot d\vec{l} \tag{2-58}$$

Furthermore, Gauss's law ($\nabla \cdot \varepsilon \vec{E} = \rho$) states that the divergence of $\varepsilon \vec{E}$ is nonzero and equal to the charge density, which implies that the sources of the electric field are the electrical charges. Equivalently, if the electric field terminates abruptly, the termination must be an electric charge. To understand how the electric field behaves at the interface between the dielectric and a conducting surface, we use the integral form of Gauss's law to calculate the boundary conditions:

$$\oint_S \varepsilon \vec{E} \cdot d\vec{s} = \int_V \rho dV = Q_{\text{enc}} \tag{2-59}$$

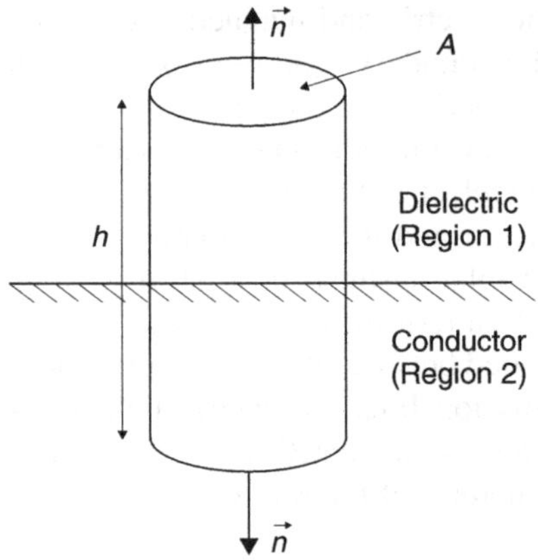

Figure 3-3 Surface used to calculate the boundary conditions at the dielectric–conductor interface.

First, we must choose a surface to integrate and compare the normal components of $\varepsilon\vec{E}$ on both sides of the conductor–dielectric boundary. A convenient surface is a cylinder, as depicted in Figure 3-3. Since we are observing the behavior of fields at the surface, the height (h) of the cylinder can be made infinitely small, reducing (2-59) to

$$\oint_S \varepsilon\vec{E}\cdot d\vec{s} = \oint_S \varepsilon_1\vec{E}_1\cdot d\vec{s}_1 + \oint_S \varepsilon_2\vec{E}_2\cdot d\vec{s}_2 = (\vec{n}\cdot\varepsilon_1\vec{E}_1)A + (-\vec{n}\cdot\varepsilon_2\vec{E}_2)A = \rho A \tag{3-2}$$

where ε_1 is the dielectric permittivity of the dielectric in region 1 and ε_2 describes the dielectric permittivity of region 2, which in this example is a perfect conductor. To interpret this equation, recall the discussion in Section 2.7.1, where it was shown that the electric field ($\vec{E}_2$ in this case) must be zero inside a perfect conductor. Setting $\vec{E}_2 = 0$ allows us to simplify (3-2) for the special case of a boundary between a dielectric and a perfect conductor:

$$\vec{n}\cdot\varepsilon\vec{E} = \rho \qquad \text{C/m}^2 \tag{3-3}$$

Equation (3-3) means that *the electric field must emanate normal from and terminate normal to the conductor surface*. Since equations (2-29) and (2-30) say that the magnetic field must be orthogonal to the electric field, we can conclude that *the magnetic field must be tangential to the conductor surface*. These two rules make it easy to visualize electric and magnetic fields on transmission-line structures with perfect electrical conductors. Simply draw the electric field lines so that they are always perpendicular to the conductor surface, emanating from the high and terminating in the low potential conductor, and then draw the magnetic field lines so that they are always perpendicular to the electric field lines

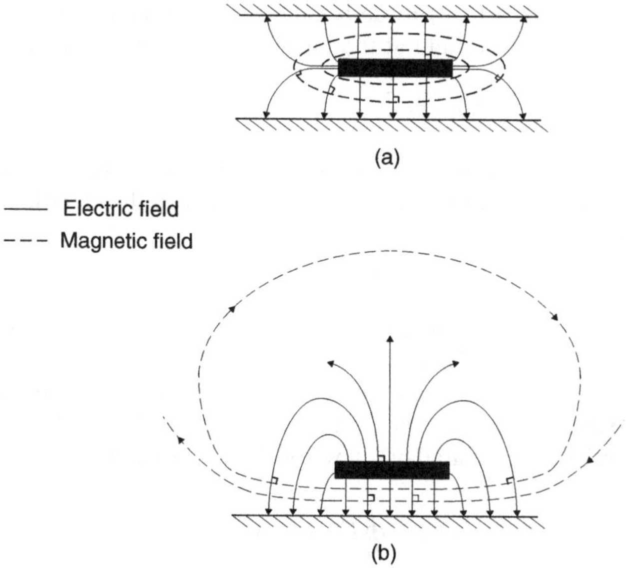

Figure 3-4 Electric and magnetic field patterns for homogeneous dielectrics in (a) a stripline and (b) a microstrip.

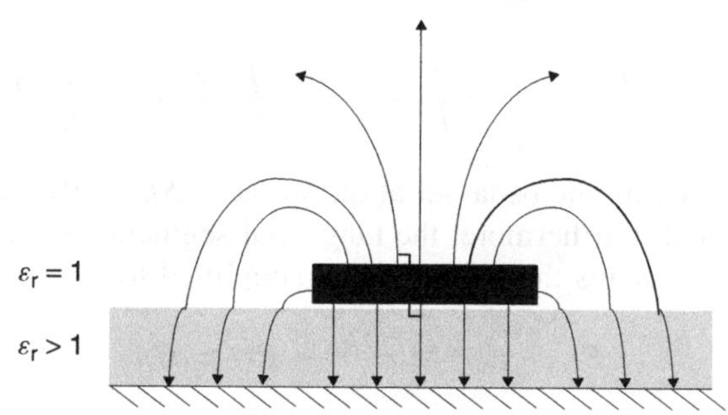

Figure 3-5 How the electric field behaves at a dielectric boundary.

and tangential to the conductor surfaces. If we assume that the voltage is applied with the positive value on the signal conductors as shown in Figure 3-4, we can draw the electric and magnetic fields for various transmission-line structures assuming a homogeneous dielectric.

When the dielectric is not homogeneous, as is almost always the case with microstrip transmission lines, the field lines are distorted as they cross dielectric boundaries. Figure 3-5 illustrates how *the electric field lines are bent away from the normal to the dielectric boundary when the relative dielectric permittivity on the top half of the structures is smaller than that on the bottom*.

To understand why the field lines are distorted, we must examine how electrostatic fields behave at the boundaries between two dielectric regions. First, we can recycle equation (3-2) to derive the boundary conditions of the normal components of the electric field at the interface between two dielectrics:

$$(\vec{n} \cdot \varepsilon_1 \vec{E}_1) - (\vec{n} \cdot \varepsilon_2 \vec{E}_2) = \rho \tag{3-4}$$

If the surface charge density between the dielectric layers is assumed to be zero (usually, a valid assumption), the relationships between the electric fields in both regions is described by

$$\vec{n} \cdot \varepsilon_1 \vec{E}_1 = \vec{n} \cdot \varepsilon_2 \vec{E}_2 \tag{3-5}$$

Equation (3-5) means that *the normal component of the electric field is not continuous across a dielectric boundary*.

In Section 2.7.2 it was mentioned that the tangential component of the electric field must remain continuous across a dielectric boundary. This can be shown with the integral form of Faraday's law for the electrostatic case:

$$\oint_l \vec{E} \cdot d\vec{l} = 0 \tag{3-6}$$

If we integrate (3-6) around a closed differential contour that encompasses the dielectric boundary such as that shown in Figure 3-6, we can calculate the tangential components of the electric field:

$$\oint_l \vec{E} \cdot d\vec{l} = \int_a^b \vec{E} \cdot d\vec{l} + \int_b^c \vec{E} \cdot d\vec{l} + \int_c^d \vec{E} \cdot d\vec{l} + \int_d^a \vec{E} \cdot d\vec{l} \tag{3-7}$$

Since we are considering the behavior at the surface ($\Delta h \to 0$), segments *da* and *bc* can be eliminated. Furthermore, the tangential segments *ab* and *cd* are equal but opposite, which means that (3-7) can be simplified to

$$(E_{1t} - E_{2t})\Delta l = 0 \to E_{1t} = E_{2t} \tag{3-8}$$

Equation (3-8) means that *the tangential components of the electric field across a dielectric boundary must remain continuous*.

Assume that an electric field $\vec{E}_1$ is incident on a boundary between two dielectrics as shown in Figure 3-7. The change in the orientation of the electric

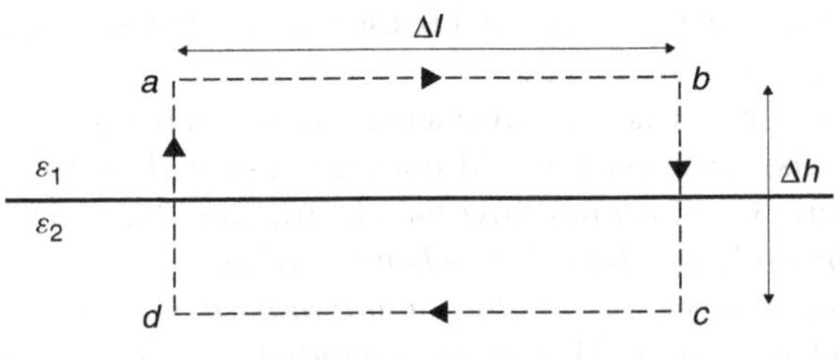

Figure 3-6 Differential contour encompassing a dielectric boundary.

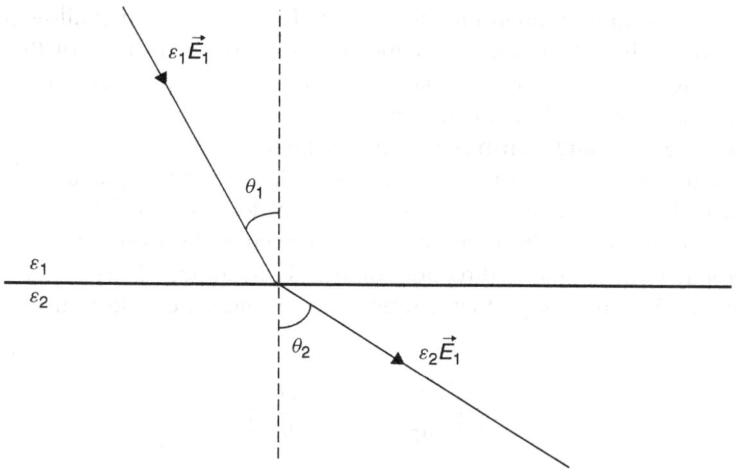

Figure 3-7 Change in the orientation of an electric flux line at a dielectric boundary.

flux lines across the interface can be calculated by using the boundary conditions for the normal and tangential components of the electric field derived in (3-5) and (3-8). When $\theta_1 = 0$, the boundary conditions of (3-5) apply. Consequently, we need a function of θ_1 that will satisfy (3-5) when the electric field is normal to the boundary. Since $\cos(0) = 1$, the following equation will satisfy the boundary conditions for the normal components of the electric field:

$$\varepsilon_1 E_1 \cos\theta_1 = \varepsilon_2 E_2 \cos\theta_2 \tag{3-9}$$

Similarly, when $\theta_1 = 90°$, the boundary conditions of (3-8) apply, and since $\sin(90°) = 1$, the following equation will satisfy the boundary conditions for the tangential components of the electric field:

$$E_1 \sin\theta_1 = E_2 \sin\theta_2 \tag{3-10}$$

If E_1 is calculated from (3-10) and substituted into (3-9), the change in orientation that the electric fields experience at a dielectric boundary can be calculated:

$$\theta_2 = \tan^{-1}\left(\frac{\varepsilon_2}{\varepsilon_1}\tan\theta_1\right) \tag{3-11}$$

Equation (3-11) means that *the field lines will be bent farther away from the normal to the dielectric interface in the medium with the higher permittivity*.

3.2.2 Telegrapher's Equations

So far, we have concentrated primarily on the derivation and calculation of the electric and magnetic fields. However, since this is a book targeted primarily at engineers, we want to express the effect of the fields in terms that allow us to use circuit theory. The telegrapher's equations express the properties of the electric and magnetic fields of transmission lines with simple equivalent circuits. This simplifies the analysis dramatically and provides a well-known mechanism for engineers to understand transmission-line behavior.

Since a transmission line is a structure intended to guide a planar electromagnetic wave down a specific path, we begin the derivation of the telegrapher's equations by observing the relationships described in Section 2.3.2 for a plane wave propagating in the z-direction in the TEM mode. First, let's examine equation (2-29), which says that a time-varying magnetic field will produce an electric field:

$$\vec{a}_y \left(\frac{\partial E_x}{\partial z} = -\mu \frac{\partial H_y}{\partial t} \right) \tag{2-29}$$

Equation (2-29) is a written in general terms assuming that the electromagnetic wave is propagating in an infinite space. In reality, the electric and magnetic fields are confined to an area dictated by the transmission-line geometry, so (2-29) can be simplified by calculating the equivalent-circuit parameters. First, we calculate the voltage between the signal conductor and the reference plane from the electric field using equation (3-1), assuming that point a is on the signal conductor and point b is the position on the reference plane directly below.

$$v = \int_a^b \vec{E}_x \cdot d\vec{l}$$

Consequently, for given transmission-line geometry, the left-hand part of (2-29) is equivalent to the partial derivative of the voltage between the signal conductor and the reference plane with respect to z:

$$\frac{\partial E_x}{\partial z} \to \frac{\partial \int \vec{E} \cdot d\vec{l}}{\partial z} = \frac{\partial v(z,t)}{\partial z}$$

Similarly, the right-hand side of equation (2-29) can be related to the inductance as discussed in Section 2.5.2. Since $\vec{B} = \mu \vec{H}$ and the magnetic flux is a function of $\vec{B}$,

$$\psi = \int \vec{B} \cdot d\vec{s}$$

(where $d\vec{s}$ is defined by the transmission-line geometry), equation (2-97) can be used to write the magnetic flux in terms of the current and inductance ($\psi = IL$):

$$L_{11} \equiv \frac{\psi_1}{I_1} \tag{2-97}$$

Therefore, for a given transmission-line geometry, the right-hand side of (2-29) can be written in terms of circuit parameters, where L is the series inductance of the transmission line per unit length:

$$\frac{\partial \mu H_y}{\partial t} \to \frac{\partial \int \vec{B}_y \cdot d\vec{s}}{\partial t} = \frac{\partial \psi}{\partial t} = L\frac{\partial i}{\partial t}$$

Don't get confused by the fact that the units of $\partial \mu H_y/\partial t$ are HA/m^2 (henry · amperes per square meter) and those of $L(\partial i/\partial t)$ are H · A/m. Equation (2-29) is in terms of generalized electric and magnetic fields. When the fields are represented by circuit parameters, the units change. Therefore, the circuit equivalent of (2-29) is

$$\frac{\partial v(z,t)}{\partial z} = -L\frac{\partial i(z,t)}{\partial t} \tag{3-12}$$

Note that equation (3-12) is simply the classic response of an inductor from circuit theory.

Similarly, equation (2-30) says that a time-varying electric field will produce a magnetic field:

$$\vec{a}_x\left(\varepsilon\frac{\partial E_x}{\partial t} = -\frac{\partial H_y}{\partial z}\right) \tag{2-30}$$

Note that the form of the numerator on the left-hand side of equation (2-30) (εE) has units of F · V/m^2 (farad · volts per square meter), which indicates that the equivalent-circuit form should be in terms of a capacitance and a voltage. From the integral form of Gauss's law (2-59) we can calculate the left-hand side of (2-30) in terms of the charge, assuming that the signal conductor is an infinite thin strip:

$$\oint_S \varepsilon\vec{E}\cdot d\vec{s} = \int_V \rho dV \to \varepsilon E_x A = \frac{QA}{A} \to \varepsilon E_x = \frac{Q}{A} \tag{3-13a}$$

where dV refers to volume. For a given transmission-line geometry, the voltage between the conductors from point a on the signal conductor to point b on the reference plane is calculated with (3-1), which has units of volts:

$$v = \int_a^b \vec{E}_x \cdot d\vec{l} = \frac{Q}{\varepsilon A}d = E_x d \tag{3-13b}$$

where A is the area of the conductors where the electric field is established and d is the distance between the signal conductor and the reference plane. Therefore, since $Q = Cv$,

$$\varepsilon E_x = \frac{Q}{A} \to \frac{Cv}{A} = C\left(\int_a^b \vec{E}_x \cdot d\vec{l}\right)\frac{1}{A} = \frac{CE_x d}{A} = \frac{C}{l}v \to Cv \tag{3-14}$$

where C is normalized to length with units of farads per meter. As a result, for a specific transmission-line geometry, we can write the left-hand side of (2-30) in terms of the capacitance:

$$\varepsilon \frac{\partial E_x}{\partial t} \rightarrow C \frac{\partial v}{\partial t} \tag{3-15}$$

To derive the circuit equivalent of the right side of (2-30), recall from equation (2-79) that current is the rate of charge flow per second:

$$i = \frac{dQ}{dt} \tag{3-16}$$

Substituting $Q = Cv$ into equation (3-16) produces the current in terms of the voltage and capacitance:

$$i = C \frac{dv}{dt} \tag{3-17}$$

which is equal to equation (3-15). Therefore, (2-30) can be rewritten in terms of circuit parameters:

$$\frac{\partial i(z,t)}{\partial z} = -C \frac{\partial v(z,t)}{\partial t} \tag{3-18}$$

Note that (3-18) is the classic response of a capacitor from circuit theory. Equations (3-12) and (3-18) are the *loss-free forms of the telegrapher's equations, which describe the electrical characteristics of a transmission line*.

3.2.3 Equivalent Circuit for the Loss-Free Case

Although signal integrity is largely a study in electromagnetic theory, application of the discipline is performed almost entirely using circuit parameters because they are more intuitive to most engineers. Consequently, it is necessary to derive a model for the transmission line in terms of the equivalent inductance L and the capacitance C per unit length. In this section the equivalent circuit of a transmission line is developed for the loss-free case. The model is refined in Chapters 5 and 6 to include loss from nonperfect dielectrics and finite conductivity conductors.

To begin, consider a differential element of transmission line with a length of Δz as shown in Figure 3-8, which represents a section of a transmission line with a signal conductor and a reference conductor. If we assume that current is traveling down the signal conductor and returning on the reference conductor (remember from circuit theory that current must always complete a loop along a return path), it can be represented by a series of differential current elements,

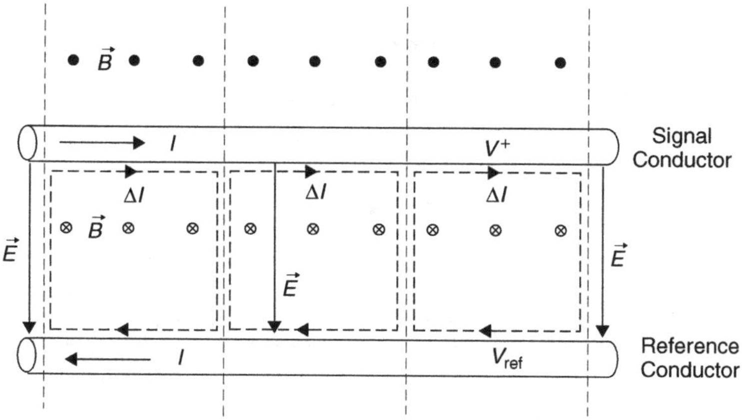

Figure 3-8 Segment of a transmission line showing the magnetic and electric fields, the associated differential current loops, and the voltage potentials at each conductor.

as was shown in Figure 2-18. When representing the total current as a series of small loops, the adjacent vertical components of the current elements cancel each other out, leaving a net current of ΔI on the signal conductor and $\Delta(-I)$ on the reference conductor. As described in equation (2-97), a current change in a loop will change the magnetic flux and will thus induce a self-inductance. Consequently, *the magnetic field of a transmission line is represented by a series inductor in the circuit model*. The value of the equivalent-circuit inductance is given by

$$L_{\Delta z} = \Delta z L \tag{3-19}$$

where Δz is the length of the differential section of transmission line and L is the inductance per unit length.

Similarly, when a voltage is applied to the transmission line between the signal conductor and the reference plane as shown in Figure 3-8 ($v = V^+ - V_{\text{ref}}$), an electric field is established in units of volts per meter:

$$v(b) - v(a) = -\int_a^b \vec{E} \cdot d\vec{l} \tag{2-58}$$

The presence of an electric field implies that charges exist on the conductors and consequently implies the existence of a capacitance (as described in Section 2.4.3). Therefore, *the electric field of a transmission line is represented by a shunt capacitor in the circuit model*. The value of the equivalent-circuit capacitance is given by

$$C_{\Delta z} = \Delta z C \tag{3-20}$$

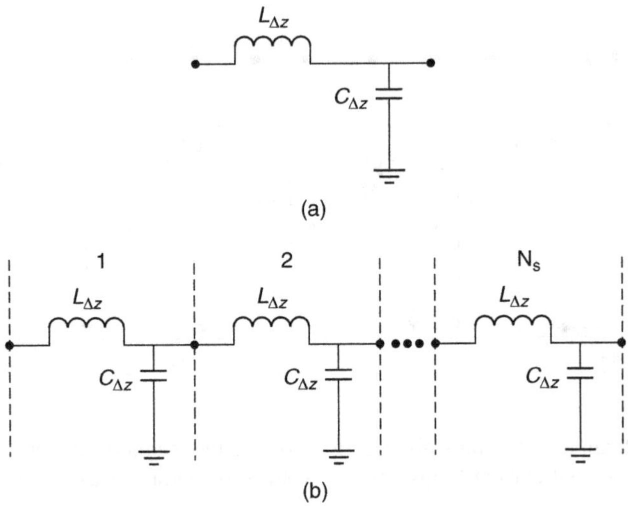

Figure 3-9 (a) Model for a differential element of a transmission line; (b) full model.

where Δz is the length of a differential section of transmission line and C is the capacitance per unit length.

Figure 3-9a shows the equivalent-circuit model of a differential element of transmission line. However, to make use of the model, we need a practical methodology to represent transmission lines that are much longer than Δz. Simply increasing Δz to scale the inductance and capacitor values for longer lengths will not produce a legitimate model unless the resonance of the *LC* circuit is much higher than the maximum frequency to be simulated:

$$\frac{1}{2\pi\sqrt{C_{\Delta z}L_{\Delta z}}} \gg f_{\text{simulation}} \tag{3-21}$$

If equation (3-21) is not satisfied, the equivalent circuit will behave like an *LC* filter circuit instead of like a transmission line. *The equivalent circuit of a transmission line can be represented by a series inductance and a shunt capacitor only if the LC resonant frequency is significantly greater than the maximum frequency of interest.*

The correct method to scale the transmission-line model for long lengths is to cascade a sufficient number of small *LC* segments together until the correct overall length is achieved, as shown in Figure 3-9b. Each *LC* segment represents a small transmission-line section of length Δz. The trick is to choose the correct value of Δz to achieve adequate model accuracy. If the value of Δz is too small, it will take an inordinate amount of time to perform SPICE simulations. However, if it is too large, the model will not exhibit realistic transmission-line properties. A good "rule of thumb" is to choose Δz so that the delay of each segment is approximately one-tenth the signal rise time if working in the time domain,

$$\Delta z \leq \frac{t_r c}{10\sqrt{\varepsilon_r}} \tag{3-22a}$$

or one-tenth the wavelength that corresponds to the maximum frequency of interest if working in the frequency domain,

$$\Delta z \leq \frac{\lambda_{f,\max}}{10} \tag{3-22b}$$

where t_r is the signal rise or fall time, c the speed of light in a vacuum (3×10^8 m/s), ε_r the dielectric permittivity, and $\lambda_{f,\max} = c/(f_{\max}\sqrt{\varepsilon_r})$ the wavelength that corresponds to the highest frequency of interest in the simulation.

When using a distributed LC model for modeling transmission lines, the number of segments for time-domain simulations is determined by

$$N_s = \frac{l}{\Delta z} = \frac{10l\sqrt{\varepsilon_r}}{t_r c} \tag{3-23a}$$

For frequency-domain simulations,

$$N_s = \frac{l}{\Delta z} = \frac{10l}{\lambda_{f,\max}} \tag{3-23b}$$

where N_s is the minimum number of segments required to model a transmission line of length l. Therefore, the capacitance and inductance per segment are given by

$$C_{\Delta z} = \frac{lC}{N_s} \tag{3-24a}$$

$$L_{\Delta z} = \frac{lL}{N_s} \tag{3-24b}$$

where C and L are the per unit length values of the capacitance and inductance.

Example 3-1 Create a transmission-line model for the 20-cm transmission line shown in Figure 3-10a assuming the following inductance and capacitance values and a dielectric permittivity of $\varepsilon_r = 4.5$.

$$L = 3.54 \times 10^{-7}\ \text{H/m}$$

$$C = 1.41 \times 10^{-10}\ \text{F/m}$$

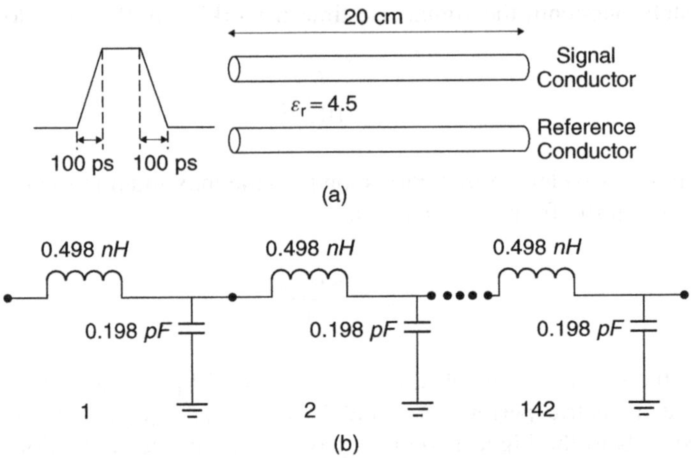

Figure 3-10 Equivalent circuit of a transmission line.

SOLUTION Since the digital waveform has rise and fall times of 100 ps ($t_r = t_f = 100$ ps), equations (3-22) and (3-23) give the required values for the model:

$$\Delta z = \frac{100 \text{ ps}(3.0 \times 10^8)}{10\sqrt{4.5}} = 1.41 \times 10^{-3} \text{ m}$$

$$N_s = \frac{l}{\Delta z} = \frac{0.2}{1.41 \times 10^{-3}} = 141.8 \text{ segments}$$

Since it is inconvenient to build an equivalent-circuit model with 141.8 segments, N_s is rounded up to 142. The inductance and capacitance values for each segment are calculated with equations (3-24):

$$C_{\Delta z} = \frac{(0.2)(1.41 \times 10^{-10})}{142} = 1.98 \times 10^{-13} \text{ F}$$

$$L_{\Delta z} = \frac{(0.2)(3.54 \times 10^{-7})}{142} = 4.98 \times 10^{-10} \text{ H}$$

Figure 3-10b shows the equivalent circuit for this transmission line.

3.2.4 Wave Equation in Terms of *LC*

The wave equations, which were used as the basis for analyzing propagating electromagnetic fields, were derived in Section 2.3.1. To analyze transmission lines in terms of circuit parameters, we rederive the wave equation from the telegrapher's equations (3-12) and (3-18):

$$\frac{\partial v(z,t)}{\partial z} = -L\frac{\partial i(z,t)}{\partial t} \tag{3-12}$$

$$\frac{\partial i(z,t)}{\partial z} = -C\frac{\partial v(z,t)}{\partial t} \tag{3-18}$$

Assuming that the digital signals can be decomposed into sinusoidal harmonics using the Fourier transform, the telegrapher's equations can be expressed in time-harmonic form where the voltage and current have the forms $v(t) = V_0 e^{j\omega t}$ and $i(t) = I_0 e^{j\omega t}$. Similar representations for the time-harmonic fields were discussed in Section 2.3.3. Consequently, the time-harmonic forms of the telegrapher's equations are

$$\frac{dv(z)}{dz} = -j\omega L i(z) \tag{3-25}$$

$$\frac{di(z)}{dz} = -j\omega C v(z) \tag{3-26}$$

Taking the derivative of (3-25) with respect to z produces

$$\frac{d^2v(z)}{dz^2} = -j\omega L\frac{di(z)}{dz} \tag{3-27}$$

and substituting (3-26) into (3-27) allows us to write an equation only in terms of voltage,

$$\frac{d^2v(z)}{dz^2} + \omega^2 LCv(z) = 0 \tag{3-28}$$

which is *the loss-free transmission-line wave equation for voltage*. Equation (3-28) is a second-order differential equation with the general solution given by

$$v(z) = v(z)^+ e^{-jz\omega\sqrt{LC}} + v(z)^- e^{jz\omega\sqrt{LC}} \tag{3-29}$$

The term $v(z)^+ e^{-jz\omega\sqrt{LC}}$ describes the voltage propagating down the transmission line in the $+z$-direction and $v(z)^- e^{jz\omega\sqrt{LC}}$ describes the voltage propagating in the $-z$-direction. Note the similarity of (3-29) and (2-41), which is the solution equivalent to the wave equation for the electric field. In Section 2.3.4, a propagation constant for a wave traveling in an infinite medium was defined that completely describes the medium where the electromagnetic wave is propagating:

$$\gamma = \alpha + j\beta \tag{2-42}$$

Comparing the solution to the electromagnetic wave equation (2-41) to equation (3-29) allows us to draw direct parallels for the propagation constant of a wave propagating on a loss-free transmission line:

$$\gamma = \alpha + j\beta = 0 + j\omega\sqrt{LC} \rightarrow \beta = \omega\sqrt{LC} \tag{3-30}$$

Equation (3-30) is the *phase constant of a transmission line*. Notice that for the loss-free transmission line, the attenuation constant (α) is zero. In Chapters 5 and 6 we describe how to calculate the losses for a transmission line and subsequently, α.

3.3 TRANSMISSION-LINE PROPERTIES

With the phase constant defined by (3-30), we can calculate the transmission-line properties using the techniques described in Section 2.3.4 for the propagation of a time-harmonic plane wave.

3.3.1 Transmission-Line Phase Velocity

Substitution of the propagation constant defined in equation (3-30) into equation (2-46) allows us to calculate the phase velocity of a harmonic component traveling on a transmission line in terms of L and C:

$$\nu_p = \frac{\omega}{\beta} = \frac{1}{\sqrt{LC}} \qquad \text{m/s} \tag{3-31}$$

were L and C are the per unit length values. Note that phase velocity can also be calculated from the dielectric properties:

$$\nu_p = \frac{c}{\sqrt{\mu_r \varepsilon_r}} \tag{2-52}$$

where c is the speed of light in a vacuum (3×10^8 m/s), μ_r the relative magnetic permeability of the dielectric (it is almost always unity for practical dielectrics), and ε_r the relative permittivity of the dielectric.

3.3.2 Transmission-Line Characteristic Impedance

In Chapter 2 we showed that the electric and magnetic fields were interdependent. Therefore, if the values of one field were known, the other could be calculated using Maxwell's equations. Similarly, for particular transmission-line geometry, if the voltage is known, the current can be calculated using the telegrapher's equations. For example, if we assume that a voltage is propagating along an infinitely long transmission line along the z-axis, the current can be calculated by differentiating the voltage and substituting into (3-25):

$$
\begin{aligned}
v(z) &= v(z)^+ e^{-jz\omega\sqrt{LC}} \\
\frac{dv(z)}{dz} &= -j\omega\sqrt{LC}\, e^{-jz\omega\sqrt{LC}}\, v(z)^+ = -j\omega L i(z)^+ \\
i(z)^+ &= \frac{\sqrt{LC}\, e^{-jz\omega\sqrt{LC}}}{L} v(z)^+ = \sqrt{\frac{C}{L}} v(z)^+ e^{-jz\omega\sqrt{LC}}
\end{aligned}
\tag{3-32}
$$

Since the units of C are F/m = A · S/m and L is in units of H/m = V · S/A, the term $\sqrt{C/L}$ has units of A/V, which is the same as siemens. Therefore, the quantity $\sqrt{L/C}$ must be similar to the intrinsic impedance defined in Section 2.3.4. The *characteristic impedance of a loss-free transmission line in ohms* is given by

$$
Z_0 = \sqrt{\frac{L}{C}} \tag{3-33}
$$

Just as the intrinsic impedance describes the relationship between the electric and magnetic fields, *the characteristic impedance describes the relationship between the voltage and current on a transmission line*. Thus, we can write the transmission-line equation for the current in terms of the voltage equation:

$$
i(z) = \frac{1}{Z_0} v(z)^+ e^{-jz\omega\sqrt{LC}} - \frac{1}{Z_0} v(z)^- e^{jz\omega\sqrt{LC}} \tag{3-34}
$$

3.3.3 Effective Dielectric Permittivity

As described in Section 3.3.1, electrical signals propagating along a transmission line will travel at a speed that is dependent on the dielectric permittivity of the surrounding medium and the geometry of the transmission-line cross section. Although this concept is relatively simple, it becomes complicated when a transmission line is built with a nonhomogeneous dielectric. The most common example of this is a microstrip transmission line. For example, consider Figure 3-11a, which shows the electric field of a signal on a microstrip transmission line built on an FR4 dielectric, which has a relative dielectric permittivity of approximately $\varepsilon_r = 4.0$. Note that a percentage the electric field is propagating simultaneously in the dielectric and in the air, where the latter has a relative dielectric permittivity of approximately $\varepsilon_r = 1.0$. Consequently, the propagation velocity will depend on an "effective" dielectric permittivity, which is a weighted average of how much electric field is fringing through the air versus how much stays contained within the dielectric material. When the electric field is contained completely within the board, as in the case of a stripline as shown in Figure 3-11b, the effective dielectric constant will be equal to the dielectric permittivity of the insulating material, and the signals will propagate more slowly than microstrip traces. When signals are routed on the external layers of the board, as in the case

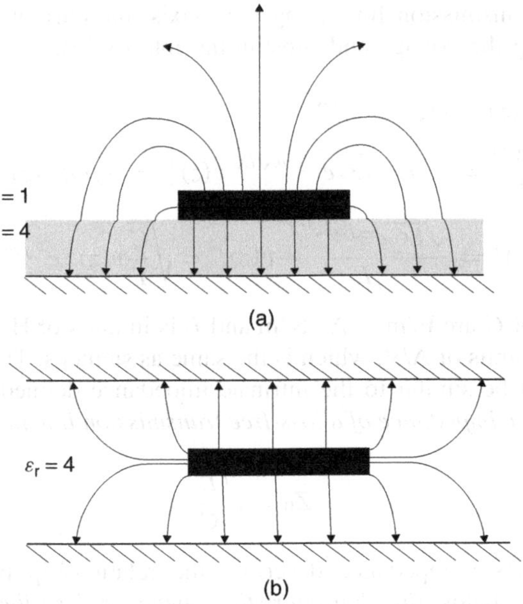

Figure 3-11 Electric fields of (a) a microstrip and (b) a stripline.

of a microstrip line, the electric field fringes through both the dielectric material and the air, which lowers the effective dielectric constant; thus, the signals will propagate more quickly than those on an internal layer. Note that this description assumes that the relative magnetic permeability of the dielectric material under consideration is unity ($\mu_r = 1$).

In Sections 3.4.3 and 3.4.4 we describe how to derive the effective dielectric permittivity of a microstrip using quasistatic approximations of Maxwell's equations. Furthermore, there are many commercially available two-dimensional electromagnetic field solvers available which will produce accurate results. However, in the absence of a field solver, equation (3-35) has been shown to produce results of reasonable accuracy [Hammerstad and Jensen, 1980] for structures where the conductor thickness t is much smaller than the dielectric thickness h.

$$a = 1 + \frac{1}{49}\ln\left[\frac{u^4 + (u/54)^2}{u^4 + 0.432}\right] + \frac{1}{18.7}\ln\left[1 + \left(\frac{u}{18.1}\right)^3\right]$$

$$b = 0.564\left(\frac{\varepsilon_r - 0.9}{\varepsilon_r + 3}\right)^{0.053} \tag{3-35}$$

$$\varepsilon_{\text{eff}}(u, \varepsilon_r) = \frac{\varepsilon_r + 1}{2} + \frac{\varepsilon_r - 1}{2}\left(1 + \frac{10}{u}\right)^{-ab}$$

where $u = w/h$ and the dimensions are defined in Figure 3-12a.

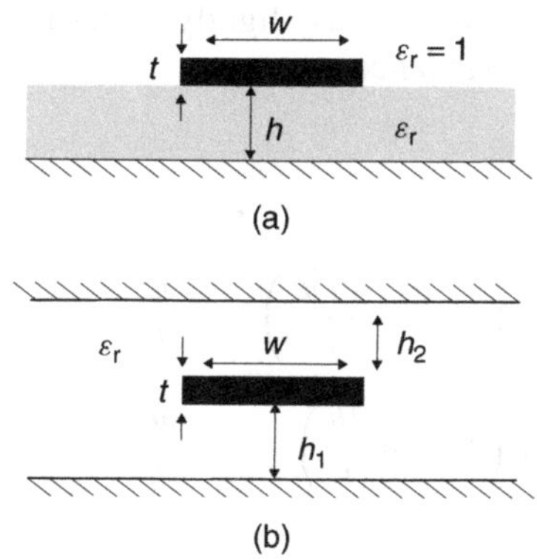

Figure 3-12 Dimensions used in the impedance and effective dielectric constant formulas: (a) microstrip line; (b) stripline.

3.3.4 Simple Formulas for Calculating the Characteristic Impedance

For maximum accuracy it is necessary to use one of the many commercially available two-dimensional electromagnetic field solvers to calculate the impedance of the PCB or MCM traces for design purposes. The solvers will typically provide the impedance, propagation velocity, and the L and C elements per unit length using many of the concepts that will be presented in Section 3.4. In the absence of a field solver, the formulas presented here will provide approximations to the impedance values of typical transmission lines as a function of the trace geometry and the dielectric constant ε_r, where the dimensions are as shown in Figure 3-12.

Microstrip: Infinitely Thin Conductors (t ≪ h) [Hammerstad and Jensen, 1980]

$$Z_0 = \frac{\eta}{2\pi}\ln\left[\frac{\xi h}{w} + \sqrt{1 + \left(\frac{2h}{w}\right)^2}\right]$$

$$\xi = 6 + (2\pi - 6)e^{-(30.666h/w)^{0.7528}} \qquad \text{(3-36a)}$$

$$\eta = \frac{377}{\sqrt{\varepsilon_{\text{eff}}}}$$

Microstrip: Finite Thickness [Collins, 1992] The formulas below are accurate for $1 < \varepsilon_r \leq 16$ and $0.25 \leq w/h \leq 6$. Note that ε_{eff} in this equation set accounts for the finite thickness of the signal conductor when calculating the effective dielectric constant for the microstrip. The term w_e is an effective width that accounts for the extra capacitance caused by the finite thickness of the signal conductor. Since electric field lines will be established between the edge of the conductor and the reference plane, thicker signal conductors will exhibit increased capacitance. The effective width w_e is slightly wider than the physical width w, to account for the extra capacitance.

$$
\begin{aligned}
Z_0 &= \sqrt{\frac{\varepsilon_0\mu_0}{\varepsilon_{\text{eff}}}}\frac{1}{C_a} \\
w_e &= \begin{cases} w + 0.398t\left(1+\ln\dfrac{4\pi w}{t}\right) & \dfrac{w}{h} \le \dfrac{1}{2\pi} \\[2ex] w + 0.398t\left(1+\ln\dfrac{2h}{t}\right) & \dfrac{w}{h} > \dfrac{1}{2\pi} \end{cases} \\
C_a &= \begin{cases} \dfrac{2\pi\varepsilon_0}{\ln(8h/w_e + w_e/4h)} & \dfrac{w_e}{h} \le 1 \\[2ex] \varepsilon_0\left[\dfrac{w_e}{h} + 1.393 + 0.667\ln\left(\dfrac{w_e}{h} + 1.444\right)\right] & \dfrac{w_e}{h} > 1 \end{cases} \\
\varepsilon_{\text{eff}} &= \frac{\varepsilon_r+1}{2} + \frac{\varepsilon_r-1}{2}\left(1+12\frac{h}{w_e}\right)^{-1/2} + \xi - 0.217(\varepsilon_r-1)\frac{t}{\sqrt{w_e h}} \\
\xi &= \begin{cases} 0.02(\varepsilon_r-1)\left(1-\dfrac{w}{h}\right)^2 & \dfrac{w}{h} < 1 \\[2ex] 0 & \dfrac{w}{h} > 1 \end{cases}
\end{aligned}
\tag{3-36b}
$$

Symmetrical Stripline ($h_1 = h_2$) ***[IPC, 1995]***

$$
Z_o = \frac{60}{\sqrt{\varepsilon_r}}\ln\frac{1.9(2h+t)}{0.8w+t}
\qquad 0.1 < \frac{w}{h} < 2.0 \qquad \frac{t}{h} < 0.25 \qquad 1 < \varepsilon_r < 15
\tag{3-36c}
$$

Asymmetrical Stripline ($h_1 > h_2$) ***[IPC, 1995]***

$$
Z_0 = \frac{80}{\sqrt{\varepsilon_r}}\ln\frac{1.9(2h_2+t)}{0.8w+t}\left(1-\frac{h_2}{4h_1}\right)
\qquad 0.1 < \frac{w}{h_2} < 2.0 \qquad \frac{t}{h_2} < 0.25 \qquad 1 < \varepsilon_r < 15
\tag{3-36d}
$$

3.3.5 Validity of the TEM Approximation

The assumptions of how the electric and magnetic fields are related for a propagating electromagnetic wave were discussed in Section 2.3.2. One of

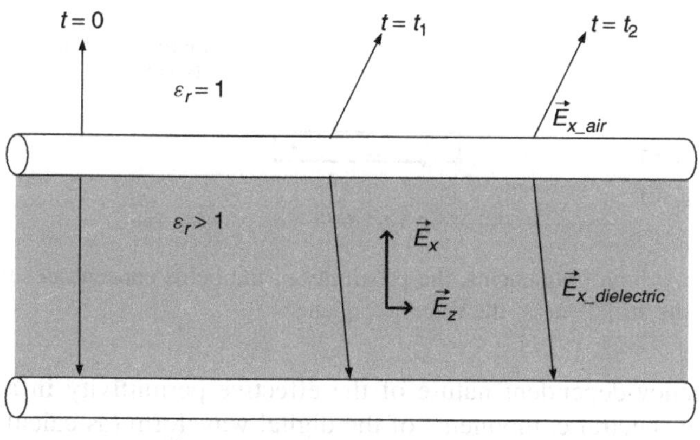

Figure 3-13 The electric field develops a z-component when propagating down a microstrip transmission line, due to the nonhomogeneous dielectric.

the fundamental concepts used throughout signal integrity analysis is that *the electric and magnetic fields are orthogonal and there are no components in the z-direction*. When waves propagate in this manner, it is called the *transverse electromagnetic mode* (TEM). However, when discussing the concept of an effective dielectric permittivity in Section 3.3.3, where the component of the electric field propagating through the air travels faster than the component propagating in the board material, it becomes obvious that the electric field is no longer restricted to a single component. For example, consider Figure 3-13, which depicts the side view of an electric field established in the x-direction at $t = 0$ on a microstrip transmission line between the signal conductor and the reference plane. As the signal begins to propagate down the line, the electric field lines in the air will travel at a faster speed than those in the board, effectively tilting the electric field in the z-direction. Consequently, the electric field develops a component in the z-direction which violates the assumption of the TEM approximation.

Furthermore, as the frequencies increase, the electric field will become more confined to the region between the microstrip and the reference plane, resulting in less fringing through the air, causing the effective dielectric permittivity to increase. To understand this, refer to Figure 3-14. When a dc voltage is applied between the signal conductor and the reference plane, the charge will be distributed uniformly across the cross section of the signal conductor. As the frequency of the signal is increased, the charge will tend to concentrate at the bottom of the signal conductor closest to the reference plane because that is the area of highest field concentration. This means that for a microstrip transmission line, *the electric field will tend to concentrate in the board material, which increases the effective dielectric permittivity with increasing frequency*. The charge distribution in transmission lines is discussed in more detail in Sections 3.4.4 and 5.1.2.

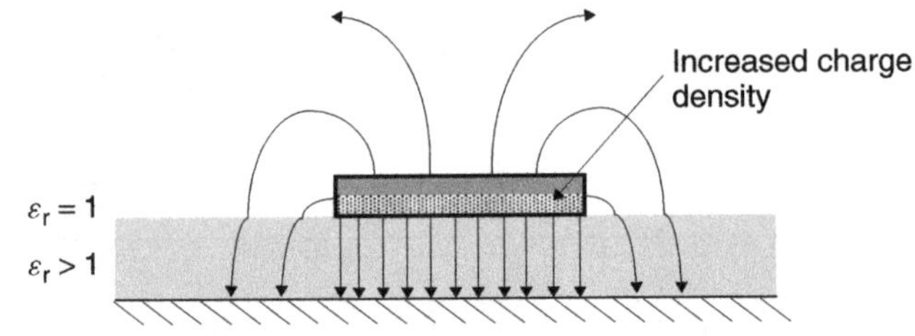

Figure 3-14 At high frequencies, the proximity of the fields concentrates the charge in the bottom of the strip nearest the reference plane.

The frequency-dependent nature of the effective permittivity in a microstrip will cause the spectral components of the digital waveform (as calculated with a Fourier transform) to travel at different speeds, which will distort the waveform. This is known as *dispersion*. A relatively simple formula for calculating how the effective dielectric permittivity for a microstrip changes with frequency due to the nonhomogeneous nature of the dielectric was developed empirically by [Collins 1992], and is given by

$$\varepsilon_{\text{eff}}(f) = \varepsilon_r - \frac{\varepsilon_r - \varepsilon_{\text{eff}}(f=0)}{1 + (f/f_a)^m} \tag{3-37}$$

where

$$f_a = \frac{f_b}{0.75 + (0.75 - 0.332\varepsilon_r^{-1.73})(w/h)}$$

$$f_b = \frac{47.746}{h\sqrt{\varepsilon_r - \varepsilon_{\text{eff}}(f=0)}}\tan^{-1}\left[\varepsilon_r\sqrt{\frac{\varepsilon_{\text{eff}}(f=0) - 1}{\varepsilon_r - \varepsilon_{\text{eff}}(f=0)}}\right]$$

$$m = m_0 m_c \le 2.32$$

$$m_0 = 1 + \frac{1}{1+\sqrt{w/h}} + 0.32\left(1 + \sqrt{\frac{w}{h}}\right)^{-3}$$

$$m_c = \begin{cases} 1 + \dfrac{1.4}{1 + w/h}\left(0.15 - 0.235e^{-0.45(f/f_a)}\right) & \dfrac{w}{h} \le 0.7 \\ 1 & \dfrac{w}{h} > 0.7 \end{cases}$$

where $\varepsilon_{\text{eff}}(f=0)$ is calculated with (3-35), f is in gigahertz, and the units of w and h are millimeters.*

*Fortunately, a frequency-dependent effective dielectric permittivity does not pose significant obstacles to modeling transmission lines. In Chapter 10, techniques that employ frequency-dependent equivalent circuits using tabular SPICE models are described that allow a unique value of the transmission-line parameters to be described at every frequency point.

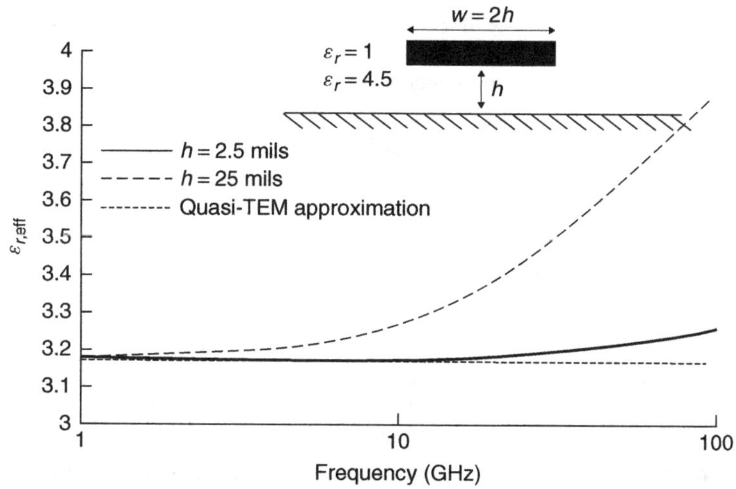

Figure 3-15 Effective dielectric permittivity compared to the quasi-TEM approximation for a 12-in. microstrip.

For most practical applications encountered in high-speed digital design, the TEM approximations are valid and the frequency-dependent nature described by (3-37) is ignored. For microstrip lines and other nonhomogeneous structures, when TEM propagation is assumed, it is referred to as the *quasi-TEM approximation*. To demonstrate the conditions when the quasi-TEM approximation breaks down, equation (3-37) was used to calculate the frequency variation of the relative effective dielectric permittivity ε_{eff} for two cases, a thick and a thin dielectric, as shown in Figure 3-15. The thin dielectric example ($h = 2.5$ mils) represents the transmission-line dimensions typically used to design buses on conventional motherboards for personal computers. The thick dielectric example ($h = 25$ mils) is an exaggerated case chosen to demonstrate when the TEM approximation breaks down (however, similar dimensions are sometimes used in radio-frequency (RF) applications where density is not as much of a concern). Equation (3-35) was used to calculate the frequency-invariant quasistatic TEM value for ε_{eff}, and equation (3-37) was used to estimate the variation of ε_{eff} with frequency. Note that for the thin case, the deviation from the quasi-TEM approximation is small up to very high frequencies, but the thick case begins to deviate much earlier.

To evaluate when the quasi-TEM approximation breaks down, we choose a metric of 1% error in total delay. Since the changes in ε_{eff} will alter the velocity as shown in equation (2-52), the errors will accumulate for longer line lengths. If we choose a line length, the valid frequency range of the TEM approximation can be calculated. Figure 3-16 shows the percent error in propagation delay caused by the quasi-TEM approximation for a 12-in. microstrip. Note that for the thick case, the quasi-TEM will induce a 1% error in delay at about 8 GHz, and the thin case remains accurate to about 80 GHz. Consequently, for typical transmission-line

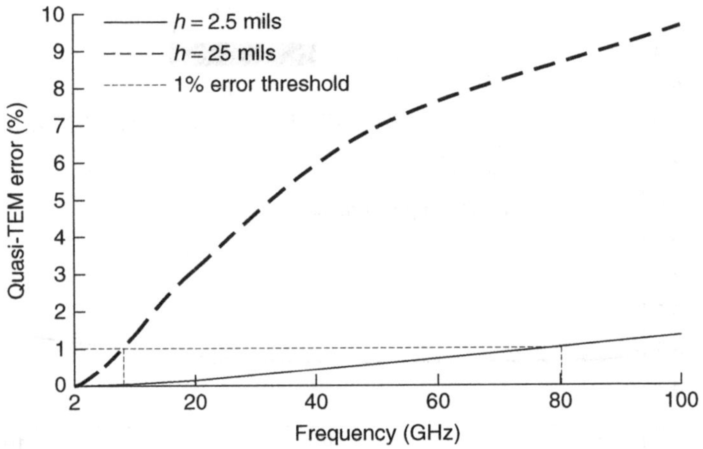

Figure 3-16 Bandwidth where the quasi-TEM approximation induces a 1% error in the delay for a 12-in. microstrip.

dimensions used in contemporary digital design, the quasi-TEM approximation for the transmission-line parameters is valid.

3.4 TRANSMISSION-LINE PARAMETERS FOR THE LOSS-FREE CASE

Earlier, we discussed how an electromagnetic wave propagates on a transmission line by drawing parallels to a plane wave in free space and developed an equivalent circuit that was constructed from inductance and capacitance elements. In this section we discuss a variety of methods to calculate the equivalent inductance and capacitance values per unit length for various transmission-line structures. Although in practice, a commercial two-dimensional field solver is usually employed to calculate the equivalent-circuit parameters, it is vital for the engineer to understand the approximations and methodologies typically used.

Note that the solutions in this chapter assume that the electrostatic solutions to Maxwell's equations are adequate approximations for high-speed digital design purposes. This is known as the *quasistatic approximation*. This technique allows significant simplification over the full-wave solution, where no approximations are assumed. Indeed, at high frequencies (above a few hundred megahertz), the quasistatic approximation breaks down because the equivalent-circuit parameters have significant frequency dependence. Nonetheless, virtually all two-dimensional field solvers used in the industry are quasistatic in nature. Full-wave solvers, which do not assume quasistatic fields are in theory more accurate. However, in practice they require too many computational resources and too much time to be of practical use for modeling transmission lines in a digital design. Fortunately, the errors caused by the quasistatic approximation can be corrected relatively painlessly with excellent accuracy up to very high frequencies (at least 50 GHz), as discussed in Chapters 5, 6, 8, and 10. Consequently, the methodologies in this book for modeling transmission lines are based on a quasistatic approach, and corrections are applied in later chapters to recover the frequency-dependent behavior.

3.4.1 Laplace and Poisson Equations

In Section 2.4 we show that the behavior of an electrostatic field can be described by the differential equations

$$\nabla \cdot \vec{D} = \rho \tag{2-3}$$

$$\vec{E} = -\nabla\Phi \tag{2-65}$$

the latter coming from the fact that Ampère's law is zero for an electrostatic field ($\nabla \times \vec{E} = 0$). If (2-65) is substituted into (2-3),

$$\nabla \cdot (-\nabla\Phi) = \frac{\rho}{\varepsilon}$$

we arrive at *Poisson's equation*:

$$\nabla^2\Phi = -\frac{\rho}{\varepsilon} \tag{3-38}$$

In a medium that lacks any charge density, Poisson's equation is reduced to *Laplace's equation*,

$$\nabla^2\Phi = 0 \tag{3-39}$$

Poisson's and Laplace's equations are particularly useful for solving quasistatic transmission-line problems, as demonstrated in the remainder of this chapter.

3.4.2 Transmission-Line Parameters for a Coaxial Line

Assume that a pair of long, coaxial, circular conductors are statically charged with the inner conductor at the potential $\Phi = V$ relative to the outer conductor, which is held at zero potential (ground), where the cross section is as shown in Figure 3-17. The region between the conductors has a dielectric permittivity of $\varepsilon = \varepsilon_0\varepsilon_r$ and a relative magnetic permeability of $\mu_r = 1$.

The first step is to calculate the capacitance per unit length. If we make the assumption that the dielectric is charge free (very reasonable; see Chapter 6), the capacitance can be derived by solving Laplace's equation (3-39) in cylindrical coordinates. From Appendix A we get

$$\nabla\Phi = \vec{a}_r\frac{\partial\Phi}{\partial r} + \vec{a}_\phi\frac{1}{r}\frac{\partial\Phi}{\partial\phi} + \vec{a}_z\frac{\partial\Phi}{\partial z}$$

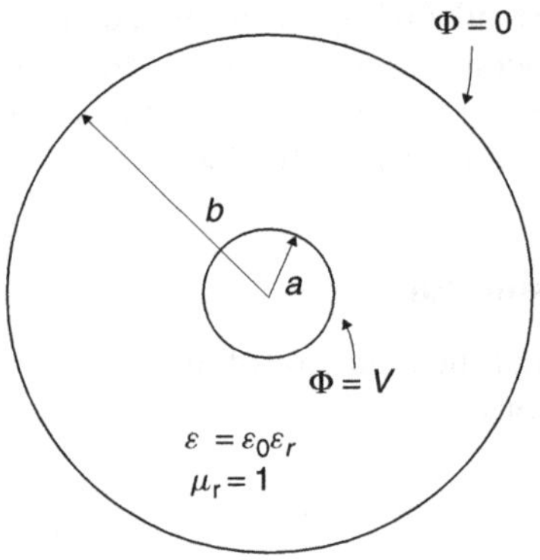

Figure 3-17 Cross section of a coaxial transmission line.

Assuming TEM, there will be no variation of the electric field in the z-direction, and because of the symmetry, there should be no variation with ϕ. Therefore,

$$\nabla^2\Phi = \nabla \cdot (\nabla\Phi) = \frac{1}{r}\frac{\partial}{\partial r}\left(r\frac{\partial\Phi}{\partial r}\right) = 0$$

Integrating $\partial[(r(\partial\Phi/\partial r)]/\partial r$ produces $r(\partial\Phi/\partial r) = C_1$, and a second integration yields the solution:

$$\Phi(r) = C_1 \ln(r) + C_2 \tag{3-40}$$

The boundary conditions must be applied to (3-40) to determine the values of C_1 and C_2. The boundary conditions are:

- When $r = b$, $\Phi = 0$.
- When $r = a$, $\Phi(a) = V$.

Applying the first boundary condition yields

$$\Phi(b) = 0 = C_1 \ln(b) + C_2$$
$$C_2 = -C_1 \ln(b)$$

which is substituted back into (3-40):

$$\Phi(r) = C_1 \ln(r) - \ln(b)C_1 = C_1 \ln\frac{r}{b}$$

Applying the second boundary condition produces the following:

$$\Phi(a) = V = C_1 \ln\frac{a}{b} \rightarrow C_1 = \frac{V}{\ln(a/b)} \rightarrow C_1 = -\frac{V}{\ln(b/a)}$$

Note that $\ln(a/b) = -\ln(b/a)$, which is done here as a convenience:

$$\Phi(r) = -\frac{V}{\ln(b/a)}\ln(r) + \frac{V}{\ln(b/a)}\ln(b) = \frac{V}{\ln(b/a)}\ln\frac{b}{r}$$

The electric field is computed using (2-65):

$$\vec{E} = -\nabla\Phi = -\nabla\left[\frac{V}{\ln(b/a)}\ln\frac{b}{r}\right] = -\vec{a}_r\frac{\partial}{\partial r}\left[\frac{V}{\ln(b/a)}\ln\frac{b}{r}\right] = \vec{a}_r\frac{V}{r\ln(b/a)} \tag{3-41}$$

The next step is to find the total surface charge on the signal conductor using (3-3), which describes the boundary conditions between a dielectric and a perfect conductor:

$$\vec{n}\cdot\varepsilon\vec{E} = \rho \qquad \text{C/m}^2 \tag{3-3}$$

At $r = a$ (the inner conductor surface), the surface charge is calculated with

$$\varepsilon E = \frac{\varepsilon V}{a\ln(b/a)} = \rho \qquad \text{C/m}^2 \tag{3-42}$$

Multiplying (3-42) by the circumference of the inner conductor will give the equivalent surface charge per unit length:

$$\frac{Q}{l} = \rho 2\pi a = \frac{2\pi\varepsilon V}{\ln(b/a)} \qquad \text{C/m} \tag{3-43}$$

Using (2-76) we can get the *capacitance per unit length of a coaxial transmission line*:

$$C = \frac{Q/l}{V} = \frac{2\pi\varepsilon}{\ln(b/a)} \qquad \text{F/m} \tag{3-44}$$

To find the inductance per unit length, we exploit the relationship between the magnetic and electric fields:

$$\nu_p = \begin{cases} \dfrac{1}{\sqrt{\mu_r\mu_0\varepsilon_r\varepsilon_0}} = \dfrac{c}{\sqrt{\mu_r\varepsilon_r}} & \text{m/s} \quad (2\text{-}52) \\[2ex] \dfrac{\omega}{\beta} = \dfrac{1}{\sqrt{LC}} & \text{m/s} \quad (3\text{-}31) \end{cases}$$

which allows us to write (assuming $\mu_r = 1$)

$$\frac{c}{\sqrt{\varepsilon_r}} = \frac{1}{\sqrt{LC}} \tag{3-45}$$

Since $\mu_r = 1$, the dielectric properties do not influence the inductance. Consequently, since the speed of light in a vacuum is constant, the inductance can be calculated from (3-45) if the capacitance is calculated with $\varepsilon_r = 1$:

$$L = \frac{1}{c^2 C_{\varepsilon_r=1}} \tag{3-46}$$

where c is the speed of light in a vacuum and $C_{\varepsilon_r=1}$ is the capacitance with the relative dielectric permittivity ε_r set to 1. Substituting (3-44) into (3-46) with $\varepsilon_r = 1$ gives the value of the *inductance per unit length for a coaxial line*:

$$L = \frac{\ln(b/a)}{c^2 2\pi \varepsilon_0} \qquad \text{H/m} \tag{3-47}$$

3.4.3 Transmission-Line Parameters for a Microstrip

To derive the impedance of a microstrip transmission line, we begin with the solution to Laplace's equation. To begin the derivation, consider Figure 3-18, which shows the boundary conditions that we must satisfy when solving the differential equations. Note that the microstrip is surrounded by a box with conducting walls at $x = \pm d/2$, $y = 0$, and $y = \infty$. This is a valid placement of the boundary conditions only if $d \gg h$, where h is the height of the dielectric. In this case, the partial differential equations are solved using a method called *separation of variables*, as described by Jackson [1999]. The rectangular nature of the microstrip geometry allows us to remain in rectangular coordinates, which gives the two-dimensional Laplace equation the form

$$\nabla^2 \Phi = \frac{\partial^2 \Phi}{\partial x^2} + \frac{\partial^2 \Phi}{\partial y^2} = 0 \tag{3-48}$$

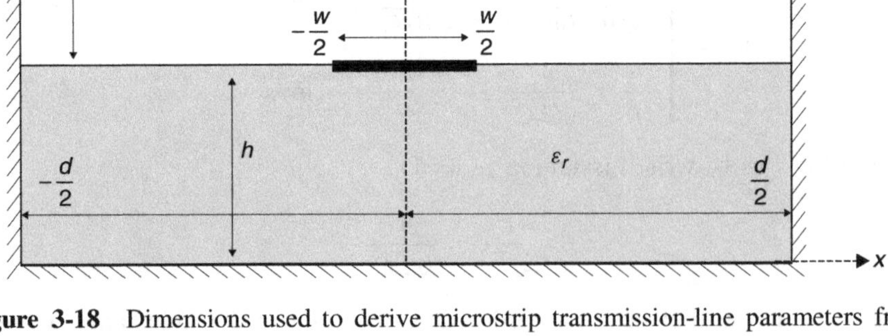

Figure 3-18 Dimensions used to derive microstrip transmission-line parameters from Laplace's equation.

The solution to this partial differential equation can be found in terms of two ordinary differential equations, assuming that the potential can be represented by the product of a function for each coordinate [Jackson, 1999]:

$$\Phi(x, y) = X(x)Y(y) \tag{3-49}$$

If equation (3-49) is substituted back into (3-48) and then divided by Φ, the result is

$$\frac{1}{XY}\frac{d^2XY}{dx^2} + \frac{1}{XY}\frac{d^2XY}{dy^2} = \frac{1}{X}\frac{d^2X}{dx^2} + \frac{1}{Y}\frac{d^2Y}{dy^2} = 0 \tag{3-50}$$

where the partial derivatives are replaced by total derivatives because each term involves only one variable. This allows us to write two separate ordinary differential equations:

$$\frac{1}{X}\frac{d^2X}{dx^2} = -\beta^2 \tag{3-51}$$

$$\frac{1}{Y}\frac{d^2Y}{dy^2} = \beta^2 \tag{3-52}$$

The constant β must satisfy the conditions $\beta^2 + (-\beta^2) = 0$ to satisfy Laplace's equation. The differential equations in (3-51) and (3-52) are solved by finding the roots of the characteristic equations:

$$\begin{aligned} X(x) &= C_{1n}e^{j\beta x} + C_{2n}e^{-j\beta x} = C_{1n}(\cos\beta x + j\sin\beta x) \\ &\quad + C_{2n}(\cos\beta x - j\sin\beta x) \end{aligned} \tag{3-53}$$

$$\begin{aligned} Y(y) &= C'_{1n}e^{\beta y} + C'_{2n}e^{-\beta y} = C'_{1n}(\cosh\beta y + \sinh\beta y) \\ &\quad + C'_{2n}(\cosh\beta y - \sinh\beta y) \end{aligned} \tag{3-54}$$

The potential is then calculated by substituting (3-53) and (3-54) into (3-49):

$$\begin{aligned} \Phi(x, y) = X(x)Y(y) &= [C_{1n}(\cos\beta x + j\sin\beta x) + C_{2n}(\cos\beta x - j\sin\beta x)] \\ &\quad \cdot [C'_{1n}(\cosh\beta y + \sinh\beta y) + C'_{2n}(\cosh\beta y - \sinh\beta y)] \end{aligned} \tag{3-55}$$

Note that since the solution is periodic, β must be an integer. The boundary conditions that must be applied to solve (3-55) are

$$\Phi(x, y) = 0 \qquad \text{when } x = \pm\frac{d}{2} \tag{3-56}$$

which is the potential at the sidewalls and

$$\Phi(x, y) = 0 \qquad \text{when } y = 0, \infty \tag{3-57}$$

which is the potential at the ground plane and at a point infinitely far away. This means that we must consider two different solutions: region 1, which exists from $y = 0$ to h (in the dielectric), and region 2, which exists from $y = h$ to infinity (in the air), where the potentials must be equal at the boundary of these regions.

Looking at $X(x)$ in (3-55), the boundary condition in (3-56) is met if $C_{1n} = C_{2n}$ and β is chosen so that the cosine term equals zero when $x = \pm d/2$. When $C_{1n} = C_{2n}$, $X(x)$ is reduced to

$$X(x) = C_{1n} \cos \beta x$$

and when $\beta = n\pi/d$ for odd n and $x = \pm d/2$,

$$X(x) = 0$$

yielding an appropriate form for $X(x)$:

$$X(x) = C_{1n} \cos\left(\frac{n\pi}{d}x\right) \tag{3-58}$$

The second term $Y(y)$ in (3-55) will satisfy the boundary condition in (3-57) at $y = 0$ if $C'_{1n} = -C'_{2n}$, yielding

$$Y(y) = C'_{2n} \sinh \beta y = C'_{2n} \sinh\left(\frac{n\pi}{d}y\right) \tag{3-59}$$

At $y = \infty$, the boundary condition (3-57) is satisfied when $C'_{1n} = 0$:

$$Y(y) = C'_{2n}(\cosh \beta y - \sinh \beta y) = C'_{2n}e^{-\beta y} \tag{3-60}$$

This allows us to write equations for the potential that satisfies the boundary conditions by multiplying $X(x)$ and $Y(y)$ for the appropriate boundary conditions, where the product of the constants have been renamed A_n and B_n for clarity.

$$\Phi(x, y) = \begin{cases} \displaystyle\sum_{\substack{n=1 \\ \text{odd}}}^{\infty} A_n \cos\left(\frac{n\pi}{d}x\right) \sinh\left(\frac{n\pi}{d}y\right) & \text{when } 0 \le y < h \quad \text{(region 1)} \quad \text{(3-61a)} \\ \displaystyle\sum_{\substack{n=1 \\ \text{odd}}}^{\infty} B_n \cos\left(\frac{n\pi}{d}x\right) e^{-(n\pi/d)y} & \text{when } h \le y < \infty \quad \text{(region 2)} \quad \text{(3-61b)} \end{cases}$$

The potential at the signal conductor ($y = h$) must be continuous, so

$$A_n \cos\left(\frac{n\pi}{d}x\right) \sinh\left(\frac{n\pi}{d}h\right) = B_n \cos\left(\frac{n\pi}{d}x\right) e^{-(n\pi/d)h}$$

yielding

$$A_n e^{(n\pi/d)h} \sinh\left(\frac{n\pi}{d}h\right) = B_n$$

which allows the equations for the potential to be written in terms of A_n alone:

$$\Phi(x, y) = \begin{cases} \sum_{\substack{n=1 \\ \text{odd}}}^{\infty} A_n \cos\left(\frac{n\pi}{d}x\right) \sinh\left(\frac{n\pi}{d}y\right) & \text{when } 0 \leq y < h \quad \text{(3-62a)} \\ \sum_{\substack{n=1 \\ \text{odd}}}^{\infty} A_n \sinh\left(\frac{n\pi}{d}h\right) \cos\left(\frac{n\pi}{d}x\right) e^{-(n\pi/d)(y-h)} & \text{when } h \leq y < \infty \quad \text{(3-62b)} \end{cases}$$

To get the electric field between the signal conductor and the ground plane, we apply equation (2-65), $E_y = -\nabla\phi = -\partial\Phi/\partial y$. Since $d(\sinh ax)/dx = a\cosh ax$ and $d(e^{ax})/dx = ae^{ax}$, the electric fields become

$$E_{yn} = -\frac{\partial}{\partial y} A_n \cos\left(\frac{n\pi}{d}x\right) \sinh\left(\frac{n\pi}{d}y\right) = -\frac{n\pi A_n}{d} \cos\left(\frac{n\pi}{d}x\right) \cosh\left(\frac{n\pi}{d}y\right)$$

for region 1 and

$$\begin{aligned} E_{yn} &= -\frac{\partial}{\partial y} A_n \sinh\left(\frac{n\pi}{d}h\right) \cos\left(\frac{n\pi}{d}x\right) e^{-(n\pi/d)(y-h)} \\ &= \frac{n\pi A_n}{d} \sinh\left(\frac{n\pi}{d}h\right) \cos\left(\frac{n\pi}{d}x\right) e^{-(n\pi/d)(y-h)} \end{aligned}$$

for region 2, yielding

$$E_y(x, y) = \begin{cases} -\sum_{\substack{n=1 \\ \text{odd}}}^{\infty} \frac{n\pi A_n}{d} \cos\left(\frac{n\pi}{d}x\right) \cosh\left(\frac{n\pi}{d}y\right) & \text{when } 0 \leq y < h \quad \text{(3-63a)} \\ \sum_{\substack{n=1 \\ \text{odd}}}^{\infty} \frac{n\pi A_n}{d} \sinh\left(\frac{n\pi}{d}h\right) \times \cos\left(\frac{n\pi}{d}x\right) e^{-(n\pi/d)(y-h)} & \text{when } h \leq y < \infty \quad \text{(3-63b)} \end{cases}$$

To calculate the coefficient A_n, we must assume a charge distribution on the signal conductor. As a first-order approximation, it can be assumed that the charge (ρ) is spread out uniformly across the signal conductor at $y = h$. Consequently, there will be no variation of ρ with x on the strip, but it will be zero off the strip.

$$\rho(x) = \begin{cases} 1 & \text{when } |x| < w/2 \\ 0 & \text{when } |x| > w/2 \end{cases} \tag{3-64}$$

From equation (3-4), the electric fields above and below the signal conductor are equated to the charge density:

$$\rho(x) = (\varepsilon_0 E_{y1}) - (\varepsilon_0 \varepsilon_r E_{y2}) = \sum_{\substack{n=1 \\ \text{odd}}}^{\infty} \frac{\varepsilon_0 n\pi A_n}{d} \cos\left(\frac{n\pi}{d}x\right) \times \left[\sinh\left(\frac{n\pi}{d}h\right) + \varepsilon_r \cosh\left(\frac{n\pi}{d}h\right)\right] \tag{3-65}$$

If we temporarily assign all terms in (3-65) that are not a function of x (except A_n) to a constant term k as in (3-66), it is easy to see that it takes the form similar to the Fourier series shown in (3-68) when a_0 and b_m are zero:

$$\rho(x) = \sum_{\substack{n=1 \\ \text{odd}}}^{\infty} A_n k \cos\left(\frac{n\pi}{d}x\right) \tag{3-66}$$

$$k = \varepsilon_0 \frac{n\pi}{d}\left[\sinh\left(\frac{n\pi}{d}h\right) + \varepsilon_r \cosh\left(\frac{n\pi}{d}h\right)\right] \tag{3-67}$$

$$f(x) = \frac{1}{2}a_0 + \sum_{n=1}^{\infty} a_m \cos\left(\frac{m\pi}{d}x\right) + \sum_{n=1}^{\infty} b_m \cos\left(\frac{m\pi}{d}x\right) \tag{3-68}$$

This allows the use of Fourier series techniques to solve for the coefficient, A_n:

$$\int_{-w/2}^{w/2} \rho(x) \cos\left(\frac{m\pi}{d}x\right) dx = \int_{-d/2}^{d/2} A_n k \cos\left(\frac{n\pi}{d}x\right) \cos\left(\frac{m\pi}{d}x\right) dx$$

Setting $m = n$, we arrive at

$$\int_{-w/2}^{w/2} \rho(x) \cos\left(\frac{n\pi}{d}x\right) dx = \int_{-d/2}^{d/2} A_n k \cos^2\left(\frac{n\pi}{d}x\right) dx \tag{3-69}$$

After substituting (3-67) for k, integration of both sides of (3-69) yields

$$\frac{2d}{n\pi}\sin\frac{n\pi w}{2d} = \frac{A_n k d}{2}$$

$$A_n = \frac{4\sin(n\pi w/2d)}{\varepsilon_0[(n\pi)^2/d]\left[\sinh((n\pi/d)h) + \varepsilon_r\cosh((n\pi/d)h)\right]} \tag{3-70}$$

The voltage of the signal conductor can now be calculated with respect to the ground plane at $x = 0$ using (3-1) and (3-63a):

$$v = -\int_a^b \vec{E}\cdot d\vec{l} = -\int_0^h E_y(x=0, y)\,dy = \sum_{\substack{n=1\\ \text{odd}}}^{\infty} A_n \sinh\left(\frac{n\pi}{d}h\right) \tag{3-71}$$

The total charge is calculated with

$$Q = \int_{-w/2}^{w/2} \rho(x)dx = w \tag{3-72}$$

So the capacitance per unit length is calculated from (3-71) and (3-72) using (2-76):

$$C = \frac{Q}{v} = \frac{w}{\sum_{\substack{n=1\\ \text{odd}}}^{\infty} A_n \sinh[(n\pi/d)h]} \tag{3-73}$$

and the inductance per unit length is calculated using (3-46) by first calculating the capacitance with (3-73) with a relative dielectric permittivity of unity ($\varepsilon_r = 1$):

$$L = \frac{1}{c^2 C_{\varepsilon_r=1}} \tag{3-46}$$

With (3-73) and (3-46), we can calculate several useful quantities, such as the phase constant (3-30), the characteristic impedance (3-33), and the effective relative dielectric permittivity (ε_{eff}). To understand how to calculate ε_{eff}, we can examine the formulation of a parallel-plate capacitor, as derived in Example 2-3:

$$\frac{C_{\varepsilon_{r,\text{eff}}}}{C_{\varepsilon_r=1}} = \frac{\varepsilon_0\varepsilon_{r,\text{eff}}A/d}{\varepsilon_0 A/d} = \varepsilon_{r,\text{eff}} \tag{3-74}$$

Consequently, the effective permittivity is determined by calculating the capacitance of the microstrip with (3-73) using the correct value of the dielectric constant and dividing by the capacitance calculated with (3-73) when $\varepsilon_r = 1$. Since the dimensions that determine the capacitance remain identical, all that remains is the effective dielectric permittivity.

The accuracy of (3-73) and (3-74) is reasonable, but two approximations made during the derivation degrade the accuracy. First, we assumed that the signal conductor is infinitely thin, which is not realistic because microstrip transmission lines manufactured on printed circuit boards often have conductor thicknesses of dimensions similar to the dielectric height. There is little benefit of deriving a finite thickness formula here because the most useful method would employ numerical solutions that are not covered in this book. Second, the charge distribution on the signal conductor is not uniform. The charge distribution near sharp edges is derived in Section 3.4.4 and applied to this formulation.

3.4.4 Charge Distribution near a Conductor Edge

The approximation made in the analysis of Section 3.4.3 assumed that charge distribution on the signal conductor on the microstrip was uniform. Realistic charge distributions tend to increase significantly near the corners on the conductors, which will alter the capacitance and subsequently the effective permittivity of the transmission line. To begin the derivation of the charge distribution, consider Figure 3-19, which shows two intersecting planes at an angle of θ. The two cases of interest are when $\theta = 270^\circ$, which will yield the charge distribution near a sharp corner, and when $\theta = 360^\circ$, which will yield the charge distribution near the edge of a very thin strip. The derivation assumes that the conducting planes are kept at a potential V with respect to a point far away. Since we are interested

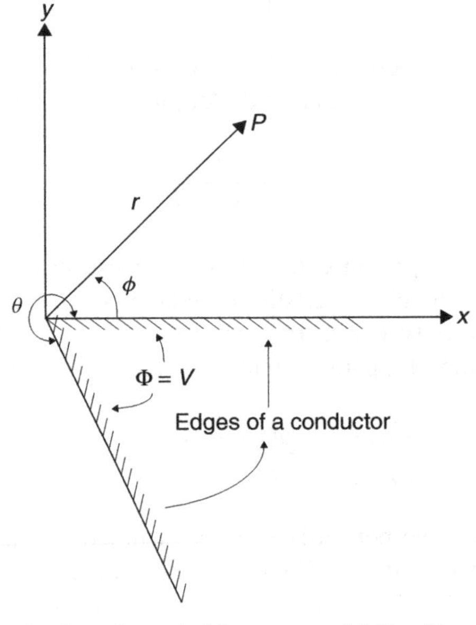

Figure 3-19 Two conducting planes held at a potential V with respect to a far-off point intersecting at an angle.

in the behavior of the fields only close to the edge, the details of the reference conductor far away are not considered.

We begin the derivation with Laplace's equation in cylindrical coordinates:

$$\nabla^2\Phi = \frac{1}{r}\frac{\partial}{\partial r}\left(r\frac{\partial\Phi}{\partial r}\right) + \frac{1}{r^2}\frac{\partial^2\Phi}{\partial\phi^2} = 0 \tag{3-75}$$

Similar to the solution of the microstrip transmission line, the solution to this partial differential equation is found using separation of variables. The problem is reduced to two ordinary differential equations, assuming that the potential can be represented by the product of a function for each coordinate [Jackson, 1999]:

$$\Phi(r, \phi) = R(r)Y(\phi) \tag{3-76}$$

Substitution of (3-76) into (3-75) and dividing each side by $R(r)Y(\phi)/r^2$ produces

$$\frac{r^2}{RYr}\frac{\partial}{\partial r}\left(r\frac{\partial RY}{\partial r}\right) + \frac{r^2}{RYr^2}\frac{\partial^2 RY}{\partial\phi^2} = \frac{r}{R}\frac{\partial}{\partial r}\left(r\frac{\partial R}{\partial r}\right) + \frac{1}{Y}\frac{\partial^2 Y}{\partial\phi^2} = 0 \tag{3-77}$$

Since the sum of the ordinary differential equations must equal zero, we can solve each one separately:

$$r\frac{d}{dr}\left(r\frac{dR}{dr}\right) = \beta^2 R \tag{3-78}$$

$$\frac{d^2Y}{d\phi^2} = -\beta^2 Y \tag{3-79}$$

For the solution of (3-78), try $R = kr^l$:

$$r\frac{d}{dr}\left(r\frac{d(kr^l)}{dr}\right) = r\frac{d}{dr}(rklr^{l-1}) = r\frac{d}{dr}klr^l = kl^2r^l$$

Setting the solution above to $\beta^2 R$ yields

$$\beta^2 R = \beta^2(kr^l) = l^2(kr^l)$$
$$l = \pm\beta$$

Hence, the solution to (3-78) is shown:

$$R(r) = ar^\beta + br^{-\beta} \tag{3-80}$$

The solution to (3-79) is found by solving the roots of the characteristic equation

$$Y(D^2 + \beta^2) = 0$$

yielding

$$Y(\phi) = A\cos\beta\phi + B\sin\beta\phi \tag{3-81}$$

Since (3-81) is composed of periodic functions, β must be an integer. Substituting (3-80) and (3-81) into (3-76) yields the general solution for the potential when $\beta \neq 0$:

$$\Phi(r,\phi) = ar^{\beta}A\cos\beta\phi + ar^{\beta}B\sin\beta\phi + br^{-\beta}A\cos\beta\phi + br^{-\beta}B\sin\beta\phi \tag{3-82}$$

The boundary conditions that must be satisfied are:

- $\Phi = V$ when $\phi = \theta$.
- $\Phi = V$ when $\phi = 0$.

If $\beta = n\pi/\theta$, the sin terms in (3-82) become zero when n is odd. Since $\sin(0) = 0$, the choice of β allows us to write a function for the potential that satisfies the boundary conditions by setting the constant $A = 0$, and combining the constants $\alpha_n = aB$ and $\alpha_n' = bB$:

$$\Phi(r,\phi) = V + \sum_{\substack{n=1\\ \text{odd}}}^{\infty} \alpha_n r^{n\pi/\theta}\sin\left(\frac{n\pi}{\theta}\phi\right) + \alpha_n' r^{-n\pi/\theta}\sin\left(\frac{n\pi}{\theta}\phi\right) \tag{3-83}$$

The ϕ component of the electric field is calculated from equation (2-65) ($\vec{E} = -\nabla\Phi$). From Appendix A we get

$$\nabla\Phi = \vec{a}_r\frac{\partial\Phi}{\partial r} + \vec{a}_\phi\frac{1}{r}\frac{\partial\Phi}{\partial\phi} + \vec{a}_z\frac{\partial\Phi}{\partial z}$$

yielding

$$E_\phi = -\frac{1}{r}\frac{\partial\Phi}{\partial\phi} = \sum_{\substack{n=1\\ \text{odd}}}^{\infty}\left[-\frac{n\pi\alpha_n}{\theta}r^{(n\pi/\theta)-1}\cos\left(\frac{n\pi}{\theta}\phi\right)\right.$$
$$\left. + \frac{n\pi\alpha_n'}{\theta}r^{-(n\pi/\theta+1)}\cos\left(\frac{n\pi}{\theta}\phi\right)\right] \tag{3-84}$$

Finally, the charge can be calculated with (3-3):

$$\vec{n} \cdot \varepsilon \vec{E} = \rho$$

$$\rho(r, \phi) = \varepsilon E_\phi = \sum_{\substack{n=1 \\ \text{odd}}}^{\infty} \left[-\frac{n\pi \varepsilon \alpha_n}{\theta} r^{(n\pi/\theta)-1} \cos\left(\frac{n\pi}{\theta}\phi\right) + \frac{n\pi \varepsilon \alpha'_n}{\theta} r^{-(n\pi/\theta+1)} \cos\left(\frac{n\pi}{\theta}\phi\right) \right]$$

Since we are interested in the behavior of the charge in the immediate vicinity of the edge, r is small and therefore only the first term, the summation, is important ($n = 1$). Also, since for a flat plane the dependence on r should vanish and the last term shows dependence on r for $\theta = \pi$ and $n = 1$, it is necessary that $\alpha'_n = 0$.

$$\rho(r, \phi) = -\frac{\pi \varepsilon \alpha_1}{\theta} r^0 \cos\left(\frac{\pi}{\theta}\phi\right) + \frac{\pi \varepsilon \alpha'_1}{\theta} r^{-2} \cos\left(\frac{\pi}{\theta}\phi\right)$$

This yields the surface charge density as a function of the angle between two conducting planes in the near vicinity of the corner:

$$\rho(r, \phi) = -\frac{\pi \varepsilon \alpha_1}{\theta} r^{(\pi/\theta)-1} \cos\left(\frac{\pi}{\theta}\phi\right) \tag{3-85}$$

For a sharp corner, $\theta = 3\pi/2$, as shown in Figure 3-20, 3.85 reduces to

$$\rho(r, \phi = 3\pi/2) = \frac{2\varepsilon\alpha_1}{3} r^{-1/3} \tag{3-86}$$

Since we are concerned with the charge density on the metal planes, ϕ will be either θ or 0, meaning that the cosine term will equal 1 or -1.

Equation (3-86) means that for a transmission line built with a square conductor, the charge density increases dramatically near the corner, as shown in Figure 3-20. In fact, for a perfectly sharp edge, the charge density becomes singular as r approaches zero. For the edge of a thin strip, $\theta = 2\pi$ and (3-85) reduces to

$$\rho(r, \phi = 2\pi) = \frac{\varepsilon\alpha_1}{2} r^{-1/2} \tag{3-87}$$

Equation (3-87) means that for an infinitely thin metal strip, as was assumed in the derivation for the microstrip line in Section 3.4.3, the charge density also approaches infinity near the edge. Since infinitely thin sheets or infinitely sharp corners do not really exist in nature, the trends of (3-86) and (3-87) are what must be understood. *The charge density increases dramatically near sharp corners on metal conductors*. The coefficients α_n are dependent on the initial conditions remote from the corner, which are not solved here.

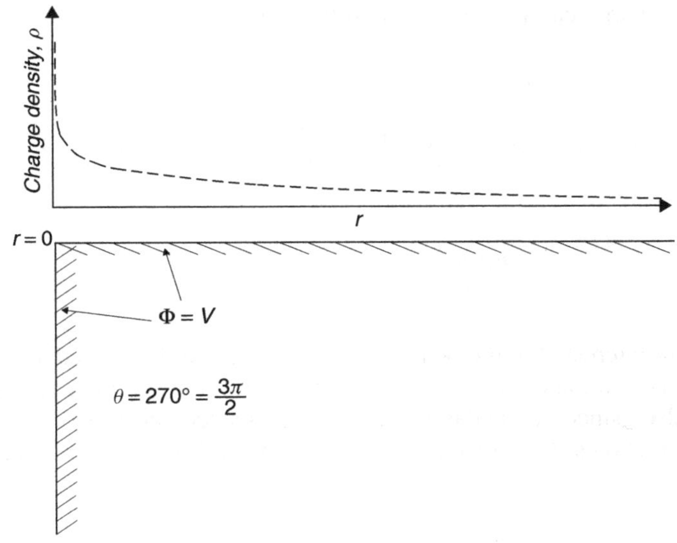

Figure 3-20 Charge distribution near a sharp corner.

3.4.5 Charge Distribution and Transmission-Line Parameters

Now that the behavior of the charge in the near vicinity of the edge is understood, it is easy to imagine what the total charge distribution should look like across the signal conductor of a transmission line. The charge density should be almost uniform near the center of the strip but will increase dramatically near the edges, as predicted by (3-85). A useful formula to predict the current distribution for a microstrip, which works well for a variety of dimensions, was derived by Collins [1992] using conformal mapping techniques:

$$\rho(x) = \frac{2Q}{\pi w \sqrt{1 - \dfrac{x^2}{\left(\dfrac{w}{2}\right)^2}}} \tag{3-88}$$

where Q is the total charge, x the distance from the conductor center, and w the width of the signal conductor. Figure 3-21 shows an example of a realistic current distribution in the signal conductor of a microstrip transmission line, as calculated with (3-88) and normalized so that the center of the conductor has a charge density of 1.

To understand how the charge distribution will influence the transmission-line parameters such as delay and impedance, the equations for the microstrip transmission line derived in Section 3.4.3 can be improved by applying a more realistic charge distribution to the signal conductor and solving (3-69) for A_n:

$$\int_{-w/2}^{w/2} \rho(x) \cos\left(\frac{n\pi}{d}x\right) dx = \int_{-d/2}^{d/2} A_n k \cos^2\left(\frac{n\pi}{d}x\right) dx \tag{3-69}$$

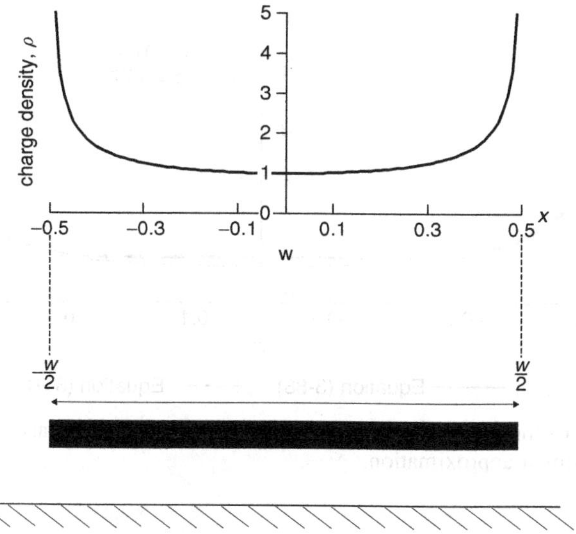

Figure 3-21 Charge distribution across a microstrip signal conductor.

The problem arises that (3-88) is singular at $x = w/2$, so it cannot be integrated. Fortunately, we can circumvent this problem by choosing a well-behaved function to approximate (3-88) that can be integrated:

$$\rho(x) = \frac{2Q}{\pi w\sqrt{1 - \dfrac{x^2}{\left(\dfrac{w}{2}\right)^2}}} \cong Q(ax^m + c) \tag{3-89}$$

where a, m, and c should be chosen so that the polynomial approximates the realistic charge distribution. If the total charge is to remain the same as in (3-72), and the width of the transmission line is normalized to $w = 1$, the charge distribution must satisfy

$$Q = \int_{-w/2}^{w/2} \rho(x)dx = 1 \tag{3-90}$$

Furthermore, the order of the polynomial must be high to approximate the sharp increase in the charge density near the edges. If the order of (3-89) is chosen to get a reasonable fit ($m = 6$), the polynomial will satisfy (3-90) when $a = 100$ and $c = 0.77$ for a strip width of 1,

$$\rho(x) \approx 100x^6 + 0.77 \tag{3-91}$$

where

$$\int_{-0.5}^{0.5} (100x^6 + 0.77)\, dx \approx 1.$$

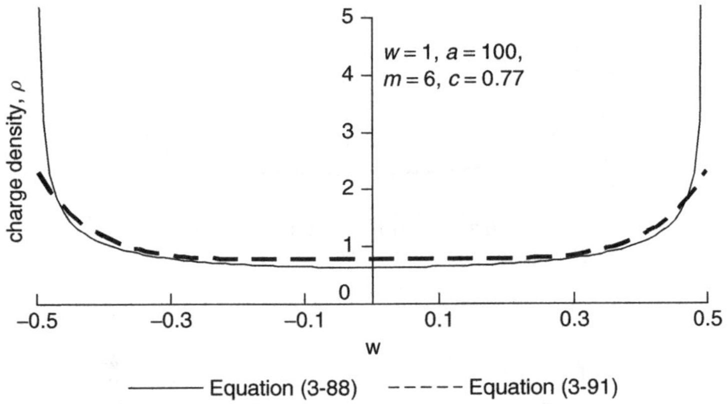

Figure 3-22 Comparison of the charge distribution calculated analytically and the non-singular polynomial approximation.

TABLE 3-1. Effect of Charge Distribution on the Impedance and Effective Relative Permittivity[a]

	$Z_0(\Omega)/\varepsilon_{\text{eff}}$		
w/h	$\rho(x)=\begin{cases}1 & \text{when } \lvert x\rvert < w/2\\ 0 & \text{when } \lvert x\rvert > w/2\end{cases}$	$\rho(x) = 100x^6 + 0.77$	Two-Dimensional Numerical Field Solver Results
4	37.6/3.26	31.6/3.37	32.5/3.38
2	59.7/3.02	52.6/3.08	51.4/3.13
1	84.8/2.86	77.0/2.90	74.6/2.96
0.667	99.9/2.81	92.1/2.83	89.3/2.89
0.5	110.8/2.78	99.1/2.76	99.9/2.84

[a] $d = 100h$, widths normalized to 1, $\varepsilon_r = 4.0$, $n = 5000$.

Figure 3-22 shows a comparison between the charge distribution approximated with (3-91), which is not singular, and (3-88). Plugging (3-91) into (3-69) and solving for A_n will provide a formula for the microstrip transmission line that accounts for a realistic charge distribution on the signal conductor. Unfortunately, the integration gets messy and the final form of A_n is ungainly and therefore is not shown here. Mathematica was programmed to perform the integration and solve (3-69) for A_n when the charge distribution was approximated with (3-91). Comparisons between the calculated impedance and the effective dielectric constant assuming a uniform charge distribution using (3-64) and a realistic distribution calculated with (3-88) and approximated with (3-91) is shown in Table 3-1. Note that when a realistic charge distribution is used, the quasistatic approximations are significantly more accurate when compared to a commercial two-dimensional numerical field solver.

3.4.6 Field Mapping

So far, we have used Laplace's equation to calculate the transmission-line parameters. Next we explore a graphical method for solving general two-dimensional electrostatic problems. The chief merit of this method is that it provides unique insight into the field patterns for complex geometries. Additionally, the method is applicable to systems where the analytical approach is either extremely difficult or impossible. It also demonstrates a methodology where the structure is meshed, which is similar to techniques used by many numerical field solving techniques. A full understanding of this method is useful because it will allow the signal integrity engineer to develop an intuitive understanding of how fields are distributed and how transmission-line geometries affect impedance and the effective dielectric permittivity.

For any two-dimensional system possessing a uniform charge distribution along the z-axis, the same electrostatic field sketch can be applied to every cross section. An example of the field sketch for an arbitrary cross section possessing the charges q and $-q$ in every length l is shown in Figure 3-23. Equation (3-3) tells us that the electric field lines must terminate on a perfect conductor normal to the surface. This means that if a line is drawn on a two-dimensional field sketch so that it is always perpendicular to the electric fields, and that line is extended into the page, the surface created must be a constant potential, as shown in Figure 3-23. These constant potential surfaces can be treated as

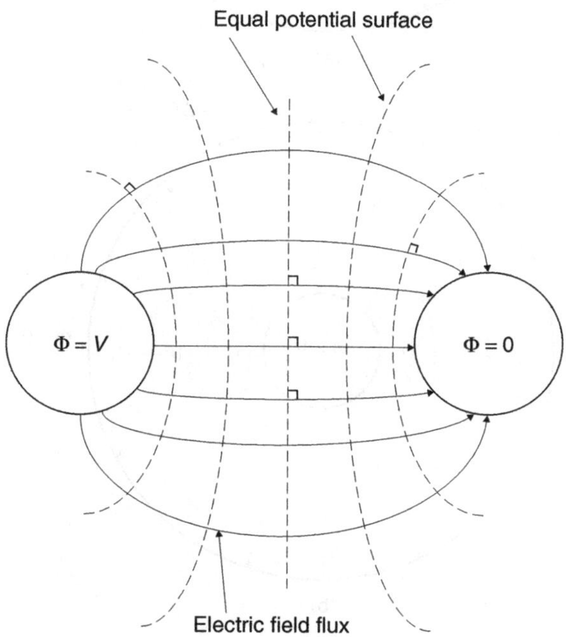

Figure 3-23 Electrostatic field sketch of two arbitrary conductors.

infinitely thin sheets of "virtual" conducting foil. Using the constant potential surfaces, the capacitance per unit length of a two-dimensional system can be found with reasonable accuracy from a careful field sketch. For example, consider the cross section of a coaxial line as shown in Figure 3-24, where electric field flux lines originate from a charge q distributed over a length l on the inner conductor and terminate on $-q$ in the same length on the outer conductor. If equal potential lines are drawn between the inner and outer conductors, the total

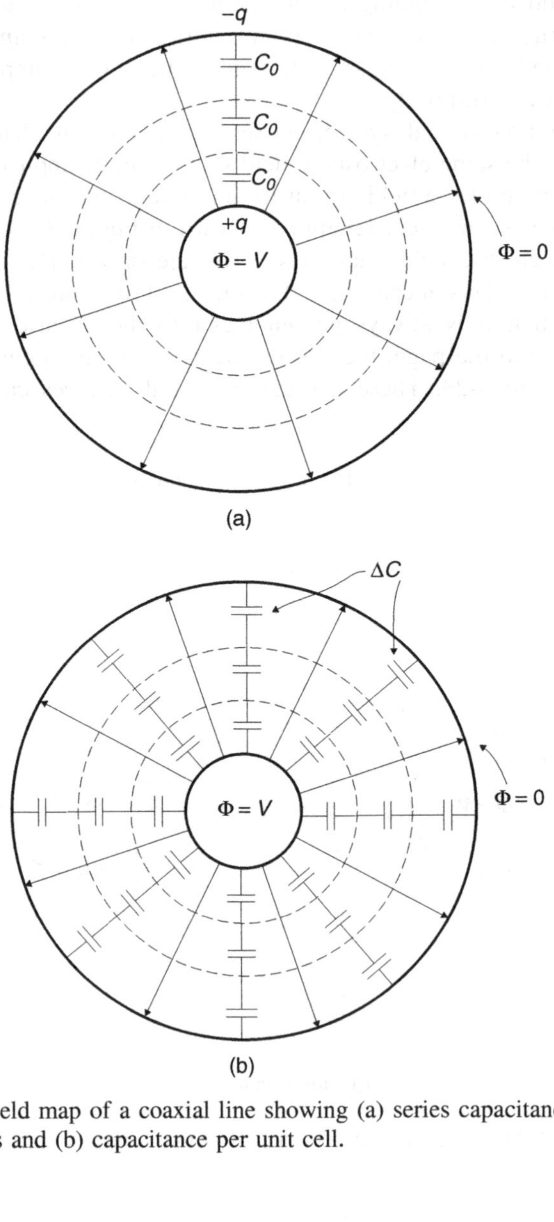

Figure 3-24 Field map of a coaxial line showing (a) series capacitance between equal potential surfaces and (b) capacitance per unit cell.

capacitance is simply the series combination of the capacitance between equal potential lines as shown in Figure 3-24a. If the equal potential lines are chosen so that each capacitance is the same and n_s denotes the number of elements in series, the total capacitance is

$$C = \frac{C_0}{n_s} \tag{3-92}$$

Furthermore, each of the series capacitors in Figure 3-24a can be subdivided into parallel capacitance values with a value of ΔC for each cell, as shown in Figure 3-24b. Assuming n_p parallel elements, $C_0 = n_p \Delta C$, yielding a total capacitance

$$C = \frac{n_p}{n_s} \Delta C \tag{3-93}$$

In order to use (3-93), we need to calculate the cell capacitance ΔC. Assuming that the charge Δq and $-\Delta q$ are present on the top and bottom of each cell wall, we can write the cell capacitance in terms of the potential difference between the boundaries using (2-76) and (3-1):

$$\Delta C = \frac{\Delta q}{\int \vec{E} \cdot d\vec{l}} \tag{3-94}$$

If the problem is simplified by drawing the field lines such that each cell is approximately the same size, the capacitance can be written in terms of the average electric fields,

$$\Delta C = \frac{\Delta q}{E_{\text{ave}} \Delta h_{\text{ave}}} \tag{3-95}$$

where Δh_{ave} is the average height of a cell and E_{ave} is the average electric field for a cell.

Equation (3-95) can be simplified to eliminate the electric field using the integral form of Gauss's law, (2-59), $\oint_S \varepsilon \vec{E} \cdot d\vec{s} = \oint_S \vec{D} \cdot d\vec{s} = q$:

$$\Delta C = \frac{\Delta q}{E_{\text{ave}} \Delta h_{\text{ave}}} = \frac{\varepsilon \Delta q}{D_{\text{ave}} \Delta h_{\text{ave}}} = \frac{\varepsilon \Delta q}{(\Delta q / l \Delta w_{\text{ave}}) \Delta h_{\text{ave}}} = \frac{\varepsilon \Delta w_{\text{ave}}}{\Delta h_{\text{ave}}} l \tag{3-96}$$

where Δw_{ave} is the average cell width and $ds = l \Delta w_{\text{ave}}$ is the area of the cell surface for the transmission line length l. If the cells can be drawn so that $\Delta w_{\text{ave}} \approx \Delta h_{\text{ave}}$, the total capacitance is calculated by substituting (3-96) into (3-93):

$$\frac{C}{l} = \varepsilon \frac{n_p}{n_s} \quad \text{F/m} \tag{3-97}$$

Example 3-2 Use field mapping techniques to calculate the impedance of a coaxial transmission line shown in Figure 3-25, where $b/a = 2$ and the permittivity of the dielectric is $\varepsilon_r = 2.3$.

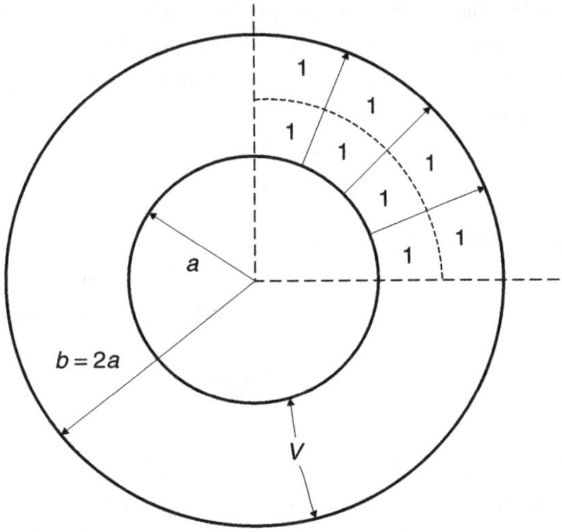

Figure 3-25 Field map used to calculate the capacitance in Example 3-2, showing one quadrant divided up into two series sections and four parallel sections.

SOLUTION Since the cross section of the coaxial line is symmetrical, we can take advantage of symmetry and draw the field lines for only one quadrant, as shown in Figure 3-25. In this sketch, the quadrant was divided into two series sections using a single equal potential line, and four parallel sections yielding $n_s = 2$ and $n_p = (4)(4) = 16$. Therefore, equation (3-97) yields

$$C = \varepsilon_r \varepsilon_0 \frac{n_p}{n_s} = (2.3)(8.85 \times 10^{-12}) \left(\frac{16}{2} \right) = 162.8 \times 10^{-12} \text{ F/m}$$

To calculate the impedance of the coaxial line, we must first recalculate the capacitance for $\varepsilon_r = 1$ and calculate the inductance from (3-46),

$$C_{\varepsilon_r = 1} = \varepsilon_0 \frac{n_p}{n_s} = 8.85 \times 10^{-12} \left(\frac{16}{2} \right) = 70.78 \times 10^{-12} F/m$$

$$L = \frac{1}{c^2 C_{\varepsilon_r = 1}} = \frac{1}{(3 \times 10^8)^2 (70.78 \times 10^{-12})} = 156.9 \times 10^{-9} H/m$$

yielding the characteristic impedance calculated from (3-33):

$$Z_{0,\text{fieldmap}} = \sqrt{\frac{156.9 \times 10^{-9}}{162.8 \times 10^{-12}}} = 31.0\ \Omega$$

This can be compared directly to the results derived in Section 3.4.2.

$$C = \frac{2\pi\varepsilon}{\ln(b/a)} = \frac{2\pi(2.3)(8.85 \times 10^{-12})}{\ln(2)} = 184.5 \times 10^{-12} \text{ F/m}$$

Again, to calculate the impedance, we must first recalculate the capacitance for $\varepsilon_r = 1$.

$$C = \frac{2\pi\varepsilon}{\ln(b/a)} = \frac{2\pi(1)(8.85\times10^{-12})}{\ln(2)} = 80.2\times10^{-12}\ \text{F/m}$$

$$L = \frac{1}{c^2 C_{\varepsilon_r=1}} = \frac{1}{(3\times10^8)^2(80.2\times10^{-12})} = 138.5\times10^{-9}\ \text{H/m}$$

$$Z_0 = \sqrt{\frac{138.5\times10^{-9}}{184.5\times10^{-12}}} = 27.3\ \Omega$$

Comparing this result to $Z_{0,\text{fieldmap}}$ shows reasonable accuracy but demonstrates the inherent errors that are inevitable when using hand sketches.

Example 3-3 Use field mapping techniques to calculate the impedance and the effective dielectric permittivity of the microstrip transmission line shown in Figure 3-26, where $w/h = 1$ and the permittivity of the dielectric material is $\varepsilon_r = 4.0$.

SOLUTION It is inherently more difficult to apply mapping techniques to a microstrip because the fields are difficult to draw accurately and the dielectric is not homogeneous. Figure 3-26 shows a drawing of the field lines. Note

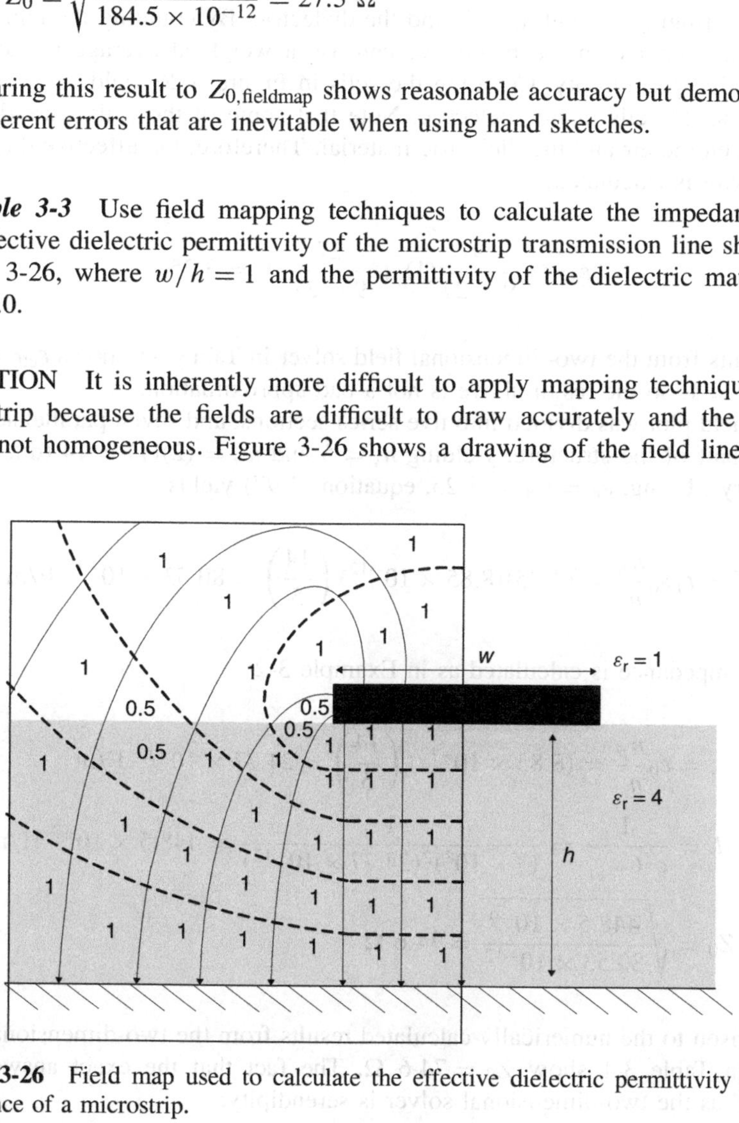

Figure 3-26 Field map used to calculate the effective dielectric permittivity and the impedance of a microstrip.

that the authors were not able to draw the cells to be all the same size; however, we will still assume an average electric field strength for each cell, which will induce some error. Since some cells are small and some cells are large, the errors induced by the cell size discrepancy will partially average out and should give reasonable results. For this example, symmetry was used and only half of the field lines were drawn. The distortion of the electric field lines at the dielectric boundary was ignored, which will induce an additional, although small error.

Initially, we must calculate the effective dielectric permittivity. In Section 3.3.3 we explained that the effective permittivity was due to the weighted average of the fields fringing through the air and the dielectric. By counting the number of cells in the air and in the board, we can use a weighted average to calculate the effective permittivity. Counting the cells in Figure 3-26 yields nine cells in the air and 27 cells in the dielectric. Note that some of the cells were divided up between the air and the dielectric material. Therefore, the effective dielectric permittivity is calculated:

$$\varepsilon_{\text{eff}} \approx \frac{9}{9+27}(1) + \frac{27}{9+27}(4) = 3.25$$

The results from the two-dimensional field solver in Table 3-1 shows $\varepsilon_{\text{eff}} = 2.96$ for $w/h = 1$, so the result above is not a bad approximation.

The field plot was divided into five series sections, and seven parallel sections for one-half of the structure, yielding $n_s = 5$ and $n_p = (2)(7) = 14$ (due to the symmetry). Using, $\varepsilon_r = \varepsilon_{\text{eff}} = 3.25$, equation (3-97) yields

$$C = \varepsilon_r \varepsilon_0 \frac{n_p}{n_s} = (3.25)(8.85 \times 10^{-12}) \left(\frac{14}{5}\right) = 80.53 \times 10^{-12} \text{ F/m}$$

and the impedance is calculated as in Example 3-2:

$$C_{\varepsilon_r=1} = \varepsilon_0 \frac{n_p}{n_s} = (8.85 \times 10^{-12}) \left(\frac{14}{5}\right) = 24.77 \times 10^{-12} \text{ F/m}$$

$$L = \frac{1}{c^2 C_{\varepsilon_r=1}} = \frac{1}{(3 \times 10^8)^2 (24.77 \times 10^{-12})} = 448.5 \times 10^{-9} \text{ H/m}$$

$$Z_0 = \sqrt{\frac{448.5 \times 10^{-9}}{80.53 \times 10^{-12}}} = 74.6\ \Omega$$

Comparison to the numerically calculated results from the two-dimensional field solver in Table 3-1 show $Z_0 = 74.6\ \Omega$. The fact that the exact answer was obtained as the two-dimensional solver is serendipity.

3.5 TRANSMISSION-LINE REFLECTIONS

So far, we have considered the propagation of signals on a single transmission line. However, most practical problems involve signals propagating on a series of cascaded transmission lines, such as those built on a motherboard, a daughtercard, and a chip package. In addition, digital designs often utilize transmission-line topologies that branch out from one driver to multiple receivers. Since each transmission line will have different electric characteristics, and topologies will affect signal integrity dramatically, it is essential to understand how a propagating electromagnetic wave will behave when it enters a region where the properties of the transmission line and/or the topology of the interconnection change.

3.5.1 Transmission-Line Reflection and Transmission Coefficient

Generally, when an electromagnetic plane wave propagating on transmission line A transitions to line B, where the characteristic impedance changes, two things happen: (1) a portion of the wave is reflected away from the impedance discontinuity back toward the source, and (2) a portion of the wave is transmitted onto transmission line B, as shown in Figure 3-27. The simultaneous existence of both the transmitted and reflected waves is a direct result of the boundary conditions that must be satisfied when solving Maxwell's equations at the interface between the two regions. These boundary conditions are identical to those derived in Section 2.7 for a plane wave impinging on a region with different properties. The parallels between a plane wave propagating in free space and a wave propagating on a transmission line were shown in Section 3.2, where the wave equation was derived in terms of the transmission-line circuit parameters (L and C).

If we assume that the values are represented by phasors, we can separate the total plane wave into three parts: the incident, the reflected, and the transmitted.

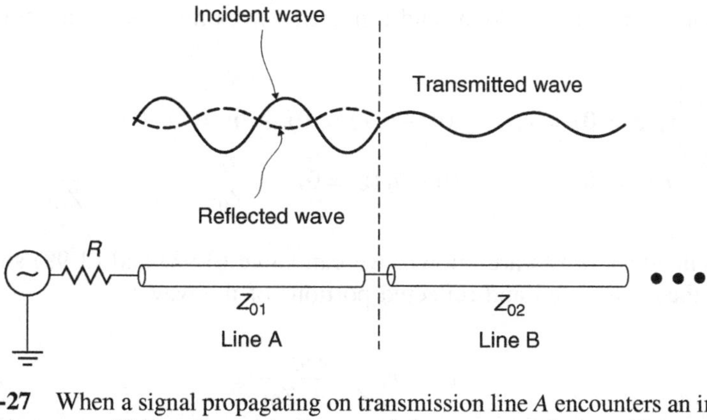

Figure 3-27 When a signal propagating on transmission line A encounters an impedance discontinuity, part will be reflected back toward the source and part will be transmitted onto transmission line B.

1. Incident wave:

$$v_i(z) = v_i^+ e^{-j\beta_1 z}$$
$$i_i(z) = \frac{1}{Z_{01}} v_i^+ e^{-j\beta_1 z}$$

2. Reflected wave:

$$v_r(z) = v_r^- e^{+j\beta_1 z}$$
$$i_r(z) = -\frac{1}{Z_{01}} v_r^- e^{+j\beta_1 z}$$

3. Transmitted wave:

$$v_t(z) = v_t^+ e^{-j\beta_2 z}$$
$$i_t(z) = \frac{1}{Z_{02}} v_t^+ e^{-j\beta_2 z}$$

where $\beta_1 = \omega\sqrt{L_1C_1}$, $\beta_2 = \omega\sqrt{L_2C_2}$, $Z_{01} = \sqrt{L_1/C_1}$, and $Z_{02} = \sqrt{L_2/C_2}$ are the phase constants and characteristic impedances of transmission line A and region B, respectively.

When the wave intersects a boundary between the transmission lines, the tangential components of both the electric and magnetic fields across the interface must remain continuous, as shown with equation (3-8). In other words, the tangential component of the fields cannot change instantaneously. In our particular scenario, the plane wave is propagating in the TEM mode in the z-direction, so both the electric and magnetic fields are oriented parallel (i.e., tangent) to the boundary of the dielectric interface. Subsequently, we can say that at the interface ($z = 0$), the sum of the incident and reflected waves must equal the transmitted wave.

$$v_t(z = 0) = v_i(z = 0) + v_r(z = 0) \rightarrow v_t = v_i + v_r \tag{3-98}$$

$$i_t(z = 0) = i_i(z = 0) + i_r(z = 0) \rightarrow \frac{i_t}{Z_{02}} = \frac{i_i}{Z_{01}} - \frac{i_r}{Z_{01}} \tag{3-99}$$

Since the incident waves are known, we can solve (3-98) and (3-99) simultaneously for the transmitted and reflected portions of the wave:

$$v_t = v_i \frac{2Z_{02}}{Z_{02} + Z_{01}} \tag{3-100}$$

$$v_r = v_i \frac{Z_{02} - Z_{01}}{Z_{02} + Z_{01}} \tag{3-101}$$

Following the method used to derive equations (2-132) and (2-133) leads to the definition of the reflection and transmission coefficients:

$$\Gamma \equiv \frac{v_r}{v_i} = \frac{Z_{02} - Z_{01}}{Z_{02} + Z_{01}} \tag{3-102}$$

$$T \equiv \frac{v_t}{v_i} = \frac{2Z_{02}}{Z_{02} + Z_{01}} = 1 + \Gamma \tag{3-103}$$

The reflection coefficient is a measure of how much is reflected back off the intersection between the two impedance regions, and the transmission coefficient tells how much of the wave is transmitted. If the reflection coefficient is zero, it means that the characteristic impedances in the two regions are identical. If the characteristic impedances are not equal, the reflection coefficient will be finite. If the impedance discontinuity is infinite, such as an open circuit, the signal propagating on transmission line A will be reflected 100%, as shown in Figure 3-28a. This is easy to show simply by taking the limit of (3-102) as Z_{02} goes to infinity:

$$\Gamma_{\text{open}} = 1 \tag{3-104}$$

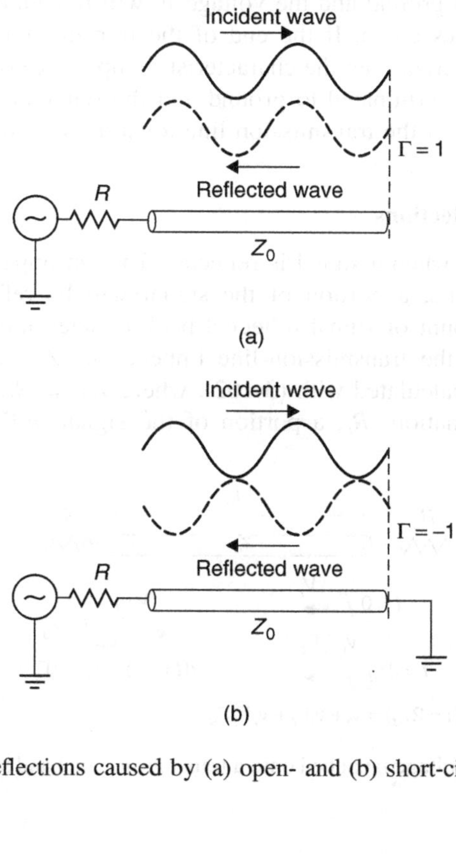

Figure 3-28 Reflections caused by (a) open- and (b) short-circuit termination.

If the impedance discontinuity is shorted to ground, as shown in Figure 3-28b, the signal propagating on transmission line A will also be reflected 100%, but the reflected wave will be out of phase from the incident by 180°, yielding

$$\Gamma_{\text{short}} = -1 \tag{3-105}$$

The reason for the phase shift is explained in Section 2.7.1, for a plane wave incident on a perfect conductor.

3.5.2 Launching an Initial Wave

When a driver launches a signal onto a transmission line, the magnitude of the initial voltage (v_i) seen on the transmission line at $t = 0$ will be governed by the voltage divider between the source resistance and the line impedance (see Figure 3-29):

$$v_i = v_s \frac{Z_0}{Z_0 + R_s} \tag{3-106}$$

If the end of the transmission line is terminated with an impedance that exactly matches the characteristic impedance of the line, the signal with amplitude v_i will be terminated to ground and the voltage v_i will remain on the line until the signal source switches again. If the end of the transmission line is terminated with a resistance different than the characteristic impedance of the line, a portion of the signal will be terminated to ground and the remainder of the signal will be reflected back down the transmission line toward the source.

3.5.3 Multiple Reflections

As described above, when a signal is reflected from an impedance discontinuity at the end of the line, a portion of the signal will be reflected back toward the source. The amount of signal reflected back is determined by the reflection coefficient between the transmission-line impedance (Z_0) and the termination resistance (R_t) and calculated with (3-102), where $Z_{02} = R_t$. When the incident wave hits the termination, R_t, a portion of the signal, $v_i\Gamma_t$, is reflected back

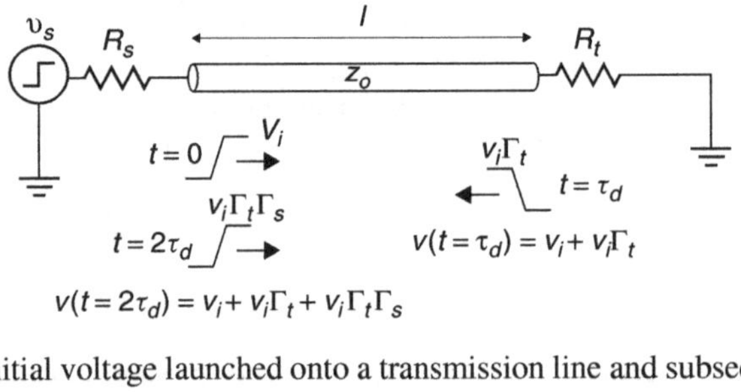

Figure 3-29 Initial voltage launched onto a transmission line and subsequent reflections.

toward the source and is added to the incident wave to produce a total magnitude on the line of $v_i\Gamma_t + v_i$, as shown in Figure 3-29. The reflected component ($v_i\Gamma_t$) will then travel back to the source and will generate another reflection if $Z_0 \neq R_s$. Obviously, if $Z_0 = R_s$, there will be no reflection at the source.

If an impedance discontinuity exists on both sides of the transmission line, the signal will bounce back and forth between the driver and receiver until the reflections eventually reach steady state at the dc solution. For example, consider Figure 3-29, which shows one example for a time interval of a few τ_d, where τ_d is the time delay for the signal to propagate from one end of the transmission line to the other, which is derived from equation (3-31) and calculated by

$$\tau_d = \frac{l}{\nu_p} = l\sqrt{LC} \qquad \text{seconds} \tag{3-107}$$

where l is the length of the transmission line and L and C are the inductance and capacitance values per unit length.†

When the source transitions from 0 volts to v_s, the initial voltage on the line at $t = 0$ is v_i, determined by (3-106). At time $t = \tau_d$, the incident voltage v_i arrives at the load R_t. At this time a reflected component is generated with a magnitude of $v_i\Gamma_t$, where Γ_t is the reflection coefficient looking into the terminating load):

$$\Gamma_t = \frac{R_t - Z_0}{R_t + Z_0}$$

The reflected component is added to the incident voltage v_i, creating a total voltage at the load of $v_i\Gamma_t + v_i$. The reflected portion of the wave $v_i\Gamma_t$ then travels back to the source and at time $t = 2\tau_d$ generates a reflection off the source determined by $v_i\Gamma_t\Gamma_s$, where

$$\Gamma_s = \frac{R_s - Z_0}{R_s + Z_0}$$

is the reflection coefficient looking into the source. At this time the voltage seen at the source will be the previous voltage (v_i) plus the voltage from the reflection ($v_i\Gamma_t$) plus the reflected wave ($v_i\Gamma_t\Gamma_s$). This reflecting and counter reflecting will continue until the line voltage has approached the steady-state dc value. As the reader can see, the reflections could take a long time to reach equilibrium if the termination and source resistors do not have values similar to the characteristic impedance of the transmission line.

†Note that the phase velocity, v_p, calculated in (3-31) is a frequency-domain variable that is valid only at a single frequency. The time delay will change with frequency because of non-TEM propagation (Section 3.3.5) and realistic dielectric behavior, as discussed in Chapter 6. However, for the purposes here, it is valid to assume frequency-invariant dielectric properties, which allows (3-107) to be applied over the broad range of frequencies contained in a trapezoidal digital waveform.

It is apparent from the example above that hand calculation of multiple reflections can be rather tedious. An easier way to predict the effect of reflections on a signal is to use a lattice diagram.

3.5.4 Lattice Diagrams and Over- or Underdriven Transmission Lines

A *lattice diagram* (sometimes called a *bounce diagram*) is a graphical technique used to solve the multiple reflections on a transmission line with linear loads. Figure 3-30 shows a sample lattice diagram for the transmission line shown in Figure 3-29. The left- and right-hand vertical lines represent the source end ($z = 0$) and load end ($z = l$) of the transmission line. The diagonal lines contained between the vertical lines represent the signal bouncing back and forth between the source and the load. The diagram progressing from top to bottom represents increasing time. Notice that the time increment is equal to the time delay τ_d of the transmission line as calculated with (3-107), and the reflection coefficients looking into the source and into the load are labeled at the top of the vertical bars. The lowercase letters represent the magnitude of the reflected signal traveling on the line. The uppercase letters represent the voltages seen at the source, and the primed uppercase letters represent the voltage seen at the load end of the line. For example, referring to Figure 3-30, the near end of the line will be held at

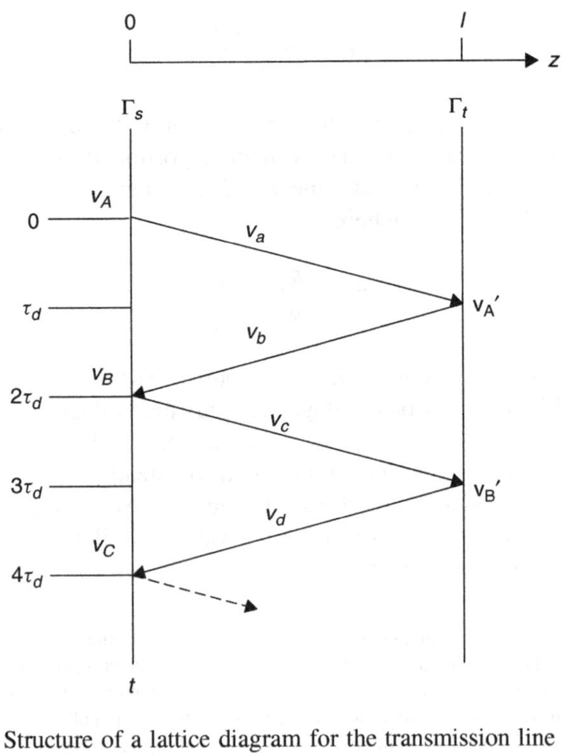

Figure 3-30 Structure of a lattice diagram for the transmission line in Figure 3-29.

a voltage of v_A volts for a duration of $2\tau_d$. The voltage v_a is simply the initial voltage v_i, which will remain constant until the reflection from the load reaches the source and is calculated with (3-106):

$$v_A = v_a = v_s \frac{Z_o}{Z_o + R_s}$$

The voltage v_a will propagate down the transmission line toward the source.

The voltage v'_A is the voltage v_a plus the voltage reflected off the load, v_b, where

$$v_b = v_a \Gamma_t$$

$$v'_A = v_a + v_b$$

The voltage v_B is the sum of the incident voltage v_a plus the signal reflected from the load v_b and the signal reflected off the source v_c:

$$v_B = v_a + v_b + v_c$$

where

$$v_c = v_b \Gamma_s$$

The reflections on the line eventually reach the steady-state voltage of the source, v_s, if the line is open. However, if the line is terminated with a resistor R_t, the steady-state voltage is computed with the voltage divider between the source and the load (assuming that the transmission line is loss-free):

$$v_{\text{steady state}} = v_s \frac{R_t}{R_t + R_s}$$

Example 3-4 Multiple Reflections for $R_s > Z_0$ Consider the transmission-line system and lattice diagram shown in Figure 3-31. The initial voltage launched onto the transmission line will be governed by the voltage divider between the source impedance R_s and the line impedance Z_0:

$$v_i = v_s \frac{Z_0}{Z_0 + R_s} = 2 \cdot \frac{50}{50 + 75} = 0.8 \text{ V}$$

The initial signal, $v_a = 0.8$ V, will travel down the line until it reaches the load. In this particular case, the load is open and thus has a reflection coefficient of 1. Subsequently, the entire signal is reflected back toward the source and is added to the incident signal of 0.8 V. So at time $t = \tau_d$, or 250 ps in this specific example, the signal seen at the load is $v_{A'} = v_a + v_b = 0.8 + 0.8 = 1.6$ V. The 0.8-V reflected signal v_b will then propagate down the line toward the source.

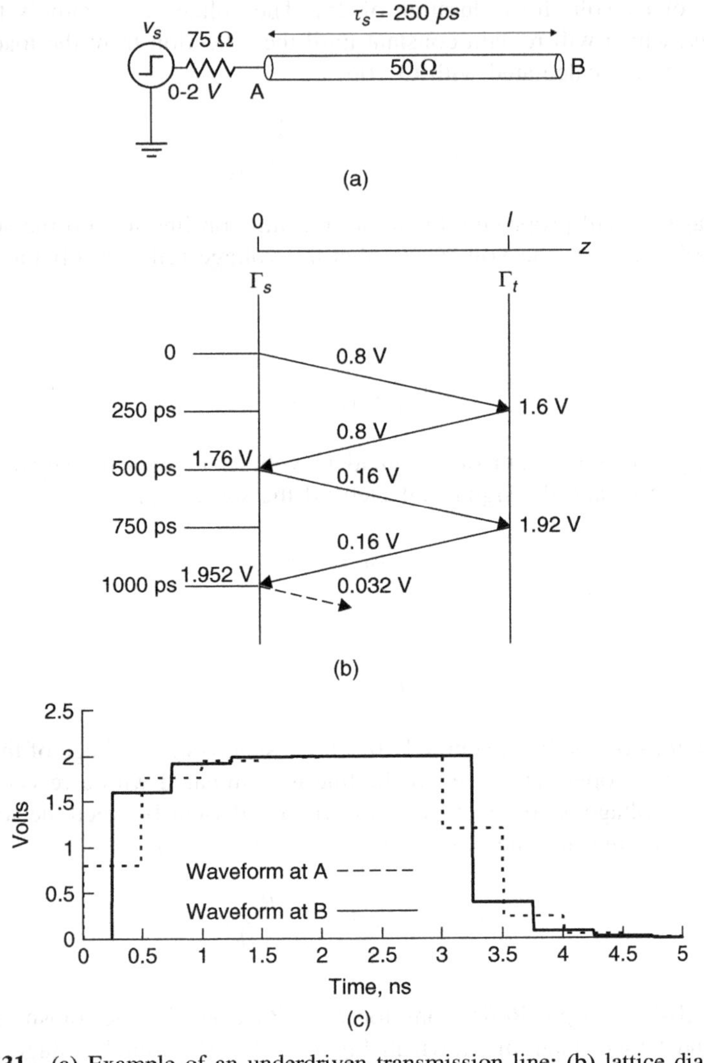

Figure 3-31 (a) Example of an underdriven transmission line; (b) lattice diagram; (c) digital waveform.

When v_b reaches the source, part of the signal will be reflected back toward the load, as determined with the reflection coefficient looking into the source using (3-102):

$$\Gamma_{\text{source}} = \frac{75 - 50}{75 + 50} = 0.2$$

The value reflected back toward the load is $v_c = v_b \Gamma_s = (0.2)(0.8) = 0.16$ V. The reflected signal will be added to the signal already present on the line, which will give a total magnitude at the source end $v_B = v_a + v_b + v_c = 0.8 + 0.8 + 0.16 = 1.76$ V, with the reflected portion v_c of 0.16 V traveling toward the load. This process is repeated until the voltage reaches a steady-state value of 2 V. If the same procedure is applied to the falling edge of a digital waveform, the signal integrity of a digital pulse propagating on this system can be calculated, as shown in Figure 3-31c. Notice how the reflections give the waveform a "stair-step" appearance at the receiver (node B), even though the unloaded output of the voltage source is a square wave. This effect occurs when the source impedance (R_s) is larger than the characteristic impedance (Z_0) and is referred to as an *underdriven transmission line*.

Example 3-5 Multiple Reflections for $R_s < Z_0$ When the characteristic impedance of the transmission line is greater than the source impedance, as shown in Figure 3-32a, the reflection coefficient looking into the source will be negative:

$$\Gamma_{\text{source}} = \frac{25 - 50}{25 + 50} = -\frac{1}{3}$$

When the lattice diagram is solved, as shown in Figure 3-32b, it is easy to show that a negative reflection at the source will produce a "ringing" effect. This is known as an *overdriven transmission line*. The resulting distorted digital waveform is shown in Figure 3-32c. Since the procedure for solving the lattice diagram is identical to Example 3-3, the exercise is left to the reader.

3.5.5 Lattice Diagrams for Nonideal Topologies

Real bus designs rarely employ only a single transmission line. For example, even in point-to-point designs, the silicon driver is connected to the main bus through a package, which often employs transmission lines anywhere from 0.25 to 1.0 in. long. Also, many high-speed designs use add-in cards, where two separately manufactured printed circuit boards are interfaced through a connector. Furthermore, it is not uncommon to encounter designs where one driver is sending data to multiple receivers, such as a front-side bus of a multiprocessor system, necessitating the solution of multiple reflections in parallel. Consequently, it is important to explore techniques for understanding and solving systems with multiple transmission-line segments in a variety of topologies.

Cascaded Topologies Consider the transmission-line structure depicted in Figure 3-33, which consists of two transmission-line segments cascaded in series. The first section is of length l_1 and has a characteristic impedance of Z_{01} ohms. The second section is of length l_2 and has an impedance of Z_{02} ohms. Finally, the structure is terminated with a value of R_t. When the signal encounters the Z_{01}: Z_{02} impedance junction, part of the signal (v_c) will be

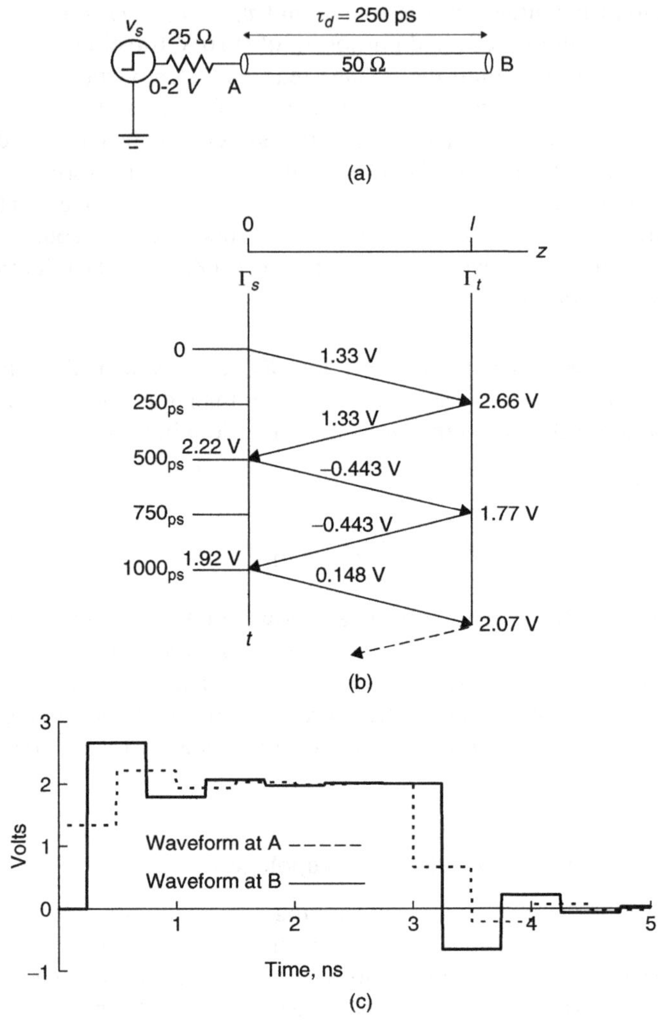

Figure 3-32 (a) Example of an overdriven transmission line; (b) lattice diagram; (c) digital waveform.

reflected, as governed by the reflection coefficient (3-102) looking into line 2 from line 1,

$$\Gamma_2 = \frac{Z_{02} - Z_{01}}{Z_{02} + Z_{01}}$$

and part of the signal (v_b) will be transmitted, as governed by the transmission coefficient, as defined in equation (3-103):

$$T_2 = \frac{2Z_{02}}{Z_{02} + Z_{01}} = 1 + \Gamma_2$$

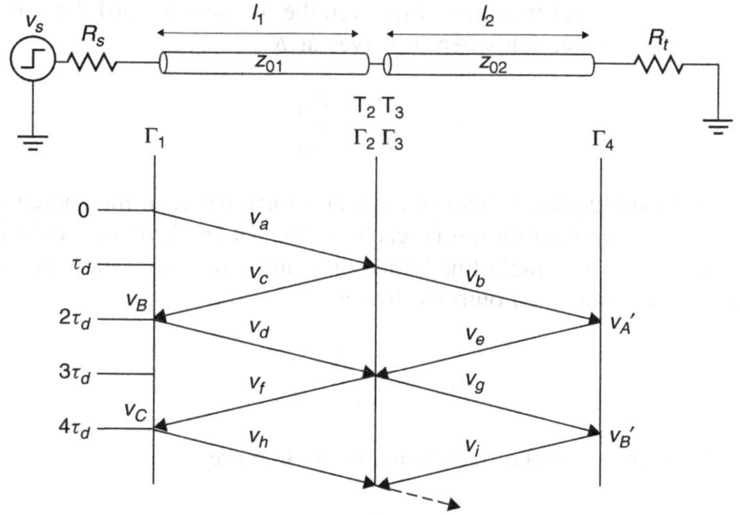

Figure 3-33 Lattice diagram for two cascaded transmission lines with different impedance values and identical lengths ($l_1 = l_2$).

Figure 3-33 also depicts how a lattice diagram can be used to solve for multiple reflections on a transmission-line system with a series of transmission lines with more than one characteristic impedance. Note that the transmission lines in this example are of equal length ($l_1 = l_2$), which simplifies the problem because the reflections from each section will be in phase. For example, in Figure 3-33, the transmitted portion of v_e adds directly to the reflection, v_f. When the two transmission lines are of different lengths, the reflections from one section will not be in phase with the reflections from the other section, which complicates the diagram drastically. When the signal reaches the termination, the reflection is governed by the reflection coefficient looking into the termination resistance at the load (Γ_4):

$$\Gamma_4 = \frac{R_t - Z_{02}}{R_t + Z_{02}}$$

The portion of the signal reflected off the termination resistor will travel back toward the source and experience another reflection when it reaches the junction between transmission lines,

$$\Gamma_3 = \frac{Z_{01} - Z_{02}}{Z_{01} + Z_{02}}$$

where Γ_3 is the reflection looking into line 1 from line 2. Part of the signal will be reflected back toward the load as calculated by Γ_3 and part transmitted toward the source, as dictated by the transmission coefficient T_3:

$$T_3 = 1 + \Gamma_3$$

The portion of the signal transmitted through the junction toward the source will experience another reflection when it arrives at R_s:

$$\Gamma_1 = \frac{R_s - Z_{01}}{R_s + Z_{01}}$$

The signal will subsequently bounce back and forth between the source and the termination load until equilibrium is reached. The voltage levels are calculated in the same manner as the single-line lattice diagram with a little more accounting. The initial voltage launched onto the line is

$$v_a = v_s \frac{Z_{01}}{Z_{01} + R_s}$$

and the voltage levels of the reflections on the line are

$$\begin{aligned} v_b &= v_a T_2 \\ v_c &= v_a \Gamma_2 \\ v_d &= v_c \Gamma_1 \\ v_e &= v_b \Gamma_4 \\ v_f &= v_d \Gamma_2 + v_e T_3 \\ v_g &= v_e \Gamma_3 + v_d T_2 \\ v_h &= v_f \Gamma_1 \\ v_i &= v_g \Gamma_4 \end{aligned}$$

giving source-side voltages of

$$\begin{aligned} v_B &= v_a + v_c + v_d \\ v_C &= v_a + v_c + v_d + v_f + v_h \end{aligned}$$

and load voltages of

$$\begin{aligned} v_{A'} &= v_b + v_e \\ v_{B'} &= v_b + v_e + v_g + v_i \end{aligned}$$

where the remaining reflections are left for the reader to calculate.

Multireceiver Topologies So far in this book we have covered many issues that deal with an interconnect connecting two components. However, this is not always the case. Often, it is required that a single driver be connected to two or

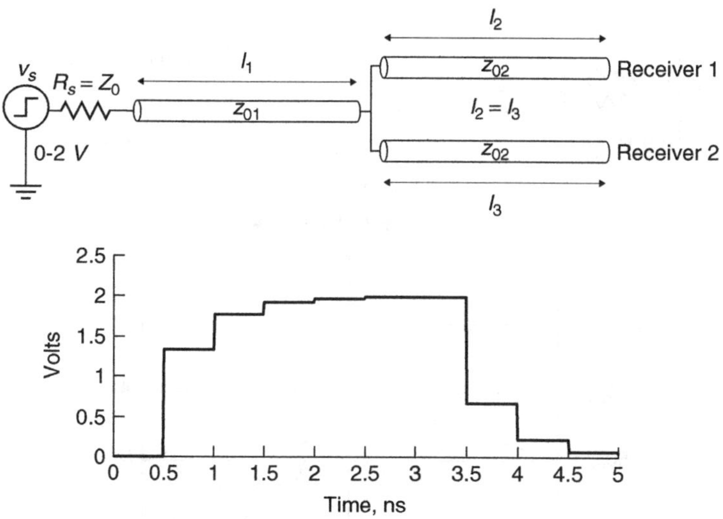

Figure 3-34 Signal integrity of a T-topology when the leg lengths and characteristic impedances are equal.

more receivers. In these cases, the topology of the interconnect can affect system performance dramatically. For example, consider Figure 3-34, which is a case where one driver is connected to two receivers. In this example, the impedance of the base (Z_{01}) is equal to the impedance of the two legs (Z_{02}), and the legs are of equal length ($l_2 = l_3$). When the signal propagates to the junction, it will see an effective impedance of $Z_{02}/2$, resulting in the waveform in the figure that steps up toward the final value in the same manner as an underdriven transmission line as calculated in Example 3-4. When the impedance of the legs is twice that of the base ($Z_{02} = 2Z_{01}$), the effective impedance the signal will see at the junction will be equal to the base, so that no reflections are generated.

When the structure is unbalanced, as in the case where one leg is longer than the other, the signal integrity will deteriorate dramatically because the reflections will arrive at the junction at different times. To gain an intuitive understanding of how the multiple reflections from different legs interact, it is useful to solve a multiple-legged lattice diagram at least once, which is demonstrated in the following example.

Example 3-6 Calculate the first few reflections of the unbalanced T-topology shown in Figure 3-35 assuming that $Z_0 = R_s = 50\ \Omega$, l_1 and l_3 are lengths that corresponds to a propagation delay of 250 ps, and l_2 has a delay of 125 ps.

SOLUTION Referring to the lattice diagram in Figure 3-35, the first and second vertical lines represents the electrical pathway between the driver and the junction, the third vertical line represents the pathway between the junction and the end of the short line (receiver 1), and the fourth vertical line represents the

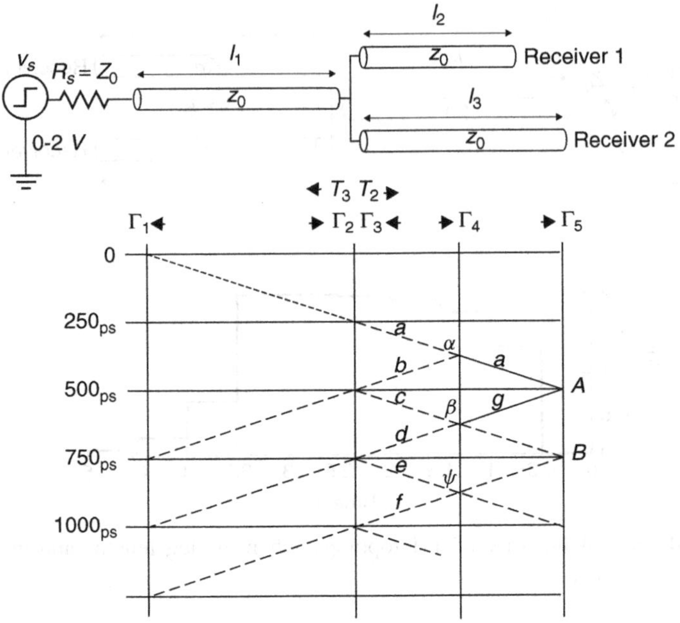

Figure 3-35 Lattice diagram of a T-topology when the leg lengths are not equal.

end of the long line (receiver 2). The initial voltage step launched onto line 1 is

$$v_i = v_s \frac{Z_0}{Z_0 + R_s} = 2 \cdot \frac{50}{50 + 50} = 1$$

The reflection and transmission coefficients looking from line 1 into the junction is

$$\Gamma_2 = \frac{(Z_0/2) - Z_0}{(Z_0/2) + Z_0} = \frac{25 - 50}{25 + 50} = -\frac{1}{3}$$

$$T_2 = 1 + \Gamma_2 = \frac{2}{3}$$

Consequently, the initial voltage launched into both legs (lines 2 and 3) is

$$v_a = T_2 v_{\mathrm{i}} = \frac{2}{3}$$

This voltage (v_a) travels down each leg and doubles when it arrives at the open circuit ($\Gamma_4 = \Gamma_5 = 1$). Therefore, the voltage at receiver 1 (v_α) occurs at $t = 375$ ps, which is the delay of line 1 plus line 2 (the short leg).

$$v_b = v_a \Gamma_4 = \tfrac{2}{3}$$

$$v_\alpha = v_a + v_b = \tfrac{4}{3}$$

The voltage at receiver 2 (v_A) occurs at $t = 500$ ps, which is the delay of line 1 plus line 2 (the long leg):

$$v_A = v_a + v_g = v_a + v_a\Gamma_5 = \tfrac{4}{3}$$

For both legs, a reflection of $v_b = v_g = \frac{2}{3}$ is reflected from the open circuit at the receivers; however, they will arrive at the junction at different times. As seen by the lattice diagram, at $t = 500$ ps, v_b will arrive at the junction. Part of v_b will be reflected back toward receiver 1,

$$\Gamma_3 = \frac{(Z_0/2) - Z_0}{(Z_0/2) + Z_0} = \frac{25 - 50}{25 + 50} = -\frac{1}{3}$$

$$v_c = v_b\Gamma_3 = \frac{2}{3}\left(-\frac{1}{3}\right) = -\frac{2}{9}$$

and part will be transmitted onto line 1 toward the source and onto line 3 toward receiver 2:

$$T_3 = 1 + \Gamma_3 = \frac{2}{3}$$

The voltage at receiver 1 at $t = 625$ ps (v_β) is calculated:

$$v_d = \Gamma_4 v_c = -\frac{2}{9}$$

$$v_\beta = v_\alpha + v_c + v_d = \frac{4}{3} - \frac{2}{9} - \frac{2}{9} = \frac{8}{9}$$

To calculate the voltage at receiver 1 at $t = 875$ ps (v_ψ), it is necessary to account for the portion of the signal reflected from receiver 2 at 500 ps and transmitted into the junction at 750 ps that travels toward receiver 1. This combination of reflections can be seen by observing the lattice diagram, where reflection g arrives at the junction simultaneously with reflection d.

$$v_e = \Gamma_3 v_d = -\frac{2}{9}\left(-\frac{1}{3}\right) = \frac{2}{27}$$

$$v_f = \Gamma_4 v_e = \frac{2}{27}$$

$$v_g = \frac{2}{3} \quad \text{(from above)}$$

$$v_\psi = v_\beta + v_e + v_f + v_g T_3 + v_g T_3 \Gamma_4$$

$$= \frac{8}{9} + \frac{2}{27} + \frac{2}{27} + \frac{2}{3}\left(\frac{2}{3}\right) + \frac{2}{3}\left(\frac{2}{3}\right)(1) = \frac{52}{27}$$

By observing the lattice diagram, we see that the voltage at receiver 2 at $t = 750$ ps is calculated by accounting for the voltage reflected from receiver 1 at $t = 375$ ps and transmitted into the junction at $t = 500$ ps:

$$v_B = v_A + T_3 v_b + T_3 v_b \Gamma_5 = \frac{4}{3} + \frac{2}{3}\left(\frac{2}{3}\right) + \frac{2}{3}\left(\frac{2}{3}\right)(1) = \frac{20}{9}$$

This process can be continued until the waveform has reached steady state. The complete waveforms for this example are shown in Figure 3-36, with the first few reflections (just calculated) labeled. Note that the complicated interactions between the reflections from each leg severely degrade the integrity of the signal. As more legs are added to the topology, it becomes more sensitive to differences in the electrical length of the legs. Furthermore, a mismatch between the source resistance and the characteristic impedance of line, differences between receiver loads, and impedance deltas between each leg will cause similar instabilities.

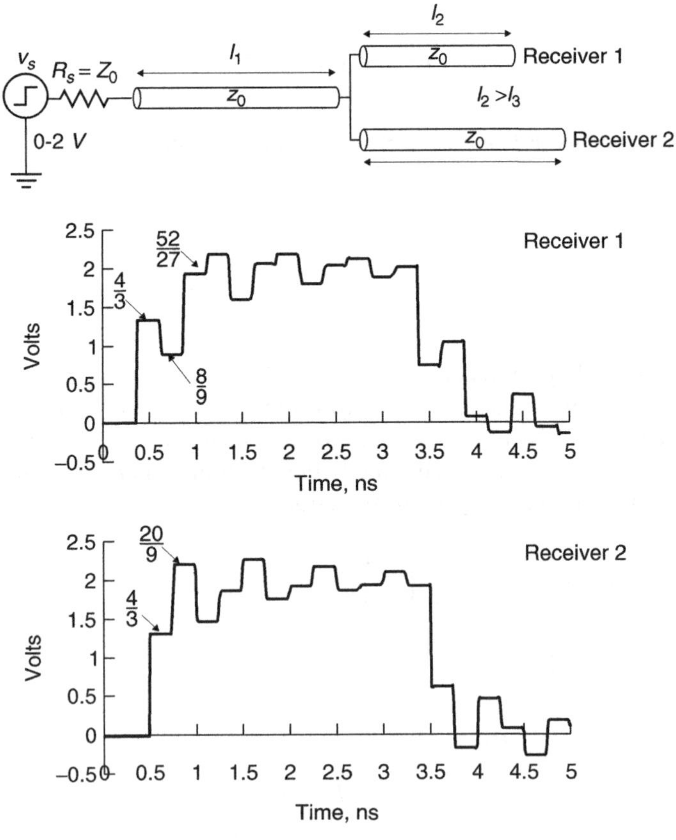

Figure 3-36 Signal integrity of a T-topology when the leg lengths are not equal.

So what can we learn form this? The answer is: *symmetry*. Whenever a topology is considered, the primary area of concern is symmetry. Make certain that the topology looks symmetrical from the point of view of any driving agent. This is usually accomplished by ensuring that the lengths, impedances, and loading are identical for each leg of the topology. The secondary concern is to minimize the impedance discontinuities at the topology junctions, although this may be impossible in some designs.

3.5.6 Effect of Rise and Fall Times on Reflections

The rise and fall times of real digital waveforms begin to have a significant effect on the wave shape when they become less than twice the delay ($2\tau_d$) of the transmission line. Figures 3-37 and 3-38 show the effect that finite rise and fall times have on over- and underdriven transmission lines. Notice how significantly the wave shape changes as the rise time exceeds twice the delay of the line. When the edge rate exceeds twice the line delay, the reflections are masked because the amount of time that it takes to transition from a low state to a high state (or vice versa) exceeds the period of the reflections.

3.5.7 Reflections from Reactive Loads

In real systems, there are rarely cases where the loads are purely resistive. The input to a CMOS gate, for example, tends to be capacitive. Additionally, bond wires, vias, lead frames of the chip packages, chip sockets, and daughtercard

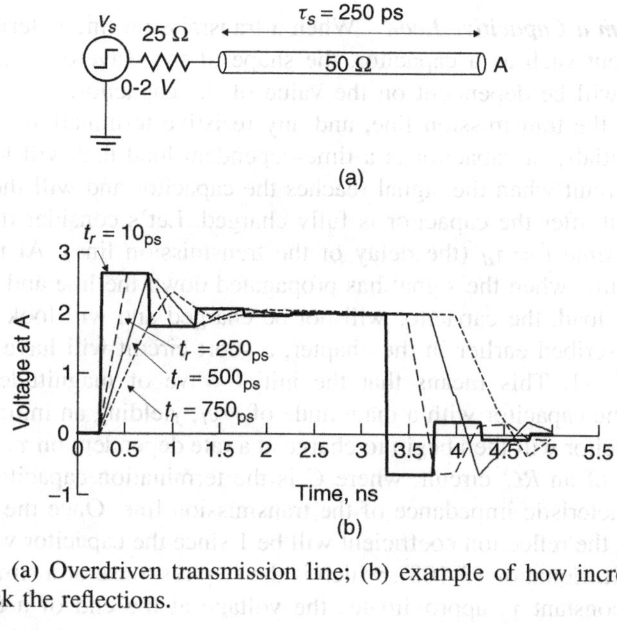

Figure 3-37 (a) Overdriven transmission line; (b) example of how increased rise and fall times mask the reflections.

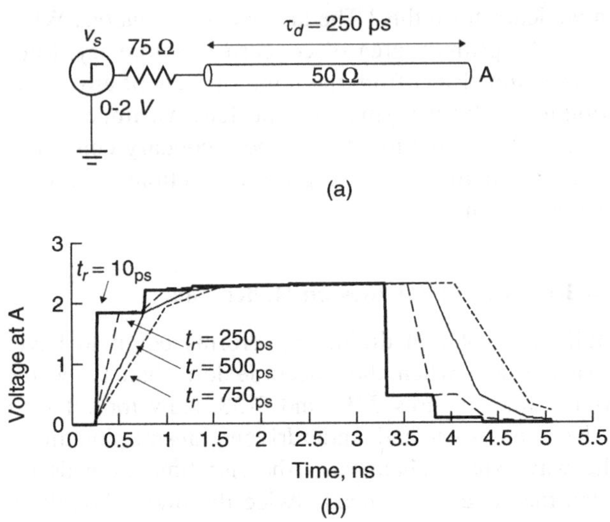

Figure 3-38 (a) Underdriven transmission line; (b) example of how increased rise and fall times mask the reflections.

connectors tend to be inductive. This makes it necessary to understand how these reactive elements affect the reflections in a transmission-line system. In this section we briefly introduce the effect that capacitors and inductors have on reflections.

Reflection from a Capacitive Load When a transmission line is terminated in a reactive element such as a capacitor, the shape of the waveforms at the driver and the load will be dependent on the value of the capacitor, the characteristic impedance of the transmission line, and any resistive terminations that may be present. Essentially, a capacitor is a time-dependent load that will look initially like a short circuit when the signal reaches the capacitor and will then look like an open circuit after the capacitor is fully charged. Let's consider the reflection coefficient at time $t = \tau_d$ (the delay of the transmission line). At time $t = \tau_d$, which is the time when the signal has propagated down the line and has reached the capacitive load, the capacitor will not be charged and will look like a short circuit. As described earlier in the chapter, a short circuit will have a reflection coefficient of -1. This means that the initial wave of magnitude v_i will be reflected off the capacitor with a magnitude of $-v_i$, yielding an initial voltage of 0 V. The capacitor will then begin to charge at a rate dependent on τ, which is the time constant of an RC circuit, where C is the termination capacitor value and R is the characteristic impedance of the transmission line. Once the capacitor is fully charged, the reflection coefficient will be 1 since the capacitor will resemble an open circuit. Equation (3-108), which is the step response of a simple network with a time constant τ, approximates the voltage at the end of a transmission line terminated with a capacitor beginning at time $t = \tau_d$,

$$v_{\text{capacitor}} = v_{\text{ss}} \left(1 - e^{-(t-\tau_d)/\tau]}\right) \qquad t > \tau_d \tag{3-108}$$

where $\tau = CZ_0$ is the time constant, τ_d the time delay of the transmission line as given by (3-107), and v_{ss} the steady-state voltage determined by the voltage source v_s and the voltage divider between the source resistance R_s and the termination resistance R_t. Note that (3-108) is an approximation because it assumes a step function with an infinite edge rate (i.e ., the rise time is infinitely fast). Figure 3-39 shows the response of a line terminated with a capacitive load. The waveform shape at node B follows equation (3-108). Notice that the waveform at the source (node A) dips toward zero at $t = 500$ ps (which is $2\tau_d$) because the capacitor initially looks like a short circuit, so the reflection coefficient is -1. Note that the shape of this wave at node A is also dictated by (3-108) when the exponent term is $e^{-[(t-2\tau_d)/\tau]}$, which simply shifts the time. The voltage reflected back toward the source is initially v_i, where v_i is the initial voltage launched onto the transmission line. After the capacitor is fully charged, it will look like an open and have a reflection coefficient of $+1$. Consequently, the reflected waveform at the receiver will double. As seen in Figure 3-39b, the waveform at the receiver (B) reaches a steady-state value of 2 V after about three time constants [$3\tau = 3(50\ \Omega)(2$ pF $= 300$ ps] after arriving at the receiver, just as circuit theory predicts.

If the line is terminated with a parallel resistor and capacitor, as depicted in Figure 3-40, the voltage at the capacitor will be dependent on the time constant

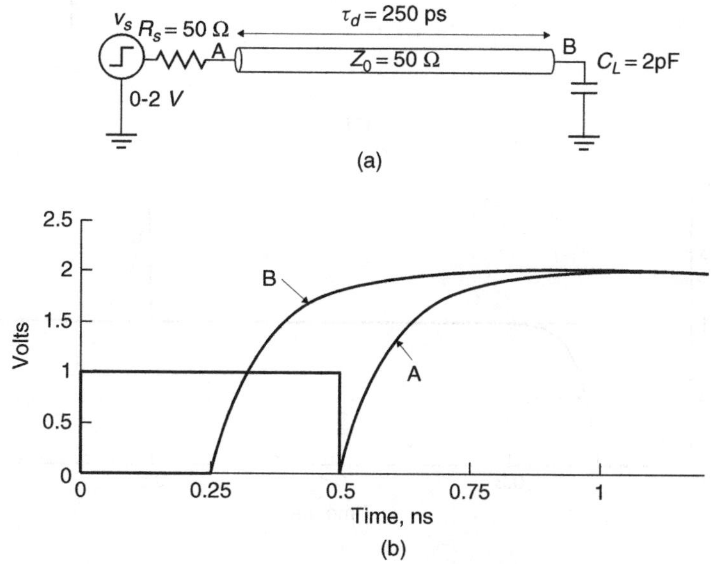

Figure 3-39 (a) Transmission line terminated with a capacitive load; (b) step response showing reflections from the capacitor.

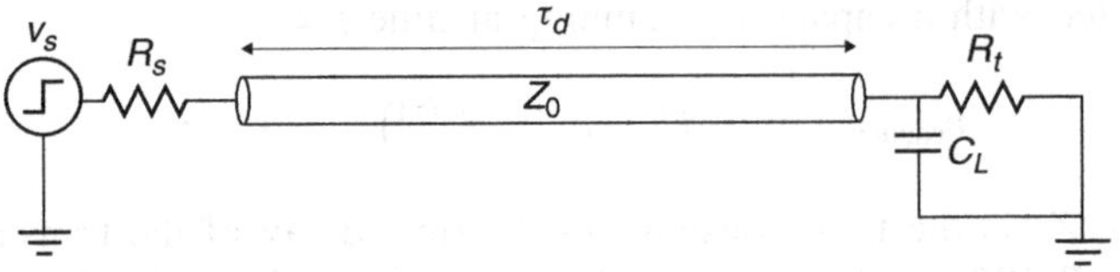

Figure 3-40 Transmission line terminated in a parallel *RC* network.

between C_L and the parallel combination of R_t and Z_0:

$$\tau_1 = \frac{C_L Z_0 R_t}{R_t + Z_0} \tag{3-109}$$

Reflection from an Inductive Load In the real world, when a transmission line is terminated, there is usually a series inductance caused by the physical connection between the transmission line and the resistor. Some common examples of this inductive connection are bond wires, lead frames, and vias. When a series inductor appears in the electrical pathway of a transmission-line termination, as depicted in Figure 3-41a, it will also act as a time-dependent load. Initially, at

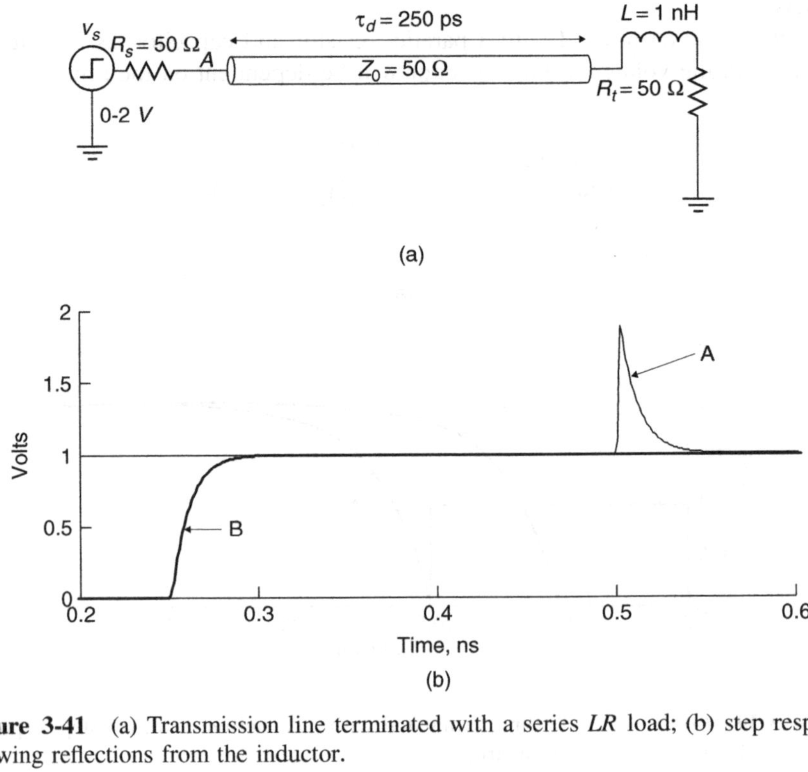

Figure 3-41 (a) Transmission line terminated with a series *LR* load; (b) step response showing reflections from the inductor.

time $t = \tau_d$, the inductor will resemble on open circuit. When a voltage step is applied initially, no current flows across the inductor. This produces a reflection coefficient of 1, causing an inductive spike seen as a reflection at node A in Figure 3-41b. The value of the inductor will determine how long the reflection coefficient will remain 1. If the inductor is large enough, the signal will double in magnitude. Eventually, the inductor will discharge its energy at a rate dependent on the time constant τ of an *LR* circuit. For the circuit depicted in Figure 3-41a, the wave shape of the rising edge at node B is calculated:

$$v_{\text{inductor}} = v_{\text{ss}}\left(1 - \exp\left[-\frac{(t-\tau_d)(Z_0+R_t)}{L}\right]\right) \qquad t > \tau_d \qquad \text{(3-110a)}$$

Note that the wave shape calculated by (3-110a) will also be valid for the falling edge of the inductive spike, shown at node B in Figure 3-41b, if τ_d is adjusted to shift the waveform to the correct position in time ($2\tau_d$), the waveform is inverted, and dc is shifted to the appropriate level:

$$v_{\text{inductor}} = v_{\text{ss}}\left(1 + \exp\left[-\frac{(t-2\tau_d)(Z_0+R_t)}{L}\right]\right) \qquad \text{(3-110b)}$$

Filtering Effects of Reactive Components Figures 3-39b and 3-41b show how the series inductor and the shunt capacitor affect the signal integrity. The series inductance will cause an inductive spike, which is seen as a positive reflection, the capacitance will cause a capacitive dip, which is seen as a negative reflection, and both will smooth the rising and falling edges seen at the receiver (node B). To understand why the edges are smoothed, we must explore how an inductor or a capacitor will filter the harmonics of a digital waveform. In Chapter 8 it will be shown that high-frequency harmonics are associated with the rising and falling edges of a digital waveform. Consequently, if the higher-frequency harmonics are filtered out by the capacitive or inductive loads, the rising and falling times will be increased. Equation (3-111) shows that the impedance of the shunt capacitor will decrease with frequency, which means that the higher-order harmonics of the digital waveform will be shunted to ground, increasing the rise and fall times:

$$Z_{\text{cap}} = \frac{1}{j\omega C} \qquad \text{(3-111)}$$

where $\omega = 2\pi f$.

Similarly, (3-112) shows that the series impedance of an inductor will increase with frequency, which will also tend to filter out the higher harmonics because they will experience larger impedances than the lower-frequency harmonics.

$$Z_{\text{ind}} = j\omega L \qquad \text{(3-112)}$$

Consequently, for a digital pulse, reactive components such as inductors and capacitors will low-pass-filter the waveform, resulting in increased rise and fall times. The exception to this statement is when specific filters are constructed using reactive components to equalize a channel, which is described is Chapter 12.

3.6 TIME-DOMAIN REFLECTOMETRY

In practice, it is often necessary to measure a system of transmission lines in a digital system to ensure compliance with design guidelines, to understand the equivalent circuit of a component, or simply to verify that the simulations are predicting the transient response correctly. One measurement method often used is called *time-domain reflectrometry* (TDR). A TDR measurement uses the premise of multiple reflections discussed in Section 3.5 to derive an impedance profile of the *device under test* (DUT). Figure 3-42 depicts a generic TDR setup where a step function is driven onto the DUT through a 50-Ω cable. A sampling oscilloscope is used to observe the waveform at A, which depicts the voltage profile of the reflected waves. The voltage profile is converted to an impedance profile, which can be used to measure the characteristic impedance and propagation delay of a transmission line, estimate inductance and capacitance values of structures such as vias, bond wires, and lead frames, and deduce the form of an equivalent circuit for many applications.

3.6.1 Measuring the Characteristic Impedance and Delay of a Transmission Line

One of the most common uses for TDR measurements is to determine the characteristic impedance of transmission lines. The impedance profile is calculated

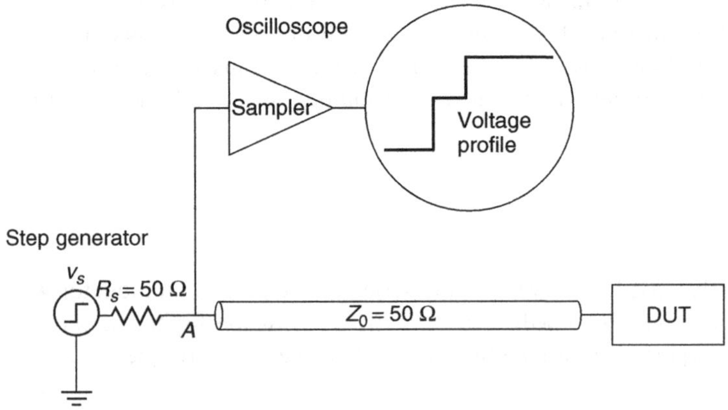

Figure 3-42 Generic time-domain reflectometry measurement setup.

from the voltage by rearranging equation (3-102),

$$\Gamma \equiv \frac{v_r}{v_i} = \frac{Z_{\text{DUT}} - Z_0}{Z_{\text{DUT}} + Z_0} \tag{3-113}$$

$$Z_{\text{DUT}} = Z_0 \frac{v_i + v_r}{v_i - v_r}$$

where v_i is the incident voltage transmitted and v_r is the voltage reflected from the DUT. Figure 3-43 depicts the TDR waveforms of a 60-Ω transmission line with a length that corresponds to a delay of 250 ps. The impedance of the transmission line under test is calculated from the first voltage step that occurs at 500 ps using equation (3-113):

$$Z_{\text{DUT}} = 50 \cdot \frac{1 + 0.091}{1 - 0.091} = 60\ \Omega$$

We can also calculate the delay of the transmission line. Since a TDR is essentially a reflected voltage at the driver, the reflection of the DUT remains for the time it takes the signal to propagate to the end of the DUT and return to the source from the open circuit. Consequently, the duration of the reflection will correspond to twice the delay of the transmission line under test, as shown in Figure 3-43.

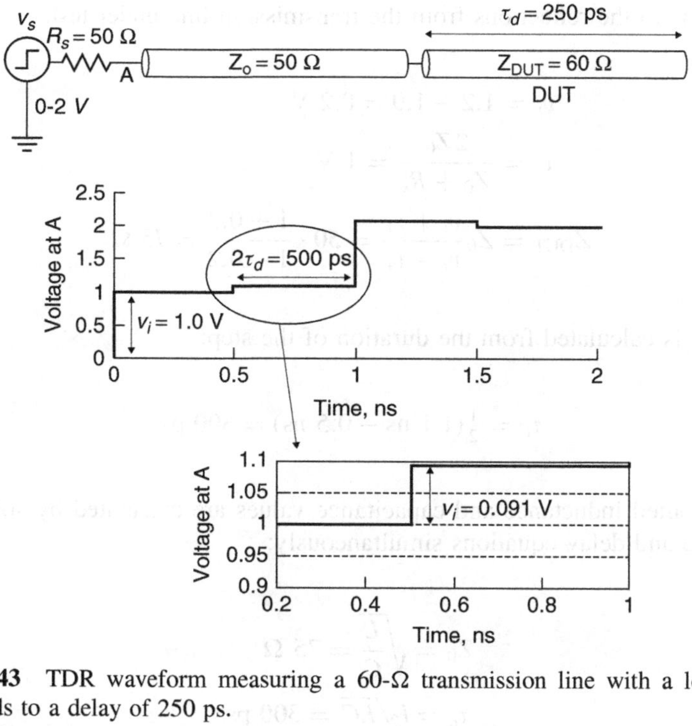

Figure 3-43 TDR waveform measuring a 60-Ω transmission line with a length that corresponds to a delay of 250 ps.

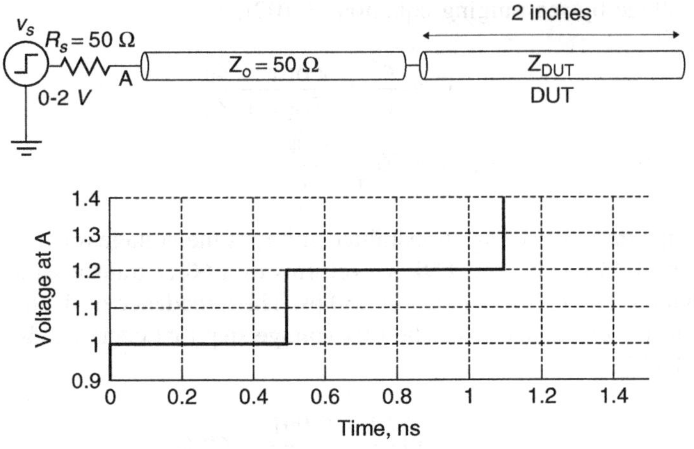

Figure 3-44 TDR profile for Example 3-7.

Example 3-7 Calculate the equivalent L and C per unit length of the 2-in. transmission line measured with TDR profile shown in Figure 3-44.

SOLUTION The impedance is calculated from the step voltage at 1.2 V, which corresponds to the reflections from the transmission line under test.

$$v_r = 1.2 - 1.0 = 0.2 \text{ V}$$

$$v_i = \frac{2Z_0}{Z_0 + R_s} = 1 \text{ V}$$

$$Z_{\text{DUT}} = Z_0 \frac{v_i + v_r}{v_i - v_r} = 50 \cdot \frac{1 + 0.2}{1 - 0.2} = 75 \ \Omega$$

The delay is calculated from the duration of the step:

$$\tau_d = \tfrac{1}{2}(1.1 \text{ ns} - 0.5 \text{ ns}) = 300 \text{ ps}$$

The distributed inductance and capacitance values are calculated by solving the impedance and delay equations simultaneously:

$$Z_0 = \sqrt{\frac{L}{C}} = 75 \ \Omega$$

$$\tau_d = l\sqrt{LC} = 300 \text{ ps}$$

The inductance per unit length is calculated by normalizing the delay to line length of the transmission line under test and multiplying by the impedance:

$$\frac{\tau_d}{2\text{ in.}} = \sqrt{LC} = 150\text{ ps/in.}$$

$$L = \sqrt{\frac{L}{C}}\sqrt{LC} = 75(150 \times 10^{-12}) = 11.25\text{ nH/in.}$$

The capacitance is calculated by dividing the normalized delay by the impedance:

$$C = \frac{\sqrt{LC}}{\sqrt{L/C}} = \frac{150 \times 10^{-12}}{75} = 2.0\text{ pF/in.}$$

3.6.2 Measuring Inductance and Capacitance of Reactive Structures

In Section 3.5.7 we discussed reflections from reactive loads. In this section we show how to estimate the value of the capacitance or the inductance from a measured TDR profile. The analysis presented here is somewhat idealized because it assumes an ideal step for a source. In reality, ideal steps are impossible to generate. However, modern TDR measurement equipment allows the generation of very fast rise times that vary anywhere between 9 and 35 ps, which are fast enough to approximate a step function for a variety of applications.

Inductive Structures In a TDR measurement, narrow spikes such as those depicted in Figure 3-41 are indicative of an inductive component such as a bond wire, a connector pin, or a package lead frame. Assuming that the input step has a sufficiently fast rise time, the value of the inductor can be estimated by measuring the area under the inductive spike, as shown in Figure 3-45. The area is calculated by integrating equation (3-110b) after subtracting the dc offset and assuming that $R_t = Z_0$:

$$\begin{aligned}
v_{ss} &= v_s\frac{R_t}{R_t + R_s} = \frac{50}{50+50} = \frac{v_s}{2} \\
A_{\text{ind}} &= \int_{2\tau_d}^{\infty}\left[\frac{v_s}{2}\left(1 + e^{-\left[\frac{(t-2\tau_d)(Z_0+R_t)}{L}\right]}\right) - \frac{v_s}{2}\right]dt \\
&= \frac{v_s}{2}\int_0^{\infty} e^{-(2Z_o/L)t'}\,dt' = \frac{v_s L}{4Z_0} \qquad (3\text{-}114) \\
L &= \frac{4Z_0 A_{\text{ind}}}{v_s}
\end{aligned}$$

Therefore, through careful measurement of the area under the inductive spike, we can achieve a good approximation of the inductance. The accuracy is maximized when the rise time is very fast compared to the duration of the inductive spike.

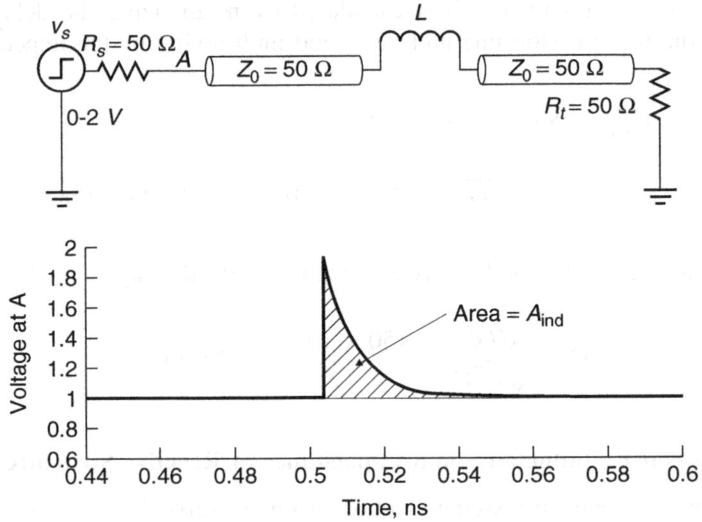

Figure 3-45 The area under the reflection can be used to estimate the inductance.

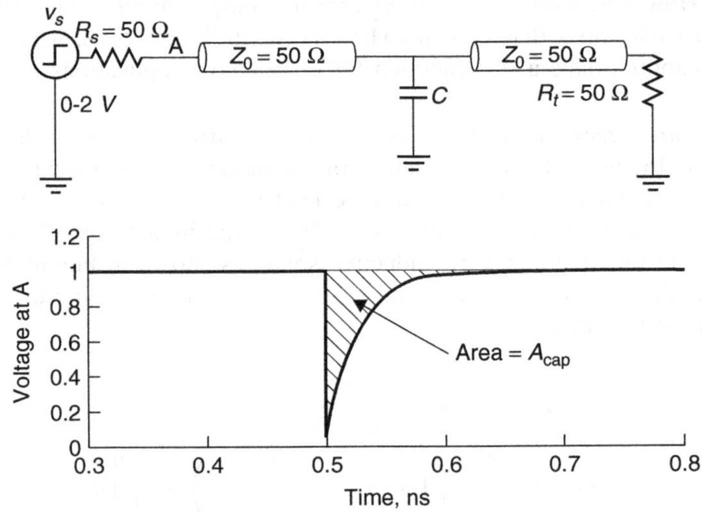

Figure 3-46 The area under the reflection can be used to estimate the capacitance.

For long rise times or very small inductance values, the area under the curve will be masked by a rise time similar to the reflections in Figure 3-37.

Capacitive Structures In a TDR measurement, narrow dips such as that depicted in Figure 3-46 are indicative of an capacitive component such as a probe pad or a via pad. Assuming that the input step has a sufficiently fast rise time, the value of the capacitance can be estimated by measuring the area under the capacitive dip. The area is calculated by integrating equation (3-108) after subtracting the dc offset, assuming that $R_t = Z_0$ where $\tau = Z_0 C/2$

$$
\begin{aligned}
v_{\text{capacitor}} &= \frac{v_s}{2}(1 - e^{-[2(t-2\tau_d)/Z_o C]}) \\
A_{\text{cap}} &= \int_{2\tau_d}^{\infty} \left[\frac{v_s}{2} - \frac{v_s}{2}(1 - e^{-[2(t-2\tau_d)/Z_0 C]}) \right] dt \\
&= \frac{v_s}{2} \int_0^{\infty} e^{-2t'/Z_0 C} dt' = \frac{v_s C Z_0}{4} \\
C &= \frac{4A_{\text{cap}}}{v_s Z_0}
\end{aligned}
\tag{3-115}
$$

Example 3-8 Calculate the value of the shunt capacitance for the waveform in Figure 3-47 assuming the circuit shown in Figure 3-46.

SOLUTION A fine grid was overlaid on the waveform in Figure 3-47 so that the area under the curve can be estimated. The total number of squares under the capacitive dip is about 60. The area per square is

$$A_{\text{square}} = (0.04)(0.01 \times 10^{-9}) = 4 \times 10^{-13}$$

The total area is therefore

$$A_{\text{tot}} = 60 A_{\text{square}} = 2.4 \times 10^{-11}$$

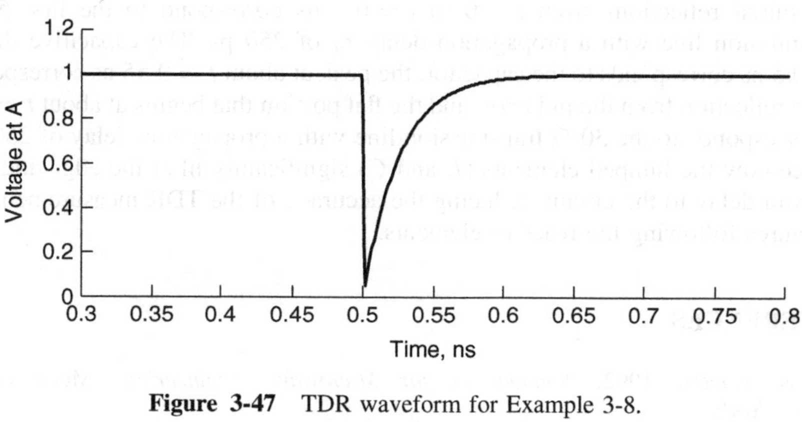

Figure 3-47 TDR waveform for Example 3-8.

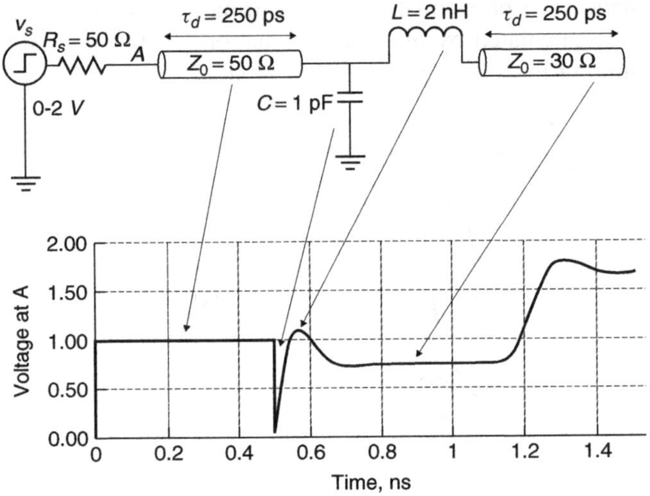

Figure 3-48 TDR profile showing the relationship between the waveform and the individual circuit components.

yielding an estimated capacitance of

$$C = \frac{4A_{\text{cap}}}{v_s Z_0} = \frac{4(2.4 \times 10^{-11})}{2(50)} = 0.96 \times 10^{-12}\ \text{F}$$

which is very reasonable because the waveform for this example was simulated using a capacitance value of 1.0 pF. More accuracy could be obtained by using a finer grid.

3.6.3 Understanding the TDR Profile

Another very useful application of a TDR measurement is to deduce the form of the equivalent circuit. For example, consider the TDR waveform in Figure 3-48. The initial reflections from $t = 0$ to $t = 0.5$ ns correspond to the first 50-Ω transmission line with a propagation delay τ_d of 250 ps. The capacitive dip at $t = 0.5$ ns corresponds to the capacitor, the peak at about $t = 0.55$ ns corresponds to the reflection from the inductor, and the flat portion that begins at about $t = 0.7$ ns corresponds to the 30-Ω transmission line with a propagation delay of 250 ps. Notice how the lumped elements (L and C) significantly filter the edge and add an extra delay to the circuit, reducing the accuracy of the TDR measurement for structures following the reactive elements.

REFERENCES

Collins, Robert, 1992, *Foundations for Microwave Engineering*, McGraw-Hill, New York.

Hall, S., G. Hall, and J. McCall, 2000, *High-Speed Digital System Design*, Wiley, New York.

Hammerstad, E., and O. Jensen, 1980, *Accurate models for microstrip computer-aided design, IEEE MTT-S International Microwave Symposium Digest*, May, pp. 407–409.

IPC, 1995, *Design Guidelines for Electronic Packaging Utilizing High-Speed Techniques*, IPC-D-317A, IPC, Chicago.

Jackson, J. D., 1999, *Classical Electrodynamics*, 3rd ed., Wiley, New York.

PROBLEMS

3-1 For the cross section in Figure 3-49, calculate the correct conductor width w to achieve 50-Ω characteristic impedance, effective dielectric permittivity, propagation delay for a 10-in.-long line, and the equivalent inductance and capacitance per meter.

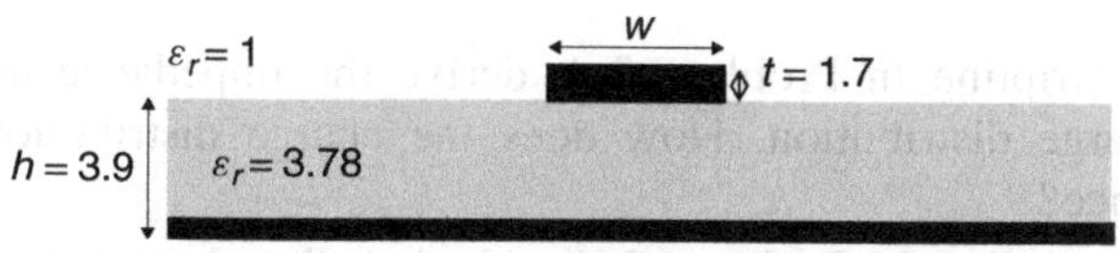

Figure 3-49 Cross section for Problem 3-1.

3-2 For the circuit shown in Figure 3-50 and the transmission line designed in Problem 3-1, use a lattice diagram to calculate the waveforms at nodes A and B when driven with a 1-V step.

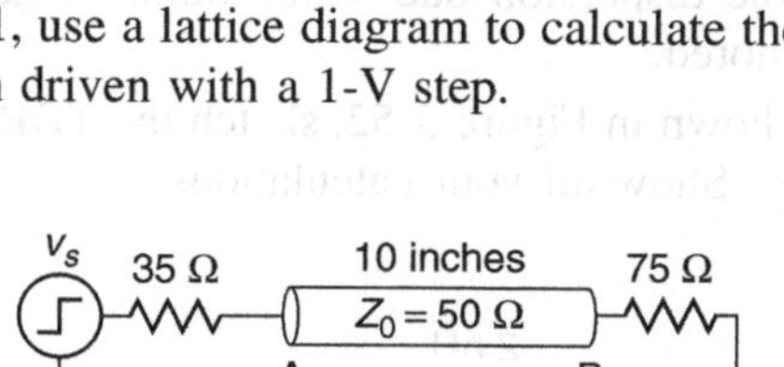

Figure 3-50 Circuit for Problem 3-2.

3-3 For the 50-Ω cross section in Problem 3-1, use field mapping techniques to calculate the effective dielectric permittivity, the propagation delay for a 10-in.-long line, and the equivalent inductance and capacitance per meter. Compare your answers to the results from Problem 3-1 to determine your accuracy.

3-4 Beginning with Laplace's equation, derive a formula for the characteristic impedance of the stripline shown in Figure 3-51. Assume an infinitely thin conductor with a uniform charge distribution. Compare the accuracy against equation (3-36c).

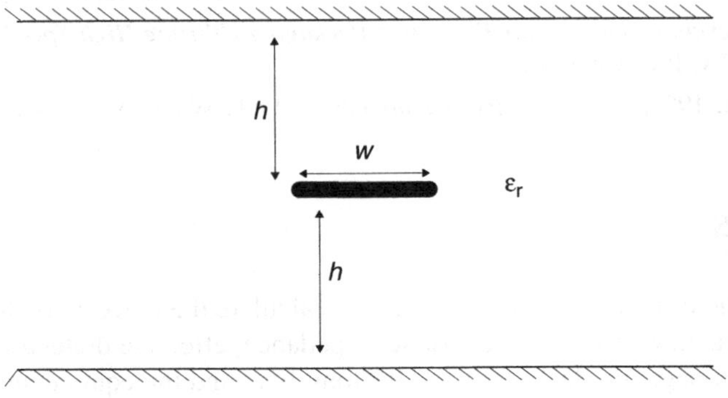

Figure 3-51 Cross section for Problem 3-4.

3-5 For the stripline in Problem 3-4, derive the impedance assuming realistic charge distribution. How does the charge distribution change the impedance?

3-6 For the stripline in Problem 3-4, calculate the characteristic impedance using field mapping techniques. Compare your answer to those obtained in Problems 3-4 and 3-5.

3-7 For the 50-Ω microstrip line designed in Problem 3-1, estimate the frequency where the dispersion due to the nonhomogeneous dielectric can no longer be ignored.

3-8 For the circuit shown in Figure 3-52, sketch the TDR waveform assuming a 2-V input step. Show all your calculations.

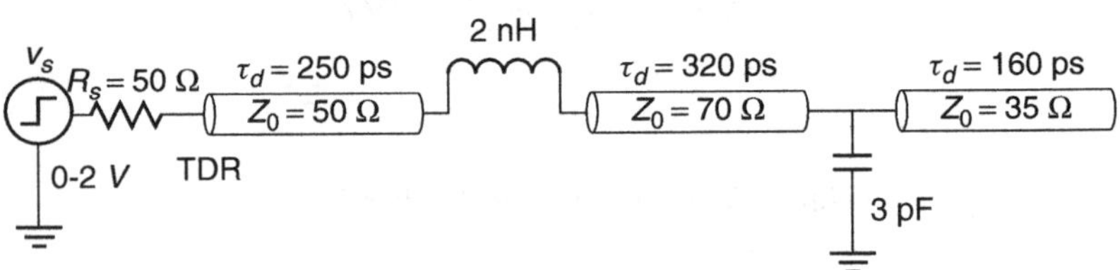

Figure 3-52 Circuit for Problem 3-8.

3-9 Draw the response of the circuit in Figure 3-53, where the device under test is (**a**) a 2-pF capacitor shunted to ground, (**b**) a series 2-pF capacitor, (**c**) a 4-nH series inductor, (**d**) a 4-nH inductor shunted to ground, and (**e**) a 1-in.-long 75-Ω transmission line with $\varepsilon_r = 4.0$.

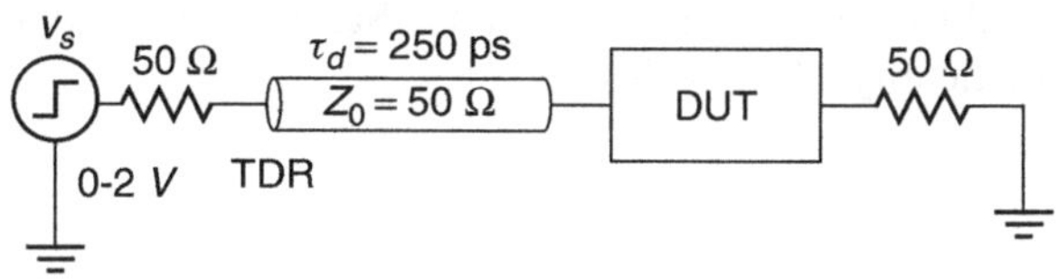

Figure 3-53 Circuit for Problem 3-9.

3-10 For the microstrip transmission line designed in Problem 3-1, develop the equivalent circuit assuming that the input pulse is 100 ps wide with 25-ps rise and fall times. Perform a virtual TDR using the circuit in SPICE and confirm that the impedance and delay match the values expected.

3-11 Calculate the first few reflections of the stub topology shown in Figure 3-54 at nodes A, B, and C using a lattice diagram. Use the equivalent circuit in Problem 3-10 to simulate the structure in SPICE. Confirm the results of your lattice diagram.

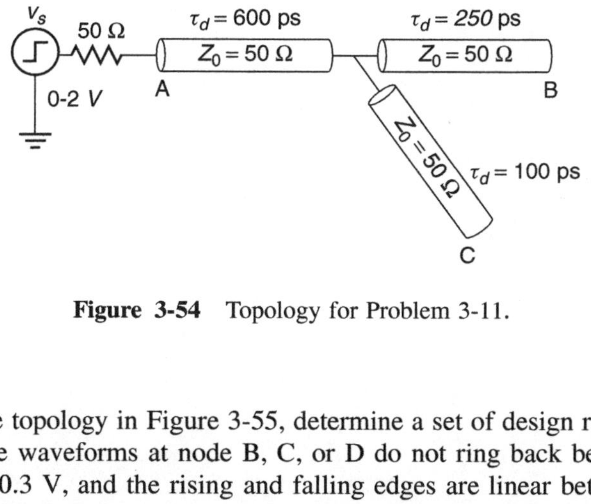

Figure 3-54 Topology for Problem 3-11.

3-12 For the topology in Figure 3-55, determine a set of design rules to ensure that the waveforms at node B, C, or D do not ring back below 0.7 V or above 0.3 V, and the rising and falling edges are linear between 0.3 and 0.7 V. Design the proper microstrip transmission-line cross sections to achieve proper impedance values for each segment, assuming a dielectric thickness of 4 mils and a relative dielectric permittivity $\varepsilon_r = 4.2$. Based on component placement requirements determined by a layout engineer, the length of line 1 (l_1) can be as short as 1.5 in., l_2 can be as short as 3 in., but it is impossible to route l_3 shorter than 5.5 in. The design guidelines should produce limits on the lengths and impedance values of each leg as well as any other conditions needed to solve the problem. Do not design a point solution. Determine a solution space that guarantees proper signal integrity. Create equivalent circuits and verify the validity of your design guidelines through SPICE simulations. State all assumptions. Comment on the practicality of your design rules.

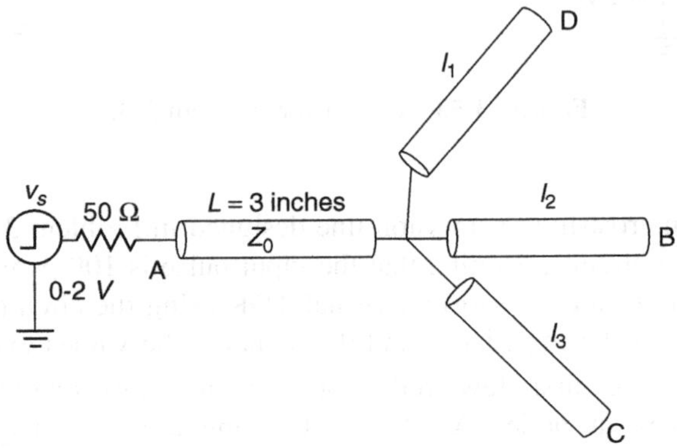

Figure 3-55 Topology for Problem 3-12.

4

CROSSTALK

As described in Section 3.2, a signal propagates along an interconnect in the form of an electromagnetic wave established between two (or more) conductors. When neighboring transmission structures are in close proximity, the electric and magnetic fields from the signal will fringe and interact with adjacent conductors. The interaction of fields induces the coupling of energy from one transmission structure to another when a stimulus is applied. This is called *crosstalk*. Since

Advanced Signal Integrity for High-Speed Digital Design, By Stephen H. Hall and Howard L. Heck

most digital systems use signaling interfaces in which large numbers of transmission lines are routed in parallel through packages, connectors, and printed circuit boards, crosstalk can play an important role in determining the performance of the system. Trends toward smaller and faster systems will drive increased crosstalk levels in the future, resulting in two major impacts. First, crosstalk will affect signal integrity and timing by modifying the propagation characteristics of the lines (characteristic impedance and propagation velocity). Second, crosstalk couples noise onto lines, which harms signal integrity and reduces noise margins.

We will see that signals on interconnects propagate in modes that are a function of the switching patterns and of the self- and mutual capacitances and inductances in the system. As was the case for isolated lines in Section 3.2.4, signals propagating in a coupled multiconductor transmission-line system are described by the wave equation. Whereas the solution to the isolated line system resulted in a pair of forward- and backward-traveling waves, for the coupled system with n signals there will be n forward- and n backward-traveling waves. Each forward–backward wave pair constitutes a *mode*, and we can analyze the behavior of the system in terms of the modes.

In this chapter we introduce the mechanism that causes crosstalk, develop a mathematical formulation of the coupled wave equations, describe techniques for analyzing and modeling coupled systems, and discuss the impact of crosstalk on system performance.

4.1 MUTUAL INDUCTANCE AND CAPACITANCE

From a circuit point of view, crosstalk is caused by mutual inductances and mutual capacitances between conductors. These two phenomena couple energy between lines via the magnetic (for mutual inductance) and electric (mutual capacitance) fields. We next examine each in more detail.

Mutual inductance L_M induces current from a driven line onto a quiet line by means of the magnetic field, as shown in Figure 4-1. Conceptually, if the "victim" trace is in close enough proximity to a driven line such that its magnetic flux lines intersect the victim trace, a current will be induced on that line. The mutual inductance creates a voltage noise on the victim in proportion to the rate of change of the current on the driven line according to

$$\Delta v_L = L_M \frac{di}{dt} \tag{4-1}$$

In equation (4-1), Δv_L is the voltage coupled by the mutual inductance L_M, in response to the transient current i.

Mutual capacitance C_M is the coupling of conductors via the electric field. Conceptually, if the victim trace is in close enough proximity to a driven line such that its electric field lines intersect the victim trace, a current is induced

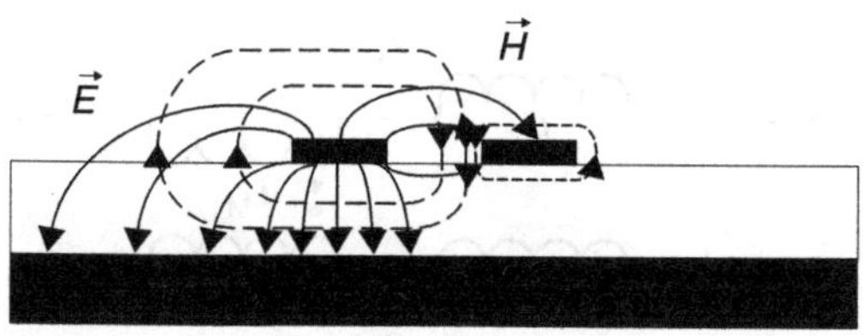

Figure 4-1 Coupled PCB transmission lines.

onto the victim line in proportion to the rate of change of voltage on the driven line:

$$\Delta i_C = C_M \frac{dv}{dt} \tag{4-2}$$

In equation (4-2), Δi_C is the amount of current coupled through the mutual capacitance C_M, when driven by the voltage signal v.

I/O circuit rise and fall times decrease as data transfer rates increase, so (4-1) and (4-2) predict that both inductive and capacitive crosstalk will play a significant role in high-speed digital applications. Thus, we must account for mutual capacitance and inductance in our modeling and analysis of coupled systems, and we now proceed to examine each in more detail.

4.1.1 Mutual Inductance

We start our discussion of mutual inductance by studying the simple inductive circuit shown in Figure 4-2. Given transient currents i_1 and i_2 injected onto lines 1 and 2, respectively, we can write expressions for v_1 and v_2 from Faraday's law,

$$v_1 = L_O \frac{di_1}{dt} + L_M \frac{di_2}{dt} \tag{4-3}$$

$$v_2 = L_O \frac{di_2}{dt} + L_M \frac{di_1}{dt} \tag{4-4}$$

where L_O is the self-inductance and L_M is the mutual inductance between lines 1 and 2. From (4-3) and (4-4) we note that the potential differences v_1 and v_2 depend on both input currents and on the self- and mutual inductances. We can better understand the impact of mutual inductance by analyzing the situations where the input currents are equal ($i_1 = i_2$) and where they are opposite ($i_1 = -i_2$). In the equal-current case we assume that their transition times are equal, so that $di_1/t = di_2/dt = di/dt$. Applying these signals leads to

$$v_1 = v_2 = (L_O + L_M)\frac{di}{dt} \tag{4-5}$$

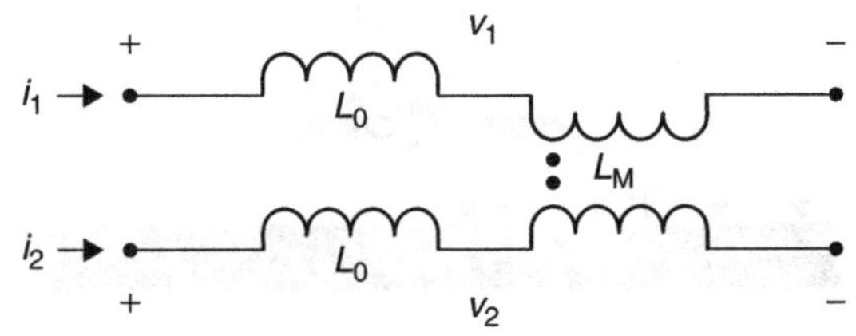

Figure 4-2 Coupled inductor circuit.

Looking next at the opposing current flow case, with the assumption that the transition time of current i_1 is equal to that of i_2, we have $di_1/dt = -di_2/dt = di/dt$, and derive

$$v_1 = -v_2 = (L_0 - L_M)\frac{di}{dt} \tag{4-6}$$

When the two inputs are equal, we say that the system is being driven in *even mode*; in the case of opposite polarity inputs, we say that the system is driven in *odd mode*. Notice that the *effective inductance* of the system as seen by our input signals is changed by the mutual inductance and is a function of the switching pattern. In particular, the even-mode inductance is increased relative to the self-inductance by an amount equal to the mutual inductance. Correspondingly, the odd-mode inductance is decreased by the mutual inductance, giving us a relative comparison of self-, even-mode, and odd-mode inductances, $L_{\text{even}} > L_0 > L_{\text{odd}}$.

As a first step toward developing a general expression for the inductance, we write the equation for the voltage across the inductive elements as a function of inductances and input currents in matrix form:

$$\begin{bmatrix} v_1 \\ v_2 \end{bmatrix} = \begin{bmatrix} L_0 & L_M \\ L_M & L_0 \end{bmatrix} \begin{bmatrix} di_1/dt \\ di_2/dt \end{bmatrix} \tag{4-7}$$

If we add a third line to our system, as shown in Figure 4-3, we can extend equation (4-7):

$$\begin{bmatrix} v_1 \\ v_2 \\ v_3 \end{bmatrix} = \begin{bmatrix} L_{11} & L_{12} & L_{13} \\ L_{21} & L_{22} & L_{23} \\ L_{31} & L_{32} & L_{33} \end{bmatrix} \begin{bmatrix} di_1/dt \\ di_2/dt \\ di_3/dt \end{bmatrix} \tag{4-8}$$

In equation (4-8), the diagonal elements L_{11}, L_{22}, and L_{33} represent the self-inductances on lines 1, 2, and 3, respectively. The mutual inductances are represented by L_{ij}, where i and j correspond to the lines coupled by the mutual inductance. The inductance matrix is symmetric. In other words, the mutual inductances between lines do not depend on the direction, so that $L_{ij} = L_{ji}$.

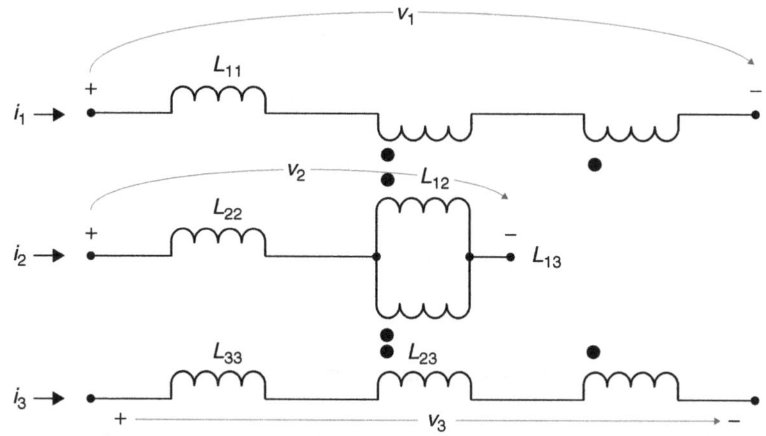

Figure 4-3 Circuit with three coupled inductors.

As a final step, we can generalize equation (4-8) to n inductively coupled lines,

$$\begin{bmatrix} v_1 \\ v_2 \\ \vdots \\ v_n \end{bmatrix} = \begin{bmatrix} L_{11} & L_{12} & \cdots & L_{1n} \\ L_{n1} & L_{22} & & L_{2n} \\ \vdots & & \ddots & \vdots \\ L_{n1} & L_{n2} & \cdots & L_{nn} \end{bmatrix} \begin{bmatrix} di_1/dt \\ di_2/dt \\ \vdots \\ di_n/dt \end{bmatrix} \tag{4-9}$$

where L_{ii} are the self-inductances and L_{ij} are the mutual inductances.

4.1.2 Mutual Capacitance

Our discussion of mutual capacitance follows a similar line of reasoning as for mutual inductance, beginning with the capacitive circuit shown in Figure 4-4. Given the input signals v_1 and v_2 on lines 1 and 2, we write expressions for the currents i_1 and i_2:

$$i_1 = C_g\frac{dv_1}{dt} + C_M\left(\frac{dv_1}{dt} - \frac{dv_2}{dt}\right) = (C_g + C_M)\frac{dv_1}{dt} - C_M\frac{dv_2}{dt} \tag{4-10}$$

$$i_2 = C_g\frac{dv_2}{dt} + C_M\left(\frac{dv_2}{dt} - \frac{dv_1}{dt}\right) = (C_g + C_M)\frac{dv_2}{dt} - C_M\frac{dv_1}{dt} \tag{4-11}$$

In analyzing the behavior of the circuit, let us first look at the case where line 1 has a transient signal dv/dt applied to it, whereas line 2 has no signal ($dv_2/dt = 0$). Applying these inputs to (4-10) and (4-11) gives

$$i_1 = (C_g + C_M)\frac{dv}{dt} \tag{4-12}$$

$$i_2 = -C_M\frac{dv_1}{dt} \tag{4-13}$$

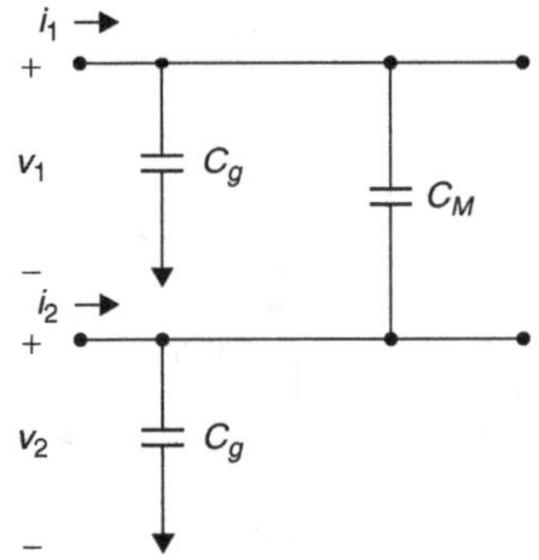

Figure 4-4 Coupled capacitor circuit.

Examination of equation (4-12) reveals that the effective capacitance of line 1 when it is driven in isolation is equal to the sum of the capacitance to ground and the mutual capacitance between lines 1 and 2. Conceptually, (4-12) says that the voltage signal applied to line 1 must charge up both the capacitance to ground and the capacitance between lines. Therefore, the sum of the capacitance to ground plus the mutual capacitance gives the total capacitance of line 1. Equation (4-13) indicates that signal on line 1 will impress an unwanted signal (i.e., crosstalk noise) on line 2 through the mutual capacitance. These effects are represented conceptually by the capacitance sketch for the three-conductor PCB (two signals plus a ground) shown in Figure 4-5.

We can also further our insight into the impact of mutual capacitance by analyzing the even-mode ($v_1 = v_2$) and odd-mode ($v_1 = -v_2$) cases. For the even mode, since $v_1 = v_2$, we assume that the rise times (or fall times) are equal, and therefore that $dv_1/dt = dv_2/dt = dv/dt$. Substitution into (4-10) and (4-11) leads to

$$i_1 = i_2 = C_g \frac{dv}{dt} \tag{4-14}$$

For the odd mode we use $dv_1/dt = -dv_2/dt = dv/dt$, and find that

$$i_1 = -i_2 = \left(C_g + 2C_M\right) \frac{dv}{dt} \tag{4-15}$$

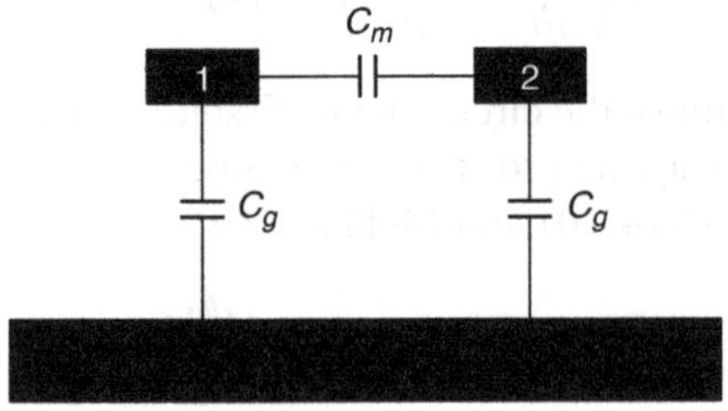

Figure 4-5 Coupled capacitances on a printed circuit board.

Equations (4-14) and (4-15) show that the *effective capacitance* of the system as seen by our input signals is changed by the mutual capacitance and is a function of the switching pattern. In particular, the even-mode capacitance is decreased relative to the total capacitance, by an amount equal to the mutual capacitance. Correspondingly, the odd-mode capacitance is increased by the mutual capacitance, giving us the relationship $C_{\text{even}} < C_{\text{total}} < C_{\text{odd}}$, where the total capacitance is $C_{\text{total}} = C_0 + C_M$.

In developing a general expression for the capacitance, we write the equation for the current flow in the capacitive elements as a function of capacitances and input voltages in matrix form:

$$\begin{bmatrix} i_1 \\ i_2 \end{bmatrix} = \begin{bmatrix} C_g + C_M & -C_M \\ -C_M & C_g + C_M \end{bmatrix} \begin{bmatrix} dv_1/dt \\ dv_2/dt \end{bmatrix} \tag{4-16}$$

If we add a third line to our system, as shown in Figure 4-6, we can extend equation (4-16):

$$\begin{bmatrix} i_1 \\ i_2 \\ i_3 \end{bmatrix} = \begin{bmatrix} C_{11} & -C_{12} & -C_{13} \\ -C_{21} & C_{22} & -C_{23} \\ -C_{31} & -C_{32} & C_{33} \end{bmatrix} \begin{bmatrix} dv_1/dt \\ dv_2/dt \\ dv_3/dt \end{bmatrix} \tag{4-17}$$

In equation (4-17), the diagonal elements C_{11}, C_{22}, and C_{33} represent the total capacitances on lines 1, 2, and 3, respectively. The total capacitance is the sum of the capacitance to ground (e.g., C_{1g} for line 1) plus the mutual capacitances between lines. The mutual capacitances are represented by C_{ij}, where i and j correspond to the lines coupled by the mutual capacitance. In other words,

$$C_{ii} = C_{ig} + \sum_{j \neq i} |C_{ij}|$$

As is the case with the inductance matrix, the capacitance matrix is symmetric since the capacitance does not depend on the polarity of the electric field.

We can generalize equation (4-17) to handle n capacitively coupled lines,

$$\begin{bmatrix} i_1 \\ i_2 \\ \vdots \\ i_n \end{bmatrix} = \begin{bmatrix} C_{11} & -C_{12} & \cdots & -C_{1n} \\ -C_{21} & C_{22} & & -C_{2n} \\ \vdots & & \ddots & \vdots \\ -C_{n1} & -C_{n2} & \cdots & C_{nn} \end{bmatrix} \begin{bmatrix} dv_1/dt \\ dv_2/dt \\ \vdots \\ dv_n/dt \end{bmatrix} \tag{4-18}$$

where the C_{ii} are the total capacitances and the C_{ij} are the mutual capacitances, with $C_{ij} = C_{ji}$. Remember that total capacitance of line i is equal to sum of the capacitance to ground for line i and the mutual capacitances between line i and all of the other lines in the system.

A feature of the capacitance matrix that immediately stands out is that the off-diagonal entries are negative. Although this may seem counterintuitive, it is

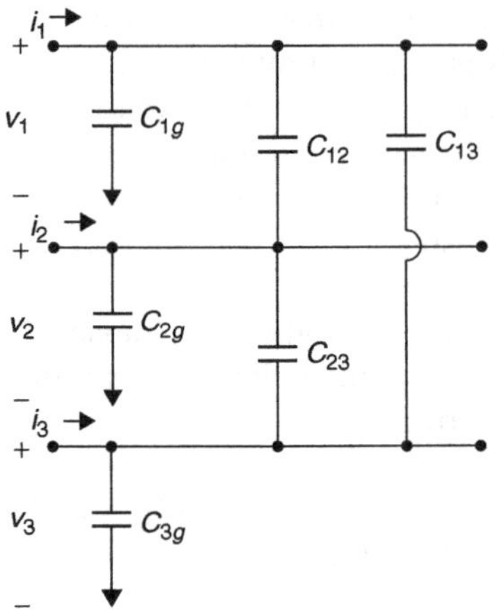

Figure 4-6 Circuit with three coupled capacitors.

simply a direct consequence of the fact that we defined the diagonal elements to be the total capacitance of the individual traces. This is necessary to simplify the mechanics of our circuit calculations while guaranteeing correct results, as we illustrate with an example.

Example 4-1 Equivalent Capacitance for a Coupled Pair We apply equation (4-18) to the two-line case, so that

$$C = \begin{bmatrix} C_{11} & -C_{12} \\ -C_{21} & C_{22} \end{bmatrix}$$

Driving the lines with $dv_1/dt = dv/dt$ and $dv_2/dt = 0$ gives $i_1 = C_{11}\ (dv/dt)$ and $i_2 = -C_{21}(dv/dt)$. Recalling that $C_{11} = C_{1g} + C_{12}$, we confirm that our solution matches the result given in equations (4-12) and (4-13). Additional analysis of the odd- and even-mode cases will also confirm our earlier results, and is left as an exercise for the reader.

4.1.3 Field Solvers

The capacitance and inductance matrices are typically obtained as the output from an electromagnetic field solver. These tools model the electromagnetic fields between transmission lines in a multiconductor system, providing the basis for equivalent circuit models and the inputs to transmission-line simulators such as HSPICE. Field simulators fall into two general categories, two-dimensional (2D) quasistatic and three-dimensional (3D) full-wave solvers. Examples of commercially available tools include Linpar [Djordjevic et al., 1999] for the 2D quasistatic case, and HFSS for 3D full-wave solvers.

Quasistatic tools use techniques similar to those outlined in Chapter 3, where Laplace's equation is solved to calculate the capacitance for a given set of boundary conditions. The inductance is then calculated from the capacitance using the speed of light as a conversion mechanism. Most 2D quasistatic tools will give the inductance and capacitance matrices (along with the resistance and conductance matrices) per unit of conductor length, under the assumption that the physical geometries and materials are uniform along the length of the transmission structures. They are typically easy to use and execute in a matter of seconds, but because of the length uniformity assumption, they cannot handle complex 3D transmission structures. Because they are static field solvers, they typically do not calculate frequency-dependent effects such as internal inductance and skin effect resistance. This is not a significant obstacle for printed circuit boards since the transmission structures are often uniform, and we can use other methods for including frequency dependence (refer to Chapters 5 and 6). The quasistatic assumption also requires that the signals propagate in transverse electromagnetic (TEM) mode, which means that the electric and magnetic fields are perpendicular and there is no field component in the direction of wave propagation, as described in Section 2.3.2. The TEM assumption is geometry dependent, but for typical PCB traces used to design high-speed digital systems (50 Ω, trace width about 5 mils), it is valid for frequencies well past 20 GHz.

Full-wave 3D solvers, on the other hand, are capable of simulating complex physical structures and will predict frequency-dependent losses, internal inductance, dispersion, and most other electromagnetic phenomena, including radiation. These tools essentially solve Maxwell's equations directly for an arbitrary geometry. Complex structures such as edge connectors and packages may require 3D tools to model their effects accurately at high data rates. The disadvantage of full-wave solvers is that they require more expertise to use, and simulations typically take hours or days rather than seconds. Additionally, the output from a full-wave simulator is often in the form of *S*-parameters, which typically require additional processing in order to use them for interconnect simulations for digital applications. As a result, design engineers typically employ 2D field solvers whenever possible, making use of 3D full-wave tools only where necessary.

4.2 COUPLED WAVE EQUATIONS

Before proceeding with the analysis of coupled systems, we first derive the wave equations, to reinforce the notion of wave propagation and to allow us to study the effects of mutual inductance and capacitance. The wave equation is central to our analysis of transmission lines, and extension to coupled systems will lend insight into the effects of crosstalk on the propagation of signals in a coupled system.

4.2.1 Wave Equation Revisited

We start our derivation of the transmission-line equations by focusing on the isolated line case shown in Figure 4-7. The circuit must satisfy Kirchhoff's laws.

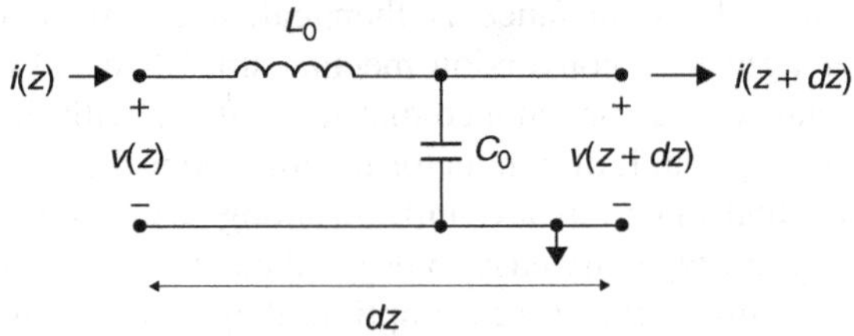

Figure 4-7 Differential circuit subsection for a lossless transmission line.

Application of Kirchhoff's voltage law (KVL) gives the voltage drop across the incremental inductance:

$$v(z) - v(z + dz) = -j\omega L_0 i(z)\, dz \tag{4-19}$$

where $j\omega L_0 dz$ is the frequency-dependent impedance of the inductor. Equation (4-19) contains a frequency term, ω, implying a sinusoidal input which has the form $v(t) = V_0 e^{j\omega t}$. A sinusoidal signal has the property that its derivative is a scaled version of the original signal, $dv(t)/dt = d(V_0 e^{j\omega t})/dt = j\omega V_0 e^{j\omega t} = j\omega v(t)$. Note that the current, $i(z,t)$, is also a sinusoid, so that $di(t)/dt = j\omega i(t)$.

Equation (4-19) is equivalent to Faraday's law for the response of an inductor to a transient current $[\Delta v = L(di/dt)]$. The analysis that follows is equally applicable for a digital input, since such a signal is composed of the superposition of multiple sinusoids, although the wave equation requires a separate solution for each frequency in the envelope of the driven signal. The next steps are to divide (4-19) through by dz, followed by differentiation with respect to z:

$$\frac{dv}{dz} = -j\omega L_0 i \tag{4-20}$$

$$\frac{d^2v}{dz^2} = -j\omega L_0 \frac{di}{dz} \tag{4-21}$$

Looking now at the incremental capacitance of the subsection, we apply Kirchhoff's current law to find the change in current, as shown in equation (4-22), where $(j\omega C_0\, dz)^{-1}$ is the impedance of the capacitor:

$$i(z + dz) - i(z) = -j\omega C_0 v(z)\, dz \tag{4-22}$$

Equation (4-22) is equivalent to $\Delta i = C_0(dv/dt)$ in the time domain. Dividing through by dz gives

$$\frac{di}{dz} = -j\omega C_0 v(z) \tag{4-23}$$

Substituting (4-23) into (4-21) to get an expression in terms of v yields the voltage wave equation:

$$\frac{d^2v}{dz^2} + \omega^2 L_0 C_0 v = 0 \tag{4-24}$$

The same approach will give the wave equation for the current:

$$\frac{d^2 i}{dz^2} + \omega^2 L_0 C_0 i = 0 \tag{4-25}$$

Equations (4-24) and (4-25) should be familiar as the wave equations for a uniform lossless transmission line.

4.2.2 Coupled Wave Equations

Our goal now is to generalize equations (4-24) and (4-25) to the n-coupled-line case. Figure 4-8 describes the circuit for a subsection of a pair of coupled lines. Our derivation follows the same method that we used above for the isolated transmission line. We start by applying KVL to develop equation (4-26). Note that the voltage drop on line 1 depends on the mutual inductance L_M and the current on the adjacent line $i_2(z)$, in addition to the line self-inductance L and driving current $i_1(z)$. We can also create a corresponding expression (4-27) for the voltage drop on line 2:

$$\begin{aligned} v_1(z) - v_1(z+dz) &= -j\omega L_0 i_1(z)\,dz - j\omega L_M i_2(z)\,dz \\ &= -j\omega[L_0 i_1(z) + L_M i_2(z)]\,dz \end{aligned} \tag{4-26}$$

$$v_2(z) - v_2(z+dz) = -j\omega[L_0 i_2(z) + L_M i_1(z)]\,dz \tag{4-27}$$

We can write the equations for the voltage drop across the coupled subcircuit in compact matrix form, where the boldface symbols represent the compact matrix:

$$\frac{d\mathbf{v}}{dz} = -j\omega \mathbf{L}\mathbf{i} \tag{4-28}$$

where

$$\frac{d\mathbf{v}}{dz} = \frac{1}{dz}\begin{bmatrix} v_1(z) - v_1(z+dz) \\ v_2(z) - v_2(z+dz) \end{bmatrix} \qquad \text{taken in the limit as } dz \to 0$$

$$\mathbf{L} = \begin{bmatrix} L_0 & L_M \\ L_M & L_0 \end{bmatrix}$$

$$\mathbf{i} = \begin{bmatrix} i_1(z,t) \\ i_2(z,t) \end{bmatrix}$$

Applying the isolated line method to the current change caused by the capacitances of the coupled line yields

$$\begin{aligned} i_1(z) - i_1(z+dz) &= -j\omega(C_g + C_M)v_1(z)\,dz + j\omega C_M v_2(z)\,dz \\ &= -j\omega[(C_g + C_M)v_1(z) - C_M v_2(z)]\,dz \end{aligned} \tag{4-29}$$

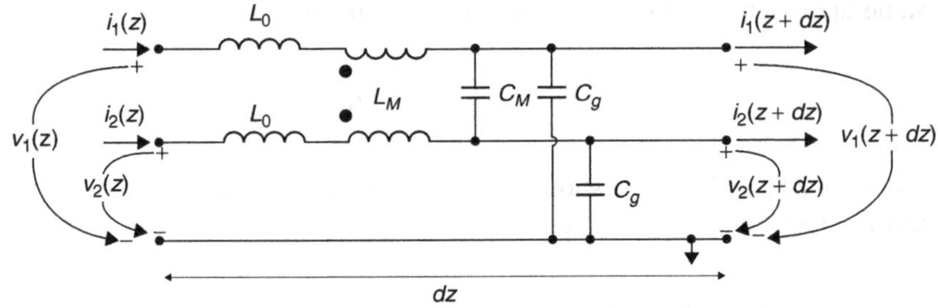

Figure 4-8 Differential circuit subsection for two lossless coupled transmission lines.

$$i_2(z) - i_2(z+dz) = -j\omega[(C_g + C_M)v_2(z) - C_M v_1(z)]\, dz \tag{4-30}$$

$$\frac{d\mathbf{i}}{dz} = -j\omega\mathbf{C}\mathbf{v} \tag{4-31}$$

where

$$\frac{d\mathbf{i}}{dz} = \frac{1}{dz}\begin{bmatrix} i_1(z) - i_1(z+dz) \\ i_2(z) - i_2(z+dz) \end{bmatrix} \quad \text{taken in the limit as } dz \to 0$$

$$\mathbf{C} = \begin{bmatrix} C_g + C_M & -C_M \\ -C_M & C_g + C_M \end{bmatrix}$$

$$\mathbf{v} = \begin{bmatrix} v_1(z) \\ v_2(z) \end{bmatrix}$$

Notice in equations (4-29) and (4-30) that the current change on line 1 caused by v_1 is proportional to the sum of the capacitance to ground C_g and the mutual capacitance between lines C_M. This is consistent with our earlier discussion of mutual capacitance. We can also reassure ourselves that this is correct by considering the response to a potential stimulus. Let us first assume that a potential, v, is applied to line 1 (relative to ground), while no potential is applied to line 2. In this case, we must charge up both the capacitance to ground and the mutual capacitance between lines, a result that (4-29) predicts. In the second case we assume that the same potential is applied to both lines. In this situation, lines 1 and 2 remain at the same potential, so that no charge is stored in the electric field between the lines. Therefore, line 1 need charge up only the capacitance to ground, a result that is also predicted by (4-29).

Returning to our derivation, we differentiate (4-28) with respect to z to get

$$\frac{d^2\mathbf{v}}{dz^2} = -j\omega L\frac{d\mathbf{i}}{dz} \tag{4-32}$$

Substituting for $d\mathbf{i}/dz$ from (4-31) results in the coupled voltage wave equation

$$\frac{d^2\mathbf{v}}{dz^2} = \omega^2\mathbf{LC}\mathbf{v} \tag{4-33}$$

Following our method, we can also derive the current wave equation for the coupled lines:

$$\frac{d^2\mathbf{i}}{dz^2} = \omega^2\mathbf{CLi} \tag{4-34}$$

Equations (4-33) and (4-34) comprehend wave propagation on each line in the system due to the source on the line itself and to sources being coupled from other lines through the electromagnetic fields. Notice also that they bear a striking similarity to equations (4-24) and (4-25). In fact, (4-33) and (4-34) reduce to (4-24) and (4-25) for the case $n = 1$.

A couple of additional observations about (4-33) and (4-34) are worth pointing out. First, in equation (4-34) the order of multiplication of the **L** and **C** matrices is reversed from that of (4-33). Since they are matrices, the product **LC** is not necessarily equal to **CL**. Second, the compact matrix equations are extensible to an arbitrary number of coupled transmission lines, providing the means for analyzing practical systems, as we demonstrate in the sections that follow.

4.3 COUPLED LINE ANALYSIS

In Section 4.1 we demonstrated that the effective capacitance and inductance will change with different switching patterns in systems where there is significant coupling between lines. This implies that the effective characteristic impedance and propagation delay will also be a function of the switching patterns. We also showed that coupling between lines can result in the appearance of noise on quiet lines. Therefore, both effects can have switching-dependent impacts on signal quality and timing characteristics of a coupled system, and we need to develop methods for quantifying them.

4.3.1 Impedance and Velocity

In this section we explain the effects that crosstalk-induced impedance and velocity changes have on signal integrity and timing, and provide a simple method for analyzing multiconductor transmission lines to provide first-order estimates of that impact. From Section 4.1, recall that the switching pattern will change the effective capacitance and inductance on a pair of couple lines, as summarized in Table 4-1. By inspection of the table we can easily construct equations for the effective impedances (Z_0) and propagation velocities (ν_p) for each case:

$$Z_{0,\text{isolated}} = \sqrt{\frac{L_0}{C_g + C_M}} \tag{4-35}$$

$$Z_{0,\text{even}} = \sqrt{\frac{L_0 + L_M}{C_g}} \tag{4-36}$$

TABLE 4-1. Summary of Effective Capacitance and Inductance for a Couple Pair

Mode	Line 1	Line 2	$C_{\text{effective}}$	$L_{\text{effective}}$
Even	lo hi hi lo	lo hi hi lo	C_g	$L_0 + L_M$
Odd	lo hi hi lo	hi lo lo hi	$C_g + 2C_M$	$L_0 - L_M$
Quiet	lo hi hi lo lo hi lo	lo lo lo lo hi hi hi hi	$C_g + C_M$	L_0

$$Z_{0,\text{odd}} = \sqrt{\frac{L_0 - L_M}{C_g + 2C_M}} \tag{4-37}$$

$$\nu_{p,\text{isolated}} = \frac{1}{\sqrt{L_0(C_g + C_M)}} \tag{4-38}$$

$$\nu_{p,\text{even}} = \frac{1}{\sqrt{(L_0 + L_M)C_0}} \tag{4-39}$$

$$\nu_{p,\text{odd}} = \frac{1}{\sqrt{(L_0 - L_M)(C_g + 2C_M)}} \tag{4-40}$$

Equations (4-35) through (4-40) are exact for the two-line case, and they give us a simple way to analyze a coupled pair via a lattice diagram or simulation of a single line using the effective characteristic impedance and effective propagation velocity. We call the models created using this method single-line equivalent models (SLEMs) [Hall et al., 2000].

It is interesting to note that the mutual inductance is always added or subtracted in the opposite manner as the mutual capacitance for odd- and even-mode propagation. The fields in Figure 4-9 help us understand why this is true. Considering odd-mode propagation as an example, the effect of mutual capacitance must be added because the conductors are at different potentials. Additionally,

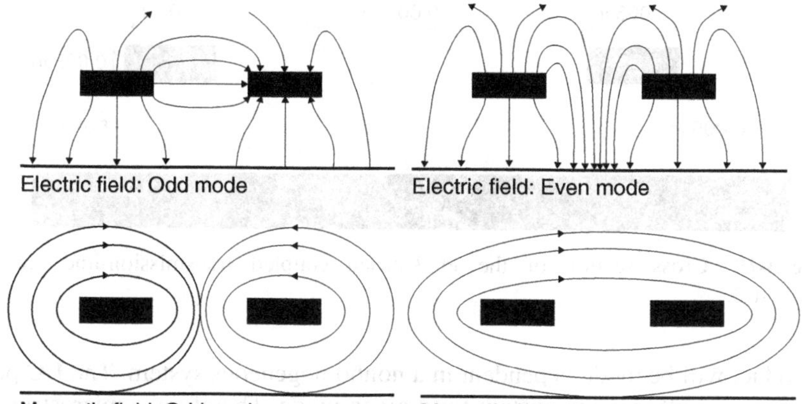

Figure 4-9 Odd- and even-mode electric and magnetic field patterns for a simple two-conductor system.

since the current in the two conductors flows in opposite directions, the currents induced on each line due to the coupling of the magnetic fields always oppose each other and cancel the effect of the mutual inductance. Therefore, the mutual inductance must be subtracted and the mutual capacitance must be added to calculate the odd-mode characteristics. These characteristics of even- and odd-mode propagation are due to the assumption that the signals are propagating only in transverse electromagnetic (TEM) mode, so that the electric and magnetic fields are always orthogonal to each other.

With homogeneous dielectrics, the product of **L** and **C** remains constant, since fields are confined within the uniform dielectric:

$$\mathbf{LC} = \frac{1}{\nu_p^2}\mathbf{I} \tag{4-41}$$

where **I** is the identity matrix. Thus, in a multiconductor homogeneous system such as a stripline array, if **L** is increased by the mutual inductance, **C** must be decreased by the mutual capacitance such that **LC** remains constant. Subsequently, a stripline, or buried microstrip, which is embedded in a homogeneous dielectric, should not exhibit velocity variations due to different switching modes. It will, however, exhibit pattern-dependent impedance variation.

In a nonhomogeneous dielectric where the electric fields fringe through more than one dielectric material, such as an array of microstrip lines, **LC** is not constant for different propagation modes because the electromagnetic fields travel partially in air and partially in the board's dielectric material. In a microstrip, the effective dielectric constant is a weighted average between air and the dielectric material of the board. Because the field patterns change with different propagation modes, the effective dielectric constant will change depending on the field densities contained within the board's dielectric material and the air. Thus, the

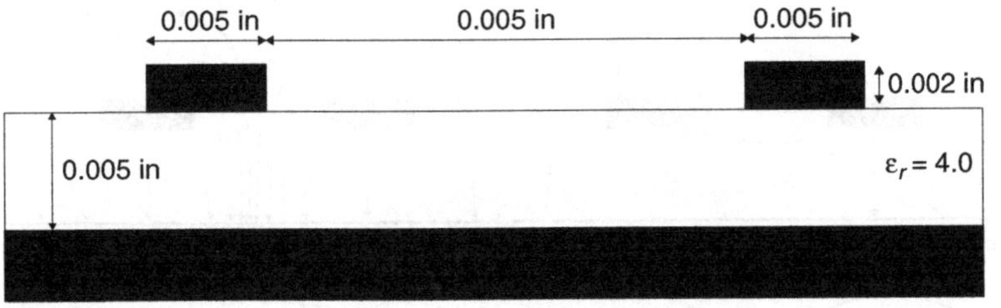

Figure 4-10 Cross section of the PCB-based coupled transmission-line pair for Example 4-2.

LC product will be mode dependent in a nonhomogeneous system. The **LC** product, however, will remain constant for a given mode. Therefore, a microstrip will exhibit both a velocity and an impedance change, due to different switching patterns. It should be noted that the description above holds for a single frequency. The product of **LC** varies with frequency but remains constant at each frequency point for a given mode.

Example 4-2 The PCB transmission lines depicted in Figure 4-10 have the following inductances and capacitances:

$$\mathbf{L} = \begin{bmatrix} 3.592 \times 10^{-7} & 3.218 \times 10^{-8} \\ 3.218 \times 10^{-8} & 3.592 \times 10^{-7} \end{bmatrix} \quad \text{H/m}$$

$$\mathbf{C} = \begin{bmatrix} 8.501 \times 10^{-11} & -2.173 \times 10^{-12} \\ -2.173 \times 10^{-12} & 8.501 \times 10^{-11} \end{bmatrix} \quad \text{F/m}$$

Assume that the waveform is driven into the line at $t = 1$ ns.

We have designed the PCB traces to have a typical (isolated) characteristic impedance of approximately 65 Ω with a length of 0.2794 m (11 in.). They are driven by a 1-V 65-Ω source, and are terminated to ground in 65 Ω at the far end. The rise and fall times are 0.1 ns. Compare the analytical results with those from a fully coupled simulation for even- and odd-mode propagation.

SOLUTION

Step 1: Calculate the impedances and velocities for all of the switching patterns.

$$Z_{0,\text{even}} = \sqrt{\frac{3.592 \times 10^{-7} + 3.218 \times 10^{-8}\ \text{H/m}}{8.501 \times 10^{-11} - 2.173 \times 10^{-12}\ \text{F/m}}} = 68.7\ \Omega$$

$$Z_{0,\text{odd}} = \sqrt{\frac{3.592 \times 10^{-7} - 3.218 \times 10^{-8}\ \text{H/m}}{8.501 \times 10^{-11} + 2.173 \times 10^{-12})\ \text{F/m}}} = 61.2\ \Omega$$

$$Z_{0,\text{isolated}} = \sqrt{\frac{3.592 \times 10^{-7}\ \text{H/m}}{8.501 \times 10^{-11}\ \text{F/m}}} = 65.0\ \Omega$$

$$\begin{aligned} \nu_{p,\text{even}} &= \frac{1}{\sqrt{[(35.92 + 3.218) \times 10^{-8}\ \text{H/m}][(85.01 - 2.173) \times 10^{-12}\ \text{F/m}]}} \\ &= 1.756 \times 10^{8}\ \text{m/s} \end{aligned}$$

$$\begin{aligned} \nu_{p,\text{odd}} &= \frac{1}{\sqrt{[(35.92 - 3.218) \times 10^{-8}\ \text{H/m}][(85.01 + 2.173) \times 10^{-12}\ \text{F/m}]}} \\ &= 1.873 \times 10^{8}\ \text{m/s} \end{aligned}$$

$$\begin{aligned} \nu_{p,\text{isolated}} &= \frac{1}{\sqrt{3.592 \times 10^{-7}\ \text{H/m}\ 8.501 \times 10^{-11}\ \text{F/m}}} \\ &= 1.810 \times 10^{8}\ \text{m/s} \end{aligned}$$

Step 2: Calculate the even-mode waveform. Calculate the values for the initial voltage and current waves, reflection coefficients, final voltage and current levels, and propagation delay as preparation for a lattice diagram analysis.

$$\begin{aligned} v(t = 0, z = 0) &= \frac{Z_{0,\text{even}}}{R_S + Z_{0,\text{even}}} V_S = \frac{68.7\ \Omega}{65\ \Omega + 68.7\ \Omega}(1\ \text{V}) = 0.514\ \text{V} \\ i(t = 0, z = 0) &= \frac{v(t = 0, z = 0)}{Z_{0,\text{even}}} = \frac{0.514\ \text{V}}{68.7\ \Omega} = 7.48\ \text{mA} \\ \Gamma(z = 0) &= \frac{R_S - Z_{0,\text{even}}}{R_S + Z_{0,\text{even}}} = \frac{65\ \Omega - 68.7\ \Omega}{65\ \Omega - 68.7\ \Omega} = -0.028 \\ \Gamma(z = l) &= \frac{R_T - Z_{0,\text{even}}}{R_T + Z_{0,\text{even}}} = \frac{65\ \Omega - 68.7\ \Omega}{65\ \Omega + 68.7\ \Omega} = -0.028 \\ v(t = \infty) &= \frac{R_t}{R_S + R_t} V_S = \frac{65\ \Omega}{65\ \Omega + 65\ \Omega}(1\ \text{V}) = 0.500\ \text{V} \\ i(t = \infty) &= \frac{v_S}{R_S + R_t} = \frac{1.000}{65\ \Omega + 65\ \Omega} = 7.69\ \text{mA} \\ t_{d,\text{even}} &= \frac{l}{\nu_{p,\text{even}}} = \frac{11\ \text{in}}{1.756 \times 10^{8}\ \text{m/s}} \frac{\text{m}}{39.37\ \text{in.}} = 1.592\ \text{ns} \end{aligned}$$

The corresponding lattice diagram is shown in Figure 4-11.

Step 3: Calculate the odd-mode waveform. Repeat the analysis for the odd-mode propagation, with the lattice diagrams shown in Figure 4-12.

$$v(t = 0, z = 0) = \frac{Z_{0,\text{odd}}}{R_S + Z_{0,\text{odd}}} V_S = \frac{61.2\ \Omega}{65\ \Omega + 61.2\ \Omega}(1\ \text{V}) = 0.485\ \text{V}$$

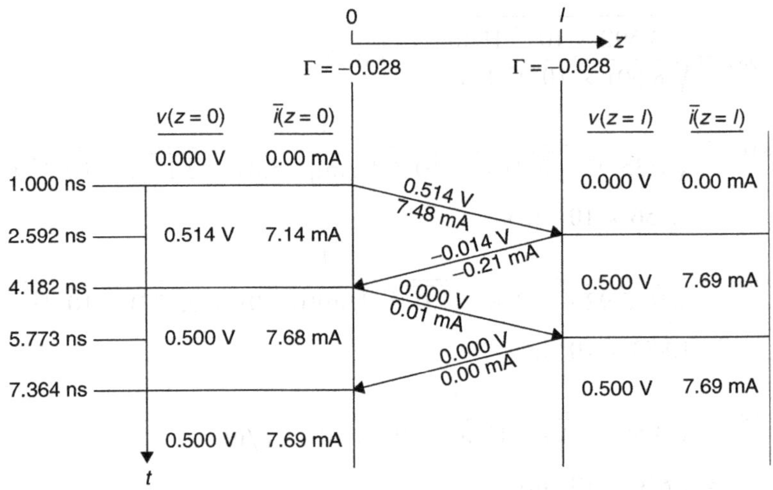

Figure 4-11 Lattice diagram for even-mode propagation of the coupled line pair for Example 4-2.

Note that the rising edge wave will be 0.485 V, starting from 0.000 V, while the falling edge wave will be −0.485 V, starting from 0.500 V.

$$i(t=0, z=0) = \frac{v(t=0, z=0)}{Z_{\text{odd}}} = \frac{0.485 \text{ V}}{61.2 \ \Omega} = 7.92 \text{ mA}$$

$$\Gamma(z=0) = \frac{R_S - Z_{0,\text{odd}}}{R_S + Z_{0,\text{odd}}} = \frac{65 \ \Omega - 61.2 \ \Omega}{65 \ \Omega + 61.2 \ \Omega} = 0.030$$

$$\Gamma(z=l) = \frac{R_T - Z_{0,\text{odd}}}{R_T + Z_{0,\text{odd}}} = \frac{65 \ \Omega - 61.2 \ \Omega}{65 \ \Omega + 61.2 \ \Omega} = 0.030$$

$$\text{Rising edge} \begin{cases} v(t=\infty) = \dfrac{R_t}{R_S + R_t} V_S = \dfrac{65 \ \Omega}{65 \ \Omega + 65 \ \Omega}(1 \text{ V}) = 0.500 \text{ V} \\ i(t=\infty) = \dfrac{v_S}{R_S + R_t} = \dfrac{1.000}{65 \ \Omega + 65 \ \Omega} = 7.69 \text{ mA} \end{cases}$$

$$\text{Falling edge} \begin{cases} v(t=\infty) = 0.000 \text{ V} \\ i(t=\infty) = 0.00 \text{ mA} \end{cases}$$

$$t_{d,\text{odd}} = \frac{l}{v_{p,\text{odd}}} = \frac{0.2794 \text{ m}}{1.873 \times 10^8 \text{ m/s}} \left(\frac{10^9 \text{ ns}}{\text{s}} \right) = 1.492 \text{ ns}$$

Step 4: Compare the calculated results to simulated results. From Figure 4-13 we see that the results from SLEM analysis match results from fully coupled SPICE time-domain simulations using the **L** and **C** matrices.

Having developed the SLEM approach for the two-line case, we want to generalize it to deal with an arbitrary number of coupled lines, since real systems

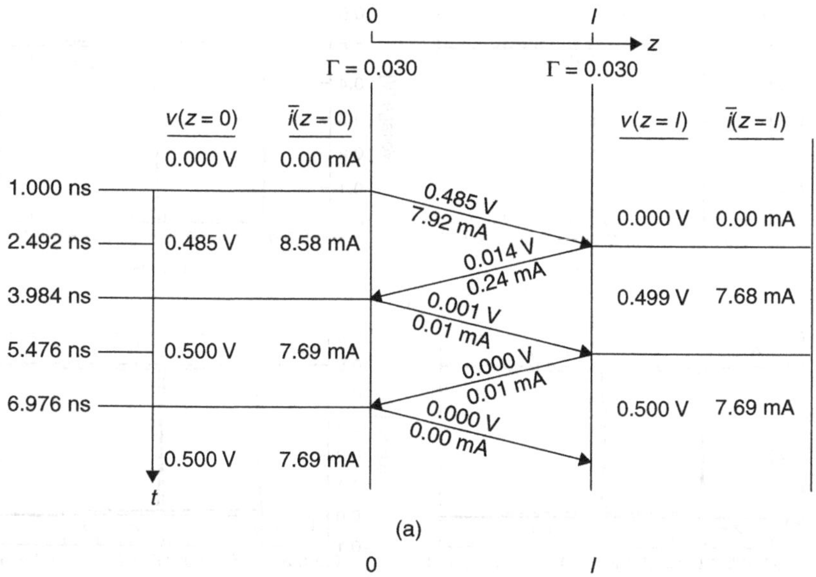

Figure 4-12 (a) Rising- and (b) falling-edge lattice diagrams for odd-mode propagation of the coupled line pair for Example 4-2.

typically contain larger numbers of coupled lines. In our earlier discussion we noted that for even-mode switching, the effective inductance is increased by the mutual inductances between lines, while the effective capacitance is decreased by the mutual capacitance. We also know that for odd-mode switching, the effective inductance is decreased by the mutual inductances between lines, while the effective capacitance is increased by the mutual capacitances. Finally, the mutual inductances and capacitances for quiet lines do not change the effective

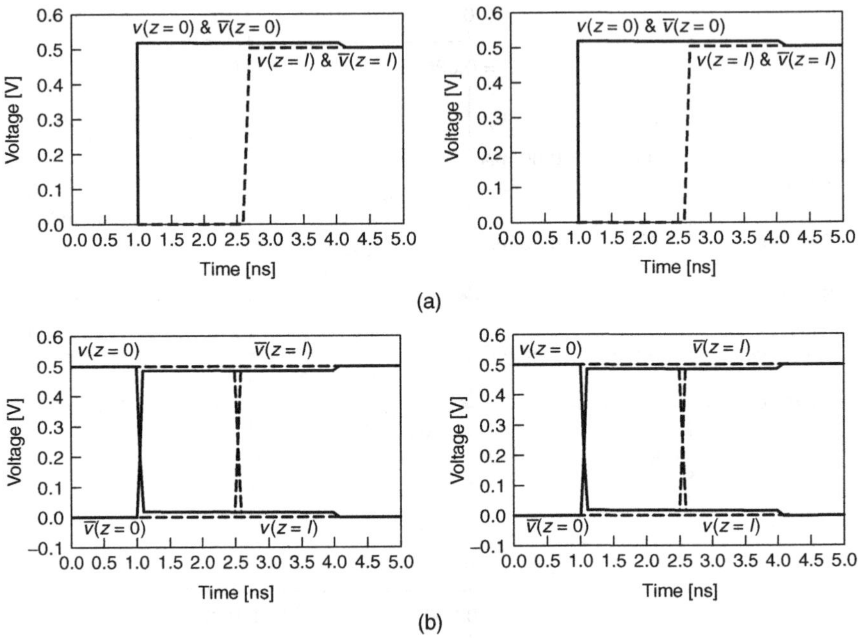

Figure 4-13 Comparison of (a) even- and (b) odd-mode calculated (left) and simulated (right) results for Example 4-2.

inductance and capacitance. With this knowledge, we can write generalized approximations for the effective inductance and capacitance, adopting the matrix notation that we developed in Section 4.1 .

$$L_{\text{eff},n} = L_{nn} + \sum L_{ne} - \sum L_{no} \tag{4-42}$$

$$C_{\text{eff},n} = C_{nn} - \sum \left|C_{ne}\right| + \sum \left|C_{no}\right| \tag{4-43}$$

where $L_{\text{eff},n}$ = effective inductance of line n

L_{nn} = self-inductance of line n

$\sum L_{ne}$ = sum of the mutual inductances for the lines that switch in phase with line n (which we approximate as "even" mode)

$\sum L_{no}$ = sum of the mutual inductances for the lines that switch out of phase with line n (which we approximate as "odd" mode)

$C_{\text{eff},n}$ = effective capacitance of line n

C_{nn} = total capacitance of line n

$\sum \left|C_{ne}\right|$ = sum of the absolute values of the mutual capacitances for the lines that switch in phase with line n

$\sum \left|C_{no}\right|$ = sum of the absolute values of the mutual capacitances for the lines that switch out of phase with line n

Recall from Section 4.1 that the elements on the matrix diagonals represent the self-inductance and total capacitance, respectively, while the off-diagonal elements represent the mutual terms. Once we obtain the effective inductance and capacitance, we can calculate the effective impedance and propagation velocity:

$$Z_{0,\text{eff},n} = \sqrt{\frac{L_{\text{eff},n}}{C_{\text{eff},n}}} \tag{4-44}$$

$$v_{p,\text{eff},n} = \frac{1}{\sqrt{L_{\text{eff},n} C_{\text{eff},n}}} \tag{4-45}$$

We can use equations (4-44) and (4-45) along with the physical length of the transmission line to analyze or simulate the behavior of any line as a simple noncoupled line while accounting for the coupling to the other lines in the system. This technique is best used during the early design phase of a bus, when I/O transceiver impedances and line-to-line spacing are being chosen. In addition, it is useful only for signals traveling in the same direction. For signals traveling in opposite directions, fully coupled simulations are required to comprehend the effects of crosstalk.

At this point it is important to note that although the SLEM method gives correct results for systems with two coupled lines, it is an approximation that will not exactly match the actual modal impedances and velocities for three or more coupled lines. As such, its use should be restricted to early design exploration aimed at narrowing down the solution space. Final simulations should always be done with fully coupled models. The accuracy of the SLEM model (for three lines) is reasonable for cross sections in which the spacing/height ratio is greater than 1. When this ratio is less than 1, the SLEM approximation should not be used. In Section 4.4 we introduce a technique for producing exact solutions to systems of three or more coupled lines.

4.3.2 Coupled Noise

Before describing the coupled noise mechanism and developing methods to quantify the noise, we need to provide some motivation for doing so. By now it should be clear that our goal is to transmit interchip digital data signals successfully. In doing so, we use an active transmitter circuit to drive the data onto the interconnect, where it propagates as an electromagnetic wave to a receiver circuit that senses the signal and restores it to the appropriate logic level. It is this signal restoration operation that can be affected by crosstalk noise (as well by any other sources of noise).

In restoring the signals, the receivers have thresholds for distinguishing between logic levels. Crossing over a logic threshold will cause the receiver to switch the output state. Referring to Figure 4-14, we see a voltage transfer characteristic for an inverting receiver. When a rising input signal crosses v_{IH}, it will cause the output of the receiver to switch from high to low. However, noise

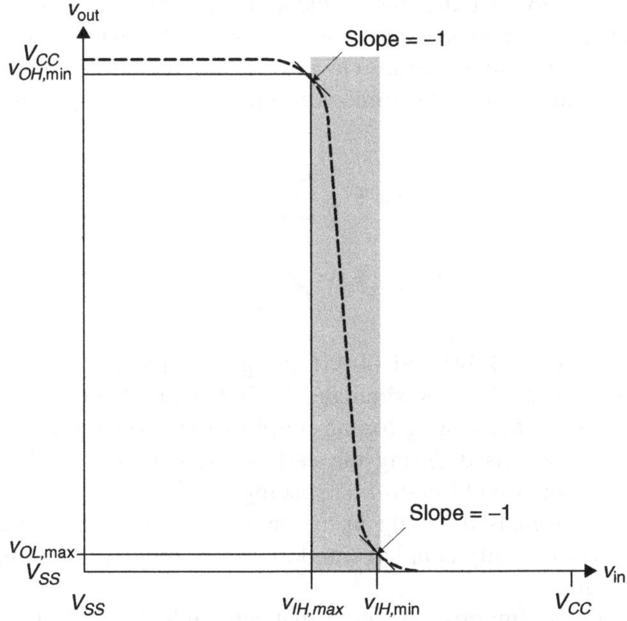

Figure 4-14 Receiver transfer characteristic and switching thresholds.

on the signal that causes an excursion back below v_{IL} will cause erroneous switching (Figure 4-15). It is the job of the system designer to make sure that this does not happen.

Refer to Chapter 11 for additional details on the operation of receivers for high-speed links.

Qualitative Description Before developing a quantitative treatment of crosstalk-induced noise, we describe the behavior qualitatively by looking at the propagation of the aggressor and coupled noise wave. From our discussion of mutual inductance and capacitance it is apparent that energy is coupled from one line to another only during signal transitions (i.e., the rising and falling edges). Subsequently, we look at the propagation of a rising edge on the aggressor line.

Figure 4-16 shows a pair of coupled lines terminated at both ends that represent a typical system. As the incident wave on the aggressor line is launched, it immediately begins coupling over to the victim line through the mutual capacitance and inductance. Current that couples through the mutual capacitance (i_C) splits into forward-traveling (i_f) and backward-traveling (i_b) components in the victim line, as shown in Figure 4-17. Current that couples through the mutual inductance (i_L) travels back toward the source end. As a result, we have a forward-coupled wave that is a function of the difference between the capacitively coupled current and the inductively coupled current. Since it is based on the difference between capacitive and inductive coupling, the amplitude may have the same polarity as

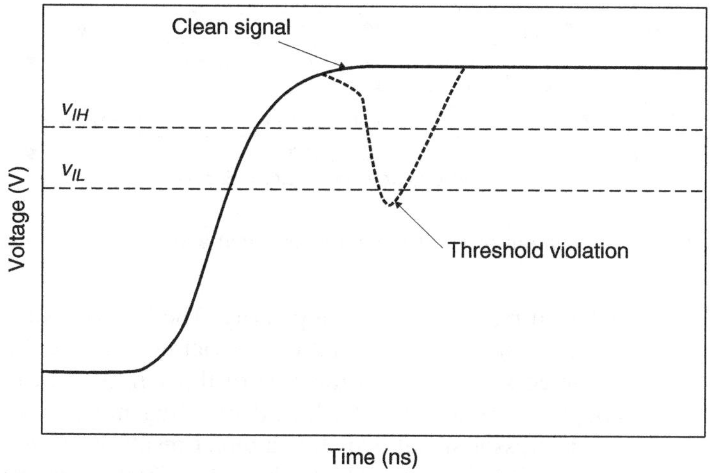

Figure 4-15 Receiver threshold violation.

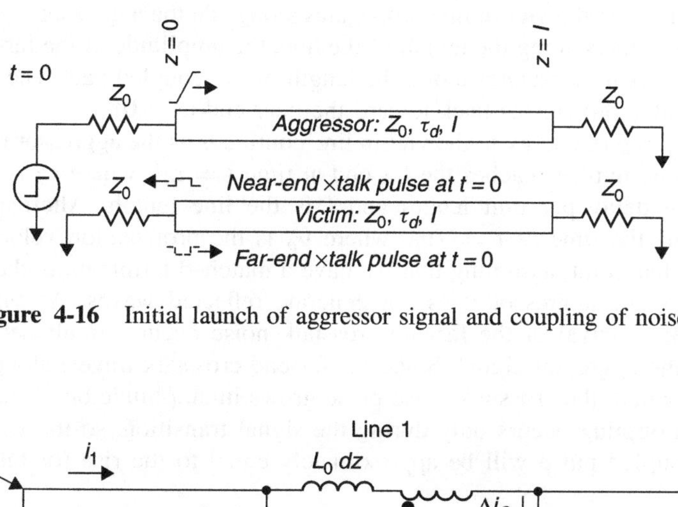

Figure 4-16 Initial launch of aggressor signal and coupling of noise.

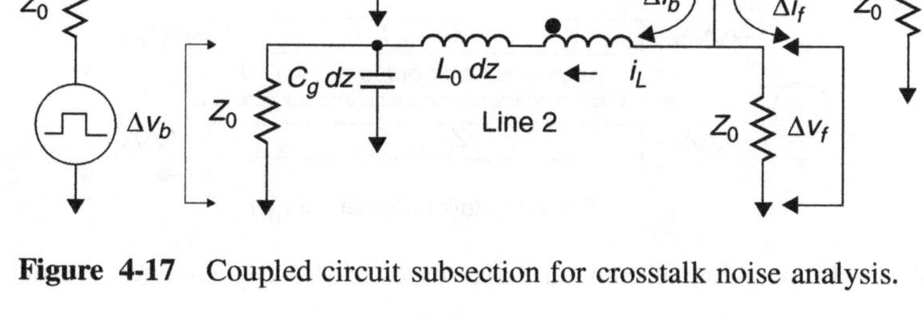

Figure 4-17 Coupled circuit subsection for crosstalk noise analysis.

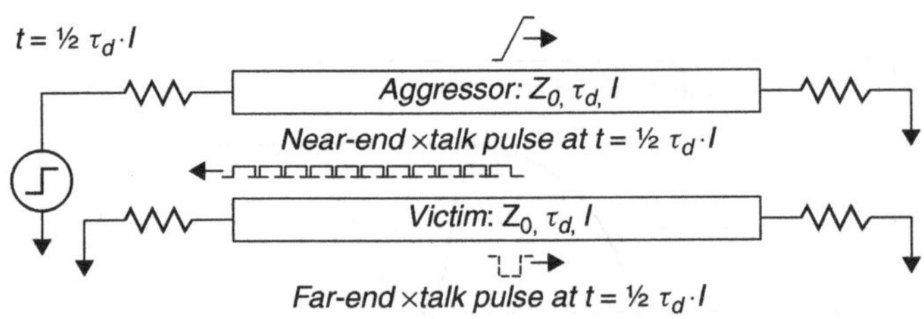

Figure 4-18 Propagation of incident aggressor signal and coupled noise pulses.

the aggressor signal, or it may have opposite polarity. The forward coupled wave then begins to propagate toward the far end of the victim line ($z = l$). We also have a backward coupled wave that is a function of the sum of the capacitively and inductively coupled currents. The backward traveling noise will have the same polarity as the aggressor signal, which is a direct result of the summing of the inductive and capacitive coupling. The backward crosstalk noise propagates toward the near end of the victim line, where it is immediately detectable.

As the incident wave on the aggressor propagates toward the far end, it continues to couple energy over to the victim line. As Figure 4-18 shows, the forward crosstalk pulse on the victim line propagates alongside the aggressor signal. Since coupling continues along the length of the line, the amplitude of the far-end noise pulse grows as it propagates along the length of the coupled pair. The backward crosstalk pulse propagates back toward the near end ($z = 0$).

The coupling of energy to the victim line continues as the aggressor propagates along the line until it reaches the far end at time $t = \tau_d l$, where τ_d is the signal propagation delay per unit length and l is the line length. Alternatively, we could define the time as $t = l/v_p$, where v_p is the propagation velocity of the signal. At that point, assuming that we have a matched termination, the coupling ceases since the aggressor does not generate reflected waves. As Figure 4-19 demonstrates, arrival of the far-end crosstalk noise occurs simultaneously with arrival of the aggressor signal. Since the far-end crosstalk travels along with the aggressor signal, the crosstalk noise pulse grows in amplitude but does not grow in width. Coupling occurs only during the signal transition, so the width of the forward coupled pulse will be approximately equal to the rise (or fall) time of

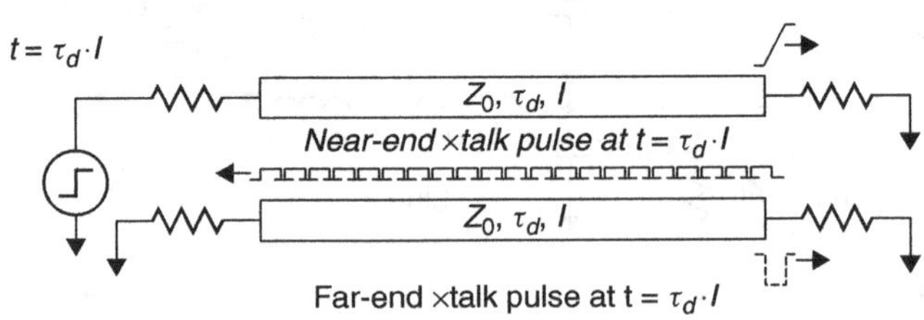

Figure 4-19 Propagation of coupled noise pulses as the aggressor reaches the far end ($z = l$).

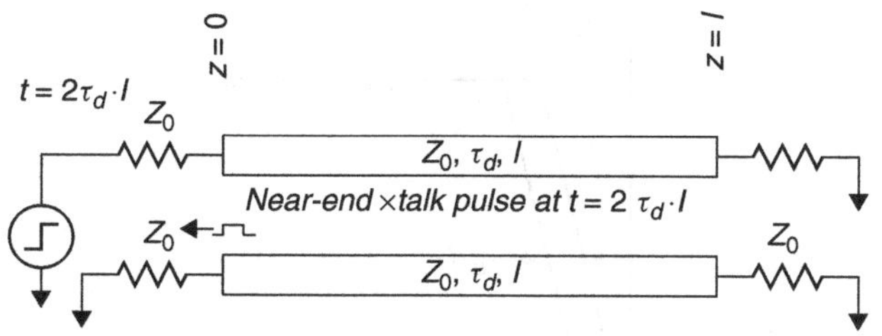

Figure 4-20 Completion of the noise pulse at the near end ($z = 0$).

the aggressor signal, as the figure illustrates. Even though additional energy is no longer being coupled at this point, the backward crosstalk wave must still travel back to the near end. This takes a full propagation delay ($t_d = \tau_d l$) to complete and is shown in Figure 4-20.

The shape of the near-end pulse depends on the electrical length of the coupled lines relative to the transition time of the aggressor signal. Consider the case for which the coupled length is less than one-half of the signal rise time. The beginning of the signal transition will reach the far end before the rising edge at the driving end has completed one-half of the transition. As the coupled noise propagates back to the near end, the signal at the far end continues to change, and therefore to couple more energy to the victim. Not until the rising-edge transition has finished propagating all the way to the far end does coupling stop. In that case, the near-end crosstalk pulse will have a similar shape (but different amplitude) to the far-end pulse, as shown in Figure 4-21a.

On the other hand, if the coupled length is greater than one-half the signal rise time, the near-end crosstalk pulse will reach maximum amplitude and will then begin to spread in time, a phenomenon known as *saturation*, which is shown in Figure 4-21b. To understand this effect, consider a case where the coupled electrical length is equal to several rise times. For this situation, we can visualize the signal edge as a traveling wave on the aggressor line. Recalling that the lines couple only during the transition, we can imagine the backward crosstalk as a train of pulses of equal magnitude with a width equal to the rise time that propagate back to the near end. As a result, the backward noise does not grow in magnitude, but instead, spreads out in time. Figure 4-22 illustrates the coupled pulse propagation as just described.

As a final note, we must realize that the crosstalk pulse magnitudes and shapes that we just described, and for which we will develop quantitative models in the next section, are specific to the matched termination case. Terminating the lines simplifies the analysis by eliminating the need to deal with reflected crosstalk and crosstalk from reflected aggressor signals. In general, the characteristics of the crosstalk noise are heavily dependent on the amount of coupling and the termination. For cases with imperfect termination and/or complex topologies, we recommend using a simulator to analyze the behavior of the system. Hall et al. [2000] describe crosstalk shapes for some general nonperfect termination schemes.

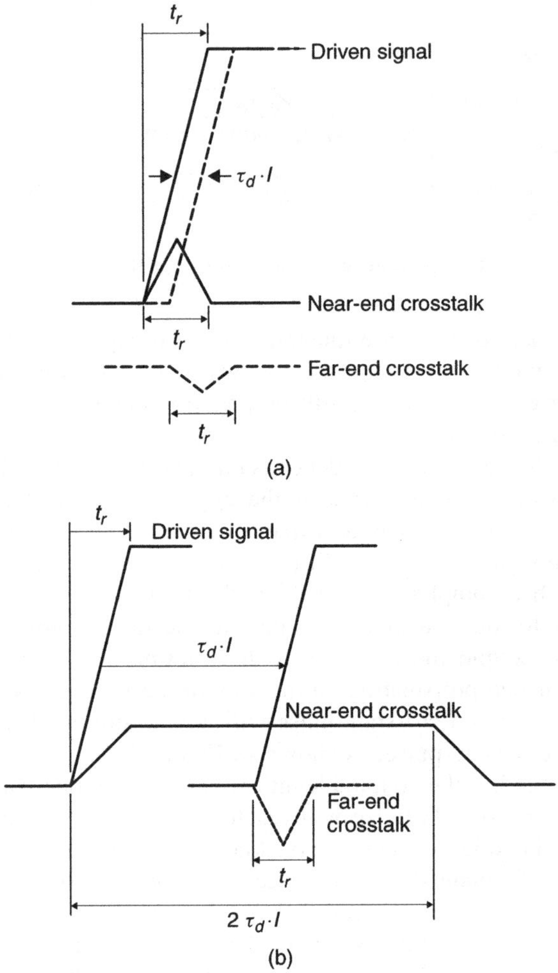

Figure 4-21 Forward- and backward-coupled noise pulses: (a) nonsaturated; (b) saturated.

Quantitative Development Having developed an intuitive understanding of crosstalk and the characteristics of crosstalk noise, we now derive the equations for noise at each end of a quiet transmission line (line 2 in Figure 4-17) that is induced by coupling from an adjacent coupled line (line 1) that is actively driven. Noise will be coupled from line 1 to line 2 through the mutual inductance $L_M\,dz$ and mutual capacitance $C_M\,dz$. Line 1, whose characteristic impedance is Z_0, has an incident pulse of magnitude v_1 (voltage) and i_1 (current), and is terminated in its characteristic impedance at both ends. We want to come up with expressions for the backward (Δv_b) and forward (Δv_f) noise pulses on line 2, which also has a characteristic impedance of Z_0 with termination at both ends. Our derivation follows the method presented by Seraphim et al. [1989].

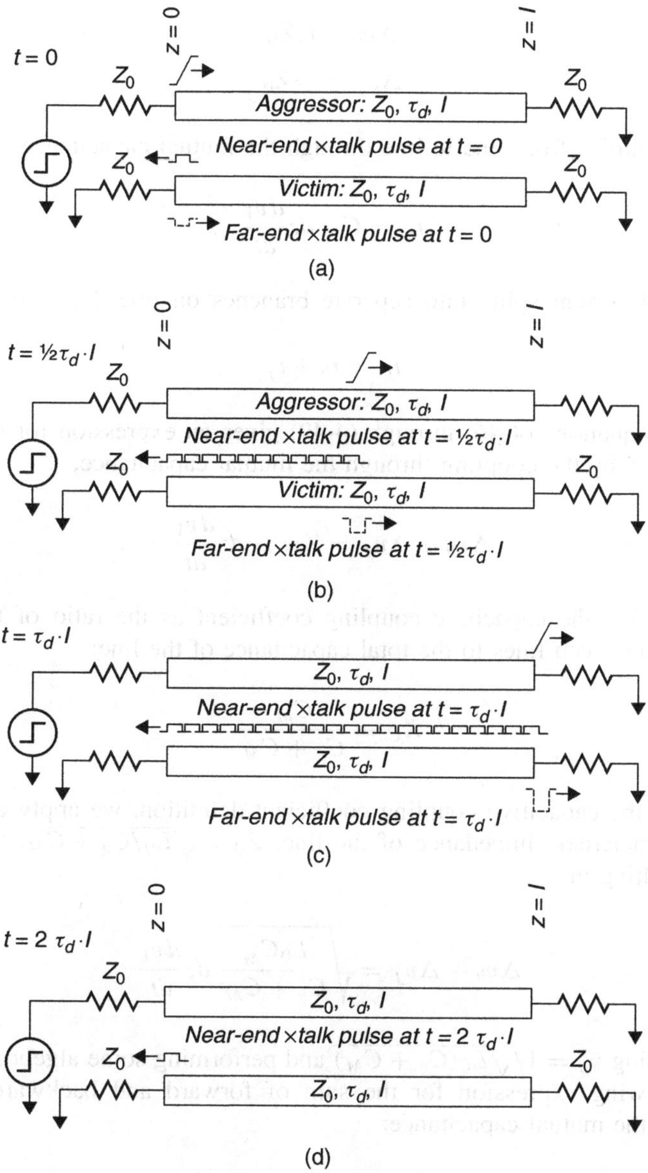

Figure 4-22 Summary of propagation of forward- and backward-coupled noise: (a) initial wave launch; (b) halfway down the line; (c) one full trip down the line; (d) round trip.

We begin the development by applying Ohm's law at each end of line 2 to get the amplitudes of the noise pulses at the near end (Δv_b) and far end (Δv_f):

$$\Delta v_b = i_b Z_0 \tag{4-46}$$

$$\Delta v_f = i_f Z_0 \tag{4-47}$$

Current is coupled from line to line through the mutual capacitance:

$$i_C = C_M \, dz \frac{dv_1}{dt} \tag{4-48}$$

The coupled current splits into separate branches on line 2 that flow in both directions:

$$i_C = i_b + i_f \tag{4-49}$$

Combining equations (4-46) through (4-49) gives an expression for the voltage pulses created by the coupling through the mutual capacitance,

$$\Delta v_b + \Delta v_f = Z_0 C_M \, dz \frac{dv_1}{dt} \tag{4-50}$$

Next we define the capacitive coupling coefficient as the ratio of the mutual capacitance between lines to the total capacitance of the line:

$$K_C \equiv \frac{C_M}{C_g + C_M} \tag{4-51}$$

Along with the capacitive coupling coefficient definition, we apply expressions for the characteristic impedance of the line, $Z_0 = \sqrt{L_0/C_g + C_M}$, to equation (4-50), resulting in

$$\Delta v_b + \Delta v_f = \sqrt{\frac{L_0 C_M^2}{C_g + C_M}} \, dz \frac{dv_1}{dt} \tag{4-52}$$

By substituting $v_p = 1/\sqrt{L_0(C_g + C_M)}$ and performing some algebra, we arrive at the following expression for the sum of forward and backward crosstalk induced by the mutual capacitance:

$$\Delta v_b + \Delta v_f = \frac{1}{v_p} K_C \, dz \frac{dv_1}{dt} \tag{4-53}$$

Turning now to the inductance, we note that the mutual inductance acts as a coupling transformer. Current on line 1 induces a voltage on line 2 that travels in the direction opposite that of the incident signal on line 1. As a result, it creates a voltage difference across the differential segment (dz) which travels back toward the source:

$$\Delta v_b - \Delta v_f = L_M \, dz \frac{di_1}{dt} \tag{4-54}$$

In the same way that we defined a capacitive coupling coefficient, we define the inductive coupling coefficient as the ratio of the mutual inductance between lines to the self-inductance of the line:

$$K_L \equiv \frac{L_M}{L_0} \tag{4-55}$$

Application of Ohm's law at the driven end of line 1 ($i_1 = v_1/Z_0$) yields

$$\begin{aligned} \Delta v_b - \Delta v_f &= dz \frac{L_M}{Z_0} \frac{dv_1}{dt} = dz \sqrt{\frac{L_M^2 (C_g + C_M)}{L_0} \frac{L_0}{L_0}} \frac{dv_1}{dt} \\ &= dz \sqrt{L_0 (C_g + C_M) \frac{L_M^2}{L_0^2}} \frac{dv_1}{dt} \end{aligned}$$

which reduces to another expression relating the forward and backward coupled noise to a coupling coefficient, in this case the inductive coupling coefficient:

$$\Delta v_b - \Delta v_f = dz \frac{K_L}{v_p} \frac{dv_1}{dt} \tag{4-56}$$

Since (4-53) and (4-56) give us two expressions with two unknowns, we can solve them for Δv_b and Δv_f and take $dz \to 0$ in the limit, to get

$$\frac{dv_f}{dz} = \frac{K_C - K_L}{2v_p} \frac{dv_1}{dt} \tag{4-57}$$

$$\frac{dv_b}{dz} = \frac{K_C + K_L}{2v_p} \frac{dv_1}{dt} \tag{4-58}$$

Forward Crosstalk Integrating (4-57) from $z = 0$ to $z = l$ gives an expression for forward crosstalk:

$$v_f = \frac{1}{2}(K_C - K_L) \frac{l}{v_p} \frac{dv_1}{dt} \tag{4-59}$$

By approximating dv_1/dt as the ratio of the voltage swing v and a 10 to 90% rise time t_r, we have our final expression for the forward crosstalk:

$$v_f = \frac{1}{2}(K_C - K_L) \frac{l}{v_p} \frac{v}{t_r} \tag{4-60}$$

Noting again that the forward crosstalk is a function of the difference between capacitive and inductive coupling, we see that the coupled pulse will have the same polarity as the aggressor signal if there is more capacitive coupling than inductive coupling in the system, and vice versa for the case where inductive coupling dominates. Equation (4-60) also suggests that it is possible to have no forward crosstalk if we can find a case where $K_C = K_L$. In fact, this is always true for coupled lines in a homogeneous dielectric (proof is left for the reader as Problem 4-9). On the other hand, for typical transmission lines in inhomogeneous dielectrics, such as microstrip PCB traces, the inductive coupling is generally greater than the capacitive coupling, so that the forward crosstalk pulse has the opposite magnitude from that of the aggressor signal.

As we described earlier, the width of the forward crosstalk pulse is

$$t_{pw,f} \cong t_r \tag{4-61}$$

where t_r is the rise time of the signal. Note that (4-60) and (4-61) are equally applicable for a falling-edge transition.

Reverse Crosstalk To get an expression for the reverse (near-end) crosstalk, we must take into account the fact that the coupling region travels in the direction opposite to the coupled waved. The output wave at the left in Figure 4-17 is a superposition of the waves coupled at earlier times that propagate and sum at the near end. This requires that we integrate from $z = 0$ to $z = l$ while accounting for the travel time of the wave:

$$v_b = \frac{K_C + K_L}{2v_p} \int_{z=0}^{l} \frac{dv(t - 2z/v_p)}{dt} dz \tag{4-62}$$

After integration we have an expression for the coupled noise at the near end.

$$v_b(t) = \frac{K_C + K_L}{4} \left[v_1(t) - v_1 \left(t - \frac{2l}{v_p} \right) \right] \tag{4-63}$$

The apparent reduction of the effect of the coupling coefficient after the integration of (4-62) is caused by the fact that the energy coupling of the backward crosstalk is spread out over a pulse width of $2l/v_p$. The width of the backward coupled pulse is calculated with

$$t_{pw,b} = 2\tau_d l \tag{4-64}$$

where τ_d is the propagation delay per unit length and l is the coupled length. Equations (4-63) and (4-64) assume that the backward crosstalk has saturated, which is realistic for multi-Gb/s links.

Finally, we note that the equations that we derived in this section apply to situation in which both lines are terminated at each end. Other configurations, such as when the near end is not terminated, will have different equations to describe the amplitudes and shapes of the crosstalk pulses. The modified crosstalk equations can be derived by considering the effect of reflections as described in Section 3.5.

Example 4-3 We now analyze the coupling from an active line to a quiet line for the PCB transmission lines from Example 4-2. Recall that the lines had the following inductances and capacitances:

$$\mathbf{L} = \begin{bmatrix} 3.592 \times 10^{-7} & 3218 \times 10^{-8} \\ 3218 \times 10^{-8} & 3.592 \times 10^{-7} \end{bmatrix} \quad \text{H/m}$$

$$\mathbf{C} = \begin{bmatrix} 8.501 \times 10^{-11} & -2.173 \times 10^{-12} \\ -2.173 \times 10^{-12} & 8.533 \times 10^{-11} \end{bmatrix} \quad \text{F/m}$$

The 0.2794-m-long traces have a typical (isolated) characteristic impedance of approximately 65 Ω and are terminated to ground in 65 Ω at the far end. They are driven by a 1-V 65-Ω source with a 100-ps rise time. Compare the analytical results with those from a fully coupled simulation.

SOLUTION

Step 1: We start by calculating the impedance and propagation velocity:

$$Z_{0,\text{isolated}} = \sqrt{\frac{3.592 \times 10^{-7}\ \text{H/m}}{8.501 \times 10^{-11}\ \text{F/m}}} = 65.0\ \Omega$$

$$v_{p,\text{isolated}} = \frac{1}{\sqrt{3.592\times 10^{-7}\ \text{H/m}\ 8.501 \times 10^{-11}\ \text{F/m}}} = 1.810 \times 10^{8}\ \text{m/s}$$

Step 2: Since we plan to analyze the coupled noise, we need the coupling coefficients.

$$K_C = \frac{2.173 \times 10^{-11}\ \text{F/m}}{8.501 \times 10^{-11}\ \text{F/m}} = 0.0256$$

$$K_L = \frac{3.218 \times 10^{-8}\ \text{H/m}}{3.593 \times 10^{-7}\ \text{H/m}} = 0.0896$$

Step 3: Analysis for a rising edge:

$$v(t = 0, z = 0) = \frac{Z_0}{R_S + Z_0} V_S = \frac{65.0\ \Omega}{65\ \Omega + 65.0\ \Omega}(1\ \text{V}) = 0.500\ \text{V}$$

$$i(t = 0, z = 0) = \frac{v(t = 0, z = 0)}{Z_0} = \frac{0.500\ \text{V}}{65.0\ \Omega} = 7.69\ \text{mA}$$

$$\Gamma(z = 0) = \frac{R_S - Z_0}{R_S + Z_0} = \frac{65\ \Omega - 65.0\ \Omega}{65\ \Omega + 65.0\ \Omega} = 0.000$$

$$\Gamma(z = 0.2794 \text{ m}) = \frac{R_T - Z_0}{R_T + Z_0} = \frac{65\ \Omega - 65.0\ \Omega}{65\ \Omega + 65.0\ \Omega} = 0.000$$

$$t_d = \frac{l}{\nu_p} = \left(\frac{0.2794 \text{ m}}{1.810 \times 10^8 \text{ m/s}}\right)\left(\frac{10^9 \text{ ns}}{\text{s}}\right) = 1.544 \text{ ns}$$

$$\begin{aligned} v_f &= \frac{1}{2}(K_C - K_L)\frac{l}{\nu_p}\frac{v}{t_r} \\ &= \frac{1}{2}(0.0256 - 0.0896)\left(\frac{0.2794 \text{ m}}{1.820 \times 10^8 \text{ m/s}}\right) \\ &\quad \times \left(\frac{0.500 \text{ V}}{100 \text{ ps}}\right)\left(\frac{10^{12} \text{ ps}}{\text{s}}\right) \\ &= -0.247 \text{ V} \\ v_b &= \frac{K_C + K_L}{4}\left[v(t) - v\left(t - 2\frac{l}{\nu_p}\right)\right] \\ &= \left(\frac{0.0256 + 0.0896}{4}\right)(0.500 \text{ V}) = 0.014 \text{ V} \\ t_{pw,f} &= t_r = 100 \text{ ps} \\ t_{pw,b} &= 2\tau_d l = 2\left(\frac{11 \text{ in.}}{1.810 \times 10^8 \text{ m/s}}\right)\left(\frac{\text{m}}{39.37 \text{ in.}}\right)\left(\frac{10^9 \text{ ns}}{\text{s}}\right) \\ &= 3.088 \text{ ns} \end{aligned}$$

Since the reflection coefficients are zero, a lattice diagram is not necessary, as we can construct the waveform directly from our calculations.

Step 4: Comparison to simulation. Figure 4-23 compares our calculated results with those from SPICE time-domain simulations. We see that although waveforms nearly match, they are not identical. In particular, the rising edge of the active signal in the SPICE simulation has grown to approximately 200 ps at the receiver end of the transmission line. The degradation of the rising edge of the active signal can be attributed to the crosstalk mechanism using an energy conservation argument. To conserve energy, the active line must give up an amount of energy that is equal to the amount coupled to the quiet line. Recall that the reactive nature of the coupling mechanism means that energy is coupled to the quiet line only during signal transition, so that the rising edge is degraded as a direct result of the coupling. In addition, the increase in rise time at the receiver causes the width of the crosstalk pulse at the receiver end to be approximately 200 ps rather than the 100 ps predicted by our calculation.

As the example demonstrates, the crosstalk model presented here is an approximation. This model begins to break down when the coupled length is long enough such that the difference in propagation delay between the even and odd modes

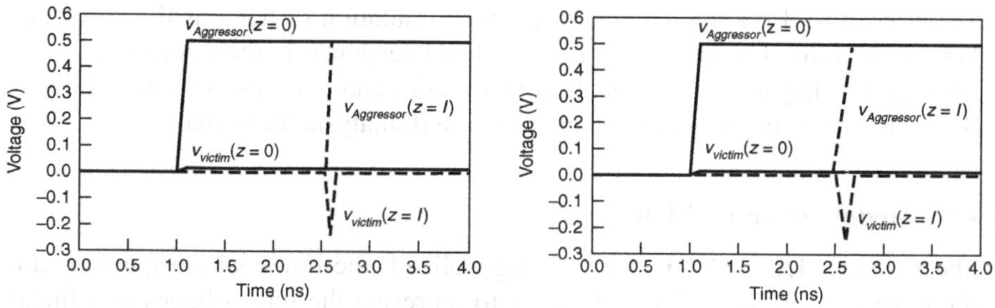

Figure 4-23 Comparison of (a) calculated and (b) simulated results for Example 4-3.

exceeds the rise time:

$$t_{d,\text{even}} - t_{d,\text{odd}} > t_r$$

When this occurs, the forward crosstalk saturates. At that point the amplitude ceases to grow; instead, the pulse width increases. In the next section we show how modal analysis allows us to explain the noise coupling mechanism in terms of modal propagation velocities and gives us the means to calculate accurately the behavior of the system.

4.4 MODAL ANALYSIS

Earlier we introduced the notion of propagation modes in the even- and odd-mode cases for a two-line system. When we have more than two lines, the notion of even- and odd-mode signals does not apply directly. A system with n coupled lines can propagate n distinct modes, each of which is a function of the electric and magnetic field strengths (or the capacitances and inductances) and of the driven signals on each of the lines. Each mode can have a different effective impedance and propagation velocity. We can calculate the modal impedances and velocities, although the mathematics is somewhat complicated.

Modal decomposition and analysis provides us with the means to model the behavior of a coupled system using multiple single line simulations, which dramatically simplifies the analysis without sacrificing accuracy. To do so, we must first express the behavior of the lines in terms of the propagation modes. The technique of modal decomposition transforms the $n \times n$ inductance and capacitance matrices into sets of n vectors, known as *eigenvectors*, that are each weighted by a constant value called an *eigenvalue*. What makes this technique interesting is that the eigenvectors are orthogonal, which allows us to diagonalize the **L** and **C** matrices so that the off-diagonal matrix entries are zero. Once we decompose the inductance and capacitances into diagonal matrices, we can analyze the behavior of the coupled line system as a set of n isolated transmission lines, each propagating modal voltage and modal current waves. In this approach we use the modal inductance and capacitance matrices $\mathbf{L_m}$ and $\mathbf{C_m}$ to transform the line voltages and currents to modal voltages and currents. The modal voltages and currents are then used to drive the isolated lines in a simulation or hand analysis such as a lattice diagram. The simulated or calculated response is then converted back from modal voltages and currents to line voltages and currents. The material that follows describes the modal decomposition and analysis technique.

4.4.1 Modal Decomposition

If the matrices **LC** and **CL** can be diagonalized, the diagonal components are orthogonal to each other. This allows us to represent the line voltages as a linear combination of modal voltages. To do this it is first necessary to determine the relationship between line voltages and modal voltages. We can use the coupled voltage wave equation (4-33) to derive a voltage transform matrix $\mathbf{T_v}$;

$$\mathbf{v} = \mathbf{T_v v_m} \tag{4-65}$$

where $\mathbf{T_v}$ is a matrix that contains the eigenvectors of **LC**, which translate normal line voltages in $\mathbf{v}$ to modal voltages in $\mathbf{v_m}$. Once we have done so, we can simulate the system as n isolated lines using $\mathbf{v_m}$ as input rather than as n coupled lines with $\mathbf{v}$ as input. Similarly, equation (4-34) for the coupled current wave is used to derive $\mathbf{T_i}$:

$$\mathbf{i} = \mathbf{T_i i_m} \tag{4-66}$$

where $\mathbf{T_i}$ is a matrix that contains the eigenvectors of **CL**, which translate normal line currents in $\mathbf{i}$ to modal currents in $\mathbf{i_m}$. From (4-28) and (4-31), the matrix equations for an $n \times n$ system become

$$\frac{d}{dz}\begin{bmatrix}\mathbf{v}\\\mathbf{i}\end{bmatrix} = \begin{bmatrix}\mathbf{0} & -j\omega\mathbf{L}\\ -j\omega\mathbf{C} & \mathbf{0}\end{bmatrix}\begin{bmatrix}\mathbf{v}\\\mathbf{i}\end{bmatrix} \tag{4-67}$$

To derive the modal inductances, we start with equation (4-28) and substitute the modal voltages, currents, and transform matrices for the line voltage and current:

$$\frac{d}{dz}(\mathbf{T_v v_m}) = -j\omega\mathbf{L T_i i_m} \tag{4-68}$$

Multiplying both sides by $\mathbf{T_v^{-1}}$ gives

$$\begin{aligned}\frac{d}{dz}(\mathbf{T_v^{-1} T_v v_m}) &= -j\omega\mathbf{T_v^{-1} L T_i i_m}\\ \frac{d\mathbf{v_m}}{dz} &= -j\omega\mathbf{T_v^{-1} L T_i i_m}\end{aligned} \tag{4-69}$$

The modal inductance matrix is defined as

$$\mathbf{L_m} = \mathbf{T_v^{-1} L T_i} \tag{4-70}$$

By following a similar approach, we can also derive the equation for modal capacitance:

$$\frac{d}{dz}(\mathbf{T_i i_m}) = -j\omega \mathbf{C T_v v_m} \tag{4-71}$$

$$\frac{d\mathbf{i_m}}{dz} = -j\omega \mathbf{T_i^{-1} C T_v v_m} \tag{4-72}$$

$$\mathbf{C_m} = \mathbf{T_i^{-1} C T_v} \tag{4-73}$$

$\mathbf{L_m}$ and $\mathbf{C_m}$ are diagonal,

$$\mathbf{L_m} = \begin{bmatrix} L_{m11} & 0 & 0 & 0 & 0 \\ 0 & L_{m22} & 0 & 0 & 0 \\ 0 & 0 & \cdots & 0 & 0 \\ 0 & 0 & 0 & L_{m(n-1)(n-1)} & 0 \\ 0 & 0 & 0 & 0 & L_{mnn} \end{bmatrix} \tag{4-74}$$

$$\mathbf{C_m} = \begin{bmatrix} C_{m11} & 0 & 0 & 0 & 0 \\ 0 & C_{m22} & 0 & 0 & 0 \\ 0 & 0 & \cdots & 0 & 0 \\ 0 & 0 & 0 & C_{m(n-1)(n-1)} & 0 \\ 0 & 0 & 0 & 0 & C_{mnn} \end{bmatrix} \tag{4-75}$$

where the diagonal elements are defined by (4-70) and (4-71).

Another useful relationship that we will find useful is [Paul, 1994]

$$\mathbf{T_i} = (\mathbf{T_v^{-1}})^{\mathrm{T}} \tag{4-76}$$

To carry out the modal transmission line analysis, we need to put the modal quantities into the wave equation, starting with (4-33):

$$\frac{d^2 \mathbf{T_v v_m}}{dz^2} = \omega^2 \mathbf{T_v L_m T_i^{-1} T_i C_m T_v^{-1} T_v v_m} = \omega^2 \mathbf{T_v L_m C_m v_m}$$

Multiplying through by $\mathbf{T_v^{-1}}$ gives the wave equation expressed in terms of the modal voltage, inductance, and capacitance:

$$\frac{d^2 \mathbf{v_m}}{dz^2} = \omega^2 \mathbf{L_m C_m v_m} \tag{4-77}$$

Following the same process with (4-15) gives

$$\frac{d^2 \mathbf{i_m}}{dz^2} = \omega^2 \mathbf{C_m L_m i_m} \tag{4-78}$$

Again, since the modal quantities are orthogonal, we can use them to simulate the system as n isolated lines rather than n coupled lines.

4.4.2 Modal Impedance and Velocity

Transmission-line analysis requires that we determine the modal impedance and propagation velocity of the lines. Recall that for an isolated line,

$$v_p = \frac{\omega}{\beta} \tag{4-79}$$

with the phase shift per unit length for the lossless case:

$$\beta = Im\left\lfloor \sqrt{(R + j\omega L)(G + j\omega C)} \right\rfloor = \omega\sqrt{LC} \tag{4-80}$$

leading to

$$v_p = \frac{1}{\sqrt{LC}} \tag{4-81}$$

By analogy, we can write an expression for the modal velocity matrix in terms of the modal inductance and capacitance matrices:

$$\mathbf{v_{pm}} = \sqrt{\mathbf{L_m^{-1} C_m^{-1}}} \tag{4-82}$$

Because $\mathbf{L_m}$ and $\mathbf{C_m}$ are square, diagonal matrices, their inverses will be, too, as will be the product of their inverses. Since the product is diagonal, the square root in (4-82) can be evaluated to yield the result

$$v_{pm,ii} = \frac{1}{\sqrt{L_{m,ii} C_{m,ii}}} \tag{4-83}$$

The subscript i represents the row and column in each matrix. By similar means, we find an equation to express the modal impedances:

$$Z_{m,ii} = \sqrt{\frac{L_{m,ii}}{C_{m,ii}}} \tag{4-84}$$

4.4.3 Reconstructing the Signal

The observable line voltage and currents that compose the signal propagating on the interconnect are linear combinations of the modal values. After the system is decomposed into its orthogonal values and a set of n single lines that correspond to each mode is analyzed to determine the modal voltage and currents, the observable line voltage and currents must be reconstructed using equations (4-65) and (4-66). For example, expanding (4-65) for an $n = 2$ signal conductor system yields

$$\begin{bmatrix} v_1 \\ v_2 \end{bmatrix} = \begin{bmatrix} T_{v1} & T_{v2} \\ T_{v3} & T_{v4} \end{bmatrix} \begin{bmatrix} v_{m1} \\ v_{m2} \end{bmatrix}$$

$$v_1 = T_{v1} v_{m1} + T_{v2} v_{m2}$$

$$v_2 = T_{v3} v_{m1} + T_{v4} v_{m2}$$

where v_1 is the line voltage observed on line 1, v_2 is the line voltage observed on line 2, and v_{m1} and v_{m2} are the modal voltages calculated from the analysis of the single transmission lines with modal impedance and velocities. Notice that the line voltages are reconstructed from linear combinations of scaled modal voltages.

4.4.4 Modal Analysis

In this section we outline the method for using the modal quantities.

1. Find, $\mathbf{T_v}$, the eigenvectors of **LC**.
2. Find, $\mathbf{T_i}$, the eigenvectors of **CL**. $\mathbf{T_v}$ and $\mathbf{T_i}$ are related by $\mathbf{T_i} = (\mathbf{T_v^{-1}})^{\mathrm{T}}$.
3. Use $\mathbf{T_v}$ and $\mathbf{T_i}$ to calculate the modal inductances, capacitances, and voltages and/or currents.
4. Use the modal inductance and capacitance to calculate the modal impedance and propagation velocity.
5. Carry out a conventional transmission-line analysis, such as a lattice diagram, or a single-line simulation for each mode.
6. Convert the modal quantities back into observable line quantities.

The following examples illustrate the application of modal analysis to a coupled transmission-line pair.

Example 4-4 Using the inductance and capacitance matrices and the interconnect design from Example 4-3, calculate the line voltages for signals switching in the odd mode using modal analysis. Compare the results to those of a fully coupled simulation.

SOLUTION Our analysis follows the method described above.

Step 1: Start by calculating the voltage eigenvector:

$$\mathbf{LC} = \begin{bmatrix} 3.0466 \times 10^{-17} & 1.9551 \times 10^{-18} \\ 1.9551 \times 10^{-18} & 3.0466 \times 10^{-17} \end{bmatrix} \quad \mathrm{s^2/m^2}$$

In finding the eigenvector of **LC**, we will make use of its symmetry, which has the form

$$\mathbf{LC} = \begin{bmatrix} a & b \\ b & a \end{bmatrix}$$

Following the method of O'Neil [1983], we subtract $\lambda\mathbf{I}$:

$$\mathbf{LC} - \lambda\mathbf{I} = \begin{bmatrix} a & b \\ b & a \end{bmatrix} - \lambda \begin{bmatrix} 1 & 0 \\ 0 & 1 \end{bmatrix} = \begin{bmatrix} a - \lambda & b \\ b & a - \lambda \end{bmatrix}$$

The eigenvalue $\mathbf{T_v}$ of $\mathbf{LC}$ satisfies the following relationship:

$$(\mathbf{LC} - \lambda\mathbf{I})\mathbf{T_v} = \begin{bmatrix} a - \lambda & b \\ b & a - \lambda \end{bmatrix} \begin{bmatrix} T_{v1} \\ T_{v2} \end{bmatrix} = \begin{bmatrix} 0 \\ 0 \end{bmatrix}$$

Carrying out the multiplication yields

$$(a - \lambda)T_{v1} + bT_{v2} = 0$$

$$bT_{v1} + (a - \lambda)T_{v2} = 0$$

From inspection of the equations above, we observe that $\mathbf{T_v}$ will satisfy the equation if $T_{v1} = -T_{v2}$, or $T_{v1} = T_{v2}$. As a result, we have a pair of eigenvectors

$$\mathbf{T_{v1}} = \begin{bmatrix} 1 \\ -1 \end{bmatrix} \quad \text{and} \quad \mathbf{T_{v2}} = \begin{bmatrix} 1 \\ 1 \end{bmatrix}$$

Combining them gives the $\mathbf{T_v}$ matrix:

$$\mathbf{T_v} = \begin{bmatrix} \mathbf{T_{v1}} & \mathbf{T_{v2}} \end{bmatrix} = \begin{bmatrix} 1 & 1 \\ -1 & 1 \end{bmatrix}$$

As a final step, we normalize the voltage eigenvector, $\mathbf{T_v}$:

$$\mathbf{T_v} = \frac{\begin{bmatrix} 1 & 1 \\ -1 & 1 \end{bmatrix}}{\sqrt{1^2 + 1^2}} = \begin{bmatrix} 0.707 & 0.707 \\ -0.707 & 0.707 \end{bmatrix}$$

Step 2: The current eigenvector is equal to the transpose of the inverse of the voltage eigenvector [O'Neil, 1983]:

$$\mathbf{T_i} = (\mathbf{T_v^{-1}})^{\mathrm{T}} = \begin{bmatrix} 0.707 & 0.707 \\ -0.707 & 0.707 \end{bmatrix}$$

Note that because we derived $\mathbf{T_v}$ and $\mathbf{T_i}$ from a general symmetric 2×2 matrix, they are applicable to any other 2×2 matrix, so we do not need to recalculate them for any other system of two coupled lines. This is true only for the 2×2 case. Whereas the matrices will be symmetrical for larger numbers of coupled lines, the eigenvectors will change as a function of the values of matrix entries.

Step 3: The modal inductances and capacitances are calculated from equations (4-70) and (4-73).

$$\mathbf{L_m} = \mathbf{T_v^{-1} L T_i}$$

$$= \begin{bmatrix} 0.707 & 0.707 \\ -0.707 & 0.707 \end{bmatrix}^{-1} \begin{bmatrix} 3.592 \times 10^{-7} & 3.218 \times 10^{-8} \\ 3.218 \times 10^{-8} & 3.592 \times 10^{-7} \end{bmatrix} \text{H/m}$$

$$\times \begin{bmatrix} 0.707 & 0.707 \\ -0.707 & 0.707 \end{bmatrix}$$

$$= \begin{bmatrix} 3.270 \times 10^{-7} & 0 \\ 0 & 3.914 \times 10^{-7} \end{bmatrix} \text{H/m}$$

$$\mathbf{C_m} = \mathbf{T_i^{-1} C T_v}$$

$$= \begin{bmatrix} 0.707 & 0.707 \\ -0.707 & 0.707 \end{bmatrix}^{-1} \begin{bmatrix} 8.501 \times 10^{-11} & -2.173 \times 10^{-11} \\ -2.173 \times 10^{-11} & 8.501 \times 10^{-11} \end{bmatrix} \text{F/m}$$

$$\times \begin{bmatrix} 0.707 & 0.707 \\ -0.707 & 0.707 \end{bmatrix}$$

$$= \begin{bmatrix} 8.718 \times 10^{-11} & 0 \\ 0 & 8.284 \times 10^{-11} \end{bmatrix} \text{F/m}$$

To calculate the modal voltage, we must first have the voltage inputs. For the odd mode, the input on line 1 is 1 V and line 2 is -1 V. Therefore, the input voltage matrix is

$$\mathbf{v_{in}} = \begin{bmatrix} 1 \\ -1 \end{bmatrix}$$

and solving (4-65) for the modal voltages yields

$$\mathbf{v_m} = \begin{bmatrix} 0.707 & 0.707 \\ -0.707 & 0.707 \end{bmatrix}^{-1} \begin{bmatrix} -1 \\ 1 \end{bmatrix} = \begin{bmatrix} v_{\text{odd}} \\ v_{\text{even}} \end{bmatrix} = \begin{bmatrix} 1.414 \\ 0 \end{bmatrix}$$

Note that there is no voltage propagating in the even mode, which is expected since we drove the system in the odd mode.

Step 4: The modal impedance and propagation velocity are calculated from (4-83) and (4-84).

$$Z_m = \begin{bmatrix} \sqrt{\dfrac{L_{m11}}{C_{m11}}} \\ \sqrt{\dfrac{L_{m22}}{C_{m22}}} \end{bmatrix} = \begin{bmatrix} \sqrt{3.270 \times 10^{-7} \text{ H/m}/8.718 \times 10^{-11} \text{ F/m}} \\ \sqrt{3.914 \times 10^{-7} \text{ H/m}/8.284 \times 10^{-11} \text{ F/m}} \end{bmatrix}$$

$$= \begin{bmatrix} 61.25 \\ 68.74 \end{bmatrix} \quad \text{ohms}$$

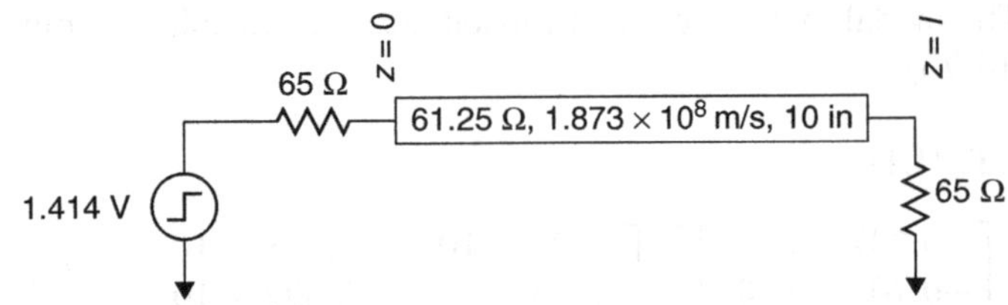

Figure 4-24 Equivalent modal circuit for the transmission-line pair in Example 4-4.

$$v_{pm} = \begin{bmatrix} \dfrac{1}{\sqrt{L_{m11}C_{m11}}} \\ \dfrac{1}{\sqrt{L_{m22}C_{m22}}} \end{bmatrix} = \begin{bmatrix} \dfrac{1}{\sqrt{(3.270 \times 10^{-7}\ \text{H/m})(8.718 \times 10^{-11}\ \text{F/m})}} \\ \dfrac{1}{\sqrt{(3.914 \times 10^{-7}\ \text{H/m})(8.284 \times 10^{-11}\ \text{F/m})}} \end{bmatrix}$$

$$= \begin{bmatrix} 1.873 \times 10^8 \\ 1.756 \times 10^8 \end{bmatrix} \quad \text{m/s}$$

This gives us the information that we require to construct the modal circuit for the odd mode, as shown in Figure 4-24.

Step 5: Transmission-line analysis. Beginning with the low-to-high transition at the voltage source, we use the odd-mode equivalent circuit in Figure 4-24 to calculate the voltage and current waves that are launched, assuming that the rising edge occurs at time $t = 1$ ns.

$$v(t = 1, z = 0) = (1.414\ \text{V})\frac{61.25\ \Omega}{61.25\ \Omega + 65\ \Omega} = 0.686\ \text{V}$$

$$i(t = 1, z = 0) = \frac{v(t = 0, z = 0)}{Z_m} = \frac{0.686\ \text{V}}{61.25\ \Omega}\left(\frac{\text{A}}{\text{V}/\Omega}\right)\left(\frac{\text{mA}}{\text{A}}\right) = 11.20\ \text{mA}$$

The propagation delay to the far end ($z = l = 0.254$ m) is

$$t_{d,\text{odd}} = \frac{\text{length}}{v_{pm,\text{odd}}} = \frac{0.254\ \text{m}}{1.873 \times 10^8\ \text{m/s}}\left(\frac{10^9\ \text{ns}}{\text{s}}\right) = 1.356\ \text{ns}$$

The voltage reflection coefficient at the far end is

$$\Gamma = \frac{65\ \Omega - 61.25\ \Omega}{65\ \Omega + 61.25\ \Omega} = 0.030$$

The voltage reflection coefficient at the source end is

$$\Gamma = \frac{65\ \Omega - 61.25\ \Omega}{65\ \Omega + 61.25\ \Omega} = 0.030$$

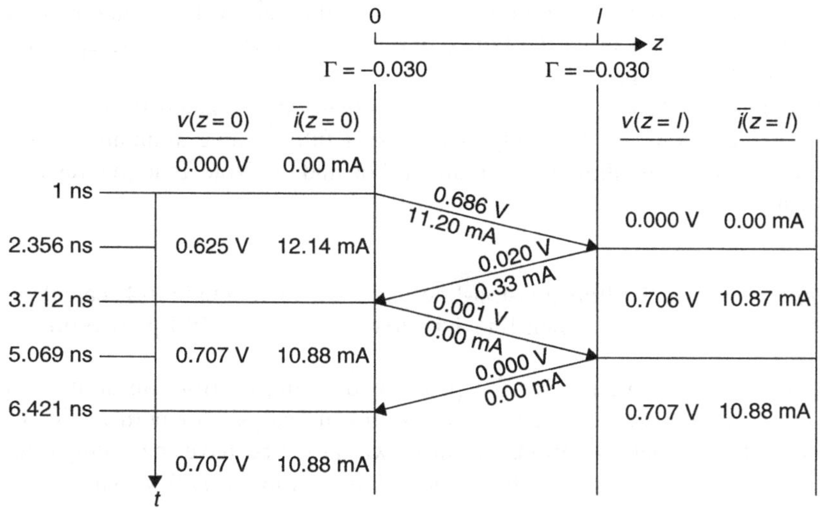

Figure 4-25 Odd-mode lattice diagram for Example 4-4.

The final steady-state voltage and current for the odd mode are

$$v(t=\infty) = (1.414\text{ V})\frac{65\ \Omega}{65\ \Omega + 65\ \Omega} = 0.707\text{ V}$$

$$i(t=\infty) = \frac{1.414}{65\ \Omega + 65\ \Omega} = 10.88\text{ mA}$$

We now have enough information to construct the odd-mode lattice diagram analysis shown in Figure 4-25.

Step 6: We now convert the modal voltages and currents back to line voltages and currents. In converting back to line voltages and currents, we use (4-65) and (4-66) along with the values calculated in the odd-mode lattice diagram analysis.

$$\mathbf{v} = \mathbf{T_v v_m} \qquad \text{with } \mathbf{v_m} = \begin{bmatrix} \text{from lattice} \\ 0 \end{bmatrix}$$

$$\mathbf{i} = \mathbf{T_i i_m} \qquad \text{with } \mathbf{i_m} = \begin{bmatrix} \text{from lattice} \\ 0 \end{bmatrix}$$

For example, the first voltage level on the odd-mode lattice diagram is 0.686 V. The equivalent observable line voltages are calculated as

$$v_1 = T_{v1}v_{m1} + T_{v2}v_{m2} = 0.707(0.686) + 0.707(0) = 0.485\text{ V}$$

$$v_2 = T_{v3}v_{m1} + T_{v4}v_{m2} = -0.707(0.686) + 0.707(0) = -0.485\text{ V}$$

Translating each point on the odd-mode lattice diagram to line voltage and currents allows us to construct the lattice diagrams for lines 1 and line 2, as shown in Figure 4-26a and b.

As a final check, we compare waveforms that we calculated from lattice diagrams created using modal analysis against a fully couple simulation, using a 100-ps rise time, as shown in Figure 4-27a and b. Note that the results are identical.

Example 4-5 For the transmission-line pair used in Example 4-4, calculate the crosstalk noise on line 2 when line 1 is driven with a IV, 100 ps rise time step.

SOLUTION The solution for the problem of coupling from an active line to a quiet line requires that we carry out the modal analysis for both modes (even and odd). To calculate the modal voltage, we must first have the voltage inputs. For this example, the input on line 1 is 1 V and on line 2 is 0 V. Therefore, the input voltage matrix is

$$\mathbf{v_{in}} = \begin{bmatrix} 1 \\ 0 \end{bmatrix}$$

and solving (4-65) for the modal voltages using the matrix $\mathbf{T_v}$ from Example 4-4 yields

$$\mathbf{v_m} = \begin{bmatrix} 0.707 & 0.707 \\ -0.707 & 0.707 \end{bmatrix}^{-1} \begin{bmatrix} -1 \\ 1 \end{bmatrix} = \begin{bmatrix} v_{\text{odd}} \\ v_{\text{even}} \end{bmatrix} = \begin{bmatrix} 0.707 \\ 0.707 \end{bmatrix}$$

Note that there is voltage propagating in both the even and odd modes. Table 4-2 summarizes the relationship between the modal and line voltages for the odd, even, and quiet cases.

Since the modal voltages for the case where line 1 is switching and line 2 is held low are

$$\mathbf{v_m} = \begin{bmatrix} 0.707 \\ 0.707 \end{bmatrix}$$

TABLE 4-2. Modal Voltages for the Coupled PCB Transmission-Line Pair of Example 4-5

Case	Line 1	Line 2	v_{in} (V)	v_m (V)
Even mode	Low → high	Low → high	$\begin{bmatrix} 1 \\ 1 \end{bmatrix}$	$\begin{bmatrix} 0.707 & 0.707 \\ -0.707 & 0.707 \end{bmatrix}^{-1} \begin{bmatrix} 1 \\ 1 \end{bmatrix} = \begin{bmatrix} 0 \\ 1.414 \end{bmatrix}$
Odd mode	Low → high	High → low	$\begin{bmatrix} 1 \\ -1 \end{bmatrix}$	$\begin{bmatrix} 0.707 & 0.707 \\ -0.707 & 0.707 \end{bmatrix}^{-1} \begin{bmatrix} -1 \\ 1 \end{bmatrix} = \begin{bmatrix} 1.414 \\ 0 \end{bmatrix}$
Quiet	Low → high	Low → low	$\begin{bmatrix} 1 \\ 0 \end{bmatrix}$	$\begin{bmatrix} 0.707 & 0.707 \\ -0.707 & 0.707 \end{bmatrix}^{-1} \begin{bmatrix} 1 \\ 0 \end{bmatrix} = \begin{bmatrix} 0.707 \\ 0.707 \end{bmatrix}$

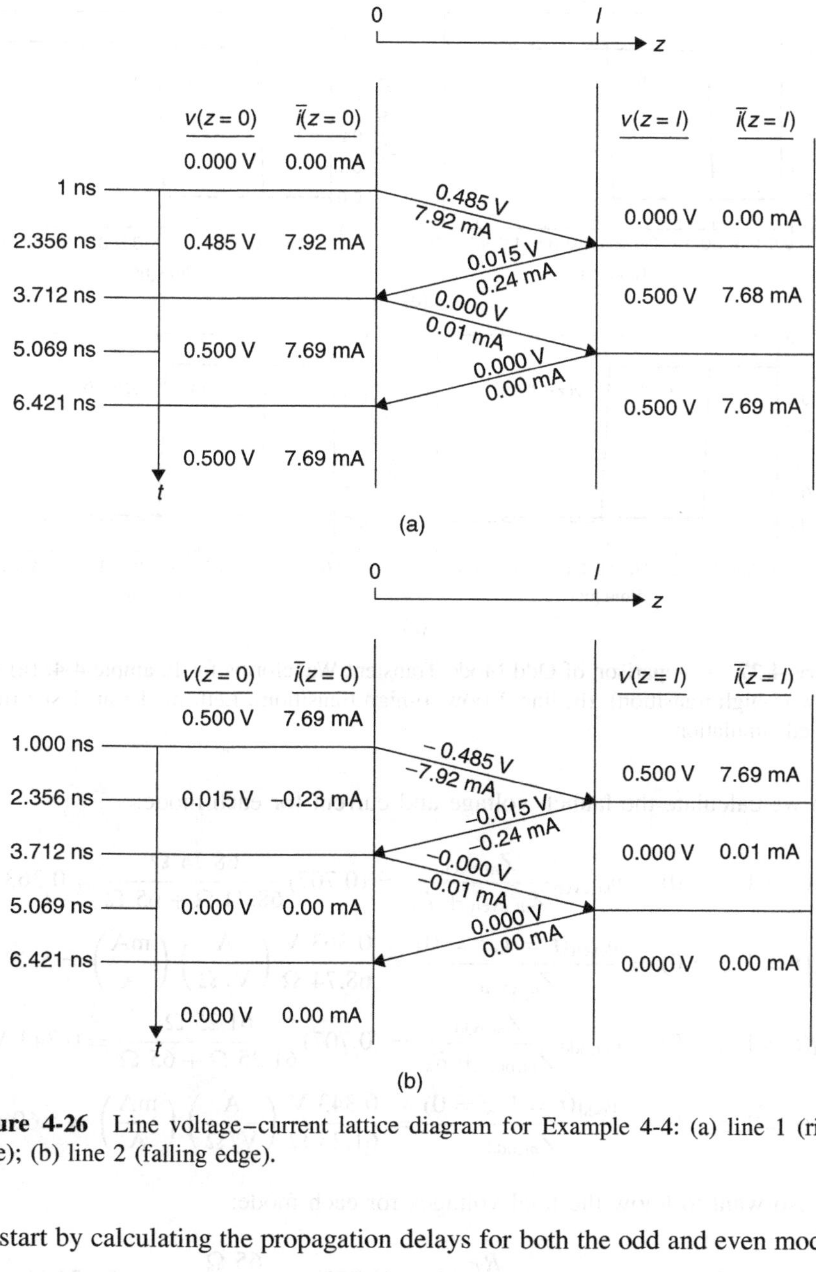

Figure 4-26 Line voltage–current lattice diagram for Example 4-4: (a) line 1 (rising edge); (b) line 2 (falling edge).

we start by calculating the propagation delays for both the odd and even modes:

$$t_{d,\text{even}} = \frac{l}{v_{pm,\text{even}}} = \frac{0.2794 \text{ m}}{1.756 \times 10^8 \text{ m/s}} = 1.446 \text{ ns}$$

$$t_{d,\text{odd}} = \frac{l}{v_{pm,\text{odd}}} = \frac{0.2794 \text{ m}}{1.873 \times 10^8 \text{ m/s}} = 1.356 \text{ ns}$$

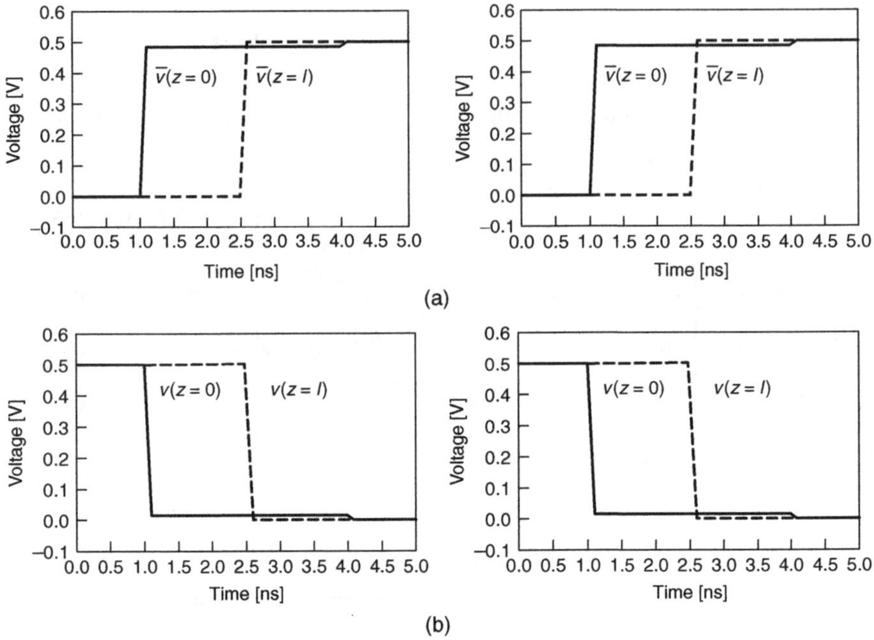

Figure 4-27 Comparison of Odd Mode Transient Waveforms for Example 4-4: (a) line 1 (low-to-high transition); (b) line 2 (low-to-high transition). Left, model analysis; right, coupled simulation.

Next we calculate the launch voltage and current for each mode:

$$v_{\text{even}}(t=1, z=0) = v_{m,\text{even}} \frac{Z_{m,\text{even}}}{Z_{m,\text{even}} + R_s} = (0.707) \frac{68.74\ \Omega}{68.74\ \Omega + 65\ \Omega} = 0.363\ \text{V}$$

$$i_{\text{even}}(t=1, z=0) = \frac{v_{\text{even}}(t=1, z=0)}{Z_{m,\text{even}}} = \frac{0.363\ \text{V}}{68.74\ \Omega} \left(\frac{\text{A}}{\text{V}/\Omega}\right) \left(\frac{\text{mA}}{\text{A}}\right) = 5.29\ \text{mA}$$

$$v_{\text{odd}}(t=1, z=0) = v_{m,\text{odd}} \frac{Z_{m,\text{odd}}}{Z_{m,\text{odd}} + R_s} = (0.707) \frac{61.25\ \Omega}{61.25\ \Omega + 65\ \Omega} = 0.343\ \text{V}$$

$$i_{\text{odd}}(t=1, z=0) = \frac{v_{\text{odd}}(t=1, z=0)}{Z_{m,\text{odd}}} = \frac{0.343\ \text{V}}{61.25\ \Omega} \left(\frac{\text{A}}{\text{V}/\Omega}\right) \left(\frac{\text{mA}}{\text{A}}\right) = 5.60\ \text{mA}$$

We also want to know the final voltages for each mode:

$$v_{\text{even}}(t=\infty) = v_{m,\text{even}} \frac{R_T}{R_T + R_S} = (0.707) \frac{65\ \Omega}{65\ \Omega + 65\ \Omega} = 0.354\ \text{V}$$

$$v_{\text{odd}}(t=\infty) = v_{m,\text{odd}} \frac{R_T}{R_T + R_S} = (0.707) \frac{65\ \Omega}{65\ \Omega + 65\ \Omega} = 0.354\ \text{V}$$

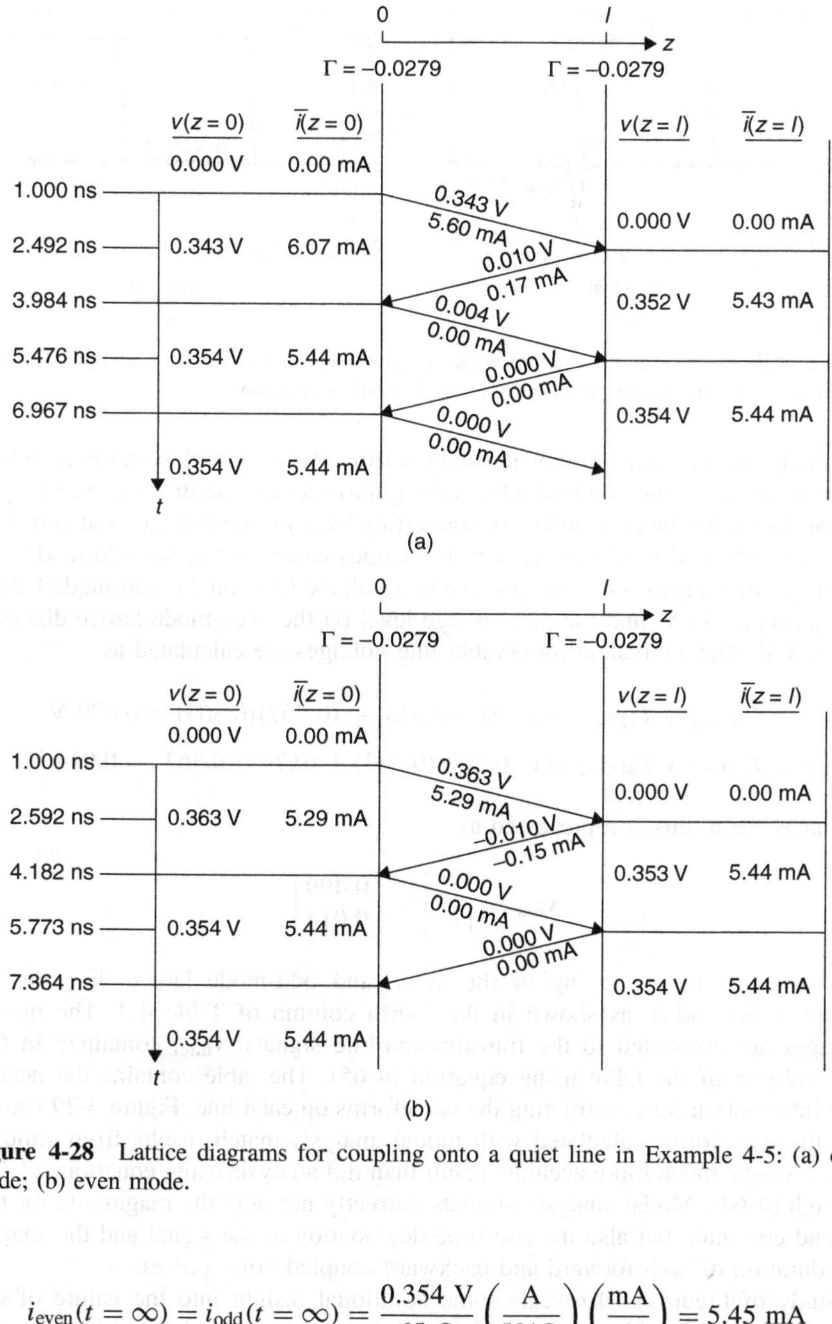

Figure 4-28 Lattice diagrams for coupling onto a quiet line in Example 4-5: (a) odd mode; (b) even mode.

$$i_{\text{even}}(t=\infty) = i_{\text{odd}}(t=\infty) = \frac{0.354\ \text{V}}{65\ \Omega}\left(\frac{\text{A}}{\text{V}/\Omega}\right)\left(\frac{\text{mA}}{\text{A}}\right) = 5.45\ \text{mA}$$

We now perform lattice diagram analysis for each of the modes, as shown in Figure 4-28a and b.

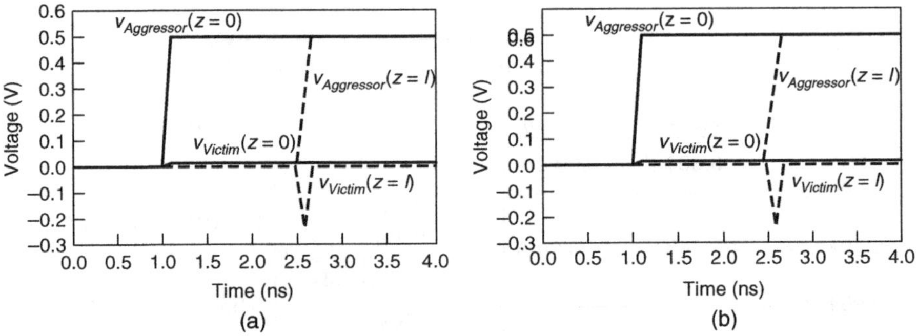

Figure 4-29 Comparison of coupling to a quiet line for the pair in Example 4-5: (a) model analysis calculation; (b) HSPICE W-element simulation.

Finally, we convert the modal voltages from the odd- and even-mode lattice diagrams back to line voltages while taking into account the difference in propagation delays for the two modes. In converting back to line voltages and currents, we use (4-65) and (4-66) along with the values calculated in the odd-mode lattice diagram analysis. For example, the first voltage level on the odd-mode lattice diagram is 0.343 V, and the first voltage level on the even-mode lattice diagram is 0.363 V. The equivalent observable line voltages are calculated as

$$v_1 = T_{v1}v_{m1} + T_{v2}v_{m2} = (0.707)(0.343) + (0.707)(0.363) = 0.499 \text{ V}$$

$$v_2 = T_{v3}v_{m1} + T_{v4}v_{m2} = (-0.707)(0.343) + (0.707)(0.363) = 0.014 \text{ V}$$

In matrix form, this is represented as

$$\mathbf{v_{line}} = \begin{bmatrix} v_1 \\ v_2 \end{bmatrix} = \begin{bmatrix} 0.499 \\ 0.014 \end{bmatrix}$$

This amounts to a merging of the even- and odd-mode lattice diagrams in Figure 4-28a and b, as shown in the fourth column of Table 4-3. The modal voltages are converted to the transmission-line signals, $\mathbf{v_{line}}$, contained in the fifth column of the table using equation (4-65). The table contains the necessary information for constructing the waveforms on each line. Figure 4-29 shows that the waveforms calculated with modal analysis match results from simulation, giving a much more accurate result than did analysis using equations (4-60) through (4-64). Modal analysis predicts correctly not only the magnitude for the far-end crosstalk, but also the rise-time degradation of the signal and the shapes and duration of both forward and backward coupled noise pulses.

Study of Figure 4-30 reveals some additional insight into the nature of the forward coupled noise. In particular, the far-end noise pulse exists over the time interval between the arrival of the odd-mode signal and the arrival of the even-mode signal. In effect, *far-end crosstalk noise is a function of the difference in propagation velocity between odd mode and even mode*. The effect is illustrated

TABLE 4-3. Modal and Line Voltages for the Coupled PCB Transmission-Line Pair of Example 4-5

Event	Location	t (ns)	v_m (V)	$v_{line} = T_V v_m$(V)	Notes
Waveform launch	$z = 0$	1.000	$\begin{bmatrix}0.000\\0.000\end{bmatrix}$	$\begin{bmatrix}0.000\\0.000\end{bmatrix}$	$t = 0$
	$z = 0$	1.100	$\begin{bmatrix}0.343\\0.363\end{bmatrix}$	$\begin{bmatrix}0.499\\0.014\end{bmatrix}$	$t = t_r$
Odd-mode incident wave	$z = l$	2.492	$\begin{bmatrix}0.000\\0.000\end{bmatrix}$	$\begin{bmatrix}0.000\\0.000\end{bmatrix}$	$t = t_{d,\text{odd}}$
	$z = l$	2.592	$\begin{bmatrix}0.352\\0.000\end{bmatrix}$	$\begin{bmatrix}0.249\\-0.249\end{bmatrix}$	$t = t_{d,\text{odd}} + t_r$
Even-mode incident wave	$z = l$	2.592	$\begin{bmatrix}0.352\\0.000\end{bmatrix}$	$\begin{bmatrix}0.249\\-0.249\end{bmatrix}$	$t = t_{d,\text{even}}$
	$z = l$	2.692	$\begin{bmatrix}0.352\\0.353\end{bmatrix}$	$\begin{bmatrix}0.499\\-0.0007\end{bmatrix}$	$t = t_{d,\text{even}} + t_r$
Odd-mode first reflection	$z = 0$	3.984	$\begin{bmatrix}0.343\\0.363\end{bmatrix}$	$\begin{bmatrix}0.499\\0.014\end{bmatrix}$	$t = 2t_{d,\text{odd}}$
	$z = 0$	4.084	$\begin{bmatrix}0.354\\0.363\end{bmatrix}$	$\begin{bmatrix}0.507\\0.006\end{bmatrix}$	$t = 2t_{d,\text{odd}} + t_r$
Even-mode first reflection	$z = 0$	4.182	$\begin{bmatrix}0.354\\0.363\end{bmatrix}$	$\begin{bmatrix}0.507\\0.006\end{bmatrix}$	$t = 2t_{d,\text{even}}$
	$z = 0$	4.282	$\begin{bmatrix}0.354\\0.354\end{bmatrix}$	$\begin{bmatrix}0.499\\0.000\end{bmatrix}$	$t = 2t_{d,\text{even}} + t_r$
Odd-mode second reflection	$z = l$	5.476	$\begin{bmatrix}0.352\\0.353\end{bmatrix}$	$\begin{bmatrix}0.499\\-0.0007\end{bmatrix}$	$t = 3t_{d,\text{odd}}$
	$z = l$	5.576	$\begin{bmatrix}0.354\\0.353\end{bmatrix}$	$\begin{bmatrix}0.500\\-0.0007\end{bmatrix}$	$t = 3t_{d,\text{odd}} + t_r$
Even-mode second reflection	$z = l$	5.773	$\begin{bmatrix}0.354\\0.353\end{bmatrix}$	$\begin{bmatrix}0.500\\-0.0007\end{bmatrix}$	$t = 3t_{d,\text{even}}$
	$z = l$	5.873	$\begin{bmatrix}0.354\\0.354\end{bmatrix}$	$\begin{bmatrix}0.501\\0.000\end{bmatrix}$	$t = 3t_{d,\text{even}} + t_r$

for this example in Figure 4-30. In the figure, the line voltage contributions from each mode are calculated at the far end of line 2 (the quiet line).
Odd-mode component:

$$v_{\text{odd}} = T_{v3} v_{m1} = -0.707 v_{m1}$$

Even-mode component:

$$v_{\text{even}} = T_{v4} v_{m2} = 0.707 v_{m2}$$

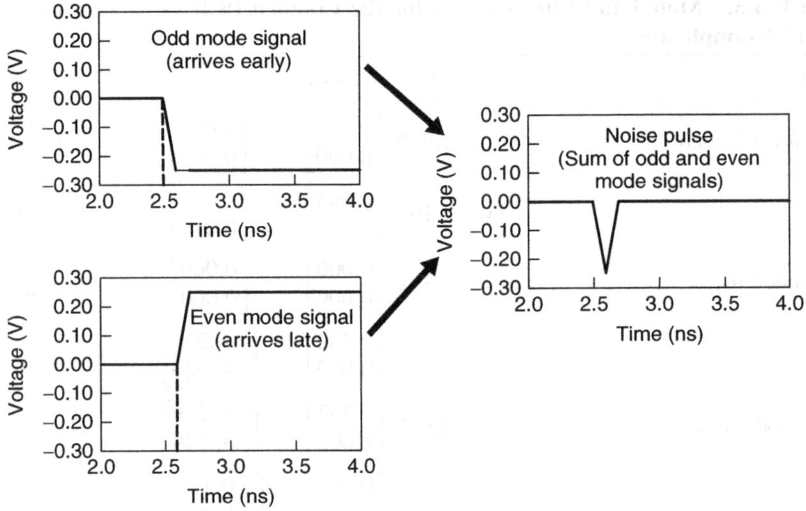

Figure 4-30 Far-end crosstalk generation due to even- and odd-mode propagation velocity differences.

Voltage on line 2:

$$v_2 = T_{v3}v_{m1} + T_{v4}v_{m2} = -0.707v_{m1} + 0.707v_{m2} = v_{\text{even}} + v_{\text{odd}}$$

When the even- and odd-mode components are summed to produce the observable line voltage, a pulse occurs at the far end when the modal velocities differ because each component arrives at a different time. Note that this same phenomenon will induce a ledge in the received waveform on the driving line that can be seen when the rise time is short compared to the difference between even- and odd-mode propagation delays. We also note that the description of crosstalk as a function of the difference in even and odd propagation velocities correctly predicts zero far-end crosstalk for coupled lines in homogeneous dielectrics, since the even and odd velocities will be equal.

As a final note on modal analysis, we acknowledge the difficulty of solving for the eigenvectors by hand. We encourage the use of computer tools in finding the eigenvectors wherever possible. Commercially available math packages such as Mathcad and Mathematica contain routines for calculating eigenvectors and eigenvalues. For readers who are interested in developing their own code, a book by Press et al. [1989] provides fully documented and widely used routines that are easily capable of handling the relatively small matrices that are part of coupled transmission-line analysis.

4.4.5 Modal Analysis of Lossy Lines

Although our discussion has focused on the application of modal analysis to lossless coupled lines, it is equally valid for analyzing lossy lines. When including the loss terms, equations (4-65) and (4-66) still apply, but we start from the lossy coupled line equations [Paul, 1994]:

$$\frac{d}{dx}\begin{bmatrix}\mathbf{v}(z)\\ \mathbf{i}(z)\end{bmatrix} = \begin{bmatrix}0 & -\mathbf{Z}\\ -\mathbf{Y} & 0\end{bmatrix}\begin{bmatrix}\mathbf{v}(z)\\ \mathbf{i}(z)\end{bmatrix} \tag{4-85}$$

$$\mathbf{Z} = \mathbf{R} + j\omega\mathbf{L} \tag{4-86}$$

$$\mathbf{Y} = \mathbf{G} + j\omega\mathbf{C} \tag{4-87}$$

In essence, we calculate the eigenvectors $\mathbf{T_v}$ and $\mathbf{T_i}$ from the product $(\mathbf{R} + j\omega\mathbf{L})(\mathbf{G} + j\omega\mathbf{C})$ for the lossy case. Recalling from our example that we calculated $\mathbf{T_v}$ and $\mathbf{T_i}$ from the product $\mathbf{LC}$, we realize that it is equivalent to using $(\mathbf{R} + j\omega\mathbf{L})(\mathbf{G} + j\omega\mathbf{C})$ with $\mathbf{R} = \mathbf{G} = 0$.

4.5 CROSSTALK MINIMIZATION

Since all of the major components in an interconnect system (i.e., PCB, packages, connectors) can have enough crosstalk to harm system performance, we present some crosstalk reduction guidelines in this section. Because it is often not possible to reduce crosstalk without affecting system cost, we include discussion of trade-offs along with reduction techniques in Table 4-4. In particular, we note that in cost-sensitive applications such as desktop personal computers, adding layers in the printed circuit board represents significant added cost to the system.

Another technique that sometimes finds use is the placement of *guard traces* between signals. These are connected to the ground return layers using plated via holes in the board. This technique requires careful attention to the design to provide the desired crosstalk benefit. Inductance of the traces will tend to create a potential difference at points that are a significant distance from the ground vias. When this occurs, the guard traces can radiate the coupled energy, thereby

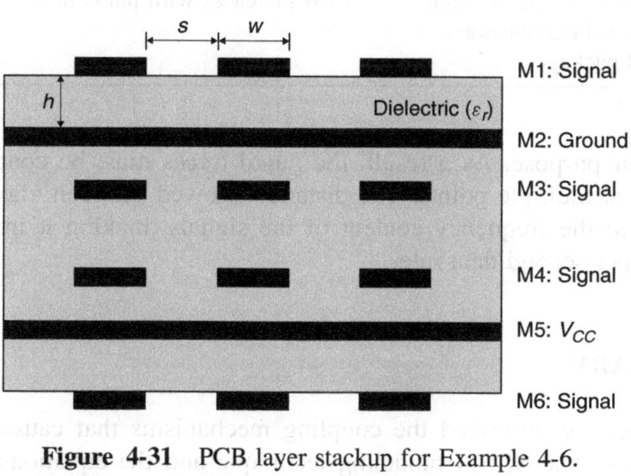

Figure 4-31 PCB layer stackup for Example 4-6.

TABLE 4-4. Crosstalk Reduction Techniques and Trade-offs

Approach	Trade-off
Increase the spacing, *s*, between the PCB and package traces.	Can add cost by requiring additional layers in PCBs and/or packages to route all traces.
Decrease the thickness of the dielectric, *h*, between the ground (return) layer and the signal layer in PCBs and packages to couple the transmission lines more tightly to the return layer and reduce the coupling to adjacent signals.	Operating limits for characteristic impedance may limit the minimum acceptable thickness (recall that Z_0 decreases as *h* decreases). The minimum manufacturable dielectric thickness may also limit the effectiveness of this approach.
Use differential signaling.	Can add cost by requiring additional layers in PCBs and/or packages to route all traces, since each signal requires two traces. Can also add cost to packages, sockets, and connectors, due to increased pin counts.
Route traces on adjacent PCB signal layers orthogonally to each other.	Can add cost by requiring additional layers in PCBs and/or packages to route all traces, since restricting signal routing direction may decrease routing efficiency.
Route the signals in PCBs and packages using striplines or embedded microstrips to eliminate velocity variations.	Striplines require at least six layers in the PCB. Refer to Figure 4-31.
Minimize parallel run lengths between signals in PCBs and packages.	Can add cost by requiring additional layers in PCBs and/or packages to route all traces, since restricting signal routing direction may decrease routing efficiency.
Reduce signal edge rates.	May limit the maximum performance, since rise and fall times typically must scale with data rate.
Insert power/ground pins between signal I/O pins in connectors, sockets, and packages.	Cost increases with pin count.

defeating their purpose. As a result, the guard traces must be connected to the ground layer at multiple points. The distance allowed between vias is inversely proportional to the frequency content of the signals, making it impractical for multigigabit per second data rates.

4.6 SUMMARY

In this chapter we described the coupling mechanisms that cause crosstalk in digital systems. The SLEM modeling technique and the equations for coupled noise provide the means to make first-order engineering estimates of the impacts of crosstalk on high-speed systems, while modal decomposition provides a more accurate analytical technique that does not require the use of coupled simulations. Guidelines for crosstalk reduction provide designers with a toolbox for managing coupled noise.

REFERENCES

In addition to the references cited in the text, the reader may find that others provide useful background on the topic of crosstalk. In particular, a book by Paul [1994] is the classic text on the analysis of coupled transmission lines; books by Bakoglu [1990] and Poon [2002] also contain general discussions of crosstalk. For an alternative treatment of modal decomposition, refer to Young's [2001] book.

Bakoglu, H. Brian, 1990, *Circuits, Interconnections, and Packaging for VLSI*, Addison-Wesley, Reading, MA.

Djordjevic, Antonije, Miodrag Bazdar, Tapan Sarkar, and Roger Harrington, 1999, *LINPAR for Windows: Matrix Parameters for Multiconductor Transmission Lines, Software and User's Manual, Version 2.0*, Artech House, Norwood, MA.

Hall, Stephen, Garrett Hall, and James McCall, 2000, *High Speed Digital System Design*, Wiley-Interscience, New York.

O'Neil, P., 1983, *Advanced Engineering Mathematics*, Wadsworth, Belmont, CA.

Paul, Clayton, 1994, *Analysis of Multiconductor Transmission Lines*, Wiley-Interscience, New York.

Poon, Ron, 1995, *Computer Circuits Electrical Design*, Prentice Hall, Upper Saddle River, NJ.

Press, William, Saul Teukolsky, William Vetterling, and Brian Flannery, 2002, *Numerical Recipes in C++: The Art of Scientific Computing*, 2nd ed., Cambridge University Press, Cambridge, UK.

Seraphim, Don, Ron Lasky, and Che-Yu Li, eds., 1989, *Principles of Electronic Packaging*, McGraw-Hill, New York.

Young, Brian, 2001, *Digital Signal Integrity*, Prentice Hall, Upper Saddle River, NJ.

PROBLEMS

4-1 Use the SLEM method to calculate the effective even- and odd-mode impedances and propagation velocities for the coupled striplines whose capacitance and inductance matrices are shown below. Estimate the impacts of crosstalk on the propagation delay for a 0.5-m coupled length.

$$L = \begin{bmatrix} 3.480 \times 10^{-7} & 1.951 \times 10^{-8} \\ 1.951 \times 10^{-8} & 3.480 \times 10^{-7} \end{bmatrix} \quad \text{H/m}$$

$$C = \begin{bmatrix} 1.271 \times 10^{-10} & -7.213 \times 10^{-12} \\ -7.213 \times 10^{-12} & 1.271 \times 10^{-10} \end{bmatrix} \quad \text{F/m}$$

4-2 For coupled striplines from Problem 4-1, calculate the near- and far-end noise for the isolated switching case, and compare your results against simulation with each line terminated at both ends in the characteristic impedance calculated.

4-3 Use the SLEM method to calculate the effective impedance and propagation velocity for the coupled striplines shown whose capacitance and inductance matrices are shown below when all three lines switch in the same direction. Compare the propagation delay for a 0.5-m coupled length when only the middle line switches.

$$L = \begin{bmatrix} 3.480 \times 10^{-7} & 5.268 \times 10^{-8} & 1.687 \times 10^{-8} \\ 5.268 \times 10^{-8} & 3.461 \times 10^{-7} & 5.268 \times 10^{-8} \\ 1.687 \times 10^{-8} & 5.268 \times 10^{-8} & 3.480 \times 10^{-7} \end{bmatrix} \quad \text{H/m}$$

$$C = \begin{bmatrix} 1.087 \times 10^{-10} & -1.172 \times 10^{-11} & -7.918 \times 10^{-11} \\ -1.172 \times 10^{-11} & 1.105 \times 10^{-10} & -1.172 \times 10^{-11} \\ -7.918 \times 10^{-11} & -1.172 \times 10^{-11} & 1.087 \times 10^{-10} \end{bmatrix} \quad \text{F/m}$$

4-4 Estimate crosstalk pulse amplitudes and pulse widths on the middle lines of the three-line system from Problem 4-3 when the two outer lines are switching from low to high. Assume that each line is terminated at both ends in its characteristic impedance. Compare your results against a fully coupled simulation.

4-5 Sketch the far-end crosstalk pulse for a two-line case with no termination at the near end, and matched termination at the far end.

4-6 Use the inductance and capacitance matrices below to determine the switching activity on lines 1 and 2, given the waveform on line 3 in Figure 4-32.

$$L = \begin{bmatrix} 3.544 \times 10^{-7} & 1.914 \times 10^{-8} & 5.161 \times 10^{-9} \\ 1.914 \times 10^{-8} & 3.826 \times 10^{-7} & 1.914 \times 10^{-8} \\ 5.161 \times 10^{-9} & 1.914 \times 10^{-8} & 3.544 \times 10^{-7} \end{bmatrix} \quad \text{H/m}$$

$$C = \begin{bmatrix} 8.266 \times 10^{-11} & -1.108 \times 10^{-11} & -2.354 \times 10^{-11} \\ -1.108 \times 10^{-11} & 1.001 \times 10^{-10} & -1.108 \times 10^{-11} \\ -2.354 \times 10^{-11} & -1.108 \times 10^{-11} & 8.266 \times 10^{-11} \end{bmatrix} \quad \text{F/m}$$

4-7 Use the inductance and capacitance matrices from Problem 4-6 to determine the switching activity on lines 1 and 3 given the waveform on line 2 in Figure 4-33.

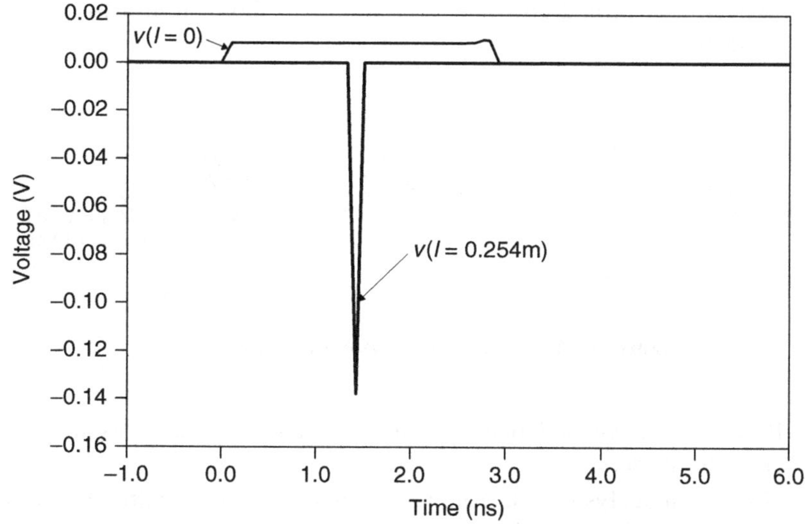

Figure 4-32 Transient response for Problem 4-6.

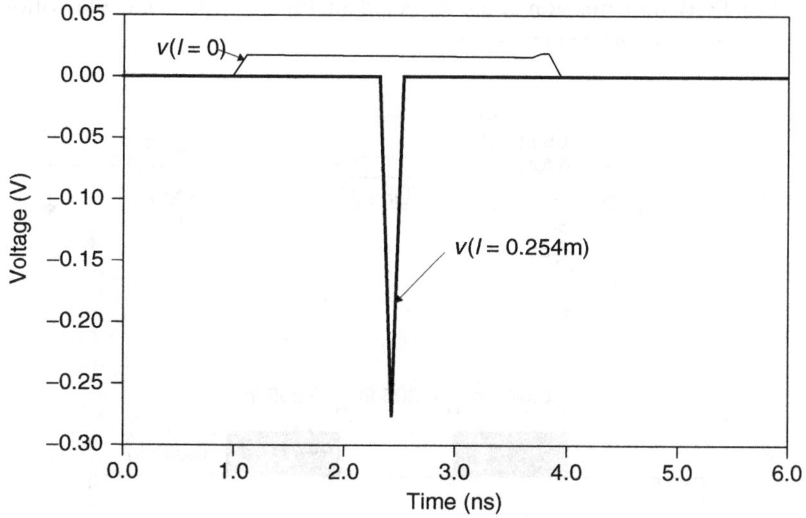

Figure 4-33 Transient response for Problem 4-7.

4-8 The circuit in Figure 4-34 uses transmission lines that have the inductance and capacitance matrices given in Problem 4-1. In this problem, line 1 undergoes a rising-edge transition at $z = 0$; simultaneously, line 2 undergoes a falling-edge transition at the other end of the pair ($z = l$). Calculate the waveform on line 1 at the receiving end ($z = l$) for a 0.254-m coupled length, and compare your results to those from a fully coupled simulation.

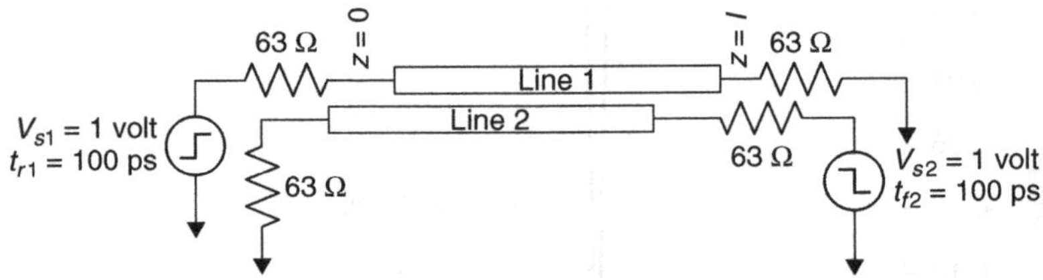

Figure 4-34 Coupled line system for Problem 4-8.

4-9 Prove that a coupled pair in a homogeneous medium has no forward crosstalk noise.

4-10 Use modal analysis to determine the resistor values required to terminate both even and odd modes using a pi network for a two-transmission-line system.

4-11 Calculate modal reflection coefficients for the two lines in Problem 4-1 when they are terminated in 50 Ω at each end.

4-12 The PCB transmission lines depicted in Figure 4-35 have the following inductances and capacitances:

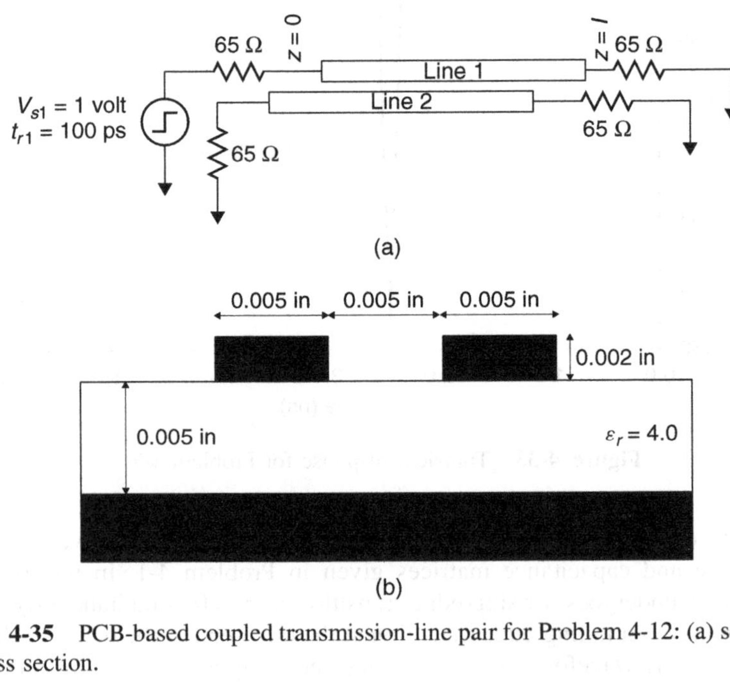

(b)

Figure 4-35 PCB-based coupled transmission-line pair for Problem 4-12: (a) schematic; (b) cross section.

$$\mathbf{L} = \begin{bmatrix} 3.537 \times 10^{-7} & 9.559 \times 10^{-8} \\ 9.559 \times 10^{-8} & 3.537 \times 10^{-7} \end{bmatrix} \quad \text{H/m}$$

$$\mathbf{C} = \begin{bmatrix} 8.533 \times 10^{-11} & -1.205 \times 10^{-11} \\ -1.205 \times 10^{-11} & 8.533 \times 10^{-11} \end{bmatrix} \quad \text{F/m}$$

The traces have a length of 0.254 m (10 in.) and are terminated to ground with 65 Ω. They are driven by a 1-V 65-Ω source, with 100-ps rise and fall times. Analyze the crosstalk noise using the approximate model from Section 4.3.2 and modal analysis, and compare your results with simulation. Do all of the results agree? If not, why not?

4-13 Use modal analysis to:

(a) Calculate the modal impedance and velocity for the three-coupled-line case in Problem 4-3.

(b) Simulate a 0.5-m coupled length and compare your results to those obtained using SLEM analysis.

4-14 Develop expressions for the forward and backward crosstalk amplitudes for the circuit shown in Figure 4-17, in which the victim line is open circuited at the near end.

4-15 Use modal analysis to derive equations for the near- and far-end crosstalk amplitudes for the terminated coupled pair shown in Figure 4-22.

4-16 *Project*: Develop a tool (Matlab, C++, etc.) to perform modal analysis for an arbitrary number of coupled lines, using the inductance and capacitance matrices as input.

5

NONIDEAL CONDUCTOR MODELS

As digital systems evolve and technology pushes for smaller and faster designs, the geometric dimensions of the physical platform are shrinking. Smaller dimensions and higher data transmission rates necessitate the use of proper techniques to model both the frequency dependent resistive losses and inductance. Without proper models that accurately predict these quantities, simulation-based bus design for multigigabit data rates is not possible. Frequency-dependent resistive losses, for example, will affect bus performance by decreasing the signal amplitude and slowing edge rates, which in turn affects voltage and timing margins, respectively. In addition, frequency dependent inductance models are required to

Advanced Signal Integrity for High-Speed Digital Design, By Stephen H. Hall and Howard L. Heck

preserve causality , which is discussed in Chapter 8 and Appendix E. In prior days it was possible to utilize simpler conductor models for digital designs because bandwidth demands were much lower. However, as digital data rates increase, the assumptions and approximations of traditional conductor models begin to break down. Consequently, the signal integrity engineer is now required to learn new techniques that compensate for variables that were insignificant in past designs.

So far in this book we have covered the topics of electromagnetic theory for signal integrity engineers, transmission-line fundamentals, and crosstalk. Up to this point, the conductors were assumed to be infinitely conductive and the dielectric was assumed to be a perfect insulator. In this chapter we develop modeling techniques to predict properly the electrical behavior of conductors used to design transmission lines on printed circuit boards, multichip modules, and chip packages. First, classic electromagnetic theory will be used to derive the frequency dependence of resistance and inductance for smooth conductors with finite conductivity. Next, three different methodologies for modeling the effects of rough copper on the electrical parameters of the transmission lines will be introduced. Detailed analysis of how currents flow on a rough surface will give physical insights into the mechanisms of surface roughness losses. Finally, a new circuit model for a transmission line that accounts for realistic conductors will be introduced along with a modified version of the telegrapher's equations that account for realistic conductor losses.

5.1 SIGNALS PROPAGATING IN UNBOUNDED CONDUCTIVE MEDIA

The topic of uniform plane waves propagating in a lossless media was discussed in Section 2.3, where the influences of the material properties μ and ε were observed. In Chapter 3 we described how the waves propagated when confined to the physical dimensions of a transmission line, yet the problem was still idealized because it was assumed that the dielectric was a perfect insulator and the conductor was infinitely conductive. To derive the equations that govern the propagation of waves on realistic transmission lines, it is first necessary to comprehend how an electromagnetic wave propagates in unbounded, conductive media.

5.1.1 Propagation Constant for Conductive Media

To derive the equations that govern electromagnetic waves propagating in conductive or lossy media, we begin with the loss-free forms of Maxwell's differential equations presented in Chapter 2 and modify them appropriately to obtain a wave equation that accounts for loss. To begin, the time-harmonic forms of Maxwell's equations are repeated here:

$$\nabla \times \vec{E} + j\omega\vec{B} = 0 \tag{2-33}$$

$$\nabla \times \vec{H} = \vec{J} + j\omega\vec{D} \tag{2-34}$$

$$\nabla \cdot \vec{D} = \rho \tag{2-35}$$

$$\nabla \cdot \vec{B} = 0 \tag{2-36}$$

where $\vec{J} = \sigma \vec{E}$, $\vec{D} = \varepsilon \vec{E}$, and $\vec{B} = \mu \vec{H}$, as derived from equations (2-6) through (2-9).

Simplification of Ampère's law (2-34) by replacing the current density term with $\sigma \vec{E}$ and $\vec{D} = \varepsilon \vec{E}$ yields

$$\nabla \times \vec{H} = \sigma \vec{E} + j\omega\varepsilon\vec{E} = j\omega\left(\frac{\sigma}{j\omega} + \varepsilon\right)\vec{E} = j\omega\left(\varepsilon - j\frac{\sigma}{\omega}\right)\vec{E} \tag{5-1}$$

Comparison to the solution of Ampère's law in a loss-free medium ($\nabla \times \vec{H} = j\omega\varepsilon\vec{E}$) allows us to define the complex permittivity for a conductive or lossy media by analogy:

$$\varepsilon = \varepsilon - j\frac{\sigma}{\omega} = \varepsilon' - j\varepsilon'' \tag{5-2}$$

where the real component is the dielectric permittivity discussed in Chapters 2 and 3 ($\varepsilon' = \varepsilon_0\varepsilon_r$) and the imaginary component accounts for the losses in the medium where the wave is propagating. The term σ can be thought of as the conductivity of the material, which will be quite high for a metal and quite low for a dielectric. If (5-2) is inserted into the time-harmonic solution of the electric field derived in Section 2.3.4,

$$E_x(z,t) = E_x^+ e^{-\gamma z} + E_x^- e^{\gamma z} \tag{2-54}$$

then the complex propagation constant for a lossy media can be derived. The complex propagation constant for a plane wave was derived in Section 2.3.4:

$$\gamma = \alpha + j\beta \tag{2-42}$$

As discussed extensively in Chapter 2, if the wave is propagating in a loss-free medium (where $\alpha = 0$), (2-42) reduces to

$$j\beta = \omega\sqrt{\mu\varepsilon} \qquad \text{rad/m} \tag{5-3}$$

where $\varepsilon = \varepsilon_r\varepsilon_0$ and $\mu = \mu_r\mu_0$. Substitution of the complex permittivity (5-2) into (5-3) provides the form of the complex propagation constant of an electromagnetic wave traveling in a conductive medium:

$$\gamma = \omega\sqrt{\mu\left(\varepsilon' - j\frac{\sigma}{\omega}\right)} \tag{5-4}$$

Setting (5-4) equal to (2-42) and separating into real and imaginary components yields the general form of the attenuation constant α and the phase constant β for an electromagnetic wave propagating in a conductive media with a conductivity of σ:

$$\alpha = \frac{\omega\sqrt{\mu\varepsilon'}}{\sqrt{2}}\left[\sqrt{1+\left(\frac{\sigma}{\varepsilon'\omega}\right)^2}-1\right]^{1/2} \qquad \text{Np/m} \tag{5-5}$$

$$\beta = \frac{\omega\sqrt{\mu\varepsilon'}}{\sqrt{2}}\left[\sqrt{1+\left(\frac{\sigma}{\varepsilon'\omega}\right)^2}+1\right]^{1/2} \qquad \text{rad/m} \tag{5-6}$$

Note that $\mu = \mu_0$ for virtually every practical digital design because the conductors are almost always constructed from copper, which is not magnetic. Both α and β have units of 1/m; however, the dimensionless terms neper (Np) and radian (rad) are used to communicate the attenuation and phase meanings in the wave equation.

5.1.2 Skin Depth

As described in Section 2.3.4, α is the attenuation constant, which will modify a wave propagating in the z-direction as described by

$$\begin{aligned} E(z,t) &= \text{Re}\left(E_x^+ e^{-\gamma z} e^{j\omega t}\right) = \text{Re}\left(E_x^+ e^{-\alpha z} e^{-j\beta z} e^{j\omega t}\right) \\ &= e^{-\alpha z} E_x^+ \cos(\omega t - \beta z) \end{aligned} \tag{2-43}$$

The factor $e^{-\alpha z}$ is known as the *wave decay* for a wave propagating in the $+z$-direction. The wave attenuation in a conductive region is governed by the term $1 + (\sigma/\varepsilon'\omega)^2$, as shown in equation (5-5). As the conductivity of the medium is increased, the attenuation constant α becomes larger and the wave decay increases with distance and time. Consequently, for a good conductor such as copper, the wave will decay very rapidly. The decay of an electromagnetic wave propagating into a conductor is measured in terms of the *skin depth*. The skin depth, denoted δ, is simply the distance of penetration where the settling exponent $-\alpha z$ of the wave decay factor is -1 ($e^{-\alpha z} = e^{-1}$). The skin depth is therefore given in units of meters:

$$\delta = \frac{1}{\alpha} \tag{5-7}$$

Since the term $\vec{J} = \sigma\vec{E}$ in Ampère's law (2-34) is no longer neglected, a current density $\vec{J}$ must accompany the electric field $\vec{E}$ in the conductive region:

$$J_x^+(z,t) = e^{-\alpha z}\sigma E_x^+ \cos(\omega t - \beta z) \tag{5-8}$$

Therefore, at a penetration distance of one skin depth ($1/\alpha$), the field intensity and the current density have been attenuated by a factor of e^{-1}, or approximately 36.7%, meaning that approximately 63.6% of the current density exists within a distance of δ from the conductor surface. Note that for a perfect conductor, the

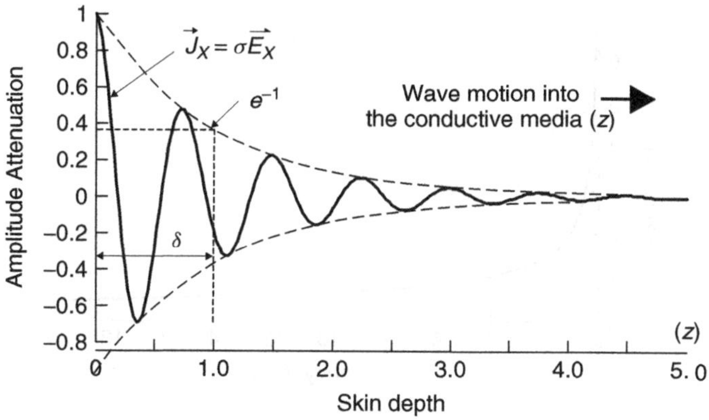

Figure 5-1 Penetration depth δ associated with the amplitude attenuation of the current density for a plane wave propagating into a conductive region.

conductivity σ is infinite and therefore α is also infinite. If α is infinite, equation (5-7) says that δ must be infinitely small. Therefore, *for a perfect conductor, the current only flows on the surface and the wave cannot penetrate the conductor*. Figure 5-1 shows how the current density decays as the wave propagates into a conductive medium. Note that although the magnitude of the current density is oscillatory, it remains within the envelope defined by the exponential decay of the skin depth.

If a *good conductor* is defined such that $\sigma/\varepsilon\omega \gg 1$, (5-5) reduces to

$$\alpha = \sqrt{\frac{\omega\mu\sigma}{2}} \tag{5-9}$$

Therefore, the skin depth for a metal with conductivity σ at a frequency ($\omega = 2\pi f$) is given by (5-10) in units of meters:

$$\delta = \frac{1}{\alpha} = \sqrt{\frac{2}{\omega\mu\sigma}} \tag{5-10}$$

Figure 5-2 shows the magnitude of the skin depth in copper as a function of frequency. Note that even at 1 GHz, the skin depth is only 2 μm, meaning that most of the current is flowing in a very small area.

5.2 CLASSIC CONDUCTOR MODEL FOR TRANSMISSION LINES

Classic conductor models are derived on the assumption of perfectly smooth surfaces. Although realistic conductors used to construct printed circuit boards (PCBs), packages, and multichip modules for high-speed digital designs rarely employ smooth conductors, a study of classical transmission-line losses will

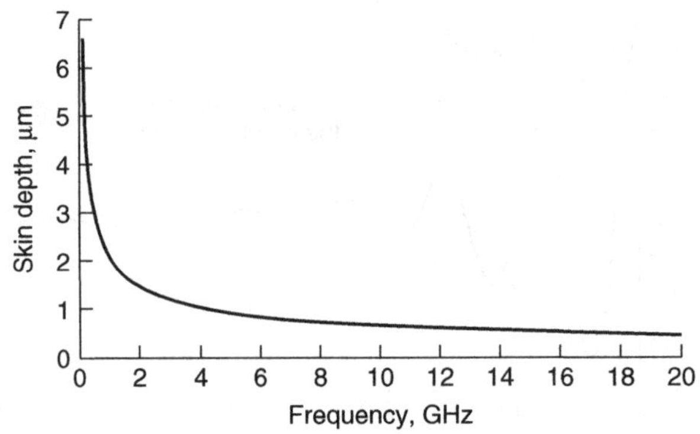

Figure 5-2 Skin depth δ in copper as a function of frequency.

provide the theoretical basis needed to derive frequency-dependent physically consistent models of real-world conductors, which are often purposely roughened to promote adhesion to dielectric layers during the manufacturing process. The resistive loss induced by general transmission-line conductors can be broken down into two components: low frequency, or dc, and high frequency, or ac. First, the dc losses will be derived and the formulas will be modified to include the frequency-dependent effects of ac resistance at high frequencies.

5.2.1 DC Losses in Conductors

Dc losses are of particular concern in small-geometry conductors, very long lines, and *multiload* (also known as *multidrop*) *buses*. Long copper telecommunication lines, for example, must have repeaters every few miles to receive and retransmit the data because of signal degradation. Additionally, designs of multiprocessor computer systems with long buses experience resistive drops that can encroach on the logic threshold levels and reduce the noise margins.

The dc loss of a transmission line depends primarily on two factors: the conductivity of the metal and the cross-sectional area of the conductor where the current is flowing. Figure 5-3 shows the current distribution in a transmission line at dc. Traditionally, dc resistance is defined to be the value at 0 Hz. However, for the purposes of this chapter, dc will be assumed to be valid for all frequencies where the skin depth is larger than the conductor thickness t, which

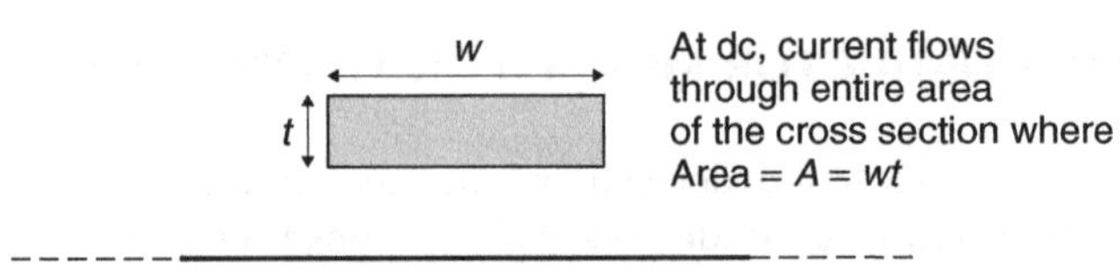

Figure 5-3 Current distribution in a microstrip at dc.

ensures almost uniform current density through the cross section of the signal conductor.

At dc the current will spread out as much as possible and flow through the entire cross section of the conductor, and the resistive loss can be found with

$$R_{\text{dc}} = \frac{l}{\sigma A_{\text{cross section}}} = \frac{l}{\sigma wt} \quad \text{ohms} \tag{5-11}$$

where l is the length, w the width, t the thickness of the signal conductor, and σ the conductivity of the metal. Note that (5-11) has neglected the dc losses of the current return path in the reference plane. This is an adequate approximation because at dc, the current will spread out and flow through the entire plane, which is several orders of magnitude larger than the signal conductor. Consequently, the cross-sectional area where the current flows in the return path will be much larger and the associated resistance will be much smaller.

5.2.2 Frequency-Dependent Resistance in Conductors

By extending the dc equation (5-11), the frequency dependence of the resistance in a transmission line can be approximated. Frequency-dependent resistance will be referred to as ac resistance or skin effect resistance in the remainder of the book. At low frequencies, the ac resistance will be identical to the dc resistance because the skin depth will be much greater than the thickness of the conductor. The ac resistance will remain equal to the dc resistance until the frequency increases to a point where the skin depth is smaller than the conductor thickness.

Microstrip Conductor Losses (Smooth Conductors) Figure 5-4 depicts the current distribution on a microstrip line at high frequencies. Notice that the current distribution is concentrated on the bottom edge of the transmission line. This is because the fields between the signal line and the ground plane pull the charge to the bottom edge, and the skin depth is much smaller than the conductor thickness. Also notice that the current density is greater near the corners of the conductor. This is because the charge density increases significantly in the proximity of a sharp edge, as described in Sections 3.4.4 and 3.4.5, and the current density along the conductor will vary in the same way. Furthermore, there is still significant field concentration along the thickness (the t dimension in Figure 5-4) of the conductor.

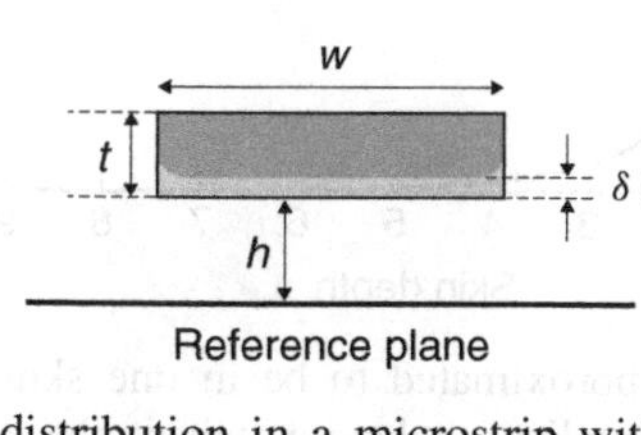

Figure 5-4 Current distribution in a microstrip with an ideal reference plane at high frequencies where the skin depth δ is small compared to the thickness t.

The skin effect will cause the cross-sectional area where the current is flowing to decrease as the frequency increases. Consequently, the frequency-dependent losses in the conductor can be approximated using the dc resistance formula by setting $t = \delta$:

$$R_{ac} = \frac{l}{\sigma w \delta} = \frac{l}{\sigma w \sqrt{2/\omega\mu\sigma}} = \frac{l}{w}\sqrt{\frac{\pi\mu f}{\sigma}} \quad \text{ohms} \tag{5-12}$$

Note that the approximation is valid only when the skin depth is smaller than the conductor thickness. Notice that the ac resistance is directly proportional to the square root of the frequency f and inversely proportional to the conductivity σ.

Equation (5-12) assumes that all the current is flowing in the first skin depth, which is not correct. Section 5.1.2 defines the skin depth such that only about 63% of the current density is contained in this depth. To test the validity of equation (5-12), the effective area of an exponential decay can be calculated by integrating $e^{-\alpha z}$ from $z = 0\delta$ to $z = \infty\delta$ and comparing it to the case where all the current is confined to one skin depth. To visualize the differences, refer to Figure 5-5, which plots penetration depth into a conductive medium in terms of skin depths versus the total current density. If 100% of the current is assumed to flow within one skin depth, the area under the curve is $J\delta = 1$, where J is the current density and δ is the skin depth. Integrating the wave decay term $e^{-\alpha z}$ from $z = 0\delta$ to $z = \infty\delta$, the area under the curve also yields $J\delta$:

$$J\int_{z=0\delta}^{z=\infty\delta} e^{-\alpha z}\,dz = \frac{J}{\alpha} = J\delta$$

Since the effective areas under each curve are identical, *it is a valid approximation to assume that all the current is flowing in an area confined by the conductor width*

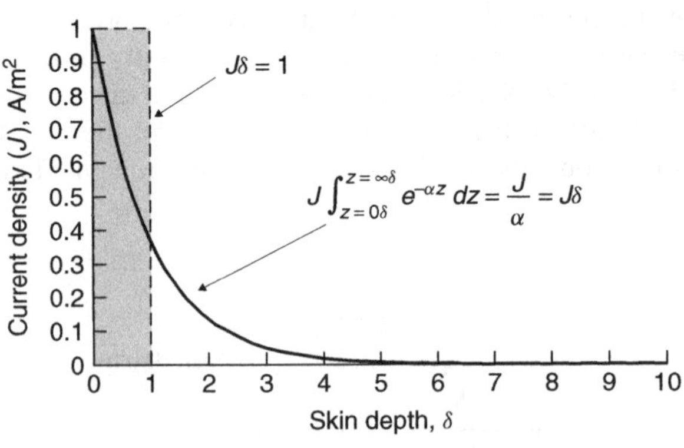

Figure 5-5 If all the current is approximated to be in one skin depth, the total area under the curve is identical to the realistic behavior, where the current density decays exponentially with increasing skin depths.

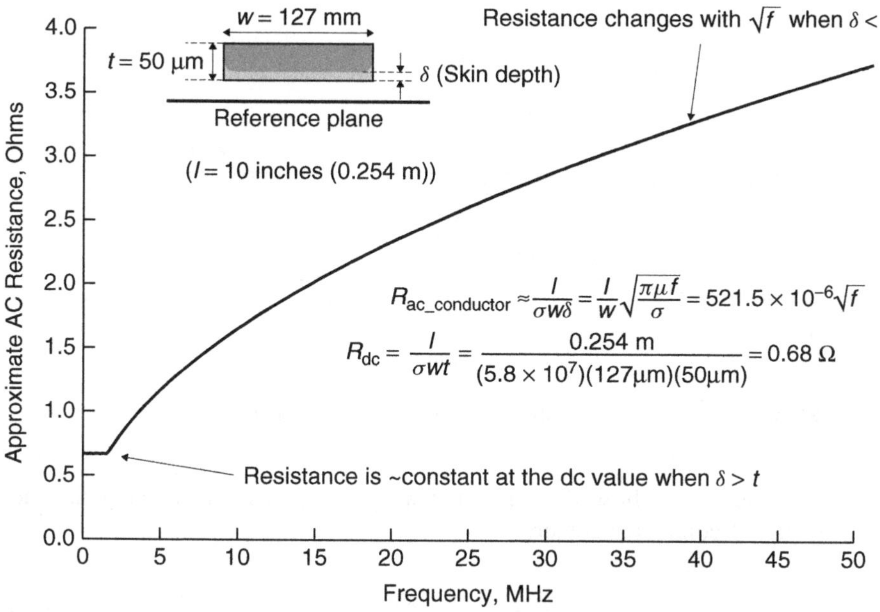

Figure 5-6 Frequency-dependent resistance of a microstrip transmission line with an ideal reference plane.

and a single skin depth. As this section progresses, more accurate methods of calculating the losses will be presented.

Figure 5-6 plots the total resistance of a transmission line over a wide frequency range. Note that the resistance will remain at approximately the dc value until the frequency where the skin depth is smaller than the thickness of the conductor, after which the ac resistance begins to take effect and increases proportional to $\sqrt{f}$. Note that the discontinuity shown in Figure 5-6 is artificial and drawn for instructional purposes. In reality, current is not completely confined to a single skin depth, so the transition from R_{dc} to R_{ac} is more gradual and not discontinuous. The discontinuity generally will not have any ill effects on simulated waveforms; however, if it is desirable to smooth the curve to provide more realistic behavior, a root-sum-square function can be used:

$$R_{\text{total}} \approx \sqrt{R_{dc}^2 + R_{ac}^2} \tag{5-13}$$

The resistance of the signal conductor, however, is only one part of the total ac resistance. The portion that is not included in equation (5-12) is the resistance of the return current on the reference plane. The return current will flow underneath the signal line in the reference plane, will be largely concentrated in one skin depth, and will spread out perpendicular to the trace direction with the highest amount of current concentrated directly beneath the signal conductor. Equation (5-14) was derived by Collins [1992] using conformal mapping

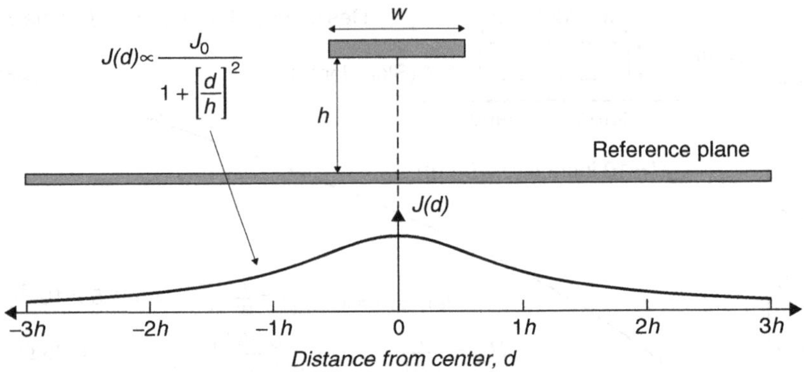

Figure 5-7 Current distribution in the reference plane of a microstrip.

techniques and shows how the current density varies with increasing distance from the signal conductor center:

$$J(d) \propto \frac{J_0}{1 + (d/h)^2} \tag{5-14}$$

where d is the distance from the conductor center and h is the height above the ground plane and J_0 is the total current density. Figure 5-7 is a graphical representation of this current density distribution.

An approximation of the ground plane resistance can be derived using a technique similar to that used to find the ac resistance of the signal conductor. First, assume that all the current will be confined to one skin depth δ. Next, an effective width w_{eff} where the current will flow must be determined. Integrating (5-14) from $-\infty$ to $+\infty$ with $h = 1$,

$$\int_{-\infty}^{\infty} \frac{1}{1 + (d/1)^2} dd = \tan^{-1}(d)|_{-\infty}^{\infty} = \pi$$

allows us to determine that (5-14) must be normalized by π so that the total current density is unity for an infinite plane. If the effective width is chosen somewhat arbitrarily at $\pm 3h$ from the center of the conductor and the normalized current density function is integrated,

$$\frac{J_0}{\pi} \int_{-3}^{3} \frac{1}{1 + (d/1)^2} dd = \frac{J_0}{\pi} 2 \tan^{-1}(3) = 0.795 J_0 \tag{5-15}$$

it can be shown that about 80% of the total current density is contained within a distance of $\pm 3h$ of the signal conductors center. Using this approximation,

$w_{\text{eff}} = 6h$ and an approximate formula can be derived for the ground-plane resistance in a microstrip transmission line in units of ohms:

$$R_{\text{ac, ground}} \approx \frac{l}{\sigma w_{\text{eff}} \delta} = \frac{l}{6h}\sqrt{\frac{\pi \mu f}{\sigma}} \tag{5-16}$$

The total resistance is the sum of (5-12) and (5-16) in ohms:

$$R_{\text{ac, micro}} = \frac{l}{w}\sqrt{\frac{\pi \mu f}{\sigma}} + \frac{l}{6h}\sqrt{\frac{\pi \mu f}{\sigma}} = \sqrt{\frac{\pi \mu f}{\sigma}}\left(\frac{1}{w} + \frac{1}{6h}\right) \tag{5-17}$$

Equation (5-17) should be considered a good "back of the envelope" estimation of the ac resistance for a microstrip transmission line [Hall et al., 2000] A more exact formula for the ac resistance of a microstrip was derived using conformal mapping techniques by Collins [1992] and is shown in equation set (5-18). This formula is significantly more cumbersome than (5-17), but should yield the most accurate results.

$$R_{\text{trace}} = \text{LR}\left(\frac{1}{\pi} + \frac{1}{\pi^2}\ln\frac{4\pi w}{t}\right)\frac{R_s}{w}$$

where LR is given by

$$\text{LR} = \begin{cases} 1 & \text{when } \dfrac{w}{h} \le \dfrac{1}{2} \\[2ex] 0.94 + 0.132\dfrac{w}{h} - 0.0062\left(\dfrac{w}{h}\right)^2 & \text{when } \dfrac{1}{2} < \dfrac{w}{h} \le 10 \end{cases}$$

$$R_{\text{ground}} = \left(\frac{w/h}{(w/h) + 5.8 + 0.03(h/w)}\right)\frac{R_s}{w} \qquad \text{when } \frac{1}{10} \le \frac{w}{h} \le 10 \tag{5-18}$$

where

$$R_s = \sqrt{\frac{\omega\mu}{2\sigma}}$$

$$R_{\text{ac, micro}} = R_{\text{trace}} + R_{\text{ground}}$$

For practical micristrip lines, formulas based on smooth conductors should simply be used as an approximation because realistic conductor surfaces are generally rough, which will increase the conductor losses significantly at frequencies where the skin depth begins to approach the magnitude of the roughness profile. The extra losses caused by surface roughness are calculated in Section 5.3.

Example 5-1 Calculate the approximate frequency where ac resistance must be used to calculate the ohmic losses of a microstrip transmission line constructed with a copper conductor with a conductivity of $\sigma = 5.8 \times 10^7 (\Omega \cdot \text{m})^{-1}$ and the following cross-sectional dimensions: $w = 5$ mils, $h = 3$ mils, $t = 2.1$ mils.

SOLUTION

The ac resistance will exist when the frequency gets high enough so that the skin depth is smaller than the conductor thickness. Above this frequency, dc resistance ceases to exist; only ac (skin effect) resistance is present. The frequency can be calculated by setting a skin depth equal to the conductor thickness using (5-10) and solving for the frequency:

$$\delta = 2.1 \text{ mils} = \sqrt{\frac{2}{\omega\mu\sigma}}$$

Since $\omega = 2\pi f$, $\mu = \mu_0 = 12.56 \times 10^{-7}$ H/m and 2.1 mils $= 55.3 \times 10^{-6}$ m:

$$f = \frac{2}{2\pi\sigma\mu_0(53.3 \times 10^{-6})}^{2} = 1.53 \times 10^{6} \text{ Hz}$$

Therefore, at 1.53 MHz, dc resistance does not exist and ac resistance begins to increase with $\sqrt{f}$.

Stripline Losses (Smooth Conductors) In a stripline transmission line, the currents of a high-frequency signal are concentrated in the upper and lower edges of the conductor. The current density will be dependent on the proximity of the local reference planes. If the stripline is referenced equidistant from both planes, the current will be divided equally in the upper and lower portions of the conductor as depicted in Figure 5-8. In an offset transmission line, the current densities on the upper and lower edges of the transmission line will be dependent on the relative distances between the ground planes and the conductor (h_1 and h_2 in Figure 5-8). The current density distributions in each stripline reference plane will be governed by an equation similar to equation

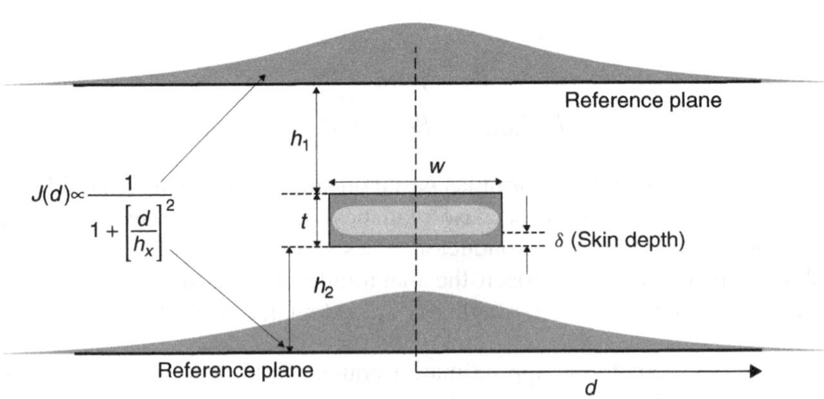

Figure 5-8 Current distribution in the signal; conductor and reference planes of a stripline.

(5-14) and will differ only in the magnitude, which is a function of the respective distances between the reference planes and the strip (h_1 and h_2). Thus, the resistance of a stripline can be approximated by the parallel combination of the resistance in the top and bottom portions of the conductor. The resistance equations for the upper and lower sections of the stripline may be obtained by applying equation (5-17) or (5-18) for the appropriate value of h. These two resistance values must then be put in parallel to obtain the total resistance for a stripline:

$$R_{\text{ac, strip}} = \frac{(R_{(h_1)\text{ac, micro}})(R_{(h_2)\text{ac, micro}})}{R_{(h_1)\text{ac, micro}} + R_{(h_2)\text{ac, micro}}} \tag{5-19}$$

5.2.3 Frequency-Dependent Inductance

Section 5.1 describes how the skin effect will force high-frequency current to flow primarily in a small layer near the periphery of the conductor, and Section 5.2 describes how this translates into frequency-dependent resistance. Another consequence of the skin effect is a frequency-dependent inductance. To conceptualize where this frequency dependence comes from, consider two filaments of current that form a loop with the return plane immediately below the conductor over a differential length of transmission line Δz, as depicted in Figure 5-9. Loop (a) passes through the center of the signal conductor, and loop (b) exists only on the conductor surfaces. As described in Section 2.5.2, the inductance is proportional to the loop area,

$$L_{11} \equiv \frac{\psi_1}{I_1} \tag{2-97}$$

where ψ_1 is the magnetic flux, which depends on the loop area. Therefore, loop (a) will have a higher inductance than loop (b) simply because the loop is larger. At low frequencies, the skin depth will be large compared to the conductor thickness and there will be significant current flowing in the interior of the conductor,

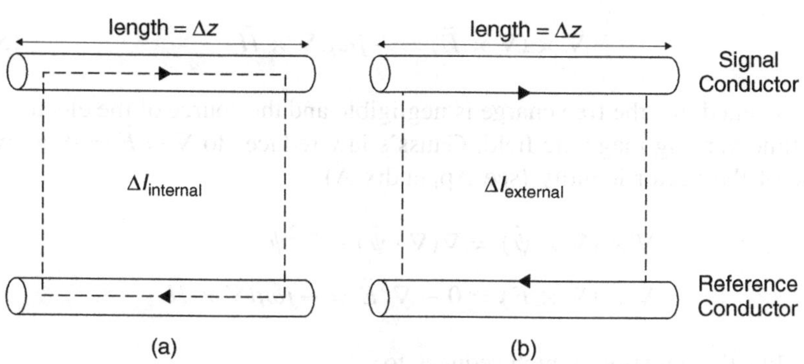

Figure 5-9 Loops for a filament of current: (a) in the center of the conductor; (b) on the surface of the conductor.

similar to loop (a). As frequency increases, the skin effect will begin to force the current to flow on the periphery of the conductor, so internal currents will decrease. At very high frequencies, there will be almost no current flowing in loop (a) because the skin effect has forced all the current to flow on the surface, causing the inductance of loop (b) to dominate. The inductance associated with loop (a), where the currents are flowing internal to the conductor, is commonly known as the *internal inductance*. The inductance associated with loop (b), where all currents are flowing on the surface of the conductor, is commonly known as the *external inductance*. The total inductance is the sum of the external and internal parts:

$$L_{\text{total}} = L_{\text{internal}} + L_{\text{external}} \tag{5-20}$$

As frequency increases, the skin effect causes the current to flow very close to the surface, with minimal current flowing interior to the conductor. At infinite frequencies, all the current is flowing on the surface and the internal contribution is zero. Consequently, as frequency increases, the effect of L_{internal} will decrease and L_{total} will asymptote to L_{external}.

The external inductance is calculated with quasistatic techniques assuming that all the charge is on the surface of the conductor and solving Laplace's equation as shown in Sections 3.4.2, 3.4.3, and 3.4.6. The internal inductance is derived by observing Ampère's (2-34) and Faraday's (2-33) laws for a good conductor. For a *good conductor*, the conduction current $\vec{J}$ will be much larger than the displacement current $j\omega\vec{D}$, and therefore Ampère's law reduces to

$$\nabla \times \vec{H} \approx \vec{J} = \sigma \vec{E} \tag{5-21}$$

Faraday's law is repeated here:

$$\nabla \times \vec{E} + j\omega\mu\vec{H} = 0 \tag{2-33}$$

Taking the curl of (2-33) yields the following equation:

$$\nabla \times (\nabla \times \vec{E}) = -j\omega\mu\nabla \times \vec{H} \tag{5-22}$$

If it is assumed that the free charge is negligible and the source of the electric field is the time-varying magnetic field, Gauss's law reduces to $\nabla \cdot \varepsilon\vec{E} = 0$, allowing the use of the vector identity (see Appendix A)

$$\nabla \times (\nabla \times \vec{\psi}) = \nabla(\nabla \cdot \vec{\psi}) - \nabla^2\vec{\psi}$$
$$\nabla \times (\nabla \times \vec{E}) = 0 - \nabla^2\vec{E} = -j\omega\mu\nabla \times \vec{H}$$

to simplify the equation, which reduces to

$$\nabla^2\vec{E} = j\omega\mu\nabla \times \vec{H} \tag{5-23}$$

Substituting Ampère's law (5-21) into (5-23) and multiplying each side by σ allows us to write (5-23) in terms of the current density (since $\vec{J} = \sigma\vec{E}$):

$$\nabla^2\sigma\vec{E} = j\omega\mu\sigma\sigma\vec{E} \rightarrow \nabla^2\vec{J} = j\omega\mu\sigma\vec{J} \tag{5-24}$$

Equation (5-24) is known as the *diffusion equation* for the current density. Similar equations govern the electric and magnetic fields. Assuming a z-directed current, the solution to (5-24) is found using standard techniques.

$$\frac{d^2\vec{J}_z}{dx^2} = j\omega\mu\sigma\vec{J}_z$$

$$\left(\frac{d^2}{dx^2} - j\omega\mu\sigma\right)\vec{J}_z = 0$$

The roots of the characteristic equation yield

$$\pm\frac{\sqrt{-4(-j\omega\mu\sigma)}}{2} = \pm\left(\sqrt{\frac{\omega\mu\sigma}{2}} + j\sqrt{\frac{\omega\mu\sigma}{2}}\right) = \pm(1+j)\sqrt{\frac{\omega\mu\sigma}{2}}$$

Comparing to equation (5-10), note that $\sqrt{\omega\mu\sigma/2} = 1/\delta$, giving a solution to the diffusion equation for current density of

$$\vec{J}_z = \vec{J}_0 e^{-(1+j)x/\delta} \tag{5-25}$$

where only the negative exponent is a valid solution because the current density decreases as the electromagnetic wave penetrates into the conductor. Equation (5-25) says that the current density will decrease very rapidly into the surface of the conductor and is essentially confined to a layer at the surface equal to a few skin depths, as depicted in Figure 5-10.

The total current can be found by integrating (5-25):

$$\vec{J}_T = \int_0^{\infty} \vec{J}_0 e^{-(1+j)x/\delta} dx = -\frac{1}{2}\vec{J}_0\delta(j-1) \tag{5-26}$$

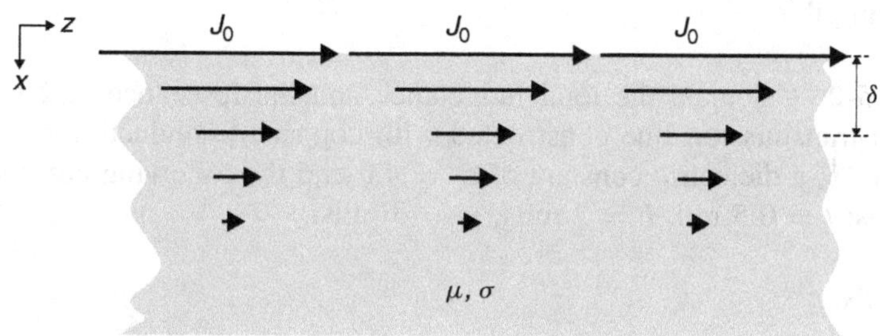

Figure 5-10 Diffusion of currents into a semi-infinite conductive material.

The surface impedance can be defined as the ratio of the electric field at the surface and the total current density:

$$\vec{Z}_s = \frac{\vec{E}_0}{\vec{J}_T} \tag{5-27}$$

Substituting (5-26) into (5-27) and expressing the current density in terms of the conductivity and the electric field ($\vec{J}_0 = \sigma \vec{E}_0$) yields

$$\vec{Z}_s = \frac{\vec{E}_0}{-\frac{1}{2}\sigma \vec{E}_0 \delta (j-1)} = (1+j)\frac{1}{\sigma\delta} \quad \text{ohms/square} \tag{5-28}$$

The surface impedance is expressed in terms of an area of unit width and unit length, so the term *ohms/square* is used. Note the similarity of the real part of (5-28) to equation (5-12), which was the series resistance caused by the skin effect of a transmission line. For the case where $l = w$, the real part of equations (5-28) and (5-12) are identical. If the real part of (5-28) is resistance, the imaginary part must therefore be reactance (the impedance of inductance). Since the impedance of inductance is $j\omega L$, (5-28) can be expressed in terms of a series resistance (due to the skin effect) and a series inductance (the internal inductance):

$$Z_s = R_{\text{ac}} + j\omega L_{\text{internal}} \tag{5-29}$$

Therefore, *the internal inductance can be calculated directly from the ac resistance*:

$$L_{\text{internal}} = \frac{R_{\text{ac}}}{\omega} \tag{5-30}$$

Equation (5-30) highlights an important relationship between the skin effect resistance (ac resistance) and the internal inductance. As the skin effect forces the current to the periphery of the conductor, the resistance increases; however, since current ceases to flow in the interior of the conductor, the inductance must decrease. Figure 5-11 plots both the internal inductance and the ac resistance. Note that when the ac resistance becomes significant, the internal inductance is almost negligible.

Example 5-2 Calculate the total inductance and the resistance at 2 GHz of a microstrip transmission line constructed with copper of conductivity $\sigma = 5.8 \times 10^7 (\Omega \cdot \text{m})^{-1}$, a dielectric constant of $\varepsilon_r = 4.0$ and the following cross-sectional dimensions: $t = 0.5$ mil, $h = 2$ mils, $w = 3$ mils.

SOLUTION

Step 1: Determine if the ac or dc resistance should be used. The methodology presented in Example 5-1 could be used, however, it is easier simply to calculate

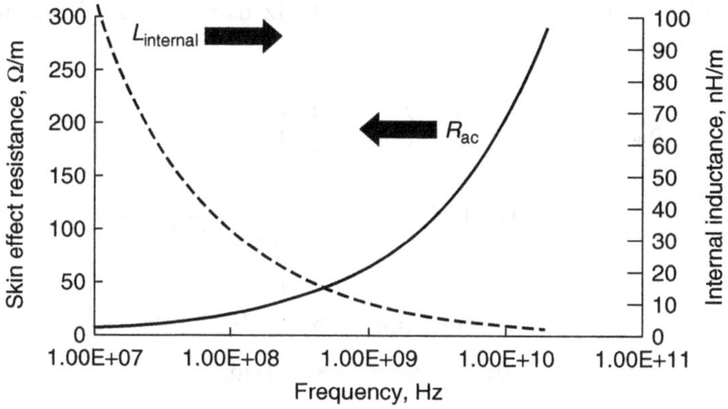

Figure 5-11 Example of how the skin effect changes resistance and internal inductance with frequency for a copper microstrip as shown in Figure 5-6.

the skin depth in copper at 2 GHz using (5-10) and compare it to the conductor thickness t:

$$\delta = \sqrt{\frac{2}{2\pi f \mu_0 \sigma}} \approx 1.41\ \mu\text{m}$$

Since $t = 0.5$ mil $= 12.7\ \mu$m, $\delta < t$, so the ac resistance must be used.

Step 2: The impedance and effective dielectric permittivity must be calculated using equation (3-36b), which assumes a perfect conductor. If it is unknown whether a particular impedance formula includes the effect of a realistic conductor, the lack of a metal conductivity or a magnetic permeability variable indicates the assumption of infinite conductivity. From (3-36b), $Z_0 \approx 55\ \Omega$ and $\varepsilon_{\text{eff}} \approx 2.95$.

Step 3: Calculate the phase velocity with (2-52):

$$\nu_p = \frac{c}{\sqrt{\mu_r \varepsilon_{\text{eff}}}} = \frac{3 \times 10^8\ \text{m/s}}{\sqrt{(1)(2.95)}} = 1.75 \times 10^8\ \text{m/s}$$

Step 4: The external inductance is solved with (3-31) and (3-33). Since (3-36b) assumes a perfect conductor, the inductance is the external value.

$$\nu_p = \frac{1}{\sqrt{LC}} = 1.75 \times 10^8 \rightarrow \sqrt{LC} = 5.73 \times 10^{-9}\ \text{s/m}$$

$$\approx 146 \times 10^{-12}\ \text{s/in.}$$

$$Z_0 = \sqrt{\frac{L}{C}} = 55\ \Omega$$

$$L_{\text{external}} = L = \sqrt{\frac{L}{C}} \cdot \sqrt{LC} = 55(146 \times 10^{-12}) = 8.03\ \text{nH/in.}$$

Step 5: Calculate the ac resistance at 2 GHz using (5-17). Note that (5-18) could be used as well.

$$
\begin{aligned}
R_{\text{ac, micro}} &= \sqrt{\frac{\pi\mu_0 f}{\sigma}}\left(\frac{1}{w}+\frac{1}{6h}\right)\\
&= 0.0117\left(\frac{1}{(3\ \text{mils})(25.4\times10^{-6}\ \text{m/mil})}\right.\\
&\qquad\left.+\frac{1}{6(2\ \text{mils})(25.4\times10^{-6}\ \text{m/mil})}\right)\\
&= 191.3\ \Omega/\text{m} = 4.86\ \Omega/\text{in.}
\end{aligned}
$$

Step 6: Calculate the internal inductance using (5-30).

$$
L_{\text{internal}} = \frac{R_{\text{ac}}}{\omega} = \frac{4.86\ \Omega/\text{in.}}{2\pi(2\times10^{9}\ \text{Hz})} = 0.387\ \text{nH/in.}
$$

Step 7: Calculate the total inductance using (5-20).

$$
L_{\text{total}} = L_{\text{internal}} + L_{\text{external}} = 8.03 + 0.387 = 8.42\ \text{nH/in.}
$$

5.2.4 Power Loss in a Smooth Conductor

In high-speed digital design, surface treatment of the copper foils used to construct printed circuit boards (PCBs) significantly affects the power losses experienced by a signal propagating on a transmission line. In this section, the power losses of an electromagnetic wave impinging on a flat, smooth plane are examined. In later sections we explore the consequences of rough conductor surfaces.

First, we assume that the fields in the vicinity of a good but not perfect conductor will behave approximately the same as for a perfect conductor. In Section 3.2.1 it was shown that the electric fields terminate normal to a perfect conducting surface and the magnetic fields are tangential to the surface. Furthermore, in Section 5.1.2 it was shown that fields inside a conductor will attenuate exponentially and are measured in terms of the skin depth δ. At high frequencies, the boundary condition shown in (3-3) is true for a good conductor, except for a thin transitional layer.

To derive an equation to predict the loss for a smooth plane, we first assume that just outside the conductor there exists only a normal component of the electric field ($\vec{E}_\perp$) and a tangential component of the magnetic field ($\vec{H}_\parallel$), which are the identical boundary conditions used for perfect conductors. Following the approach outlined by Jackson [1999], Maxwell's equations are then used to calculate the fields within the transition layer.

If a tangential $\vec{H}_\parallel$ exists just outside the surface, the same $\vec{H}_\parallel$ must exist just inside the conductor surface. With the neglect of the displacement current, (2-33) and (2-34) become

$$\nabla \times \vec{E}_c + j\omega\mu\vec{H}_c = 0 \tag{5-31}$$

$$\nabla \times \vec{H}_c = \vec{J} = \sigma\vec{E}_c \tag{5-32}$$

where $\vec{E}_c$ and $\vec{H}_c$ denote the field values inside the conductor. If it is assumed that $\vec{n}$ is the normal vector pointing outward from the conductor surface and z is the normal coordinate inward into the conductor, the gradient operator is $\nabla \approx -\vec{n}(\partial/\partial z)$, reducing Maxwell's equations to

$$\vec{H}_c = j\frac{\vec{n} \times \partial\vec{E}_c/\partial z}{\mu\omega} \tag{5-33}$$

$$\vec{E}_c = -\frac{\vec{n} \times \partial\vec{H}_c/\partial z}{\sigma} \tag{5-34}$$

These equations can be solved to yield the fields inside the conductor. The first step is to take the partial derivative of (5-34):

$$\frac{\partial\vec{E}_c}{\partial z} = -\frac{1}{\sigma}\vec{n} \times \frac{\partial^2\vec{H}_c}{\partial z^2} \tag{5-35}$$

Next is to take the cross product of (5-33) with the unit vector,

$$\vec{n} \times \vec{H}_c = \frac{j}{\mu\omega}\vec{n} \times \left(\vec{n} \times \frac{\partial\vec{E}_c}{\partial z}\right)$$

so the vector identity from Appendix A can be used to simplify the math,

$$a \times (b \times c) = (a \cdot c)b - (a \cdot b)c$$

$$\vec{n} \times \vec{H}_c = \frac{j}{\mu\omega}\left\{\left(\vec{n} \cdot \frac{\partial\vec{E}_c}{\partial z}\right)\vec{n} - (\vec{n} \cdot \vec{n})\frac{\partial\vec{E}_c}{\partial z}\right\}$$

where $\vec{n} \cdot \partial\vec{E}_c/\partial z = 0$ because when (5-35) is substituted, the form becomes proportional to $\vec{n} \cdot \left(\vec{n} \times \partial^2\vec{H}_c/\partial z^2\right)$, which is zero. Furthermore, $\vec{n} \cdot \vec{n} = 1$, yielding

$$n \times \vec{H}_c = -\frac{j}{\mu\omega}\frac{\partial\vec{E}_c}{\partial z}$$

Next, equation (5-35) is substituted for $\partial\vec{E}_c/\partial z$, yielding

$$\vec{n} \times \vec{H}_c = \frac{j}{\mu\omega}\left(\frac{1}{\sigma}\vec{n} \times \frac{\partial^2\vec{H}_c}{\partial z^2}\right)$$

$$= \frac{j}{\mu\omega\sigma}\frac{\partial^2}{\partial z^2}\left(\vec{n} \times \vec{H}_c\right)$$

After rearranging the equation, we get a more manageable form:

$$-j\mu\omega\sigma\left(\vec{n}\times\vec{H}_c\right)=\frac{\partial^2}{\partial z^2}\left(\vec{n}\times\vec{H}_c\right)$$

Since $\mu\omega\sigma = 2/(2/\mu\omega\sigma) = 2/\delta^2$, the equation can be written in terms of the skin depth,

$$-j\frac{2}{\delta^2}\left(n\times\vec{H}_c\right)=\frac{\partial^2}{\partial z^2}\left(\vec{n}\times\vec{H}_c\right)$$

which yields the differential equation

$$\frac{\partial^2}{\partial z^2}\left(\vec{n}\times\vec{H}_c\right)+\frac{2j}{\delta^2}\left(\vec{n}\times\vec{H}_c\right)=0 \tag{5-36}$$

To solve (5-36), we reasonably assume that the field inside the conductor is a function of the external applied field,

$$\vec{H}_c=\vec{H}_{||}f(z) \tag{5-37}$$

where $f(z=0)=1$, which says that at the surface of the conductor, $\vec{H}_c=\vec{H}_{||}$. This allows us to substitute (5-37) into (5-36) and solve the equation for $f(z)$.

$$\frac{\partial^2}{\partial z^2}f(z)+\frac{2j}{\delta^2}f(z)=0 \tag{5-38}$$

The solution to (5-38) is

$$f(z)=Ae^{-z/\delta}e^{-jz/\delta}$$

where $f(z=0)=1=A$, yielding

$$\vec{H}_c=\vec{H}_{||}e^{-z/\delta}e^{-jz/\delta} \tag{5-39}$$

where $\vec{H}_{||}$ is the tangential magnetic field applied to the surface of the conductor.

The electric field inside the conductor is calculated from (5-34):

$$\begin{aligned}\vec{E}_c&=-\frac{\vec{n}\times\partial\vec{H}_c/\partial z}{\sigma}=-\frac{1}{\sigma}\vec{n}\times\left(-\frac{1}{\delta}(1+j)\vec{H}_{||}e^{-z/\delta}e^{-jz/\delta}\right)\\&=\frac{1+j}{\sigma\delta}\left(\vec{n}\times\vec{H}_{||}\right)e^{-z/\delta}e^{-jz/\delta}\\&=(1+j)\sqrt{\frac{\omega\mu}{2\sigma}}\left(\vec{n}\times\vec{H}_{||}\right)e^{-z/\delta}e^{-jz/\delta}\end{aligned} \tag{5-40}$$

Since the tangential component of the electric field must remain continuous, as shown in equation (3-8), the electric field just outside the conductor surface can be calculated with (5-40) evaluated at $z = 0$:

$$\vec{E}_{||} = (1 + j)\sqrt{\frac{\omega\mu}{2\sigma}}\left(\vec{n} \times \vec{H}_{||}\right) \tag{5-41}$$

Section 3.2.1 says the electric field must terminate normal to a perfectly conducting surface, however, equation (5-41) shows that *for a good conductor, a tangential component of* $\vec{E}$ *must exist just outside the conductor*. Since (5-40) describes the electric field decaying with increasing depth z into the surface, there must be power flow into the conductor. The time-averaged value of the Poynting vector described in Section 2.6.1 is used to calculate the power absorbed per unit area:

$$\vec{S}_{\text{ave}} = \vec{a}_z \frac{(E^+)^2}{2\eta} \tag{2-121}$$

The intrinsic impedance (2-53) at the surface of the conductor, η_s, is calculated with (5-37) and (5-41):

$$\eta_s(z = 0) = \frac{\vec{E}_c}{\vec{H}_c} = (1 + j)\sqrt{\frac{\omega\mu}{2\sigma}} \tag{5-42}$$

It is interesting to note that since $\sqrt{\mu\omega/2\sigma} = 1/\sigma\delta$, (5-42) reduces to equation (5-28), which is the series impedance of a transmission line:

$$(1 + j)\sqrt{\frac{\omega\mu}{2\sigma}} = (1 + j)\frac{1}{\sigma\delta}$$

Finally, the power flow per unit area into the conductor is calculated using equation (2-121):

$$\vec{S}_{\text{ave}} = \frac{\left|\vec{E}_c\right|^2}{2\eta} = \frac{\omega\mu}{4\sigma}\left|\vec{H}_{||}\right|^2\sqrt{\frac{2\sigma}{\omega\mu}} = \frac{\omega\mu}{4\sigma}\left|\vec{H}_{||}\right|^2\sqrt{\frac{2\sigma^2}{\omega\mu\sigma}}$$

From (5-10), $\delta = \sqrt{2/\omega\mu\sigma}$, yielding (5-43), which is the time-averaged power absorbed by a flat conducting plane per unit area:

$$\vec{S}_{\text{ave}} = P_{\text{plane}} = \frac{\omega\mu\delta\left|\vec{H}_{||}\right|^2}{4} \quad \text{W/m}^2 \tag{5-43}$$

Equation (5-43) will allow for the solution of the resistive power losses of a good conducting plane provided that the applied magnetic field $\vec{H}_{||}$ has been solved for the idealized case of a perfect conducting plane.

The power loss from a plane can also be put in more intuitive form. Equation (2-7), which is simply Ohm's law, can be used to calculate the current density from (5-40):

$$\vec{J} = \sigma \vec{E}_c = \sigma \sqrt{\frac{\mu\omega}{2\sigma}}(1 + j)\left(\vec{n} \times \vec{H}_{||}\right) e^{-z/\delta} e^{-jz/\delta}$$

Again, since $\sqrt{\mu\omega/2\sigma} = 1/\sigma\delta$, the equation can be rewritten as

$$\vec{J} = \sigma \vec{E}_c = \frac{1}{\delta}(1 + j)\left(\vec{n} \times \vec{H}_{||}\right) e^{-z/\delta} e^{-jz/\delta} \qquad (5\text{-}44)$$

Equation (5-44) simply says that most of the current density will be confined to a very small thickness, as described in Section 5.1.2. Note that the cross product in (5-44) implies that the current will flow perpendicular to the magnetic field.

To derive a more intuitive form of the power dissipated by a flat plane, it is first necessary to define an effective surface current. If the current density (5-44) is integrated to get the total current, an equivalent surface current can be calculated for use in the classic time-averaged power equation $P = \frac{1}{2}RI^2$:

$$\vec{J}_{\text{eff}} = \int_0^\infty \vec{J}\,dz \qquad \text{amperes} \qquad (5\text{-}45)$$

Equation (5-45) simply calculates the total current that is decaying exponentially into the conductor surface. To calculate the power per unit area, we make the approximation that all the current exists on the surface and use the real part of the surface impedance in (5-28):

$$P_{\text{plane}} = \frac{1}{2}RI^2 = \frac{1}{2\sigma\delta}\left|\vec{J}_{\text{eff}}\right|^2 \qquad \text{W/square} \qquad (5\text{-}46)$$

5.3 SURFACE ROUGHNESS

To account properly for the frequency variation of L_{internal} and R_{ac}, the nonideal effects of the copper surface must be considered. The problem is that most (if not all) commercial 2D field solvers calculate the resistance and inductance assuming smooth conductors. Real copper surfaces, however, are purposely roughened to promote adhesion to the dielectric when manufacturing printed circuit boards. The resulting copper surfaces have a "tooth structure" as depicted in Figure 5-12. When the tooth height is comparable to the skin depth, the smooth copper assumptions break down. The root-mean-square (RMS) tooth height of common copper foils used to manufacture printed circuit boards range from approximately 0.3 to 5.8 μm, with peak heights exceeding 11 μm [Brist et al., 2005]. The skin depth in copper at 1 GHz is about 2 μm, indicating that for many copper foils, most of the current will be flowing in the tooth structure for multigigabit designs

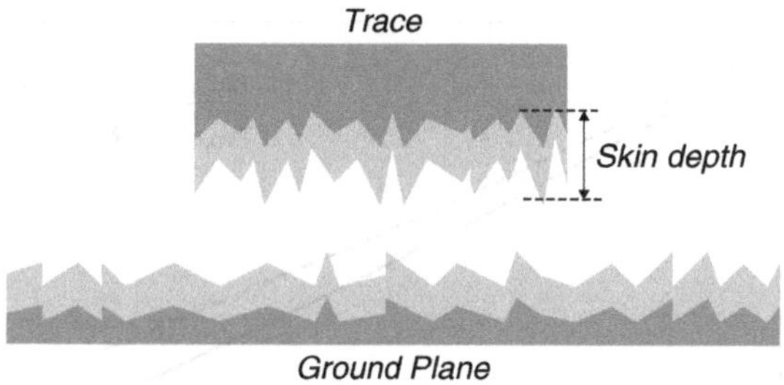

Figure 5-12 Realistic conductors used to manufacture transmission lines exhibit a rough surface often called the "tooth structure." When the skin depth is similar to the tooth size, power dissipation is increased.

[Hall et al., 2000]. Since the rough copper surface affects current flow, it will also affect power dissipation, and thus insertion loss.

Insertion loss, which is described extensively in Chapter 9, is a common way to measure the frequency-dependent losses in the form of a transfer function by injecting a sinusoidal waveform at port 1 (such as the input to a transmission line) and measuring at port 2 (such as the output). Expressed in terms of power, the insertion loss is

$$S_{21}(f) = 20 \log \sqrt{\frac{P_2(f)}{P_1(f)}} \qquad \text{dB}$$

where S_{21} is the insertion loss in decibels, P_2 is the power measured at the output of the transmission line, and P_1 is the power injected into the input of the transmission line. Insertion loss is a convenient method to evaluate the power loss of a transmission line. Note that the ratio of powers reduce to a ratio of voltages when the port impedances are identical, as shown in equation (9-21), which is why the form 20 log is used instead of 10 log to calculate the magnitude in decibels [Hall et al., 2000].

Figure 5-13 shows the measured results of two identical transmission lines built with rough and relatively smooth copper. The copper foil used to construct the test boards was characterized with an optical profilometer prior to lamination yielding an RMS tooth height of $h_{\text{RMS}} = 1.2$ μm for the smoother copper and $h_{\text{RMS}} = 5.8$ μm for the rough copper. Note the significant increase in insertion losses, and therefore power losses, due to increased roughness profile. At high frequencies, *surface roughness will increase the ohmic losses of a transmission-line conductor significantly*.

5.3.1 Hammerstad Model

The traditional way to account for surface roughness losses in a transmission-line model is to use the Hammerstad equation:

$$R_{\text{ac}} = K_{\text{H}} R_s \sqrt{f} \tag{5-47}$$

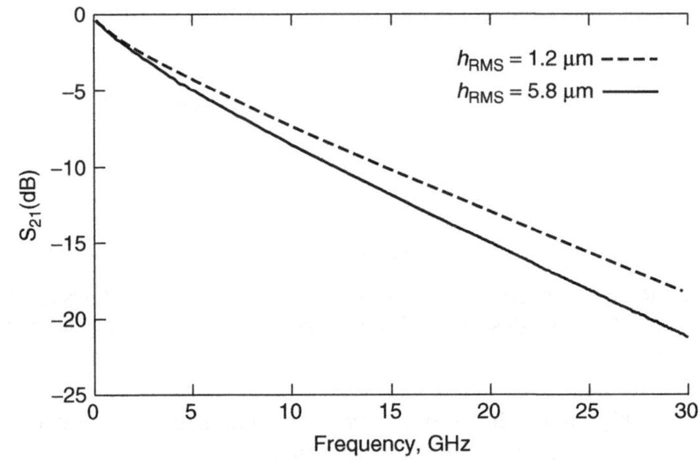

Figure 5-13 Measured results of identical 7-in. transmission lines with relatively smooth ($h_{\text{RMS}} = 1.2\ \mu\text{m}$) and rough ($h_{\text{RMS}} = 5.8\ \mu\text{m}$) copper showing how surface roughness affects losses.

where $R_s\sqrt{f}$ is the classic skin resistance for a smooth conductor as calculated in (5-17) and (5-18) and K_{H} is the Hammerstad coefficient:

$$K_{\text{H}} = 1 + \frac{2}{\pi}\arctan\left[1.4\left(\frac{h_{\text{RMS}}}{\delta}\right)^2\right] \tag{5-48}$$

where h_{RMS} is the root-mean-square value of the surface roughness height and δ is the skin depth [Hammerstad and Jensen, 1980; Brist et al., 2005]. The Hammerstad coefficient is used to model the extra losses caused by the copper surfaces on a transmission line that are often purposely roughened to promote adhesion to the dielectric.

The frequency dependence of the skin effect resistance and total inductance using the Hammerstad correction for surface roughness is implemented with†

$$R_{\text{H}}(f) = \begin{cases} K_{\text{H}} R_s\sqrt{f} & \text{when } \delta < t \\ R_{\text{dc}} & \text{when } \delta \geq t \end{cases} \tag{5-49a}$$

$$L_{\text{H}}(f) = \begin{cases} L_{\text{external}} + \dfrac{R_{\text{H}}(f)}{2\pi f} & \text{when } \delta < t \\ L_{\text{external}} + \dfrac{R_{\text{H}}(f_{\delta=t})}{2\pi f_{\delta=t}} & \text{when } \delta \geq t \end{cases} \tag{5-49b}$$

†Note that the method shown here for calculating the internal portion of the inductance ($L_{\text{internal}} = R_{\text{ac}}/\omega$) for a rough conductor is an approximation based on the result for a smooth conductor. The approximation will induce causality errors that tend to be small enough to ignore, so this method is generally acceptable. For the interested reader, Appendix E derives the internal inductance using a more rigorous approach based on the discussion in Chapter 8.

where K_H is calculated with (5-48), t is the conductor thickness, δ the skin depth, and $f_{\delta=t}$ the frequency where the skin depth equals the thickness of the conductor. When the skin depth (δ) is larger than the conductor thickness, the dc value of the resistance and low-frequency inductance where the skin depth is equal to the conductor thickness should be used.

This approach has been shown to be accurate by comparison to vector network analyzer measurements of insertion loss for copper surface roughness profiles less than approximately 2 μm (RMS). Figure 5-14 shows the accuracy of the Hammerstad model by comparing measured transmission-line structures constructed with relatively smooth and very rough copper to simulations using (5-49). Notice that the Hammerstad model is considerably less accurate for the rough copper case for frequencies greater than about 5 GHz.

To understand why the accuracy breaks down for some copper types, it is useful to explore the assumptions behind (5-48), which assumes a 2D corrugated surface similar to that shown in Figure 5-15 [Pytel, 2007]. The first published work providing a theoretical investigation into the power loss from copper roughness was in 1948 by Samuel Morgan (Bell Laboratories), who studied the effects up to 10 GHz providing loss equations for current flow transverse and parallel to corrugated structures similar to that depicted in Figure 5-15. He concluded

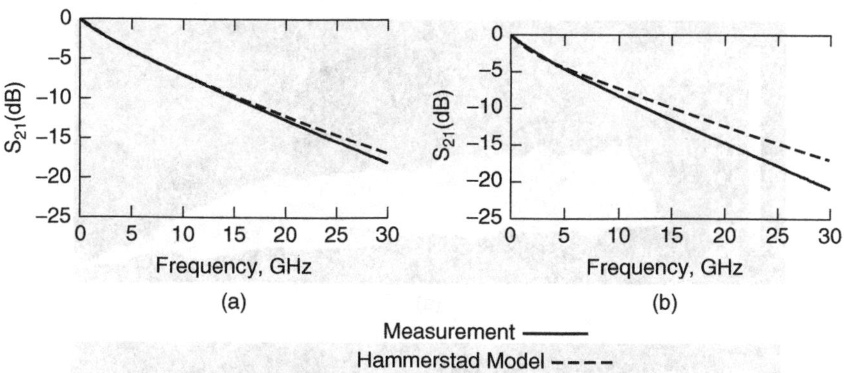

Figure 5-14 Example of the Hammerstad model accuracy for (a) relatively smooth ($h_{RMS} = 1.2$ μm) and (b) rough ($h_{RMS} = 5.8$ μm) copper foils; 7-in. microstrip line.

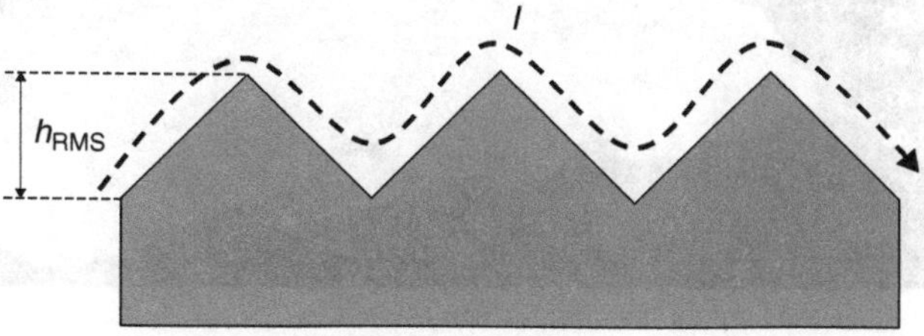

Figure 5-15 Two-dimensional corrugated surface assumption behind the Hammerstad equation.

that *power loss is proportional to the surface area of the roughness structure*, current flow transverse to the corrugated surface could increase the power losses by 100%, and current flow parallel to the grooves increases losses by about 30%. In 1975, a Norwegian scientist named Erik Hammerstad used Morgan's work to fit the data to an arctan function, producing the Hammerstad equation shown in (5-48), which became the standard equation in industry to account for the effects of surface roughness [Pytel, 2007]. The model assumes that at high frequencies, when the skin depth becomes small compared to the tooth height, the current will begin to follow the contour of the corrugated surface, which will increase the losses.

To understand why the Hammerstad equation breaks down for some roughness profiles, the surfaces of relatively smooth and rough copper surfaces were measured using an optical profilometer. The surface shown in Figure 5-16a [Hall et al., 2007] can be described as corrugated with sparse protrusions on the surface, suggesting that the Hammerstad equation (5-48) might be adequate for approximating the surface roughness losses for a transmission line manufactured with this copper foil. Figure 5-14a depicts a measurement of a transmission line that was constructed with the copper foil depicted in Figure 5-16a compared to a model created using (5-47) and (5-48). Note that the model and measurement

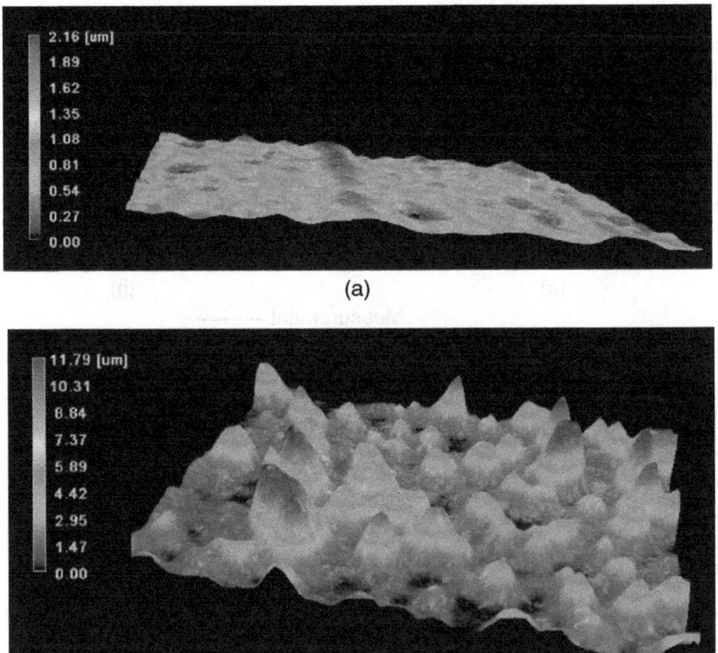

(b)

Figure 5-16 Surface profile measurement of (a) relatively smooth and (b) rough copper foil used to construct PCBs.

correlate very nicely until about 15 GHz, indicating that the Hammerstad model is adequate to model the losses for this case.

Conversely, Figure 5-16b depicts an optical profilometer measurement of a very rough copper foil. Note the significant difference between the corrugated surface assumed by the Hammerstad equation and the surface depicted in Figure 5-16b, suggesting that (5-48) will not work well for this type of 3D surface profile. Figure 5-14b depicts the measurement of a transmission line constructed with the rough copper sample shown in Figure 5-16b compared to a model created with (5-47) and (5-48). Note that the accuracy breaks down after about 5 GHz, indicating that an alternative modeling methodology is needed for losses induced by this surface profile.

Example 5-3 Assuming the transmission line in Example 5-2, calculate the frequency where surface roughness begins to affect the losses for an RMS tooth height of 1.8 μm, calculate the ac resistance and total inductance at an operating frequency of 2 GHz, and determine how surface roughness changes the resistance and inductance compared to the smooth case.

SOLUTION

Step 1: To determine the approximate frequency where the surface roughness will begin to affect the ac losses, the skin depth must be set equal to the RMS roughness height, and the frequency is calculated using (5-10):

$$f_r = \frac{2}{2\pi\sigma\mu_0(1.8 \times 10^{-6})^2} = 1.34 \times 10^9 \text{ Hz}$$

Since this is close to the operating frequency of 2 GHz, the surface roughness will be significant and cannot be ignored. Note that the roughness will influence the losses at frequencies below f_r; however, if $f_r \gg f$, the surface roughness will have no significant effect, and it can be ignored.

Step 2: Calculate the ac resistance at 2 GHz using (5-47) and (5-48). The ac resistance at 2 GHz for a smooth surface was calculated in Example 5-2.

$$R = K_{\text{H}} R_s \sqrt{f} = K_{\text{H}} \sqrt{\frac{\pi\mu_0 f}{\sigma}} \left(\frac{1}{w} + \frac{1}{6h}\right) = K_{\text{H}} \cdot 4.86\ \Omega/\text{in.}$$

Since the RMS roughness height is less than 2 μm, (5-48) is suitable to correct the smooth formula to estimate the extra losses due to the rough surface.

$$\delta(2\text{ GHz}) = \sqrt{\frac{2}{2\pi f \mu_0 \sigma}} \approx 1.41\ \mu\text{m}$$

$$K_H = 1 + \frac{2}{\pi} \arctan\left[1.4\left(\frac{1.8 \times 10^{-6}}{1.41 \times 10^{-6}}\right)^2\right] = 1.73$$

Therefore, the ac resistance including the surface roughness is

$$R = K_H \cdot 4.86\ \Omega/\text{in.} = (1.73)(4.86) = 8.4\ \Omega/\text{in.}$$

Step 3: Calculate the internal inductance using (5-30).

$$L_{\text{internal}} = \frac{R_{\text{ac}}}{\omega} = \frac{8.4\ \Omega/\text{in.}}{2\pi(2 \times 10^9\ \text{Hz})} = 0.669\ \text{nH/in.}$$

Step 4: Calculate the total inductance using (5-20), where L_{external} is as calculated in Example 5-2.

$$L_{\text{total}} = L_{\text{internal}} + L_{\text{external}} = 8.03 + 0.669 = 8.7\ \text{nH/in.}$$

Step 5: Compare the values to the smooth case in Example 5-2. Note that the surface roughness increases the resistance and internal portion of the inductance significantly compared to the identical values for a smooth conductor from Example 5-2.

Smooth conductor	$L_{\text{total}} = 8.42$ nH/in.	$R_{\text{ac}} = 4.86\ \Omega/\text{in.}$
Rough conductor ($h_{\text{RMS}} = 1.8\ \mu\text{m}$)	$L_{\text{total}} = 8.7$ nH/in.	$R_{\text{ac}} = 8.4\ \Omega/\text{in.}$

5.3.2 Hemispherical Model

The rough surface depicted in Figure 5-16b can be characterized as random protrusions sitting on a flat plane, which precludes rigorous derivation of an analytical formula to calculate the extra losses due to current flowing in the tooth structure. Subsequently, for the rough copper, an approximation of the tooth structure is required so that an analytical solution can be derived. As a first approximation, a hemispherical boss sitting on a plane can be used to represent the individual surface protrusions as shown in Figure 5-17 [Hall et al., 2007]. The complete surface is modeled using N hemispheres randomly distributed on a flat plane. A TEM (transverse electromagnetic) wave is assumed incident on the hemisphere at a grazing angle of 90° with respect to the flat plane and with the H field (the magnetic field intensity) tangential to the surrounding plane as shown in Figure 5-17. To find the power dissipated by the structure, the absorption and scattering of the incident TEM wave on the hemisphere must be calculated. The problem of scattering of a plane wave from the hemispherical protrusion on the flat surface can be approximated using superposition. First the power losses of a sphere are calculated and then divided in half because the

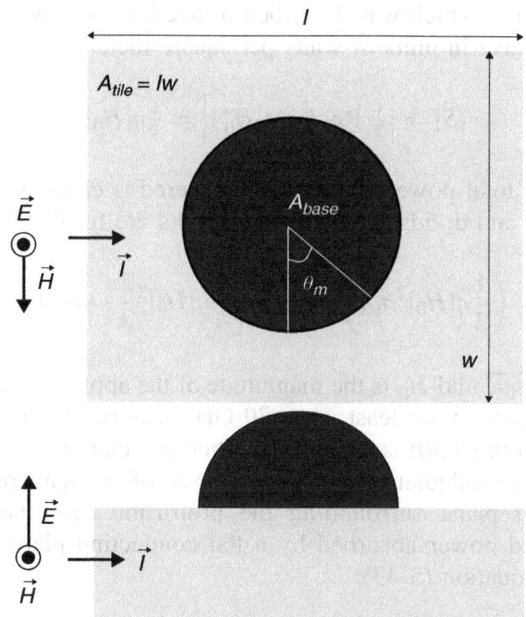

Figure 5-17 Simplified hemispherical shape approximating a single surface protrusion: top and side views; current direction and applied TEM field orientations shown.

structure is a hemisphere, and then the power loss of the flat plane surrounding the hemisphere is calculated. The total power dissipated is then calculated simply by adding the results.

For a good conducting protrusion (as opposed to a perfect electrical conductor protrusion) a plane electromagnetic wave incident on a conducting sphere will be partly scattered and partly absorbed. The total power scattered and absorbed from a sphere divided by the incident flux is known as the *total cross section* and is calculated by Jackson [1999] (with units of square meters) as

$$\sigma_{\text{tot}} = -\frac{\pi}{k^2} \sum_m (2m+1)\ \text{Re}[\alpha(m) + \beta(m)] \tag{5-50}$$

where $k = 2\pi/\lambda$, $\lambda = c/f\sqrt{\varepsilon'}$, c is the speed of light, and the scattering coefficients are approximated assuming that $kr \ll 1$, where r is the sphere radius, and are given by [Jackson, 1999]

$$\alpha(1) = -\frac{2j}{3}(kr)^3 \frac{1 - (\delta/r)(1+j)}{1 + (\delta/2r)(1+j)} \tag{5-51a}$$

$$\beta(1) = -\frac{2j}{3}(kr)^3 \frac{1 - (4j/k^2 r\delta)[1/(1-j)]}{1 + (2j/k^2 r\delta)[1/(1-j)]} \tag{5-51b}$$

The Poynting vector, which was described in Section 2.6, gives the power of an electromagnetic wave in units of watts per square meter:

$$|\vec{S}| = \left|\tfrac{1}{2}\,\mathrm{Re}(\vec{E}_0 \times \vec{H}_0^*)\right| = \tfrac{1}{2}\eta|H_0|^2 \tag{5-52}$$

Subsequently, the total power absorbed or scattered is calculated by multiplying (5-50) and (5-52) and dividing by 2 so the results are for that of a hemisphere:

$$P_{\text{hemisphere}} = \frac{1}{2}\left(\frac{1}{2}\eta|H_0|^2\sigma_{\text{tot}}\right) = -\mathrm{Re}\left[\frac{1}{4}\eta|H_0|^2\frac{3\pi}{k^2}(\alpha(1)+\beta(1))\right] \tag{5-53}$$

where $\eta = \sqrt{\mu_0/\varepsilon_0\varepsilon'}$ and H_0 is the magnitude of the applied magnetic field. Note that reasonable accuracy (at least up to 30 GHz) can be obtained when only the first term ($m = 1$) in (5-50) is considered when calculating (5-53).

Equation (5-53) calculates the power loss of a hemisphere. Now, the losses of the flat plane surrounding the protrusion must be accounted for. The time-averaged power absorbed by a flat conducting plane per unit area is calculated from equation (5-43):

$$\frac{dP_{\text{plane}}}{da} = \frac{\mu_0\,\omega\delta}{4}|H_0|^2 \tag{5-54}$$

To approximate the losses of a single hemispherical boss sitting on a flat plane of finite conductivity, the power loss of the hemisphere is added to the loss of the plane less the base area of the hemisphere:

$$P_{\text{tot}} = \left|-\mathrm{Re}\left[\frac{1}{4}\eta|H_0|^2\frac{3\pi}{2k^2}(\alpha(1)+\beta(1))\right]\right| + \frac{\mu_0\,\omega\delta}{4}|H_0|^2(A_{\text{tile}} - A_{\text{base}}) \tag{5-55}$$

where A_{tile} is the tile area of the plane surrounding the protrusion (see Figure 5-17) and A_{base} is the base area of the hemisphere. Note that (5-55) is an approximation because it assumes that the magnetic field (H_0) on the tile is not affected by the presence of the hemisphere and the loss of the surrounding plane is simply a function of the area.

To gain an intuitive understanding of how a propagating electromagnetic wave behaves in the presence of a protrusion, it is useful to observe the fields and solve for the surface currents on a PEC (perfect electrical conducting) sphere, which is a good approximation of how the current will flow at very high frequencies when the skin depth is small compared to the sphere. In Section 3.2.1 the boundary conditions for a PEC were described; the electric field must emanate from and terminate normal to a perfectly conducting surface, and the magnetic field must be tangential to the conductor surface. First consider Figure 5-18a, which depicts the front cross-sectional view of a hemispherical protrusion sitting on a conducting plane where the current flow is out of the page. Note that the electric fields are drawn perpendicular to the conductor surface and the magnetic fields are tangent

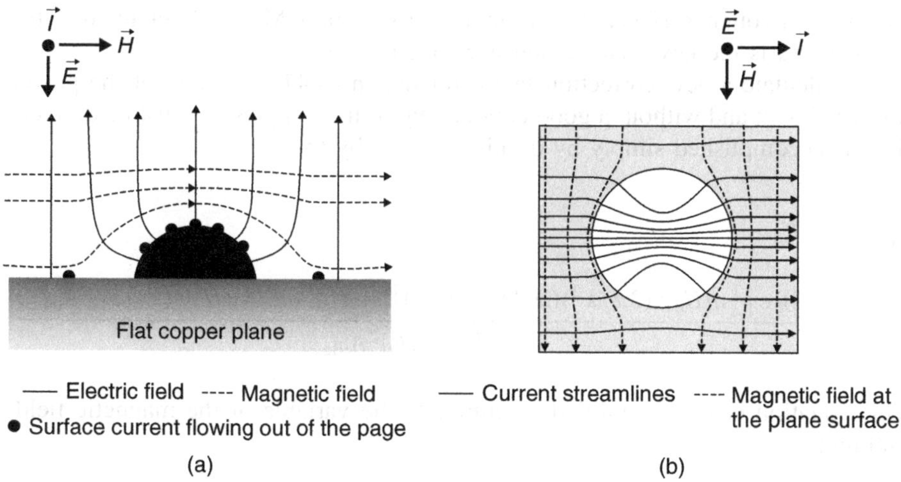

Figure 5-18 (a) Front view of fields and current showing the normal electric field and tangential magnetic field; (b) current streamlines flowing over the top of a single surface protrusion.

to the surface. Figure 5-18b shows the top view of the same protrusion with the current flowing from the left to the right. If TEM is assumed, the magnetic field will be perpendicular to the electric field as described is Section 2.3.2. The magnetic field intensity lines depicted in Figure 5-18b are those near the surface of the plane and not on the sphere. Note that these fields bend around the protrusion because they must satisfy the boundary conditions and remain tangent to the PEC hemisphere. Furthermore, equation (5-44) says that lines of constant magnetic field intensity are orthogonal to the lines of surface current flow, indicating that if the magnetic field bends around the hemisphere, an area of low current density perpendicular to the current flow will be induced. The surface current density on the hemisphere (J_{eff}) (in A/m^2) can be derived by defining the magnetic field in terms of a magnetic scalar potential [Orlando and Delin, 1991; Huray et al., 2007; Huray, 2008]:

$$J_{\text{eff}} = -\text{Re}\left(\tfrac{3}{2} H_0 \sin\theta_m e^{j\omega t}\right) \tag{5-56}$$

where θ_m is the angle between the applied magnetic field and the current flow (see Figure 5-17). If the uniform current streamlines on the plane are matched with those calculated with (5-56), the influence of a hemisphere sitting on a plane can be observed. Figure 5-18b depicts how current will flow in the presence of a spherical protrusion. Note that the current on the flat portion of the plane is drawn toward the protrusion with the highest density on the top and minimal current density on the side perpendicular to the current flow. The current crowding at the top of the protrusion effectively decreases the area where current flows, increases the path length, and thus helps explain the physical mechanisms that cause extra losses on a rough surface. It also appears to support Morgan's claim that the surface area is the key factor to surface roughness losses.

To calculate a new correction factor for use in (5-47), the ratio of the power absorbed with and without a good conducting protrusion present must be derived. This is accomplished simply by dividing (5-55) by (5-54):

$$\frac{P_{\text{tot}}}{P_{\text{plane}}} = K_s$$

$$= \frac{\left|\text{Re}\left[\frac{1}{4}\eta|H_0|^2(3\pi/k^2)(\alpha(1)+\beta(1))\right]\right| + (\mu_0\ \omega\delta/4)|H_0|^2(A_{\text{tile}} - A_{\text{base}})}{\frac{\mu_0\ \omega\delta}{4}|H_0|^2 A_{\text{tile}}}$$

This equation can be simplified to eliminate the variable of the magnetic field, yielding

$$K_s = \frac{\left|\text{Re}\left[\eta(3\pi/4k^2)(\alpha(1)+\beta(1))\right]\right| + (\mu_0\ \omega\delta/4)(A_{\text{tile}} - A_{\text{base}})}{(\mu_0\omega\delta/4)A_{\text{tile}}} \tag{5-57}$$

Note that (5-57) becomes invalid when the skin depth is greater than the surface protrusion height. At these frequencies the power dissipated by a flat plane with an area equal to the base of the protrusion will be greater than the power dissipated by the protrusion. Subsequently, a knee frequency can be defined when $K_s = 1$, where the roughness begins to affect the losses significantly. Below the knee frequency, the correction factor, K_s, is unity. Subsequently, implementation of this correction factor is shown as

$$K_{\text{hemi}} = \begin{cases} 1 & \text{when } K_{\text{s}} \leq 1 \\ K_s & \text{when } K_{\text{s}} > 1 \end{cases} \tag{5-58}$$

To implement (5-58) accurately, the surface shown in Figure 5-16b must somehow be represented by equivalent hemispheres. The additional surface area of simple hemisphere bosses are insufficient to account for the measured surface roughness losses. This is not surprising when one compares the additional hemisphere model area to the 3D surface of the rough copper sample shown in Figure 5-16b. To account for additional surface area the root mean square (RMS) volume of the rough surface must be calculated, and volume equivalent hemispheres are created to determine A_{base}. The RMS distance between peaks in the roughness profile is used to calculate the tile area, A_{tile}. To obtain these input parameters, the surface is measured using a profilometer, as shown in Figure 5-19. To facilitate the volume equivalent model, the tooth shape is approximated as one-half of a prolate spheroid instead of a hemisphere, because it more closely resembles the shape of the protrusion. A prolate spheroid is a surface of revolution obtained by rotating an ellipse about its major axis. A symmetrical egg (i.e., with the same shape at both ends) would approximate a prolate spheroid

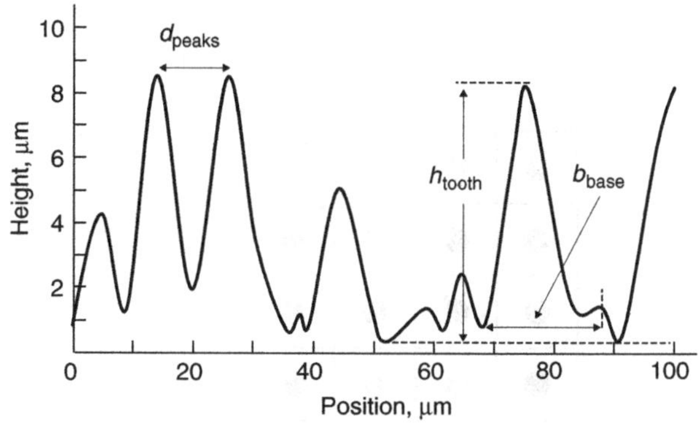

Figure 5-19 Example of a surface profilometer measurement of a rough copper sample showing peak heights that range from 0.7 to 8.5 μm. The flat surface is assumed to be at 0.5 μm.

[Mathworld, n.d.]. The spheroid volume is in turn equal to one-half the volume of a sphere to calculate the radius of a hemisphere with the same volume as the hemispheroid-shaped surface protrusion:

$$r_e = \sqrt[3]{h_{\text{tooth}} \left(\frac{b_{\text{base}}}{2} \right)^2} \tag{5-59}$$

where b_{base} is the tooth base width, h_{tooth} the tooth height, and r_e the radius of a hemisphere with equivalent tooth volume. The base area of the hemisphere, A_{base}, is then calculated:

$$A_{\text{base}} = \pi \left(\frac{b_{\text{base}}}{2} \right)^2 \tag{5-60}$$

The square tile area of the surrounding flat plane is calculated based on the distance between peaks:

$$A_{\text{tile}} = d_{\text{peaks}}^2 \tag{5-61}$$

If the RMS values of d_{peaks}, h_{tooth}, and b_{base} values are calculated, the surface shown in Figure 5-16b and measured in Figure 5-19 can be represented as the equivalent surface in Figure 5-20.

A comparison between the correction factor calculated from the Hammerstad model (5-48) and the hemispherical model (5-58) with the modified equivalent volume is shown in Figure 5-21. Note that the hemisphere model saturates at a much higher value than Hammerstad, which will always saturate at a value of 2. The implementation shown in (5-58) causes a nonphysical discontinuity at the frequency where the model transitions from dc to ac behavior, which is the point where the surface of the flat plane with an area equal to the base of the hemisphere has more loss than the hemisphere. Some engineers may be concerned that the discontinuity may induce nonphysical glitches into the time-domain responses. However, simulations of pulses as fast as 30 Gb/s were

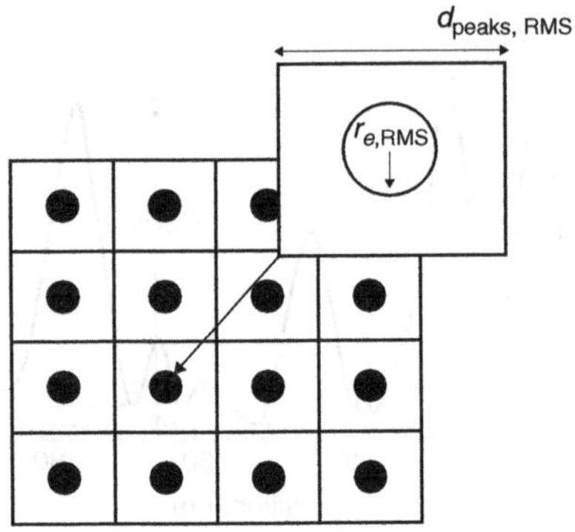

Figure 5-20 Equivalent surface represented by hemispheres with the same RMS volume as that of the measured surface profile.

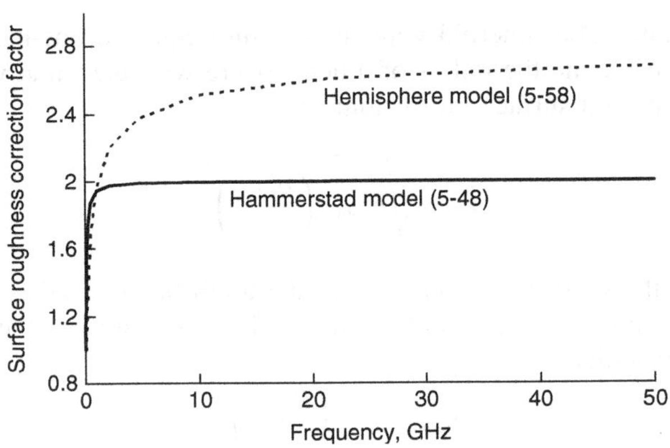

Figure 5-21 Hammerstad correction factor (5-48) compared to the hemisphere model (5-58). RMS roughness: $h_{\text{RMS}} = 5.8\ \mu\text{m}$, $d_{\text{peaks,RMS}} = 9.4\ \mu\text{m}$.

performed using this method in HSPICE and Nexxim with no apparent nonphysical aberrations observed.

The frequency dependence of the skin effect resistance and total inductance using the hemisphere correction for surface roughness is implemented with (5-62a) and (5-62b).† When the skin depth δ is larger than the conductor thickness (which includes the roughness profile), the dc value of the resistance and low-frequency inductance where the skin depth is equal to the total conductor thickness should be used:

†Just as with equation (5.49b), the method shown here for calculating the internal portion of the inductance ($L_{\text{internal}} = R_{\text{ac}}/\omega$) for a rough conductor is an approximation based on the result for a smooth conductor. The approximation will induce causality errors that tend to be small enough to ignore, so this method is generally acceptable. Appendix E derives the internal inductance using a more rigorous approach based on the discussion in Chapter 8.

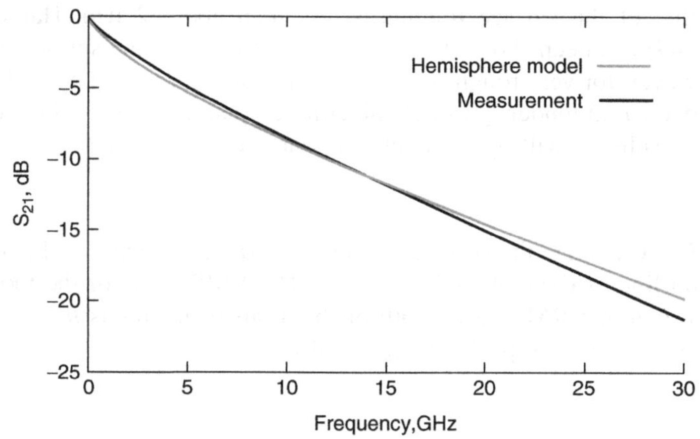

Figure 5-22 Accuracy of the hemisphere surface roughness model (5-58); 7 in. microstrip; $\varepsilon_r/\tan\delta = 3.9/0.0073$ at 1 GHz; RMS roughness of copper foil $h_{\text{RMS}} = 5.8$ μm, $d_{\text{peaks,RMS}} = 9.4$ μm. (Adapted from Hall et al. [2007].)

$$R_{\text{hemi}}(f) = \begin{cases} K_{\text{hemi}} R_s \sqrt{f} & \text{when } \delta < t \\ R_{\text{dc}} & \text{when } \delta \geq t \end{cases} \tag{5-62a}$$

$$L_{\text{hemi}}(f) = \begin{cases} L_{\text{external}} + \dfrac{R_{\text{hemi}}(f)}{2\pi f} & \text{when } \delta < t \\ L_{\text{external}} + \dfrac{R_{\text{hemi}}(f_{\delta=t})}{2\pi f_{\delta=t}} & \text{when } \delta \geq t \end{cases} \tag{5-62b}$$

where K_{hemi} is calculated with (5-58), t is the conductor thickness, δ is the skin depth, and $f_{\delta=t}$ is the frequency where the skin depth equals the thickness of the conductor.

Figure 5-22 depicts the accuracy of the hemisphere model for very rough copper. The small deviation that occurs between the model and the measurement in Figure 5-22 is because (5-57) essentially uses superposition to combine the losses of the tile and protrusion and does not account for the interaction between the hemisphere and the plane. Consequently, the formula is very accurate (1) at low frequencies where the skin depth is large compared to the protrusion and the loss is due primarily to the plane, and (2) at high frequencies where the skin depth is small compared to the protrusion and the loss is due primarily to the roughness. At intermediate frequencies, where the skin depth is on the same order as the roughness height, an error is introduced. Additionally, the roughness shape was approximated with a spheroid and the interactions between spheres were neglected. Nonetheless, for very rough copper surfaces, Figure 5-22 shows that the methodology produces very reasonable accuracy over a wide bandwidth with a simple, easy-to-use formula. More important, it provides valuable intuition into the mechanisms that cause surface roughness losses.

Note that the method of calculating the surface roughness correction factor K should be chosen carefully. If relatively smooth copper is being used, with a RMS value of the surface roughness less than about 2 μm, Hammerstad's formula (5-48) has been shown to adequately approximate the surface roughness losses. However, for very rough copper (which is often preferred by PCB vendors due to its decreased tendency of delaminating), equation (5-58) using equivalent volume hemispheres will approximate the surface roughness losses with more accuracy.

Example 5-4 Calculate the surface roughness correction factor of the microstrip transmission line in Example 5-2 at 5 GHz. The RMS value of the tooth height $h_{\text{tooth}} = 5.8\ \mu\text{m}$, the RMS base width of the tooth structures is $b_{\text{base}} = 9.4\ \mu\text{m}$, and the distance between peaks $d_{\text{peaks}} = 9.4\ \mu\text{m}$.

SOLUTION

Step 1: Use equations (5-59) through (5-61) to calculate a sphere with the same volume as a spheroid-shaped surface protrusion:

$$r_e = \sqrt[3]{h_{\text{tooth}}\left(\frac{b_{\text{base}}}{2}\right)^2} = 5\ \mu\text{m}$$

$$A_{\text{base}} = \pi\left(\frac{b_{\text{base}}}{2}\right)^2 = 69.4\ \mu\text{m}^2$$

$$A_{\text{tile}} = d_{\text{peaks}}^2 = 88.4\ \mu\text{m}^2$$

Note that the diameter of the equivalent volume hemisphere (10 μm) is actually larger than the edge length of the tile (9.4 μm), meaning that the hemispheres will overlap. This is not a concern since it is understood that the surface roughness shape is not hemispherical in nature and the shape of the protrusion is assumed to be a spheroid. However, the equivalent volume hemisphere allows for a much simpler solution to the electromagnetic fields without sacrificing much accuracy.

Step 2: Use (5-57) to calculate the correction factor at 5 GHz. To calculate the intrinsic impedance $\eta = \sqrt{\mu_0/\varepsilon_0\varepsilon'}$, the value of the dielectric permittivity directly under the rough surface should be used and not the effective value for the microstrip as calculated in Example 5-2.

$$\eta = \sqrt{\frac{\mu_0}{\varepsilon_0\varepsilon'}} = \frac{377}{\sqrt{\varepsilon'}} = \frac{377}{\sqrt{4}} = 188.5\ \Omega$$

$$\delta(5\ \text{GHz}) = \sqrt{\frac{2}{2\pi f \mu_0 \sigma}} \approx 0.935\ \mu\text{m}$$

$$\lambda = \frac{c}{f\sqrt{\varepsilon'}} = \frac{3\times 10^8}{5\times 10^9\sqrt{4}} = 0.03 \text{ m}$$

$$k = \frac{2\pi}{\lambda} = 209.3 \text{ m}^{-1}$$

$$\alpha(1) = -\frac{2j}{3}(kr_e)^3 \frac{1-(\delta/r_e)(1+j)}{1+(\delta/2r_e)(1+j)}$$

$$= -\frac{2j}{3}[209.3(5\times 10^{-6})]^3 \left[\frac{1-\dfrac{0.935\times 10^{-6}}{5\times 10^{-6}}(1+j)}{1+\dfrac{0.935\times 10^{-6}}{2\cdot 5\times 10^{-6}}(1+j)}\right]$$

$$= -j0.764\times 10^{-9}\left[\frac{1-0.187(1+j)}{1+0.0935(1+j)}\right]$$

$$= -1.78\times 10^{-10} - j5.53\times 10^{-10}$$

For this example, it can be shown that $\beta(1)$ is negligible, so it is ignored.

$$K_s = \frac{\left|\mathrm{Re}\left[\eta(3\pi/4k^2)\alpha(1)\right]\right| + (\mu_0\omega\delta/4)(A_{\text{tile}} - A_{\text{base}})}{(\mu_0\omega\delta/4)A_{\text{tile}}}$$

$$= \frac{1.8\times 10^{-12} + 1.75\times 10^{-13}}{8.15\times 10^{-13}} = 2.42$$

In this case, the series resistance of a transmission line manufactured with a copper conductor with this roughness profile would be approximately 2.42 times as high at 5 GHz as the same transmission line constructed with a smooth conductor.

5.3.3 Huray Model

In 2006 at the University of South Carolina, Paul G. Huray was researching new wideband modeling techniques for surface roughness that would provide better accuracy than both the Hammerstad and hemisphere models. Upon observation of scanning electron microscope (SEM) photographs of copper foil samples used to manufacture printed circuit boards (PCBs), he observed that the structures appeared to be constructed of conducting "snowballs," as shown in Figure 5-23. Subsequently, he formed a material and physical basis of a theoretical model that is composed of a distribution of spherical shapes [Olufemi, 2007; Hurray, 2009].

Printed circuit boards are a "stackup" of layers of copper conductors and intervening layers of an insulating propagating medium such as FR4 joined under heat and pressure. To assure that the copper sheets do not delaminate from the dielectric layers, manufacturers typically electrodeposit an additional surface layer of copper on a relatively smooth copper foil that creates irregular features as large as 11 μm, to promote good adhesion. The electroplated copper produces a surface

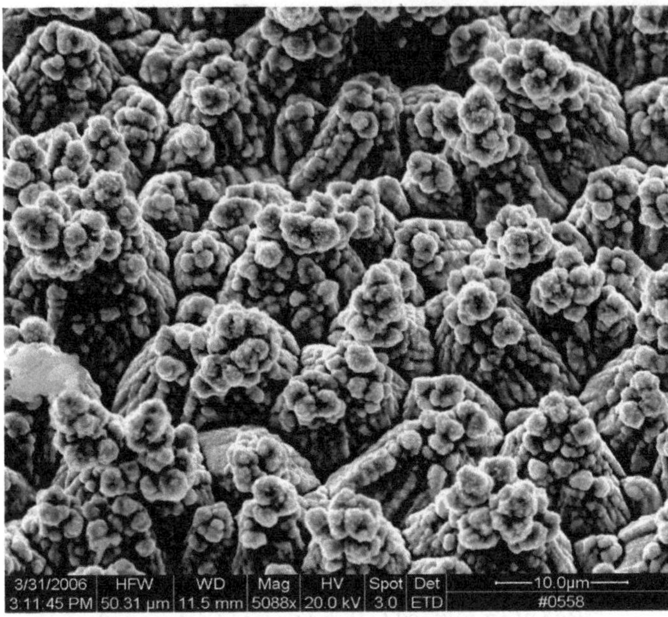

Figure 5-23 SEM photograph of rough copper at 5000× magnification at a 30° angle.

roughness profile that can be described as spherical particles, joined together in a network to form a distorted surface, as shown in Figure 5-24. An individual snowball is located a distance x_i below a flat copper surface, has radius a_i, and, when it experiences an external electromagnetic field intensity, as it would in the region below the copper trace when a signal is propagating on a transmission line, it will dissipate power similar to the hemisphere described in equation

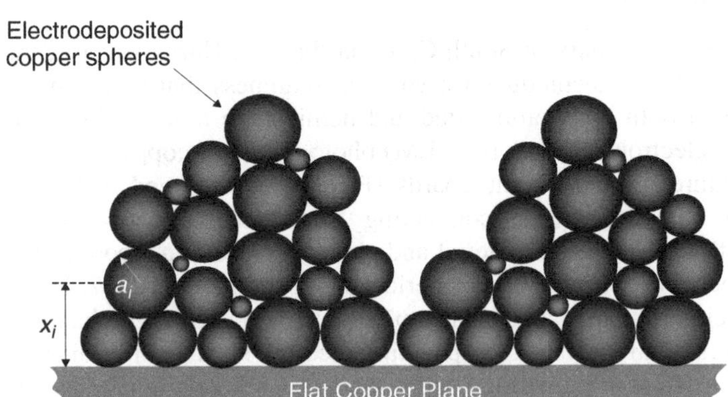

Figure 5-24 Cross section of a distribution of copper spheres that create a 3D rough surface in the form of copper "pyramids" on a flat conductor.

(5-53) with twice the magnitude since it is a whole sphere:

$$P_{\text{sphere}} = \frac{1}{2}\eta|H_0|^2\sigma_{\text{tot}} = -\text{Re}\left[\frac{1}{2}\eta|H_0|^2\frac{3\pi}{k^2}(\alpha(1)+\beta(1))\right] \tag{5-63}$$

For a distribution of copper spheres such as the pyramidal configuration that creates 3D copper roughness profiles shown in Figures 5-23 and 5-24, superposition of the sphere losses can be used to calculate the total losses of the structure [Olufemi, 2007; Huray, 2009]. Since roughness losses are proportional to the surface area of the tooth structure, the number and size of the spheres can be chosen to approximate the surface roughness correction factor very accurately. To do this, the general tooth shape must be approximated, the average snowball size must be measured, and the number of spheres that correspond to the surface area of the protrusion must be calculated. If SEM photographs are analyzed extensively, it may be possible to build a model with multiple snowball sizes that very closely replicated the actual tooth shape. However, since electron microscopes are typically not available, profilometers will provide approximate tooth shapes, and a reasonable model can be constructed.

The total power dissipated by the tooth structure is simply the sum of the power dissipated from the total number of spheres (N) required to replicate the surface area of the tooth structure:

$$P_{N,\text{spheres}} = -\sum_{n=1}^{N}\text{Re}\left[\frac{1}{2}\eta|H_0|^2\frac{3\pi}{k^2}(\alpha(1)+\beta(1))\right]_n \tag{5-64}$$

where $\alpha(1)$ and $\beta(1)$ are calculated with (5-51a) and (5-51b), $\eta = \sqrt{\mu_0/\varepsilon_0\varepsilon'}$, and H_0 is the magnitude of the applied magnetic field.

To calculate a new surface roughness correction factor for use in (5-47), the ratio of the power absorbed with and without surface roughness must be derived using (5-65) and (5-54):

$$K_{\text{Huray}} = \frac{P_{\text{flat}} + P_{N,\text{spheres}}}{P_{\text{flat}}} = \frac{(\mu_0\omega\delta/4)A_{\text{tile}} + \left|\sum_{n=1}^{N}\text{Re}\left[\frac{1}{2}\eta(3\pi/k^2)(\alpha(1)+\beta(1))\right]_n\right|}{(\mu_0\omega\delta/4)A_{\text{tile}}} \tag{5-65}$$

To calculate the total number of spheres (N) needed to represent the surface area of the roughness profile, it is useful to choose a geometric shape that resembles a typical tooth. If one-half of a spheroid (a hemispheroid) is chosen as the approximate tooth shape, laboratory data indicate that reasonable accuracy can be obtained. However, the SEM photograph in Figure 5-23 clearly shows that a spheroid is only an approximation because the real tooth shape has more surface area. Consequently, this assumption is expected to slightly underpredict the surface roughness losses at high frequencies. A detailed statistical analysis of the

SEM photograph could provide a more accurate representation of the specific size and number of spheres; however, the spheroid assumption produces reasonable results compared to measured surfaces.

To construct a Huray surface roughness model, the surface is measured as in Figure 5-19 and the RMS values of b_{base} and h_{tooth} are calculated. Next, the surface area of a hemispheroid is calculated with a base b_{base} and a height h_{tooth}. An appropriate snowball radius is chosen (usually between 0.5 and 1 μm), and the total number of copper spheres (N) are calculated so that the total surface area of all N spheres is equal to that of the hemispheroid constructed using the RMS values measured with the profilometer. The procedure is demonstrated with the following example.

Example 5-5 Calculate the surface roughness correction factor using the Huray equation at 5 GHz. The RMS value of the tooth height $h_{\text{tooth}} = 5.8$ μm, the RMS base width of the tooth structures $b_{\text{base}} = 9.4$ μm, and the distance between peaks $d_{\text{peaks}} = 9.4$ μm.

SOLUTION

Step 1: Calculate the number of spheres required. If profiliometer measurements indicate that the RMS value of the tooth height $h_{\text{tooth}} = 5.8$ μm with an RMS spacing between protrusions of $b_{\text{base}} = 9.4$ μm and the tooth shape is approximated as a hemispheroid, the number of spheres can be calculated. The lateral surface area of a hemispheroid is given by

$$A_{\text{lat}} = \pi \frac{b_{\text{base}}}{2}\left[h_{\text{tooth}} \frac{\arcsin\left(\sqrt{1-(b_{\text{base}}/2)^2/(h_{\text{tooth}})^2}\right)}{\sqrt{1-\frac{(b_{\text{base}}/2)^2}{(h_{\text{tooth}})^2}}} + \frac{b_{\text{base}}}{2}\right]$$

Plugging in the values for b_{base} and h_{tooth} yields the surface area:

$$A_{\text{lat}} = 161\ \mu\text{m}^2$$

Assuming a 0.8-μm sphere radius, the surface area of a single sphere is calculated:

$$A_{\text{sphere}} = 4\pi a^2 = 8\ \mu\text{m}^2$$

The number of spheres that would have the same surface area as the hemispheroid is

$$\frac{A_{\text{lat}}}{A_{\text{sphere}}} = \frac{161}{8} = 20.125$$

Step 2: Calculate the correction factor at 5 GHz. Since the peak–peak distance is 9.4 μm, $A_{\text{tile}} = (9.4\ \mu\text{m})^2$ is chosen for the tile area. However, any tile size could be used as long as the total number of spheres that equals the surface area of the profile within the tile size is calculated. The sphere radius $a = 0.8$ μm. From Example 5-4, $k = 209.3\ \text{m}^{-1}$, $\eta = \sqrt{\mu_0/\varepsilon_0\varepsilon'} = 377/\sqrt{\varepsilon'} = 377/\sqrt{4} = 188.5\ \Omega$, and $\delta(5\ \text{GHz}) = \sqrt{2/2\pi f \mu_0 \sigma} \approx 0.935$ μm.

The scattering coefficient is calculated from equation (5-51a). For this example, it can be shown that $\beta(1)$ is negligible, so it is ignored.

$$\alpha(1) = -\frac{2j}{3}(ka)^3 \frac{1 - (\delta/a)(1+j)}{1 + (\delta/2a)(1+j)}$$

$$= -\frac{2j}{3}[209.3(0.8 \times 10^{-6})]^3 \left[\frac{1 - \dfrac{0.935 \times 10^{-6}}{0.8 \times 10^{-6}}(1+j)}{1 + \dfrac{0.935 \times 10^{-6}}{2 \cdot 0.8 \times 10^{-6}}(1+j)} \right]$$

$$= -1.93 \times 10^{-12} - j1.05 \times 10^{-12}$$

The Huray surface roughness correction factor is calculated with (5-66) at 5 GHz:

$$K_{\text{Huray}} = \frac{P_{\text{flat}} + P_{N\_\text{spheres}}}{P_{\text{flat}}}$$

$$= \frac{(\mu_0 \omega \delta/4) A_{\text{tile}} + \sum_{n=1}^{N} \text{Re}\left[\frac{1}{2}\eta(3\pi/k^2)(\alpha(1) + \beta(1))\right]_n}{(\mu_0 \omega \delta/4) A_{\text{tile}}}$$

$$= \frac{(\mu_0 \omega \delta/4) A_{\text{tile}} + 20\ \text{Re}\left[\frac{1}{2}\eta(3\pi/k^2)\alpha(1)\right]}{(\mu_0 \omega \delta/4) A_{\text{tile}}}$$

$$= \frac{8.15 \times 10^{-13} + 7.8 \times 10^{-13}}{8.15 \times 10^{-13}} = 1.95$$

Therefore, the Huray model predicts that the series resistance of a transmission line manufactured according to the roughness profile defined above would be approximately 1.95 times higher than the same transmission line constructed with a smooth conductor at a frequency of 5 GHz.

Figure 5-25 shows a comparison of the insertion losses of a transmission line constructed with rough copper modeled using the Huray equation compared to measured results. Note that the Huray model correctly predicts the shape of the insertion loss curve with less than 1.5 dB of error at 30 GHz. Simulations show that if 23 spheres are used in this example, the model fits the measured results almost exactly [Olufemi, 2007]. However, aside from detailed statistical analysis of the SEM photographs, there is no known deterministic methodology to arrive at the perfect fit. The methodology presented allows the use of a profilometer and produces reasonable wideband results.

Figure 5-26 shows the Huray equation (5-65) constructed with twenty 0.8-μm spheres as calculated in Example 5-5, compared to the Hammerstad and hemisphere models for the roughness profile assumed in Examples 5-4 and 5-5. Note that the Hammerstad equation saturates at 2, which does not provide enough loss

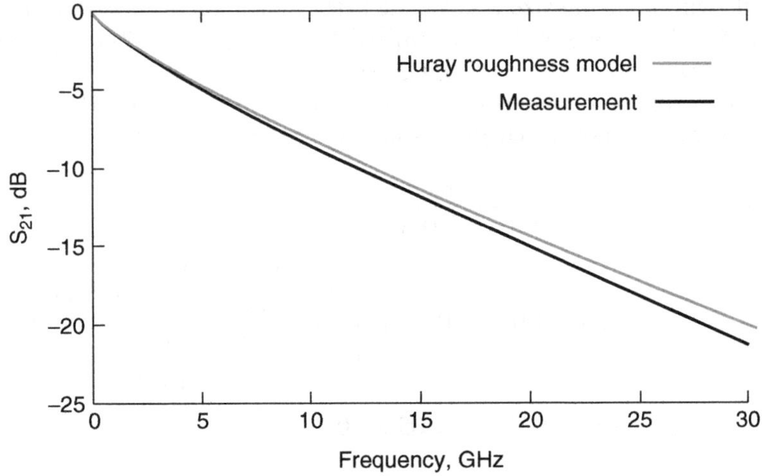

Figure 5-25 Accuracy of the Huray surface roughness model (5-65) constructed with $N = 20$ spheres with a radius of $a = 0.8\ \mu\text{m}$, assuming a hemispheroid tooth shape; 7-in. microstrip; $\varepsilon_r / \tan\delta = 3.9/0.0073$ at 1 GHz; RMS roughness of copper foil $h_{\text{RMS}} = 5.8\ \mu\text{m}$, $d_{\text{peaks,RMS}} = 9.4\ \mu\text{m}$.

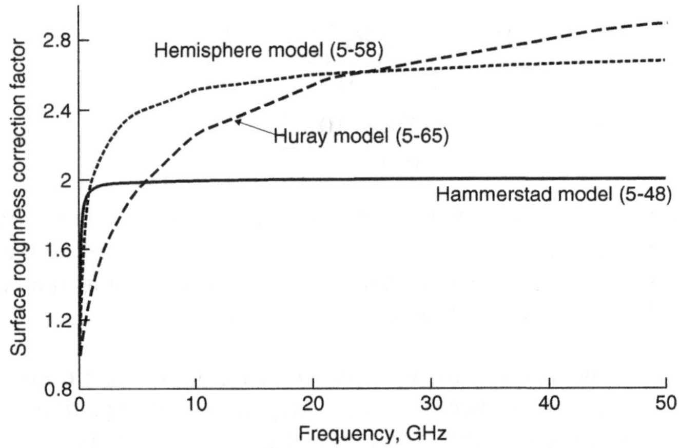

Figure 5-26 Surface roughness correction factors for rough copper; Huray model (5-65), hemisphere model (5-58), and Hammerstad model (5-48); RMS roughness: $h_{\text{RMS}} = 5.8\ \mu\text{m}$, $d_{\text{peaks,RMS}} = 9.4\ \mu\text{m}$.

for rough copper. The hemisphere model overpredicts at middle frequencies and slightly underpredicts at high frequencies. As demonstrated by Figure 5-25, a properly constructed Huray model predicts a realistic correction curve.

The frequency dependence of the skin effect resistance and total inductance using the Huray equation for surface roughness is implemented with (5-66a) and

(5-66b).* When the skin depth (δ) is larger than the conductor thickness, the dc value of the resistance and low-frequency inductance where the skin depth is equal to the conductor thickness should be used:

$$R_{\text{Huray}}(f) = \begin{cases} K_{\text{Huray}} R_s \sqrt{f} & \text{when } \delta < t \\ R_{\text{dc}} & \text{when } \delta \geq t \end{cases} \tag{5-66a}$$

$$L_{\text{Huray}}(f) = \begin{cases} L_{\text{external}} + \dfrac{R_{\text{Huray}}(f)}{2\pi f} & \text{when } \delta < t \\ L_{\text{external}} + \dfrac{R_{\text{Huray}}(f_{\delta=t})}{2\pi f_{\delta=t}} & \text{when } \delta \geq t \end{cases} \tag{5-66b}$$

where K_{Huray} is calculated with (5-65), t is the conductor thickness, δ is the skin depth, and $f_{\delta=t}$ is the frequency where the skin depth equals the thickness of the conductor.

5.3.4 Conclusions

For rough copper, Figure 5-14b shows that the Hammerstad model slightly overpredicts surface roughness losses at low frequencies and significantly underpredicts losses at high frequencies. However, Figure 5-14a shows that the Hammerstad method works well for copper profiles that are relatively smooth with a corrugated surface profile. The Hammerstad model should be used only for RMS copper roughness profiles less than about 2 μm.

The hemisphere model is an improvement over Hammerstad for rough copper but still overpredicts the losses at middle frequencies and underpredicts the losses at high frequencies, which causes the "belly" in the simulated results in Figure 5-22. The major benefit gained by studying the hemisphere model is a physical understanding of how the fields and surface currents behave in the presence of a protrusion. The hemisphere is the simplest geometry that can be solved analytically to predict the impact of a rough surface compared to a smooth surface. Furthermore, the hemisphere model is not appropriate for smoother copper profiles that are corrugated in nature. The hemisphere model is valid only for copper profiles that can be characterized by a distribution of distinct protrusions, such as the surfaces shown in Figures 5-16b and 5-23.

The Huray approach is the most diverse and accurate modeling methodology for surface roughness. It can be used for any type of copper as long as a detailed image of the surface profile can be obtained with a profilometer or a scanning electron microscope. The surface area of the roughness profile prior to lamination must be determined, and the number of appropriately sized spheres less than 1 μm

*Just as with equations (5.49b) and (5.62b), the method shown here for calculating the internal portion of the inductance ($L_{\text{internal}} = R_{\text{ac}}/\omega$) for a rough conductor is an approximation based on the result for a smooth conductor. The approximation will induce causality errors that tend to be small enough to ignore, so this method is generally acceptable. Appendix E derives the internal inductance using a more rigorous approach based on the discussion in Chapter 8.

in radius must be calculated so that the total surface area is equal to that of the roughness profile.

5.4 TRANSMISSION-LINE PARAMETERS FOR NONIDEAL CONDUCTORS

As bus data rates increase and physical implementations of high-speed digital designs shrink, the transmission-line losses become more important. Consequently, the engineer must have the ability to calculate the response of a transmission line successfully and account for realistic conductor behavior. The next two sections will (1) describe how to include ac resistance and internal inductance in an equivalent circuit, and (2) modify telegrapher's equations to comprehend realistic conductors.

5.4.1 Equivalent Circuit, Impedance, and Propagation Constant

In deriving the equivalent circuit for a transmission line in Section 3.2.3, which is shown in Figure 3-9, the conductor was considered to be infinitely conductive, meaning that all the current flows only on the surface because the skin depth $\delta = 0$, as shown by taking the limit of (5-10):

$$\lim_{\sigma \to \infty}[\delta] = \lim_{\sigma \to \infty}\left[\sqrt{\frac{2}{\omega\mu\sigma}}\right] = 0$$

Furthermore, the assumption of perfect conductors did not allow the calculation of a resistive term or an internal inductance term because there was no field penetration into the conductor. As described in Section 5.1.2, physical conductors manufactured with metals of finite (although good) conductivity behave very similar to perfect conductors except for a small transition region where internal currents exist that are mostly confined to a few skin depths. As described in Sections 5.2.2 and 5.2.3, the skin effect leads directly to frequency-dependent resistance and internal inductance terms that must be comprehended in the equivalent circuit.

Fortunately, the form of the equivalent circuit derived in Section 3.3 is also applicable to a line whose conductors have finite conductivity. To begin this derivation, the series impedance of an ideal transmission line with infinite conductivity is calculated in units of ohms.

$$\vec{Z}_{\text{external}} = j\omega L_{\text{external}} \tag{5-67}$$

The idealized parameters must be modified to include the surface impedance (also known as the internal impedance):

$$\vec{Z}_s = R_{\text{ac}} + j\omega L_{\text{internal}} \tag{5-29}$$

The total inductance was calculated in (5-20):

$$L_{\text{total}} = L_{\text{internal}} + L_{\text{external}} \tag{5-20}$$

indicating that the series impedance contribution from the inductance is simply $j\omega(L_{\text{internal}} + L_{\text{external}})$. To calculate the total series impedance of a transmission-line segment, the resistive component is added:

$$\begin{aligned} Z_{\text{series}} &= R_{\text{ac}} + j\omega(L_{\text{internal}} + L_{\text{external}}) \\ &= R_{\text{ac}} + j\omega L_{\text{total}} \qquad \text{ohms} \end{aligned} \tag{5-68}$$

leading to the equivalent circuit for a transmission line with a conductor of finite conductivity, as shown in Figure 5-27, where N_s is the number of segments, $C_{\Delta z} = \Delta z\, C$ and $L_{\Delta z} = \Delta z\, L_{\text{total}}$, as calculated in Section 3.2.3, and $R_{\Delta z} = \Delta z\, R_{\text{ac}}$, where Δz is the length of the differential section of transmission line and C, L_{total}, and R_{ac} are the capacitance, inductance, and resistance per unit length.

The characteristic impedance, which was defined in equation (3-33), can be calculated by dividing the series impedance as defined by (5-68) by the parallel admittance of the capacitance, $Y_{\text{shunt}} = j\omega C$, for a short section of transmission line of length Δz.

$$Z_0 = \sqrt{\frac{Z_{\text{series}}}{Y_{\text{shunt}}}} = \sqrt{\frac{R_{\text{AC}} + j\omega L_{\text{total}}}{j\omega C}} \qquad \text{ohms} \tag{5-69}$$

Note that the units in (5-69) are $\sqrt{\text{ohms}/(1/\text{ohms})} = \sqrt{(\text{ohms})^2} = \text{ohms}$.

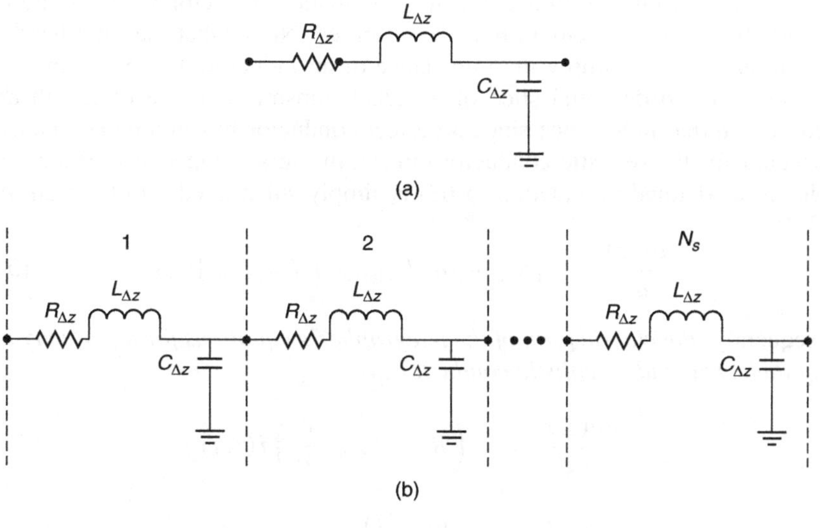

Figure 5-27 (a) Model for a differential element of a transmission line; (b) full model.

The propagation constant can be derived by inserting the complex values of the series impedance and shunt admittance into the loss-free formula that was derived in Section 3.2.4, which takes the form

$$\gamma = \alpha + j\beta = 0 + j\omega\sqrt{LC} = \sqrt{(j\omega L)(j\omega C)} = \sqrt{Z_{\text{lossless}} Y_{\text{lossless}}}$$

Substitution of Z_{series} and Y_{shunt} in place of the loss-free values of the series impedance and the parallel admittance yields the propagation constant for a transmission line with an ideal dielectric and a conductor with a finite conductivity, as shown in

$$\gamma = \alpha + j\beta = \sqrt{Z_{\text{series}} Y_{\text{shunt}}} = \sqrt{(R_{\text{AC}} + j\omega L_{\text{total}})j\omega C} \qquad (5\text{-}70)$$

Equations (5-69) and (5-70) account for realistic conductor effects, such as skin effect resistance, internal inductance, and surface roughness, as outlined throughout this chapter.

5.4.2 Telegrapher's Equations for a Real Conductor and a Perfect Dielectric

The time-harmonic forms of the telegrapher's equations for a transmission line with a perfect dielectric and perfect conductor were shown in equations (3-25) and (3-26):

$$\frac{dv(z)}{dz} = -j\omega L i(z) \qquad (3\text{-}25)$$

$$\frac{di(z)}{dz} = -j\omega C v(z) \qquad (3\text{-}26)$$

To adjust these formulas to account for a realistic conductor with a finite conductivity, the equivalent circuit must be observed. Notice that the right-hand side of equation (3-25) is simply the impedance of an inductor. Therefore, the series impedance of a differential slice of an ideal transmission line of length dz is based on external inductance since a perfect conductor has infinite conductivity. To account for the realistic conductor effects, the series impedance of a realistic conductor as defined in equation (5-68) is simply substituted into the right hand of (3-25):

$$\frac{dv(z)}{dz} = -[R_{\text{ac}} + j\omega(L_{\text{internal}} + L_{\text{external}})]i(z) \qquad (5\text{-}71)$$

Consequently, *the classic form of the telegrapher's equations for a perfectly insulating dielectric and a realistic conductor are*

$$\frac{\partial v(z,t)}{\partial z} = -\left(R_{\text{ac}} + L_{\text{total}}\frac{\partial}{\partial t}\right) i(z,t) \qquad (5\text{-}72\text{a})$$

$$\frac{\partial i(z,t)}{\partial z} = -C\frac{\partial v(z,t)}{\partial t} \qquad (5\text{-}72\text{b})$$

REFERENCES

Brist, G., S. Hall, S. Clouser, and T. Liang, 2005, Non-classical conductor losses due to copper foil roughness and treatment, *CircuiTree*, May.

Collins, Robert, 1992, *Foundations for Microwave Engineering*, McGraw-Hill, New York.

Hall, S., G. Hall, and J. McCall, 2000, *High-Speed Digital System Design*, Wiley, New York.

Hall, Stephen, Steven G. Pytel, Paul G. Huray, Daniel Hua, Anusha Moonshiram, Gary A. Brist, and Edin Sijercic, 2007, Multi-GHz causal transmission line modeling using a 3-D hemispherical surface roughness approach, *IEEE Transactions on Microwave Theory and Techniques*, vol. 55, no. 12, Dec.

Hammerstad, E., and O. Jensen, 1980, Accurate models for microstrip computer-aided design, *IEEE MTT-S International Microwave Symposium Digest*, May, pp. 407–409.

Huray, Paul, 2008, *Foundations of Signal Integrity*, Wiley, Hoboken, NJ.

Huray, P. G., S. Hall, S. G. Pytel, F. Oluwafemi, R. Mellitz, D. Hua, and P. Ye, 2007, Fundamentals of a 3D "snowball" model for surface roughness power losses, *Proceedings of the IEEE Conference on Signals and Propagation on Interconnects*, Genoa, Italy, May 14.

Jackson, J. D., 1999, *Classical Electrodynamics*, 3rd ed., Wiley, New York.

Mathworld, n.d., http://mathworld.wolfram.com/ProlateSpheroid.html.

Olufemi, Oluwafemi, 2007, *Surface Roughness and Its Impact on System Power Losses*, ProQuest, Ann Arbor, MI.

Orlando, Terry P., and Kevin A. Delin, 1991, *Foundations of Applied Superconductivity*, Addison-Wesley, Reading, MA.

Pytel, Steven G., 2007, Multi-gigabit data signaling rates for PWBs including dielectric losses and effects of surface roughness, Ph.D. dissertation, Department of Electrical Engineering, University of South Carolina.

PROBLEMS

5-1 Draw the resistance, inductance, and impedance curves from 0 to 10 GHz for a microstrip transmission line constructed with copper that has the parameters $w = 5$ mils, $h = 3$ mils, $t = 2.1$ mils, $\varepsilon_r = 4.1$, $b_{\text{base}} = 10.2\ \mu\text{m}$, $h_{\text{tooth}} = 4.5\ \mu\text{m}$, $d_{\text{peaks}} = 12\ \mu\text{m}$, and $\sigma = 5.8 \times 10^8$ S/m. Assume a general tooth shape that is hemispheroid in nature. Create the model with the hemisphere model and the Huray model. Discuss the physical mechanisms that cause the differences between the curves.

5-2 Create an equivalent-circuit model valid at 15 GHz of the transmission lines defined in Problem 5-1.

5-3 For the transmission-line description in Problem 5-1, under what conditions is the Hammerstad approximation valid?

5-4 For the transmission line defined in Problem 5-1, when do ac losses overtake dc losses?

5-5 What causes internal inductance? Is internal inductance larger or smaller for a thick conductor line versus a thin conductor? What is the relationship between inductance and resistance? In physical terms, explain why the relationship must hold.

5-6 For a stripline transmission line with the parameters $w = 5$ mils, $h_1 = h_2 = 6$ mils (refer to Figure 5-8), $t = 0.7$ mil, $\varepsilon_r = 4.1$, $b_{\text{base}} = 10.2\ \mu\text{m}$, $h_{\text{tooth}} = 0.9\ \mu\text{m}$, and $d_{\text{peaks}} = 15\ \mu\text{m}$, calculate the resistance and inductance at 1 GHz.

5-7 For the transmission-line models defined in Problem 5-1, when do the surface roughness losses become significant?

5-8 For a periodic roughness pattern such as that shown in Figure 5-20, how will the current flow between and on top of the surface protrusions? Assuming TEM fields, draw the magnetic and electric fields and the associated current. What is significant about the periodic pattern? What assumptions break down for a periodic pattern?

5-9 For a 10-in. transmission line with the parameters $w = 5$ mils, $h = 3$ mils, $t = 2.1$ mils, $\varepsilon_r = 4.1$, and $\sigma = 5.8 \times 10^8$ S/m, sourced with 5 mA of current, what is the total power per unit area dissipated by the reference plane?

5-10 For the transmission line described in Problem 5-1, how much error would be introduced by neglecting surface roughness at 10 GHz?

5-11 For the transmission line described in Problem 5-1, how much surface roughness would be required to double the power loss of the transmission line?

5-12 For a plane wave impinging onto an infinitely thick slab of a good-conducting medium, draw the electric field outside the conductor and the magnetic fields inside the conductor. Compare and contrast this to the boundary conditions for a perfect conductor.

6

ELECTRICAL PROPERTIES OF DIELECTRICS

Advanced Signal Integrity for High-Speed Digital Design, By Stephen H. Hall and Howard L. Heck

As the speed of digital systems continues to increase with Moore's law, the electrical performance of the dielectric layers of the printed circuit board, package, or multichip modules becomes significantly more important. Dielectric materials that worked perfectly well for slower designs become increasingly difficult to design with because new phenomena, such as frequency-dependent dielectric permittivity and loss tangents, environmental factors, and localized interactions between the electromagnetic signal and the fiber weave reinforcement of the board, become significant and can no longer be ignored. Without properly accounting for the high-speed dielectric phenomena, it becomes impossible to properly predict phase delay and signal losses, leading to nonphysical behavior of the transmission line model. In short, simulation-based digital bus designs exceeding 3 to 5 GHz are not possible without accounting for the effects covered in this chapter.

Dielectrics, more commonly called *insulators*, are substances whose charges in the molecules and atoms are bound and therefore cannot move over macroscopic distances under the influence of an applied field. Ideal dielectrics do not contain any free charge (such as in conductors), and chemical structure is macroscopically neutral. When a field is applied to a dielectric, the bound charges do not move to the surface of the material as they would in a conductor, but the electron clouds associated with the atomic and molecular structures of the dielectric can be distorted, reoriented, or displaced, inducing electric dipoles. When this happens, the dielectric is said to be *polarized*. The *polarizability* of a dielectric leads directly to the definition of the relative permittivity, dielectric losses, and the relationship between energy propagation and losses.

6.1 POLARIZATION OF DIELECTRICS

For a metal, conductivity is caused by the redistribution of free charges over a macroscopic distance. For example, Figure 5-4 shows how the current, and therefore the charge are largely contained within one skin depth on the bottom of the conductor. For dielectrics, the applied field only displaces a few electrons per atom over very small, subatomic distances. In a dielectric, the electrons are tightly bound to the atoms, and only a negligible number of electrons are available for conduction of electric current. The difference in electrical behavior of a conductor versus a dielectric is essentially the difference between free and bound charges.

6.1.1 Electronic Polarization

When a dielectric material constructed of nonpolar molecules is exposed to an external electric field, the electrons react by shifting with respect to the nucleus opposite the applied field. This establishes numerous small electric dipoles that align with the electric field. When the external electric field is removed, the electric dipoles return to the neutral position, as shown in Figure 6-1a. Essentially,

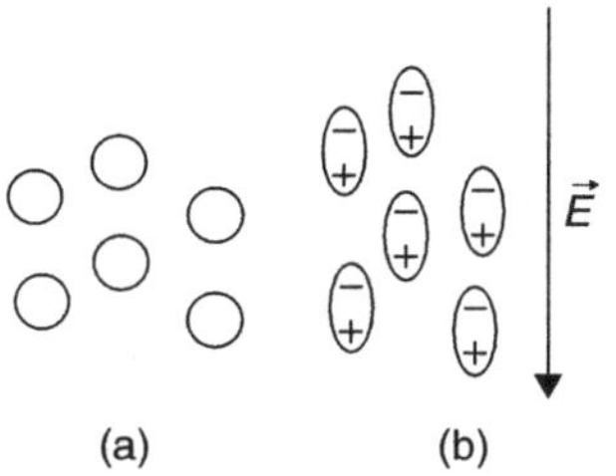

Figure 6-1 (a) Atoms with no external electric field applied; (b) electric dipoles induced as the externally applied electric field distorts the electron cloud.

the electron clouds of the atoms are distorted by the applied electric field, similar to stretching a spring as depicted in Figure 6-1b. This gives the material the ability to store electric energy (as potential energy).

To demonstrate the relationship between the applied electric field and the distortion of the electron cloud, a very simplified model of an atom can be employed by assuming a point positive charge of q_{e+} representing the nucleus encompassed by a sphere, with a charge of q_{e-} representing the electron cloud. When an electric field is applied, the electron cloud is displaced until the attractive force between the positively charged nucleus and the negatively charged sphere is equal to the force of the applied electric field. By equating these two forces, the displacement of the electron cloud can be estimated. Note that this model assumes that the electron cloud remains spherical when it is displaced. In reality, when the electron cloud is moved by the external electric field, it is elongated and stretched. However, this simple derivation will provide valuable insight into the mechanisms that cause electronic polarization.

Consider the spherical cloud of uniform charge density as in Figure 6-2a. The charge density per unit volume is

$$\rho = \frac{Q}{V} = \frac{Q}{\frac{4}{3}\pi r_e^3}$$

where Q is the total charge distributed over the volume of the sphere V and r_e is the radius of the electron cloud. The electric field is solved using the integral form of Gauss's law (2-59):

$$\oint_S \varepsilon \vec{E} \cdot d\vec{s} = \int_V \rho \, dV = Q_{\text{enclosed}}$$

In this case, only the electric field inside the sphere is relevant because the force required to move the electron cloud away from the positive nucleus must be calculated. Assuming that the electron cloud encompasses free space, $\varepsilon = \varepsilon_0$, Gauss's law reduces to

$$\varepsilon_0 E_r 4\pi r^2 = \tfrac{4}{3}\pi r^3 \rho$$

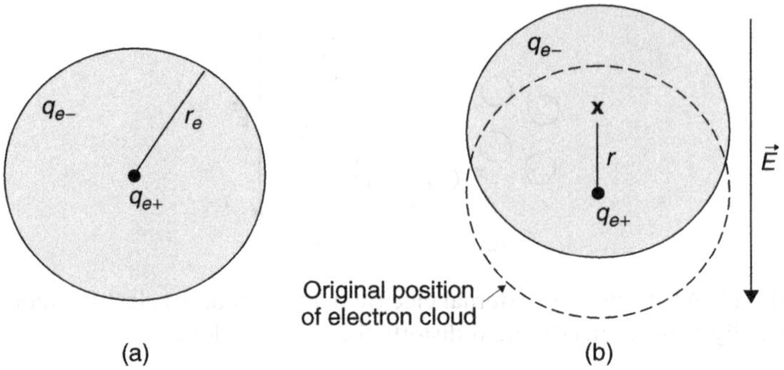

Figure 6-2 (a) Atom with no external electric field applied; (b) the applied electric field displaces the electron cloud a distance r, inducing an electric dipole.

and for surfaces with a radius of less than r_e,

$$\varepsilon_0 E_r 4\pi r^2 = Q\left(\frac{r}{r_e}\right)^3 \rightarrow E_r = \frac{Qr}{4\pi\varepsilon_0 r_e^3}$$

Assuming the total charge $Q = q_e$, the force required to move the electron cloud a distance of r is calculated with equation (2-57):

$$\vec{F} = q_e\vec{E}_r = \frac{q_e^2 r}{4\pi\varepsilon_0 r_e^3} \tag{6-1}$$

Consequently, if an external electric field with a magnitude equal but opposite to E_r is applied to the atom, it will move the electron cloud center a distance of r with respect to the nucleus, polarizing the atom, as shown in Figure 6-2b.

The *electric dipole moment* (or *electric dipole* for short) is a measure of the polarity of a system of electric charges. In the simple case of two point charges, one with charge $+q_e$ and one with charge $-q_e$, the electric dipole moment is

$$\vec{p} = q_e\vec{r} \tag{6-2}$$

where $\vec{r}$ is the displacement vector pointing from the negative charge to the positive charge. This implies that the electric dipole moment vector points in the same direction as the electric field.

Solving (6-1) for q_e and multiplying by r produces the dipole moment of a single atom in units of coulomb-meters per atom:

$$\vec{p} = q_e\vec{r} = 4\pi\varepsilon_0 r_e^3 E_r \tag{6-3}$$

Note that (6-3) indicates that the polarization is directly proportional to the electric field strength. For a volume of dielectric that has N atoms, the dipole moment per unit volume is

$$\vec{P} = N\vec{p} \tag{6-4}$$

where $\vec{P}$ is the *polarization vector*. To simplify the notation, (6-3) is often denoted as

$$\vec{p} = \alpha_e E_r \tag{6-5}$$

where $\alpha_e = 4\pi\varepsilon_0 r_e^3$ and is called the *electronic polarizability* of the material.

6.1.2 Orientational (Dipole) Polarization

If the material contains polar molecules, such as liquid water, which possess a permanent dipole moment, the orientation of the molecules is usually randomized by thermal agitation so that the dipoles are oriented in random directions, as shown in Figure 6-3a. When the material is exposed to an applied electric field, the polar molecules will tend to orient themselves with the electric field, as shown in Figure 6-3b. The *orientational polarizability* of the material is denoted α_o. An analysis similar to that shown in section 6.1.1 can be performed to derive α_o, but it is more involved and generally not practical because a detailed analysis of the intermolecular distances must be known. A more detailed discussion may be found in a book by Elliott [1993].

6.1.3 Ionic (Molecular) Polarization

Polarization of a dissolved ionic substance may be considered to be a special case, because the solution, unlike the ideal dielectric described above, consists of mobile, dissociated positive and negative charged particles, or ions. For example, NaCl dissolved in water exists principally as individual Na^+ and Cl^- ions rather than as neutral ion pairs. Each ion is associated with several of the neutral but polar water molecules, where the latter are oriented in a "sphere of hydration" surrounding each ion, as illustrated in Figure 6-4a for a positive sodium ion. Note that the partial negative charge on the oxygen of the water molecule is

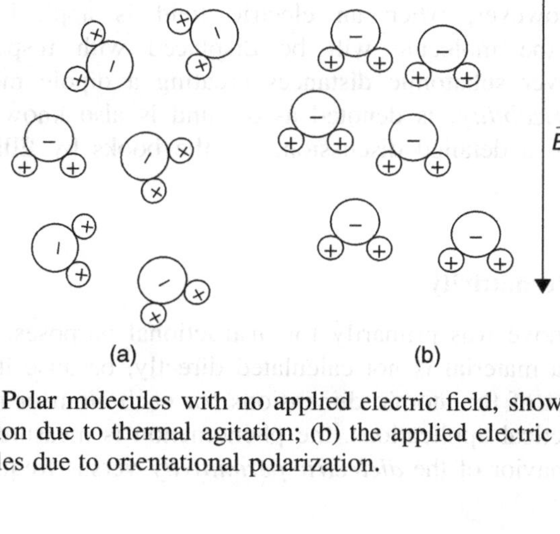

Figure 6-3 (a) Polar molecules with no applied electric field, showing somewhat randomized orientation due to thermal agitation; (b) the applied electric field tends to align the polar molecules due to orientational polarization.

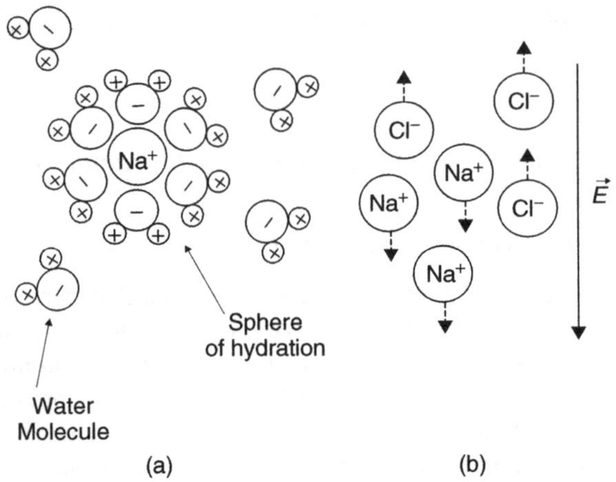

Figure 6-4 (a) In the absence of an external electric field, polar molecules and ions will tend to arrange themselves so that the total polarity is neutral. Example depicts the positive sodium ion associated with and surrounded by the polar water molecule. (b) When an external electric field is applied, the positive and negative ions will be displaced relative to each other, inducing ionic or molecular polarization.

oriented toward the positive sodium and that the hydrated ion maintains a net positive charge. When this solution is exposed to an external electric field, the positive ions are displaced with respect to the negative ones, resulting in physical migration—and separation—of the oppositely charged hydrated ions, as shown in Figure 6-4b. If the poles of the applied field represent an electron sink and source, and if the applied potential is sufficiently high, oxidation and reduction (redox) reactions occur at the poles and there is net current in the circuit. In the absence of the physical requirements for electron transfer (i.e., redox reactions), only displacement occurs in response to the applied field. In a solid material (such as NaC1), the individual ions are not able to dissociate. However, when an electric field is applied, the electrons associated with the molecule will be displaced with respect to the Na and C1 nuclei over subatomic distances creating a dipole moment. This is called *ionic polarizability*, is denoted as α_i, and is also known as *molecular polarizability*. For a detailed discussion, see the books by Elliott [1993] and Pauling [1948].

6.1.4 Relative Permittivity

The discussion above was primarily for instructional purposes. In general, the polarizability of a material is not calculated directly, because it would require precise knowledge of the atomic dimensions for each atom or molecule in the material. For practical applications, the phenomenon is measured indirectly by looking at the behavior of the *dielectric permittivity* versus frequency.

To derive the relationship between the *polarizability* of a material and the *dielectric permittivity*, the total polarization per unit volume is observed:

$$\vec{P} = N(\alpha_e + \alpha_i + \alpha_o)\vec{E}_{\text{tot}} \tag{6-6}$$

where N is the number of dipoles per unit volume. The electric field $\vec{E}_{\text{tot}}$ in (6-6) consists of the externally applied field plus the molecular field created by the alignment of the electric dipoles. If the dielectric material were removed, the value of the applied electric field $\vec{E}_0$ could be measured and the electric flux density would be given by

$$\vec{D}_0 = \varepsilon_0 \vec{E}_0 \tag{6-7}$$

When the dielectric material is included, the effect of the polarization on the electric field is added and the electric flux density becomes

$$\vec{D} = \varepsilon_0 \vec{E}_0 + \vec{P} \tag{6-8}$$

Because $\vec{P}$ and therefore $\vec{E}_{\text{tot}}$ in (6-6) is proportional to the applied electric field, $\vec{E}_0$, as implied by (6-3), a constant of proportionality can be chosen to relate $\vec{D}$ and $\vec{E}_0$. This value, the dielectric permittivity ε, has been used throughout the book:

$$\vec{D} = \varepsilon \vec{E}_0 \tag{6-9}$$

Comparing (6-8) and (6-9) allows us to choose another constant to relate $\vec{P}$ to the applied electric field, $\vec{E}_0$:

$$\vec{D} = \varepsilon_0 \vec{E}_0 + \vec{P} = \varepsilon_0(1 + \chi)\vec{E}_0 \tag{6-10}$$

where

$$\chi = \frac{\vec{P}}{\varepsilon_0 \vec{E}_0} \tag{6-11}$$

where χ is known as the *electric susceptibility* and is dimensionless. Comparing (6-9) and (6-10) shows that the dielectric permittivity can be expressed in terms of χ:

$$\varepsilon = \varepsilon_0(1 + \chi) \tag{6-12}$$

Furthermore, in Section 2.1 the dielectric permittivity is defined as

$$\varepsilon = \varepsilon_r \varepsilon_0 \tag{2-10b}$$

indicating that the relative dielectric permittivity ε_r can be defined in terms of χ and therefore the polarizability of the material:

$$\varepsilon_r = 1 + \chi \tag{6-13}$$

This means that the relative dielectric permittivity is a direct result of how the material reacts to an applied electric field.

6.2 CLASSIFICATION OF DIELECTRIC MATERIALS

In Section 5.1.1 it was shown that when a plane wave propagates in a conductive medium such as a metal, Ampère's law reduces to

$$\nabla \times \vec{H} = j\omega\left(\varepsilon - j\frac{\sigma}{\omega}\right)\vec{E} \tag{6-14}$$

In the case of a conductor, which was the topic of Chapter 5, the conductivity term σ represented the metal losses. Conductivity implies the movement of free charges within a material. In the case of a good dielectric, the charges are bound. However, as discussed in Section 6.1, the interaction between the molecular or atomic structure of the dielectric and the applied electric field changes the orientation of the bound charges within the material. As an example, equation (6-1) calculated the force required to displace the electron cloud of an atom. Consequently, as the electric dipoles within a material attempt to remain aligned with the time-varying electric field, energy is consumed, which manifests itself as dielectric losses. Subsequently, the term σ in (6-14) can be thought of as the *equivalent conductivity* of the dielectric, which represents the losses due to the polarization of the material. This allows the definition of the complex permittivity for a lossy dielectric, as was done for conductors in Chapter 5:

$$\varepsilon = \varepsilon' - j\frac{\sigma_{\text{dielectric}}}{\omega} = \varepsilon' - j\varepsilon'' \tag{6-15}$$

The imaginary portion of the complex dielectric permittivity represents the dielectric losses, and the real portion represents the dielectric permittivity $\varepsilon' = \varepsilon_r\varepsilon_0$, discussed throughout this book. For most practical purposes, materials are classified by the real part of (6-15) divided by the permittivity of free space:

$$\varepsilon_r = \frac{\varepsilon'}{\varepsilon_0} \tag{6-16a}$$

which is known as the relative dielectric permittivity, and the loss tangent,

$$\tan|\delta| = \frac{\varepsilon''}{\varepsilon'} \tag{6-16b}$$

which is simply the ratio of the imaginary and real components of (6-15).

6.3 FREQUENCY-DEPENDENT DIELECTRIC BEHAVIOR

In Section 6.1 we showed that when a dielectric is subjected to an external electric field, the positive and negative charges bound to the atoms and molecules are displaced relative to their average positions, causing electric dipoles to be formed which are quantified using the polarization vector $\vec{P}$. Furthermore, the relative dielectric permittivity ε_r was introduced to account for the presence of $\vec{P}$ in a dielectric. When the applied fields begin to alternate in polarity, the polarization vector $\vec{P}$, and subsequently the dielectric permittivity, is affected. Since the electric dipoles in a material will not align instantaneously with the applied time-varying fields, the polarization and relative permittivity are a function of the frequency of the alternating field. In this section we explore the frequency dependence of dielectric materials and derive useful models that can be used to simulate these effects.

6.3.1 DC Dielectric Losses

Dc losses imply a dielectric with conduction electrons that are free to move according to Ohm's law:

$$\vec{J} = \sigma_d \vec{E} \tag{6-17}$$

where $\vec{J}$ is the current density and σ_d is the conductivity of the dielectric. Assuming a time-harmonic field, substitution of (6-17) into Ampère's equation gives

$$\begin{aligned} \vec{\nabla} \times \vec{H} &= \vec{J}(\vec{x}, t) + \frac{\partial \vec{D}(\vec{x}, t)}{\partial t} \\ &= \sigma_d \vec{E}(\vec{x}) + j\omega[\varepsilon' - j\varepsilon'']\vec{E}(\vec{x}) \\ &= j\omega\varepsilon_0 \left(\varepsilon_r' - j\varepsilon_r'' - j\frac{\sigma_d}{\varepsilon_0 \omega}\right) \vec{E}(\vec{x}) \end{aligned} \tag{6-18}$$

Rearranging (6-18) yields the dependence of Ampère's equation on permittivity by adding the term $\sigma_d/\varepsilon_0\omega$ to (6-15) to account for the conduction electrons. Do not get σ_d in (6-18) confused with the term $\sigma_{\text{dielectric}}$ in (6-15), which is the effective conductivity of the dielectric due to the energy it takes to polarize the electric dipoles in the material. The term σ_d used in (6-18) is true conductivity, similar to that of a metal where electrons are not bound and are free to move. Note that a pole is created at $\omega = 0$ based on the rearrangement of (6-18), rendering the equation invalid at dc. Since the dielectric conductivity σ_d is very small in practical dielectrics ($\sigma_d/\varepsilon_0\omega \ll 1$), the dc term is almost always neglected [Huray, 2009].

6.3.2 Frequency-Dependent Dielectric Model: Single Pole

Following the derivation in Section 6.1.1, assume that an atom in the absence of an applied electric field is represented by a positive nucleus and a negative electric cloud with centers that coincide. Since protons and neutrons are much heavier than electrons, we assume that when the external electric field is applied, the nucleus remains stationary and the electron cloud moves. Therefore, when the electric field is applied, the electron cloud is displaced, and when the field is removed, the electron cloud returns to its original position, similar to the

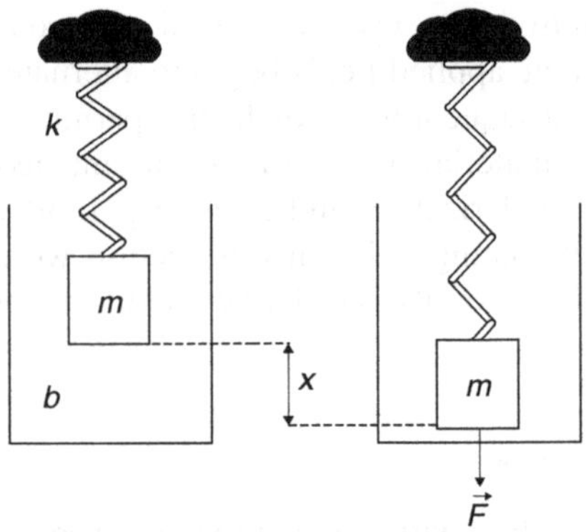

Figure 6-5 Analogy to an electron cloud in the presence of an electric field: When the mass (electron) is pulled downward by a force (electric field), it stretches the spring (displaces the electron cloud), displacing the mass (electron) by x through a medium with damping factor b. The spring constant is k.

spring and mass system shown in Figure 6-5. When the electric field is applied (the mass is pulled downward, stretching the spring by $\vec{F}$), the electron (the mass) is displaced by a distance of x and an electric dipole is created. When the electric field is removed (the mass is no longer being pulled downward), the electron cloud returns to its neutral position (the spring is no longer stretched). This analogy suggests that a simple mechanical spring model can be used to model the polarization and therefore the dielectric permittivity as a function of frequency [Huray, 2009].

The classic differential equation that governs the solution of a damped driven harmonic oscillator (and therefore a damped spring, as shown in Figure 6-5) is

$$\left(m\frac{d^2x}{dt^2} + b\frac{dx}{dt} + kx\right) = Fe^{j\omega t}$$

where m is the mass, b the damping factor, k the spring constant, and $Fe^{j\omega t}$ the driving force. The spring equation can be rewritten in terms of the applied electric field and the electron mass. From equation (2-57), the force can be written in terms of the electric field and the charge, $\vec{F} = q\vec{E}$, where q is the charge of the dipole, m the mass of the negative charge relative to the stationary nucleus, and x the displacement distance:

$$\frac{d^2x}{dt^2} + \frac{b}{m}\frac{dx}{dt} + \frac{k}{m}x = \frac{qE_0}{m}e^{j\omega t} \tag{6-19}$$

The homogeneous solution to (6-19) is uninteresting because it is heavily damped and will not contribute to the steady-state solution. The particular solution is solved assuming that x takes the form $x = Ae^{j\omega t}$. Therefore,

$$\frac{dx}{dt} = jA\omega e^{j\omega t}$$
$$\frac{d^2x}{dx^2} = -\omega^2 Ae^{j\omega t}$$

which are substituted back into (6-19), so the coefficient A can be solved for:

$$-\omega^2 Ae^{j\omega t} + j\frac{b}{m}A\omega e^{j\omega t} + \frac{k}{m}Ae^{j\omega t} = \frac{qE_0}{m}e^{j\omega t}$$
$$A\left(-\omega^2 + j\frac{b}{m}\omega + \frac{k}{m}\right) = \frac{qE_0}{m}$$
$$A = \frac{qE_0/m}{-\omega^2 + j(b/m)\omega + k/m}$$

Therefore,

$$x = Ae^{j\omega t} = \frac{(qE_0/m)e^{j\omega t}}{-\omega^2 + j(b/m)\omega + k/m}$$

Dropping the time-harmonic function and rearranging yields

$$x = \frac{qE_0/m}{(k/m - \omega^2) + j(b/m)\omega} \tag{6-20}$$

The natural (resonant) frequency of the oscillator is defined when, $\omega^2 = k/m$, which allows the definition of ω_0:

$$\omega_0^2 = \frac{k}{m} \tag{6-21}$$

Therefore, (6-20) can be simplified:

$$x = \frac{qE_0/m}{\omega_0^2 - \omega^2 + j(b/m)\omega} \tag{6-22}$$

From equation (6-2), the electric dipole moment is

$$\vec{p} = q\vec{x} = \frac{q^2E_0/m}{\omega_0^2 - \omega^2 + j(b/m)\omega} \tag{6-23}$$

and subsequently, the polarization vector is

$$\vec{P} = N\vec{p} = \frac{N(q^2E_0/m)}{\omega_0^2 - \omega^2 + j(b/m)\omega}$$

and the electric susceptibility is defined from (6-11):

$$\chi = \frac{\vec{P}}{\varepsilon_0 E_0} = \frac{N(q^2/\varepsilon_0 m)}{\omega_0^2 - \omega^2 + j(b/m)\omega}$$

leading to an expression for the relative dielectric constant from (6-13) [Huray, 2009]:

$$\varepsilon_r = 1 + \chi = 1 + \frac{N(q^2/\varepsilon_0 m)}{\omega_0^2 - \omega^2 + j(b/m)\omega} \tag{6-24}$$

The real and imaginary parts are shown as [Balanis, 1989]

$$\varepsilon_r' = 1 + \frac{N(q^2/\varepsilon_0 m)(\omega_0^2 - \omega^2)}{(\omega_0^2 - \omega^2)^2 + (\omega(b/m))^2} \tag{6-25a}$$

$$\varepsilon_r'' = N\frac{q^2}{\varepsilon_0 m}\frac{\omega(b/m)}{(\omega_0^2 - \omega^2)^2 + \left(\omega\frac{b}{m}\right)^2} \tag{6-25b}$$

Equations (6-24) and (6-25) calculate the frequency response for a material that exhibits an atomic or molecular structure with only one natural or resonant frequency, an example of which is shown in Figure 6-6. At the natural (i.e., resonant) frequency of the harmonic oscillator, the imaginary portion of the complex permittivity will peak, which in turn dramatically increases the dielectric losses, which are quantified with the loss tangent (6-16). Also note that the real portion of the dielectric permittivity is almost constant until the operating frequency approaches the resonant frequency of the oscillator, which in this case is an atomic structure. In the vicinity of the natural frequency, the real portion begins

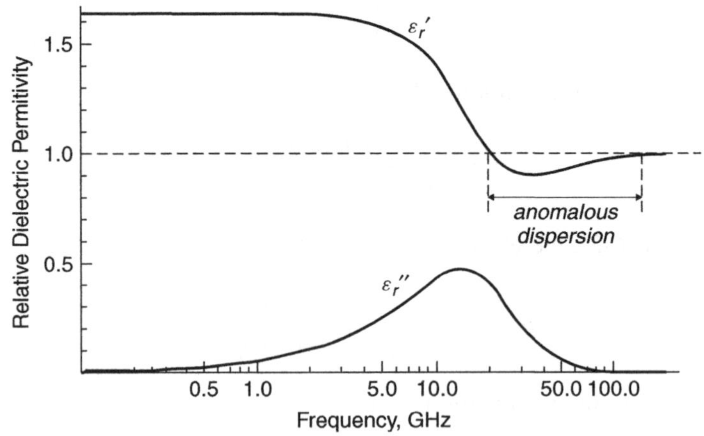

Figure 6-6 Frequency response for a pure material that exhibits an atomic or molecular structure with only one natural or resonant frequency.

to drop and the imaginary portion begins to increase, implying a relationship between the real and imaginary parts of the complex permittivity. This relationship is discussed in detail in Section 6.4. Finally, the value of the permittivity settles out at a new level after the operating frequency ω has surpassed the natural frequency ω_0.

6.3.3 Anomalous Dispersion

Perhaps the most interesting part of the curve shown in Figure 6-6 is the area where the relative dielectric permittivity drops below 1. This area is often referred to as *anomalous dispersion*. When this is first encountered, it often causes great confusion, as demonstrated by observing the definition of the speed of light in a vacuum, which was introduced in Section 2.3.4 in units of meters per second:

$$c \equiv \frac{1}{\sqrt{\mu_0 \varepsilon_r \varepsilon_0}} \qquad \text{when } \varepsilon_r = 1$$

This implies that if $\varepsilon_r < 1$, the velocity will exceed the speed of light in a vacuum ($c \approx 3 \times 10^8$ m/s). One of the consequences of Einstein's theory of special relativity is that speeds greater than c are not attainable, yet the phase velocity of a wave traveling at frequencies where $\varepsilon_r < 1$ in Figure 6-6 appears to break this fundamental law of physics.

The apparent conflict with the laws of special relativity comes from a widely mistaken assumption that all quantities with units of velocity must obey this rule. In fact, special relativity only places an upper value on the speed of material bodies that include signals, or information. Since a single-frequency harmonic plane wave is not a material body and is not a signal, it cannot be used to transmit information by itself. To understand this, the definition of phase velocity must be examined from Chapter 2. To determine how fast the wave is propagating, it is necessary to observe the cosine term for a small duration of time, Δt. Since the wave is propagating, a small change in time will be proportional to a small change in distance Δz, which means that an observer moving with the wave will experience no phase change because she is moving at the phase velocity (v_p). However, the only way to measure the velocity of a signal is to turn on the transmitter, time how long it takes for a response to arrive at the receiver, and divide by the distance. When velocity is measured in this way, it cannot exceed c. To understand why, consider the propagation of information, such as a digital pulse. If the pulse is decomposed into its Fourier components, each will propagate with its own velocity, some slower than c and some faster than c; *however, when all components are combined, the total velocity of the pulse cannot exceed the speed of light*. A question that often arises when this subject is discussed is the possibility of modulating a narrowband signal with a frequency that coincides with an area of anomalous dispersion to sidestep the laws of special relativity and transmit information faster than c. However, a single-frequency plane wave still cannot carry information unless another signal is combined with it. For example, if a narrowband signal was simply turned on and off as a simplistic form of modulation, this is equivalent to convolving the narrowband signal with a step response, which contains an infinite number of Fourier components.

The proof that the velocity of a signal used to transmit information cannot exceed the speed of light is too complicated to address here. However, Brillouin [1960] showed that a signal propagating in a realistic dielectric medium cannot exceed c.

6.3.4 Frequency-Dependent Dielectric Model: Multipole

As discussed in Section 6.1, the polarization of a realistic dielectric may consist of any number of ionic, orientational, and electronic mechanisms, depending on the molecular and atomic structure of the material. In general, ε' and ε'' depend on frequency in complicated ways, exhibiting several resonances over a wide range. To account for this, a separate harmonic oscillator model is constructed for each molecular or atomic resonance, and the results are combined using superposition. Thus, for a material with n natural frequencies, the relative dielectric permittivity is represented by summing the responses from n independent harmonic oscillator models:

$$\varepsilon_r = 1 + \sum_{i=1}^{n} \frac{N_i(q^2/\varepsilon_0 m)}{\omega_i^2 - \omega^2 + j(b_i/m)\omega} \tag{6-26}$$

An example is shown in Figure 6-7, where three harmonic oscillator models were chosen arbitrarily with natural frequencies $\omega_1 = 20$ GHz, $\omega_2 = 100$ GHz, and $\omega_3 = 400$ GHz. Note how the permittivity tends to "stair-step" down to smaller values as the frequency increases, separated by areas of anomalous dispersion, producing a change in the steady-state value $(\Delta\varepsilon_r')$ at frequencies higher than

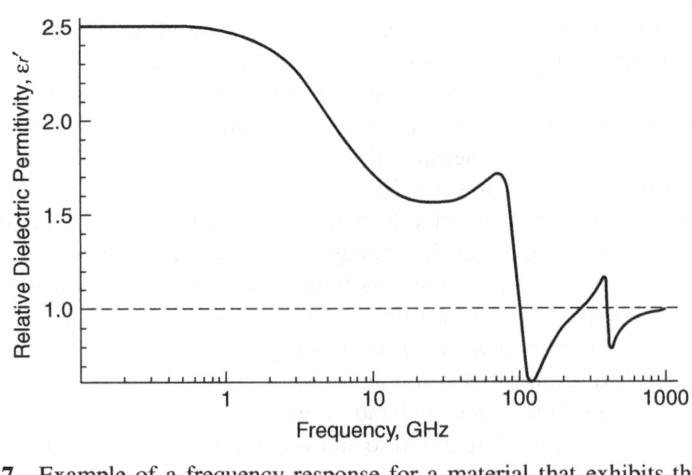

Figure 6-7 Example of a frequency response for a material that exhibits three atomic or molecular resonances at 20, 100, and 400 GHz.

the resonance. The summation indicates that even low-frequency values of the permittivity depend on the high-frequency resonances.

Although (6-26) is quite useful for developing an understanding of the physical mechanisms that cause the dielectric to vary with frequency, it does not lend itself well to simulating dielectrics for digital designs because minuscule details of the molecular structure are required. A more pragmatic approach is to represent (6-26) in terms of measured permittivity. If (6-26) is multiplied by ε_0 and the right-hand term is simplified by dividing the top and bottom by $1/\omega_0^2$, the top part, $(1/\omega_0^2)[N(q^2/m)]$, has units of farads per meter:

$$1\ \mathrm{F} = \frac{\mathrm{s}^4 \cdot \mathrm{A}^2}{\mathrm{m}^2 \cdot \mathrm{kg}}$$

$$1\ \mathrm{C} = \mathrm{A} \cdot \mathrm{s}$$

$$\frac{1}{\omega_0^2}\left(N\frac{q^2}{m}\right) \to \mathrm{s}^2\left(\frac{1}{\mathrm{m}^3}\frac{\mathrm{A}^2 \cdot \mathrm{s}^2}{\mathrm{kg}}\right) = \mathrm{F/m}$$

which are the same units as ε', allowing (6-26) to be rewritten as

$$\varepsilon = \varepsilon'_\infty + \sum_{i=1}^{n} \frac{\Delta\varepsilon'_i}{1 - \omega^2/\omega_{1i}^2 + j\omega/\omega_{2i}} - j\frac{\sigma_d}{\varepsilon_0\omega} \tag{6-27}$$

where ω_{1i} and $\omega_{2i}[(1/\omega_i^2)(b/m)\omega = \omega/\omega_{2i}]$ are the frequencies where the dielectric variations are occurring, $\omega = 2\pi f$ is the operating frequency, $\Delta\varepsilon'$ the variation of dielectric permittivity over the frequency of interest, σ_d the true conductivity of the dielectric material, ε'_∞ the dielectric permittivity value at very high frequencies in the area of interest, and ε_0 the permittivity of free space. The term $j(\omega/\omega_{2i})$ accounts for the damping of the molecular dipoles (orientational polarization) in the mid-frequency ranges, the term ω^2/ω_{1i}^2 accounts for resonance of induced atomic and molecular dipoles (ionic and electronic polarization), and the final term, $j(\sigma_d/\varepsilon_0\omega)$ (derived in Section 6.3.1), accounts for the low-frequency loss of the dielectric, which is usually ignored.

Figure 6-8 shows a conceptualized plot of the real and imaginary dielectric permittivity as a function of frequency. It should be noted that the figure does not represent any particular dielectric material; rather, it is simply a guide to help the reader conceptualize when the different forms of polarization begin to become significant. For most dielectric materials used in digital design, laboratory measurements show that the permittivity is dominated by the damping factor in (6-27)(the $j\omega/\omega_{2i}$ term) and not resonances (the ω^2/ω_{1i}^2 term). For orientational polarization (where polar molecules attempt to remain aligned with the time varying electric field), the damping factor tends to be high. Consequently, the classic model is derived from the concept of orientational polarization even though other forms may affect the damping. When the high frequency resonance term is ignored, the dielectric equation reduces to (6-28), which is applicable for

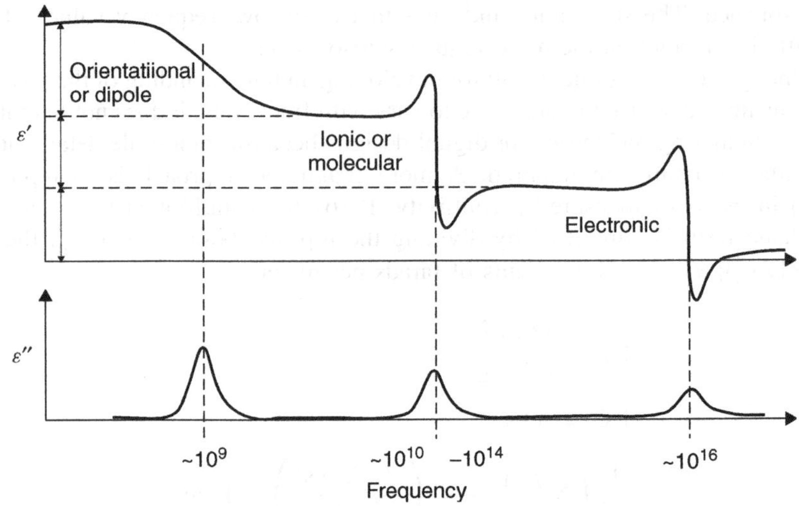

Figure 6-8 Conceptual complex dielectric permittivity variations as a function of frequency, showing approximate regions where each polarization mechanism exists.

most practical high-speed digital platforms built on commonly available dielectric materials.

$$\varepsilon = \varepsilon'_{\infty} + \sum_{i=1}^{n} \frac{\Delta\varepsilon'_i}{1 + j(\omega/\omega_{2i})} \tag{6-28}$$

Note that equation (6-28) is identical to the well-known *Debye equation* and is often curve-fit to measured data to build a very accurate dielectric model library, used for designing high-speed digital systems.

There is evidence based on laboratory measurements that dielectric resonances can exist as low as 30 GHz in FR4 dielectrics, which may affect some very high frequency designs. When designing digital systems with significant harmonics past about 20 GHz, care should be taken to examine carefully the measured phase delay and loss characteristics of transmission lines built on a representative dielectric so that any resonance can be accounted for. In Chapter 9 we describe methodologies for extracting the loss characteristics and the phase velocity from S-parameter measurements. A dramatic narrowband increase in the loss accompanied by a simultaneous increase in the phase velocity is a telltale sign that a dielectric resonance exists in the frequency of interest.

Equations (6-27) and (6-28) are much more useful than (6-26) for one very important reason: *They can be fit empirically*. Since equation (6-26) requires intimate knowledge of the atomic substructure, it is not very useful for practical applications. However, the dielectric permittivity can be measured using the phase delay (see Chapter 9 for details), the results of which can be fit to (6-27) by choosing the appropriate poles (ω_{1i} and ω_{2i}), ε'_{∞} and $\Delta\varepsilon'$, to essentially curve-fit the dielectric behavior to a physically consistent model. The particular implementation of (6-27) is dependent on the characteristics of the material.

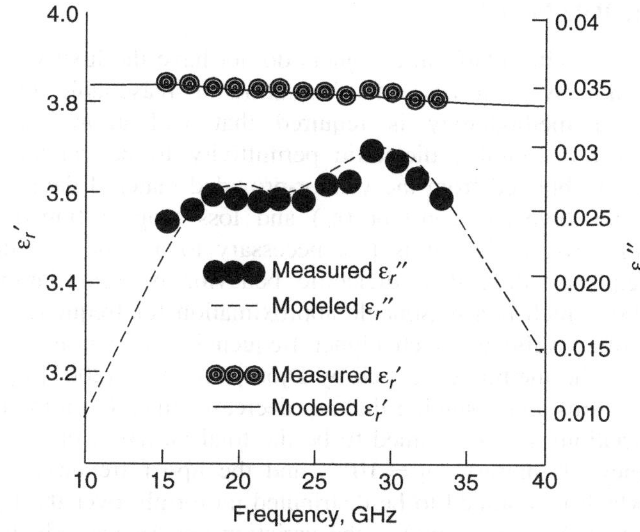

Figure 6-9 Measured dielectric response curve fit to equation (6-27).

Subsequently, for a generic material, the most straightforward usage requires a measured response of ε_r and $\tan\delta$ so the damping poles and resonant peaks can be identified.

Example 6-1 Use equation (6-27) to create an empirical model for the dielectric measured in Figure 6-9.

SOLUTION Figure 6-9 depicts the measured value of the real and imaginary permittivity of a dielectric sample between 15 and 35 GHz. Note that the measured data shows small ε''_r peaks in the vicinity of 19 and 32 GHz, indicating the presence of two resonant induced dipoles, which are probably ionic or molecular in nature. Consequently, ignoring the dc term, the implementation of (6-27) for this material over the frequency range 15 to 35 GHz is

$$\varepsilon = \varepsilon_\infty + \frac{\Delta\varepsilon_1}{1 + j(\omega/\omega_1) - \omega^2/\omega_2^2} + \frac{\Delta\varepsilon_2}{1 + j(\omega/\omega_3) - \omega^2/\omega_4^2}$$

where $\omega_1 = 2\pi(20\ \text{GHz})$, $\omega_2 = 2\pi(19\ \text{GHz})$, $\omega_3 = 2\pi(63\ \text{GHz})$, $\omega_4 = 2\pi$ (32 GHz), $\Delta\varepsilon_1 = 0.0163$, $\Delta\varepsilon_2 = 0.012$, and $\varepsilon_\infty = 3.8$. The resonant terms ω_2 and ω_4 were chosen at the peaks, $\Delta\varepsilon_1$ and $\Delta\varepsilon_2$ are the variations in the real part of the dielectric constant in the vicinity of the peaks, and ω_1 and ω_3 are damping terms that were varied until the width and the height of the peaks match the data measured. In this case, the dielectric model is valid only between 15 and 35 GHz, which are the limits of the measured data [Hall et al., 2007].

6.3.5 Infinite-Pole Model

Most high-speed digital platform designers do not have the luxury of wideband complex permittivity measurements that allow a reasonable fit to (6-27). Consequently, a methodology is required that will allow the frequency dependence of the complex dielectric permittivity to be calculated from a single data point obtained from the vendor-provided material data sheet, which usually lists the dielectric constant (ε_r) and loss tangent ($\tan\delta$) at 1 GHz and/or 10 GHz. To do this, it is first necessary to assume that the primary source of frequency dependent dielectric behavior is from damping, as in equation (6-28), which is a reasonable approximation for frequencies below 20 GHz, and is often valid to much higher frequencies. Equation (6-28) can be simplified using the method described by Djordjevic et al. [2001] by increasing the number of terms and assuming that ω_{2i} decreases linearly on the logarithmic scale. The variation $\Delta\varepsilon'$ is assumed to be the total variation of ε' between the lower frequency of interest, $\omega_1 = 10^{m1}$, and the upper frequency of interest, $\omega_2 = 10^{m2}$, which is assumed to be distributed uniformly over the logarithm of frequency so that $\Delta\varepsilon'/(m_2 - m_1)$ is the variation per decade, which is a linear decay of ε' on a logarithmic scale. By taking an infinite number of terms, the sum in (6-28) becomes (6-29), which is referred to as the *infinite pole model*:

$$\sum_{i=1}^{n} \frac{\Delta\varepsilon'}{1 + j(\omega/\omega_{2i})} \to \frac{\Delta\varepsilon'}{m_2 - m_1} \int_{x=m_1}^{m_2} \frac{dx}{1 + j(\omega/10^x)}$$

$$= \frac{\Delta\varepsilon'}{m_2 - m_1} \frac{\ln[(\omega_2 + j\omega)/(\omega_1 + j\omega)]}{\ln(10)}$$

$$\varepsilon' - j\varepsilon'' = \varepsilon_\infty + \frac{\Delta\varepsilon'}{m_2 - m_1} \frac{\ln[(\omega_2 + j\omega)/(\omega_1 + j\omega)]}{\ln(10)} \tag{6-29}$$

By comparing the results to laboratory measurements, it has been shown that (6-29) is reasonably accurate for typical dielectrics used to design modern digital platforms under dry and nominal environmental conditions (0 to about 50% relative humidity). However, for dielectric materials that have an affinity to absorb water (such as FR4) when exposed to very humid environmental conditions (i.e., 95% relative humidity), the model breaks down for frequencies greater than 10 GHz and (6-28) must be curve fit to measured data to achieve realistic dielectric behavior. More details on how dielectric properties change with the environment are discussed in Section 6.6.

To calculate the wideband dielectric permittivity and the loss tangent from a single-frequency data point, it is necessary to know the value of ε_r and $\tan\delta$ at that frequency, which allows the calculation of ε' and ε'' using

$$\varepsilon = \varepsilon - j\frac{\sigma_{\text{dielectric}}}{\omega} = \varepsilon' - j\varepsilon'' \tag{6-15}$$

$$\tan|\delta| = \frac{\varepsilon''}{\varepsilon'} \tag{6-16b}$$

Equating the real and imaginary terms of the permittivity to the real and imaginary components of (6-29) allows the calculation of $\Delta\varepsilon'$ and ε'_∞, which can then be inserted into (6-29) so the response at all other frequencies can be calculated.

Recall that for a complex number, $z = a + jb = re^{j\theta}$, where $r = \sqrt{a^2 + b^2}$ and $\ln(z) = \ln(r) + j(\theta + 2\pi k)$, where k is an integer. This allows the log term in (6-29) to be simplified assuming that $\omega_1 \ll \omega \ll \omega_2$:

$$\mathrm{Re}\left(\ln \frac{\omega_2 + j\omega}{\omega_1 + j\omega}\right) = \ln \sqrt{\frac{\omega^2 + \omega_2^2}{\omega^2 + \omega_1^2}} \approx \ln \frac{\omega_2}{\omega}$$

allowing the real part of (6-29) to be approximated as

$$\varepsilon' \approx \varepsilon'_\infty + \frac{\Delta\varepsilon'}{m_2 - m_1} \frac{\ln(\omega_2/\omega)}{\ln(10)} \tag{6-30a}$$

The imaginary part is simplified by noting that for the assumption $\omega_1 \ll \omega \ll \omega_2$, the logarithm function in (6-29) is almost constant at a value of $-\pi/2$, allowing the imaginary part of (6-29) to be approximated as [Pytel, 2007]

$$\varepsilon'' \approx \frac{\Delta\varepsilon'}{m_2 - m_1} \frac{-\pi/2}{\ln(10)} \tag{6-30b}$$

Figure 6-10 depicts an example of the modeled and measured relationship between the loss tangent and the dielectric permittivity as calculated with (6-29) and measured with a Fabry–Perot open-cavity resonator. The modeled response was calculated from a known data point at 1 GHz, where $\varepsilon_r / \tan\delta = 3.9/0.0073$, which was measured using a split post resonator. Note the deviation between

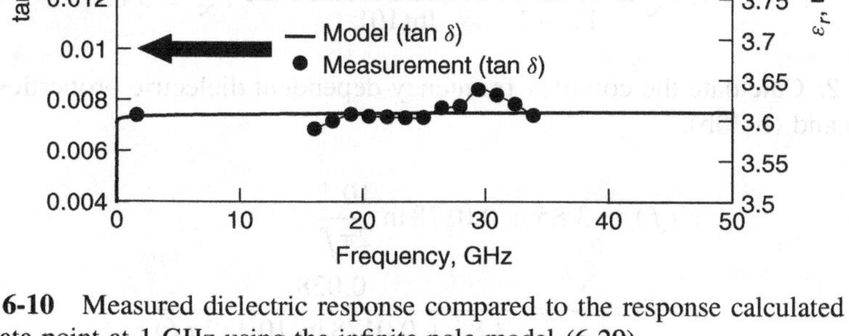

Figure 6-10 Measured dielectric response compared to the response calculated from a single data point at 1 GHz using the infinite-pole model (6-29).

the measured $\tan\delta$ data points and that the model has a maximum error of only 0.0009 at 30 GHz with a maximum error less than 0.01 for ε_r at 20 GHz [Hall et al., 2007].

Example 6-2 Calculate the frequency dependence of the dielectric permittivity using the infinite-pole Debye model (6-29) assuming a nominal relative humidity of 30% and a measured data point at 1 GHz where $\tan\delta = 0.0073$ and $\varepsilon_r = 3.9$ to 10 GHz.

SOLUTION

Step 1: Calculate the infinite-pole variables ε'_∞ and $\Delta\varepsilon'$. Choose a lower and an upper frequency limit that are well beyond the frequency of interest. We choose 10 rad/s (1.6 Hz) for the lower limit and 100 Grads/s (16 GHz) for the upper limit:

$$\omega_1 = 10^1 \to m_1 = 1 \quad \text{and} \quad \omega_2 = 10^{11} \to m_2 = 11$$

The real part of the complex permittivity at 1 GHz is $\varepsilon' = \varepsilon_r\varepsilon_0 = 3.9\varepsilon_0$. The imaginary part of the complex permittivity is calculated with (6-16):

$$\tan|\delta| = \frac{\varepsilon''}{\varepsilon'}$$

$$\varepsilon'' = \varepsilon' \tan|\delta| = (3.9)(0.0073) = 0.028$$

The imaginary part of the permittivity is substituted into (6-30b) and $\Delta\varepsilon'$ is calculated:

$$0.028 = \frac{\Delta\varepsilon'}{11-1}\frac{-\pi/2}{\ln(10)} \to \Delta\varepsilon' = -0.417$$

where the negative sign is ignored because it is already included in (6-15). Next, ε'_∞ is calculated using (6-30a):

$$3.9 \approx \varepsilon'_\infty + \frac{0.417}{11-1}\frac{\ln(10^{11}/2\pi(10^9))}{\ln(10)} \to \varepsilon'_\infty = 3.85$$

Step 2: Calculate the complex frequency-dependent dielectric properties using (6-30a) and (6-30b):

$$\varepsilon'(f) \approx 3.85 + 0.0178 \ln\frac{10^{11}}{2\pi f}$$

$$\tan|\delta| = \frac{\varepsilon''}{\varepsilon'} = \frac{0.028}{3.85 + 0.0178\ln(10^{11}/2\pi f)}$$

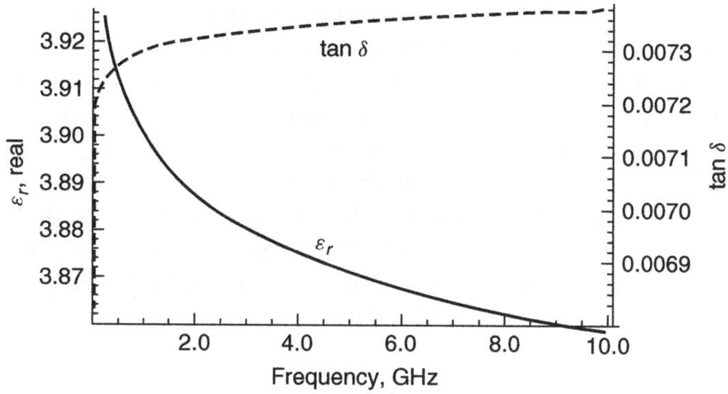

Figure 6-11 Dielectric properties versus frequency from Example 6-2.

Figure 6-11 shows a plot of the dielectric properties versus frequency. As a double check, it can be confirmed that at 1 GHz, the relative permittivity is 3.9 and the loss tangent is 0.0073.

6.4 PROPERTIES OF A PHYSICAL DIELECTRIC MODEL

For a dielectric model to remain physical, it must predict correctly the frequency-dependent behavior observed in real life. A common mistake in digital design is to use transmission-line models that have frequency-invariant dielectric properties. Although this approximation works fine for low frequencies, as system data rates increase past about 2 Gb/s, such methods will introduce large errors into the simulations, rendering the analysis almost useless. In this section we describe the properties that must be satisfied to ensure a physically realizable model. Although the focus here is on dielectric models, any model is required to pass these tests to ensure behavior consistent with nature. First, the relationship between the dielectric permittivity and the loss tangent is discussed, and then specific mathematical tests are introduced to ensure a model that obeys the laws of physics.

6.4.1 Relationship Between ε' and ε''

In Section 5.2.3, specific relationships were shown to exist between the real and imaginary parts of the series impedance of a transmission line in the form of ac resistance and internal inductance. There is a similar relationship between the real and imaginary portions of the complex permittivity, which in turn implies a relationship between the relative dielectric permittivity ε_r and the loss tangent $\tan\delta$. To conceptualize the relationship, assume that the frequency range is below 20 GHz, where measurements have shown that damping is the dominant dielectric response. In this region, large polar molecules are attempting to align themselves with the time-varying electric field (e.g., orientational polarization), which consumes energy and therefore dissipates power. As the frequency increases, it becomes more difficult for the polar molecules to follow the electric field, meaning that more energy is consumed at higher frequencies. Think of pulling a paddle through the water when rowing a boat. If you increase the rate of paddling, it becomes more difficult to move the oar though the water, increasing the amount of energy used. Following this analogy, the dielectric loss, and therefore the loss tangent, should tend to increase with frequency. If the definition of the volume energy density from Section 2.4.2 is observed, it shows that the energy stored in a charge distribution is proportional to ε:

$$w_e = \frac{\varepsilon}{2}E^2 \qquad \text{J/m}^3 \tag{2-75}$$

If energy (and therefore power) is being dissipated by the polarization of the molecules, the total energy stored (power transmitted through the dielectric) must decrease proportionally. Since ε dictates how much energy can be stored in an electric field, the permittivity must therefore decrease with a corresponding increase in the loss tangent. This general trend is demonstrated in Figures 6-10 and 6-11. Note that as the frequency of operation approaches the frequency of a pole or a resonance in the dielectric models derived earlier, this analogy breaks down. However, for the majority of high-speed digital designs that are built on common PCB materials, this analogy holds true. The analogy above hints toward a specific relationship between the real and imaginary parts of the dielectric permittivity that must be accounted for properly during the design of high-speed buses. To demonstrate this, the single-pole Debye model (6-31) can be separated into real and imaginary parts:

$$\varepsilon' - j\varepsilon'' = \varepsilon'_\infty + \frac{\Delta\varepsilon'}{1 + j(\omega/\omega_0)} \tag{6-31a}$$

$$\varepsilon' = \varepsilon'_\infty + \frac{\Delta\varepsilon'}{1 + (\omega/\omega_0)^2} \tag{6-31b}$$

$$\varepsilon'' = \frac{\Delta\varepsilon'(\omega/\omega_0)}{1 + (\omega/\omega_0)^2} \tag{6-31c}$$

Next, (6-31b) and (6-31c) are combined to yield

$$\varepsilon' = \varepsilon'_\infty + \frac{\varepsilon''\omega_0}{\omega} \tag{6-32a}$$

$$\varepsilon'' = (\varepsilon' - \varepsilon'_\infty)\sqrt{\frac{\Delta\varepsilon'}{\varepsilon' - \varepsilon'_\infty} - 1} \tag{6-32b}$$

Note that the real part of the permittivity (ε') is a function of the imaginary part (ε''), and vice versa. Equations (6-32) show that *a specific relationship governs*

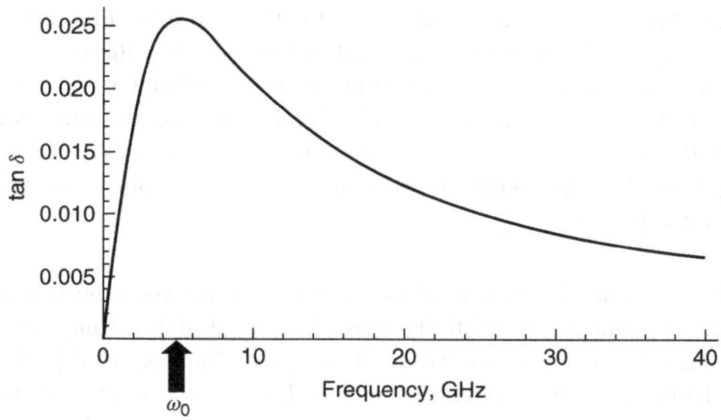

Figure 6-12 When a harmonic oscillator is being driven at a frequency much larger than its natural or resonant frequency ($\omega_0 < \omega$), the losses vanish. This example plots the loss generated from the Debye equation (6-28) with a pole at $\omega_0 = \omega_{2i} = 5$ GHz.

the real and imaginary parts of the complex permittivity. Furthermore, if $\omega_0 \gg \omega$, the loss tangent reduces to

$$\tan|\delta| \approx \frac{\omega}{\omega_0}\left(1 - \frac{\varepsilon_\infty}{\varepsilon_\infty + \Delta\varepsilon'}\right) \tag{6-33}$$

which indicates an increase in the loss tangent with frequency. So for dielectrics that can be described by the Debye equation, equation (6-32) shows that ε' (and subsequently ε_r) will decrease with frequency, and (6-33) shows that the loss tangent will increase with frequency.

A question arises regarding the very high frequency response of the Debye model. If the model is subjected to a very high frequency oscillatory forcing function, where $\omega_0 < \omega$, there will be no time for the system to respond before the forcing function has switched direction, so the losses will vanish as ω becomes large, as is demonstrated by Figure 6-12, which has a pole at $\omega_0 = 5$ GHz.

6.4.2 Mathematical Limits

For a transmission-line model to be physically consistent with the laws of nature, certain mathematical limits must be obeyed. Presently, the vast majority of engineers utilize models to design high-speed digital systems that employ frequency-invariant values of the dielectric permittivity and loss tangent. This assumption, although perfectly valid at low frequencies, induces amplitude errors and phase miscalculations for digital data rates faster than 1 to 2 Gb/s propagating on transmission lines because realistic dielectric materials have frequency-dependent properties that must be modeled correctly. As data rates increase, the spectral content of the digital pulse trains also increase and errors induced by the frequency-invariant approximation are amplified, especially for longer line lengths. When models are used that do not obey these rules, incorrect solutions are determined, lab correlation becomes difficult or impossible, and the design time is increased significantly. In this section we introduce specific limitations that dielectric models must obey to remain physically consistent with nature. Specifically, the conditions of *causality, analytic functions, reality*, and *passivity* are explored.

Causality An in-depth analysis of the relationship between the real and imaginary parts of complex permittivity was done by Ralph Kronig and Hendrik Anthony Kramers in the early twentieth century [Balanis, 1989]. They developed the Kramers–Kronig relations, which describe the relationships between the real and imaginary parts of any complex function that is analytic in the upper half-plane:

$$\varepsilon'(\omega) = 1 + \frac{2}{\pi} \int_0^\infty \frac{\omega' \varepsilon_r''(\omega')}{(\omega')^2 - \omega^2}\, d\omega' \tag{6-34a}$$

$$\varepsilon''(\omega) = \frac{2\omega}{\pi} \int_0^\infty \frac{1 - \varepsilon_r'(\omega')}{(\omega')^2 - \omega^2}\, d\omega' \tag{6-34b}$$

The Kramers–Kronig formalism is applied to response functions. In physics, a response function $\chi(t - t')$ describes how a property $P(t)$ of a physical system responds to an applied force $F(t')$. For example, $P(t)$ could be the angle of a pendulum and $F(t')$ the applied force of a motor driving the pendulum motion. The response $\chi(t - t')$ must be zero for $t < t'$ since a system cannot respond to a force before it is applied. Such a function is said to *be causal*. From a commonsense point of view, it makes sense that an effect cannot precede its cause, which is the fundamental principle of causality that every physical model must respect, as expressed mathematically by

$$h(t) = 0 \qquad \text{when } t < 0 \tag{6-35}$$

The causality requirement is described mathematically in Chapter 8.

Analytic Functions The response of a dielectric to an applied electric field is quantified in terms of ε, and therefore a causal dielectric model cannot respond prior to the electric field being applied, which makes sense for a physical system. It can be shown that this causality condition implies that *the Fourier transform of the complex dielectric permittivity $\varepsilon(\omega)$ is analytic* [Jackson, 1998], *which in turn implies a specific relationship between the real (ε') and imaginary (ε'') parts*.

For a complex function $f(x + jy) = u(x, y) + jv(x, y)$, if u and v have continuous first partial derivatives and satisfy the Cauchy–Riemann equations, f is said to be *analytic* [LePage, 1980]. The *Cauchy–Riemann* equations dictate specific relationships between the real and imaginary parts of such a complex function:

$$\frac{\partial u}{\partial x} = \frac{\partial v}{\partial y} \tag{6-36a}$$

$$\frac{\partial u}{\partial y} = -\frac{\partial v}{\partial x} \tag{6-36b}$$

It can be shown that the dielectric models presented earlier in the book satisfy the Cauchy–Riemann equations assuming that the dielectric response is a function of complex frequency. For example, if $\omega = \omega_R + j\omega_I$ is inserted into the Debye model in (6-31),

$$\varepsilon(\omega_R + j\omega_I) = \varepsilon'_\infty + \frac{\Delta\varepsilon'}{1 + j[(\omega_R + j\omega_I)/\omega_0]} = \frac{\omega_0 \Delta\varepsilon'(\omega_0 - \omega_I)}{(\omega_0 - \omega_I)^2 + \omega_R^2}$$
$$- j\frac{\omega_0 \Delta\varepsilon' \omega_R}{(\omega_0 - \omega_I)^2 + \omega_R^2}$$

it can be shown that

$$\frac{\partial \mathrm{Re}(\varepsilon)}{\partial \omega_R} = \frac{\partial \mathrm{Im}(\varepsilon)}{\partial \omega_I}$$

$$\frac{\partial \mathrm{Re}(\varepsilon)}{\partial \omega_I} = -\frac{\partial \mathrm{Im}(\varepsilon)}{\partial \omega_R}$$

Therefore, the real and imaginary parts of a realistic dielectric model are interrelated, meaning that the imaginary response can be calculated from the real response, and vice versa.

Reality Another constraint is that a *physical model must produce a response that is real in the time domain*, as discussed briefly in Section 2.3.3. If a function is real in the time domain, its Fourier transform must satisfy the complex-conjugate rule [LePage, 1980]:

$$F(-\omega) = F(\omega)^* \tag{6-37a}$$

For example, consider Gauss's law:

$$\nabla \cdot \vec{D} = \nabla \cdot \varepsilon \vec{E} = \rho$$

If $\vec{D}$ and $\vec{E}$ are real functions in the time domain, they must satisfy the complex conjugate rule in (6-37a) and therefore so must the permittivity [Jackson, 1998]:

$$\varepsilon(-\omega) = \varepsilon(\omega)^* \tag{6-37b}$$

Passivity A physical system is denoted as *passive* when it is unable to generate energy. For example, an n-port network is said to be passive if

$$\int_{-\infty}^{t} \mathbf{v}^{\mathrm{T}}(\tau) \cdot \mathbf{i}(\tau)\, d\tau \geq 0 \tag{6-38}$$

where $\mathbf{v}^{\mathrm{T}}(\tau)$ is the transpose of a matrix containing the port voltages and $\mathbf{i}(\tau)$ a matrix containing the currents. Integral (6-38) represents the cumulative net energy (power) absorbed by the system up to time t. In a passive system, this energy must be positive for all t. This requirement is satisfied if (1) the system absorbs more energy than it generates, and (2) any generation of energy happens after the absorption. A noncausal system that generates energy before it absorbs it would be considered to be nonpassive.

Summary It has been shown that a realistic dielectric model must be:

1. *Causal*, meaning that the model cannot respond to a stimulus until some time after it has been applied.
2. *Analytic*, which says that the function must satisfy the Cauchy–Riemann equations, meaning that the real and imaginary parts of the permittivity are related.
3. *Real* in the time domain, which means that its Fourier transform must satisfy the complex-conjugate rule (6-37).
4. *Passive*, which means that the model cannot generate energy.

The dielectric models derived in Section 6.3 satisfy these mathematical constraints.

6.5 FIBER-WEAVE EFFECT

As described briefly in Section 3.1, printed circuit boards (PCBs) are most commonly constructed from an FR4-type material. Many dielectric materials used in the electronics industry, including FR4, have been treated historically as a homogeneous dielectric medium for the propagation of digital signals in PCBs. As data rates for system buses push into the multi-Gb/s range, the composition of FR4, in which woven fiberglass bundles are embedded in epoxy resin, causes this assumption to break down. At high frequencies, local variations in dielectric constants create spatially dependent values of the time delay τ_d and the characteristic impedance Z_0. If not controlled properly, the spatially dependent transmission-line parameters can severely degrade voltage and timing margins, especially in differential signal-based buses, described in detail in Chapter 7. In this section we demonstrate how PCB physical structures create local, systematic dielectric constant variation, and describe material and design options for mitigating the effect.

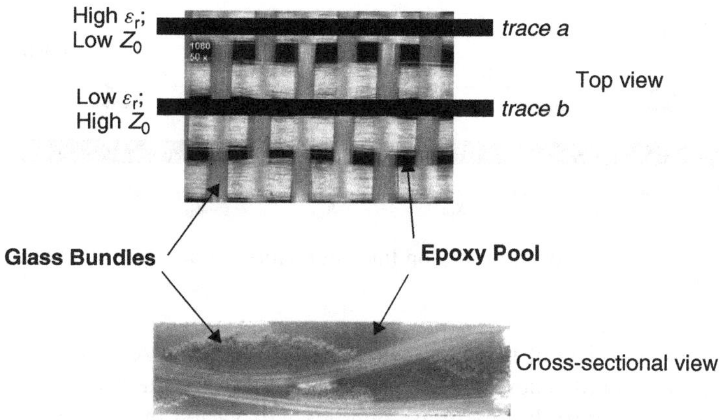

Figure 6-13 Physical structure of FR4 dielectric; note how the physical position of the transmission line with respect to the fiberglass weave dictates the impedance.

6.5.1 Physical Structure of an FR4 Dielectric and Dielectric Constant Variation

FR4 dielectric is a composite material made from a matrix of woven bundles of fiberglass embedded in an epoxy resin. The physical structure of FR4 is illustrated in Figure 6-13, which demonstrates how local material variation makes the homogeneous dielectric assumption inaccurate when propagating signals with multi-GHz frequency content. The reinforcing fiberglass bundles have a dielectric constant (ε_r) of approximately 6, whereas ε_r is close to 3 for the epoxy resin in which the bundles are embedded. The bulk dielectric constant is dependent on the glass/resin volume ratio:

$$\varepsilon_r = \varepsilon_{\text{rsn}} V_{\text{rsn}} + \varepsilon_{\text{gls}} V_{\text{gls}} \tag{6-39}$$

where ε_{rsn} and ε_{gls} are the dielectric permittivities and V_{rsn} and V_{gls} are the volume ratios of the epoxy resin and glass.

From Figure 6-13 we see that there is a significant gap between bundles over which a signal trace may be routed. This creates the opportunity for a differential pair to be routed such that one trace (a) is located over a fiberglass bundle, while the other (b) is routed above the gap between bundles. The result is that a signal on *trace a* will see a higher effective dielectric permittivity ($\varepsilon_{r,\text{eff}}$) and a lower impedance then *trace b*. Furthermore, it should be noted that this is a statistically likely scenario, since the fiber weave is usually manufactured so that the bundles are parallel to the edge of the board. The most common layout methodology is to route the transmission lines of the system buses at either 0° or 90° with respect to the board edges, which maximizes the chance that spatially dependent effective dielectric permittivity ($\varepsilon_{r,\text{eff}}$) will be observed.

The effective dielectric permittivity variability can be shown experimentally by using a test board structure as shown in Figure 6-14. If the structures are designed with a center–center trace distance (x_t) slightly larger than the expected

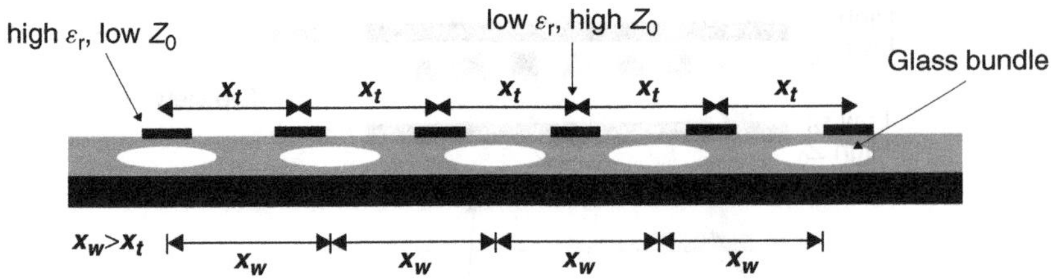

Figure 6-14 Depiction of transmission lines in relation to the fiber bundles in a printed circuit board.

distance between adjacent fiberglass bundles (x_w), routing several traces parallel to the fiberglass bundles across a large section of a PCB, the relative signal trace location with respect to the fiberglass bundles will varied in a systematic fashion to allow determination of the worst-case difference in $\varepsilon_{r,\text{eff}}$ between traces.

Figure 6-15 shows effective dielectric permittivity extracted from time-domain reflectometry (TDR) measurements on test structures similar to Figure 6-14. The effective dielectric permittivity is calculated from the measured time delay τ_d of the transmission line, which is determined from the TDR profile similar to Example 3-7. Next, equation (2-52) is used to calculate the effective permittivity where $\mu_r = 1$ and $\varepsilon_r = \varepsilon_{r,\text{eff}}$:

$$\nu_p = \frac{c}{\sqrt{\mu_r \varepsilon_{r,\text{eff}}}} \rightarrow \frac{1}{\nu_p} = \tau_d = \frac{\sqrt{\varepsilon_{r,\text{eff}}}}{c}$$
$$\rightarrow \varepsilon_{r,\text{eff}} = (c\tau_d)^2 \tag{6-40}$$

Note that the effective dielectric constant is a composite function of the glass cloth, epoxy resin, and surrounding air for microstrip transmission-line structures. Since the relative location of the traces is slowly varied across the PCB, the worst-case difference in effective dielectric constant is equal to the difference between minimum and maximum measured $\varepsilon_{r,\text{eff}}$. The example in Figure 6-15 shows a worst-case difference in the effective dielectric constant to be $\Delta\varepsilon_{r,\text{eff}} \sim 0.23$ for microstrip structures using 2116-type fiberglass cloth for the board, which corresponds to an actual dielectric permittivity difference for the material ($\Delta\varepsilon_r$) of approximately 1.5 to $2.25\Delta\varepsilon_{r,\text{eff}}$, based on empirical observations of measured data. Empirical data from numerous FR4-based test boards indicate that the differences in the measured effective dielectric permittivity for a microstrip structure can be as high as $\Delta\varepsilon_{r,\text{eff}} = 0.4$, leading to a material value of $\Delta\varepsilon_r \sim 0.8$ to 0.9. Although these spatial variations might seem small, the analysis in Chapter 7 will demonstrate that it can have a severe impact for relatively short differential transmission lengths at data rates of 5 to 10 Gb/s.

6.5.2 Mitigation

Routing pairs at angles relative to the direction of weave will average out the fiber-weave effect by constantly changing the relative position of the traces

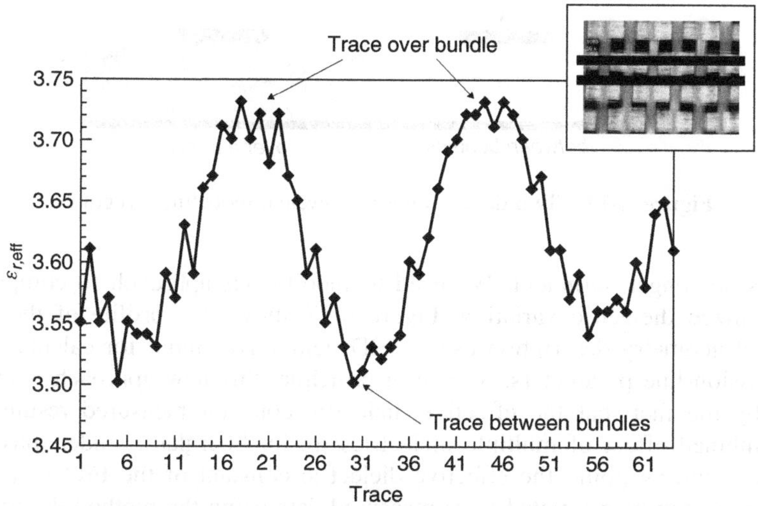

Figure 6-15 Example of $\varepsilon_{r,\,\mathrm{eff}}$ variations due to the fiber-weave effect for 64 parallel transmission lines routed parallel to a board edge constructed with FR4 dielectric material.

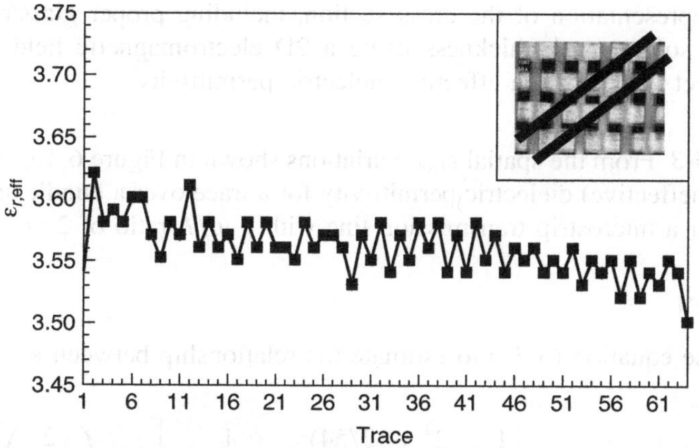

Figure 6-16 Example of the reduction in $\varepsilon_{r,\,\mathrm{eff}}$ variations from the fiber-weave effect for 64 parallel transmission lines routed 45° with respect to a board edge constructed with FR4 dielectric material.

relative to the bundles. As Figure 6-16 shows, routing at 45° angles minimizes the impact. It has been shown that offsets as small as 5 to 10° are adequate to mitigate most of the spatial effects.

6.5.3 Modeling the Fiber-Weave Effect

The preceding discussion indicates that the traditional approach of modeling a transmission line using a uniform dielectric constant throughout the dielectric

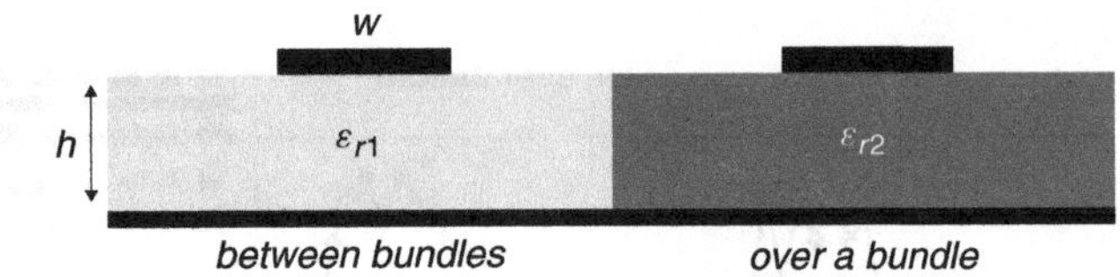

Figure 6-17 Two-dimensional fiber-weave modeling structure.

layer is no longer sufficient. We need to modify our approach to comprehend the localized dielectric variation. Figure 6-17 shows the profile of the cross-sectional geometry description used as 2D field solver input for calculating the transmission-line parameters. Simple in principle, this new approach is complicated by the fact that the effective dielectric constant measured results from the combined effect of multiple dielectrics: resin, fiberglass, solder mask, and air. As a starting point, the effective dielectric constant of the FR4 material in each region can be estimated from measured data using the method described in equation 6-39 for multicomponent dielectrics. To determine the actual values of ε_{r1} and ε_{r2} (see Figure 6-17), they should be varied until the delay of the 2D simulation is equal to the measured response from the test board. Alternatively, accurate representation of the cross section, including proper glass/resin height ratios and solder mask thickness using a 2D electromagnetic field solver will yield correct values of the effective dielectric permittivity.

Example 6-3 From the spatial $\varepsilon_{r,\text{eff}}$ variations shown in Figure 6-15, calculate the actual (noneffective) dielectric permittivity for a trace over a bundle and between bundles for a microstrip transmission line with a w/h ratio of 2 and $t \ll h$.

SOLUTION

Step 1: Use equation (3-35) to estimate the relationship between ε_r and $\varepsilon_{r,\text{eff}}$.

$$a = 1 + \frac{1}{49}\ln\frac{2^4 + (2/54)^2}{2^4 + 0.432} + \frac{1}{18.7}\ln\left[1 + \left(\frac{2}{18.1}\right)^3\right]$$

$$b = 0.564\left(\frac{\varepsilon_r - 0.9}{\varepsilon_r + 3}\right)^{0.053}$$

$$\varepsilon_{\text{eff}}(u, \varepsilon_r) = \frac{\varepsilon_r + 1}{2} + \frac{\varepsilon_r - 1}{2}\left(1 + \frac{10}{2}\right)^{-ab}$$

The relationships between ε_r and $\varepsilon_{r,\text{eff}}$ are plotted in Figure 6-18a. The value of ε_r that corresponds to the minimum and maximum values of $\varepsilon_{r,\text{eff}}$ in Figure 6-15 are chosen to create equivalent cross sections to represent the case where the trace is over a bundle and between bundles.

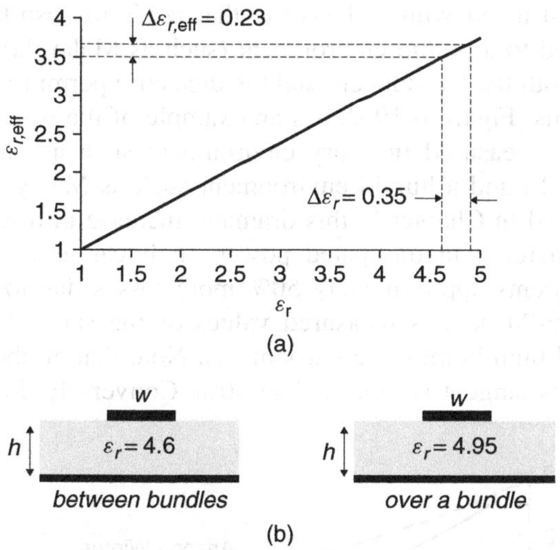

Figure 6-18 (a) Relationship between ε_r and $\varepsilon_{r,\text{eff}}$; (b) equivalent cross sections for a trace directly over a bundle and between bundles for Example 6-2.

Trace between bundles:

$$\varepsilon_{r,\text{eff}} = 3.5 \text{ (from Figure 6-15) and } \varepsilon_r \sim 4.6 \text{ (from Figure 6-18a)}$$

Trace over a bundle:

$$\varepsilon_{r,\text{eff}} = 3.72 \quad \text{and} \quad \varepsilon_r \sim 4.95$$

Step 2: Create the equivalent cross sections, as shown in Figure 6-18b.

6.6 ENVIRONMENTAL VARIATION IN DIELECTRIC BEHAVIOR

One problem often overlooked in high-speed digital design is the impact that the relative humidity (RH) of the environment has on the electrical performance of a dielectric material. The electrical properties of a dielectric are partially a function of the amount of moisture in the material. The extent to which a material absorbs moisture is characterized by its moisture diffusivity and saturated moisture concentration. *Moisture diffusivity* describes the rate of change of a material's moisture concentration. The *saturated moisture concentration* provides an expression for the limit to the amount of moisture that a material can contain. It is important to note that both are a function of temperature and relative humidity. Another measure of a material's susceptibility to moisture is its *maximum moisture uptake*. This is typically reported on material data sheets as % weight and is related to the saturated moisture concentration. If a printed circuit board (PCB) constructed with a dielectric that tends to absorb moisture (such as FR4) is exposed to a humid environment (such as Malaysia) for a significant amount of time, both the loss tangent and the dielectric permittivity will increase. To demonstrate this, Figure 6-19 shows an example of the insertion losses of an identical structure measured in a dry environment such as an Arizona winter (15% RH and 60°F) and a humid environment such as Malaysia (95% RH and 95°F). As explained in Chapter 9, this dramatic increase in insertion losses corresponds to an increase in dissipated power. In linear terms, the example in Figure 6-19 represents approximately 50% more losses due to the dielectric at 10 GHz. Figure 6-20 depicts measured values of the same dielectric material for both low- and high-humidity environments. Note that at about 7.5 GHz, the increase in the loss tangent is greater than 50%. Conversely, Figure 6-21 shows

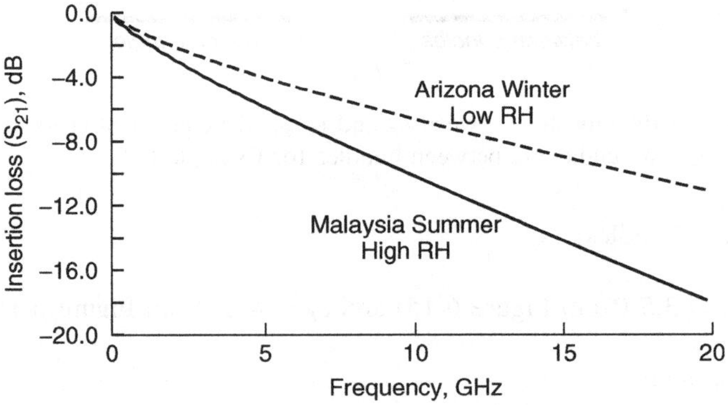

Figure 6-19 Example of the impact of the environment on PCB losses for a microstrip structure constructed on an FR4 dielectric.

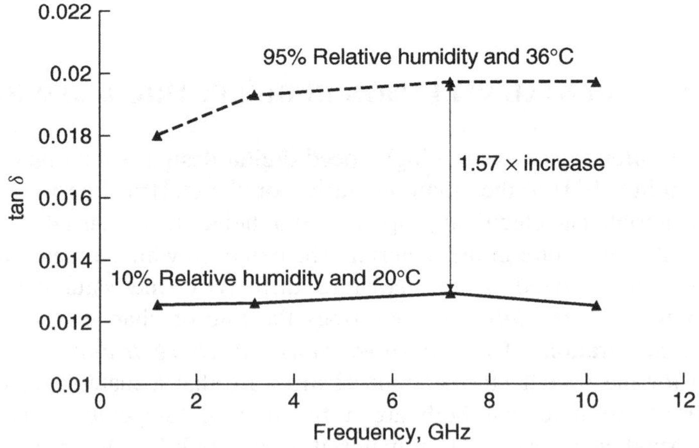

Figure 6-20 Example of the measured variation in tanδ for 7628-FR4 dielectric for different environmental extremes.

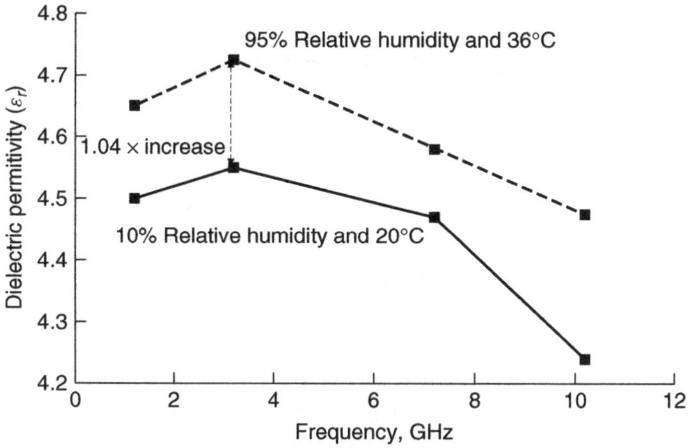

Figure 6-21 Example of the measured variation in ε_r for 7628-FR4 dielectric for different environmental extremes.

a relatively small increase of less than 5% in the dielectric permittivity for the environmental extremes [Hamilton et al., 2007].

The mechanism that explains this dramatic increase in losses and the minor increase in permittivity is probably a combination of orientational and ionic polarization. The water molecule is relatively heavy, and it therefore takes a significant amount of energy for the molecule to follow the external time-varying electric field, as shown in Figure 6-3. Furthermore, the relative dielectric permittivity of distilled water is $\varepsilon_r \sim 81$. Since the dielectric permitivity of the FR4 dielectric does not change much (~5%) between dry and humid conditions whereas the loss tangent changes dramatically (~50%), it indicates that (1) a relatively small amount of water is absorbed by the dielectric, and (2) a small amount of absorbed water (per unit volume of dielectric) has a large impact on the dielectric losses.

6.6.1 Environmental Effects on Transmission-Line Performance

The absorption of water in an FR4 dielectric can be observed with vector network analyzer (VNA) measurements. Using linear values for S_{21} and S_{11}, the power loss of a transmission line as derived in Chapter 9 is

$$\alpha_{\text{loss}} = 1 - S_{21}^2 - S_{11}^2$$

where α_{loss} is a measure of the power dissipated by the transmission line from conductor, dielectric, and radiation losses. The humidity-induced change in α_{loss} is defined as the difference between the measured loss of a transmission line in a dry state ($\alpha_{\text{loss,dry}}$) and the measured loss of a transmission line that has been exposed in a humid environment for a significant amount of time ($\alpha_{\text{loss,humid}}$). This humidity-induced change in α_{loss} from the dry state is termed $\Delta\alpha_{\text{humid}}$:

$$\Delta\alpha_{\text{humid}} = \alpha_{\text{loss,humid}} - \alpha_{\text{loss,dry}} \tag{6-41}$$

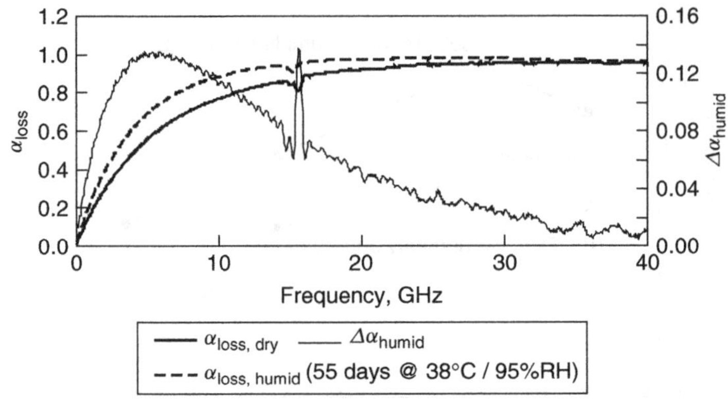

Figure 6-22 Example of how a humid environment will change the losses of 7628-FR4 dielectric. (Adapted from Hamilton et al. [2007].)

Hamilton et al. [2007] detail several experiments where the relative humidity was controlled with an environmental chamber and measurements were taken with a vector network analyzer (VNA) to determine how dry and wet environments influence dielectric losses. Figure 6-22 depicts an example of a measurement showing how humidity changes the loss characteristics of a test board built on an FR4 dielectric. The dry case ($\alpha_{loss,dry}$) was preconditioned in an environmental chamber to bake out the moisture already absorbed by the dielectric and then stored in a dry box at 20°C and 10% RH. The wet case ($\alpha_{loss,humid}$) was conditioned in an environment of 38°C and 95% RH for 55 days. Note that $\Delta\alpha_{humid}$ peaks at about 5 GHz and then decreases to almost zero at high frequencies. The peak is caused by the increase in power dissipation caused by the affinity of the dielectric material to absorb water from a humid environment as demonstrated by the increased slope of $\Delta\alpha_{humid}$ prior to the peak. After 5 GHz, the insertion loss (S_{21}) for both humid and dry conditions will increase (meaning that the transmitted power decreases), causing both $\Delta\alpha_{loss,humid}$ and $\Delta\alpha_{loss,dry}$ to approach unity. Consequently, $\Delta\alpha_{humid}$ will initially peak because $\Delta\alpha_{loss,humid}$ will increase faster than $\Delta\alpha_{loss,dry}$, then approach zero. This measurement highlights how a transmission line can exhibit a significant increase in dissipated power losses if the dielectric material has an affinity to absorb water from a humid environment.

For design purposes, the question arises of how long it takes for moisture to diffuse into the dielectric. Obviously, each dielectric material will exhibit its own moisture absorption characteristics. For example, Teflon (PTFE) generally exhibits very little or no tendency to absorb moisture, whereas common solder-mask materials exhibit a very high tendency to absorb moisture. For the purposes of this discussion, we concentrate on an FR4 dielectric sample.

Figure 6-23 shows the measured change in insertion loss (S_{21}) at 10 GHz for a microstrip transmission line built on 7628-FR4 dielectric that was preconditioned to a dry state by baking the moisture out of the sample and then exposed to humid

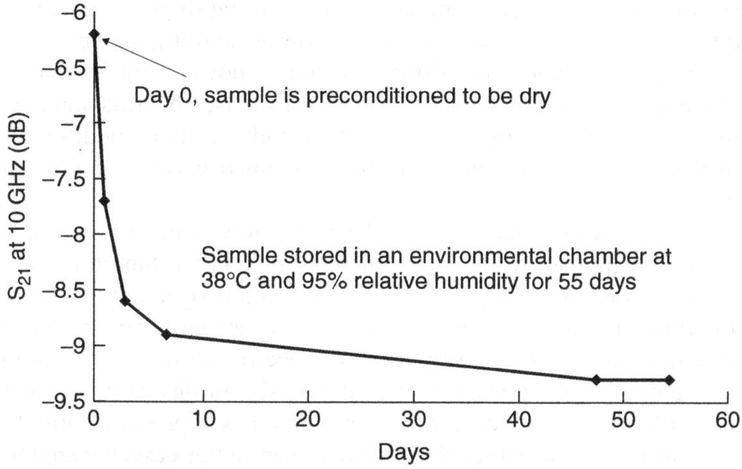

Figure 6-23 Insertion loss at 10 GHz of a transmission line built on 7628-FR4 dielectric, showing how losses change with time exposed to a humid environment. (Adapted from Hamilton et al. [2007].)

conditions in an environmental chamber and measured at specific intervals for 55 days. In this example, the dielectric transmission-line losses at 10 GHz were -6.3 dB under dry conditions at $t = 0$. When the structure was exposed to a humid environment, the dielectric became almost fully saturated within the first 7 days, increasing the loss to -8.9 dB. Finally, the transmission-line loss stabilized to a value of -9.3 dB of loss at $t = 48$ days.

Stripline transmission lines tend to absorb moisture at a significantly slower rate than microstrip lines because there is much less dielectric surface area exposed to the surrounding environment. Hamilton et al. [2007] report experimental data for striplines built on 7628-FR4 dielectric with 90% coverage of the dielectric material by copper planes, showing that it takes about 5 months for the stripline to reach the same loss value as a microstrip does in 7 days under identical conditions. This means that striplines are not immune to environmental conditions; however, they absorb moisture at a significantly slower rate.

6.6.2 Mitigation

Unfortunately, there are very few practical methods for mitigating the effect of the environment on transmission-line performance. In consumer products such as laptop or desktop mother boards, the full range of environmental conditions must be accounted for in the design to ensure that performance specifications are maintained in both dry and humid environments (such as a dry Arizona winter or a humid Florida summer). This means that an FR4-based transmission-line model used in the design must comprehend as much as a 50% or more variation in the loss tangent, which can significantly reduce solution spaces, particularly for long bus lengths and high data rates. Generally, the use of stripline transmission lines does not cure the effect of the environment because products are generally designed to operate under all reasonable conditions, and it must be assumed that there will be periods where the digital design is powered down long enough for the dielectric to absorb enough water to increase the loss tangent. However, striplines can help mitigate the problem under certain conditions because it takes a very long time for the dielectric contained between copper layers to absorb water.

In some cases, the product usage model may allow mitigation of the environmental effects by not allowing large changes in relative humidity for normal operation. For example, computer servers are often operated in a controlled temperature/humidity room, allowing engineers to design for minimum environmental variation, thus increasing the performance of the platform. Additionally, some computers are often powered on continuously. In this case, the heat generated by the CPU and the movement of air by the fan will produce a low-humidity environment inside the computer chassis. However, in this case, the engineer must understand that if the computer is powered down for any significant amount of time, the board dielectric material may begin to absorb moisture. The absorbed moisture will increase the loss of the transmission lines and decrease the performance of the high-speed buses until the absorbed moisture diffuses out, which could take several hours to several days, depending on the dielectric material and the environmental conditions inside the chasses during operation and transmission-line structure.

6.6.3 Modeling the Effect of Relative Humidity on an FR4 Dielectric

The modeling method employed to simulate the effect of moisture absorption in a dielectric material is entirely dependent on the dielectrics affinity to absorb water, how the absorbed moisture interacts with the chemical structure of the dielectric, and the frequency range of interest. The dielectric most commonly used in the electronics industry is FR4, which readily absorbs moisture. To understand how to account properly for environmental variations in the relative humidity, we consider how to model FR4 dielectric for two cases:

1. *Low frequencies (below 2 GHz).* Below 2 GHz, the effect of moisture absorption has very little effect on total transmission-line losses (at least for FR4 dielectrics). This behavior is demonstrated in Figure 6-19, where the difference in insertion loss is minimal until approximately 2 GHz. This is because at low frequencies, the total loss is usually dominated by the skin effect resistance of the signal conductor, as described in Section 5.2. For typical transmission lines, the losses of the dielectric begin to dominate above about 1 to 2 GHz.
2. *High frequencies (2 to 50 GHz).* For this frequency range, the dry and nominal environmental corners (up to approximately 50% relative humidity for FR4) can be modeled with the infinite-pole Debye model, as described in Section 6.3.5 and shown in equation (6-29). However, for

high-humidity conditions, the assumptions used to derive the infinite-pole model break down at frequencies greater than about 10 GHz. Specifically, the approximation that ω_{2i} in equation (6-28) decreases linearly on the logarithmic scale is no longer valid for a dielectric that has absorbed a significant amount of moisture. Therefore, for humid conditions (above 50% relative humidity) and frequencies above approximately 10 GHz, FR4 dielectric must be fit empirically to a multipole model as described in Section 6.3.4 with equation (6-27) and demonstrated in Example 6-1. It is sometimes necessary to account for induced dipole resonances with the squared term in equation (6-27) (ω^2/ω_{1i}^2).

6.7 TRANSMISSION-LINE PARAMETERS FOR LOSSY DIELECTRICS AND REALISTIC CONDUCTORS

As bus data rates increase and physical implementations of high-speed digital designs shrink, transmission-line losses become more important. Consequently, the engineer must have the ability to calculate the response of a transmission line successfully and account for realistic conductor and dielectric behavior. In the next two sections we build on the equivalent circuit derived in Section 5.4, which included nonideal conductor effects, and describe how to include dielectric losses in the equivalent circuit and modify the telegrapher's equations to comprehend dielectric losses.

6.7.1 Equivalent Circuit, Impedance, and Propagation Constant

Building directly on Section 5.4.1, where we developed the equivalent circuit, impedance, and propagation constant for a transmission line with a realistic conductor, we can complete the transmission-line model by adding the effects of dielectric losses. The assumption of a perfect dielectric did not allow calculation of a loss term. Fortunately, the form of the equivalent circuit derived in Section 3.2 is also applicable to a line whose dielectrics have an imaginary term as in equation (6-15), and therefore losses. To begin this derivation, the shunt admittance of an ideal transmission line with a loss-free dielectric is calculated as

$$Y_{\text{shunt}} = j\omega C \tag{6-41}$$

which has units of siemens. The equivalent circuit is easily modified to include a shunt resistor with a conductance G with units of siemens, in parallel with the capacitance to account for the dielectric losses, as shown in Figure 6-24, where N_s is the number of segments, $C_{\Delta z} = \Delta z\, C$ and $L_{\Delta z} = \Delta z\, L_{\text{total}}$, as calculated from Section 3.23 and $R_{\Delta z} = \Delta z\, R_{\text{ac}}$ from Section 5.4.1 and $G_{\Delta z} = \Delta z\, G$, where Δz is the length of the differential section of transmission line and C, L_{total}, R_{ac}, and G are the capacitance, inductance, resistance, and conductance per unit length.

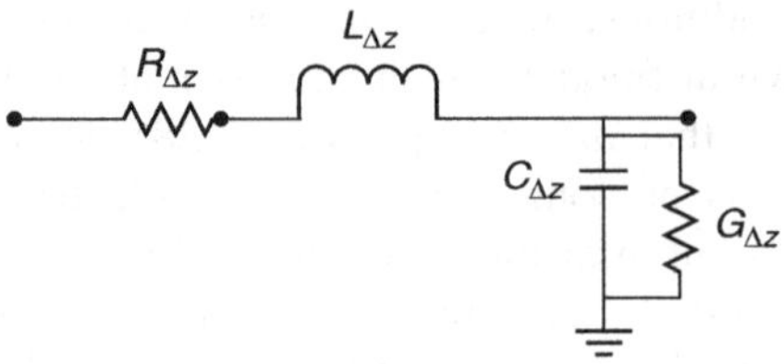

(a)

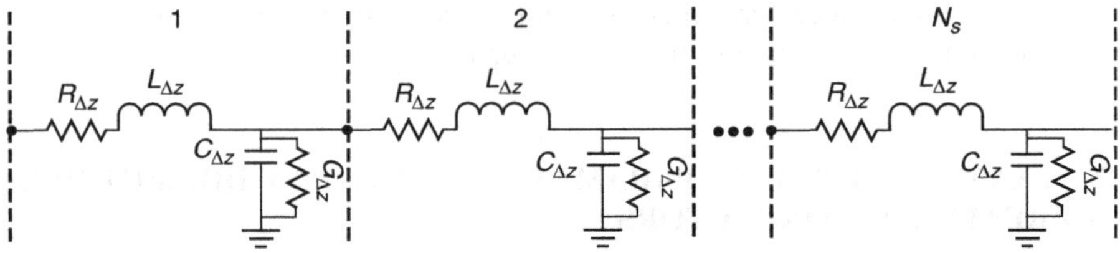

(b)

Figure 6-24 (a) Model for a differential element of a transmission line; (b) full model.

For this equivalent circuit, equation (6-42) calculates the shunt admittance by adding a conductance term G:

$$Y_{\text{shunt}} = G + j\omega C \tag{6-42}$$

The formula for G can be derived by

$$\varepsilon = \varepsilon' - j\frac{\sigma_{\text{dielectric}}}{\omega} = \varepsilon' - j\varepsilon'' \tag{6-15}$$

$$\tan|\delta| = \frac{\varepsilon''}{\varepsilon'} \tag{6-16b}$$

which show clearly that the dielectric losses will be proportional to ε'' and $\tan|\delta|$, and therefore G.

Furthermore, we know from Section 6.4.1 that the real and imaginary parts of the dielectric permittivity must be related. Consequently, there must also be a relationship between the conductance G and the capacitance C. If the dielectric losses are treated as an equivalent conductivity, we can say that the dielectric carries a current of $\vec{J} = \sigma_{\text{dielectric}}\vec{E}$ [equation (2-7)]. Equation (3-1) says that the voltage between the signal conductor and the reference plane is $v = -\int_a^b \vec{E} \cdot d\vec{l}$ and that the total current is calculated from (2-20) as $i = \int_S \vec{J} \cdot d\vec{s}$. Therefore, in circuit terms, the conductance G can be written

$$G = \frac{i}{v} = \frac{\int_S \vec{J} \cdot d\vec{s}}{-\int_a^b \vec{E} \cdot d\vec{l}} = \frac{\sigma_{\text{dielectric}} \int_S \vec{E} \cdot d\vec{s}}{-\int_a^b \vec{E} \cdot d\vec{l}} \tag{6-43a}$$

Similarly, from equations (2-76), (2-59) and (3-1), we can write the capacitance in the form

$$C = \frac{Q}{v} = \frac{\varepsilon \int_S \vec{E} \cdot d\vec{s}}{-\int_a^b \vec{E} \cdot d\vec{l}} \tag{6-43b}$$

Dividing (6-43a) by (6-43b) yields the following ratio:

$$\frac{G}{C} = \frac{\sigma_{\text{dielectric}}}{\varepsilon} \tag{6-44}$$

From (6-15) and (6-16b),

$$\sigma_{\text{dielectric}} = \tan|\delta|\varepsilon'\omega$$

which is substituted into (6-44) to yield

$$G(\omega) = \tan|\delta|\omega C \tag{6-45}$$

where $G(\omega)$ is the frequency-dependent conductance in units of siemens per unit length, $\omega = 2\pi f$, and C is the capacitance per unit length for the transmission line.

The characteristic impedance, which was defined in equation (3-33), can be calculated by dividing the series impedance as defined by (5-68) by the parallel admittance defined by (6-42) for a short section of transmission line of length Δz:

$$Z_0 = \sqrt{\frac{Z_{\text{series}}}{Y_{\text{shunt}}}} = \sqrt{\frac{R_{\text{ac}} + j\omega L_{\text{total}}}{G + j\omega C}} \quad \text{ohms} \tag{6-46}$$

Note that the units in (6-46) are $\sqrt{\text{ohms}/(1/\text{ohms})} = \sqrt{(\text{ohms})^2} = \text{ohms}$.

The propagation constant can be derived by inserting the complex values of the series impedance and shunt admittance into the loss-free formula derived in Section 3.2.4 in (3-30), which takes the form

$$\gamma = \alpha + j\beta = 0 + j\omega\sqrt{LC} = \sqrt{(j\omega L)(j\omega C)} = \sqrt{Z_{\text{lossless}} Y_{\text{lossless}}}$$

Substitution of Z_{series} and Y_{shunt} in place of the loss-free values of the series impedance and the parallel admittance yields the propagation constant for a transmission line with a lossy dielectric and a conductor with a finite conductivity:

$$\gamma = \alpha + j\beta = \sqrt{Z_{\text{series}} Y_{\text{shunt}}} = \sqrt{(R + j\omega L)(G + j\omega C)} \tag{6-47}$$

If the loss parameters of equation (6-47) are *small* compared to $j\omega L_{\text{total}}$ and $j\omega C$, which is the case for most practical transmission lines, (6-47) can be approximated [Johnk, 1988]:

$$\gamma \approx \alpha + j\beta = \frac{1}{2}\left(R\sqrt{\frac{C}{L}} + G\sqrt{\frac{L}{C}}\right) + j\omega\sqrt{LC} \qquad \text{(6-48a)}$$

$$\alpha \approx \frac{1}{2}\left(R\sqrt{\frac{C}{L}} + G\sqrt{\frac{L}{C}}\right) \qquad \text{(6-48b)}$$

$$\beta \approx \omega\sqrt{LC} \qquad \text{(6-48c)}$$

As described in Section 3.2.4, the voltage propagating on a loss-free transmission ($\alpha = 0$) line can be written

$$v(z) = v(z)^+ e^{-jzw\sqrt{LC}} + v(z)^- e^{jzw\sqrt{LC}} \qquad \text{(3-29)}$$

However, for a lossy transmission line, the voltage is described by multiplying equation (3-29) by the decay factor $e^{-\alpha z}$:

$$v(z) = v(z)^+ e^{-\left(\alpha + jw\sqrt{LC}\right)z} + v(z)^- e^{\left(\alpha + jw\sqrt{LC}\right)z} = v(z)^+ e^{-\gamma z} + v(z)^- e^{\gamma z} \qquad \text{(6-49)}$$

Equation (6-49) describes the voltage propagating on a lossy transmission line.

Example 6-4 Calculate the frequency-dependent voltage at the output of a 0.5-m transmission line that is perfectly terminated in its characteristic impedance with the following properties: $L_{\text{ext}} = 2.5 \times 10^{-7}$ H/m, $C_{\text{quasistatic}} = 1.5 \times 10^{-10}$ F/m, $\sigma = 5.8 \times 10^{7}$ S/m, $\mu = 4\pi \times 10^{-7}$ H/m, $l = 0.5$ m, and $w = 100 \times 10^{-6}$ m, $\varepsilon_{r,\text{eff}} = 3.32$, and $\tan\delta = 0.0205$ at 1 GHz.

SOLUTION

Step 1: Calculate the frequency-dependent parameters of the conductor. The resistance can be calculated approximately using equation (5-12), and is plotted in Figure 6-25a.

$$R_{\text{ac}} = \frac{l}{w}\sqrt{\frac{\pi\mu f}{\sigma}}$$

The inductance is calculated with equations (5-20) and (5-30) and plotted in Figure 6-25b.

$$L_{\text{total}} = L_{\text{internal}} + L_{\text{external}} \qquad \text{(5-20)}$$

$$L_{\text{internal}} = \frac{R_{\text{ac}}}{\omega} \qquad \text{(5-30)}$$

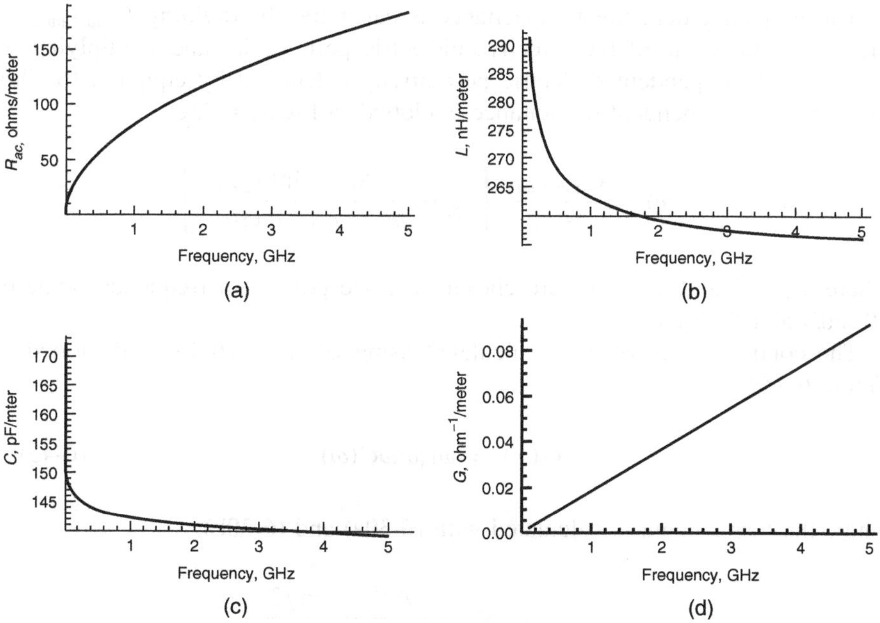

Figure 6-25 Frequency-dependent transmission-line parameters for Example 6-4.

Step 2: Calculate the frequency-dependent parameters of the dielectric. First, the real and imaginary parts of the dielectric constant must be found:

$$\varepsilon' = \varepsilon_r = 3.32$$

$$\tan\delta = \frac{\varepsilon''}{\varepsilon'} = 0.0205$$

$$\varepsilon'' = 0.068$$

$$\varepsilon = \varepsilon' - j\varepsilon'' = 3.32 - j0.068$$

Equations (6-30a) and (6-30b) are solved simultaneously to get $\Delta\varepsilon'$ and ε'_∞:

$$3.32 = \varepsilon'_\infty + \frac{\Delta\varepsilon'}{m_2 - m_1}\frac{\ln(\omega_2/\omega)}{\ln(10)}$$

$$0.068 = \frac{\Delta\varepsilon'}{m_2 - m_1}\frac{-\pi/2}{\ln(10)}$$

where $m_1 = 1$ and $m_2 = 11$ are chosen to correspond to a frequency range of 10 rad/s to 100 Grad/s, ω is $2\pi(1\text{ GHz})$, and $\omega_2 = 10^{11}$.

$$\Delta\varepsilon' = 0.997$$

$$\varepsilon'_\infty = 3.2$$

The frequency-dependent capacitance is calculated by dividing $C_{\text{quasistatic}}$ by the quasistatic value of the effective dielectric permittivity and multiplying by the frequency-dependent dielectric permittivity calculated by equation (6-30a). The frequency-dependent capacitance is plotted in Figure 6-25c:

$$C(\omega) = \frac{C_{\text{quasistatic}}}{\varepsilon_{r,\text{eff}}} \left[\varepsilon'_{\infty} + \frac{\Delta\varepsilon'}{m_2 - m_1} \frac{\ln(\omega_2/\omega)}{\ln(10)} \right]$$

where $m_1 = 1$ and $m_2 = 11$ are chosen to correspond to a frequency range of 10 rad/s to 100 Grad/s.

The conductance, $G(\omega)$, is calculated using equation (6-45) and plotted in Figure 6-25d:

$$G(\omega) = \tan|\delta|\omega C(\omega) \tag{6-45}$$

where the loss tangent is calculated with (6-30a) and (6-30b):

$$\tan|\delta = |\frac{\varepsilon''}{\varepsilon'} \approx \frac{\dfrac{\Delta\varepsilon'}{m_2 - m_1} \dfrac{-\pi/2}{\ln(10)}}{\varepsilon'_{\infty} + \dfrac{\Delta\varepsilon'}{m_2 - m_1} \dfrac{\ln(\omega_2/\omega)}{\ln(10)}}$$

Step 3: Calculate the frequency-dependent voltage. Since the transmission line is perfectly terminated, no reflections will be generated. Therefore, the loss-free voltage wave will behave as described by equation (6-49) with $v(z)^- = 0$.

$$v_{\text{out}} = v_{\text{in}} e^{-\alpha z} e^{-j\beta z} = v_{\text{in}} e^{-\gamma z}$$

Therefore, using equation (2-31) to simplify,

$$v_{\text{out}} = v_{\text{in}} e^{-\alpha z} [\cos(-\beta z) + j\sin(-\beta z)]$$

where γ is defined by equation (6-47) and z is the line length:

$$\gamma(\omega) = \alpha + j\beta = \sqrt{(R + j\omega L)(G + j\omega C)} \tag{6-47}$$

Next, the magnitude of the voltage is calculated:

$$v_{\text{out,mag}} = \sqrt{\text{Re}(v_{\text{out}})^2 + \text{Im}(v_{\text{out}})^2}$$

and is plotted in Figure 6-26 assuming that $v_{\text{in}} = 1$ V.

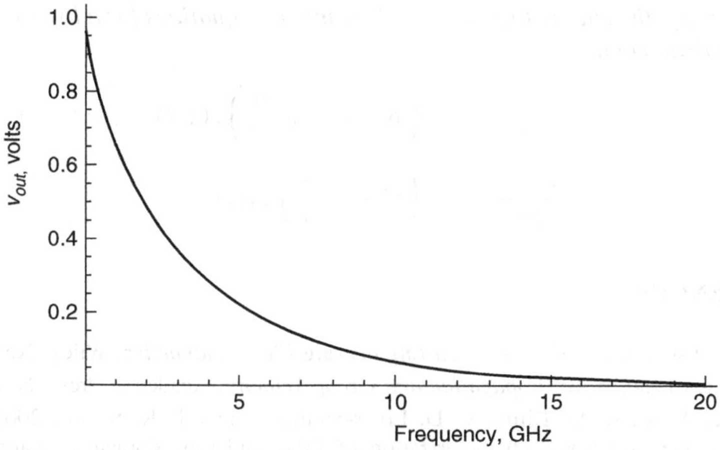

Figure 6-26 Magnitude of voltage at the output of a 0.5-m transmission line of Example 6-4.

6.7.2 Telegrapher's Equations for Realistic Conductors and Lossy Dielectrics

The time-harmonic forms of the telegrapher's equations for a transmission line with a perfect dielectric and perfect conductor were shown in Section 3.2.4:

$$\frac{dv(z)}{dz} = -j\omega L i(z) \tag{3-25}$$

$$\frac{di(z)}{dz} = -j\omega C v(z) \tag{3-26}$$

These formulas were adjusted in Section 5.4.2 to account for a realistic conductor with a finite conductivity. The classic form of the telegrapher's equations for a perfectly insulating dielectric and a realistic conductor are

$$\frac{\partial v(z,t)}{\partial z} = -\left(R_{\text{ac}} + L_{\text{total}}\frac{\partial}{\partial t}\right) i(z,t) \tag{5-72a}$$

$$\frac{\partial i(z,t)}{\partial z} = -C\frac{\partial v(z,t)}{\partial t} \tag{5-72b}$$

To include the effects of a lossy dielectric, equation (5-72) must be modified to include the conductance G, as defined in equation (6-45). Notice that the right-hand side of equation (3-26) is simply the inverse impedance of a capacitor (i.e., the conductance). To account for the dielectric losses, G is simply added into the right-handside of (3-26):

$$\frac{di(z)}{dz} = -[G + j\omega(C)]v(z) \tag{6-50}$$

Consequently, *the classic form of the telegrapher's equations for a lossy dielectric and a realistic conductor are*

$$\frac{\partial v(z,t)}{\partial z} = -\left(R_{\text{ac}} + L_{\text{total}}\frac{\partial}{\partial t}\right) i(z,t) \tag{5-72a}$$

$$\frac{\partial i(z,t)}{\partial z} = -\left(G + C\frac{\partial}{\partial t}\right) v(z,t) \tag{6-51}$$

REFERENCES

Balanis, Constantine, 1989, *Advanced Engineering Electromagnetics*, Wiley, New York.

Brillouin, L., 1960, *Wave Propagation and Group Velocity*, Academic Press, New York.

Djordjevic, A. R., R. M. Biljic, V. D. Likar-Smiljanic, and T. K. Sarkar, 2001, Wide-band frequency domain characterization of FR-4 and time domain causality, *IEEE Transactions on Electromagnetic Compatability*, vol. 43, no. 4, Nov., pp. 662–667.

Elliot, R., 1993. *Electromagnetics*, IEEE Press, Piscataway, NJ.

Hall, Stephen, Steven G. Pytel, Paul G. Huray, Daniel Hua, Anusha Moonshiram, Gary A. Brist, and Edin Sijercic, 2007, Multi-GHz causal transmission line modeling using a 3-D hemispherical surface roughness approach, *IEEE Transactions on Microwave Theory and Techniques*, vol. 55, no. 12, Dec.

Hamilton, Paul, Gary Brist, Guy Barnes, Jr., and Jason Schrader, 2007, Humidity dependent losses in FR4 dielectric, presented at the IPC Apex Conference.

Huray, Paul, 2009, *Foundations of Signal Integrity*, Wiley, Hoboken, NJ.

Jackson, J. D., 1998, *Classical Electrodynamics*, 3rd ed., Wiley, New York.

Johnk, Carl T. A., 1988, *Engineering Electromagnetic Fields and Waves*, Wiley, New York.

LePage, Wilbur P., 1980, *Complex Variables and the Laplace Transform*, Dover Publications, New York.

Pauling, Linus, 1948, *The Nature of the Chemical Bond*, Cornell University Press, Ithaca, NY.

Pytel, Steven G., 2007, Multi-gigabit data signaling rates for PWBs including dielectric losses and effects of surface roughness, Ph.D. dissertation, Department of Electrical Engineering, University of South Carolina.

PROBLEMS

6-1 If the dielectric permittivity of a dielectric is known to be $\varepsilon_r = 4.2$ and $\tan\delta = 0.021$ at 1 GHz, plot the frequency dependent values $\varepsilon_r(f)$ and $\tan\delta(f)$ from 1 MHz to 20 GHz.

6-2 Given the plot of the dielectric permittivity in Figure 6-27a and the loss tangent plot in Figure 6-27b, derive an equation for a frequency-dependent dielectric model.

6-3 Given the real part of the dielectric permittivity in Figure 6-28, calculate the loss tangent.

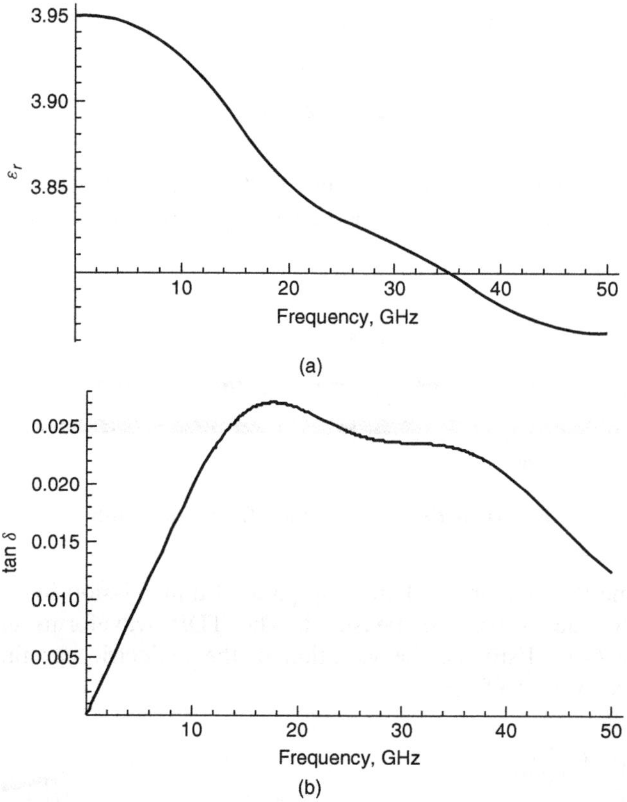

Figure 6-27 (a) Permittivity data and (b) loss tangent data for Problem 6-2.

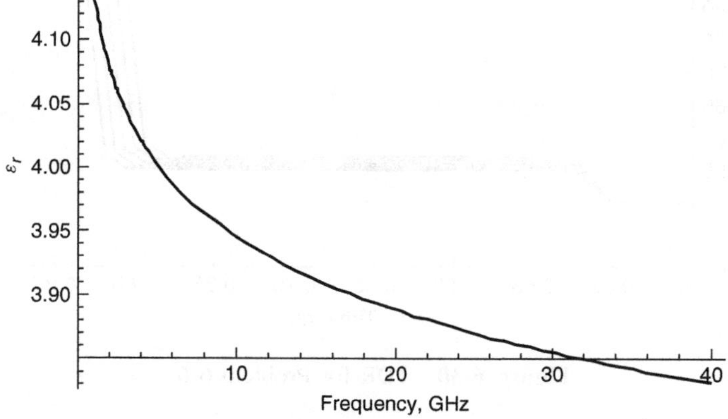

Figure 6-28 Dielectric permittivity for Problem 6-3.

6-4 Show that the Debye dielectric model [equation (6-28)] satisfies the Cauchy–Riemann equations for $i = 1$:

$$\varepsilon = \varepsilon'_\infty + \sum_{i=1}^{n} \frac{\Delta\varepsilon'_i}{1 + j(\omega/\omega_{2i})} \tag{6-28}$$

6-5 Given the cross section in Figure 6-29, estimate the bulk value of the dielectric permittivity. Assume that $\varepsilon_{r,\text{glass}} = 6.1$ and $\varepsilon_{r,\text{epoxy}} = 3.2$.

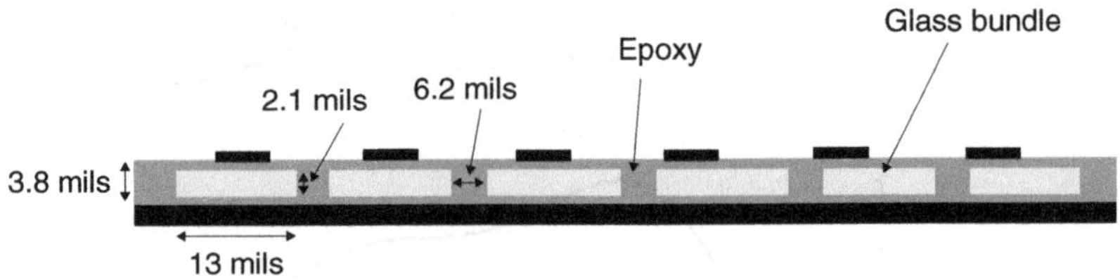

Figure 6-29 Cross section for Problem 6-5.

6-6 Assume that a series of 1-in.-long parallel transmission lines routed adjacent to each other are measured. The TDR waveforms are shown in Figure 6-30. Estimate the variation in the dielectric permittivity due to the fiber-weave effect.

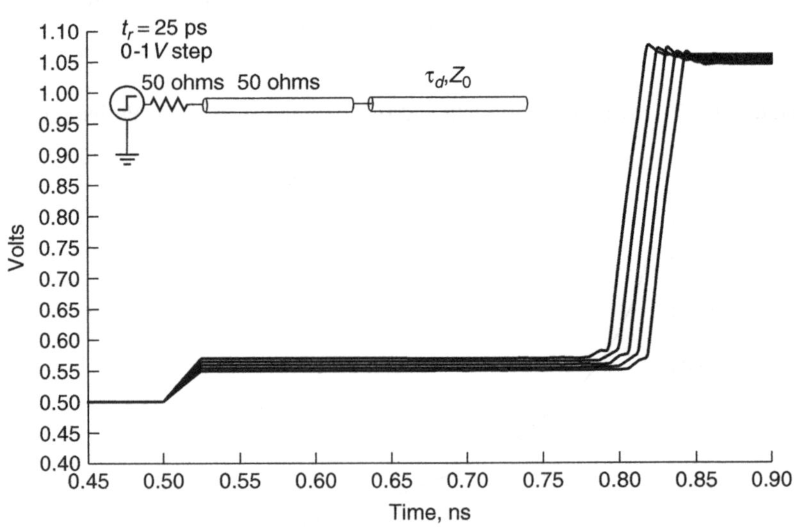

Figure 6-30 TDR for Problem 6-6.

6-7 Assuming a conductor width of 5 mils and a loss-free dielectric thickness of 4 mils, create an equivalent-circuit model of the corner cases in

Problem 6-6. Simulate them to show that the model matches the TDR measurements.

6-8 What is the best way to minimize losses for a given cross section of a transmission line with constant impedance? What is the best way to minimize losses if constant impedance is not required? Plot the voltage magnitude versus frequency for a 10-in.-long transmission line to prove your theory for each case.

6-9 Given the relationship between the fiber-weave and transmission-line dimensions shown in Figure 6-31, determine the minimum angle needed to mitigate the impedance and velocity swings caused by the fiber-weave effect. Assume that the maximum and minimum delay cases of the TDR measurements of Figure 6-30 correspond to the cross sections of Figure 6-31.

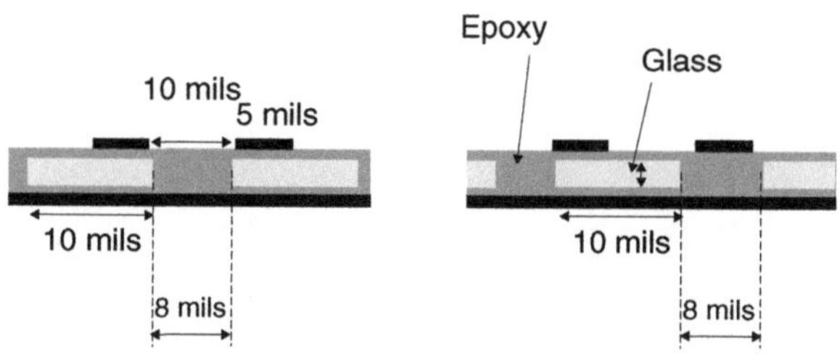

Figure 6-31 Cross sections for Problem 6-9.

6-10 For the cross section shown in Figure 6-32, plot the frequency-dependent R, L, C, G, Z_0, and τ_d from 100 MHz to 20 GHz. Assume that the surface of the copper is specified to have a RMS tooth height of 0.8μm.

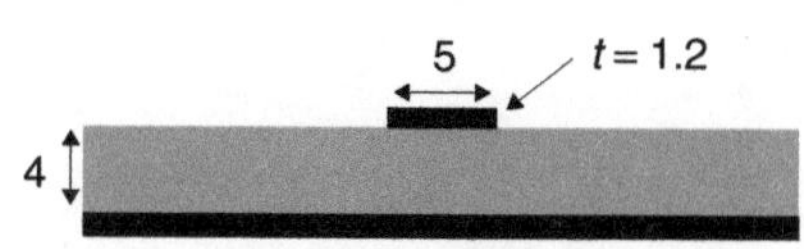

Figure 6-32 Cross section for Problem 6-10.

7

DIFFERENTIAL SIGNALING

When the interconnections between drivers and receivers on a bus are implemented with a dedicated transmission line for each bit, the signaling scheme is said to be *single ended*. Buses designed with single-ended signaling generally work well up to approximately 1 to 2 Gb/s. As data rates increase, it becomes increasingly difficult to maintain adequate signal integrity because digital systems are notoriously noisy. For example, large arrays of I/O circuits used to drive digital information onto the bus induce noise on the power and ground planes called *simultaneous switching noise* (for a complete description, see a book by Hall et al. [2000]). There are many other sources of noise that can severely distort the integrity of the digital waveform such as crosstalk (as discussed in Chapter 4) and nonideal current return paths (as discussed in Chapter 10). With single-ended signaling, each data bit is transmitted on a single transmission line and latched into the receiver with the bus clock. The decision of whether the bit is a 0 or a 1 is determined by comparing the received waveform to a reference voltage v_{ref}. If the received waveform has a voltage greater than v_{ref}, the signal is latched in as a 1, and if it is below v_{ref}, it is latched in as a logic 0. Noise coupled

Advanced Signal Integrity for High-Speed Digital Design, By Stephen H. Hall and Howard L. Heck

onto the driver, receiver, transmission lines, reference planes, or clock circuits will degrade the ideal relationship between the transmitted waveform and v_{ref}. If the magnitude of the noise is large enough, the incorrect digital states will be latched into the receiver, and bit errors will occur. Figure 7-1 depicts how noise can make the determination of a logic 0 or 1 uncertain.

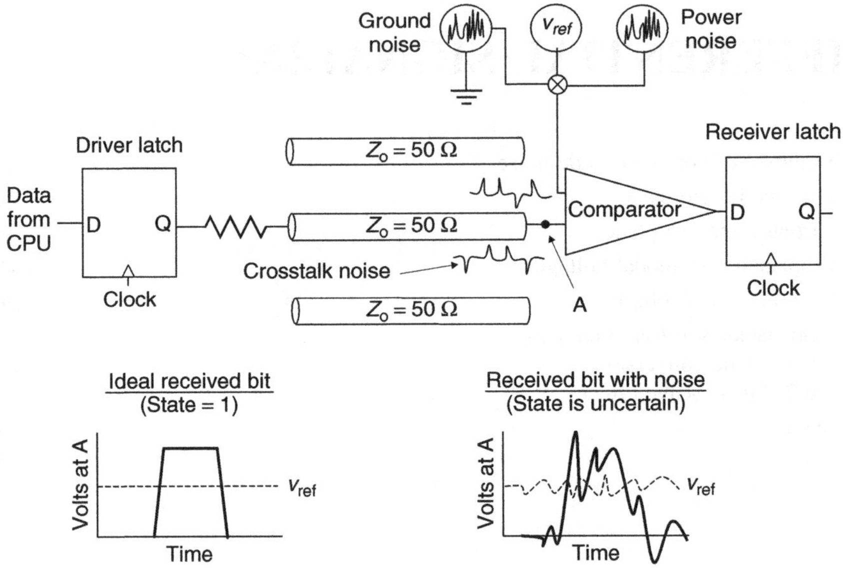

Figure 7-1 How system noise can severely degrade signal integrity on single-ended buses. The ideal versus noisy receiver voltages compared to the reference voltage.

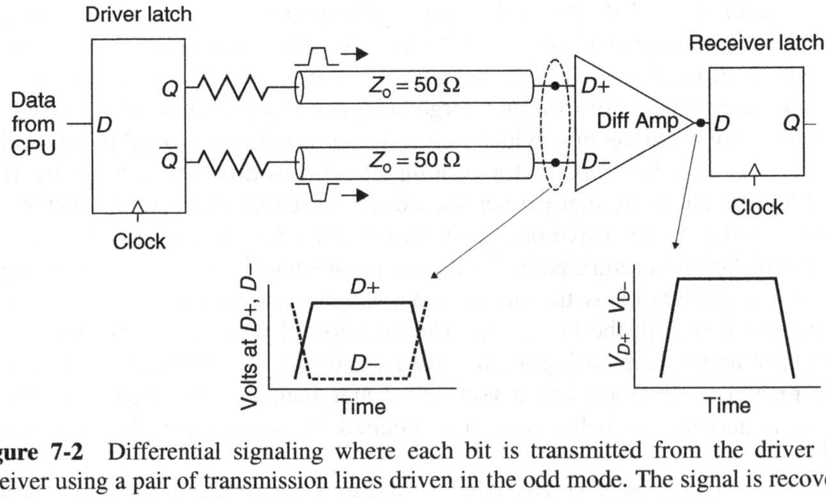

Figure 7-2 Differential signaling where each bit is transmitted from the driver to a receiver using a pair of transmission lines driven in the odd mode. The signal is recovered at the receiver with a differential amplifier.

A strategy that reduces the effect of the system noise dramatically is to dedicate a pair of transmission lines for each bit on the bus. The two transmission lines are driven 180° out of phase (in the odd mode), and the differences between the voltages are used to recover the signal at the receiver using a differential amplifier. This technique is called *differential signaling* and is depicted in Figure 7-2.

7.1 REMOVAL OF COMMON-MODE NOISE

Differential signaling is very effective for removing *common-mode noise*, which is defined as noise present on both legs of the differential pair. If the bus is designed properly so that the legs of the differential pairs are in close proximity to each other, the noise on $D+$ will be approximately equal to the noise on $D-$. Therefore, assuming that the receiver's differential amplifier has a reasonable common-mode rejection ratio, the noise will be eliminated. For example, assume that there is noise with a magnitude of v_{noise} coupled equally onto both legs of a differential pair. The output of a differential amplifier with unity gain is

$$v_{\text{diff}} = (v_{D+} + v_{\text{noise}}) - (v_{D-} + v_{\text{noise}}) = v_{D+} - v_{D-} \tag{7-1}$$

which removes the common-mode noise.

Figure 7-3 shows an example of a differential interconnect with noise present on the ground plane (v_{noise}). In this case, the noise is common to both legs of the driver. Figure 7-4a shows a simulated stream of data bits at the input to a differential amplifier with significant common-mode noise. Note that the digital state of the single-ended waveforms v_{D+} and v_{D-} is indeterminable because the noise is so large. However, since the noise is common mode, the bit stream can be recovered when the signals are subtracted by a differential amplifier ($v_{D+} - v_{D-}$), as shown in Figure 7-4b.

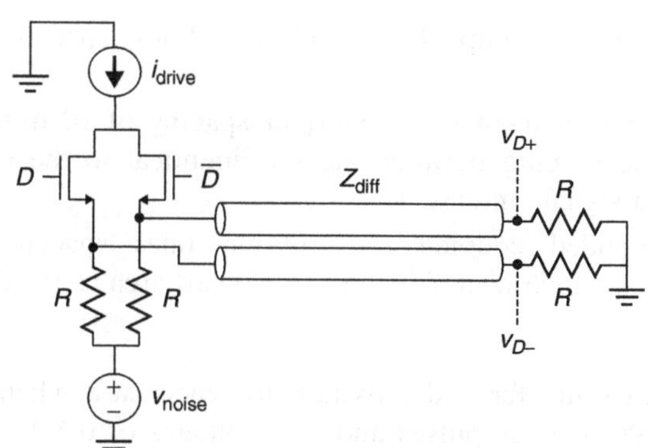

Figure 7-3 Differential driver and interconnect with common-mode noise present on the ground. Noise on both the power and ground planes is very common in digital designs.

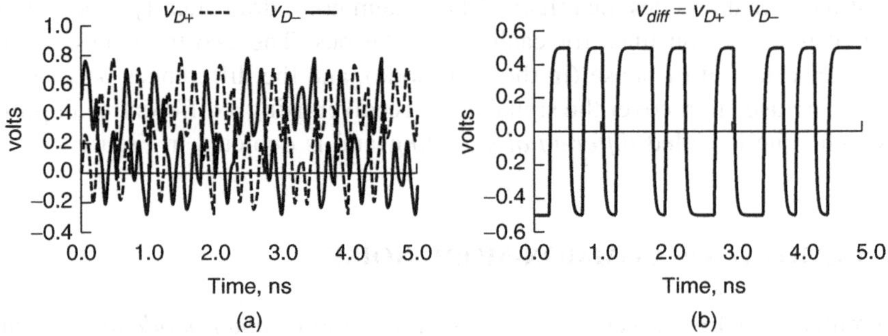

Figure 7-4 Example of how common-mode noise is eliminated with differential signaling: (a) single-ended waveforms at each leg of a differential receiver showing common-mode noise; (b) differential waveform.

7.2 DIFFERENTIAL CROSSTALK

Although crosstalk noise on a differential pair has a significant common-mode component, it also has a differential component because the distance between an aggressor and each side of the pair differs (in some cases, if the aggressor is on another layer, the crosstalk could be 100% common mode, but this is rare). Consequently, each leg will experience slightly different crosstalk, which will not be rejected by the differential amplifier. Nonetheless, under certain conditions, differential signaling can reduce crosstalk significantly.

If the crosstalk between single-ended lines is compared to the crosstalk between differential pairs, the differential crosstalk will be less if the spacing is similar. The drawback is that the differential pair will occupy significantly more board area. Additionally, if the single-ended pairs are spaced far enough apart so that the same board area as the differential pairs is occupied, the single-ended crosstalk is typically lower simply because the signals are spaced so far apart. For example, consider Figure 7-5, which shows three cases:

1. Two single-ended coupled transmission lines spaced 10 mils apart (Figure 7-5a).
2. Two differential pairs with an interpair spacing of 10 mils (Figure 7-5b). Note that the spacing between pairs is identical to the spacing between single-ended signals in case 1.
3. Two single-ended coupled transmission lines spaced 28 mils apart (Figure 7-5c), which occupies the same board area as the differential pairs of case 2.

Figure 7-6 shows the far end crosstalk for each case when the aggressors are driven with 100-ps-wide pulses and line voltages of 0.5 V. The differential crosstalk is calculated as the difference between $V+$ and $V-$:

$$v_{\text{differential}} = (V+) - (V-)$$

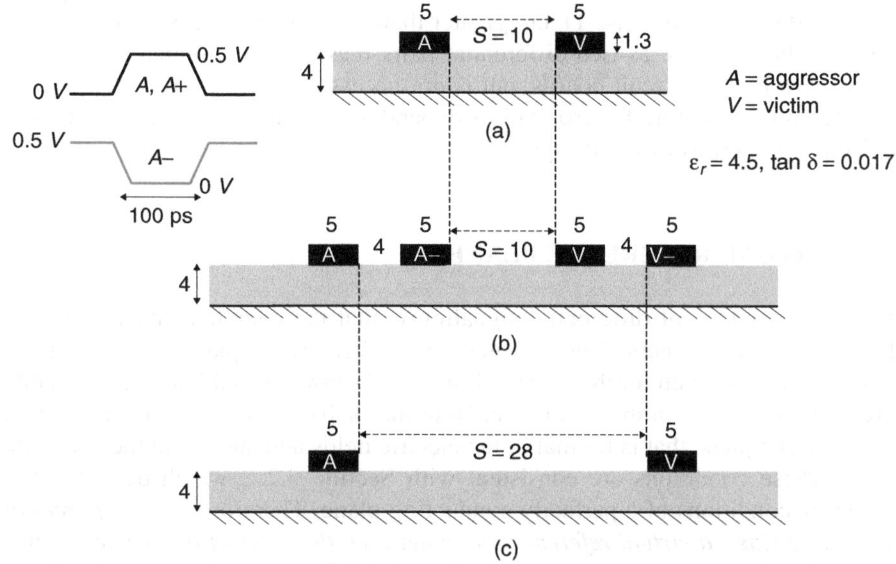

Figure 7-5 Cross sections used to compare differential versus single-ended crosstalk: (a) two single-ended signals; (b) two differential pairs with the same pair spacing as that of the single-ended signals in (a); (c) two widely spaced single-ended signals that occupy the same board area as the differential pairs (dimensions in mils).

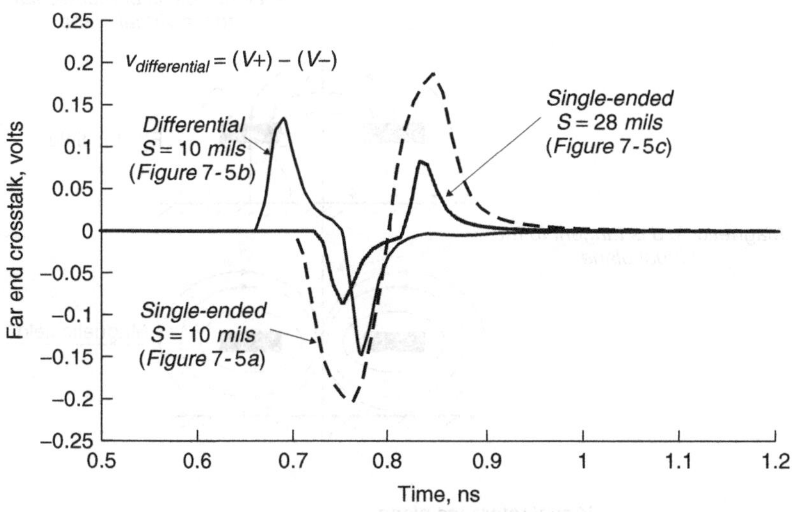

Figure 7-6 Crosstalk for the cross sections in Figure 7-5 demonstrating that differential signaling does not reduce crosstalk compared to single-ended signaling if the same board area is used. However, if the spacing between differential pairs is similar to the spacing between single-ended signals, crosstalk in reduced (5-in. microstrip).

Note that the differential crosstalk (case 2) is less than the identically spaced single-ended crosstalk (case 1), but greater than a single-ended case that occupies the same board space as two differential pairs (case 3). This scenario is typical for modern printed circuit boards, but is not necessarily always true. The benefit of differential signaling for crosstalk is dependent on the specific geometries and dielectric properties of the design.

7.3 VIRTUAL REFERENCE PLANE

Another advantage of differential signaling is that the complementary nature of the electric and magnetic fields creates a virtual reference plane that provides a continuous return path for the current. Figure 7-7 shows the field patterns for a differential pair driven in the odd mode. Note that halfway between the conductors there exists a plane that is normal to the electric fields and tangent to the magnetic fields. These conditions are consistent with Section 3.2.1, which describes the boundary conditions of a perfectly conducting plane. Therefore, *for a differential pair, there exists a virtual reference plane between the conductors*. The existence of the virtual reference plane is extremely helpful for cases when a nonideal reference exists that helps preserve signal integrity. Some common examples of a nonideal reference include connector transitions, via fields, layer transitions, and routing over a slot in the reference plane. In Chapter 10 we describe in detail the

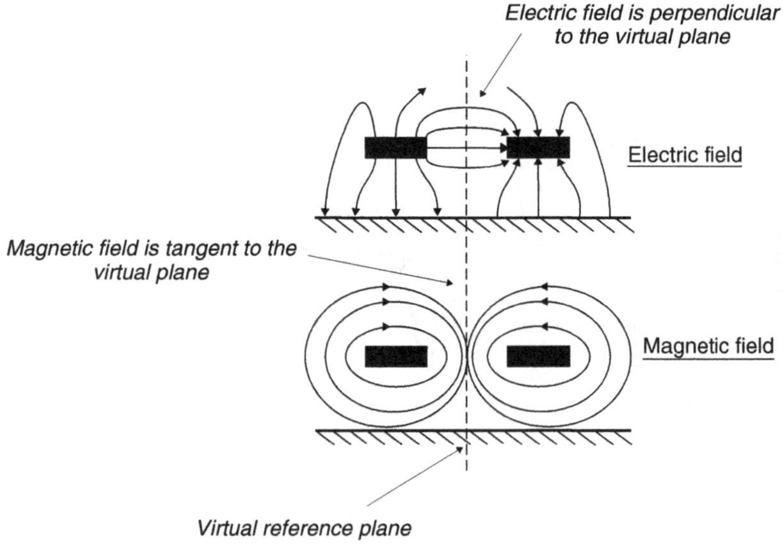

Figure 7-7 For an odd-mode (or differential) signal, the fields orient so that an ideal virtual reference plane existed between the conductors. When the physical plane is interrupted, the virtual plane provides a continuous reference for a differential signal and helps preserve signal integrity.

ill effects of nonideal reference planes. Additionally, when the fields are confined between the conductors (i.e., strongly coupled to the virtual reference plane), they are less apt to fringe out to other signals, which helps reduce crosstalk.

7.4 PROPAGATION OF MODAL VOLTAGES

For a multiconductor system, in Chapter 4 we discussed how all digital signal states are composed of linear combinations of the modal voltages. For a single differential pair, the two modes are the *even* and *odd* modes. Ideally, the differential pair is driven with lines 1 and 2 180° out of phase so that all of the energy is contained solely in the odd mode. However, if common-mode noise is coupled onto the differential pair, some energy will exist simultaneously in the even mode. This is easy to show mathematically using modal analysis, as discussed in Section 4.4. For example, consider a differential pair with line voltages v_{D+} and v_{D-}, which are composed of linear combinations of the odd- and even-mode voltages:

$$\begin{bmatrix} v_{D+} \\ v_{D-} \end{bmatrix} = [T_V] \begin{bmatrix} v_{\text{odd}} \\ v_{\text{even}} \end{bmatrix} \tag{7-2}$$

where $[T_V]$ is a matrix containing the eigenvectors of the product LC as developed in Section 4.4.1.

If the differential pair is driven exactly 180° out of phase and there is no noise present, the odd- and even-mode voltages are calculated, where

$$[T_V] = \begin{bmatrix} 0.707 & 0.707 \\ -0.707 & 0.707 \end{bmatrix}$$

from Example 4-4:

$$\begin{bmatrix} 1 \\ -1 \end{bmatrix} = \begin{bmatrix} 0.707 & 0.707 \\ -0.707 & 0.707 \end{bmatrix} \begin{bmatrix} v_{\text{odd}} \\ v_{\text{even}} \end{bmatrix} \tag{7-3}$$

resulting in

$$v_{\text{odd}} = 1.41443 \text{ V}$$

$$v_{\text{even}} = 0 \text{ V}$$

which proves that all the energy is contained on the odd mode.

However, if common-mode noise is present, energy is introduced into the even mode. If v_{noise} is the voltage noise introduced to each leg of the pair, the even and odd modes are calculated:

$$\begin{bmatrix} 1 + v_{\text{noise}} \\ -1 + v_{\text{noise}} \end{bmatrix} = \begin{bmatrix} 0.707 & 0.707 \\ -0.707 & 0.707 \end{bmatrix} \begin{bmatrix} v_{\text{odd}} \\ v_{\text{even}} \end{bmatrix} \tag{7-4a}$$

$$v_{\text{odd}} = 1.41443 \text{ V} \tag{7-4b}$$

$$v_{\text{even}} = 1.41443 v_{\text{noise}} \text{ V} \tag{7-4c}$$

Therefore, *the common-mode noise propagates in the even mode*.

The voltage present on each leg of the differential pair can be calculated from the modal voltages by solving (7-2):

$$v_{D+} = 0.707v_{\text{odd}} + 0.707v_{\text{even}} = 1 + 0.707(1.41443v_{\text{cm}}) = 1 + v_{\text{cm}}$$
$$v_{D-} = -0.707v_{\text{odd}} + 0.707v_{\text{even}} = -1 + 0.707(1.41443v_{\text{cm}}) = -1 + v_{\text{cm}}$$

In a differential bus, the line voltages are subtracted with a differential amplifier at the receiver which eliminates the noise.

$$v_{D+} - v_{D-} = (1 + v_{\text{cm}}) - (-1 + v_{\text{cm}}) = 2$$

To summarize, *a differential signaling scheme eliminates energy in the even mode*.

7.5 COMMON TERMINOLOGY

For systems with more *than two signal conductors, the terms* even mode *and* odd mode *are no longer applicable*. When analyzing differential buses, it is common to refer to a pair being driven with two signals 180° out of phase as the *differential mode* and two signals driven in phase as the *common mode*. The differential- and common-mode terms are simply naming conventions and are not technically modes at all. Analysis of the modal voltages propagating on multiconductor systems as described in Section 4.4 will show that the digital states do not correspond directly to modal voltages for systems with more than two signal conductors.

The differential-mode impedance is defined as twice the odd mode, and the common-mode impedance is one-half the even mode. The odd- and even-mode impedance values are described in Section 4.3.1.

$$Z_{\text{differential}} = 2Z_{\text{odd}} \tag{7-5a}$$

$$Z_{\text{common}} = \frac{Z_{\text{even}}}{2} \tag{7-5b}$$

It should be noted that equations (7-5a) and (7-5b) are typically used for the purpose of specifying design guidelines. The actual impedance of a differential pair may not correspond directly to $Z_{\text{differential}}$ if there is significant coupling to adjacent pairs. Equations (7-5a) and (7-5b) are representative of the true pair impedance values only if the interpair coupling is weak. Remember: In a multiconductor system with N signal conductors, there will be N unique modal impedance values.

It is also true that for a system with N signal conductors, there will be N modal propagation velocities. If the transmission lines are routed in a homogeneous dielectric (such as a stripline), all the modal velocities will be identical. However, for a nonhomogeneous dielectric (such as a microstrip), the differential- and common-mode propagation velocities are approximately equal to the odd- and even-mode velocities if the interpair coupling is weak. Odd- and even-mode velocities are defined in Section 4.3.1:

$$\nu_{\text{differential}} \approx \nu_{\text{odd}} \tag{7-6a}$$

$$\nu_{\text{common}} \approx \nu_{\text{even}} \tag{7-6b}$$

To quantify the voltage propagating in the differential and common modes, the following definitions are often used:

$$V_{\text{dm}} = V - \overline{V} \tag{7-7a}$$

$$V_{\text{cm}} = \frac{V + \overline{V}}{2} \tag{7-7b}$$

where V_{dm} is the voltage propagating in the differential mode, V_{cm} is the voltage propagating in the common mode, V represents the voltage propagating on line 1 (the same as v_{D+} used in the two-signal-conductor example), and $\overline{V}$ represents the complementary signal propagating on line 2 (v_{D-} used in the two-signal-conductor example).

7.6 DRAWBACKS OF DIFFERENTIAL SIGNALING

Differential signaling is a powerful tool used to design high-speed digital buses that dramatically reduces the amount of common-mode noise seen at the receiver, which allows higher data rates to be realized. However, differential signaling is not a panacea. Most obviously, a differential bus consumes significantly more area on the printed circuit board than does a single-ended bus. Sometimes, differential buses are so large that they force designers to use printed circuit boards with more layers, which increases cost. The increased number of signals drives package, socket, and connector pin counts to higher levels, which complicates designs and further increases system costs. Also, the impedance tolerance of differential pairs tends to be wider than single ended because the variations in etching and plating profiles of the conductors affect mutual inductance and capacitance and, consequently, differential impedance. Furthermore, small asymmetries in the differential pair can have a large impact on signal integrity if not controlled properly.

7.6.1 Mode Conversion

In Section 7.4 we showed how common-mode noise coupled onto a differential pair causes voltage to propagate in the even mode, which is rejected by the receiver assuming that the common-mode rejection ratio is high enough. Another mechanism that causes even-mode voltage to propagate are phase errors between V and $\overline{V}$. Ideally, the phase difference between the waveforms propagating on V and $\overline{V}$ is 180°, which keeps the energy in the odd mode. However, if the phase relationship between V and $\overline{V}$ deviates from 180° as the signals propagate toward the receiver, some of the energy will be converted from odd mode to even mode. This phenomenon has many names, including *mode conversion, differential-to-common mode conversion*, and *ac common-mode conversion (ACCM conversion)*. In this text we use the term *ACCM conversion*.

ACCM conversion is caused by asymmetry between V and $\overline{V}$ in the differential pair. The asymmetry can be caused by length differences, coupling differences, etching differences, proximity effects, termination differences, bends, or anything else that would make one leg of the pair look electrically different from the other. Some of these examples are shown in Figure 7-8. To demonstrate how intrapair asymmetry affects voltage and timing noise, consider the simple (but very common) example where lengths of lines 1 and 2 in the pair are not equal, as shown in Figure 7-8a. Assuming that the signal is launched differentially, the difference in propagation delay between lines 1 and 2 will change the phase relationship at the receiver because the voltage propagating on one leg of the pair will arrive early, converting part (or all) of the differential signal to common mode. Figure 7-9 shows how a perfect differential signal at the transmitter (Tx) is distorted at the receiver (Rx) when asymmetry exists in the pair. The distortion is proportional to the amount of voltage that exists in the common mode.

The amount of signal that is converted to common mode depends on the length of the pair, the difference in propagation delays, and the frequency. This

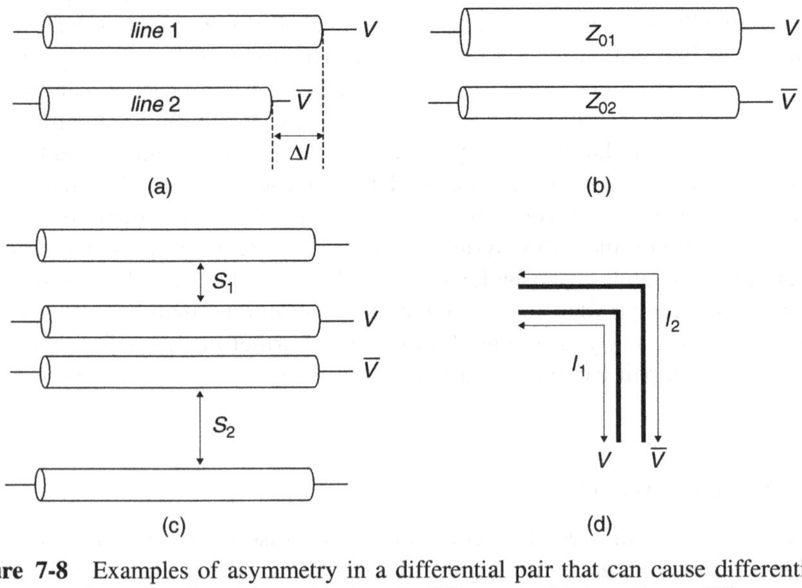

Figure 7-8 Examples of asymmetry in a differential pair that can cause differential to common-mode conversion: (a) routing-length differences; (b) impedance differences due to etching variation; (c) crosstalk differences; (d) length differences due to bends.

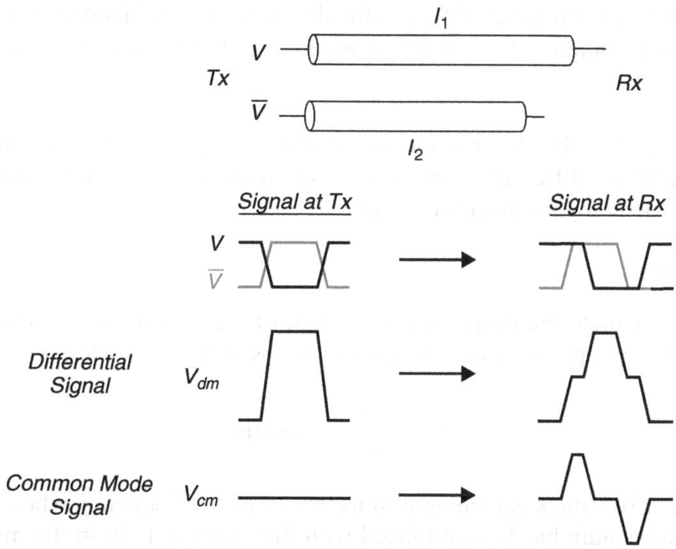

Figure 7-9 When asymmetry exists in the differential pair, part of the signal gets converted to common mode at the receiver.

can be shown by calculating the voltage on each leg of the transmission line at the receiver when the signals are launched 180° (π) out of phase:

$$V(\omega, l_1) = v_1^+ e^{-\alpha l_1} e^{j(\omega t - \beta l_1)} \tag{7-8a}$$

$$\overline{V}(\omega, l_2) = v_2^+ e^{-\alpha l_2} e^{j(\omega t + \pi - \beta l_2)} \tag{7-8b}$$

where β is the propagation constant as defined in equation (6-48c), α the attenuation constant as defined in equation (6-48b), l_1 the length of line 1, and l_2 the length of line 2. Note that since there is no backward-propagating component (v^-), all reflections are perfectly terminated in this example.

The differential-to-common mode conversion (ACCM) is calculated from (7-8a) and (7-8b) with $\alpha = 0$:

$$\text{ACCM} = \frac{V(z = l_1) + \overline{V}(z = l_2)}{V(z = 0) - \overline{V}(z = 0)} = \frac{v_1^+ e^{j(\omega t - \beta l_1)} + v_2^+ e^{j(\omega t + \pi - \beta l_2)}}{v_1^+ e^{j(\omega t)} - v_2^+ e^{j(\omega t + \pi)}}$$

$$= \frac{v_1^+ e^{-j\beta l_1} + v_2^+ e^{-j\beta l_2}}{v_1^+ - v_2^+} \tag{7-9}$$

where $V(z = l_1)$ and $\overline{V}(z = l_2)$ are the voltages at the receiver, and $V(z = 0)$ and $\overline{V}(z = 0)$ are the voltages at the driver.

At low frequencies where the wavelength is large, the phase delay difference between lines 1 and 2 is small, so the numerator of (7-9) is approximately zero. However, as the frequency increases, the phase difference becomes large. When the phase difference reaches $180^\circ(\pi)$, the differential signal launched at the driver is converted completely to a common-mode signal at the receiver, and equation (7-9) is unity.

Example 7-1 For the differential pair shown in Figure 7-10, calculate the frequency where the differential signal injected at the driver is 100% converted to a common-mode signal at the receiver.

SOLUTION

Step 1: Calculate the propagation constant of the transmission lines. Equation (2-46) defines the propagation constant in terms of the wavelength:

$$\beta = \frac{2\pi}{\lambda} \qquad \text{rad/m}$$

equation (2-45) defines wavelength in the terms of the frequency where the speed of light in a vacuum has been replaced with the speed of light in the media (ν_p):

$$f = \frac{\nu_p}{\lambda} \qquad \text{Hz}$$

and equation (2-52) calculates the speed of light in the media (assuming that $\mu_r = 1$):

$$\nu_p = \frac{c}{\sqrt{\varepsilon_r}} \qquad \text{m/s}$$

Therefore, the propagation constant is calculated as a function of frequency:

$$\beta = \frac{2\pi f\sqrt{\varepsilon_r}}{c} = (41.866 \times 10^{-9})f \qquad \text{rad/s}$$

Step 2: Use equation (7-9) to plot the differential-to-common mode conversion. Since $V(z=0) = 1$ and $\overline{V}(z=0) = -1$, the terms $v_1^+ = 1$ and $v_2^+ = -1$. The plot is shown in Figure 7-11. When ACCM = 1, the phase error due to the length mismatch equals 180° and the differential signal launched at the driver shows

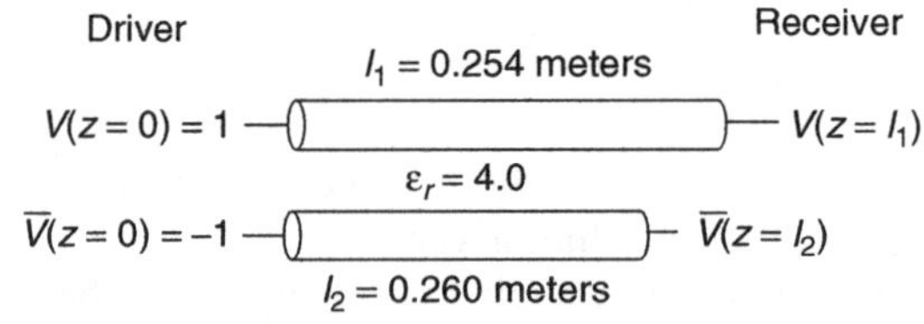

Figure 7-10 Figure for Example 7.1.

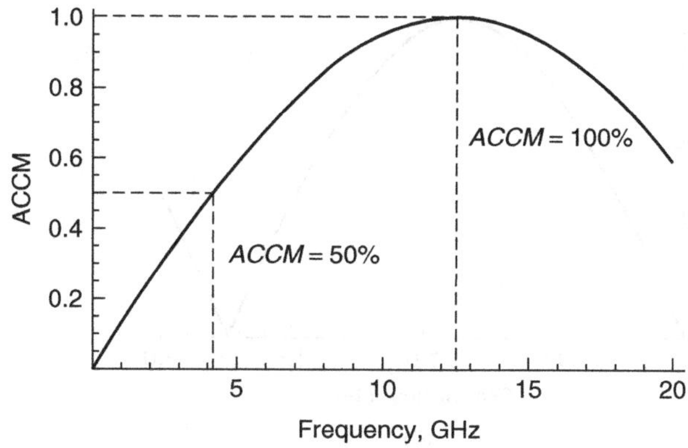

Figure 7-11 Differential common-mode conversion plotted for Example 7-1 showing that a differential signal launched at the driver becomes common mode at the receiver when the frequency is about 12.5 GHz.

up as a common-mode signal at the receiver. Therefore, the frequency where the differential-to-common mode conversion is 100% is 12.5 GHz.

In a differential system, only the difference between $V(z = l_1)$ and $\overline{V}(z = l_2)$ is recovered. Consequently, the percentage of differential energy converted to common mode looks like loss to the differential bus in the frequency domain. In Example 7-1 the differential receiver would see no signal at 12.5 GHz. At 4 GHz, approximately 50% of the energy has been converted to the common mode and only one-half the signal would be sensed by the differential receiver.

Note that the differential-to-common mode conversion in Example 7-1 decreases after 12.5 GHz. Do not be tempted to operate the bus in this frequency range. The phase difference between signals on the differential pairs will continue to increase until they are 360° out of phase, where (7-9) predicts zero mode conversion. If a digital bus were operated so that the phase difference was 360° at the receiver, bit 1 on line 1 would align with bit 2 on line 2 and the incorrect data would be latched into the receiver. This can easily be demonstrated by observing the real portion of $V(z = l_1)$ and $\overline{V}(z = l_2)$ as calculated with (7-8a) and (7-8b). Figure 7-12a shows the differential-to-common mode conversion (ACCM) for the differential pair in Example 7-1 when $l_1 = 0.254$ m and $l_2 = 0.340$ m. Figure 7-12b shows the real part of $V(z = l_1)$ and $\overline{V}(z = l_2)$ at the receiver. Note that at low frequencies, the waveforms are 180° out of phase and the differential-to-common mode conversion is zero. At about 880 MHz, the differential-to-common mode conversion is 100%, and the waveforms in Figure 7-12b are in phase. At ~1.75 GHz, the ninth peak of $V(z = l_1)$ is 180° out of phase with the seventh peak of $\overline{V}(z = l_2)$, and equation (7-9) predicts that ACCM = 0. Although the magnitude of the differential-to-common mode conversion begins to decrease after it peaks, the phase error is large, so the bus will not function properly.

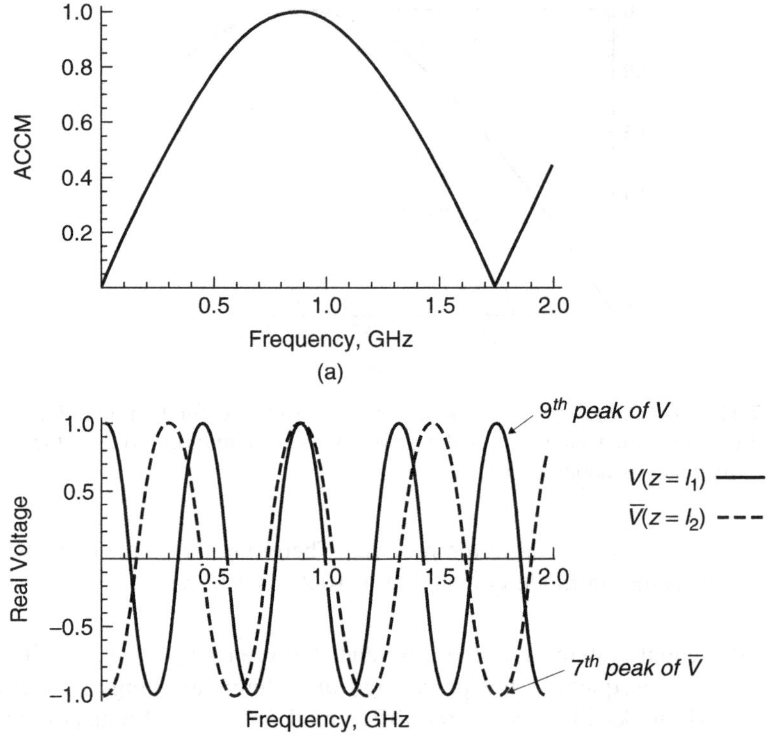

Figure 7-12 The differential-to-common mode conversion peaks when the difference in phase delay between lines 1 and 2 in a differential pair is 180°. As the phase delay difference increases with frequency, the ACCM becomes zero again, but the phase error is so large that the bus will not operate properly.

7.6.2 Fiber-Weave Effect

In Section 6.5.1 we described how FR4 and similar dielectrics are composite materials made from a matrix of woven bundles of fiberglass embedded in an epoxy resin. Figure 7-13 depicts how it is possible for one leg of the differential signal to be routed between glass bundles and the other to be routed over a glass bundle that effectively gives each trace a unique propagation velocity. The reinforcing fiberglass bundles have a dielectric permittivity ε_r of approximately 6, whereas ε_r is close to 3 for the epoxy resin in which the bundles are embedded. When a differential pair is aligned with the reinforcing fiber matrix of the dielectric in this manner, it causes an imbalance in the differential pair that causes differential-to-common mode (ACCM) conversion. Even if the lines are routed symmetrically, the difference in the dielectric permittivity will make the electrical delay shorter in one line versus the other. Figure 7-13 depicts a cross section of a differential pair that is asymmetric due to the fiber-weave effect.

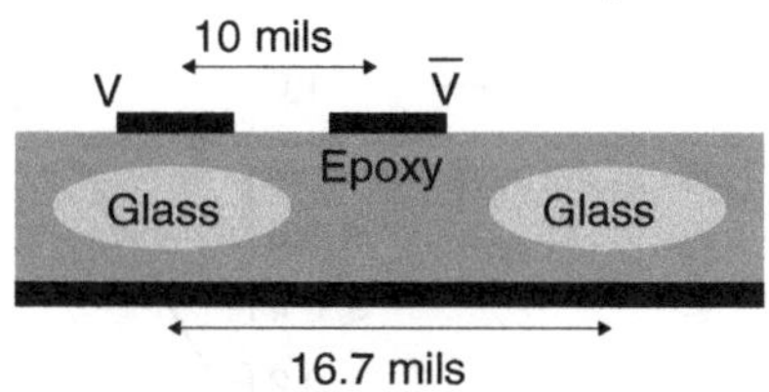

Figure 7-13 The fiber-weave matrix used to construct composite dielectrics such as FR4 can cause differential signals to be asymmetric due to the differences in permittivity of the glass weave ($\varepsilon_r \sim 6$) and the epoxy ($\varepsilon_r \sim 3$). The glass pitch shown is typical for many glass weaves used in the printed circuit board industry.

Unless specific measures are used to eliminate the alignment of the differential pair with the fiber weave, the traditional approach of modeling the transmission line using a uniform dielectric permittivity throughout the dielectric layer is no longer sufficient. The approach must be modified to comprehend the localized dielectric variation. Figure 7-14 shows the profile of the cross-section geometry description used as 2D field solver input for calculating the transmission-line parameters. Simple in principle, this approach is complicated by the fact that the measured effective dielectric permittivity results from the combined effect of multiple dielectrics: resin, fiberglass, solder mask, and air. As a starting point, the effective dielectric permittivity of the FR4 material in each region can be calculated using the measured value of the propagation delay as in Figure 6-15, and the actual dielectric differences can be estimated as described in Section 6.5.1. Alternatively, the results from a 2D field simulation with a cross section that describes the composite nature of the dielectric can be used if sufficient information is available. Many commercial field solvers have this capability, including Ansoft's Q2D. Nonetheless, once reasonable values for ε_{r1} and ε_{r2} are estimated, the fiber-weave effect can be approximated by generating the appropriate transmission-line parameters for each leg of the pair using a cross section similar to that shown in Figure 7-14.

The differential-to-common mode (ACCM) conversion due to dielectric permittivity variations can be calculated using the same procedure that was used to derive equation (7-9). The only difference is that the line length (l) stays constant

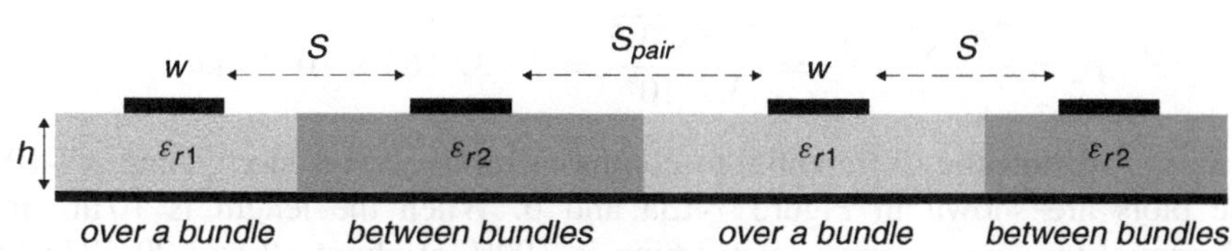

Figure 7-14 Cross section for modeling the fiber-weave effect in differential pairs. The dielectric permittivity is adjusted to account for a trace routed over a glass bundle or over an epoxy pool.

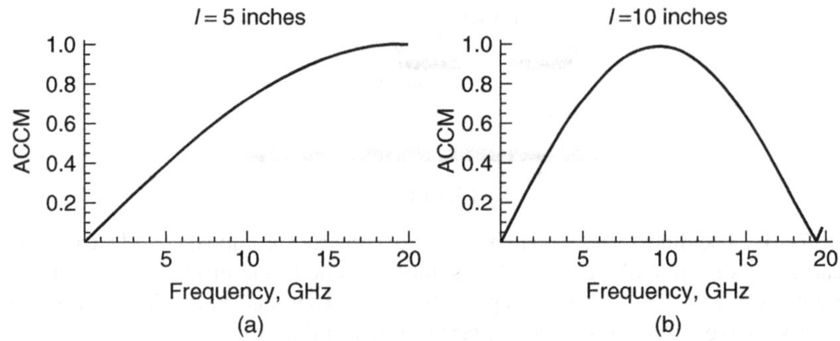

Figure 7-15 Waveforms for Example 7-2: (a) length = 5 in. (b) length = 10 in.

and the propagation constant (β) changes:

$$\text{ACCM} = \frac{V(z = l) + \overline{V}(z = l)}{V(z = 0) - \overline{V}(z = 0)} = \frac{v_1^+ e^{-j\beta_1 l} + v_2^+ e^{-j\beta_2 l}}{v_1^+ - v_2^+} \tag{7-10}$$

where $\beta_1 = 2\pi f \sqrt{\varepsilon_{r,\text{eff1}}}/c$, $\beta_2 = 2\pi f \sqrt{\varepsilon_{r,\text{eff2}}}/c$, l the differential pair length, and v_1^+ and v_1^- are the driving voltages.

Example 7-2 Determine the frequency where the differential-to-common mode conversion is 100% for a 10-in. (0.254-m) and a 5-in. (0.178-m) differential pair routed where one leg is over a bundle and one leg is between glass bundles. Use the measured data shown in Figure 6-15.

SOLUTION

Step 1: Determine the maximum spread in the effective dielectric permittivity. From Figure 6-15 the spread is 0.23:

$$\Delta\varepsilon_{\text{eff}} \approx 3.73 - 3.5 = 0.23$$

Step 2: Calculate β_1 and β_2:

$$\beta_1 = \frac{2\pi f \sqrt{\varepsilon_{r,\text{eff1}}}}{c} = \frac{2\pi f \sqrt{3.73}}{3 \times 10^8} = f \cdot 40.429 \times 10^{-9} \text{ rad/s}$$

$$\beta_2 = \frac{2\pi f \sqrt{\varepsilon_{r,\text{eff2}}}}{c} = \frac{2\pi f \sqrt{3.5}}{3 \times 10^8} = f \cdot 39.163 \times 10^{-9} \text{ rad/s}$$

Step 3: Plot the differential-to-common mode conversion using (7-10). The plots are shown in Figure 7-15a and b. When the length is 10 in., the differential-to-common mode conversion is 100% at about 10 and 20 GHz for 5 in.

Mitigation of the fiber-weave effect was discussed briefly in Section 6.5.2.

REFERENCE

Hall, Stephen, Garrett Hall, and James McCall, 2000, *High Speed Digital System Design*, Wiley-Interscience, New York.

PROBLEMS

7-1 Assume that a differential pair is routed so that the fiber-weave effect occurs similar to Figure 7-16. Assume that the line is 7 in. long and the metal is copper with a surface roughness of 0.5 μm. If the differential-to-common mode conversion is calculated with losses, how does the answer differ from the case where it is calculated with no losses? Which answer better predicts the percentage of the signal converted to the common mode?

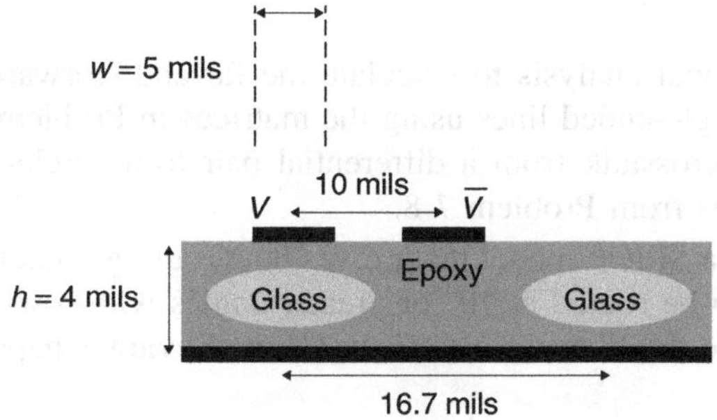

Figure 7-16 The permittivity/loss tangent for the glass weave is $\varepsilon_r = 6/\tan\delta = 0.0002$ and the epoxy $\varepsilon_r = 3/\tan\delta = 0.025$.

7-2 Derive metrics based on line lengths that will help predict the magnitude of the voltage noise on a digital signal when the fiber-weave effect is present. Use the results of Problem 7-1 for the derivation.

7-3 How would the differential pair of Problem 7-1 affect the timing of a digital signal? (*Hint:* Consider the phase noise.)

7-4 Describe three different methods for mitigating differential-to-common mode conversion.

7-5 Derive a formula to predict differential-to-common mode conversion for the case where the impedance of each leg does not match.

7-6 Use modal analysis described in Section 4.4 to calculate the differential waveform expected for a 10-in. transmission line with a cross section of Figure 7-16.

7-7 Show that the modal voltages correspond to specific digital states for two signal conductors. Use the following **L** and **C** matrices:

$$\mathbf{L} = \begin{bmatrix} 3.77 \times 10^{-7} & 5.18 \times 10^{-8} \\ 5.18 \times 10^{-8} & 3.77 \times 10^{-7} \end{bmatrix} \quad \text{H/m}$$

$$\mathbf{C} = \begin{bmatrix} 8.21 \times 10^{-11} & -4.26 \times 10^{-12} \\ -4.26 \times 10^{-12} & 8.21 \times 10^{-11} \end{bmatrix} \quad \text{F/m}$$

7-8 Show that the modal voltages <u>do not</u> correspond to specific digital states for three or more signal conductors. Use the following **L** and **C** matrices:

$$\mathbf{L} = \begin{bmatrix} 3.77 \times 10^{-7} & 5.17 \times 10^{-8} & 1.08 \times 10^{-8} \\ 5.17 \times 10^{-8} & 3.77 \times 10^{-7} & 3.21 \times 10^{-8} \\ 1.08 \times 10^{-8} & 3.21 \times 10^{-8} & 3.77 \times 10^{-7} \end{bmatrix} \text{H/m}$$

$$\mathbf{C} = \begin{bmatrix} 8.21 \times 10^{-11} & -4.25 \times 10^{-12} & -3.74 \times 10^{-12} \\ -4.25 \times 10^{-12} & 8.21 \times 10^{-11} & -4.25 \times 10^{-12} \\ -3.74 \times 10^{-12} & -2.04 \times 10^{-12} & 8.19 \times 10^{-11} \end{bmatrix} \text{F/m}$$

7-9 Use modal analysis to calculate the far-end (forward) crosstalk between two single-ended lines using the matrices in Problem 7-7, and compare it to the crosstalk from a differential pair to a single-ended line using the matrices from Problem 7-8.

7-10 Create a SPICE model that proves how routing guidelines that force signal lines to be routed at 10° or greater angles with respect to the board edge mitigate the fiber-weave effect. (*Hint*: Review Chapter 6.)

8

MATHEMATICAL REQUIREMENTS FOR PHYSICAL CHANNELS

Modern high-speed digital design requires extensive signal integrity simulations to assess the electrical performance of the system prior to fabrication of prototypes. To ensure accurate results from the simulations, careful attention must be given to system component models such as transmission lines, vias, connectors, and packages. For a model to be physically consistent with the laws of nature, certain mathematical constraints must be obeyed to ensure proper balance between signal propagation, energy storage, and losses. For example, the vast majority of engineers designing high-speed digital systems today utilize simplified modeling techniques that employ frequency-invariant values of dielectric permittivity, loss tangent, and inductance for transmission-line models. A review of Chapters 5 and 6 will remind the reader that transmission-line models have frequency-dependent properties that must be modeled correctly if a

Advanced Signal Integrity for High-Speed Digital Design, By Stephen H. Hall and Howard L. Heck

realistic response is expected. Such assumptions, although valid at low frequencies or for very short electrical structures, induce amplitude and phase errors for digital data rates faster than 1 to 2 Gb/s. In fact, the computer currently being used to write this chapter was designed using traditional modeling techniques, which assume frequency-invariant electrical properties of dielectrics and conductors. Such assumptions, however, cease to produce valid results for higher data rates. As data rates increase, bandwidth demands skyrocket, form factors shrink, and new phenomenan that were insignificant in past designs become significant. When the correct model assumptions are not used, incorrect solution spaces are determined, lab correlation becomes difficult or impossible, and the design time is increased significantly. In this chapter we outline some of the most important techniques used to determine if a physical channel model is adequate for the design at hand. First, the fundamentals of calculating the channel response and methods of implementing frequency-domain phenomena in time-domain simulations are explored. Next, the mathematical requirements for a channel that is consistent with nature are explained, and methodologies for testing these requirements are defined. It is not always necessary for a model to obey these physical rules if the error is small enough; however, it is an important concept for the modern-day digital designer to understand to ensure full comprehension of the modeling assumptions.

8.1 FREQUENCY-DOMAIN EFFECTS IN TIME-DOMAIN SIMULATIONS

Although high-speed digital design is focused largely on the signal integrity of time-domain digital waveforms, many of the phenomena that heavily influence the propagation of signals on interconnects are best described in the frequency domain. Examples covered in previous chapters include skin effect resistance, surface roughness, internal inductance, and frequency-dependent dielectric properties. Consequently, it is important for the digital engineer to understand the relationship between a time-domain waveform and its equivalent frequency-domain representation. In fact, many modern buses have component specifications in terms of frequency-domain parameters because it is the most convenient way to describe the wideband behavior. In this section we outline some of the fundamental principles for linear time-invariant systems that will allow the engineer to translate between the frequency and time domains.

8.1.1 Linear and Time Invariance

A system is *linear* if the relationship between the input and output of the system satisfies the superposition property. For example, if the input to the system is the sum of two component signals,

$$x(t) = c_1 x_1(t) + c_2 x_2(t)$$

where c_1 and c_2 are constants, the output of the system will be

$$y(t) = c_1 y_1(t) + c_2 y_2(t)$$

where $y_n(t)$ is the output resulting from the input $x_n(t)$ ($n = 1$ and 2). To generalize, a linear system with input

$$x(t) = \sum_n c_n x_n(t)$$

will have output

$$y(t) = \sum_n c_n y_n(t)$$

for any constants c_n and where the output $y_n(t)$ results from the input $x_n(t)$.

Time invariance means that whether an input to the system is applied at $t = 0$ or $t = \tau$, the output will be identical except for a time delay of τ. For example, if the output due to input $x(t)$ is $y(t)$, the output due to input $x(t - \tau)$ is $y(t - \tau)$. Simply put, *a time delay at the input should produce a corresponding time delay at the output*.

8.1.2 Time- and Frequency-Domain Equivalencies

Any time-domain waveform in an LTI system has an equivalent spectrum in the frequency domain. This means that any time-domain signal, such as a digital waveform, can also be described fully in terms of its frequency-domain parameters. This concept is important because it allows the frequency-dependent nature of electromagnetic models to be integrated into time-domain waveforms so that the signal integrity of digital bits propagating on a bus can be analyzed. The relationship between a time-domain signal and its frequency-domain equivalent is described with the *Fourier transform*:

$$F(\omega) = \sqrt{\frac{|b|}{(2\pi)^{1-a}}} \int_{-\infty}^{\infty} f(t) e^{jb\omega t}\, dt \tag{8-1a}$$

$$f(t) = \sqrt{\frac{|b|}{(2\pi)^{1+a}}} \int_{-\infty}^{\infty} F(\omega) e^{-jb\omega t}\, d\omega \tag{8-1b}$$

where (8-1a) translates a time-domain signal into a frequency-domain representation and (8-1b) translates the frequency-domain representation into a time-domain waveform. When using the Fourier transform, it is important to ensure that the correct conventions are used. Different technical and scientific fields have different conventions that can confuse the analysis if consistency is not maintained. The conventions are based on the choices for a and b. Some popular conventions are ($a = 0$, $b = -1$) for modern physics, ($a = 1$, $b = -1$) for systems engineering, ($a = -1$, $b = 1$) for classical physics, and ($a = 0$, $b = -2\pi$) for signal processing [Wolfram, 2007]. In this chapter we utilize both the systems engineering and signal processing conventions, depending on the application.

The Fourier transform simply relates a time-domain waveform to a frequency response by superimposing an infinite number of sinusoidal functions to reconstruct the original waveform. For example, if the Fourier expansion of a 50% duty cycle square wave is calculated,

$$f(t) = \frac{A}{2} + \frac{2A}{\pi} \sum_{n=1,3,5\ldots} \frac{1}{n} \sin(n2\pi f t) \tag{8-2}$$

the wave shape can be approximated by superimposing several individual harmonics, as demonstrated in Figure 8-1. As the harmonic n is increased, the reconstructed waveform better approximates the original.

The Fourier transform shown in equation (8-1a) can be used to calculate the real and imaginary parts of the individual sinusoids needed to reconstruct the time-domain waveform, which is known as the *spectrum of the waveform*. For example, the frequency spectrum of a perfect square pulse, with a width of 2 time units as shown in Figure 8-2a, is calculated using the systems engineering convention ($a = 1$, $b = -1$).

$$F(\omega) = \int_{-\infty}^{\infty} f(t)e^{-j\omega t}\, dt$$

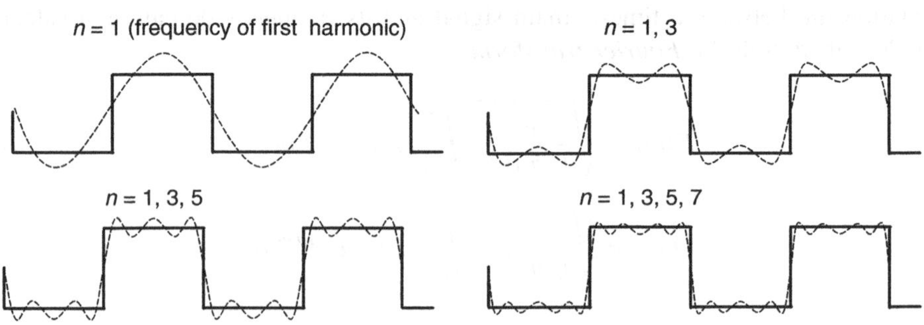

Figure 8-1 Fourier expansion for a 50% duty cycle square wave.

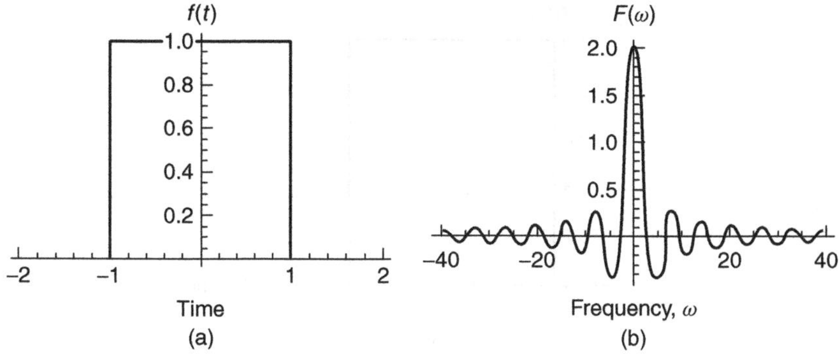

Figure 8-2 (a) Square wave $f(t)$, which is the most primitive approximation of a digital bit; (b) frequency-domain equivalent spectrum of the square wave.

$$= \int_{-1}^{1} e^{-j\omega t}\, dt$$

$$= j\, \frac{e^{-j\omega t}}{\omega}\bigg|_{t=-1}^{t=1} = -\frac{j}{\omega}(e^{j\omega} - e^{-j\omega})$$

However, since $(e^{j\omega} - e^{-j\omega})/2j = \sin\omega$, the frequency spectrum can be simplified:

$$F(\omega) = -\frac{j}{\omega}\sin\omega\; 2j$$

$$= 2\frac{\sin\omega}{\omega} = 2\;\text{sinc}\;\omega \tag{8-3}$$

Equation (8-3) is the frequency spectrum of a square-wave pulse and is plotted in Figure 8-2b. *This means that the frequency-domain representation of a square wave takes the form of a sinc function, and vice versa*. Note that for a perfect step or pulse, an infinite number of harmonics is required to reproduce the waveform.

The centering of the square wave around $t = 0$ in Figure 8-2a simplifies the mathematics by eliminating the imaginary portion of the transform. For a realistic digital pulse with only positive time values, the spectrum becomes complex. For example, the Fourier transform of a square pulse with a width of 2 time units is shown in Figure 8-3 calculated with system engineering conventions ($a = 1$, $b = -1$). Note that the spectrum contains both real and imaginary components.

The examples shown in Figures 8-2 and 8-3 demonstrate three very important concepts that link the time-domain waveform to its frequency-dependent equivalent.

1. The spectrum of the time-domain waveform has both positive and negative frequency components.

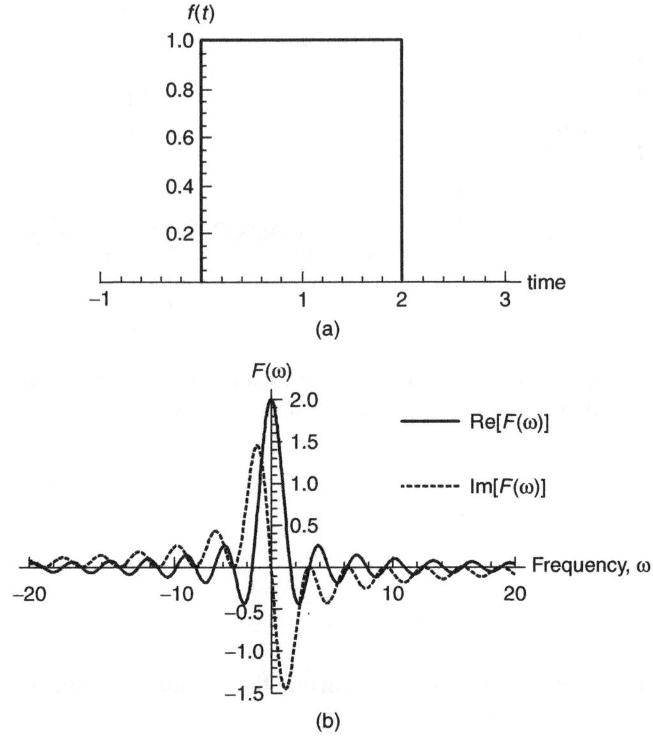

Figure 8-3 (a) Square wave $f(t)$ with only positive time values; (b) frequency-domain equivalent spectrum of the square wave. Note that the spectrum has both real and imaginary values.

2. The spectrum of a time-domain waveform with only positive time values is complex.
3. Since time-domain signals that are physically observable have no imaginary component, they must be real. Reality in the time domain is guaranteed when the positive frequencies of the Fourier transform are the complex conjugate of the negative frequencies [LePage, 1980]:

$$F(-\omega) = F(\omega)^* \tag{8-4}$$

The behavior of equation (8-4) can be observed in Figure 8-3b. For example, at the frequency $\omega = -2$ rad/s, the spectrum has a value of

$$F(-2) = -0.4 + j0.74$$

and at $\omega = 2$ rad/s, the spectrum has a value of

$$F(2) = -0.4 - j0.74$$

Since $F(-2) = F(2)^*$, the spectrum is for a real time-domain waveform.

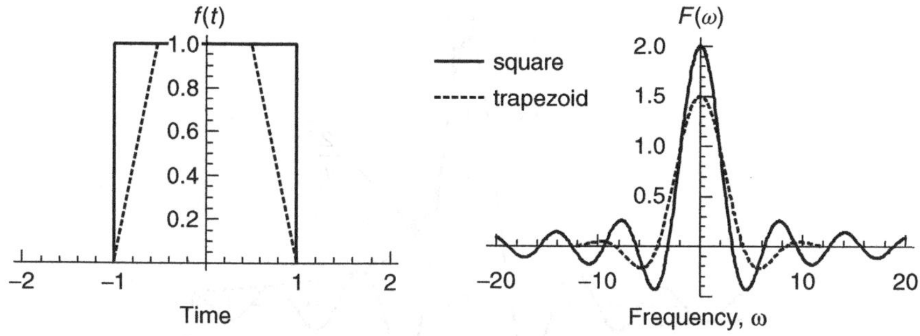

Figure 8-4 Frequency response of a trapezoidal wave compared to an ideal square wave.

8.1.3 Frequency Spectrum of a Digital Pulse

Although a square pulse is a first-order approximation of a digital bit, a better approximation is a trapezoid. Figure 8-4 shows the frequency response of a trapezoidal wave compared to an ideal square wave. Notice the shape of the trapezoid's frequency response is very similar to that of a square wave, except that the magnitude of the harmonics are smaller, especially at high frequencies. This demonstrates another important concept: *The rise and fall times of a digital waveform determine the magnitude of the high-frequency harmonics in the frequency-domain spectrum*.

Figure 8-4 also indicates that the spectrum of a trapezoid can be derived by applying a low-pass filtering function to the spectrum of a square wave. This allows a relationship to be defined between the spectral content of a digital waveform and the rise and fall times. To begin the derivation of this relationship, some unique properties of the square-wave spectrum need to be observed. Equation (8-3) shows that the harmonics of a square wave are a sinc function. Taking the limits of the sinc function allows the frequency spectrum to be generalized:

$$\frac{\sin\omega}{\omega} \approx \begin{cases} 1 & \text{when } \omega \text{ is small} \\ 1/\omega & \text{when } \omega \text{ is large} \end{cases} \tag{8-5}$$

The quantity $2/\omega$ is compared to the spectrum calculated with equation (8-3) in Figure 8-5. Note that the quantity $1/\omega$ is equivalent to a slope of 20 dB/decade on a log-log plot:

$$20\log\frac{1/10\omega}{1/\omega} = -20\ \text{dB/decade}$$

meaning that the spectral content of a square wave will fall off at a rate of −20 dB/decade.

One way to approximate the spectrum of a trapezoidal wave is to apply a low-pass filtering function to the harmonics of a square wave until the desired

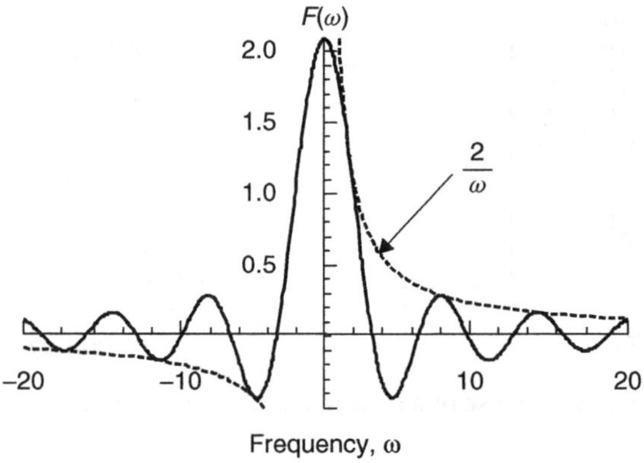

Figure 8-5 The spectrum of a square wave fits within the envelope of $2/\omega$ at high frequencies.

rise and fall times are realized. The easiest way to apply this filtering is to use a simple low-pass one-pole filtering response, such as an *RC* network. The step response of a simple one-pole filter is

$$\frac{v_{out}}{v_{input}} = 1 - e^{-t/\tau} \tag{8-6}$$

where v_{input} is the input voltage to the filter, v_{out} the output voltage, and τ the time constant. If the rise times are defined with the 10% and 90% voltage magnitude points, the time constant required to degrade a step to a specific $t_{10\text{–}90\%}$ can be calculated. The rise time of a unit step after it passes though a one-pole filter with a time constant of τ is calculated as

$$t_{10\text{–}90\%} = t_{90\%} - t_{10\%} = 2.3\tau - 0.105\tau = 2.195\tau \tag{8-7}$$

Note that $t_{10\%}$ *and* $t_{90\%}$ *are calculated from* $0.1 = 1 - e^{-t_{10\%}/\tau}$ and $0.9 = 1 - e^{-t_{90\%}/\tau}$. An example is shown in Figure 8-6, where the time constant was calculated assuming an *RC* network with $R = 50\ \Omega$ and $C = 5$ pf.

The 3-dB bandwidth of a one-pole filter is

$$f_{3dB} = \frac{1}{2\pi\tau}$$
$$\rightarrow \tau = \frac{1}{2\pi f_{3dB}}$$

Solving for τ and substituting into (8-7) produces the well-known relationship between the spectral content of an edge and the rise time:

$$t_{10\text{–}90\%} = 2.195\tau = \frac{2.195}{2\pi f_{3dB}} \approx \frac{0.35}{f_{3dB}} \tag{8-8}$$

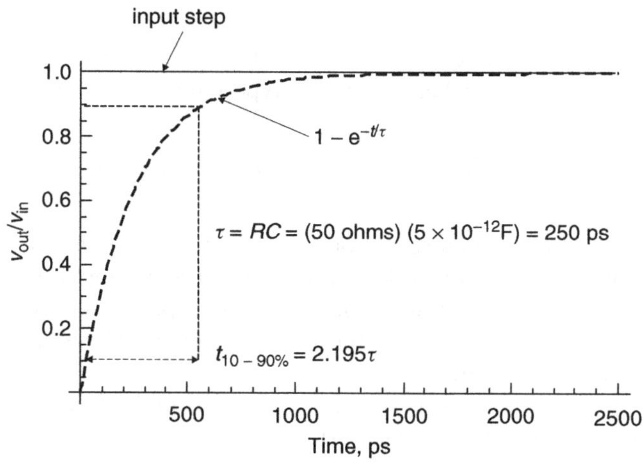

Figure 8-6 If a unit step is driven into a single-pole network with time constant τ, the resulting 10 to 90% rise time can be calculated.

Equation (8-8) is a good "back of the envelope" calculation that estimates the spectral bandwidth of a digital signal with a rise/fall time of $t_{10-90\%}$.

Equations (8-5) and (8-8) can be used to estimate the spectral envelope of a trapezoidal digital signal. The spectrum of a square wave will fall off at -20 dB/decade, as described by (8-5). When the frequency ($f_{3\text{dB}}$) that corresponds to the bandwidth of an edge with rise and fall times of $t_{10-90\%}$ is reached, the low-pass filtering function becomes significant, and also falls off at a rate of -20 dB/decade. This allows us to draw the approximate spectral envelope of a digital pulse, as shown in Figure 8-7.

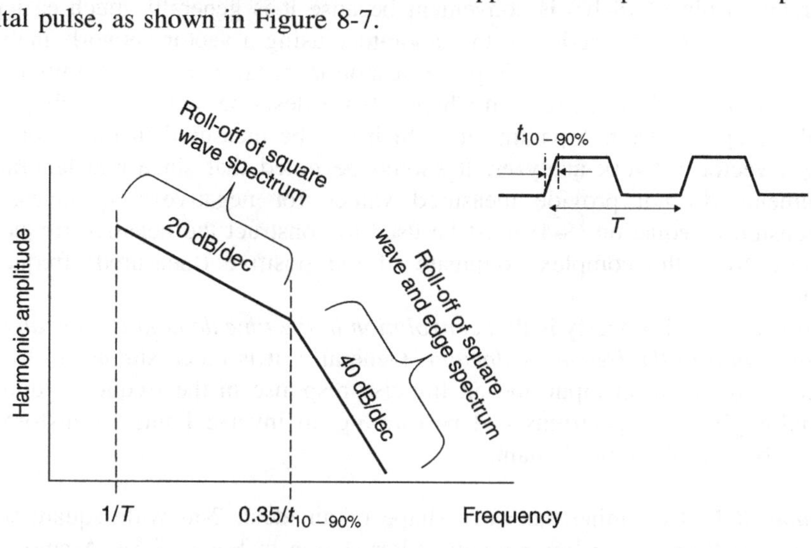

Figure 8-7 Approximation of the spectral envelope of a digital pulse.

8.1.4 System Response

When a system is linear and time invariant, the output and input can be represented with a *convolution*, where bold face type indicates a matrix:

$$\mathbf{y}(t) = \mathbf{h}(t) * \mathbf{x}(t) = \int_{-\infty}^{\infty} \mathbf{h}(t-\tau)\mathbf{x}(\tau)\,d\tau \tag{8-9a}$$

where $\mathbf{h}(t)$ is the system impulse response matrix, and each element $h_{ij}(t)$ is the response at port i when Dirac's delta function (an ideal impulse) is applied at port j with all other inputs set to zero. This is an important concept, because if the impulse response of a system is known, the response of the system with any input, $\mathbf{x}(\tau)$, can be determined.

When the system impulse response is represented in the frequency domain, it is referred to as the *transfer function* $H(\omega)$:

$$\mathbf{Y}(\omega) = \mathbf{H}(\omega)\mathbf{X}(\omega) \tag{8-9b}$$

where $\mathbf{Y}(\omega)$ is the frequency response matrix of the of the system output and $\mathbf{X}(\omega)$ is the frequency response of the system input. It is usually much more convenient to analyze systems using transfer functions in the frequency domain rather than impulse responses in the time domain. However, due to the equivalency between a time-domain waveform and a frequency-domain spectrum, *the inverse Fourier transform of the transfer function is the impulse response*:

$$h(t) = \mathsf{F}^{-1}\{H(\omega)\} \tag{8-10}$$

The relationship of (8-10) is convenient because it is generally much easier to measure the transfer function in the laboratory using a vector network analyzer (VNA) than to measure the impulse response because an ideal impulse is impossible to produce. In fact, in Chapter 9 we describe methods to obtain the impulse response from S-parameters which can be measured in the laboratory using a vector network analyzer. It should be noted that since real laboratory instruments do not provide measured values for negative frequencies, the relationship of equation (8-4) must be used to construct the negative frequency response from the complex conjugate of the positive (measured) frequency response.

Another useful property is that *convolution in the time domain is equivalent to multiplication in the frequency domain*. Generally, it is much simpler to handle the convolution of an input and an impulse response in the frequency domain by multiplying the spectrums and performing an inverse Fourier transform to convert back to the time domain.

Example 8-1 Determine the wave shape of an ideal, 2-ns-wide square wave propagating through the low-pass RC filter shown in Figure 8-8a. Assume that $R = 50\ \Omega$, $C = 5$ pF, and the amplitude of the square wave is 1 V.

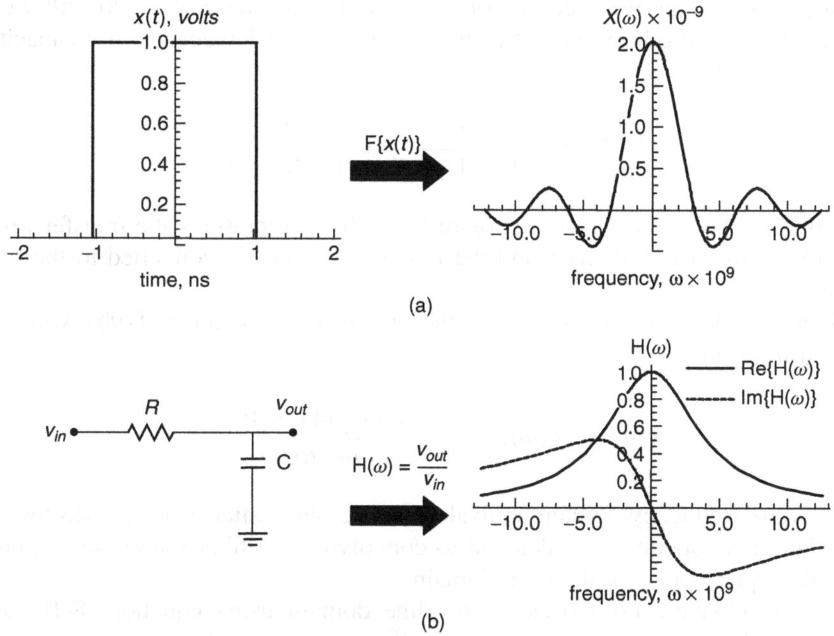

Figure 8-8 (a) Spectrum of a 2-ns-wide square wave for example 8-1; (b) transfer function of an *RC* filter with $R = 50\ \Omega$ and $C = 5$ pF.

SOLUTION

Step 1: Calculate the spectrum of the square wave using equation (8-1a). For this example, system engineering conventions ($a = 1$, $b = -1$) were chosen.

$$
\begin{aligned}
X(\omega) &= \int_{-\infty}^{\infty} x(t)e^{-j\omega t}\,dt \\
&= \int_{-\infty}^{\infty} u_s(t + 1 \times 10^{-9})u_s(-t + 1 \times 10^{-9})e^{-j\omega t}\,dt \\
&= \int_{-1\times10^{-9}}^{1\times10^{-9}} e^{-j\omega t}\,dt \\
&= j\left.\frac{e^{-j\omega t}}{\omega}\right|_{t=-1\times10^{-9}}^{t=1\times10^{-9}} \\
&= \frac{2\sin[\omega\left(1 \times 10^{-9}\right)]}{\omega}
\end{aligned}
$$

$X(\omega)$ is plotted in Figure 8-8a.

Step 2: Calculate the spectrum of the filter. The response of the *RC* filter can be calculated using basic circuit techniques where the impedance of a capacitor is $Z_c = 1/j\omega C$:

$$H(\omega) = \frac{1/j\omega C}{R + 1/j\omega C} = \frac{1}{j\omega RC + 1}$$

The filter response is plotted in Figure 8-8b. $H(\omega)$ represents the transfer function in the frequency domain and the impulse response if converted to the time domain.

Step 3: Calculate the response of the output using equation (8-9b), which is plotted in Figure 8-9a:

$$Y(\omega) = X(\omega)H(\omega) = \frac{2\sin\left[\omega(1 \times 10^{-9})\right]}{j\omega^2 RC + \omega}$$

$Y(\omega)$ is the frequency-domain equivalent of the pulse after it has passed though the filter. This operation is identical to convolving the filter's impulse response with the square pulse in the time domain.

Step 4: Convert $Y(\omega)$ back to the time domain using equation (8-1b) and system engineering conventions:

$$y(t) = \mathsf{F}^{-1}\{Y(\omega)\}$$

The inverse Fourier transform was evaluated using Mathematica and plotted in Figure 8-9b.

An alternative to the impulse response of a system is the step response. Often, the step response is more conducive to laboratory measurements because time-domain reflection (TDR) instruments are widely available and are capable

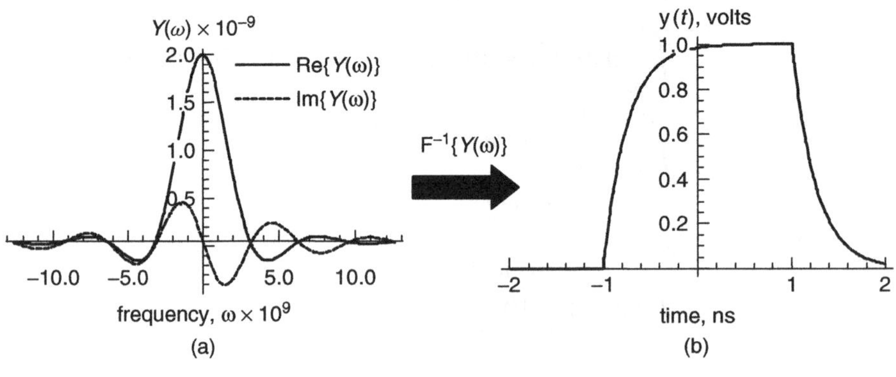

Figure 8-9 (a) Spectrum of the driving function for example 8-1 convolved with the filter response; (b) output of the filter when driven with a square wave.

of injecting rise times as fast as 9-25 ps into a test structure, which can approximate a step function. The unit step function is the integral of the impulse function (Dirac delta function),

$$u_s(\tau) = \int_{-\infty}^{t} \delta(t)\, dt \tag{8-11a}$$

$$\frac{du_s(t)}{dt} = \delta(t) \tag{8-11b}$$

where $\delta(t)$ is the impulse function and $u_s(t)$ is the unit step function.

8.1.5 Single-Bit (Pulse) Response

Perhaps the most useful means to characterize a channel in the time domain is to use the *single-bit response*, also known as the *pulse response*. The utility of the pulse response, as opposed to the impulse response, is that it is directly measurable in the laboratory. The pulse response is obtained by driving a system with a waveform that corresponds to a single digital bit of information for the system being designed.

The *data rate* (DR) is defined as the maximum number of bits per seconds the system will support. This means that the maximum data rate is determined by the width of a single bit:

$$\text{DR} = \frac{1}{\Delta t_{\text{bit}}} \tag{8-12a}$$

where Δt_{bit} is the width of a single data bit, as shown in Figure 8-10a. Sometimes, Δt_{bit} is referred to as a *unit interval* (UI). This means that the maximum fundamental frequency of the digital pulse train, where alternating bits of 1 and 0 are transmitted sequentially, is half the data rate, as shown in Figure 8-10b:

$$f_{\text{fundamental}} = \frac{1}{2\Delta t_{\text{bit}}} \tag{8-12b}$$

The pulse response is more practical than the impulse response for two reasons. First, the pulse response will give the engineer an intuitive feeling of how the bus will operate because it represents how the system will respond to an actual waveform that will be propagating on the bus. Second, it can be used to calculate the worst-case eye and the worst-case bit pattern using the peak distortion analysis, which is described in Chapter 13.

Analytically, the pulse response is calculated by convolving the input waveform with the system impulse response in the time domain. It is usually more convenient to perform the convolution in the frequency domain,

$$Y(\omega) = \mathsf{F}\{x_{\text{pulse}}(t)\} \cdot H(\omega) \tag{8-13a}$$

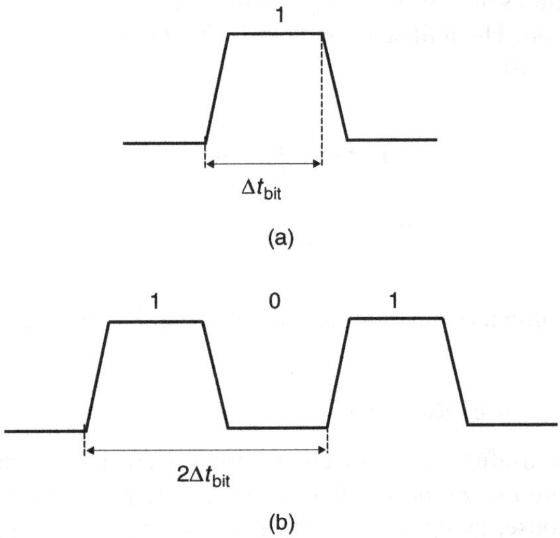

Figure 8-10 Definitions used to calculate (a) the maximum data rate and (b) the fundamental frequency of a bit stream.

where $Y(\omega)$ is the response in the frequency domain, $\mathsf{F}\{x_{\text{pulse}}(t)\}$ is the Fourier transform of the input pulse and $H(\omega)$ is the transfer function of the system. The pulse response is then computed by taking the inverse Fourier transform of $Y(\omega)$:

$$y(t) = \mathsf{F}^{-1}\{Y(\omega)\} \tag{8-13b}$$

Example 8-2 Calculate the 10-Gb/s pulse response of the 0.5-m transmission line calculated in Example 6-4. Assume that the input pulse has rise and fall times of 33 ps and a magnitude of 1 V.

SOLUTION

Step 1: Calculate the real and imaginary parts of the transmission-line transfer function in the frequency domain. This is done with the equation derived in Example 6-4:

$$v_{\text{out}} = v_{\text{in}} e^{-\alpha z}[\cos(-\beta z) + j\sin(-\beta z)]$$

where γ is defined by equation (6-47) and z is the line length,

$$\gamma(\omega) = \alpha + j\beta = \sqrt{(R + j\omega L)(G + j\omega C)}$$

and the frequency-dependent values of R, L, C, and G are as calculated in Example 6-4. Next the transfer function is calculated:

$$H(\omega) = \frac{v_{\text{out}}}{v_{\text{in}}} = e^{-\alpha z}[\cos(-\beta z) + j\sin(-\beta z)]$$

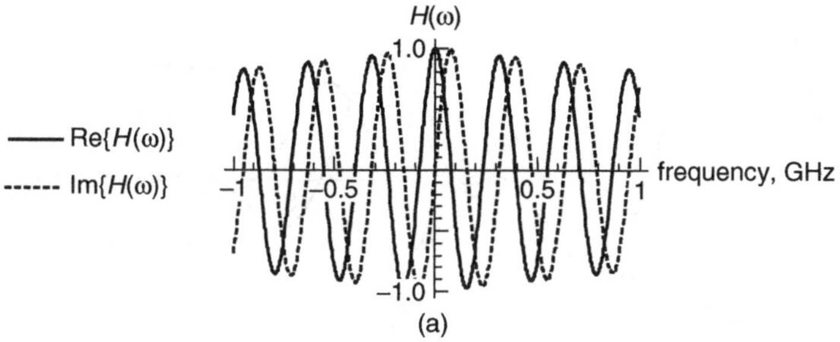

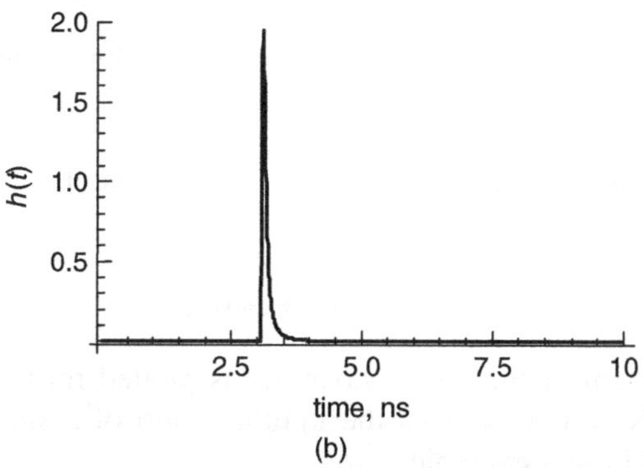

Figure 8-11 (a) Transfer function of the transmission line for example 8-2 (b) impulse response calculated with the inverse Fourier transform of the transfer function.

Using the values derived in Example 6-4, the real and imaginary parts of the transfer function are plotted in Figure 8-11a.

Step 2: Calculate the impulse response by taking the inverse Fourier transform of $H(\omega)$ using equation (8-10):

$$h(t) = \mathrm{F}^{-1}\{H(\omega)\}$$

The impulse response is plotted in Figure 8-11b. Due to the complexity of the integration, for this example the fast Fourier transform (FFT) was used instead of the Fourier integral to evaluate the spectral response of the waveforms. The FFT is a numerical method used to calculate the Fourier transform of an arbitrary waveform. The FFT was equated by sampling the positive and negative frequency response 2000 times in steps of 100 MHz, which gives a time-domain resolution of

$$\Delta t = \frac{1}{N\Delta f} = \frac{1}{2000(100 \times 10^6)} = 5 \times 10^{-12}\ \mathrm{s}$$

where $\Delta f = 100 \times 10^6$ Hz and $N = 2000$. The FFT is described by Press et al. [2005] and is built into many commercial tools, such as Mathematica, Matlab, Mathcad, and even Microsoft Excel.

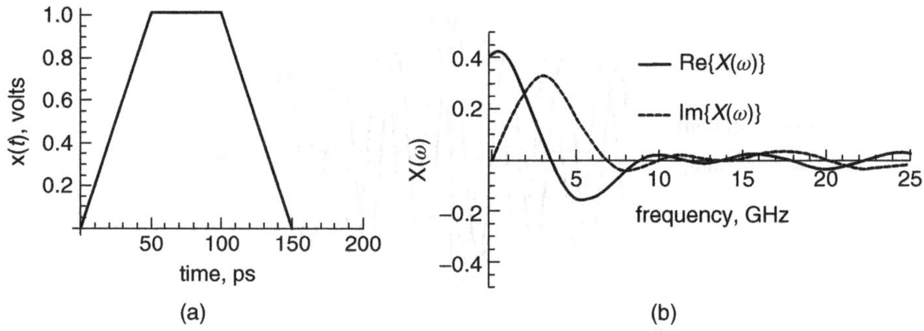

Figure 8-12 (a) Waveform driving the transmission line for example 8-2; (b) Fourier transform of the input waveform.

Step 3: Calculate the Fourier transform of the input waveform shown in Figure 8-12a:

$$X(\omega) = \mathsf{F}\{x_{\text{pulse}}(t)\}$$

The Fourier transform of the input waveform is plotted for positive frequencies in Figure 8-12b. Note that it takes the familiar form of a sinc function. Again, the FFT was used in this example.

Step 4: Calculate the pulse response. First, the spectrum of the pulse response $Y(\omega)$ is calculated using equation (8-9b):

$$Y(\omega) = X(\omega)H(\omega)$$

Finally, the pulse response is calculated by taking the inverse Fourier transform of $Y(\omega)$:

$$y(t) = \mathsf{F}^{-1}\{Y(\omega)\}$$

The pulse response, which depicts the input waveform after it has propagated from one end of the transmission line to the other, is plotted in Figure 8-13.

Note that the pulse arrives at approximately 3.1 ns. This result can be verified by estimating the total delay from the quasistatic values of the inductance given in Example 6-4 and capacitance using equation (3-107):

$$\tau_d = l\sqrt{LC} = 0.5\sqrt{(2.5 \times 10^{-7})(1.5 \times 10^{-10})} = 3.06 \times 10^{-9} \text{ s}$$

The quasistatic approximation will not be identical to the delay calculated with frequency-dependent parameters, but it should be close. Therefore, the delay of the waveform in Figure 8-13 passes the sanity check.

It is interesting to note how the transmission line distorts the pulse as it propagates down the transmission line. For example, the amplitude of the waveform

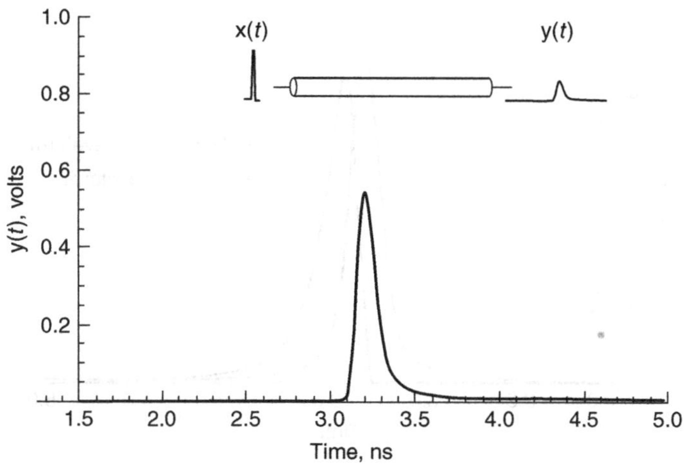

Figure 8-13 Pulse response of the transmission line for Example 8-2.

in Example 8-2 (Figure 8-13) is degraded significantly due to the conductor and dielectric losses. Furthermore, the output pulse is wider than the input pulse because the frequency components of the digital waveform (as calculated with a Fourier transform) will travel at different speeds due to the frequency-dependent dielectric permittivity used to calculate the capacitance. The velocity differences between each harmonic will distort the waveform by spreading it out in time, which is known as *dispersion*.

8.2 REQUIREMENTS FOR A PHYSICAL CHANNEL

In this section we introduce specific limitations that channel models of a LTI system must obey to remain physically consistent with nature. Specifically, the conditions of *causality, passivity*, and *stability* are described and defined by the appropriate mathematical conditions that must be met to ensure physical behavior. The analysis is restricted to linear and time-invariant electrical networks, which is appropriate for all passive components used in modern bus design, such as transmission lines, vias, packages, connectors, and so on.

8.2.1 Causality

One seemingly obvious requirement of a model that obeys the laws of nature is that an output cannot precede its input. In other words, in the real world we live in, an effect cannot precede its cause. This fundamental principle is called *causality*. Mathematically, a linear time-invariant system is causal only if for every input all the elements of its impulse response h_{ij} vanish for $t < 0$:

$$\mathbf{h}(t) = 0 \qquad \text{when } t < 0 \tag{8-14a}$$

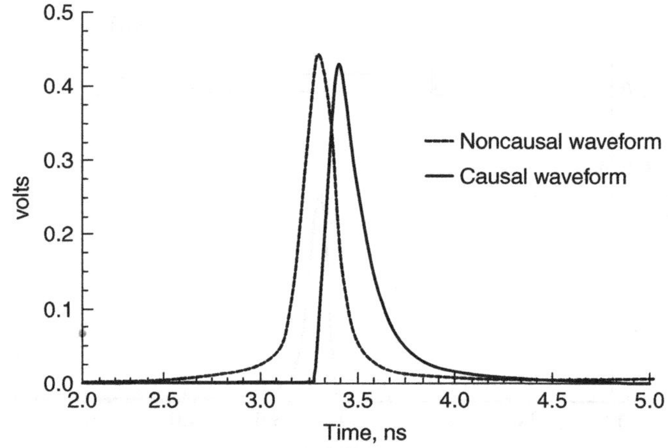

Figure 8-14 Simulated pulse response for a 20-in. transmission line, showing the outputs of a causal and a noncausal model. Note how the noncausal waveform arrives early.

But more generally, if a system has known delay $[\tau]$, the system is causal if

$$\mathbf{h}(t) = 0 \qquad \text{when } t < [\tau] \tag{8-14b}$$

An example of a causal and a noncausal pulse response is shown in Figure 8-14, where a pulse response was simulated by driving a 100-ps-wide digital pulse into a causal and a noncausal 20-in. transmission-line model.† Note that the noncausal pulse response has a component that arrives early, which is a telltale sign of a nonphysical model. In this example, the noncausal transmission-line model was created using frequency-invariant values of the loss tangent and the dielectric permittivity. The causal model was created using the infinite-pole dielectric model presented in Chapter 6.

Unfortunately, it is not always easy to ascertain whether a model is causal by observing the pulse response. Furthermore, it is not always obvious if it matters. For example, a very short transmission line may be noncausal, but the error could be so small that it would not affect the final result significantly. However, if numerous noncausal models are cascaded together to simulate a bus, the causality errors could accumulate and significant waveform miscalculations could be realized. Fortunately, it is not very difficult to create a causal model by ensuring that the dielectric and conductor properties outlined in Chapters 5 and 6 are followed.

The discussion above prompts an inevitable question: *What are the mathematical conditions that specify whether or not a system is causal?* To begin, the frequency response of the model is observed. If the Fourier transform of the

†Don't get the *pulse response* confused with the *impulse response* $[h(t)]$. Since pulse responses are more representative of a realistic digital driver, they are often used to analyze interconnects.

impulse response $h(t)$ is equated, the frequency-domain equivalent $H(\omega)$ can be used to test for the conditions of causality:

$$\mathsf{F}\{h(t)\} = H(\omega)$$

The properties of the Fourier transform can be used to make some initial conclusions about the causality of a system.

1. If $h(t)$ is real, $H(-\omega) = H(\omega)^*$, where * indicates the complex conjugate [LePage, 1980]. Since time-domain waveforms are always real, this is a necessary requirement.
2. If $h(t)$ is real and odd, $H(\omega)$ is imaginary and odd [O'Neil, 1991].
3. If $h(t)$ is real and even, $H(\omega)$ is real and even [O'Neil, 1991].

As a reminder, odd and even functions obey the following rules: Let $f(t)$ be a real-valued function of a real variable. Then f is *even* if

$$f(t) = f(-t)$$

and *odd* if

$$-f(t) = f(-t)$$

This means that if $h(t)$ is even or odd, there will exist a nonzero value for $t < 0$, which violates the definition of causality as expressed in equation (8-14a).

Another useful property is that the vector space of all real-valued functions is the direct sum of the subspaces of even and odd functions. In other words, every function can be written uniquely as the sum of an even function and an odd function:

$$f(x) = f_e(x) + f_o(x) = \frac{f(x) + f(-x)}{2} + \frac{f(x) - f(-x)}{2}$$

Therefore, a causal function [where $h(t) = 0$ for $t < 0$] must be the sum of an even function $h_e(t)$ and an odd function $h_o(t)$.

$$h(t) = h_e(t) + h_o(t) = \tfrac{1}{2}\left[h(t) + h(-t)\right] + \tfrac{1}{2}\left[h(t) - h(-t)\right] \qquad \text{(8-15)}$$

Consequently, for $h(t)$ to be causal, $h(t)$ must be composed of *both* odd and even functions. Therefore, based on Fourier transform properties 2 and 3 above, $H(\omega)$ must have both a real and an imaginary part. *For an impulse response $h(t)$ to be real and causal, $H(\omega)$ must be complex and satisfy the complex-conjugate rule shown in condition 1 above.*

However, just because $H(\omega)$ is complex obviously does not guarantee causality. Equation (8-15) shows how the negative time components of the odd and even functions must cancel each other out to ensure a causal response. Since the odd and even values of $h(t)$ are generated from imaginary and real parts of $H(\omega)$, respectively, causality requires a specific relationship between $\mathbf{Re}[H(\omega)]$ and $\mathbf{Im}[H(\omega)]$.

To demonstrate the relationship between $\mathbf{Re}[H(\omega)]$ and $\mathbf{Im}[H(\omega)]$, consider a simple causal system with the impulse response $h(t) = u_s(t)e^{-pt}$, where $u_s(t)$ is a unit step function with a value of 0 for $t \leq 0$ and a value of 1 for $t > 0$ and $p > 0$. Following equation (8-15), the odd and even functions of $h(t)$ can be written:

$$h_e(t) = \tfrac{1}{2}u_s(t)e^{-pt} + \tfrac{1}{2}u_s(-t)e^{pt}$$

$$h_o(t) = \tfrac{1}{2}u_s(t)e^{-pt} - \tfrac{1}{2}u_s(-t)e^{pt}$$

The functions $h_e(t)$, $h_o(t)$, and $h(t)$ and are plotted in Figure 8-15. Note that when $h_o(t) = h_e(t)$ for $t > 0$ and $h_o(t) = -h_e(t)$ for $t < 0$, the output $h(t)$ is zero for $t < 0$ and therefore causal. This allows the odd function to be written in terms of the even component:

$$h_o(t) = \mathrm{sgn}(t)h_e(t) \tag{8-16}$$

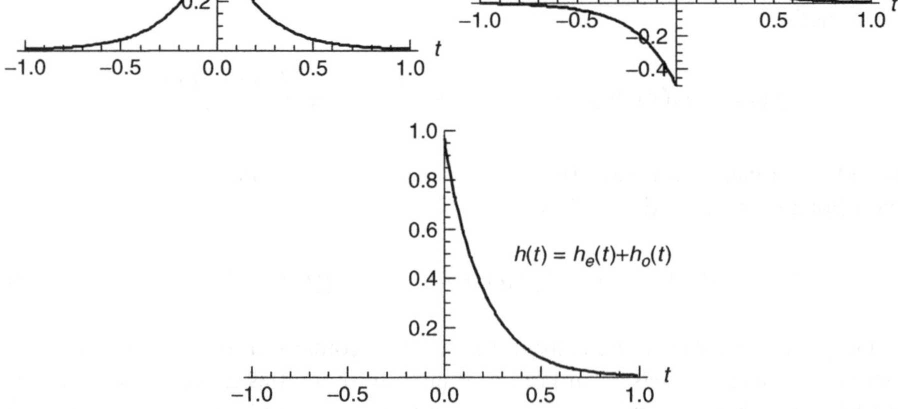

Figure 8-15 The odd and even functions are summed to produce the final causal impulse response $h(t)$.

where $\text{sgn}(t) = 1$ and $\text{sgn}(-t) = -1$. Now the causal impulse response can be written in terms of only the even component:

$$h(t) = h_e(t) + \text{sgn}(t)h_e(t) \tag{8-17}$$

In the frequency domain, $H(\omega)$ can also be written in terms of the even function using the property that multiplication in the time domain is the same as convolution in the frequency domain and $\mathsf{F}\{\text{sgn}(t)\} = -j(1/\pi\omega)$ using the signal processing convention ($a = 0$, $b = -2\pi$) of equation (8-1a):

$$H(\omega) = \mathsf{F}\{h_e(t)\} + (\mathsf{F}\{sgn(t)\} * \mathsf{F}\{h_e(t)\}) = H_e(\omega) - j\left(\frac{1}{\pi\omega} * H_e(\omega)\right) \tag{8-18}$$

Equation (8-18) can be simplified by using the definition of a *Hilbert transform*, which is the convolution of a function $g(\omega)$ and $1/\pi\omega$:

$$\hat{g}(\omega) = g(\omega) * \frac{1}{\pi\omega} = \frac{1}{\pi}\int_{-\infty}^{\infty} \frac{g(\omega')}{\omega - \omega'}\, d\omega' \tag{8-19}$$

Therefore, equation (8-18) can be written in terms of the Hilbert transform of the even function:

$$\begin{aligned} \hat{H}_e(\omega) &= \frac{1}{\pi\omega} * H_e(\omega) \\ H(\omega) &= H_e(\omega) - j\hat{H}_e(\omega) \end{aligned} \tag{8-20}$$

where $\hat{H}_e(\omega)$ denotes the Hilbert transform of $H_e(\omega)$.

Equation (8-20) demonstrates two very important properties of a system that will produce a real, linear, and causal response in the time domain.

1. The imaginary part of the frequency response is determined by the Hilbert transform of the real part. *Knowledge of the real part is sufficient to define the entire function.*
2. Causality can be tested by performing the Hilbert transform of the real part and ensuring that it is identical to the imaginary part.

Note that the real part of a causal signal can be derived from its imaginary part also. It is interesting to note that the Kramers–Kronig relations mentioned in Chapter 6 are another form of the Hilbert transforms that relate the real and imaginary parts of the complex dielectric permittivity to each other.

$$\varepsilon'(\omega) = 1 + \frac{2}{\pi}\int_0^{\infty} \frac{\omega'\varepsilon_r''(\omega')}{(\omega')^2 - \omega^2}\, d\omega' \tag{6-34a}$$

$$\varepsilon''(\omega) = \frac{2\omega}{\pi}\int_0^{\infty} \frac{1 - \varepsilon_r'(\omega')}{(\omega')^2 - \omega^2}\, d\omega' \tag{6-34b}$$

As described in Section 6.4.1, there is a specific relationship between $\varepsilon'(\omega)$ and $\varepsilon''(\omega)$ that must be enforced to ensure realistic behavior. If the dielectric models do not satisfy the Kramers–Kronig relations, the system will be noncausal.

Example 8-3 Use the Hilbert transform to prove that the waveform shown in Figure 8-16a is noncausal and the waveform depicted in Figure 8-16b is casual.

SOLUTION

Step 1a: Calculate the Fourier transform of the noncausal waveform (Figure 8-16a) using the signal processing convention ($a = 0,\ b = -2\pi$) of equation (8-1a):

$$f(t) = u_s(t+1)u_s(-t+2)$$

$$F(\omega) = \frac{\cos\pi\omega\sin 3\pi\omega}{\pi\omega} - j\frac{\sin\pi\omega\sin 3\pi\omega}{\pi\omega}$$

Step 2a: Calculate the Hilbert transform of the real part of $F(\omega)$ and compare it to the imaginary part:

$$\hat{F}_{\text{Re}}(\omega) = \text{Re}[F(\omega)] * \frac{1}{\pi\omega} = \mathsf{F}^{-1}\left(\mathsf{F}\left(\frac{\cos\pi\omega\sin 3\pi\omega}{\pi\omega}\right)\mathsf{F}\left(\frac{1}{\pi\omega}\right)\right)$$

$$= \frac{(3 + 2\cos 2\pi\omega)\sin(\pi\omega)^2}{\pi\omega}$$

Since $\text{Im}[F(\omega)] \neq -\hat{F}_{\text{Re}}(\omega)$, the waveform depicted in Figure 8-16a is noncausal. Of course, simple observation of the waveform for this example is proof of noncausality since it exhibits nonzero values for $t < 0$.

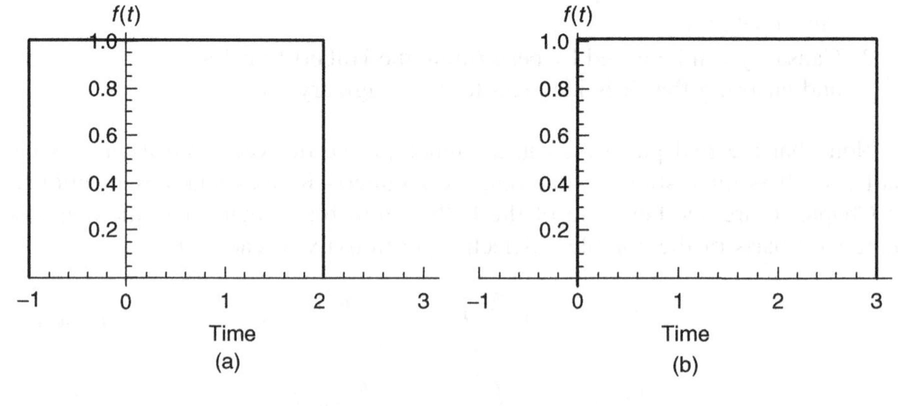

Figure 8-16 (a) Noncausal and (b) causal waveforms for Example 8-3.

Step 1b: Calculate the Fourier transform of the causal waveform (Figure 8-16b):

$$f(t) = u_s(t)u_s(-t+3)$$

$$F(\omega) = \frac{\sin 6\pi\omega}{2\pi\omega} - j\frac{\sin(3\pi\omega)^2}{\pi\omega}$$

Step 2b: Calculate the Hilbert transform of the real part of $F(\omega)$ and compare it to the imaginary part:

$$\hat{F}_{\mathrm{Re}}(\omega) = \mathrm{Re}[F(\omega)] * \frac{1}{\pi\omega} = \mathsf{F}^{-1}\left(\mathsf{F}\left(\frac{\sin 6\pi\omega}{2\pi\omega}\right)\mathsf{F}\left(\frac{1}{\pi\omega}\right)\right) = \frac{\sin(3\pi\omega)^2}{\pi\omega}$$

Since $\mathrm{Im}[F(\omega)] = -\hat{F}_{\mathrm{Re}}(\omega)$, the waveform depicted in Figure 8-16b is causal. Of course, simple observation of the waveform for this example is proof of causality since it exhibits only zero values for $t < 0$.

During bus design, commercial simulation tools are used to implement the models and generate a system response used to evaluate the signal integrity. Each simulator will have its own assumptions, approximations, and numerical methods that may affect causality. The methods detailed above should be used to verify that causal models are being generated. A practical method of judging causality is to observe the rising edge of a pulse response to see if a portion of the signal is arriving early, as demonstrated in Figure 8-14. If a portion of the pulse is arriving early, the system is noncausal. Although this method is less reliable than the rigorous methods defined above, it will provide a general idea if the model is causal.

Example 8-4 Use the Hilbert transform to determine the causality of a transmission line with a length of $z = 2.0$ in. ($\sim$0.05 m) that is terminated perfectly with a relative dielectric permittivity of $\varepsilon_r = 4.0$, $\mu_r = 1$, and a decay factor of $\alpha = 0.00000001|f|$.

SOLUTION

Step 1: Calculate the transfer function of the transmission line. Since the transmission line is perfectly terminated, no reflections will be generated. Therefore, the loss-free voltage wave will behave as described by equation (3-29). However, since this is not a loss-free network, the voltage equation must be multiplied by the decay factor $e^{-\alpha z}$:

$$v_{\mathrm{out}} = v_{\mathrm{in}}e^{-\alpha z}e^{-\beta z} = v_{\mathrm{in}}e^{-\gamma z}$$

Therefore, using equation (2-31) to simplify, the voltage is

$$v_{\mathrm{out}} = v_{\mathrm{in}}e^{-\alpha z}[\cos(-\beta z) + j\sin(-\beta z)]$$

From equations (2-45), (2-46), and (2-52), the propagation constant is calculated:

$$f = \frac{c}{\lambda}$$

$$\lambda = \frac{c}{f\sqrt{\varepsilon_r}} = \frac{3.0 \times 10^8}{2f}$$

$$\beta = \frac{2\pi}{\lambda} = \frac{4\pi f}{3.0 \times 10^8}$$

and the transfer function is calculated:

$$H(f) = \frac{v_{\text{out}}}{v_{\text{in}}} = e^{-0.00000001|f|z} \left(\cos \frac{-4\pi f}{3.0 \times 10^8} z + j \sin \frac{-4\pi f}{3.0 \times 10^8} z \right)$$

The real and imaginary parts of $H(f)$ are plotted in Figure 8-17a.

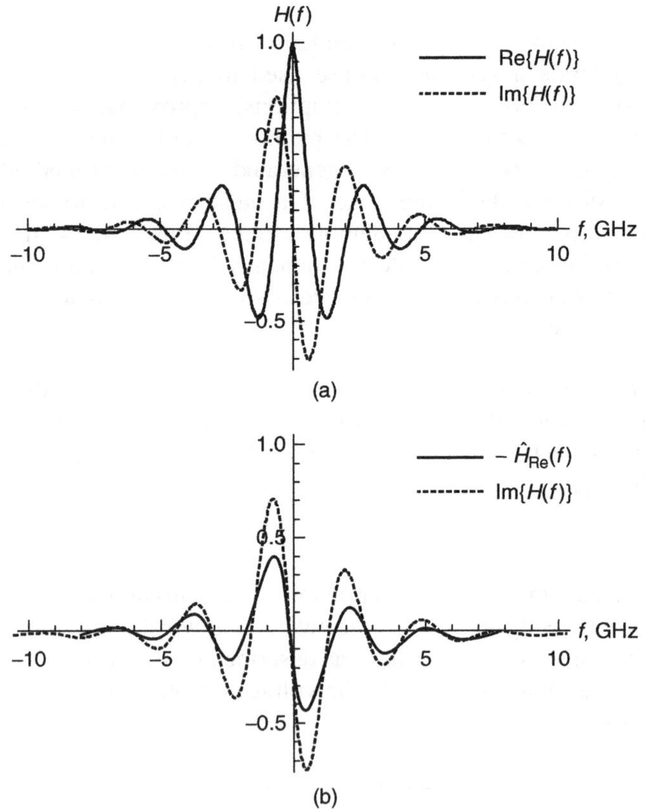

Figure 8-17 (a) Transfer function of the transmission line for Example 8-4; (b) Hilbert transform of the real part of $H(f)$ compared to the imaginary part of $H(f)$, showing that the model is noncausal.

Step 2: Calculate the Hilbert transform of the real part of $H(f)$ compare it to the imaginary part of $H(f)$.

$$\hat{H}_{\text{Re}}(f) = \text{Re}[H(f)] * \frac{1}{\pi f} = \mathsf{F}^{-1}\left(\mathsf{F}\left(e^{-0.00000001|f|z}\cos\frac{-4\pi f}{3.0\times 10^8}\right)\mathsf{F}\left(\frac{1}{\pi\omega}\right)\right)$$

The quantity $-\hat{H}_{\text{Re}}\ (f)$ and the imaginary part of $H(f)$ are plotted in Figure 8-17b. Since $\text{Im}[H(f)] \neq -\hat{H}_{\text{Re}}(\omega)$, the transmission-line model is noncausal. The noncausal nature of this model can also be observed by looking at the impulse response, which is calculated by taking the inverse Fourier transform of $H(f)$:

$$\mathsf{F}^{-1}\{H(f)\} = h(t)$$

and is plotted in Figure 8-18. Note how the impulse response rises prematurely. The theoretical delay of this transmission line can be calculated from the length, the speed of light, and the relative dielectric permittivity.

$$\frac{\sqrt{\varepsilon_r}}{c} = 6.66\times 10^{-9}\text{s/m}$$

$$\tau_d = 169.333\times 10^{-12}\text{s/in.}$$

Since the transmission line is 2 in. long, the pulse should arrive at $\tau_d = (169.333\times 10^{-12})(2.0) = 339$ ps. It is obvious that the waveform has components arriving much earlier than 339 ps, indicating that the model is noncausal and is not obeying the limits placed on it by the speed of light.

The causality problems associated with the transmission line in Example 8-4 are caused by the assumption of frequency-independent dielectric properties. In Chapter 6 we described numerous dielectric models that show how the dielectric

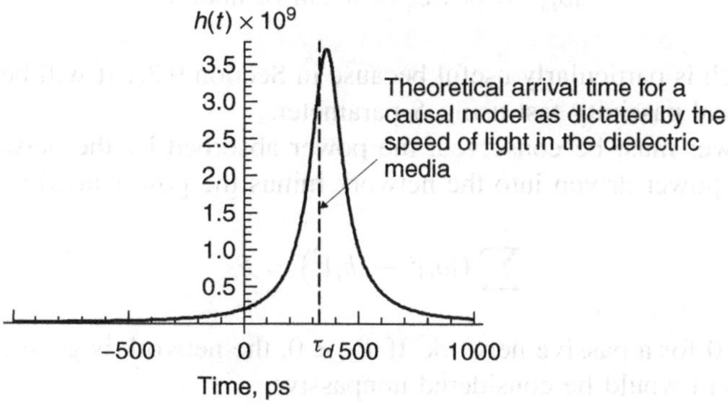

Figure 8-18 Noncausal impulse response of the transmission line of Example 8-4.

permittivity is frequency dependent. Furthermore, the relationship between the dielectric permittivity and the dielectric losses must be maintained; otherwise, energy will propagate when it should be attenuated, inducing noncausal errors. The dielectric models described in Chapter 6 will produce causal responses when used to simulate transmission lines. Unfortunately, many commercially available simulators do not account properly for the frequency dependence of the dielectric. Consequently, the engineer should be very wary of prepackaged transmission-line models to ensure that physical behavior is being observed in the simulations.

In reality, performing the Hilbert transform analytically is difficult, and numerical methods have to be used. Also, with bandlimited frequency responses, the Hilbert transform may not be a good check for causality, due to aberrations in the time-domain waveform, which is discussed briefly in Example 9-8.

8.2.2 Passivity

A physical system is *passive* when it is unable to generate energy from within. For example, an n-port network is said to be passive if

$$\int_{-\infty}^{t} \mathbf{v}^{\mathrm{T}}(\tau) \cdot \mathbf{i}(\tau)\, d\tau \geq 0 \tag{8-21}$$

where $\mathbf{v}^{\mathrm{T}}(\tau)$ is the transpose of a matrix containing the port voltages and $\mathbf{i}(\tau)$ is a matrix containing the currents. The integral (8-21) represents the cumulative net power absorbed by the system up to time t. In a passive system, this quantity must be positive for all t.

A more useful approach for digital designers would be to test the passivity in terms of the incident (a_i) and exiting (b_i) power waves at each port, as defined in Section 9.3 and shown in Figure 8-19.

$$|a_i|^2 = \text{power incident to node } i$$

$$|b_i|^2 = \text{power flow out of node } i$$

This approach is particularly useful because in Section 9.3.1 it will be developed into a practical passivity test using S-parameters.

Since power must be conserved, the power absorbed by the network (P_a) is equal to the power driven into the network minus the power flowing out:

$$\sum \left(|a_i|^2 - |b_i|^2\right) = P_a \tag{8-22}$$

where $P_a \geq 0$ for a passive network. If $P_a < 0$, the network is generating power and the system would be considered nonpassive.

When working in the frequency domain, the power waves a_i and b_i will be complex. Since power is real, equation (8-22) must be implemented using the

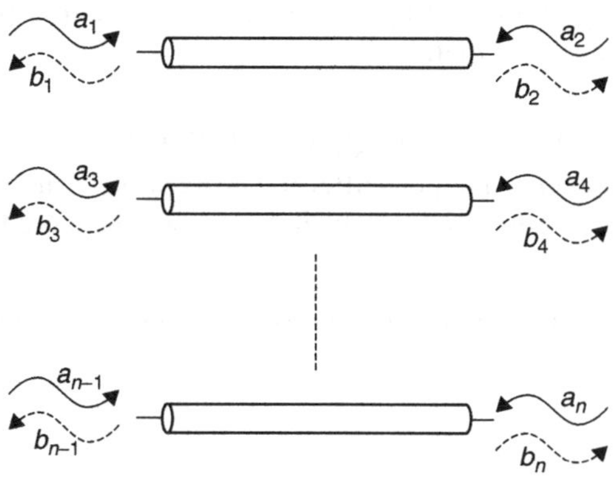

Figure 8-19 Power waves for an n-port system.

Hermitian transpose, where the complex conjugate of each term is taken, and then the matrix is transposed. For example, consider matrix **A**:

$$\mathbf{A} = \begin{bmatrix} a - jb & c + jd \\ \mathrm{e} + jf & g - jh \end{bmatrix}$$

where the Hermitian transpose is $\mathbf{A}^{\mathrm{H}}$:

$$\mathbf{A}^{\mathrm{H}} = \begin{bmatrix} a + jb & \mathrm{e} - jf \\ c - jd & g + jh \end{bmatrix}$$

This allows equation (8-22) to be written in terms of the power wave matrices that will produce a real value for the power absorbed by the network. A system is passive if

$$\mathbf{a}^{\mathrm{H}}\mathbf{a} - \mathbf{b}^{\mathrm{H}}\mathbf{b} \geq 0 \tag{8-23}$$

where **a** is a matrix that contains all the power waves incident to each port and **b** contains the power waves coming out of each port. The product of a matrix with complex values and its Hermitian transpose produces a real value. Therefore, equation (8-23) simply ensures that the total power absorbed in a network is greater than or equal to zero.

In the frequency domain, (8-23) is evaluated at each frequency point. In the time domain, the passivity requirement is essentially the same, except that the function must be integrated:

$$\int_{-\infty}^{t} \mathbf{a}(\tau)^{\mathrm{T}}\mathbf{a}(\tau) - \mathbf{b}(\tau)^{\mathrm{T}}\mathbf{b}(\tau)\, d\tau \geq 0 \tag{8-24}$$

where the transpose is used instead of the Hermitian transpose because time-domain signals are always real. Equations (8-21), (8-23), and (8-24) represent the cumulative energy absorbed by the system. A system is passive only if the following requirements are met:

1. The system absorbs more energy than it generates.
2. Generation of energy happens after the absorption. A noncausal system that generates energy before it absorbs it would be considered to be nonpassive.

Example 8-5 Determine if the transmission-line system depicted in Figure 8-20 is passive.

SOLUTION

Step 1: Calculate the power waves. The incident wave is $a_1(t)$. The reflected wave $b_1(t)$ is determined from the reflection coefficient between the termination resistor R_t and the transmission-line impedance Z_0. Using equation (3-102), the reflection coefficient is calculated:

$$\Gamma = \frac{R_t - Z_0}{R_t + Z_0}$$

Therefore, the reflected wave is calculated as

$$b_1(t) = \Gamma a_1(t - 2\tau_d)$$

which arrives at the input at $t = 2\tau_d$, which is twice the electrical length of the transmission line as defined in equation (3-107).

Step 2: Test for passivity. Since this example is in the time domain, equation (8-24) is used:

$$\int_{-\infty}^{t} [a_1(\tau)]^2 - \Gamma^2[a_1(t - 2\tau_d)]^2 \, d\tau \geq 0$$

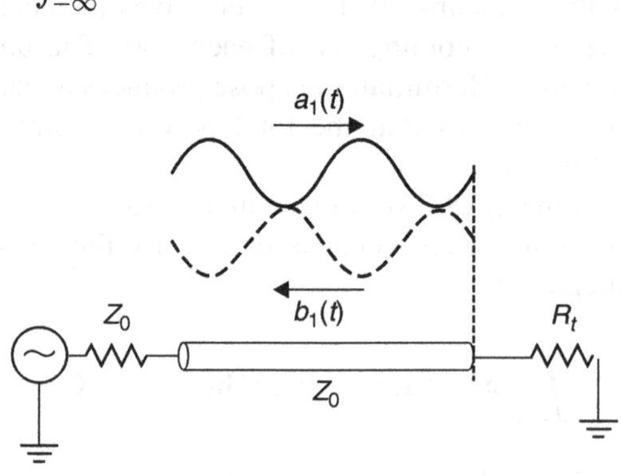

Figure 8-20 Power waves for Example 8-5.

It is easy to see that the integral above will be positive as long as the square of the reflection coefficient is less than or equal to 1 ($\Gamma^2 \leq 1$). This puts a limitation on the termination resistance R_t.

$$\Gamma^2 = \left(\frac{R_t - Z_0}{R_t + Z_0}\right)^2 \leq 1$$

The only condition where $\Gamma^2 > 1$ is when R_t is negative. Therefore, the system is passive as long as $R_t \geq 0$.

8.2.3 Stability

A model of a passive component such as a transmission line, via, connector, or package must remain stable in order to mimic real-world behavior successfully. For the purposes of this book, stability will be defined so that the output of a system $y(t)$ is stable for all bounded inputs $x(t)$. Using this definition, the stability of a linear time-invariant system is guaranteed only if all the elements in the impulse response matrix $[h(t)]$ satisfy [Triverio et al., 2007]

$$\int_{-\infty}^{\infty} |h_{ij}(t)|\, dt < \infty \tag{8-25}$$

Example 8-6 Determine the conditions of stability of a mass–spring system similar to that used to derive the frequency dependence of the dielectric permittivity in Section 6.3.2. Assume that the system is subjected to a sharp impulse at time $t = \tau$. The spring equation is given:

$$\left(m\frac{d^2x}{dt^2} + b\frac{dx}{dt} + kx\right) = \delta(t - \tau)$$

SOLUTION

Step 1: Solve the differential equation and compute the impulse response. For simplicity sake, assume initially that $m = 1$, $b = 3$, and $k = 2$. Since the inverse Laplace transform of the transfer function is the impulse response, it is easiest to solve this problem by converting to the Laplace domain:

$$s^2Y(s) + 3sY(s) + 2Y(s) = e^{-\tau s}$$

where the Laplace transform of the input is $\mathsf{L}[\delta(t - \tau)] = e^{-\tau s}$. From equation (8-9b), $Y(s)$ is solved, where $X(s) = e^{-\tau s}$:

$$Y(\omega) = H(\omega)X(\omega) \tag{8-9b}$$
$$Y(s) = H(s)e^{-\tau s}$$

and $H(s)$ is the transfer function, which is determined from the roots of the characteristic equation:

$$H(s) = \frac{1}{(s+1)(s+2)} = \frac{1}{s+1} - \frac{1}{s+2}$$

The inverse Laplace transform of $H(s)$ yields the impulse response in the time domain:

$$h(t) = \mathsf{L}^{-1}[H(s)] = e^{-t} - e^{-2t}$$

Step 2: Check the stability criteria. The system is stable of the integral of equation (8-25) converges:

$$\int_{-\infty}^{\infty} |e^{-t} - e^{-2t}|\, dt = \begin{cases} 0 & \text{if } -\infty < t < \tau \\ \left|\frac{1}{2}e^{-2(t-\tau)} - e^{-(t-\tau)}\right| & \text{if } t \geq \tau \end{cases}$$

At $t = \infty$, the integral is zero, so the integral converges and the system is stable. The impulse response to the system can be plotted by multiplying $h(t)$ by the unit step function $u_s(t - \tau)$, which occurs at time $t = \tau$:

$$y(t) = \mathsf{L}^{-1}[e^{-\tau s} H(s)] = h(t - \tau)u_s(t - \tau)$$

Assuming that $\tau = 1$, Figure 8-21a shows the response of this system.

Now consider the case where the damping coefficient b is negative, which means that the system must generate energy, which is clearly not possible for a damped mass–spring system. Assume that that $m = 1$, $b = -3$, and $k = 2$, which yields

$$s^2 Y(s) - 3sY(s) + 2Y(s) = e^{-\tau s}$$

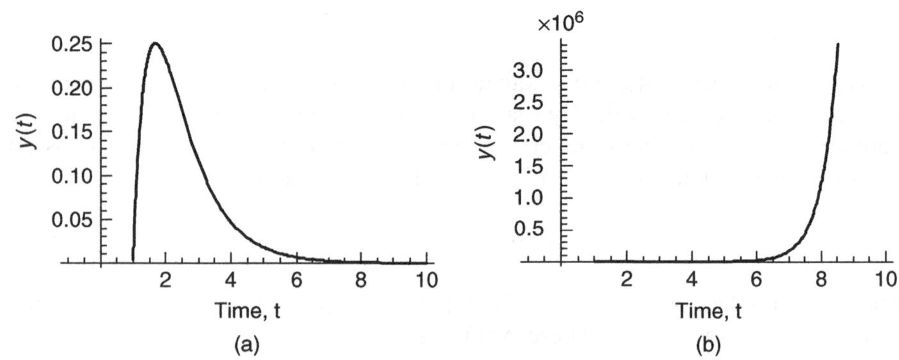

Figure 8-21 Impulse response of a mass–spring system (Example 8-6) with (a) a positive damping coefficient that is stable and (b) a negative damping coefficient that is divergent and therefore unstable.

The roots of the characteristic equation give the following transfer function:

$$H(s) = \frac{1}{(s-1)(s-2)} = \frac{1}{s-2} - \frac{1}{s-1}$$

which yields

$$h(t) = \mathsf{L}^{-1}[H(s)] = e^{2t} - e^{t}$$

When plugged into equation (8-25) to test for stability, it clearly does not converge and therefore is unstable when the damping coefficient b is negative. Assuming that $\tau = 1$, Figure 8-21b shows the response of this system.

During bus design, simulators such as HSPICE are generally used to implement the models and generate a system response used to evaluate the signal integrity. Each simulator will have its own assumptions, approximations, and numerical methods that will affect stability. Consequently, a practical method of testing stability is to generate a pulse response by driving the model with an ideal trapezoid with edge rates approximating those of realistic drivers. The transient simulation should be evaluated for a time period much longer than the time constant of the model. For example, if the propagation delay of a 10-in. bus is 1.5 ns, the reflections should have diminished to zero within two or three round trips (3 to 4.5 ns). However, the simulation should be evaluated for at least 10 round trips (30 to 45 ns) to ensure that the system remains stable.

REFERENCES

LePage, Wilbur P., 1980, *Complex Variables and the Laplace Transform*, Dover Publications, New York.

O'Neil, Peter V., 1991, *Advanced Engineering Mathematics*, Wadsworth, Belmont, CA.

Press, William H., Saul A. Teukolsy, William T. Vetterling, and Brian P. Flannery, 2005, *Numerical Recipes in C++*, 2nd ed., Cambridge University Press, New York.

Triverio, Piero, Stefano Grivet-Talocia, Michel Nakhla, Flavio Canavero, and Ramachandra Achar, 2007, Stability, causality and passivity in electrical interconnect models, *IEEE Transactions on Advanced Packaging*, vol. 30, no. 4, Nov.

Wolfram, Stephen, 2007, *Mathematica™ 6.0 On-Line Manual*, Mathematica, Champaign, IL.

PROBLEMS

8-1 Derive an equation and draw a plot that approximates the spectral bandwidth for a digital waveform with 20 to 80% rise times.

8-2 If a perfect step was driven onto a 50-Ω transmission line terminated in a 5-pF capacitor, what would the rise time be?

8-3 How many harmonics are necessary to adequately represent the bandwidth of a 10-Gb/s pulse with 25-ps rise and fall times?

8-4 For a causal waveform, derive the real part of the spectrum from the imaginary part.

8-5 For an arbitrary time-domain waveform, devise a test to determine if the waveform is causal. Implement your causality tester in a program such as Mathematica or Matlab. Prove that your causality tester works using ideal square-wave inputs.

8-6 Create a causal and a noncausal microstrip model for a 50-Ω transmission line built on FR4 (ε_r at 1 GHz $= 3.9$, $\tan\delta = 0.019$) with a 5-mil-wide smooth copper signal conductor (no surface roughness). Evaluate the causality of your models with the tester developed in Problem 8-5.

8-7 Would the noncausal model of Problem 8-6 cause significant error for a 10-Gb/s bus design with transmission-line lengths of 15 in.? If so, how did you determine the impact of the causality error?

8-8 For the transmission line described in Problem 8-6, how does surface roughness influence the causality of the system?

8-9 For the transmission line described in Problem 8-6, what has more of an effect on the causality of the model, the conductor characteristics or the dielectric characteristics? Demonstrate how each affects the causality.

8-10 Is it possible for a real, causal time-domain waveform to have only real frequency components? Show proof of your answer.

8-11 For a loss free system, do the capacitance and inductance matrices need to be frequency dependent to guarantee causality? Show proof of your answer.

9

NETWORK ANALYSIS FOR DIGITAL ENGINEERS

Historically, the techniques used to analyze signal integrity for digital designs required the use of equivalent circuits to describe components such as vias, connectors, sockets, and even transmission lines for low-data-rate applications. At low frequencies where the interconnects between the components of a digital system are small compared to the wavelength of the signal, the circuits can be described with lumped elements using resistors, capacitors, and inductors. In general, circuit theory works well for these types of problems because there is

Advanced Signal Integrity for High-Speed Digital Design, By Stephen H. Hall and Howard L. Heck

negligible phase change in the voltage and current across the circuit. In other words, the signal frequency is low enough so that the electrical delay of the circuit is small compared to the switching rate of the digital waveforms. However, as system data rates increase, the delay of the interconnects becomes significant. In fact, in many modern digital designs, such as in high-speed computers, the delay of the system interconnects is so long compared to the width of a single bit of digital information that many bits can be propagating on the bus simultaneously. In this case, the phase changes of the voltage and current across the interconnects are very significant. As a result, digital engineers have turned to new techniques for describing and analyzing circuits at high frequencies, called *network analysis*.

Network analysis is a method used traditionally by microwave and radio-frequency engineers to characterize devices such as waveguides, cables, couplers, and antennas. They are used to describe completely the behavior of linear time-invariant systems using only parameters evaluated at the input and output ports. Network analysis is a frequency domain methodology that allows discrete characterization of a linear network at each frequency. The question often arises: *Why would a digital engineer use frequency-domain analysis when a digital system uses time-domain pulses*? The answer is simple: *It is often easier to analyze and characterize systems in the frequency domain*. In Chapter 8 we discussed methods to describe a system in terms of its impulse response. Although this is a fine theoretical concept, the problem remains that it is impossible to create or measure a true impulse physically. Furthermore, as described with equation (8-10), the impulse response can be calculated from the transfer function, which is a measurable frequency-domain parameter. Additionally, many of the concepts described in the book, such as skin effect resistance and loss tangents, are best analyzed in the frequency domain. In fact, the validity of a model in the time domain is sometimes judged using frequency-domain techniques, as described in Section 8.2. In short, *network analysis is a useful tool for characterizing system interconnects, specifying component performance and creating portable, tool-independent models*.

Although general network theory is presented in this chapter, the main area of concentration will be on the derivation and use of the scattering matrix, more commonly known as *S-parameters*. S-parameters are quickly gaining acceptance in the electronics industry for specifying the performance of digital components such as transmission lines, CPU sockets, and connectors. Furthermore, methodologies have developed that allow the use of S-parameters as a portable "black box" model that can be included in the simulation environments of several commercial tools. The problem is that most digitally oriented engineers are not familiar with the concept of network theory or S-parameters because it is traditionally taught in microwave, electromagnetic interference, or radio engineering curricula. In this chapter we discuss the applications and usage models of network analysis that are most applicable to the design and validation of high-speed digital systems.

9.1 HIGH-FREQUENCY VOLTAGE AND CURRENT WAVES

As a prerequisite to describing network theory, it is required to understand how voltage and current waves propagating on an interconnect will interact with different loads. Many of these concepts were covered partially in Chapter 3 when lattice diagrams and the reflection coefficient at an impedance junction were discussed. In this chapter we build on those concepts to calculate the reflection coefficient looking into a network, such as a transmission line terminated in a load that is not equal to the characteristic impedance. Similarly, the impedance looking into a terminated network is also calculated. These concepts are important for the development of network theory.

9.1.1 Input Reflection into a Terminated Network

The reflection coefficient looking into a network with a finite electrical length is different from the reflection coefficient looking into an impedance junction because it has a phase component that will change with electrical length and frequency. Equation (3-102) from Section 3.5.1 defines the reflection coefficient looking into an impedance junction:

$$\Gamma \equiv \frac{v_r}{v_i} = \frac{Z_{02} - Z_{01}}{Z_{02} + Z_{01}} \tag{3-102}$$

where v_r and v_i are the reflected and incident voltage values, respectively. In the case of equation (3-102) the reflection occurs immediately, so there is zero phase delay between the incident and reflected waves. However, consider the case shown in Figure 9-1, where there is a significant distance between the point where the reflection is being evaluated and the impedance discontinuity. The reflection coefficient at the load, $\Gamma(z = 0)$, can be calculated with equation (3-102):

$$\Gamma_0 = \frac{R_l - Z_0}{R_l + Z_0}$$

However, consider $\Gamma(z = -l)$, which is the reflection coefficient looking into the input of the network. After a signal is driven onto the network, the reflection will not arrive back at the input until the signal propagates down the network, reflects off the impedance discontinuity at $z = 0$ (defined by Γ_0), and propagates back to the source. Depending on when the reflections arrive at the receiver, the incident and reflected waves will combine at specific frequencies and interact either constructively or destructively. If the incident and reflected waves interact destructively, the reflection coefficient will be minimized (and vice versa). This means that *the reflection coefficient looking into the network will be influenced by propagation delay, characteristic impedance, termination impedance, length, and frequency*.

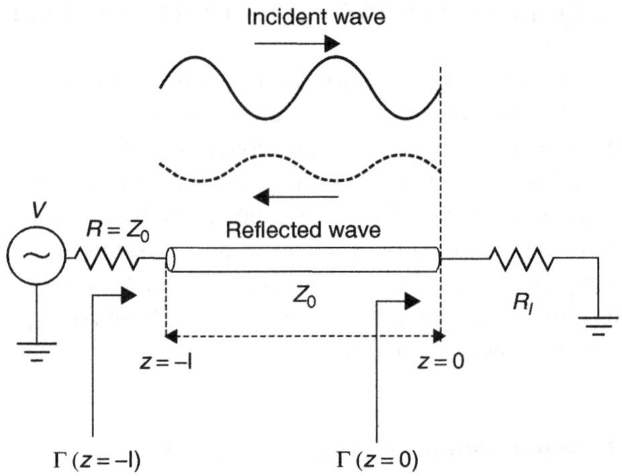

Figure 9-1 The reflection looking into a network is dependent on the distance between the point where the reflection is being evaluated and the impedance discontinuity.

The reflection coefficient looking into a network can be derived from equations (6-49) and (3-102):

$$v(z) = v(z)^{+}e^{-\gamma z} + v(z)^{-}e^{\gamma z}$$

Let $v(z)^{+} = v_i$ and $v(z)^{-} = v_r$. Then

$$v(z) = v_i e^{-\gamma z} + v_r e^{\gamma z} = v_i (e^{-\gamma z} + \Gamma_0 e^{\gamma z}) = v_i e^{-\gamma z}[1 + \Gamma(z)] \qquad (9\text{-}1)$$

where

$$\Gamma(z) \equiv \frac{v_r e^{\gamma z}}{v_i e^{-\gamma z}} = \Gamma_0 e^{2\gamma z} = \frac{R_l - Z_0}{R_l + Z_0} e^{2\gamma z}$$

Equation (9-1) describes the reflection coefficient looking into a transmission line with characteristic impedance Z_0, length z, termination impedance R_l, and propagation constant γ.

In Section 3.5 the concept of lattice diagrams was introduced to demonstrate how time-domain signals propagate on transmission lines. An important concept demonstrated was that the period of transmission-line “ringing” was dependent on the electrical length of the line. In frequency-domain analysis, the same principles apply; however, it is more useful to calculate the frequency when the reflection coefficient is either maximum or minimum, which is dependent on both the electrical length of the structure and the frequency of the input stimulus. To demonstrate this concept, consider the case of a loss-free transmission line as defined by

$$v(z) = v(z)^+ e^{-jzw\sqrt{LC}} + v(z)^- e^{jzw\sqrt{LC}} \tag{3-29}$$

$$\gamma = \alpha + j\beta = 0 + j\omega\sqrt{LC} \rightarrow \beta = \omega\sqrt{LC} \tag{3-30}$$

Beginning with equation (9-1), the reflection coefficient looking into a terminated transmission line with a length of $-l$ is calculated and expanded using equation (2-31):

$$\cos\phi + j\sin\phi = e^{j\phi} \tag{2-31}$$

$$\begin{aligned} \Gamma_0 e^{2\gamma(-l)} &= \Gamma_0 e^{j2\beta(-l)} = \Gamma_0 e^{-j2\omega l\sqrt{LC}} \\ &= \Gamma_0(\cos 4\pi f l\sqrt{LC} - j\sin 4\pi f l\sqrt{LC}) \end{aligned} \tag{9-2}$$

where a negative length convention is chosen for convenience.

Since the real and imaginary parts of (9-2) are periodic, the frequencies where the function is only real or imaginary can be calculated. The reflection looking into a nonperfectly terminated transmission line is purely imaginary when $4\pi f l\sqrt{LC} = n\pi/2$ for odd n because it will force the cosine term to be zero. Solving for the frequency where the reflection is imaginary produces

$$f_{\Gamma(\text{imaginary})} = \left.\frac{n}{8l\sqrt{LC}}\right|_{n=1,3,5,\ldots} \tag{9-3a}$$

Similarly, the frequencies where equation (9-2) is purely real can be calculated for the conditions where the sine term is zero: $4\pi f l\sqrt{LC} = n\pi$. The frequency where the reflection is real is shown by

$$f_{\Gamma(\text{real})} = \left.\frac{n}{4l\sqrt{LC}}\right|_{n=1,2,3,\ldots} \tag{9-3b}$$

Equations (9-3) demonstrate that when the real part of the input reflection is zero, the imaginary portion is maximum, and vice versa, as plotted in Figure 9-2. This periodic behavior can be used to extract out useful information about the device under test.

Example 9-1 Calculate the propagation delay, characteristic impedance, and the dielectric permittivity from the input reflection coefficient plot in Figure 9-2, assuming a circuit similar to Figure 9-1 with a length of 2.28 in. (0.058 m) and termination impedance R_l of 50 Ω.

SOLUTION

Step 1: To begin, consider the real portion of Figure 9-2. According to equation (9-3b), the frequencies will be purely real at multiples of n. Examination of

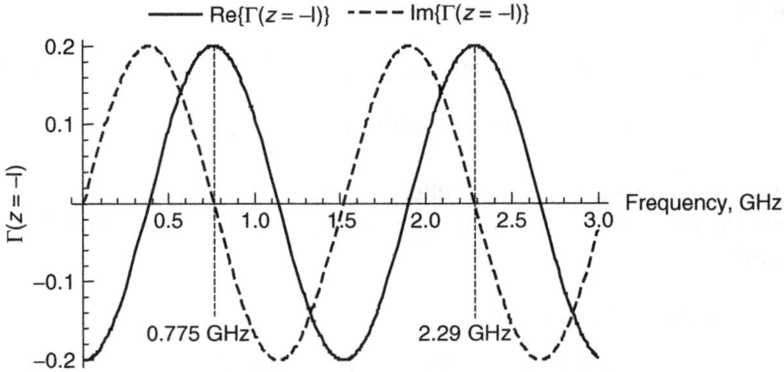

Figure 9-2 When the real part of the reflection coefficient looking into a network is maximum, the imaginary part is zero.

Figure 9-2 shows that the period of the real portion can be measured at 0.775 and 2.29 GHz. Since these peaks are the first and the third frequencies, where the reflection coefficient is purely real, they correspond to $n = 1$ and $n = 3$. Using equation (9-3b) and subtracting produces an equation in terms of the frequency difference between positive real peaks and the electrical delay:

$$f_{n=3} - f_{n=1} = \frac{3}{4l\sqrt{LC}} - \frac{1}{4l\sqrt{LC}} = \frac{1}{2l\sqrt{LC}} = 2.29 \text{ GHz} - 0.775 \text{ GHz}$$
$$= 1.515 \text{ GHz}$$

where $\tau_d = l\sqrt{LC}$, from equation (3-107). Therefore, the propagation delta can be calculated:

$$\tau_d = l\sqrt{LC} = \frac{1}{2(f_{n=3} - f_{n=1})} = \frac{1}{2(1.515 \text{ GHz})} = 330 \text{ ps}$$

Note that the propagation delay calculated using this technique is the average value between 0.775 and 2.29 GHz. Due to the frequency dependence of the dielectric permittivity described in Chapter 6, the actual value actually changes across the bandwidth.

Step 2: To calculate the dielectric permittivity, the propagation delay must first be translated into a velocity:

$$v_p = \frac{1}{l\sqrt{LC}} l = \frac{0.058}{330 \text{ ps}} = 1.75 \times 10^8 \text{ m/s}$$

Next, equation (2-52) with $\mu_r = 1$ is used to calculate the relative dielectric permittivity.

$$\nu_p = \frac{3.0 \times 10^8}{\sqrt{\varepsilon_r}} = 1.75 \times 10^8$$

$$\varepsilon_r = 2.9$$

Step 3: Calculate the characteristic impedance using the peak values of the reflection coefficient. When the imaginary term is zero, the real term will peak because the cosine term of equation (9-2) will equal 1 at frequencies predicted by (9-3b). Therefore, the easiest way to calculate the characteristic impedance is to use the value of the reflection coefficient measured at a real peak.

The first real peak at 0.775 GHz shows a maximum reflection coefficient of 0.2. The characteristic impedance can be calculated by setting the reflection coefficient equal to equation (9-2) at 0.775 GHz and solving for Z_0.

$$\cos 4\pi f l\sqrt{LC} = \cos[4\pi(0.775 \times 10^9)(330 \times 10^{-12})] \approx -1$$

$$\sin 4\pi f l\sqrt{LC} = \sin[4\pi(0.775 \times 10^9)(330 \times 10^{-12})] \approx 0$$

$$0.2 = \Gamma_0(\cos 4\pi f l\sqrt{LC} - j\sin 4\pi f l\sqrt{LC}) = \frac{R_l - Z_0}{R_l + Z_0}[-1]$$

$$= -\frac{50 - Z_0}{50 + Z_0} \rightarrow Z_0 = 75\ \Omega$$

Step 1 in Example 9-1 demonstrates a very useful relationship between the periodicity of the input reflection coefficient looking into a network and the propagation delay. If the distance between peaks $(f_{n=3} - f_{n=1})$ is represented as Δf and $\tau_d = l\sqrt{LC}$, the time delay can be calculated using

$$\tau_d = \frac{1}{2\Delta f} \tag{9-4}$$

The utility of equation (9-4) will become apparent when analyzing S-parameters in Section 9.2.2.

In summary, the reflection coefficient looking into a network is dependent on (1) the impedance discontinuities, (2) the frequency of the stimulus, and (3) the electrical length between discontinuities.

9.1.2 Input Impedance

Not surprisingly, if the reflection coefficient looking into a network is a function of length, impedance discontinuities, and frequency, the input impedance looking into the network must be a function of the same variables. Following a procedure similar to that used to derive equation (9-1), the impedance looking into a

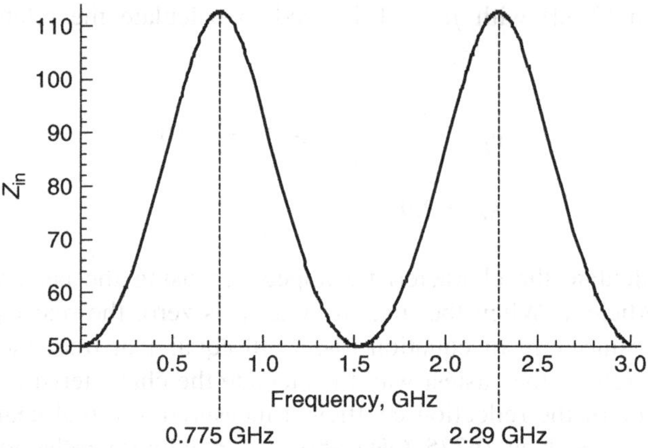

Figure 9-3 Impedance looking into a 75-Ω 2.28-in. transmission line terminated in 50 Ω. Z_{in} varies significantly with frequency due to the constructive and destructive combinations of the input stimulus and the reflected waves.

transmission line of length z terminated with R_l, as depicted in Figure 9-1, is easily derived, where $\Gamma(z)$is equated with equation (9-1):

$$v(z) = v_i e^{-\gamma z} + v_r e^{\gamma z} = v_i (e^{-\gamma z} + \Gamma_0 e^{\gamma z}) = v_i e^{-\gamma z}[1 + \Gamma(z)]$$

$$i(z) = \frac{1}{Z_0}(v_i e^{-\gamma z} - v_r e^{\gamma z}) = \frac{1}{Z_0}(v_i e^{-\gamma z}[1 - \Gamma(z)])$$

$$Z_{\text{in}} = Z(z) = \frac{v(z)}{i(z)} = \frac{v_i e^{-\gamma z}[1 + \Gamma(z)]}{1/Z_0(v_i e^{-\gamma z}[1 - \Gamma(z)])} = Z_0 \frac{1 + \Gamma(z)}{1 - \Gamma(z)} \tag{9-5}$$

Figure 9-3 shows the input impedance as a function of frequency for the terminated transmission line in Example 9-1. Note that although the characteristic impedance and the termination value are constant, the input impedance varies significantly with frequency, due to the constructive and destructive combinations of the input stimulus and the reflected waves. At frequencies where the reflection is real and the imaginary term is zero, the reflected wave is aligned with the incident wave, causing the input impedance to peak.

9.2 NETWORK THEORY

Network theory is based on the property that a linear time-invariant system can be characterized completely by parameters evaluated *only* at the input and output ports, without regard to the contents of the system. This allows the behavior of a system to be described fully in a frequency-dependent matrix that relates input stimuli to the output responses of the system. Networks can have any number of ports; however, consideration of a two-port network is sufficient to explain the theory. The discussion begins with the most intuitive method of describing a network, which is the impedance matrix.

9.2.1 Impedance Matrix

Consider the two-port network depicted in Figure 9-4. If the voltage and current are measured at the input and output ports, the system can be characterized in terms of its impedance matrix. The impedance from port 1 to port 2 is calculated by measuring the open-circuit voltage at port 2 when current is injected into port 1:

$$Z_{21} = \left.\frac{v_{\text{open,port2}}}{i_{\text{port1}}}\right|_{i_{\text{port2}}=0} \tag{9-6a}$$

Similarly, the input impedance looking into port 1 is measured by injecting current into port 1 and measuring the voltage at port 1:

$$Z_{11} = \left.\frac{v_{\text{open,port1}}}{i_{\text{port1}}}\right|_{i_{\text{port2}}=0} \tag{9-6b}$$

Using the definition shown in equations (9-6a) and (9-6b), a set of linear equations can be written to describe the network in terms of its port impedances:

$$v_1 = Z_{11}i_1 + Z_{12}i_2$$
$$v_2 = Z_{21}i_1 + Z_{22}i_2$$

which is expressed more efficiently in matrix form:

$$\begin{vmatrix} v_1 \\ v_2 \end{vmatrix} = \begin{vmatrix} Z_{11} & Z_{12} \\ Z_{21} & Z_{22} \end{vmatrix} \cdot \begin{vmatrix} i_1 \\ i_2 \end{vmatrix} \tag{9-7}$$

More generally, the elements of an impedance matrix are described in equation (9-8) for an arbitrary number of ports,

$$Z_{ij} = \frac{v_i}{i_j} = \frac{\text{open-circuit voltage measured at port } i}{\text{current injected into port } j} \tag{9-8}$$

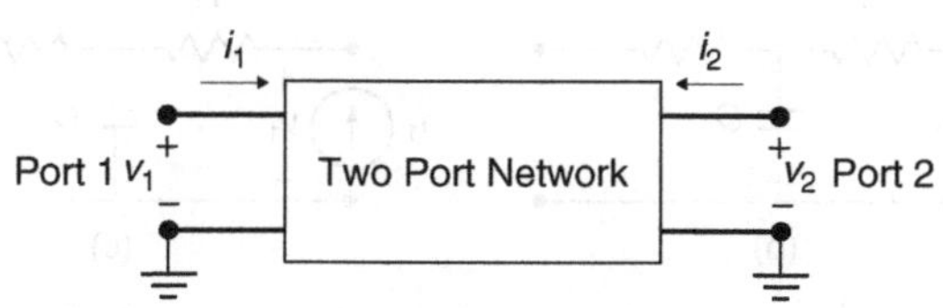

Figure 9-4 Two-port network used to generate the impedance matrix.

with $i_k = 0$ for all $k \neq j$. If the impedance matrix of a system is known, the response of the system can be predicted for any input.

Example 9-2 Calculate the impedance matrix at 1 GHz for the circuit shown in Figure 9-5a, where $R_1 = 50\ \Omega$, $R_2 = 50\ \Omega$, and $C = 5$ pF.

SOLUTION

Step 1: Calculate the input impedance (Z_{11}) by injecting a current into port 1 and measuring the voltage at port 1, as shown in Figure 9-5b. The impedance of the capacitor at 1 GHz is $Z_c = 1/j2\pi fC = 31.8\ \Omega$.

$$v_1 = i_1(R_1 + Z_c) = i_1(50 + 31.8)$$

$$Z_{11} = \frac{v_1}{i_1} = R_1 + Z_c = 81.8\ \Omega$$

Step 2: Calculate the through impedance (Z_{21}) by injecting a current into port 1 and measuring the voltage at port 2, as shown in Figure 9-5c.

$$v_2 = v_1 \frac{Z_c}{R_1 + Z_c}$$

$$= i_1(Z_c + R_1)\frac{Z_c}{R_1 + Z_c} = i_1 Z_c$$

$$Z_{21} = \frac{v_2}{i_1} = Z_c = 31.8\ \Omega$$

Step 3: Construct the impedance matrix at 1 GHz. Since the circuit is symmetrical, $Z_{12} = Z_{21}$ and $Z_{22} = Z_{11}$.

$$\mathbf{Z} = \begin{bmatrix} 81.8 & 31.8 \\ 31.8 & 81.8 \end{bmatrix}$$

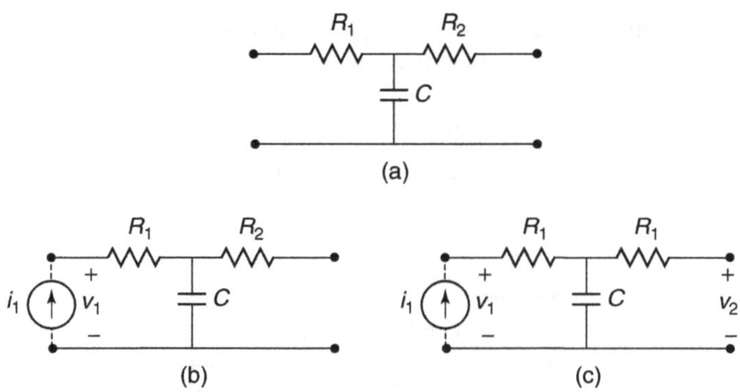

Figure 9-5 (a) General circuit to be analyzed in Example 9-2; (b) calculating Z_{11}; (c) calculating Z_{21}.

The admittance matrix is very similar to the impedance matrix except that it is characterized with short-circuit currents instead of open-circuit voltages. The admittance matrix is the inverse of the impedance matrix:

$$\mathbf{Y} = \mathbf{Z}^{-1} \tag{9-9}$$

Although the impedance and admittance matrices are intuitive and relatively easy to understand, they have severe drawbacks for the high-frequency characterization of interconnects. The problem is that at high frequencies, open and short circuits are very difficult to realize. Open circuits invariably have finite capacitances, and short circuits have inductance that can significantly affect the accuracy of the measurements. Consequently, as a measurement technique, these methods are applicable only for low frequencies.

Example 9-3 Calculate Z_{21} for the circuit analyzed in Example 9-2, assuming that the current was injected into port 1 and the voltage was measured at port 2 with a probe that has a capacitance of $C = 0.3$ pF, according to the instrument specifications.

SOLUTION

Step 1: Draw the equivalent circuit of the circuit and the probes, as depicted in Figure 9-6. From Example 9-2, $R_1 = 50\ \Omega$, $R_2 = 50\ \Omega$, $C = 5$ pF, and $Z_c = 1/j2\pi fC = 31.8\ \Omega$. The impedance of the probe at 1 GHz is

$$Z_{\text{probe}} = \frac{1}{2\pi(1\times 10^9)(0.3\times 10^{-12})} = 530.8\ \Omega$$

Step 2: Solve the circuit for Z_{11} and Z_{21}:

$$Z_{11} = \frac{v_1}{i_1} = (Z_c || (R_2 + Z_{\text{probe}}) + R_1) || Z_{\text{probe}} = 69.6\ \Omega$$

$$Z_{21} = \frac{v_2}{i_1} = \left(v_1 \frac{Z_c}{Z_c + R_1}\right) \frac{Z_{\text{probe}}}{Z_{\text{probe}} + R_2} = 24.7\ \Omega$$

where $v_1 = Z_{11} i_1$. Since the circuit is symmetrical, $Z_{12} = Z_{21}$ and $Z_{22} = Z_{11}$.

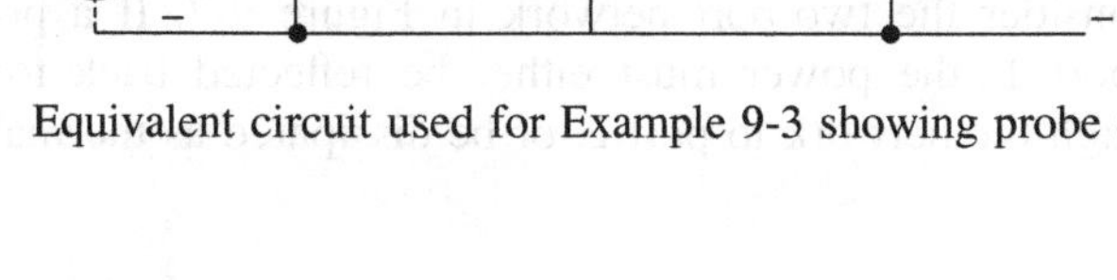

Figure 9-6 Equivalent circuit used for Example 9-3 showing probe capacitance.

Step 3: Compare the matrix measured with a realistic probe to the ideal case:

$$[Z]_{\text{ideal}} = \begin{bmatrix} 81.8 & 31.8 \\ 31.8 & 81.8 \end{bmatrix} \qquad [Z]_{\text{probed}} = \begin{bmatrix} 69.6 & 24.7 \\ 24.7 & 69.6 \end{bmatrix}$$

Note that even for a relatively low capacitance value of the probe, it makes a significant difference in the impedance matrix. This is because the voltage measured at port 2 is not across an open circuit; it is across a small capacitance. The fact is that at high frequencies, true short and open circuits do not really exist for small dimensions. There will always be a certain amount of parasitic capacitance or inductance with finite impedance.

9.2.2 Scattering Matrix

In the previous section we discussed some of the problems associated with measuring high-frequency voltage and current waves, where short and open circuits do not practically exist due to parasitic inductance and capacitance values. The scattering matrix is the most common form of network parameters used in high-speed digital design. Instead of measuring voltages and currents at the ports, it relates the power waves incident on each port to those reflected from the ports. The scattering matrix, more commonly known as *S-parameters*, can be measured in the laboratory using a *vector network analyzer* (VNA). Once the S-parameters are known, conversion to other matrix parameters, such as the impedance or admittance matrices, can be done algebraically.

Traditionally, S-parameters have been a tool used primarily by microwave and RF engineers to design antennas, waveguides, and other high-frequency narrowband applications. Higher speed data transmission on system buses is causing a convergence of two disciplines in industry today: microwave and digital engineering. Microwave engineers tend to concentrate mostly on high-frequency multi-GHz waveguides, resonators, and couplers, whereas digital engineers concentrate on binary signaling. Over the past decade, S-parameters have become much more common in the world of digital design and are often used for the dissemination of electrical models to design teams. In fact, most contemporary software suites used to design modern systems have built-in capabilities to handle S-parameter models.

S-parameters can be very confusing, as unlike the impedance matrix, they are not intuitive. In this chapter we focus on the most important aspects of S-parameters that the digital engineer needs for signal integrity analysis of modern, high-speed buses. Emphasis is given to both the theoretical development of S-parameters and intuitive techniques that will allow the engineer to interpret data quickly, share models, and estimate channel performance.

Definition Consider the two-port network in Figure 9-7. If a power wave is injected into port 1, the power must either be reflected back toward port 1, propagate through the network to port 2, or be dissipated as thermal or radiation

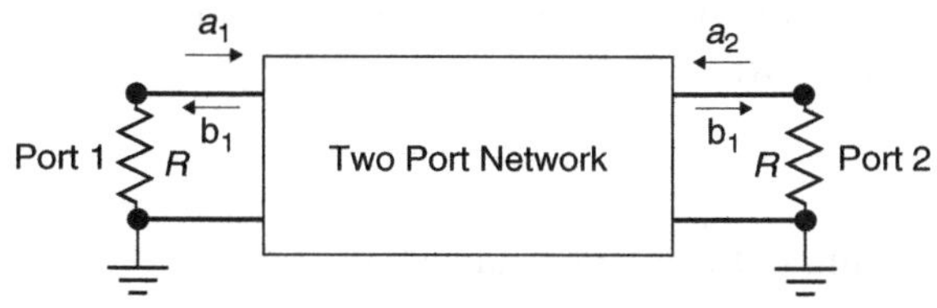

Figure 9-7 Two-port model used to define S-parameters.

losses. Similar to the derivation of the Poyting vector in Section 2.6, the power balance equation can be expressed as

$$P_{\text{out}} = P_{\text{input}} - P_{\text{loss}} - P_{\text{radiated}} \tag{9-10}$$

where P_{input} is the total power injected into all ports, P_{out} is the total power flowing out of all ports, P_{loss} is the power dissipated through ohmic losses (skin effect, loss tangent, etc.), and P_{radiated} is the power radiated into free space.

The incident and reflected power waves are calculated from voltage and current waves. The voltage waves are obtained from equation (6-49),

$$v(z) = v(z)^+ e^{-\gamma z} + v(z)^- e^{\gamma z}$$

and the current wave is derived by dividing the voltage wave by the characteristic impedance of the structure,

$$i(z) = \frac{v(z)^+ e^{-\gamma z}}{Z_0} - \frac{v(z)^- e^{\gamma z}}{Z_0}$$

where $v(z)^+$ is the voltage traveling in the $+z$-direction and $v(z)^-$ is the voltage traveling in the $-z$-direction. The power wave propagating on the network is calculated by multiplying the current and voltage waves:

$$P(z) = \frac{[v(z)^+ e^{-\gamma z}]^2}{Z_0} - \frac{[v(z)^- e^{\gamma z}]^2}{Z_0} \tag{9-11}$$

If it is defined so that at port j, $z = 0$, the voltage at a port can be calculated where $v(0) = v_i$ and $i(0) = i_i$, which are the incident voltage and current, respectively:

$$\begin{aligned} v^+ &= \tfrac{1}{2}(v_i + Ri_i) \\ v^- &= \tfrac{1}{2}(v_i - Ri_i) \end{aligned} \tag{9-12}$$

where R is the termination values at the ports of the network.

Since equation (9-10) says that the power must be balanced, the amount of power delivered to the network or radiated is defined simply as the input power minus the output power:

$$P_{\text{input}} - P_{\text{out}} = P_{\text{loss}} + P_{\text{radiated}}$$

From equation (9-12), the power going into a node can be calculated using the power relation $P = v^2/R$:

$$P_{\text{input}} = \frac{(v^+)^2}{R} \tag{9-13a}$$

and the power coming out can be calculated as

$$P_{\text{out}} = \frac{(v^-)^2}{R} \tag{9-13b}$$

meaning that the power delivered to the network is calculated as

$$P_{\text{input}} - P_{\text{out}} = P_{\text{network}} \tag{9-13c}$$

Equations (9-13) allow the definition of terms that describe the power wave propagating into port j and the power wave propagating out of port j:

$$a_j = \frac{v(z)^+}{\sqrt{R}} = \sqrt{P_{\text{input}}} \tag{9-14a}$$

$$b_j = \frac{v(z)^-}{\sqrt{R}} = \sqrt{P_{\text{out}}} \tag{9-14b}$$

where a_j is the square root of the power propagating *into* port j and b_j is the square root of the power propagating *out* of port j, as shown in Figure 9-7 for a two-port network. Equations (9-14a) and (9-14b) are known as the *scattering coefficients*. Since they are defined in terms of the square root of power, ratios of the scattering coefficients simplify into ratios of voltage as long as the termination impedance of each port (R) is the same.

S-parameters are derived from the ratios of scattering coefficients. For example, referring to Figure 9-7, the term S_{11} is calculated by the root of the reflected and incident power ratio at port 1, which is written in terms of the scattering coefficients:

$$S_{11} = \frac{b_1}{a_1} \tag{9-15a}$$

Similarly, the term S_{21} is calculated by injecting power into port 1 and measuring at port 2.

$$S_{21} = \frac{b_2}{a_1} \tag{9-15b}$$

Using the definition shown in equations (9-15a) and (9-15b), a set of linear equations can be written to describe the network in terms of its scattering coefficients:

$$b_1 = S_{11}a_1 + S_{12}a_2$$

$$b_2 = S_{21}a_1 + S_{22}a_2$$

which is more efficiently expressed in matrix form as

$$\begin{vmatrix} b_1 \\ b_2 \end{vmatrix} = \begin{vmatrix} S_{11} & S_{12} \\ S_{21} & S_{22} \end{vmatrix} \cdot \begin{vmatrix} a_1 \\ a_2 \end{vmatrix} \tag{9-16}$$

More generally, the elements of a scattering matrix are described in equation (9-17) for an arbitrary number of ports,

$$S_{ij} = \frac{b_i}{a_j} = \sqrt{\frac{\text{power measured at port } i}{\text{power injected into port } j}} \tag{9-17}$$

and an arbitrary-sized scattering matrix takes the form

$$\mathbf{b} = \mathbf{Sa} \tag{9-18}$$

If the scattering matrix of a system is known, the response of the system can be predicted for any input.

Return Loss Consider the circuit depicted in Figure 9-8. In this case there will be no reflections from the far end because the line is perfectly terminated with the characteristic impedance. However, the source impedance is not equal to the characteristic impedance, indicating that a portion of the power wave incident to port 1 will be reflected. This scenario allows the simplest definition of S_{11}, which is simply the reflection coefficient between the source resistor and the impedance of the transmission line. Note that $a_2 = 0$ because there is no source at port 2.

$$S_{11} = \left.\frac{b_1}{a_1}\right|_{a_2=0} = \frac{v_1^-/\sqrt{R}}{v_1^+/\sqrt{R}} = \frac{v_1^-}{v_1^+} = \frac{v_{\text{reflected}}}{v_{\text{incident}}} = \Gamma_0 = \frac{Z_0 - R}{Z_0 + R} \tag{9-19}$$

The term S_{11} is often referred to as the *return loss*, because it is a measure of power reflected, or returned to the source.

The calculation of S_{11} becomes more complex when the far end of the network is not perfectly terminated because the reflection arriving at the source will have

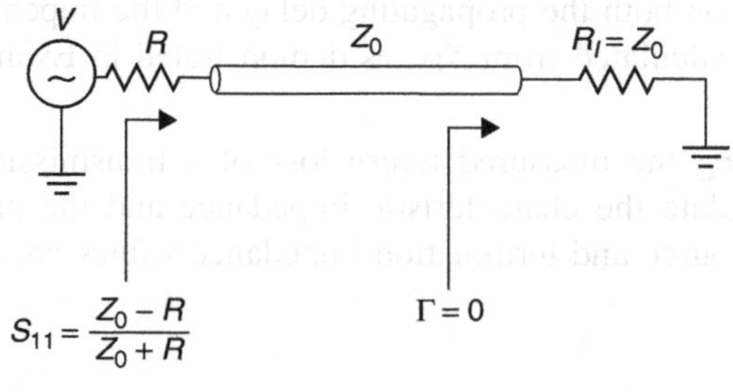

Figure 9-8 Return loss for the special case when the network is perfectly terminated in its characteristic impedance.

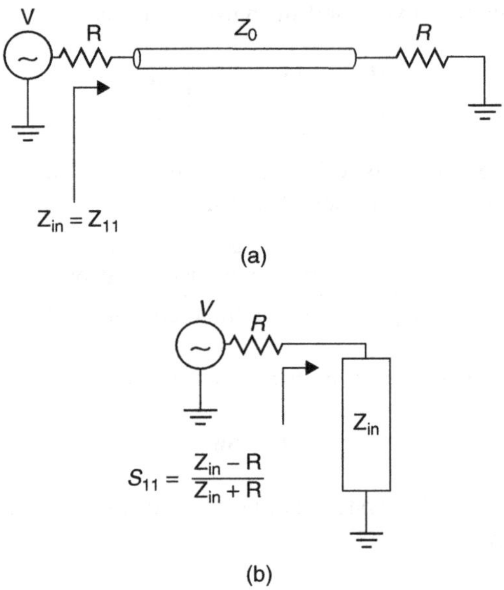

Figure 9-9 Return loss for the more general case where the network is not perfectly terminated in its characteristic impedance: (a) input impedance looking into the network; (b) equivalent circuit for the return loss.

contributions from both the impedance discontinuity at the source and at the far-end termination. This means that the input impedance looking into the network from the source will be dependent on frequency. The return loss for a nonperfectly terminated structure such as the circuit shown in Figure 9-9a is calculated as

$$S_{11}(f) = \frac{Z_{\text{in}}(f) - R}{Z_{\text{in}}(f) + R} \tag{9-20}$$

where $Z_{\text{in}}(f)$ is calculated for a transmission line with equation (9-5) for the general case. An intuitive understanding of the return loss can be achieved by constructing a simple equivalent circuit as shown in Figure 9-9b. Since Z_{in} (or Z_{11}) is dependent on both the propagating delay and the impedance of the structure, both can be calculated from S_{11}, as demonstrated in Example 9-4.

Example 9-4 Using the measured return loss of a transmission line shown in Figure 9-10, calculate the characteristic impedance and the propagation delay. Assume that the source and termination impedance values are 50 Ω.

SOLUTION

Step 1: Determine the propagation delay. This is easily calculated from the periodic behavior of S_{11} using equation (9-4). The distance between peaks is

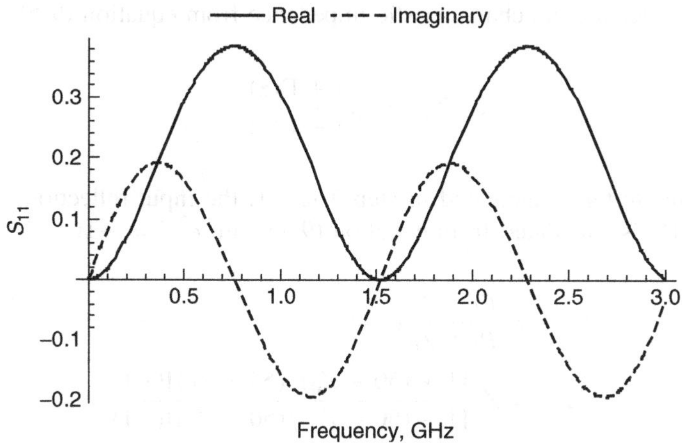

Figure 9-10 Return loss (S_{11}) for Example 9-4.

used to calculate Δf.

$$\tau_d = \frac{1}{2\Delta f} = \frac{1}{2(2.29\text{ GHz} - 0.775\text{ GHz})} = 330\text{ ps}$$

Note that τ_d is the average propagation delay over the frequency range.

Step 2: Calculate the input impedance using equation (9-20). When the return loss (S_{11}) is maximum, the imaginary part is zero. Therefore, it is convenient to measure S_{11} at a peak to simplify the analysis:

$$S_{11}(0.775\text{ GHz}) = 0.38 = \frac{Z_{\text{in}} - 50}{Z_{\text{in}} + 50}$$
$$Z_{\text{in}} = 111.3\ \Omega$$

Step 3: Determine the polarity of the phase term. The polarity of the phase term $e^{2\gamma z}$ in equation (9-1) must be determined so that the input impedance can be properly related to the characteristic impedance. Since S_{11} is being evaluated at a peak, the imaginary term of $e^{2\gamma z}$ is zero, so the real part will either be 1 or -1. Since the propagation delay has been calculated, the polarity of the phase term can be evaluated using the real part of equation (9-2):

$$\text{Re}(e^{-j2\omega l\sqrt{LC}}) = \cos 4\pi f l\sqrt{LC}$$

where $\tau_d = l\sqrt{LC} = 330$ ps and

$$\cos 4\pi f \tau_d = \cos[4\pi(0.775 \times 10^9)(330 \times 10^{-12})] \approx -1$$

Step 4: Calculate the characteristic impedance from equation (9-5):

$$Z_{\text{in}} = Z_0 \frac{1 + \Gamma(z)}{1 - \Gamma(z)}$$

Since the phase term calculated in step 3 is -1, the input reflection coefficient at 0.775 GHz is calculated from equation (9-1) with $e^{2\gamma z} = -1$:

$$\begin{aligned} \Gamma(z) &= \frac{R_l - Z_0}{R_l + Z_0}(-1) \\ Z_{\text{in}} &= Z_0 \frac{[1 + (50 - Z_0)/(50 + Z_0)](-1)}{[1 - (50 - Z_0)/(50 + Z_0)](-1)} \\ &= 111.3 = \frac{Z_0^2}{50} \\ Z_0 &= 74.6\ \Omega \end{aligned}$$

Insertion Loss When power is injected into port 1 and measured at port 2, the square root of the power ratio reduces to a voltage ratio. S_{21}, the measure of the power transmitted from port 1 to port 2, is called the *insertion loss*:

$$S_{21} = \left|\frac{b_2}{a_1}\right|_{a_2=0} = \frac{v_2^-/\sqrt{R}}{v_1^+/\sqrt{R}} = \frac{v_2^-}{v_1^+} = \frac{v_{\text{transmitted}}}{v_{\text{incident}}} \tag{9-21}$$

which is shown in Figure 9-11 for a transmission line. In digital system design, the insertion loss is the most commonly used parameter of the scattering matrix because it is a measure of both delay and amplitude as seen at the receiving agent.

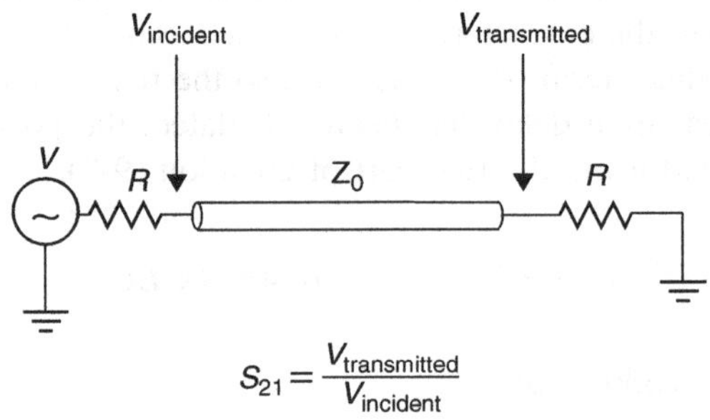

Figure 9-11 Insertion loss (S_{21}) of a transmission line.

The frequency-dependent delay can be calculated from the phase of the insertion loss,

$$\tau_p = \frac{\theta_{S_{21}}}{360^\circ f} \tag{9-22}$$

where

$$\theta_{S_{21}} = \arctan\frac{\mathrm{Im}(S_{21})}{\mathrm{Re}(S_{21})} \tag{9-23}$$

Note that equation (9-22), called the *phase delay*, has a slightly different meaning than that of the propagation delay of a digital pulse on a transmission line. Since a digital pulse is composed of numerous harmonics, and realistic dielectrics have values that vary with frequency, the propagation delay of a time-domain pulse will have numerous frequency components, each propagating with a unique phase velocity. The propagation delay of a digital pulse can be thought of as a group of harmonics propagating simultaneously and is sometimes called the *group delay*. The phase delay, as described with (9-22), is associated with only a single frequency.

Figure 9-12a shows an example of the phase of S_{21} for a 1-in. transmission-line model constructed with a realistic frequency-dependent dielectric, as described in Chapter 6. To calculate the phase delay, the phase must be unwrapped, as shown in the figure. Figure 9-12b is the phase delay calculated from the unwrapped phase using equation (9-22). Note the frequency-dependent nature of the delay, which is required for a causal transmission-line model.

For a loss-free network, the total power exiting the N ports must equal the total incident power. This means that for a two-port loss-free network, the power transmitted from port 1 to port 2 plus the power reflected from port 1 must be conserved:

$$\frac{P_{\text{reflected,port1}}}{P_{\text{incident}}} + \frac{P_{\text{transmitted,port1-port2}}}{P_{\text{incident}}} = 1 \tag{9-24}$$

Essentially, equation (9-24) says that if the power is not transmitted from port 1 to port 2, it must be reflected. This allows us to write an equation that relates the

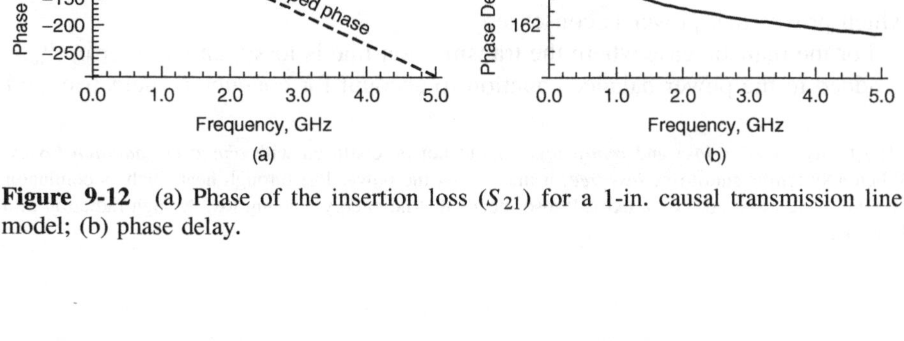

Figure 9-12 (a) Phase of the insertion loss (S_{21}) for a 1-in. causal transmission line model; (b) phase delay.

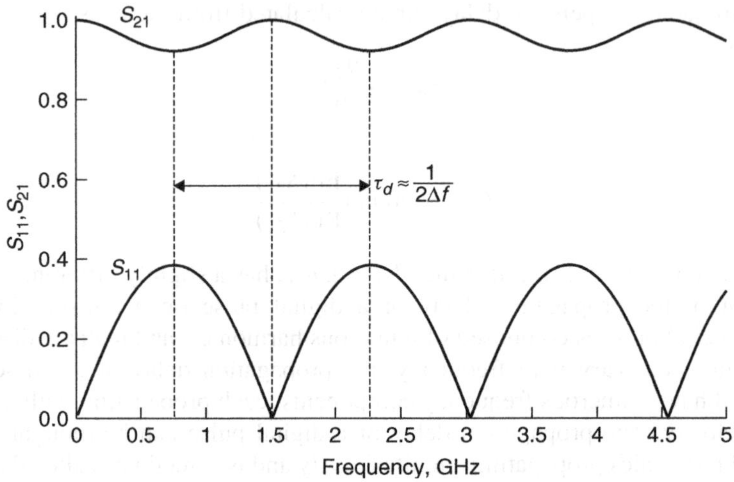

Figure 9-13 Insertion and return loss for a 2-in. loss-free transmission line.

insertion loss and the return loss for a loss-free system†:

$$S_{11}S_{11}^* + S_{21}S_{21}^* = 1 \tag{9-25}$$

where S_{ij}^* represents the complex conjugate of the term. Since S_{ij} is the square root of power measured at port i and injected at port j, the power ratio is S_{ij}^2. However, since power is a real quantity, the squaring function is replaced with the complex conjugate to ensure that the imaginary part is zero.

Figure 9-13 shows the insertion and return loss for the loss-free transmission line used in Example 9-4 terminated with 50-Ω reference impedances. Note that when the return loss (S_{11}) peaks, the insertion loss (S_{21}) dips, as described by (9-25). Since the imaginary part is zero when the terms peak and the real part is zero when the terms dip, it is easy to show that the power is conserved when the insertion loss is maximum and the return loss is minimum simply by squaring the terms, which is equivalent to the complex conjugate for these conditions. For example, at the first S_{11} peak,

$$S_{11}^2 + S_{21}^2 = (0.384)^2 + (0.923)^2 = 1$$

which proves that power is conserved.

For the realistic case where the transmission line is lossy, an extra term, P_{loss}, is added to the power balance equation to account for conductor, dielectric, and

†The terms *insertion loss* and *return loss* should not be confused with *ohmic* or *radiation losses*. When a system is said to be *loss free*, it means that the power lost through heat, such as conductor and dielectric losses, does not exist. It also means that no energy is being lost through radiation into free space.

radiation losses:

$$\frac{P_{\text{reflected,port1}}}{P_{\text{incident}}} + \frac{P_{\text{transmitted,port1-port2}}}{P_{\text{incident}}} + \frac{P_{\text{loss}}}{P_{\text{incident}}} = 1 \qquad (9\text{-}26)$$

Therefore, equation (9-25) can be rewritten to account for finite power losses:

$$S_{11}S_{11}^* + S_{21}S_{21}^* = 1 - \frac{P_{\text{loss}}}{P_{\text{incident}}} \qquad (9\text{-}27)$$

Figure 9-14 shows an example of the insertion and return losses for a lossy transmission line. The power absorbed by the network can be calculated using (9-27). At the first peak (770 MHz),

$$S_{11}^2 + S_{21}^2 = (0.379)^2 + (0.909)^2 = 0.970$$

meaning that the power absorbed by the system or radiated into space is $1 - 0.970 = 0.03$, or 3% of the total power. Of course, the absorbed power percentage will increase with frequency because both skin effect resistance and dielectric losses increase. For example, at 5.25 GHz, the total power absorbed by the network is 9.6%.

$$S_{11}^2 + S_{21}^2 = (0.35)^2 + (0.833)^2 = 0.9035$$

$$1 - 0.9035 = 0.096$$

Since properly designed transmission lines are very inefficient radiators, it is a valid assumption that very little or none of the power loss is due to radiation into free space. Of course, if there is significant coupling (crosstalk) to adjacent structures, the power will be affected, which is covered in the next section.

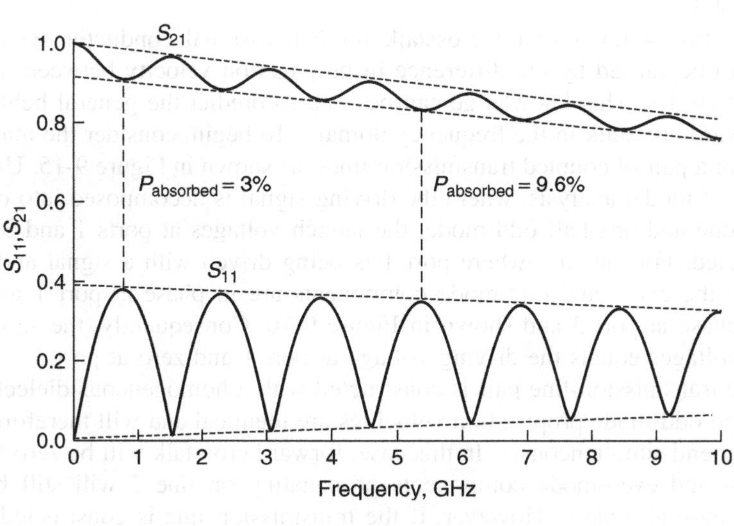

Figure 9-14 Insertion and return loss for a 2-in. lossy transmission line.

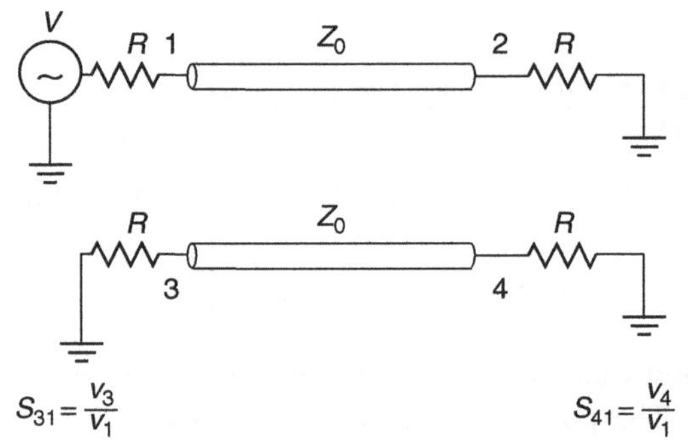

Figure 9-15 Two-transmission-line system for evaluating crosstalk using S-parameters.

Forward (Far-End) Crosstalk When power is injected into port 1 and measured at port 4, as shown in Figure 9-15, it is called *forward crosstalk*, as described in Chapter 4 and defined by

$$S_{41} = \left|\frac{b_4}{a_1}\right|_{a_2=0} = \frac{v_4^-/\sqrt{R}}{v_1^+/\sqrt{R}} = \frac{v_4^-}{v_1^+} \tag{9-28}$$

Note that *forward crosstalk* is often called *far-end crosstalk*. As detailed in Section 4.4, any bit pattern propagating on a bus with N signal conductors can be decomposed into N orthogonal modes. Section 4.4.2 describes how each mode will have a unique impedance and velocity. If a two-signal conductor system is considered, such as that shown in Figure 9-15, all digital bit patterns will be a linear superposition of the even and odd modes, which are described in Section 4.3.

In Section 4.4.4, forward crosstalk for a two-signal conductor system was shown to be caused by the difference in propagation velocity between the even and odd modes. This knowledge can be used to predict the general behavior of the forward crosstalk in the frequency domain. To begin, consider the magnitude of S_{41} for a pair of coupled transmission lines, as shown in Figure 9-15. Using the concept of modal analysis, where the driving signal is decomposed into one-half even mode and one-half odd mode, the launch voltages at ports 1 and 3 can be constructed. For the case where port 1 is being driven with a signal and port 3 is quiet, the even- and odd-mode components are in phase at port 1 and 180° out of phase at port 3 and shown in Figure 9-16. Consequently, the sum of the modal voltages equals the driving voltage at port 1 and zero at port 3.

If the transmission-line pair is constructed with a homogeneous dielectric, the even- and odd-mode propagation velocities are identical and will therefore arrive at the far end simultaneously. In this case, forward crosstalk will be zero because the odd- and even-mode components propagating on line 2 will still be 180° out of phase at node 4. However, if the transmission line is constructed with a nonhomogeneous dielectric such as a microstrip, the even and odd propagation

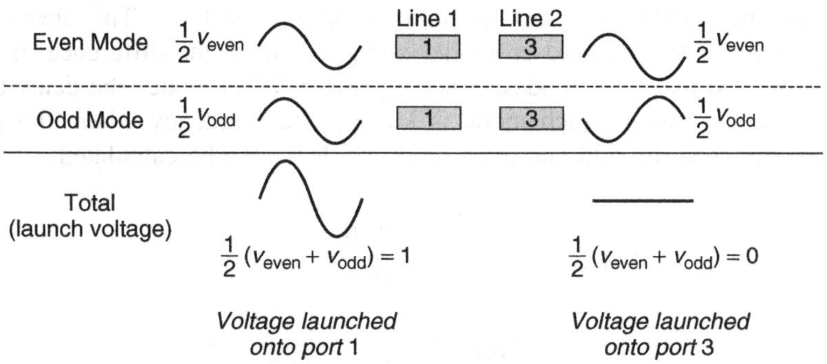

Figure 9-16 Modal decomposition of launch voltages when port 1 is driving as shown in Figure 9-15.

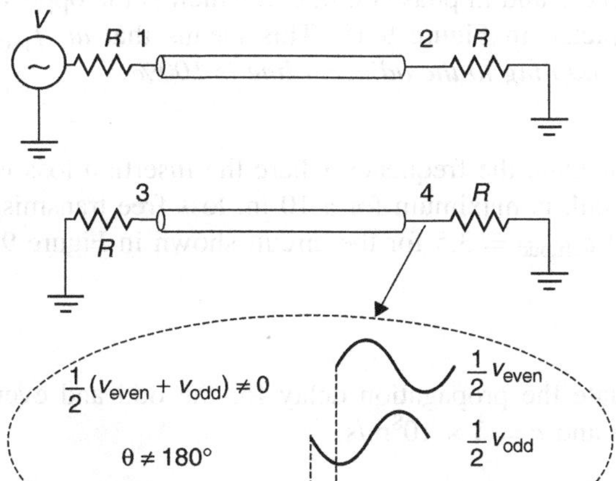

Figure 9-17 Modal decomposition of voltages at node 4 when port 1 is driving, showing that the odd- and even-mode components of the signal arrive at different times when the dielectric media is nonhomogeneous, causing forward crosstalk to be finite.

velocities will differ. Therefore, the odd and even components of the signal will arrive at node 4 at different times and will no longer be 180° out of phase. Consequently, for a nonhomogeneous dielectric, the forward crosstalk will be finite, as depicted in Figure 9-17. The propagation delay of the mode is calculated from the modal velocities using equation (4-81):

$$\nu_{p,\text{odd}} = \frac{1}{\sqrt{L_{\text{odd}}C_{\text{odd}}}} = \frac{c}{\sqrt{\varepsilon_{\text{eff,odd}}}}$$

$$\nu_{p,\text{even}} = \frac{1}{\sqrt{L_{\text{even}}C_{\text{even}}}} = \frac{c}{\sqrt{\varepsilon_{\text{eff,even}}}}$$

where the phase delay per unit length is simply $\tau_p = 1/\nu_p$. This means that the magnitude of the forward crosstalk is dependent on the difference in delay between the odd and even modes. Since equation (9-22) relates the delay to the phase, if the velocity of each mode is known, the frequency where the phase difference between the odd and even modes is 180° can be calculated:

$$\tau_{p,\text{even}} - \tau_{p,\text{odd}} = \frac{180^\circ}{360^\circ f_{180^\circ}}$$

$$f_{180^\circ} = \frac{1}{2(\tau_{p,\text{even}} - \tau_{p,\text{odd}})} \tag{9-29}$$

When $f = f_{180^\circ}$, the odd- and even-mode voltage components will be perfectly out of phase on line 1 and in phase on line 2, which is the opposite of the launch conditions as depicted in Figure 9-18. This means that *at* f_{180°, $S_{21} = 0$ *and* $S_{41} = 1$, *and the coupling to the adjacent line is 100%.*

Example 9-5 Calculate the frequency where the insertion loss is minimum and the forward crosstalk is maximum for a 10-in. loss-free transmission line where $\varepsilon_{\text{eff,even}} = 4.0$ and $\varepsilon_{\text{eff,odd}} = 3.5$ for the circuit shown in Figure 9-15.

SOLUTION

Step 1: Calculate the propagation delay for the odd and even modes where 10 in. = 0.254 m and $c = 3 \times 10^8$m/s:

$$\tau_{p,\text{odd}} = \frac{l\sqrt{\varepsilon_{\text{eff,odd}}}}{c} = \frac{0.254\sqrt{3.5}}{3 \times 10^8} = 1.58 \times 10^{-9}\ \text{s}$$

$$\tau_{p,\text{even}} = \frac{l\sqrt{\varepsilon_{\text{eff,even}}}}{c} = \frac{0.254\sqrt{4.0}}{3 \times 10^8} = 1.69 \times 10^{-9}\ \text{s}$$

Step 2: Calculate the frequency where the phase delay between odd and even modes is 180°:

$$f_{180^\circ} = \frac{1}{2(\tau_{p,\text{even}} - \tau_{p,\text{odd}})} = \frac{1}{2(1.69 \times 10^{-9} - 1.58 \times 10^{-9})} = 4.41 \times 10^9\ \text{Hz}$$

At 4.41 GHz, the insertion loss (S_{21}) will be minimum and the forward crosstalk (S_{41}) will be maximum. A simulation of this case is shown in Figure 9-19 for a 10-in. pair of coupled transmission lines with $\varepsilon_{\text{eff,even}} = 4.0$, $\varepsilon_{\text{eff,odd}} = 3.5$, $Z_{\text{odd}} = 25\ \Omega$, and $Z_{\text{even}} = 100\ \Omega$.

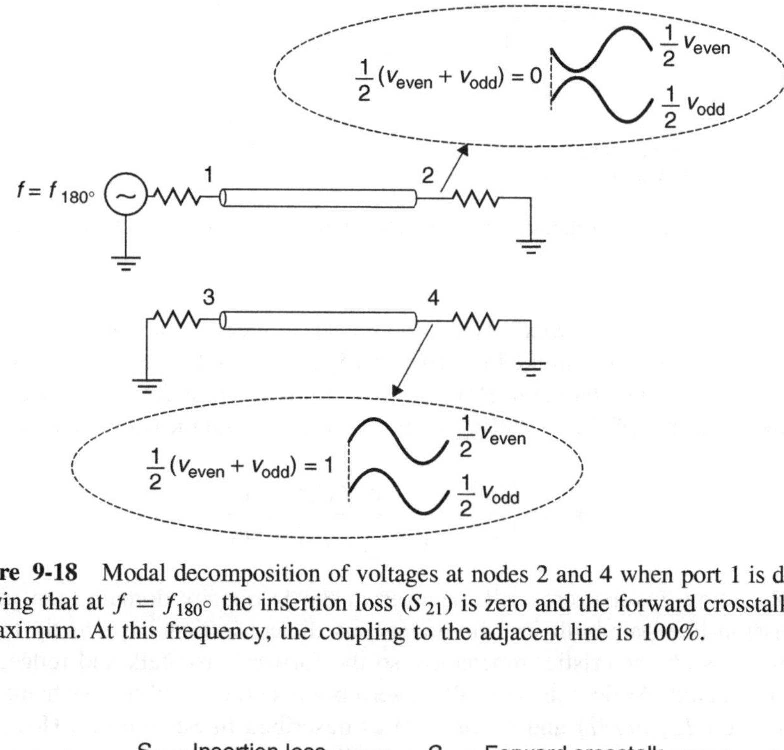

Figure 9-18 Modal decomposition of voltages at nodes 2 and 4 when port 1 is driving, showing that at $f = f_{180°}$ the insertion loss (S_{21}) is zero and the forward crosstalk (S_{41}) is maximum. At this frequency, the coupling to the adjacent line is 100%.

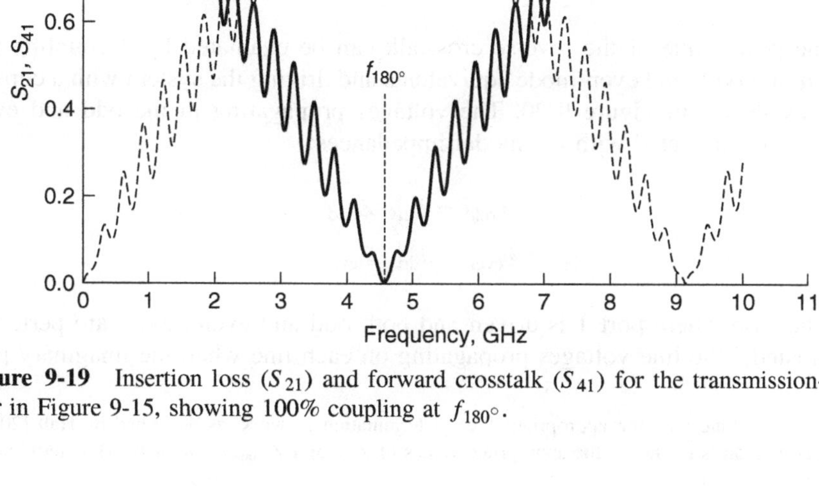

Figure 9-19 Insertion loss (S_{21}) and forward crosstalk (S_{41}) for the transmission-line pair in Figure 9-15, showing 100% coupling at $f_{180°}$.

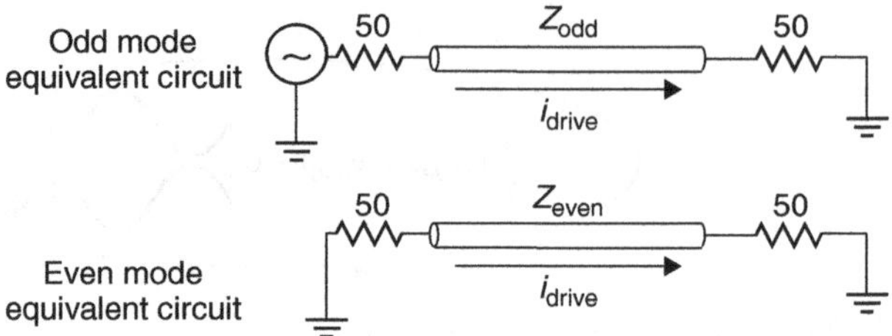

Figure 9-20 Modal decomposition of launch voltages when port 1 is driving as shown in Figure 9-15.

Reverse (Near-End) Crosstalk When the power is injected into port 1 and measured at port 3 for the circuit in Figure 9-15, it is called *reverse crosstalk*, as described in detail in Chapter 4. Reverse crosstalk is often referred to as *near-end crosstalk*. In terms of the scattering matrix, reverse crosstalk is defined as

$$S_{31} = \left|\frac{b_3}{a_1}\right|_{a_2=0} = \frac{v_3^-/\sqrt{R}}{v_1^+/\sqrt{R}} = \frac{v_3^-}{v_1^+} \tag{9-30}$$

To explain how reverse crosstalk behaves in the frequency domain, consider a transmission-line pair built in a homogeneous dielectric that is perfectly terminated with its characteristic impedance, so the forward crosstalk and reflections can be neglected. At dc, the crosstalk is zero because the coupling mechanism is dependent on $L_m(\partial i/\partial t)$ and $C_m(\partial v/\partial t)$ as described in Section 4.1. However, as the frequency starts to increase, energy will be coupled onto a victim line. As described in Section 9.1, the peak will occur when the imaginary part is zero as described by

$$f_{\Gamma(\text{real})} = \left.\frac{n}{4l\sqrt{LC}}\right|_{n=1,2,3,\ldots} \tag{9-3b}$$

The peak value of the reverse crosstalk can be evaluated by decoupling the circuit into odd- and even-mode equivalents and driving the system with a current i_{drive} as shown in Figure 9-20. The voltages propagating in the odd and even modes are calculated with the modal impedances:

$$v_{\text{odd}} = i_{\text{drive}} Z_{\text{odd}}$$

$$v_{\text{even}} = i_{\text{drive}} Z_{\text{even}}$$

For the case where port 1 is driven and both odd and even modes are perfectly terminated,† the line voltages propagating on each line when the imaginary part

†This can be done with the appropriate T or pi termination network, as described by Hall [2000]. Another method is to choose the appropriate values of Z_{odd} and Z_{even}, so the network is terminated.

is zero are given by

$$v_{\text{line1}} = \tfrac{1}{2}(v_{\text{even}} + v_{\text{odd}}) = \tfrac{1}{2}i_{\text{drive}}(Z_{\text{even}} + Z_{\text{odd}}) \quad \text{(9-31a)}$$

$$v_{\text{line2}} = \tfrac{1}{2}(v_{\text{even}} - v_{\text{odd}}) = \tfrac{1}{2}i_{\text{drive}}(Z_{\text{even}} - Z_{\text{odd}}) \quad \text{(9-31b)}$$

Since the voltage across the 50-Ω resistor at port 2 is $v_{\text{port2}} = 50i_{\text{drive}}$, the insertion loss can be calculated:

$$S_{21} = \frac{v_{\text{port2}}}{v_{\text{line1}}} = \frac{100}{Z_{\text{even}} + Z_{\text{odd}}} \quad \text{(9-32a)}$$

and the peak value of the reverse crosstalk is the ratio of the voltage coupled onto line 2 and the voltage propagating on line 1:

$$S_{31} = \frac{v_{\text{line2}}}{v_{\text{line1}}} = \frac{Z_{\text{even}} - Z_{\text{odd}}}{Z_{\text{even}} + Z_{\text{odd}}} \quad \text{(9-32b)}$$

Example 9-6 Calculate the frequency where the insertion loss will be minimum and the reverse crosstalk will be maximum for a 10-in. loss-free homogeneous transmission line where $\varepsilon_r = 4.0$, with a circuit as shown in Figure 9-15. Assume that $Z_{\text{even}} = 100\ \Omega$, $Z_{\text{odd}} = 25\ \Omega$, and all ports are terminated in 50 Ω.

SOLUTION

Step 1: Calculate the propagation delay for the odd and even modes where 10 in. = 0.254 m and $c = 3 \times 10^8$ m/s:

$$\tau_p = \frac{l\sqrt{\varepsilon_r}}{c} = \frac{0.254\sqrt{4.0}}{3 \times 10^8} = 1.69 \times 10^{-9}\ \text{s}$$

Step 2: Use equation (9-3a) to calculate the frequency of the first hump in S_{31}, where $\tau_p = l\sqrt{LC}$:

$$f_{\Gamma(\text{real})} = \left.\frac{1}{4\tau_p}\right|_{n=1} = 147 \times 10^6\ \text{Hz}$$

Step 3: Calculate the maximum value of the crosstalk:

$$S_{31} = \frac{Z_{\text{even}} - Z_{\text{odd}}}{Z_{\text{even}} + Z_{\text{odd}}} = \frac{100 - 25}{100 + 25} = 0.6$$

Step 4: Calculate the minimum value of the insertion loss. Since this is a loss-free transmission line, the insertion loss (S_{21}) will be minimum when the input reflections and crosstalk are maximum. Both the input reflections (S_{11}) and the reverse crosstalk (S_{31}) will peak when the frequency is equal to $f_{\Gamma(\text{real})}$, as

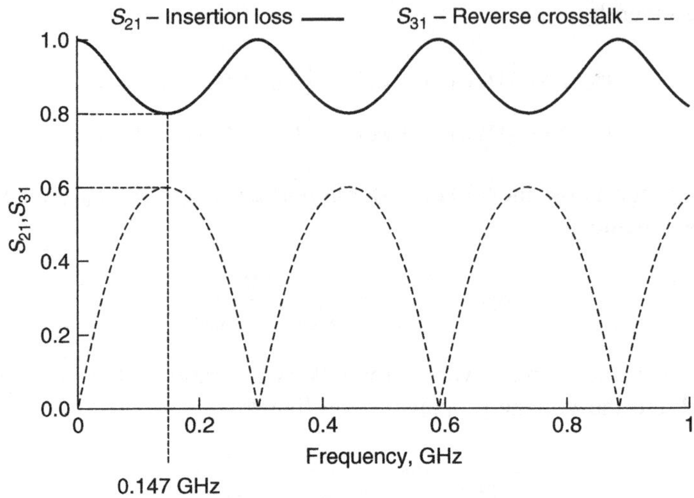

Figure 9-21 Insertion loss and reverse crosstalk for Example 9-6.

described by equation (9-3b). Consequently, when S_{11} and S_{31} are maximum, S_{21} must be minimum and is calculated with equation (9-32a):

$$S_{21} = \frac{100}{Z_{\text{even}} + Z_{\text{odd}}} = \frac{100}{100 + 25} = 0.8$$

A simulation of this case is shown in Figure 9-21.

Relationship Between S- and Z-Parameters Perhaps the most intuitive form of network analysis are the Z-parameters, which were explained in Section 9.2.1. The disadvantage of using an impedance (or admittance) matrix is simply that they are impossible to measure directly at high frequencies. Fortunately, it is relatively easy to convert Z-parameters into S-parameters and vice versa. The derivation is shown here.

In matrix form, let's begin with the voltage at port n of an N-port system, which is composed of an ingoing wave (v^+) and an outgoing wave (v^-):

$$v_n = v_n^+ + v_n^-$$

where the current is calculated from the port voltage and the port (reference) impedance Z_n:

$$i_n = i_n^+ - i_n^- = (v_n^+ - v_n^-)\frac{1}{Z_n}$$

In matrix form this becomes

$$\mathbf{v} = \mathbf{v}^+ + \mathbf{v}^- = \mathbf{Zi} = \mathbf{Zv}^+\frac{1}{Z_n} - \mathbf{Zv}^-\frac{1}{Z_n} \tag{9-33}$$

where $\mathbf{Z}$ is the impedance matrix. Assuming that the port impedance values are identical, equations (9-14) and (9-17) can be combined to show that the S-matrix can be calculated from $\mathbf{v}^+$ and $\mathbf{v}^-$:

$$S_{ij} = \frac{b_i}{a_j} = \frac{v_i^-}{v_j^+}$$

Therefore, the scattering matrix can be calculated by solving (9-33) for $(\mathbf{v}^-)(\mathbf{v}^+)^{-1}$:

$$\mathbf{v}^+ + \mathbf{v}^- = \mathbf{Zv}^+\frac{1}{Z_n} - \mathbf{Zv}^-\frac{1}{Z_n}$$

$$\mathbf{S} = (\mathbf{v}^-)(\mathbf{v}^+)^{-1} = (\mathbf{Z} + Z_n\mathbf{U})^{-1}(\mathbf{Z} - Z_n\mathbf{U}) \tag{9-34}$$

where $\mathbf{U}$ is the identity or unit matrix and Z_n is the termination impedance of each port. The derivation assumes that each port is terminated in the same value. Equation (9-34) converts Z-parameters to S-parameters.

Solving (9-34) for $\mathbf{Z}$ allows the conversion from S-parameters to Z-parameters for an arbitrary-sized matrix:

$$\mathbf{Z} = Z_n(\mathbf{U} + \mathbf{S})(\mathbf{U} - \mathbf{S})^{-1} \tag{9-35}$$

Equations (9-34) and (9-35) are solved for a two-port network and summarized in Table 9-1.

Impulse Response Section 8.1 we introduced the concept of an impulse response matrix to completely describe the behavior of a system. If the systems impulse

TABLE 9-1. Conversions Between S- and Z-Parameters for a Two-Port Network

$$\begin{bmatrix} S_{11} & S_{12} \\ S_{21} & S_{22} \end{bmatrix} = \begin{bmatrix} \dfrac{(Z_{11} - Z_n)(Z_{22} + Z_n) - Z_{12}Z_{21}}{(Z_{11} + Z_n)(Z_{22} + Z_n) - Z_{12}Z_{21}} & \dfrac{2Z_{12}Z_n}{(Z_{11} + Z_n)(Z_{22} + Z_n) - Z_{12}Z_{21}} \\ \dfrac{2Z_{21}Z_n}{(Z_{11} + Z_n)(Z_{22} + Z_n) - Z_{12}Z_{21}} & \dfrac{(Z_{11} + Z_n)(Z_{22} - Z_n) - Z_{12}Z_{21}}{(Z_{11} + Z_n)(Z_{22} + Z_n) - Z_{12}Z_{21}} \end{bmatrix}$$

$$\begin{bmatrix} Z_{11} & Z_{12} \\ Z_{21} & Z_{22} \end{bmatrix} = \begin{bmatrix} Z_n\dfrac{(1 + S_{11})(1 - S_{22}) + S_{12}S_{21}}{(1 - S_{11})(1 - S_{22}) - S_{12}S_{21}} & \dfrac{2Z_nS_{12}}{(1 - S_{11})(1 - S_{22}) - S_{12}S_{21}} \\ \dfrac{2Z_nS_{21}}{(1 - S_{11})(1 - S_{22}) - S_{12}S_{21}} & Z_n\dfrac{(1 - S_{11})(1 + S_{22}) + S_{12}S_{21}}{(1 - S_{11})(1 - S_{22}) - S_{12}S_{21}} \end{bmatrix}$$

matrix is known, it can be convolved with any arbitrary input (such as a pulse or a bit stream) and the signal integrity can be evaluated. It is impossible to measure the impulse response in the laboratory because it requires a driver capable of driving a Dirac delta function that has infinitely fast rise and fall times. Furthermore, even when a fast pulse is generated in the laboratory, inductance and capacitive loading of the probes introduces into the measured response unwanted noise, filtering, and resonances that are not associated with the electrical behavior of the device under test. Consequently, experimental evaluation of the impulse response in the laboratory using time-domain techniques is an impractical endeavor.

For most practical purposes, the impulse response of the system interconnects can be measured indirectly using a vector network analyzer (VNA), which is a device used to evaluate the scattering matrix as a function of frequency in the laboratory. Standard techniques to remove the parasitic inductance and capacitance effects of the probes and test fixtures from the measured scattering network are achieved through proper instrument calibration. Once the scattering matrix is measured, the impulse response can be calculated by taking the inverse Fourier transform, described in Section 8.1.4:

$$\mathbf{h}(t) = \mathsf{F}^{-1}\{\mathbf{S}(\omega)\} \tag{9-36a}$$

where $\mathbf{S}(\omega)$ is the scattering matrix measured with a VNA.

To calculate the impulse response, the scattering matrix must contain values for negative frequencies that obey the complex-conjugate rule to ensure a real-valued time-domain response, as described in Section 8.2.1. Since VNA measurements only provide values for the positive frequencies, the negative frequency values must be calculated from the positive values as

$$S(-f) = S(f)^* \tag{9-36b}$$

When calculating the impulse response using the fast Fourier transform (FFT), the negative values are appended to the end of the positive values, as demonstrated in Example 9-7.

Example 9-7 Calculate the impulse response from the measured values of S_{21} shown in Figure 9-22a.

SOLUTION

Step 1: Calculate the negative frequency values of S_{21} using (9-36b) as shown in Figure 9-22b.

$$S(-f) = S(f)^*$$

Step 2: Append the negative frequency values to the positive frequency values to create a continuous spectrum with both positive and negative frequencies.

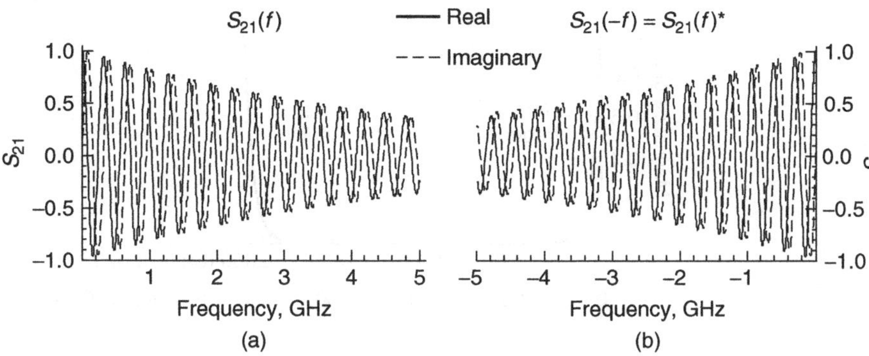

Figure 9-22 For Example 9-7: (a) measured values of S_{21} for positive frequencies; (b) negative frequency values constructed from the complex conjugate of the positive frequencies.

The magnitude of the continuous spectrum is shown in Figure 9-23a. Don't get confused by the data format required by the FFT, which assumes that the input samples are periodic, resulting in a frequency response that looks more intuitive if proper windowing of the periodic response is chosen, as shown in Figure 9-23b.

Step 3: Calculate the inverse FFT of the complex frequency spectrum to obtain the pulse response. In this case, the frequency data shown in Figure 9-22 were sampled at 100-MHz sample intervals for 1000 points for a bandwidth of 100 GHz. The final spectrum with the negative frequency values appended to the positive values has a total of 2000 sample points. Mathematica was used to perform the inverse FFT on the frequency data to produce the impulse response shown in Figure 9-24.

Although infinite bandwidth is required to produce a true impulse response with infinitely fast rise and fall times, the pulse response (i.e., single-bit response) as described in Section 8.1.5 can still be calculated with high accuracy if the harmonic bandwidth of the pulse is small compared to the bandwidth of the measurement. For example, if an 8-Gb/s single-bit response is calculated with rise and fall times of 35 ps, the bandwidth of the VNA used to evaluate the transfer function of the interconnects would need to be greater than 10 GHz, as calculated with Equation (8-8):

$$f_{\text{VNA,BW}} > \frac{0.35}{35 \times 10^{-12}} = 10 \times 10^9 \text{ Hz}$$

Furthermore, properties of the FFT place specific requirements on the frequency-domain bandwidth to generate minimum granularity in the time domain. If a minimum granularity of Δt is required for the time-domain waveform, the maximum

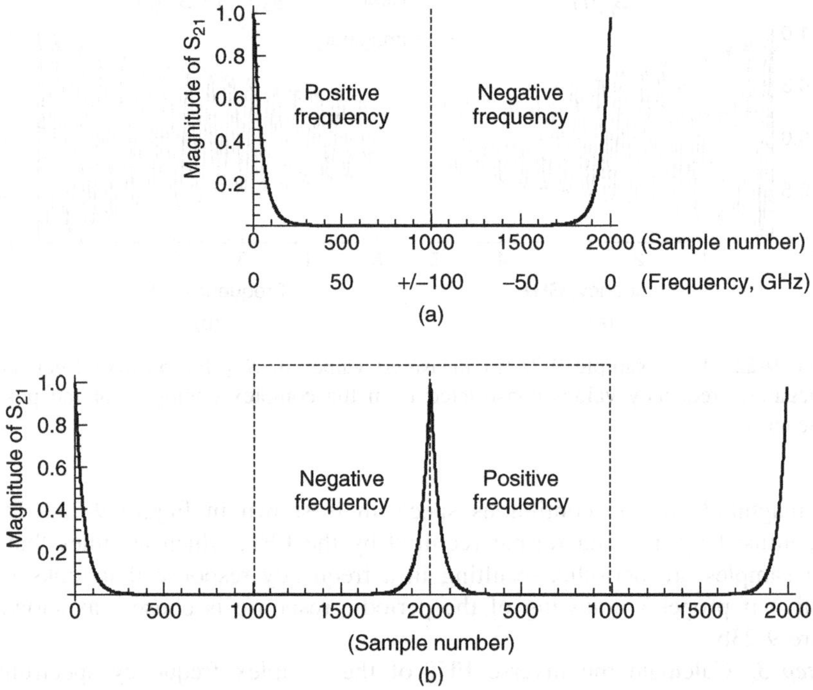

Figure 9-23 For Example 9-7: (a) magnitude of the frequency spectrum constructed by appending the measured positive frequency values of S_{21} and the complex conjugate to create the negative frequency values; (b) FFT periodic treatment of the sampled data allows a more intuitive look at the magnitude of the frequency response when windowed properly.

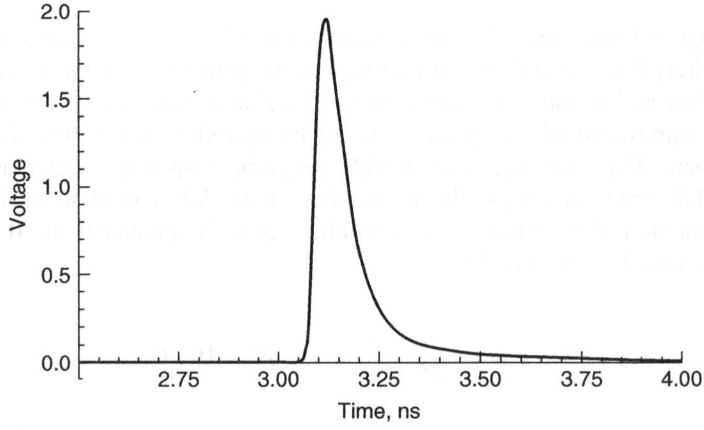

Figure 9-24 Impulse response calculated from the inverse FFT of the S_{21} data shown in Figure 9-23.

positive frequency bandwidth of the measurement is defined by

$$f_{\max} = \frac{1}{2\Delta t} \tag{9-37a}$$

To support a time-domain step size of Δt, the frequency sample size must be

$$\Delta f = \frac{1}{2n\,\Delta t} \tag{9-37b}$$

and the maximum valid time-domain signal would be

$$t_{\max} = \frac{1}{2\Delta f} \tag{9-37c}$$

where n is the number of samples for the *positive frequency values*.

For example, if a time-domain granularity of 5 ps is required, the maximum bandwidth required using 1000 positive frequency sample points would be 100 GHz:

$$f_{\max} = \frac{1}{2(5 \times 10^{-12})} = 100 \times 10^{9}\ \text{Hz}$$

Consequently, a time-domain waveform constructed from a frequency-domain measurement requires significant bandwidth if fine granularity is required. Unfortunately, it becomes very difficult to perform VNA measurements above about 20 GHz. Both the calibration techniques and equipment costs become prohibitive.

Fortunately, there are mathematic ways to sidestep the lack of high-frequency measured data without losing time-domain granularity. As long as the spectral bandwidth of the digital waveforms propagating on the system interconnects are significantly lower than $f_{\max}$, extrapolation or zero padding of the measured scattering matrix can be used to increase granularity with only a small degradation in accuracy. For example, consider a driving digital waveform with rise and fall times of 25 ps. The spectral bandwidth of this waveform is approximated by equation (8-8):

$$f_{3\text{dB}} \approx \frac{0.35}{25 \times 10^{-12}} = 14 \times 10^{9}\ \text{Hz}$$

Common bandwidths currently available on VNAs range from 20 to 110 GHz, which are equivalent to pulses with rise and fall times of about 3 to 18 ps.

$$t_{10\text{–}90\%} = \begin{cases} \dfrac{0.35}{20 \times 10^{9}} = 17.5 \times 10^{-12}\ \text{s} \\[2ex] \dfrac{0.35}{110 \times 10^{9}} = 3.18 \times 10^{-12}\ \text{s} \end{cases}$$

Therefore, measurements taken using a standard 20-GHz VNA would have adequate bandwidth to resolve rise and fall times as fast as 17.5 ps. Although a bandwidth of 20 GHz is sufficient to resolve the edge rate, the granularity of the time-domain waveform calculated using the FFT is only 25 ps, as calculated by (9-37a):

$$\Delta t = \frac{1}{2f_{\max}} = \frac{1}{2(20 \times 10^9)} = 25 \times 10^{-12} \text{ s}$$

Consequently, extrapolation or zero padding is required to ensure reasonable granularity in the time-domain waveform, as demonstrated in Example 9-8.

Example 9-8 Assume that the complex values of S_{21} used in Example 9-7 have been measured to 20 GHz. Calculate the impulse response with a resolution of 5 ps.

SOLUTION

Step 1: Calculate $f_{\max}$ using equation (9-37a):

$$f_{\max} = \frac{1}{2\Delta t} = \frac{1}{2(5 \times 10^{-12})} = 100 \times 10^9 \text{ Hz}$$

Step 2: Calculate the negative frequency values from −20 GHz to dc using equation (9-36b):

$$S(-f) = S(f)^*$$

Step 3: Calculate the sample interval of the frequency-domain data assuming 1000 samples of the measurable positive frequency values:

$$\Delta f = \frac{1}{2n\Delta t} = \frac{1}{2(1000)(5 \times 10^{-12})} = 100 \times 10^6 \text{ Hz}$$

Step 4: Calculate the number of zero points that need to be added to the spectrum. At a sample rate of 100 MHz, the number of samples up to 20 GHz is

$$\frac{20 \times 10^9 \text{ Hz}}{100 \times 10^6 \text{ Hz/sample}} = 200 \text{ samples}$$

Therefore, 800 zero points need to be added to both the positive and negative spectra to achieve ±100-GHz bandwidth.

Step 5: Append the positive and negative spectrums together as shown in Figure 9-25.

Step 6: Perform an inverse FFT on the zero-padded spectrum to get the impulse response, as shown in Figure 9-26.

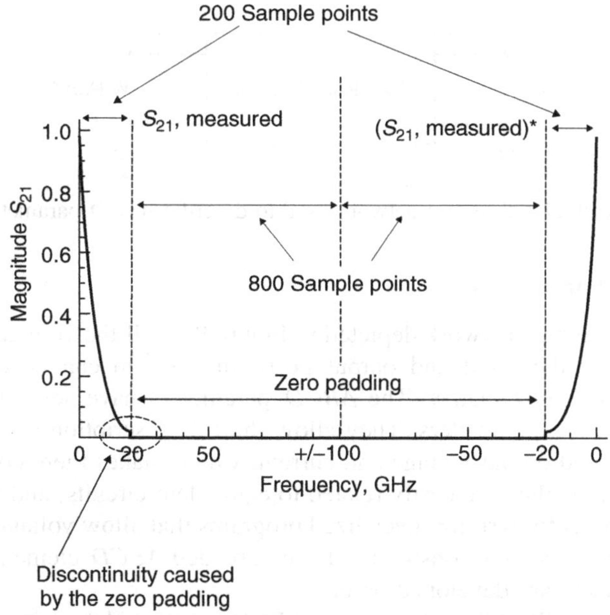

Figure 9-25 Zero-padded spectrum. Bandwidh of measured response is 20 GHz and zero-padded to 100 GHz to ensure 5-ps resolution in the time domain.

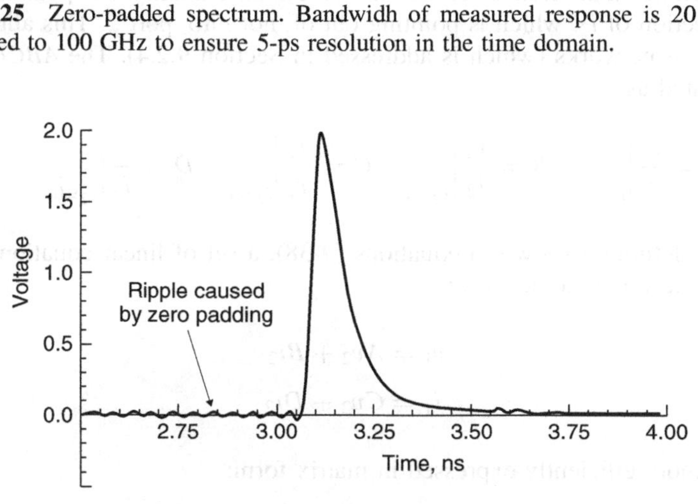

Figure 9-26 Impulse response calculated from the inverse FFT of the 20-GHz padded to 100-GHz S_{21} data shown in Figure 9-25.

Note that the zero padding has introduced a small amount of ringing into the impulse response. The error is caused by the discontinuous spectrum where the measured data stops and the zero padding begins. This error can be minimized by increasing the bandwidth of the measured data, extrapolating the real and imaginary parts of the measured data instead of zero padding or smoothing the zero-padding discontinuity.

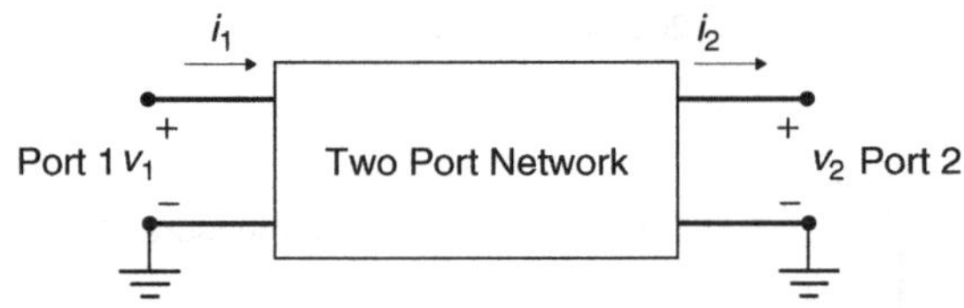

Figure 9-27 Two-port network used to describe *ABCD* parameters.

9.2.3 *ABCD* Parameters

Consider the two-port network depicted in Figure 9-27. If the voltage and current are measured at the input and output ports, the system can be characterized in terms of its *ABCD* matrix. The *ABCD* parameters have several advantages over other network parameters. They allow the full description of a network in terms of input and output voltage and current, which makes them convenient for cascading circuits; they are easily related to equivalent circuits; and they provide a convenient basis for writing specialized programs that allow voltage and current sources to drive channels constructed from cascaded *ABCD* elements. Two-port *ABCD* parameters are developed here.

One significant difference between the *ABCD* matrix and the impedance matrix is the direction of i_2, which is pointing out of, not into, port 2. This allows easy cascading of networks (which is addressed in Section 9.2.4). The *ABCD* values are evaluated as

$$A = \left.\frac{v_1}{v_2}\right|_{i_2=0} \qquad B = \left.\frac{v_1}{i_2}\right|_{v_2=0} \qquad C = \left.\frac{i_1}{v_2}\right|_{i_2=0} \qquad D = \left.\frac{i_1}{i_2}\right|_{v_2=0} \tag{9-38}$$

Using the definition shown in equations (9-38), a set of linear equations can be written to describe the network:

$$v_1 = Av_2 + Bi_2$$

$$i_1 = Cv_2 + Di_2$$

which is more efficiently expressed in matrix form:

$$\begin{vmatrix} v_1 \\ i_1 \end{vmatrix} = \begin{vmatrix} A & B \\ C & D \end{vmatrix} \cdot \begin{vmatrix} v_2 \\ i_2 \end{vmatrix} \tag{9-39}$$

Consequently, if the *ABCD* matrix of a system is known, the response of the system can be predicted for any input.

Since the *ABCD* parameters are evaluated with short and open circuits as shown in equation (9-38), they are not practical to measure directly. However, relationships exist that allow the *ABCD* matrix to be calculated directly from the *S*-parameters as will be shown later.

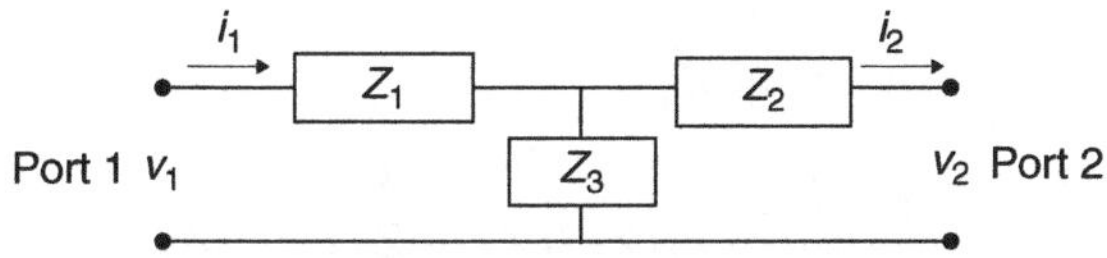

Figure 9-28 Deriving the *ABCD* parameters for a *T*-topology equivalent circuit.

Relationship to Common Circuit Parameters The *ABCD* parameters can be used to create equivalent-circuit models for common circuit topologies. For example, consider the T-circuit shown in Figure 9-28. To determine A, port 2 must be open (since $i_2 = 0$) while port 1 is driven with v_1.

$$A = \left.\frac{v_1}{v_2}\right|_{i_2=0}$$

$$v_2 = v_1 \frac{Z_3}{Z_3 + Z_1}$$

$$A = \frac{Z_3 + Z_1}{Z_3} \tag{9-40a}$$

The term B is determined by shorting port 1 (since $v_2 = 0$):

$$B = \left.\frac{v_1}{i_2}\right|_{v_2=0}$$

A current divider is used to calculate i_2, which is substituted into the equation for B:

$$i_2 = i_1 \frac{Z_3}{Z_3 + Z_2}$$

$$B = \frac{v_1}{i_1[Z_3/(Z_3 + Z_2)]} = \frac{v_1}{i_1}\frac{Z_3 + Z_2}{Z_3}$$

Since v_1/i_1 is equal to the impedance looking into port 1, B can be simplified:

$$B = (Z_1 + Z_2||Z_3)\frac{Z_3 + Z_2}{Z_3} = Z_1 + Z_2 + \frac{Z_1 Z_2}{Z_3} \tag{9-40b}$$

To determine C, port 2 must be open (since $i_2 = 0$) while port 1 is driven with v_1.

$$C = \left.\frac{i_1}{v_2}\right|_{i_2=0}$$

$$v_2 = v_1 \frac{Z_3}{Z_3 + Z_1}$$

$$i_1 = \frac{v_1}{Z_1 + Z_3}$$

$$C = \frac{i_1}{v_2} = \frac{v_1/(Z_1 + Z_3)}{v_1[Z_3/(Z_3 + Z_1)]} = \frac{1}{Z_3} \tag{9-40c}$$

To determine D, port 2 must be short (since $v_2 = 0$) while port 1 is driven:

$$D = \left.\frac{i_1}{i_2}\right|_{v_2=0}$$

$$i_2 = i_1 \frac{Z_3}{Z_3 + Z_2}$$

$$D = \frac{i_1}{i_2} = \frac{Z_3 + Z_2}{Z_3} \tag{9-40d}$$

Therefore, for a T-circuit like the one depicted in Figure 9-28, the *ABCD* matrix takes on the form

$$\begin{bmatrix} A & B \\ C & D \end{bmatrix}_{\text{T-circuit}} = \begin{bmatrix} \dfrac{Z_3 + Z_1}{Z_3} & Z_1 + Z_2 + \dfrac{Z_1 Z_2}{Z_3} \\ \dfrac{1}{Z_3} & \dfrac{Z_3 + Z_2}{Z_3} \end{bmatrix} \tag{9-41}$$

The *ABCD* parameters of a lossy transmission line can be derived for the case where the line is terminated is its characteristic impedance. This conversion is particularly useful for extracting transmission-line parameters such as Z_0 and γ from measurements. For the case where the termination impedance is not equal to the characteristic impedance, the matrix can be renormalized as described in Section 9.2.6.

To begin the derivation, consider the transmission line in Figure 9-29:

$$A = \left.\frac{v_1}{v_2}\right|_{i_2=0}$$

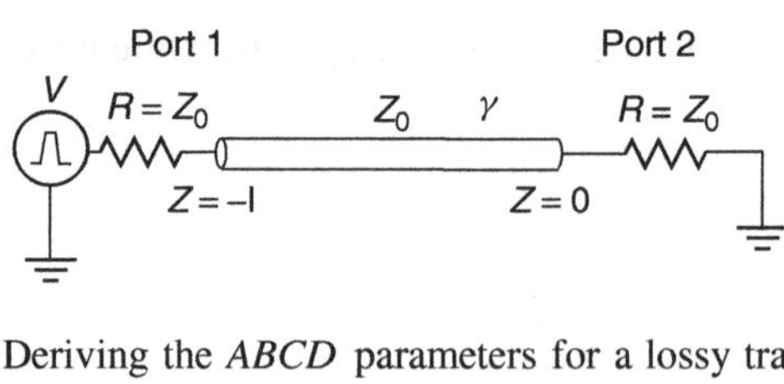

Figure 9-29 Deriving the *ABCD* parameters for a lossy transmission line.

Since $i_2 = 0$, port 2 is open. Therefore, the voltage at port 2 will have an incident and a reflected component. For convenience, the total voltage at port 2 is set equal to 1:

$$v_2 = v^+ + v^- = 1$$
$$v^+ = v^- = \tfrac{1}{2}$$

The voltage at port 1 is calculated using equation (6-49) with $z = -l$.

$$v_1 = v(z = -l) = v^+ e^{\gamma l} + v^- e^{-\gamma l} = \frac{e^{\gamma l} + e^{-\gamma l}}{2}$$

The ratio of v_1 and v_2 is used to calculate A:

$$A = \frac{e^{\gamma l} + e^{-\gamma l}}{2} = \cosh \gamma l \tag{9-42a}$$

The term C is also calculated with $i_2 = 0$.

$$C = \left.\frac{i_1}{v_2}\right|_{i_2=0}$$

The current i_1 is calculated at $z = -l$ using the characteristics impedance.

$$i_1 = i(z = l) = \frac{1}{Z_0} v^+ e^{\gamma l} - \frac{1}{Z_0} v^- e^{-\gamma l} = \frac{1}{Z_0}\frac{e^{\gamma l} - e^{-\gamma l}}{2}$$

The ratio of i_1 and v_2 is used to calculate C:

$$C = \frac{1}{Z_0}\frac{e^{\gamma l} - e^{-\gamma l}}{2} = \frac{1}{Z_0}\sinh \gamma l \tag{9-42b}$$

Terms B and D are calculated with $v_2 = 0$:

$$B = \left.\frac{v_1}{i_2}\right|_{v_2=0}$$

Since $v_2 = 0$, port 2 is shorted:

$$v_2 = v^+ + v^- = \frac{1}{2} - \frac{1}{2} = 0$$
$$i_2 = \frac{v^+}{Z_0} - \frac{v^-}{Z_0} = \frac{1/2}{Z_0} - \frac{-1/2}{Z_0} = \frac{1}{Z_0}$$
$$v_1 = v(z = -l) = v^+ e^{\gamma l} + v^- e^{-\gamma l} = \frac{e^{\gamma l} - e^{-\gamma l}}{2}$$

The ratio of v_1 and i_2 is used to calculate B:

$$B = Z_0 \frac{e^{\gamma l} - e^{-\gamma l}}{2} = Z_0 \sinh \gamma l \tag{9-42c}$$

$$D = \left. \frac{i_1}{i_2} \right|_{v_2=0}$$

$$i_2 = \frac{1}{Z_0}$$

$$i_1 = i(z = l) = \frac{1/2}{Z_0} e^{\gamma l} - \frac{-1/2}{Z_0} e^{-\gamma l} = \frac{1}{Z_0} \frac{e^{\gamma l} + e^{-\gamma l}}{2}$$

The ratio of i_1 and i_2 is used to calculate D:

$$D = \frac{e^{\gamma l} + e^{-\gamma l}}{2} = \cosh \gamma l \tag{9-42d}$$

Therefore, the $ABCD$ matrix for a lossy transmission line.

$$\begin{bmatrix} A & B \\ C & D \end{bmatrix}_{\text{lossy transmssion line}} = \begin{bmatrix} \cosh \gamma l & Z_0 \sinh \gamma l \\ \dfrac{1}{Z_0} \sinh \gamma l & \cosh \gamma l \end{bmatrix} \tag{9-43a}$$

Using the same procedure, the $ABCD$ parameters for a loss-free line can easily be derived where $\gamma = j\beta$:

$$\begin{bmatrix} A & B \\ C & D \end{bmatrix}_{\text{loss-free transmssion line}} = \begin{bmatrix} \cos \beta l & j Z_0 \sin \beta l \\ \dfrac{j}{Z_0} \sin \beta l & \cos \beta l \end{bmatrix} \tag{9-43b}$$

Table 9-2 depicts the relationship between common circuits and the $ABCD$ parameters. These common forms are useful for extracting equivalent circuits from S-parameter measurements. Of course, a methodology is needed to convert S-parameters into an $ABCD$ matrix, which is covered in the next section.

Relationship Between ABCD and S-Parameters To take advantage of the relationships between the $ABCD$ matrix and common circuit forms, it is necessary to determine the relationship between the $ABCD$ and S-parameters. The most straightforward derivation is first to define the transformation of $ABCD$ parameters into a two-port Z-matrix and then use equation (9-34) to get the S-parameters.

Beginning with the definition of Z_{11} from equation (9-8),

$$Z_{11} = \left. \frac{v_1}{i_1} \right|_{i_2=0}$$

TABLE 9-2. Relationship Between Common Circuits and the *ABCD* Parameters

Circuit	ABCD Matrix
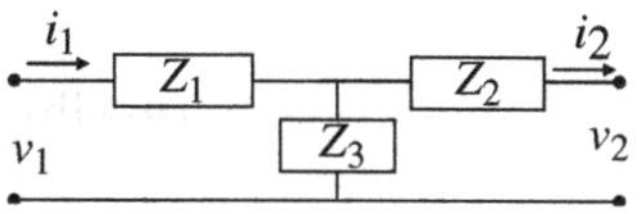	$\begin{bmatrix} \dfrac{Z_3+Z_1}{Z_3} & Z_1+Z_2+\dfrac{Z_1Z_2}{Z_3} \\ \dfrac{1}{Z_3} & \dfrac{Z_3+Z_2}{Z_3} \end{bmatrix}$
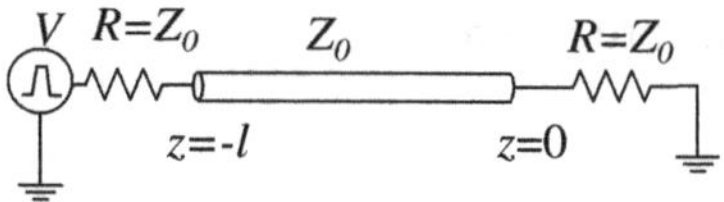	$\begin{bmatrix} \cosh\gamma l & Z_0\sinh\gamma l \\ \dfrac{1}{Z_0}\sinh\gamma l & \cosh\gamma l \end{bmatrix}$
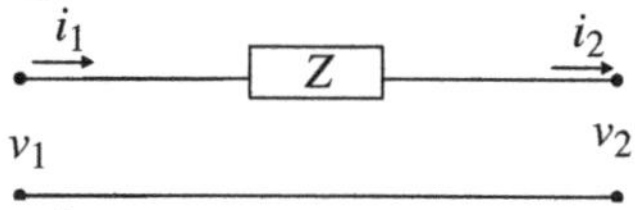	$\begin{bmatrix} 1 & Z \\ 0 & 1 \end{bmatrix}$
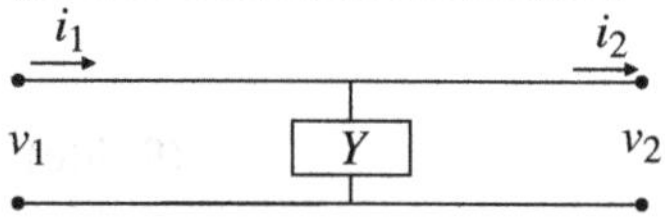	$\begin{bmatrix} 1 & 0 \\ Y & 1 \end{bmatrix}$
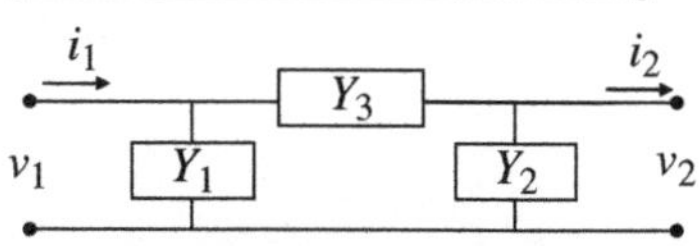	$\begin{bmatrix} 1+\dfrac{Y_2}{Y_3} & \dfrac{1}{Y_3} \\ Y_1+Y_2+\dfrac{Y_1Y_2}{Y_3} & 1+\dfrac{Y_1}{Y_3} \end{bmatrix}$

the *ABCD* equations reduce to

$$v_1 = Av_2$$
$$i_1 = Cv_2$$

resulting in

$$Z_{11} = \frac{v_1}{i_1} = \frac{A}{C} \tag{9-44a}$$

The definition of Z_{12} is

$$Z_{12} = \left.\frac{v_1}{i_2}\right|_{i_1=0}$$

so the ABCD equations reduce to

$$v_1 = Av_2 + Bi_2$$
$$0 = Cv_2 + Di_2$$

Solving the equations above for v_1/i_2 produces

$$\frac{v_1}{i_2} = \frac{BC - AD}{C}$$

However, the convention for an *ABCD* matrix assumes that i_2 is flowing out of port 2, and a Z matrix assumes that it is flowing into port 2. Consequently, the sign of i_2 must be changed, which results in the definition of Z_{12} in terms of *ABCD* parameters:

$$Z_{12} = \frac{AD - BC}{C} \tag{9-44b}$$

The terms Z_{21} and Z_{22} are derived using a similar procedure.

$$Z_{21} = \left.\frac{v_2}{i_1}\right|_{i_2=0}$$

$$v_1 = Av_2$$

$$i_1 = Cv_2$$

$$Z_{21} = \frac{1}{C} \tag{9-44c}$$

$$Z_{22} = \left.\frac{v_2}{i_2}\right|_{i_1=0}$$

$$v_1 = Av_2 + Bi_2$$

$$0 = Cv_2 + Di_2$$

$$Z_{22} = \frac{D}{C} \tag{9-44d}$$

The final relationship between a two-port Z-matrix and the *ABCD* matrix is shown as

$$\begin{bmatrix} Z_{11} & Z_{12} \\ Z_{21} & Z_{22} \end{bmatrix} = \begin{bmatrix} \dfrac{A}{C} & \dfrac{AD - BC}{C} \\ \dfrac{1}{C} & \dfrac{D}{C} \end{bmatrix} \tag{9-45}$$

To derive the transformation of the *ABCD* to S-parameters, the results of (9-45) are substituted into equation (9-34). The final solution is summarized in (9-46), where Z_n is the termination impedance at the ports, which are all assumed to be equal.

$$\begin{bmatrix} S_{11} & S_{12} \\ S_{21} & S_{22} \end{bmatrix} = \begin{bmatrix} \dfrac{B - Z_n(D - A + CZ_n)}{B + Z_n(D + A + CZ_n)} & \dfrac{2Z_n(AD - BC)}{B + Z_n(D + A + CZ_n)} \\ \dfrac{2Z_n}{B + Z_n(D + A + CZ_n)} & \dfrac{B - Z_n(A - D + CZ_n)}{B + Z_n(D + A + CZ_n)} \end{bmatrix} \tag{9-46}$$

Similar techniques are used to derive the transformation of S into *ABCD* parameters. The complete sets of transformations are summarized in Table 9-3.

TABLE 9-3. Relationships Between Two-Port S and $ABCD$ Parameters[a]

$\begin{bmatrix} S_{11} & S_{12} \\ S_{21} & S_{22} \end{bmatrix}$	$\begin{bmatrix} \dfrac{B - Z_n(D - A + CZ_n)}{B + Z_n(D + A + CZ_n)} & \dfrac{2Z_n(AD - BC)}{B + Z_n(D + A + CZ_n)} \\ \dfrac{2Z_n}{B + Z_n(D + A + CZ_n)} & \dfrac{B - Z_n(A - D + CZ_n)}{B + Z_n(D + A + CZ_n)} \end{bmatrix}$
$\begin{bmatrix} A & B \\ C & D \end{bmatrix}$	$\begin{bmatrix} \dfrac{(1 + S_{11})(1 - S_{22}) + S_{12}S_{21}}{2S_{21}} & Z_n\dfrac{(1 + S_{11})(1 + S_{22}) - S_{12}S_{21}}{2S_{21}} \\ \dfrac{1}{Z_n}\dfrac{(1 - S_{11})(1 - S_{22}) - S_{12}S_{21}}{2S_{21}} & \dfrac{(1 - S_{11})(1 + S_{22}) + S_{12}S_{21}}{2S_{21}} \end{bmatrix}$

[a]Z_n is the termination impedance at the ports.

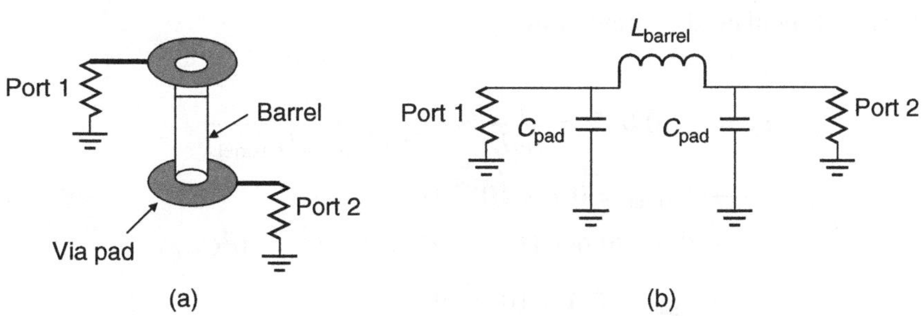

Figure 9-30 Via and equivalent circuit for Example 9-9.

Example 9-9 Extract an equivalent circuit for the via shown in Figure 9-30a from the following S-parameter matrix measured at 5 GHz assuming port impedance values (Z_n) of 50 Ω:

$$\begin{bmatrix} S_{11} & S_{12} \\ S_{21} & S_{22} \end{bmatrix} = \begin{bmatrix} -0.1235 - j0.1516 & 0.7597 - j0.6190 \\ 0.7597 - j0.6190 & -0.1235 - j0.1516 \end{bmatrix}$$

SOLUTION

Step 1: Transform the S-matrix into $ABCD$ parameters using the relations in Table 9-3.

$$\begin{bmatrix} A & B \\ C & D \end{bmatrix} = \begin{bmatrix} 0.790 & j22.22 \\ j0.01686 & 0.790 \end{bmatrix}$$

Step 2: Choose the form of the equivalent circuit. A signal propagating through the via will experience the capacitance of the via pad, the inductance of the barrel, and then the capacitance of the second pad. This configuration fits the pi model in Table 9-2 as shown in Figure 9-30b.

Step 3: Use the relations in Table 9-2 to calculate the admittance of each segment in the pi model:

$$B = \frac{1}{Y_3} = j22.22$$

$$Y_3 = -j0.045$$

$$A = 0.790 = 1 + \frac{Y_2}{Y_3} = 1 + \frac{Y_2}{-j0.045}$$

$$Y_2 = j0.00945$$

Due to symmetry in the circuit, $Y_1 = Y_2$.

Step 4: Calculate the circuit values:

$$\begin{aligned}
Y_3 &= -j0.045 = \frac{1}{j\omega L} = \frac{1}{j2\pi(5 \times 10^9 L_{\text{barrel}})} \\
&\rightarrow L_{\text{barrel}} = 0.7 \times 10^{-9}\ \text{H} \\
Y_1 &= Y_2 = j0.00945 = j\omega C = j2\pi(5 \times 10^9 C_{\text{pad}}) \\
&\rightarrow C_{\text{pad}} = 0.3 \times 10^{-12}\ \text{F}
\end{aligned}$$

9.2.4 Cascading *S*-Parameters

One of the most useful aspects of network analysis is the ability to cascade independently measured structures. For example, if the *S*-parameters are measured independently for a transmission line, via, package, and connector, the engineer has the ability to create the response of an entire channel from the individual measurements. This allows *S*-parameter files to be used as portable models, gives the designer the ability to evaluate different topologies, and provides the mechanism to deembed or calibrate out unwanted items in the measurements. The two most common methods of cascading *S*-parameters are with the *ABCD* parameters and the T-matrix.

Cascading with the ABCD Matrix For two-port networks, the most common cascading methodology is to use the *ABCD* parameters. Cascading is achieved simply by multiplying the *ABCD* matrices. This is possible because of the current convention that flows outward at port 2, as shown in Figure 9-27. For example, consider the *ABDC* parameters of the two cascaded circuits shown in Figure 9-31. The equations that describe the port voltage and currents of circuits 1 and 2 can

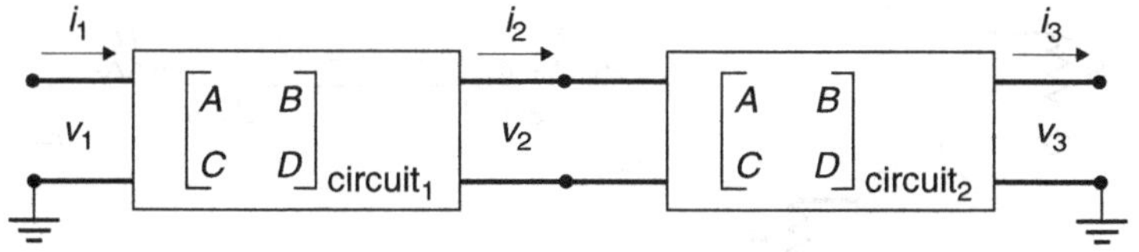

Figure 9-31 *ABCD* parameters are cascaded though multiplication.

easily be written.

$$\begin{bmatrix} v_1 \\ i_1 \end{bmatrix} = \begin{bmatrix} A & B \\ C & D \end{bmatrix}_{\text{circuit}_1} \begin{bmatrix} v_2 \\ i_2 \end{bmatrix}$$

$$\begin{bmatrix} v_2 \\ i_2 \end{bmatrix} = \begin{bmatrix} A & B \\ C & D \end{bmatrix}_{\text{circuit}_2} \begin{bmatrix} v_3 \\ i_3 \end{bmatrix}$$

Note that the output of circuit 1 is v_2 and i_2, which is the input of circuit 2. This allows the simple substitution of

$$\begin{bmatrix} v_2 \\ i_2 \end{bmatrix} \quad \text{with} \quad \begin{bmatrix} A & B \\ C & D \end{bmatrix}_{\text{circuit}_2} \begin{bmatrix} v_3 \\ i_3 \end{bmatrix}$$

resulting in

$$\begin{bmatrix} v_1 \\ i_1 \end{bmatrix} = \begin{bmatrix} A & B \\ C & D \end{bmatrix}_{\text{circuit}_1} \begin{bmatrix} A & B \\ C & D \end{bmatrix}_{\text{circuit}_2} \begin{bmatrix} v_3 \\ i_3 \end{bmatrix}$$

Therefore, two-port S-parameters can be cascaded by converting to ABCD parameters and multiplying. The cascaded scattering matrix is then calculated by converting the cascaded *ABCD* matrix back to *S*-parameters using Table 9-3. The procedure is demonstrated in Example 9-10.

Example 9-10 Using the two independently measured values of *S*-parameters for a via and a loss-free transmission line at 5 GHz, calculate the equivalent *S*-parameters that would be measured if the two circuits were cascaded as shown Figure 9-32. Assume that the termination impedance is 50 Ω.

$$\begin{bmatrix} S_{11} & S_{12} \\ S_{21} & S_{22} \end{bmatrix}_{\text{via}} = \begin{bmatrix} -0.1235 - j0.1516 & 0.7597 - j0.6190 \\ 0.7597 - j0.6190 & -0.1235 - j0.1516 \end{bmatrix}$$

$$\begin{bmatrix} S_{11} & S_{12} \\ S_{21} & S_{22} \end{bmatrix}_{\text{t-line}} = \begin{bmatrix} 0.00325 - j0.00323 & -1.00 - j0.003 \\ -1.00 - j0.003 & 0.00325 - j0.00323 \end{bmatrix}$$

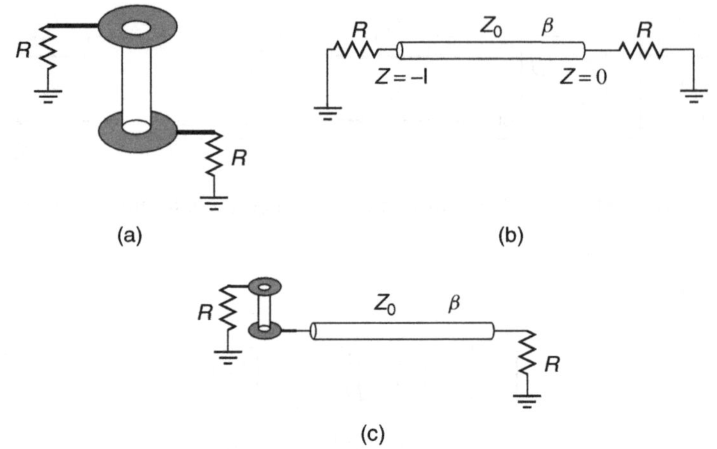

Figure 9-32 For Example 9-10: (a) configuration of the independently measured via; (b) transmission line; (c) via cascaded with the transmission line.

SOLUTION

Step 1: Convert to *ABCD* parameters using Table 9-3:

$$\begin{bmatrix} A & B \\ C & D \end{bmatrix}_{\text{via}} = \begin{bmatrix} 0.790 & j22.22 \\ j0.01686 & 0.790 \end{bmatrix}$$

$$\begin{bmatrix} A & B \\ C & D \end{bmatrix}_{\text{t-line}} = \begin{bmatrix} -1 & j0.3228 \\ j0.000129 & -1 \end{bmatrix}$$

Step 2: Multiply the *ABCD* matrices:

$$\begin{aligned} \begin{bmatrix} A & B \\ C & D \end{bmatrix}_{\text{cascade}} &= \begin{bmatrix} A & B \\ C & D \end{bmatrix}_{\text{via}} \begin{bmatrix} A & B \\ C & D \end{bmatrix}_{\text{t-line}} \\ &= \begin{bmatrix} 0.790 & j22.22 \\ j0.01686 & 0.790 \end{bmatrix} \cdot \begin{bmatrix} -1 & j0.3228 \\ j0.000129 & -1 \end{bmatrix} \\ &= \begin{bmatrix} -0.790 & -j21.965 \\ -j0.01686 & -0.795 \end{bmatrix} \end{aligned}$$

Step 3: Convert $\begin{bmatrix} A & B \\ C & D \end{bmatrix}_{\text{cascade}}$ back to *S*-parameters using Table 9-3 where $Z_n = 50\ \Omega$ (the termination impedance)

$$\begin{bmatrix} S_{11} & S_{12} \\ S_{21} & S_{22} \end{bmatrix}_{\text{cascade}} = \begin{bmatrix} -0.1259 - j0.1553 & -0.7635 + j0.6186 \\ -0.7645 + j0.6182 & -0.1200 - j0.1565 \end{bmatrix}$$

The cascaded *S*-matrix is equivalent to what would be measured if the via was cascaded with the transmission line, as shown in Figure 9-32c.

Cascading with the T-Matrix Another method commonly used to cascade S-parameters is the *T-matrix*, sometimes called *transmission parameters*. The T-parameters are derived simply by rearranging the equations for the S-parameters. Equation (9-18) describes the relationship between the incident power waves a and the exiting power waves b:

$$\begin{bmatrix} b_1 \\ b_2 \end{bmatrix} = \begin{bmatrix} S_{11} & S_{12} \\ S_{21} & S_{22} \end{bmatrix} \begin{bmatrix} a_1 \\ a_2 \end{bmatrix}$$

To facilitate the cascading of networks by simple matrix multiplication, the equation needs to be rearranged so that the ingoing and outgoing waves of port 1 can be described in terms of the waves at port 2. This is done with the T-matrix. A two-port T-matrix is

$$\begin{bmatrix} a_1 \\ b_1 \end{bmatrix} = \begin{bmatrix} T_{11} & T_{12} \\ T_{21} & T_{22} \end{bmatrix} \begin{bmatrix} b_2 \\ a_2 \end{bmatrix} \tag{9-47}$$

If the output of circuit A is attached to the input of circuit B, the total response can be calculated simply by multiplying the T-matrices, because the power wave exiting circuit A is b_2, which feeds into the input of circuit B, which is a_3. Therefore, $b_2 = a_3$ and $a_2 = b_3$, as shown in Figure 9-33.

$$\begin{bmatrix} a_1 \\ b_1 \end{bmatrix} = \begin{bmatrix} T_{11} & T_{12} \\ T_{21} & T_{22} \end{bmatrix}_A \begin{bmatrix} b_2 \\ a_2 \end{bmatrix}$$

$$\begin{bmatrix} a_3 \\ b_3 \end{bmatrix} = \begin{bmatrix} T_{11} & T_{12} \\ T_{21} & T_{22} \end{bmatrix}_B \begin{bmatrix} b_4 \\ a_4 \end{bmatrix}$$

$$\begin{bmatrix} a_1 \\ b_1 \end{bmatrix} = \begin{bmatrix} T_{11} & T_{12} \\ T_{21} & T_{22} \end{bmatrix}_A \begin{bmatrix} T_{11} & T_{12} \\ T_{21} & T_{22} \end{bmatrix}_B \begin{bmatrix} b_4 \\ a_4 \end{bmatrix}$$

Therefore, S-parameters can be cascaded by converting to T-parameters and multiplying. The cascaded scattering matrix is then calculated by converting the product of the T-matrices back to S-parameters.

Conversion between the T- and S-parameters (and vice-versa) requires simple algebraic manipulation of the equations, which can be done for any number of

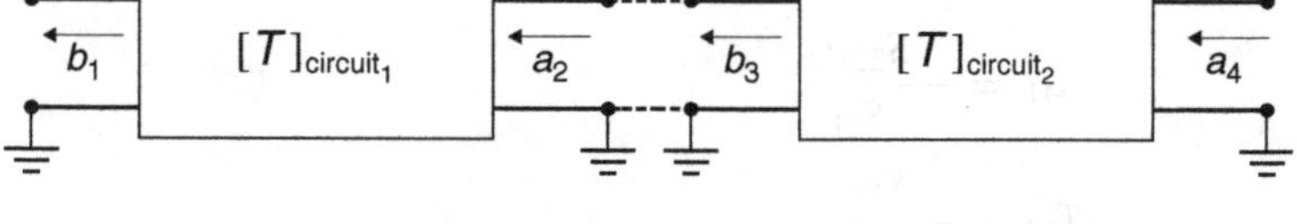

Figure 9-33 T-parameters are cascaded through multiplication.

ports. The conversion from S- to T-parameters is derived here for a two-port system to demonstrate the process. To begin, the equations that relate the incoming power waves to the outgoing are written

$$b_1 = S_{11}a_1 + S_{12}a_2$$
$$b_2 = S_{21}a_1 + S_{22}a_2$$

and equation (9-47) is expressed in algebraic form:

$$a_1 = T_{11}b_2 + T_{12}a_2$$
$$b_1 = T_{21}b_2 + T_{22}a_2$$

From the equations above, the individual T-parameters can be defined in terms of the S-parameters:

$$T_{11} = \left.\frac{a_1}{b_2}\right|_{a_2=0}$$
$$b_2 = S_{21}a_1 + S_{22}(0)$$
$$T_{11} = \frac{1}{S_{21}} \tag{9-48a}$$
$$T_{22} = \left.\frac{b_1}{a_2}\right|_{b_2=0}$$
$$0 = S_{21}a_1 + S_{22}a_2 \rightarrow a_1 = -\frac{S_{22}a_2}{S_{21}}$$
$$b_1 = S_{11}a_1 + S_{12}a_2 = \left(S_{12} - \frac{S_{11}S_{22}}{S_{21}}\right)a_2$$
$$T_{22} = S_{12} - \frac{S_{11}S_{22}}{S_{21}} \tag{9-48b}$$
$$T_{12} = \left.\frac{a_1}{a_2}\right|_{b_2=0}$$
$$0 = S_{21}a_1 + S_{22}a_2$$
$$a_1 = -\frac{S_{22}a_2}{S_{21}}$$
$$T_{12} = -\frac{S_{22}}{S_{21}} \tag{9-48c}$$

TABLE 9-4. Relationships Between T- and S-Parameters for Two-Ports

$\begin{bmatrix} T_{11} & T_{12} \\ T_{21} & T_{22} \end{bmatrix}$	$\begin{bmatrix} \dfrac{1}{S_{21}} & -\dfrac{S_{22}}{S_{21}} \\ \dfrac{S_{11}}{S_{21}} & S_{12} - \dfrac{S_{11}S_{22}}{S_{21}} \end{bmatrix}$
$\begin{bmatrix} S_{11} & S_{12} \\ S_{21} & S_{22} \end{bmatrix}$	$\begin{bmatrix} \dfrac{T_{21}}{T_{11}} & T_{22} - \dfrac{T_{21}T_{12}}{T_{11}} \\ \dfrac{1}{T_{11}} & -\dfrac{T_{12}}{T_{11}} \end{bmatrix}$

$$T_{21} = \left.\frac{b_1}{b_2}\right|_{a_2=0}$$

$$b_1 = S_{11}a_1$$

$$b_2 = S_{21}a_1$$

$$T_{21} = \frac{S_{11}}{S_{21}} \tag{9-48d}$$

A very similar analysis is used to transform the T-parameters back into S-parameters, which is left to an exercise for the reader. The transformations for two ports are summarized in Table 9-4.

The four-port T-matrix takes the form

$$\begin{bmatrix} a_1 \\ a_3 \\ b_1 \\ b_3 \end{bmatrix} = \begin{bmatrix} T_{11} & T_{12} & T_{13} & T_{14} \\ T_{21} & T_{22} & T_{23} & T_{24} \\ T_{31} & T_{32} & T_{33} & T_{34} \\ T_{41} & T_{42} & T_{43} & T_{44} \end{bmatrix} \begin{bmatrix} b_2 \\ b_4 \\ a_2 \\ a_4 \end{bmatrix} \tag{9-49}$$

for the network shown in Figure 9-34. The transformations between T- and S-parameters for n-ports is developed in an identical manner, which is shown in Appendix B.

9.2.5 Calibration and De-embedding

Very often in digital design it is desirable to measure a component such as a transmission line, a CPU socket, or a daughter-card connector using a vector network analyzer (VNA). The measurement can be used to create an equivalent-circuit model, validate the modeling methodologies used to simulate the channel, or estimate the performance of a component. Practicality dictates that the device under test (DUT) is usually mounted in a fixture and connected to the VNA with a cable using probes or SMA connectors. Consequently, methodologies are

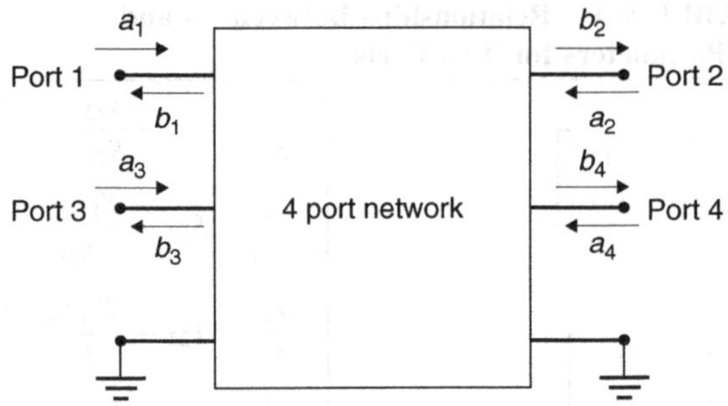

Figure 9-34 Four-port network.

needed to separate the S-parameters of the DUT from the test fixture, cables, or probes. In this section we introduce the basic calibration and deembedding methodologies so that the concept will be understood. Since each measurement setup requires a specialized calibration and deembedding procedure, the breadth of the topic is too wide to be covered here.

The terms *calibration* and *deembedding* refer to the same concept: removing an unwanted part of the measurement. More specifically, *calibration* refers to removing the effect of the VNA cables and probes, and *deembedding* refers to removing unwanted portions of the DUT, such as a via, a test fixture, or a cable connector. A VNA measures S-parameters as ratios of complex voltage amplitudes. The reference for the measurement is some place within the VNA, not at the cable ends which are attached to the DUT, so the measurement will include the losses and phase delays of the cables, connectors, and probes used to connect the DUT to the analyzer. Calibration refers to the removal of these unwanted effects from the measured response so that only the measured response of the DUT remains. Figure 9-35 demonstrates this concept.

For calibration purposes, it is convenient to work with *ABCD* parameters. For example, the measurement setup in Figure 9-35 can be represented by cascaded *ABCD* matrices:

$$\begin{bmatrix} A & B \\ C & D \end{bmatrix}_{\text{measured}} = \begin{bmatrix} A & B \\ C & D \end{bmatrix}_{\text{error}_{\text{port1}}} \begin{bmatrix} A & B \\ C & D \end{bmatrix}_{\text{DUT}} \begin{bmatrix} A & B \\ C & D \end{bmatrix}^{-1}_{\text{error}_{\text{port2}}} \tag{9-50}$$

Note that the *ABCD* matrix for port 2 is inverted because the VNA drives into the port and measures the responses at all other ports. The inversion simply ensures that it represents current flowing *from* port 2 *into* the DUT and not the other way around. Equation (9-50) suggests that if the *ABCD* matrices of the errors are known, the *ABCD* matrix (and thus the S-parameters) of the DUT can be calculated. In short, the calibrated measurement is simply the error-corrected S-parameters.

The most straightforward method to remove the errors and calibrate a VNA is to use three or more carefully controlled loads, such as a short, open, and

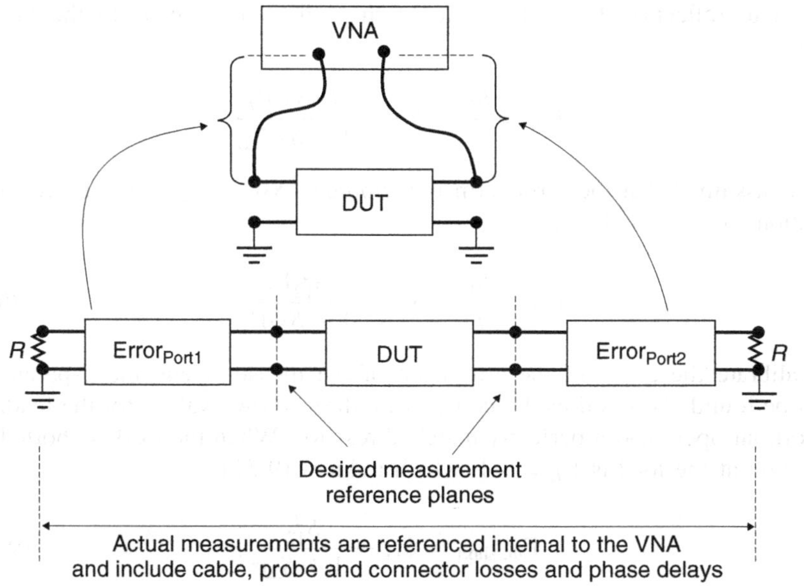

Figure 9-35 Calibration removes the unwanted parts of the measurement, such as the cable and probe effects.

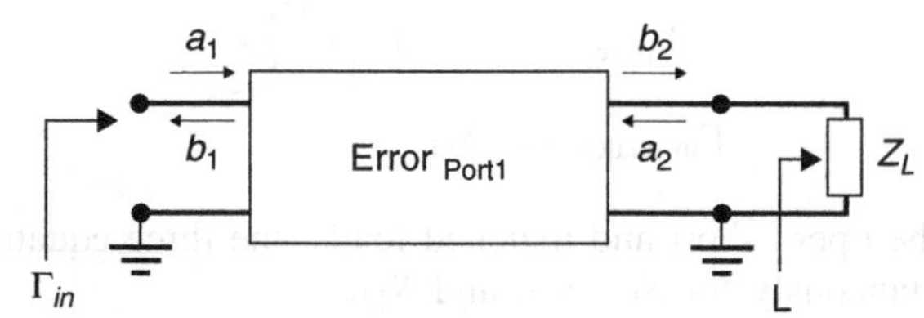

Figure 9-36 The simplest way to calibrate a VNA is to characterize the port error by driving different loads (Z_L): an open, a short, and a resistive load.

precision resistive load. For example, consider Figure 9-36, which represents one port of a VNA driving a load Z_L. The equations for the S-parameters can be written in terms of the reflection coefficient at the load where the incoming wave at port 2 (a_2) is simply the reflected portion of the exiting wave (b_2) $a_2 = b_2\Gamma_L$.

$$b_1 = S_{11}a_1 + S_{12}a_2 = S_{11}a_1 + S_{12}b_2\Gamma_L$$

$$b_2 = S_{21}a_1 + S_{22}a_2 = S_{21}a_1 + S_{22}b_2\Gamma_L$$

$$b_2(1 - S_{22}\Gamma_L) = S_{21}a_1 \rightarrow b_2 = \frac{S_{21}a_1}{1 - S_{22}\Gamma_L} \tag{9-51a}$$

$$b_1 = S_{11}a_1 + S_{12}\frac{S_{21}a_1}{1 - S_{22}\Gamma_L}\Gamma_L \tag{9-51b}$$

The input reflection Γ_{in} is the ratio of the reflected wave b_1 to the incident wave a_1:

$$\Gamma_{in} = \frac{b_1}{a_1} = S_{11} + \frac{S_{12}S_{21}\Gamma_L}{1 - S_{22}\Gamma_L}$$

If it is assumed that the error term is reciprocal, $S_{21} = S_{12}$, the input reflection equation can be simplified:

$$\Gamma_{in} = \frac{b_1}{a_1} = S_{11} + \frac{S_{12}^2\Gamma_L}{1 - S_{22}\Gamma_L} \tag{9-52}$$

To calibrate the effect of the errors out of the measurement, the S-parameters must be found. This is done by considering three known values for the load, Z_L: a short, an open, and a perfectly matched resistor. When the load is shorted, the reflection at the load is $\Gamma_L = -1$, which reduces (9-52) to

$$\Gamma_{\text{in, short}} = S_{11} - \frac{S_{12}^2}{1 + S_{22}} \tag{9-53a}$$

When the load is open and perfectly impedance matched, the reflection at the load is $\Gamma_L = 1$ and $\Gamma_L = 0$ respectively, producing

$$\Gamma_{\text{in, open}} = S_{11} + \frac{S_{12}^2}{1 - S_{22}} \tag{9-53b}$$

$$\Gamma_{\text{in, matched}} = S_{11} \tag{9-53c}$$

By measuring the open, short and matched loads, the three equations above can be solved simultaneously for S_{11}, S_{12}, and S_{22}:

$$S_{12} = S_{21} = \sqrt{(\Gamma_{\text{in, matched}} - \Gamma_{\text{in,short}})(1 + S_{22})} \tag{9-54a}$$

$$S_{22} = \frac{2\Gamma_{\text{in, matched}} - \Gamma_{\text{in,short}} - \Gamma_{\text{in, open}}}{\Gamma_{\text{in,short}} - \Gamma_{\text{in,open}}} \tag{9-54b}$$

$$S_{11} = \Gamma_{\text{in,matched}} \tag{9-54c}$$

The equations above are then used to create a scattering matrix for the errors at the ports:

$$\begin{bmatrix} S_{11} & S_{12} \\ S_{21} & S_{22} \end{bmatrix}_{\text{short,open,load}} \Rightarrow \begin{bmatrix} A & B \\ C & D \end{bmatrix}_{\text{error}_{\text{port}}} \tag{9-55}$$

Finally, the measurements of the DUT are calculated by multiplying by the inverse of the error terms:

$$\begin{bmatrix} A & B \\ C & D \end{bmatrix}_{\text{DUT}} = \begin{bmatrix} A & B \\ C & D \end{bmatrix}^{-1}_{\text{error}_{\text{port1}}} \begin{bmatrix} A & B \\ C & D \end{bmatrix}_{\text{measured}} \begin{bmatrix} A & B \\ C & D \end{bmatrix}_{\text{error}_{\text{port2}}} \tag{9-56}$$

The calibrated S-parameters are obtained by converting the $ABCD$ matrix of the DUT calculated with (9-56) using Table 9-3.

$$\begin{bmatrix} A & B \\ C & D \end{bmatrix}_{\text{DUT}} \Rightarrow \begin{bmatrix} S_{11} & S_{12} \\ S_{21} & S_{22} \end{bmatrix}_{\text{DUT}} \tag{9-57}$$

Basic deembedding principles utilize the same concepts to remove unwanted parts of the measurement. For example, Figure 9-32c shows a case where a via is cascaded with a transmission line. If it is desired to obtain only the measurements of the transmission line, the effects of the via must be deembedded. If the test board contains suitable structures to measure the S-parameters of the via in isolation, it can be effectively deembedded from the measurements using the same procedure shown in (9-56):

$$\begin{bmatrix} A & B \\ C & D \end{bmatrix}_{\text{T-line}} = \begin{bmatrix} A & B \\ C & D \end{bmatrix}_{\text{via}}^{-1} \begin{bmatrix} A & B \\ C & D \end{bmatrix}_{\text{measured}} \tag{9-58}$$

Using the concept of cascaded matrices, any number of structures can be deembedded from the measured data as long as the S-parameters for the structures are known.

9.2.6 Changing the Reference Impedance

S-parameters are dependent on the reference impedance of the VNA. If the port impedance values change, the S-parameters change. It is generally standard to measure S-parameters assuming a reference impedance of 50 Ω at each port. However, sometimes the port impedance values need to be adjusted after the measurements are performed. For example, perhaps the port impedance of the VNA was determined to be something other than 50 Ω, or the engineer wished to examine the performance of the circuit referenced to an impedance consistent with the transmission lines used in a specific board design. When an S-parameter is measured with a reference impedance at the ports of Z_n, it is said to be *normalized* to that impedance.

To renormalize the S-matrix from Z_{n1} to Z_{n2}, the definition of the Z-parameters from equation (9-35) is used:

$$\mathbf{Z} = Z_n(\mathbf{U}+\mathbf{S})(\mathbf{U}-\mathbf{S})^{-1} \tag{9-35}$$

Since the impedance matrix is not dependent on the port impedance, it can be used to renormalize the S-matrix.

$$Z_{n1}(\mathbf{U}+\mathbf{S}_1)(\mathbf{U}-\mathbf{S}_1)^{-1} = Z_{n2}(\mathbf{U}+\mathbf{S}_2)(\mathbf{U}-\mathbf{S}_2)^{-1}$$

Solving for $\mathbf{S}_2$ yields

$$\mathbf{S}_2 = \left[\frac{Z_{n1}}{Z_{n2}}(\mathbf{U}+\mathbf{S}_1)(\mathbf{U}-\mathbf{S}_1)^{-1}+\mathbf{U}\right]^{-1}\left[\frac{Z_{n1}}{Z_{n2}}(\mathbf{U}+\mathbf{S}_1)(\mathbf{U}-\mathbf{S}_1)^{-1}-\mathbf{U}\right] \tag{9-59}$$

Example 9-11 Renormalize the following S-matrix measured with a 50-Ω reference load to 75 Ω:

$$\mathbf{S}_{50\Omega} = \begin{bmatrix} 0.385 & j0.923 \\ j0.923 & 0.385 \end{bmatrix}$$

SOLUTION Using equation (9-59), the S-matrix can be manipulated so that it looks like it was measured with a 75-Ω reference impedance.

$$\mathbf{S}_{75\Omega} = \left[\frac{50}{75}\left(\begin{bmatrix} 1 & 0 \\ 0 & 1 \end{bmatrix}+\begin{bmatrix} 0.385 & j0.923 \\ j0.923 & 0.385 \end{bmatrix}\right)\left(\begin{bmatrix} 1 & 0 \\ 0 & 1 \end{bmatrix}-\begin{bmatrix} 0.385 & j0.923 \\ j0.923 & 0.385 \end{bmatrix}\right)^{-1}\right.$$
$$\left.+\begin{bmatrix} 1 & 0 \\ 0 & 1 \end{bmatrix}\right]^{-1}\cdot\left\{\frac{50}{75}\left(\begin{bmatrix} 1 & 0 \\ 0 & 1 \end{bmatrix}+\begin{bmatrix} 0.385 & j0.923 \\ j0.923 & 0.385 \end{bmatrix}\right)\right.$$
$$\left.\times\left(\begin{bmatrix} 1 & 0 \\ 0 & 1 \end{bmatrix}-\begin{bmatrix} 0.385 & j0.923 \\ j0.923 & 0.385 \end{bmatrix}\right)^{-1}-\begin{bmatrix} 1 & 0 \\ 0 & 1 \end{bmatrix}\right\}=\begin{bmatrix} 0 & -j \\ -j & 0 \end{bmatrix}$$

Therefore, the magnitude of the renormalized S-matrix is

$$|\mathbf{S}_{75\Omega}| = \begin{bmatrix} 0 & 1 \\ 1 & 0 \end{bmatrix}$$

meaning that the transmission line is loss free with a characteristic impedance of 75 Ω since there is no reflection ($S_{11} = 0$) and the insertion loss is unity ($S_{21} = 1$).

9.2.7 Multimode *S*-Parameters

In Chapter 4 differential signaling was explained. Since many of the high-speed buses being designed in modern computing systems consist of differential pairs, it is often convenient to describe the behavior of the interconnects in terms of multimode S-parameters. The multimode S-matrix breaks the signal on a differential pair into terms of differential (i.e., odd mode) and common (i.e., even mode) signaling states. A multimode matrix for two modes can be derived

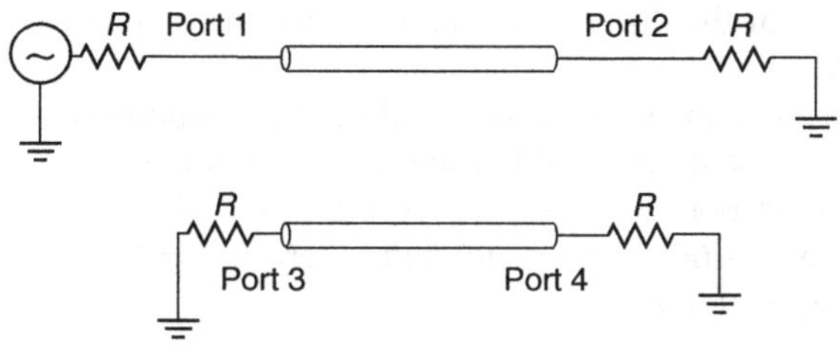

Figure 9-37 A pair of coupled transmission lines is a common example of a four-port network that is conveniently described with a multimode S-matrix.

directly from the four-port S-parameters. For example, the S-parameters of the four-port system pictured in Figure 9-34 are shown as follows:

$$\begin{bmatrix} b_1 \\ b_2 \\ b_3 \\ b_4 \end{bmatrix} = \begin{bmatrix} S_{11} & S_{12} & S_{13} & S_{14} \\ S_{21} & S_{22} & S_{23} & S_{24} \\ S_{31} & S_{32} & S_{33} & S_{34} \\ S_{41} & S_{42} & S_{43} & S_{44} \end{bmatrix} \begin{bmatrix} a_1 \\ a_2 \\ a_3 \\ a_4 \end{bmatrix} \tag{9-60}$$

An example of a four-port system is two coupled transmission lines, as shown in Figure 9-37. To derive the multimode matrix, it is first necessary to calculate the S-parameters for each mode.

The differential S-parameters, where the energy is being transmitted in the odd mode, are calculated by driving the four-port network with $+v$ on port 1 and $-v$ on port 3 when ports 2 and 4 are not being driven. This allows (9-60) to be simplified.

$$\begin{bmatrix} b_1 \\ b_2 \\ b_3 \\ b_4 \end{bmatrix} = \begin{bmatrix} S_{11} & S_{12} & S_{13} & S_{14} \\ S_{21} & S_{22} & S_{23} & S_{24} \\ S_{31} & S_{32} & S_{33} & S_{34} \\ S_{41} & S_{42} & S_{43} & S_{44} \end{bmatrix} \begin{bmatrix} v \\ 0 \\ -v \\ 0 \end{bmatrix} \tag{9-61}$$

The differential return loss at port 1 can then be calculated algebraically where $b_{d1} = b_1 - b_3$ and $a_{d1} = a_1 - a_3 = v - (-v) = 2v$.

$$S_{dd11} = \left.\frac{b_{d1}}{a_{d1}}\right|_{a_2=a_4=0} = \tfrac{1}{2}(S_{11} + S_{33} - S_{13} - S_{31}) \tag{9-62a}$$

S_{dd11} is a measure of the differential energy reflected, or returned to the source. Similarly, the differential insertion loss is calculated where $b_{d2} = b_2 - b_4$ and $a_{d1} = a_1 - a_3 = v - (-v) = 2v$.

$$S_{dd21} = \left.\frac{b_{d2}}{a_{d1}}\right|_{a_2=a_4=0} = \tfrac{1}{2}(S_{21} + S_{43} - S_{23} - S_{41}) \tag{9-62b}$$

S_{dd21} is a measure of the differential energy transmitted across the network from port 1 to port 2.

Identical analysis can be used to calculate the common-mode S-parameters, where the energy is being transmitted in the even mode, by driving the four-port network with $+v$ on port 1 and $+v$ on port 3 when ports 2 and 4 are not being driven. This allows (9-60) to be simplified where $b_{c1} = b_1 + b_3$, $b_{c2} = b_2 + b_4$, and $a_{c1} = a_1 + a_3 = v + v = 2v$.

$$\begin{bmatrix} b_1 \\ b_2 \\ b_3 \\ b_4 \end{bmatrix} = \begin{bmatrix} S_{11} & S_{12} & S_{13} & S_{14} \\ S_{21} & S_{22} & S_{23} & S_{24} \\ S_{31} & S_{32} & S_{33} & S_{34} \\ S_{41} & S_{42} & S_{43} & S_{44} \end{bmatrix} \begin{bmatrix} v \\ 0 \\ v \\ 0 \end{bmatrix} \tag{9-63}$$

The common-mode S-parameters are easily obtained algebraically from (9-63):

$$S_{cc11} = \left.\frac{b_{c1}}{a_{c1}}\right|_{a_2=a_4=0} = \tfrac{1}{2}(S_{11} + S_{33} + S_{13} + S_{31}) \tag{9-64a}$$

$$S_{cc21} = \left.\frac{b_{c2}}{a_{c1}}\right|_{a_2=a_4=0} = \tfrac{1}{2}(S_{21} + S_{23} + S_{41} + S_{43}) \tag{9-64b}$$

Equations (9-64a) and (9-64b) describe how much energy is being reflected and transmitted when the four-port system is being driven with a common-mode source (i.e., the system is being driven in the even mode).

As described in Chapter 4, for a system with two signal conductors, the voltages at the ports are combinations of odd and even modes. Consequently, for a perfectly symmetric system (where each leg of the differential pair is electrically identical), if the system is driven differentially, all the energy will be contained within the odd mode. However, if the pair exhibits any asymmetry, a portion of the energy will be flowing in the even mode. The multimode matrix also accounts for the *differential-to-common mode conversion*, which describes the amount of energy being transformed into the even mode when the system is being driven differentially and the *common mode-to-differential conversion*, which tells how much energy is being converted to the odd mode when driven commonly. Since most high-speed buses are driven differentially, the *differential-to-common mode conversion* is the most important parameter of the two.

The *differential-to-common mode S-parameters*, where the energy is being transmitted in the odd mode and received in the even mode, are calculated by driving the four-port network differentially at the driver and sensing in common mode at the receiver. The differential-to-common mode coefficients are $b_{c1} = b_1 + b_3$, $b_{c2} = b_2 + b_4$, and $a_{d1} = a_1 - a_3 = v - (-v) = 2v$.

$$\begin{bmatrix} b_1 \\ b_2 \\ b_3 \\ b_4 \end{bmatrix} = \begin{bmatrix} S_{11} & S_{12} & S_{13} & S_{14} \\ S_{21} & S_{22} & S_{23} & S_{24} \\ S_{31} & S_{32} & S_{33} & S_{34} \\ S_{41} & S_{42} & S_{43} & S_{44} \end{bmatrix} \begin{bmatrix} v \\ 0 \\ -v \\ 0 \end{bmatrix} \tag{9-65}$$

$$S_{cd11} = \left.\frac{b_{c1}}{a_{d1}}\right|_{a_2=a_4=0} = \tfrac{1}{2}(S_{11} - S_{13} + S_{31} - S_{33}) \tag{9-66a}$$

$$S_{cd21} = \left.\frac{b_{c2}}{a_{d1}}\right|_{a_2=a_4=0} = \tfrac{1}{2}(S_{21} - S_{23} + S_{41} - S_{43}) \tag{9-66b}$$

The *common mode-to-differential S-parameters*, where the energy is being transmitted in the even mode and received in the odd mode, are calculated by driving the four-port network in the common mode and sensing differentially at the receiver. The derivation is left to the reader.

The multimode S-parameters for a four-port system can be combined to create the *multimode matrix*

$$\begin{bmatrix} b_{d1} \\ b_{d2} \\ b_{c1} \\ b_{c2} \end{bmatrix} = \begin{bmatrix} S_{dd11} & S_{dd12} & S_{dc11} & S_{dc12} \\ S_{dd21} & S_{dd22} & S_{dc21} & S_{dc22} \\ S_{cd11} & S_{cd12} & S_{cc11} & S_{cc12} \\ S_{cd21} & S_{cd22} & S_{cc21} & S_{cc22} \end{bmatrix} \begin{bmatrix} a_{d1} \\ a_{d2} \\ a_{c1} \\ a_{c2} \end{bmatrix} \tag{9-67a}$$

Note that each quadrant of the multimode matrix represents unique parameters. For example, the upper left quadrant is the differential S-parameter matrix,

$$\begin{bmatrix} b_{d1} \\ b_{d2} \end{bmatrix} = \begin{bmatrix} S_{dd11} & S_{dd12} \\ S_{dd21} & S_{dd22} \end{bmatrix} \begin{bmatrix} a_{d1} \\ a_{d2} \end{bmatrix} = [S_{dd}] \begin{bmatrix} a_{d1} \\ a_{d2} \end{bmatrix} \tag{9-67b}$$

which assumes that all the energy is sourced and sensed in a differential manner. Consequently, the multimode matrix takes the form

$$\begin{bmatrix} b_{d1} \\ b_{d2} \\ b_{c1} \\ b_{c2} \end{bmatrix} = \begin{bmatrix} [\mathbf{S}_{dd}] & [\mathbf{S}_{dc}] \\ [\mathbf{S}_{cd}] & [\mathbf{S}_{cc}] \end{bmatrix} \begin{bmatrix} a_{d1} \\ a_{d2} \\ a_{c1} \\ a_{c2} \end{bmatrix} \tag{9-67c}$$

where each matrix describes a specific driving and receiving configuration:

$\mathbf{S}_{dd}$: Driver and receiver are differential.
$\mathbf{S}_{cc}$: Driver and receiver are common mode.
$\mathbf{S}_{dc}$: Driver is common mode and receiver is differential.
$\mathbf{S}_{cd}$: Driver is differential mode and receiver is common mode.

The format of (9-67c) allows the four-port S-parameters to be reduced to an equivalent simple two-port system for specific circumstances. For example, in high-speed differential buses, each pair is driven and received assuming that all

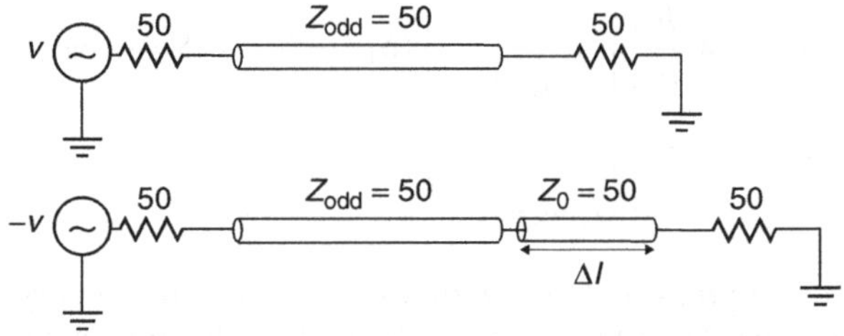

Figure 9-38 Unbalanced differential pairs cause energy to be converted from the odd mode to the even mode, which looks like additional differential insertion loss.

energy is in the odd mode, which means that for all practical purposes, equation (9-67b) is sufficient for design purposes.

Invariably, a question is asked: *If there is significant energy in the even mode, doesn't the common mode matrix need to be accounted for*? The answer is that it is already included in the differential matrix of (9-67b). If an asymmetrical four-port system is driven differentially, the energy that is converted to common mode will simply look like differential insertion loss. For example, consider the loss-free unbalanced differential transmission line pair being driven in the odd mode in Figure 9-38. Since the system is being driven differentially (in the odd mode), and the odd-mode impedance is equal to the termination, there will be no reflections. However, since it is unbalanced, it is expected that a portion of the energy will be transferred to the even mode. From a differential receiver's point of view, this will look like extra channel loss.

To demonstrate this concept, consider the case where the difference in length between the legs of the differential pair (Δl) is 200 mils with a delay of 27.77 ps (refer to Figure 9-38). If the initial phase difference between v^+ and v^- at the driver is of 180°, when the signals arrive at the receiver, the phase difference will no longer be equal to 180°, due to the extra length on the second leg. The actual phase difference at the receiver can be calculated using equation (9-22) as a function of frequency. For example, at 1 GHz the phase difference at the receiver will be

$$180^\circ + \theta = 180^\circ + (360^\circ)(27.77 \times 10^{-12})(1 \times 10^9) = 190^\circ$$

meaning that since the signals are no longer exactly 180° out of phase, and some of the energy is being dumped into the even mode.

As the frequency increases, the phase difference due to the extra delay will approach 180°. When $\theta = 180^\circ$, the differential signal at the driver has been fully transformed into a common-mode signal at the receiver. When the delay difference between the legs of the differential pair is 27.77 ps, the frequency where the signal is 100% common mode at the receiver is 18 GHz:

$$\theta = (360^\circ)(27.777 \times 10^{-12})(f_{180}) = 180^\circ$$
$$\Rightarrow f_{180} = 18 \times 10^9$$

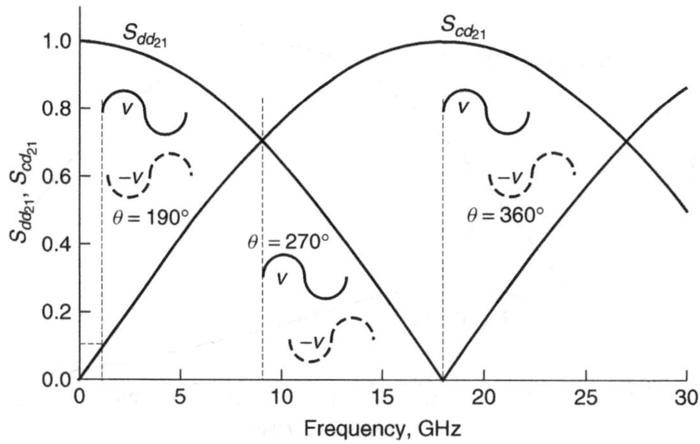

Figure 9-39 As the frequency increases, small asymmetries in a differential pair will cause larger phase differences between signals propagating on each leg, causing energy to be converted from odd to even mode. This shows up as differential insertion loss. In this case the differential signal is 100% converted to common mode at 18 GHz.

So at 18 GHz, the phase difference at the receiver will be

$$180^\circ + \theta = 180^\circ + (360^\circ)(27.77 \times 10^{-12})(18 \times 10^9) = 360^\circ$$

Therefore, at 18 GHz the differential insertion loss is zero because there is no longer any differential energy being transmitted to the receiver because all the energy has been converted to the common mode.

Figure 9-39 shows the S-parameters for the loss-free perfectly terminated asymmetrical transmission line with a interleg delay difference of 27.77 ps. Note that the differential energy transferred from port 1 to port 2 (S_{dd21}) decreases until 18 GHz, after which it begins to increase again. At 1 GHz (where $\theta = 190^\circ$), $S_{cd21} = 0.1$, which means that 10% of the differential energy is lost to the common mode. At 18 GHz, $S_{cd21} = 1.0$, so 100% of the energy is lost to the common mode. It is easy to see that the decrease in S_{dd21} corresponds with the increase in S_{cd21}.

Do not falsely conclude that differential signals can be transmitted properly at frequencies above f_{180} (18 GHz in this example). Although S_{dd21} increases, the phase difference between signals on each leg approaches 540° ($3 \cdot 180^\circ$), not the ideal value of 180°. For a digital signal, this means that the bit on line 1 is 180° out of phase with the next bit in the digital pulse train on line 2. Therefore, even though the common-mode conversion is small, the data are invalid.

When loss is included in the transmission line, the S_{cd21} curve will peak at a lower frequency that does not correspond to the point where the differential-to-common mode conversion is 100%, as shown in Figure 9-40. Be careful not to misinterpret the differential S-parameter data. The mode conversion is 100% when S_{dd21} is zero, not necessarily when S_{cd21} is maximum.

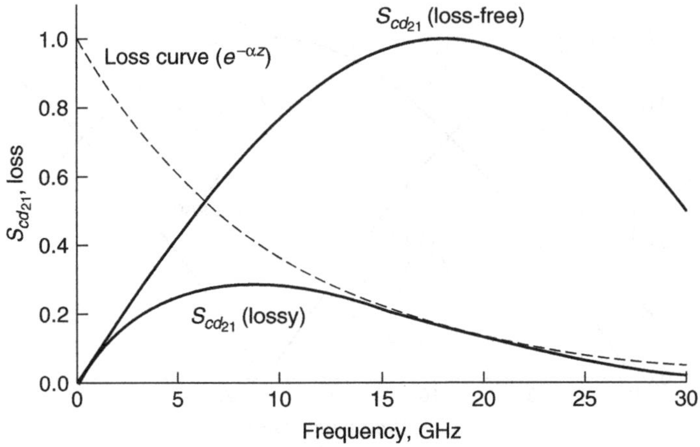

Figure 9-40 For a loss-free case, the differential-to-common mode conversion curve will be maximum when the phase difference between legs is 180°. However, for a lossy case, the S_{cd21} curve will peak at a lower frequency, due to losses from the conductors and dielectrics.

9.3 PROPERTIES OF PHYSICAL S-PARAMETERS

S-parameters are a valuable tool in digital design. They can be used to analyze the performance of a component, extract equivalent-circuit models, or be employed as a tool-independent portable model. The problem arises that S-parameters are difficult to generate accurately. For example, proper calibration of a vector network analyzer to high frequencies is both a science and an art. If probe effects are not removed or test fixtures are deembedded improperly, the S-parameters could contain significant errors that are difficult to detect. Additionally, when component models (i.e., for a connector or CPU socket) are distributed to engineers in the form of a frequency-dependent S-parameter file, the assumptions of how the model was constructed are lost, so it is impossible to judge correctness or applicability. Fortunately, even though it is impossible to judge the accuracy of S-parameters without comparison to measurements or examination of the underlying model, the concepts outlined in Chapter 8 can be used to look for gross errors that blatantly violate the laws of physics.

9.3.1 Passivity

As defined in Section 8.2.2, a physical system is *passive* when it is unable to generate energy. For digital design, S-parameters are used to analyze the physical components of the bus, such as the transmission lines, sockets, and connectors, none of which generate energy. Therefore, in digital design, if an S-parameter measurement or model is shown to be nonpassive, either the VNA was not calibrated correctly or the underlying assumptions of the modeling methodology are wrong. A practical test for the passivity of S-parameters can be developed directly from equations (8-22) and (8-23)[Ling, 2007].

Since power must be conserved, the power absorbed by the network (P_a) is equal to the power driven into the network less power flowing out:

$$\sum (|a_i|^2 - |b_i|^2) = P_a \tag{8-22}$$

where $P_a \geq 0$ for a passive network. If $P_a < 0$, the network is generating power and the system would be considered nonpassive. This allows equation (8-22) to be written in terms of the power wave matrices that will produce a real value for the power absorbed by the network. A system is passive if

$$\mathbf{a}^{\mathrm{H}}\mathbf{a} - \mathbf{b}^{\mathrm{H}}\mathbf{b} \geq 0 \tag{8-23}$$

where $\mathbf{a}$ is a matrix that contains all the power waves incident to each port, $\mathbf{b}$ contains the power waves coming out of each port, and $\mathbf{x}^{\mathrm{H}}$ indicates the *conjugate transpose* (sometimes called the *Hermitian transpose*), which is calculated by taking the transpose of $\mathbf{x}$ and then taking the complex conjugate of each entry.

Using the definition of S-parameters from equation (9-18), the passivity requirement of (8-23) can be rewritten as

$$\begin{gathered}\mathbf{b} = \mathbf{S}\mathbf{a},\ \mathbf{b}^{\mathrm{H}} = \mathbf{S}^{\mathrm{H}}\mathbf{a}^{\mathrm{H}} \\ \mathbf{a}^{\mathrm{H}}\mathbf{a} - \mathbf{S}^{\mathrm{H}}\mathbf{a}^{\mathrm{H}}\mathbf{S}\mathbf{a} \geq 0 \\ \mathbf{a}^{\mathrm{H}}(\mathbf{U} - \mathbf{S}^{\mathrm{H}}\mathbf{S})\mathbf{a} \geq 0\end{gathered} \tag{9-68}$$

where $\mathbf{U}$ is the unity (identity) matrix. Equation (9-68) leads to the general requirement for passivity:

$$\mathbf{U} - \mathbf{S}^{\mathrm{H}}\mathbf{S} \geq 0 \tag{9-69}$$

If $\mathbf{S}^{\mathrm{H}}\mathbf{S}$ is greater than 1, the requirement of (8-22) is violated and the system is not passive.

A quick test to ensure passivity can be derived based on the eigenvalues of $\mathbf{S}^{\mathrm{H}}\mathbf{S}$. The eigenvectors ζ and eigenvalues λ are determined from the solution of the equation

$$\mathbf{S}^{\mathrm{H}}\mathbf{S}\xi = \lambda\xi$$

Techniques for calculating the eigenvectors and eigenvalues are detailed by O'Neil [1991]; however, there are many software packages that can be used, such as Mathematica or Matlab.

The eigenvectors can be formed into an $N \times N$ matrix,

$$\mathbf{V} = [\xi_1 \quad \xi_2 \quad \cdots \quad \xi_N] = \begin{bmatrix} \zeta_{1_1} & \zeta_{2_1} & \cdots & \zeta_{N_1} \\ \zeta_{1_2} & \zeta_{2_2} & \cdots & \zeta_{N_2} \\ \vdots & \cdots & \ddots & \\ \zeta_{1_N} & \zeta_{2_N} & \cdots & \zeta_{N_N} \end{bmatrix}$$

with the property that $\mathbf{V}\mathbf{V}^{\mathrm{T}} = \mathbf{U}$. The matrix $\mathbf{S}^{\mathrm{H}}\mathbf{S}$ can be diagonalized into its eigenvalues with

$$\mathbf{S}^{\mathrm{H}}\mathbf{S} = \mathbf{V}\lambda\mathbf{V}^{\mathrm{T}} = \mathbf{V}\begin{bmatrix} \lambda_1 & 0 & \cdots & 0 \\ 0 & \lambda_2 & \cdots & 0 \\ \vdots & \cdots & \ddots & \vdots \\ 0 & 0 & \cdots & \lambda_N \end{bmatrix}\mathbf{V}^{\mathrm{T}} \tag{9-70}$$

The definitions above allow (9-68) to be rewritten in terms of the eigenvalues:

$$\begin{aligned} \mathbf{a}^{\mathrm{H}}(\mathbf{V}\mathbf{V}^{\mathrm{T}} - \mathbf{V}\lambda\mathbf{V}^{\mathrm{T}})\mathbf{a} &\geq 0 \\ \mathbf{a}^{\mathrm{H}}\mathbf{V}(\mathbf{U} - \lambda)\mathbf{a}\mathbf{V}^{\mathrm{T}} &\geq 0 \end{aligned} \tag{9-71}$$

Equation (9-71) leads to a simple test for passivity [Ling, 2007]. Since $\mathbf{U} - \lambda$ must be greater or equal to zero and nonnegative, the system is passive if all the eigenvalues are between 0 and 1. The test for passivity is shown as

$$0 \leq \lambda_i \leq 1 \tag{9-72}$$

Some tools employ the strategy of applying correction algorithms for the nonpassive cases. Generally, the use of a passivity-corrected model or measurement is an unwise strategy because nonpassivity indicates an incorrect model or calibration. The best strategy is to reevaluate the method for equating the S-parameters and fix the problem.

9.3.2 Reality

Time-domain signals that are used to analyze digital interconnect are real. In other words, there is no imaginary part of a time-domain signal that can exist in nature. As mentioned in Section 8.2.1, the mathematical requirement for a frequency-domain response to ensure a real time-domain waveform is

$$S_{ij}(-\omega) = S_{ij}(\omega)^* \tag{9-73}$$

where* indicates the complex conjugate and j and i represent the driving and receiving ports of the S-parameters. Most S-parameter data are presented only with positive frequency-domain data, and the tools construct the negative frequency portion from (9-73). However, if the negative frequency values of the S-parameters are available, (9-73) is a useful check.

9.3.3 Causality

As covered in Section 8.2.1, one seemingly obvious requirement of a model or measurement is that its output cannot precede its input. In other words, an effect cannot precede its cause. This fundamental principle is called *causality*.

Very often, VNA calibration problems or incorrect model assumptions generate S-parameters that are not causal. This error becomes especially critical when S-parameters are used to represent the behavior of physical channel components when simulating system performance. Causality errors will introduce delay and amplitude errors in the time-domain responses, as well as distort waveforms to such a degree that signal integrity may be falsely predicted, which can lead to false solution spaces, incorrect equalization settings, and incorrect understanding of bus performance. Many commercially available tools create models that are noncausal simply because of the assumptions of a frequency-invariant dielectric model where ε_r and $\tan\delta$ are not properly related. Figure 8.14 is a good example of how a noncausal model can distort the waveform of a single bit. The errors depicted in Figure 8.14 will be amplified for very long channel lengths and high frequencies. Many of the digital designs created in the past were designed using noncausal models that assume frequency-invariant dielectric parameters. At low data rates (below about 1 to 2 Gbit/s) and short lengths (a few inches), causality errors do not appreciably affect signal integrity. However, at high frequencies and/or long lengths, they can completely destroy the model's ability to predict realistic behavior.

Causality requirements are often difficult to judge by looking directly at S-parameters because the phase errors are small compared to the propagation delay of the structure. Furthermore, even if the causality errors of a single component such as a connector or via are small in isolation, when numerous models are cascaded in a complete channel simulation the errors become cumulative. Consequently, it is important to check the causality of each component as well as the system response for the entire channel.

As described fully in Section 8.2.1 and demonstrated in Examples 8-3 and 8-4, *causality can be tested by performing the Hilbert transform of the real part and ensuring that it is identical to the imaginary part*. If it is not identical, the model is not causal. However, care should to be taken when performing this test because for bandlimited data, the Hilbert transform can exhibit truncation errors.

Equation (8-19) is adapted to calculate the Hilbert transform of the real part of the S-parameter data,

$$\hat{S}_{\mathrm{Re},ij}(f) = \mathrm{Re}[S_{ij}(f)] * \frac{1}{\pi f} \tag{9-74}$$

where $\hat{S}_{\mathrm{Re},ij}(f)$ denotes the Hilbert transform of the real portion of the S-parameter, $S_{ij}(f)$ is an S-parameter where port j is driving and port i is receiving, and $*$ denotes convolution. The S-parameter is causal when (9-75) is satisfied:

$$\hat{S}_{\mathrm{Re},ij}(f) = \mathrm{Im}[S_{ij}(f)] \tag{9-75}$$

To implement (9-74), it is often necessary to reconstruct the negative frequency components of the response using

$$S(-f) = S(f)^* \tag{9-36b}$$

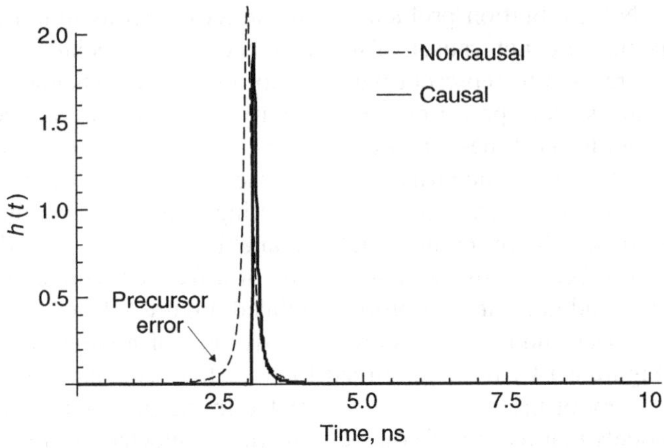

Figure 9-41 Classic impulse response behavior of a severely noncausal S-parameter. Note the precursor error, where part of the energy is arriving too early.

The most straightforward way to calculate the Hilbert transform is to use the Fourier transform (usually in the form of the fast Fourier transform), as demonstrated in Examples 8-3 and 8-4.

$$\hat{S}_{\text{Re},ij}(f) = \mathsf{F}^{-1}\left[\mathsf{F}(\text{Re}[S_{\text{Re},ij}(f)])\mathsf{F}\left(\frac{1}{\pi f}\right)\right] \tag{9-76}$$

If the fast Fourier transform is used, the negative frequency values must be appended to the end of the positive values, as demonstrated in Example 9-7. A less rigorous way to evaluate the causality is to transform the S-parameters into an impulse response, as described in 9.2.2. If there is no distinct point at which the pulse arrives, it is an indication of a noncausal response, as shown in Figure 9-41.

9.3.4 Subjective Examination of S-Parameters

Often, it is desirable to recognize the trustworthiness of S-parameter data without the rigor of the mathematical analysis above. Understanding the basic properties of S-parameter data will allow an intuitive analysis. For example, a gross passivity violation can often be seen by observing the magnitude of the insertion loss. Figure 9-42 shows an example of measured insertion loss for a transmission line. Note that the scale often used with VNA measurements is, in decibels,

$$\text{dB} = 20\log(\text{mag}) \tag{9-77}$$

where mag is the magnitude of the complex S-parameter value. Notice how S_{21} in Figure 9-42 rises above zero at the lower frequencies, indicating that the transmission line has gain and is therefore producing energy and nonpassive

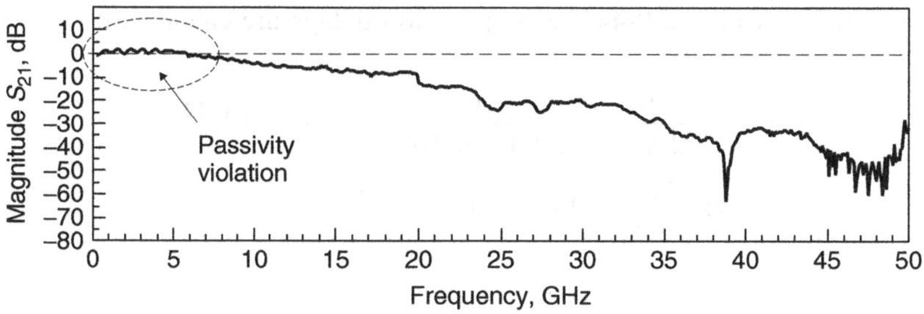

Figure 9-42 Bad S-parameter measurement of a transmission line, showing passivity violations due to incorrect VNA calibration.

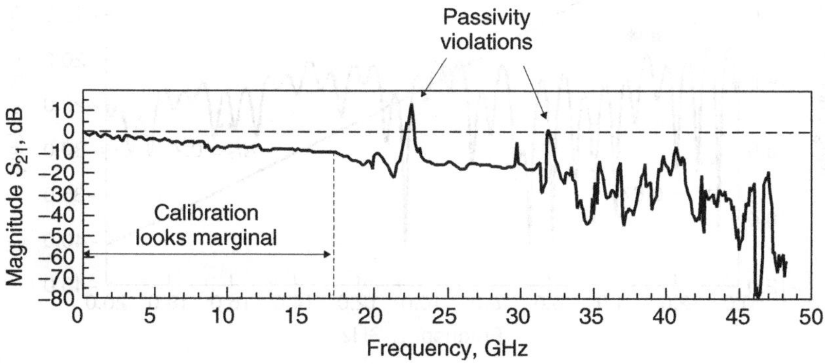

Figure 9-43 Bad S-parameter measurement of a transmission line, showing passivity violations at high frequencies where the VNA calibration breaks down.

(0 dB is equal to an insertion loss of 1 on a linear scale). In this case, the passivity violation was a result of an incorrect calibration of the vector network analyzer. Another example of a severe passivity violation is shown in Figure 9-43, where the calibration breaks down after about 17 GHz. Even below 17 GHz this measurement looks questionable, due to the significant amount of nonperiodic ripples on the waveform. Periodic ripples would probably be a result of reflections or crosstalk, but nonperiodic noise on S_{21} indicates an inaccurate measurement.

Figure 9-44 shows an example of a good measurement of a 5-in. microstrip transmission line on a FR4 substrate. Note that the insertion loss does not rise above 0 dB and that it does not contain any nonperiodic noise. It should be noted that the S-parameters could still be nonpassive even if S_{21} does not rise above 0 dB. Passivity can only be verified using the techniques outlined in Section 9.3.1; however, simple observation of S_{21} is often sufficient to catch gross errors.

The periodic behavior of S_{11} can also be used to get an intuitive feeling about the validity of the S-parameters by ensuring reasonable phase delays. If the structure under test is a transmission line, the return loss (S_{11}) should be periodic and the distance between peaks (or dips) is related to the time delay with equation (9-4). The propagation delay can be double-checked at various frequencies. From Figure 9-44 the propagation delays are calculated:

$$\tau_{d1} = \frac{1}{2\Delta f_1} = \frac{1}{2(0.67 \times 10^9)} = 746 \times 10^{-12}\ \text{s}$$

$$\Rightarrow \frac{\tau_{d1}}{5\ \text{in.}} = 149\ \text{ps/in.}$$

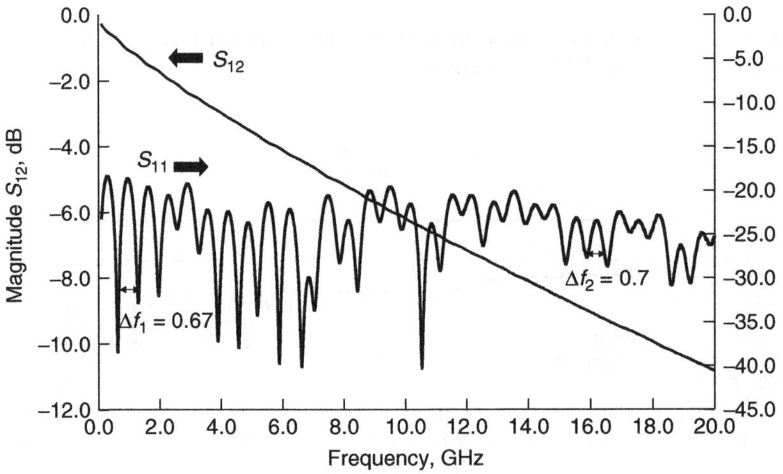

Figure 9-44 Good S-parameter measurement with no obvious problems.

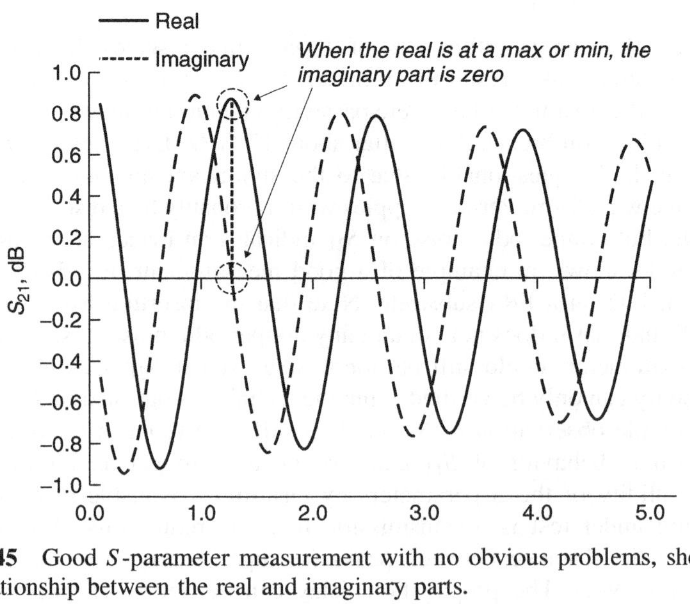

Figure 9-45 Good S-parameter measurement with no obvious problems, showing the proper relationship between the real and imaginary parts.

Since the delay of a typical microstrip line on FR4 is about 140 to 160 ps/in., the result seems reasonable.

At higher frequencies, the same test can be applied. From Figure 9-44, $\Delta f_2 = 0.7$ GHz.

$$\tau_{d2} = \frac{1}{2\Delta f_2} = \frac{1}{2(0.7 \times 10^9)} = 714 \times 10^{-12} \text{ s}$$

$$\Rightarrow \frac{\tau_{d1}}{5 \text{ in.}} = 142 \text{ ps/in.}$$

Note that the delay is less that it was at the lower-frequency test. This confirms the validity of the measurement because the dielectric constant, and therefore the propagation delay, should decrease with frequency, as described in Chapter 6.

Finally, the relationships between the real and imaginary parts of S_{21} can be examined. Figure 9-45 shows the real and imaginary parts of the insertion loss for the 5-in. transmission line measured in Figure 9-44. As discussed in Section 9.1.1, when the real part peaks, the imaginary part should be zero.

REFERENCES

Ling, Yun, 2007, Demystify *S*-parameters, I: The basics, class given at Intel Corporation.

O'Neil, Peter V., 1991, *Advanced Engineering Mathematics*, Wadsworth, Belmont, CA.

Hall, Stephen, Garrett Hall, and James McCall, 2000, *High Speed Digital System Design*, Wiley-Interscience, New York.

PROBLEMS

9-1 Consider the measured *S*-parameters of a transmission line shown in Figure 9-46. What is the propagation delay of the transmission line? What is the characteristic impedance of the transmission line assuming that the port impedances are 50 Ω? Is the measurement valid? How can you tell?

9-2 Consider the *S*-parameters shown in Figure 9-47 of a transmission line. Are these *S*-parameters valid? How can you tell?

9-3 Consider the measured *S*-parameters of a transmission line shown in Figure 9-48. Are these *S*-parameters valid? How can you tell?

9-4 Derive the conversion from *T*-parameters back into *S*-parameters for two ports.

9-5 Derive the formulas to extract *R*, *L*, *C*, and *G* from a transmission line with arbitrary impedance. Prove that the method is accurate by extracting the frequency-dependent *R*, *L*, *C*, and *G* from a simulated transmission line.

9-6 Calculate the common-mode matrix and the common mode-to-differential conversion matrix of a four-port *S*-matrix.

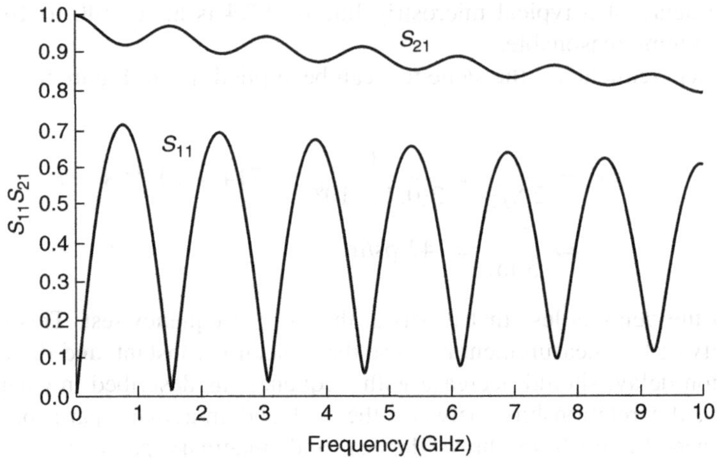

Figure 9-46 Figure for Problem 9-1.

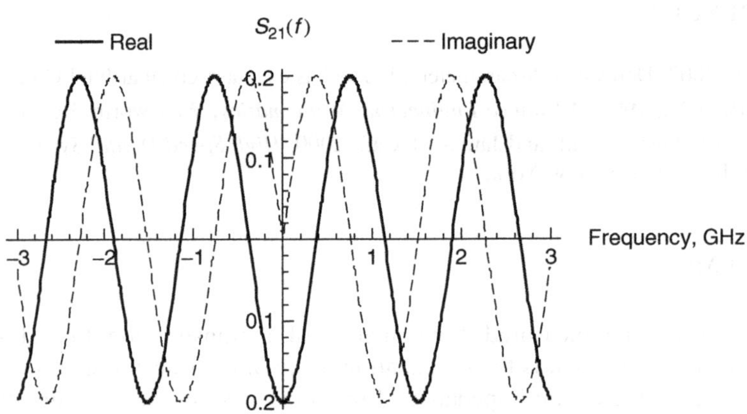

Figure 9-47 Figure for Problem 9-2.

9-7 If a differential pair has 10% of its energy transformed into the common mode when driven differentially, is 10% of its energy transformed into the differential mode when driven commonly? Show the math to support your answer.

9-8 Derive the *ABCD* matrix for the pi circuit shown in the bottom of Table 9-2.

9-9 For the differential pair in Figure 9-49, at what frequency will the ACCM be 100%? Assume that the effective permittivities are $\varepsilon_{r,\text{odd}} = 3.5$ and $\varepsilon_r = 4.0$.

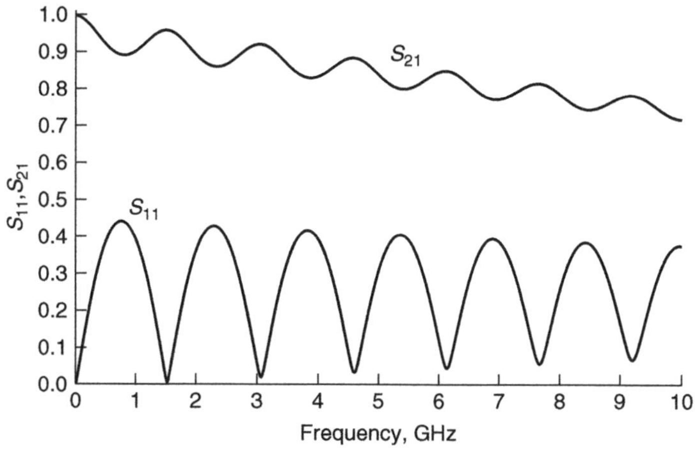

Figure 9-48 Figure for Problem 9-3.

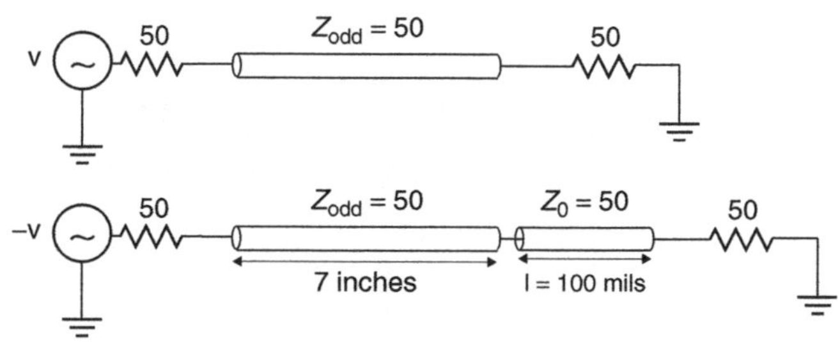

Figure 9-49 Figure for Problem 9-9.

9-10 Derive the equations to renormalize the S-matrix measured with 50-Ω loads at each port to arbitrary and different reference impedances at each port.

9-11 Derive the two-port S-parameters for a frequency-dependent, causal 10-in. transmission line from 10 Hz to 100 GHz. Using a tool such as Mathematica or Matlab, plot the S-parameters and prove that they are passive and causal.

9-12 From the S-parameters calculated from the transmission line in Problem 9-11, calculate the pulse (single-bit) response for a 15-Gbit/s data rate.

9-13 Create the S-matrices for 4- and 10-in. transmission lines using the method derived in Problem 9-11. Cascade the S-parameters to get the equivalent S-matrix for a 14-in. transmission line. Compare the cascaded S-matrix to that of a single 14-in. transmission line.

10

TOPICS IN HIGH-SPEED CHANNEL MODELING

So far, we have covered numerous topics that are required for the design of modern high-speed digital systems. There are a few important topics, however, that do not fit nicely into the flow of the other chapters. This section covers a few critical aspects of digital design that are important to understand. First, a methodology for creating frequency-dependent tabular transmission-line models is presented. The tabular format is a convenient way to model the frequency-dependent nature of the conductor and dielectrics described in Chapters 5 and 6. Next, the problem of nonideal current return paths is explored so that the engineer will understand how they affect signal integrity. Finally, the signal integrity impact of transitioning between layers using plated-through-hole vias is discussed.

Advanced Signal Integrity for High-Speed Digital Design, By Stephen H. Hall and Howard L. Heck

10.1 CREATING A PHYSICAL TRANSMISSION-LINE MODEL

Digital design is typically performed using circuit simulators. Fortunately, most commercially available tools have transmission-line models that can be integrated with the digital circuitry to perform system-level bus simulations. Problems arise, however, when the assumptions built into the models break down or are simply not known by the engineer. For example, many simulators employ transmission-line models with frequency-invariant dielectric permittivity and loss tangents. Although these models are perfectly adequate for very short bus lengths or data rates below about 1 Gbit/s the incorrect relationship between ε_r and $\tan\delta$ induces causality errors (such as the model in Example 8-4) that can render the simulations almost useless. Additionally, few commercial simulators account properly for such things as surface roughness or internal inductance. Consequently, it is usually desirable for the digital designer to create custom transmission-line models that use the methodologies presented in Chapters 3 through 6 to ensure accurate results.

10.1.1 Tabular Approach

A convenient method of implementing user-defined transmission-line models is to use a tabular approach, which is simply a lookup table that defines the transmission-line equivalent-circuit parameters at each frequency. Fortunately, user-definable tabular models are available in many commercial simulators. A tabular methodology allows calculation of the system transfer function in the frequency domain. An inverse Fourier transform can then be used to get a time-domain impulse response, which can be convolved with an arbitrary input waveform to evaluate signal integrity.

A practical and efficient approach to incorporate surface roughness, internal inductance, and wideband frequency-dependent dielectric properties into printed circuit board (PCB) transmission-line models is defined here as a two-step procedure. In the first step, quasistatic RLGC matrices in the transmission line model are generated at a single reference frequency (ω_{ref}), using a 2D transmission-line calculator or analytically with the methods outlined in Chapters 3 through 6. In this step it is important that the value of ε_r and $\tan\delta$ be known at the reference frequency. This model will reflect the single-valued frequency response of the transmission-line geometry and configuration but will not yet account for the frequency dependence of surface roughness losses, internal inductance, and dielectric properties. The matrices calculated during this step will be known as the reference values $\mathbf{R}(\omega_{\text{ref}})$, $\mathbf{L}(\omega_{\text{ref}})$, $\mathbf{C}(\omega_{\text{ref}})$, and $\mathbf{G}(\omega_{\text{ref}})$. In the second step, the reference values are modified to account for frequency-dependent material properties, internal inductance, and surface roughness [Liang et al., 2006].

10.1.2 Generating a Tabular Dielectric Model

The real portion of the dielectric permittivity, as calculated with

$$\varepsilon'(\omega) \approx \varepsilon'_{\infty} + \frac{\Delta\varepsilon'}{m_2 - m_1}\frac{\ln(\omega_2/\omega)}{\ln(10)} \tag{6-30a}$$

is used to determine the frequency-dependent capacitance $C(\omega)$ by dividing $C(\omega_{\text{ref}})$ by the value of the dielectric permittivity calculated at the reference frequency $\varepsilon'(\omega_{\text{ref}})$ and multiplying by the frequency-dependent value using equation (10-1), where boldfold type indicates a matrix for the multiconductor case:

$$\begin{aligned} C(\omega) &= C(\omega_{\text{ref}})\frac{\varepsilon'(\omega)}{\varepsilon'(\omega_{\text{ref}})} \\ &\rightarrow \mathbf{C}(\omega) = \mathbf{C}(\omega_{\text{ref}})\varepsilon'(\omega)\varepsilon'^{-1}(\omega_{\text{ref}}) \end{aligned} \tag{10-1}$$

To implement equation (10-1), the effective dielectric permittivity must be calculated from the reference values. As detailed in equation (3-74), the effective permittivity can be calculated by dividing the capacitance of each value in $\mathbf{C}$ by the corresponding value in $\mathbf{C}_{\text{air}}$, which is calculated by setting the dielectric permittivity to unity ($\varepsilon_r = 1$).

$$\frac{C_{\varepsilon_{r,\text{eff}}}}{C_{\varepsilon_r=1}} = \varepsilon_{r,\text{eff}} \tag{3-74}$$

As described in Chapter 3, the values in $\mathbf{C}_{\text{air}}$ can be calculated directly from the inductance using equation (3-46), assuming that the magnetic permeability is unity ($\mu_r = 1$), which is almost always the case because copper is usually the metal of choice.

$$L = \frac{1}{c^2 C_{\varepsilon_r=1}} \tag{3-46}$$

Equations (3-74) and (3-46) lead to equation (10-2), which is the value of the dielectric permittivity at the reference frequency:

$$\varepsilon'(\omega_{\text{ref}}) = \mathbf{C}(\omega_{\text{ref}})\mathbf{C}_{\text{air}}^{-1} = c^2[\mathbf{C}(\omega_{\text{ref}})\mathbf{L}(\omega_{\text{ref}})] \tag{10-2}$$

The frequency-dependent loss tangent is calculated as

$$\varepsilon''(\omega) \approx \frac{\Delta\varepsilon'}{m_2 - m_1}\frac{-\pi/2}{\ln(10)} \tag{6-30b}$$

$$\tan|\delta| = \frac{\varepsilon''}{\varepsilon'} \tag{6-16b}$$

The definition of $G(\omega)$ in shown in equation (6-45):

$$G(\omega) = \tan\delta\ \omega C \tag{6-45}$$

The reference value of the loss tangent is calculated by rearranging equation (6-45):

$$\tan\delta(\omega_{\text{ref}}) = \frac{G(\omega_{\text{ref}})}{C(\omega_{\text{ref}})\omega_{\text{ref}}}$$

$$\rightarrow \mathbf{tan}\ \delta(\omega_{\text{ref}}) = \frac{\mathbf{G}(\omega_{\text{ref}})\mathbf{C}^{-1}(\omega_{\text{ref}})}{\omega_{\text{ref}}} \tag{10-3}$$

For a microstrip line, the values extracted would be the effective dielectric constant and effective loss tangent. This is an easy way to handle nonhomogeneous dielectrics, such as in a microstrip configuration. For a stripline, the values extracted would be equal to the bulk properties of the material, including any nonhomogenous structures such as the fiber weave.

Next, $\mathbf{G}(\omega_{\text{ref}})$ is scaled by the frequency-dependent loss tangent and permittivity, leading to (10-4), which is the frequency-dependent conductance of a transmission line.

$$G(\omega) = G(\omega_{\text{ref}})\frac{\varepsilon'(\omega)}{\varepsilon'(\omega_{\text{ref}})}\frac{\tan\delta(\omega)}{\tan\delta(\omega_{\text{ref}})}\frac{\omega}{\omega_{\text{ref}}}$$

$$\rightarrow \mathbf{G}(\omega) = [\mathbf{tan}\,\delta^{-1}(\omega_{\text{ref}})\,\boldsymbol{\varepsilon}'^{-1}(\omega_{\text{ref}})]\mathbf{G}(\omega_{\text{ref}})\boldsymbol{\varepsilon}'(\omega)\mathbf{tan}\ \delta(\omega)\frac{\omega}{\omega_{\text{ref}}} \tag{10-4}$$

Using this procedure, equations (10-1) and (10-4) can be used to create the tabular transmission-line parameters that represent a frequency-dependent causal dielectric model that is valid for the conditions described during the derivation of the infinite-pole model in Section 6.3.5. However, it should be noted that a transmission line using any dielectric model can be implemented using this technique, providing the frequency-dependent behavior of ε' and ε'' are known.

10.1.3 Generating a Tabular Conductor Model

When creating a frequency-dependent conductor model for a transmission line, Chapter 5 will remind the reader of the four effects that must be accounted for properly:

1. External inductance
2. Internal inductance
3. Dc resistance
4. Ac (skin effect) resistance

Quasistatic techniques, which are used by most commercial transmission-line calculators, provide reasonable values of the external inductance. Furthermore, many tools also provide realistic values of the skin effect and dc resistance as well as the internal inductance for ideal smooth conductors. Problems arise when creating models of realistic conductors, which are usually purposely roughened during the PCB manufacturing process to ensure adhesion of the metal traces to the dielectric substrate. Fortunately, the shortcomings of frequency-invariant quasistatic techniques can easily be overcome by using a tabular transmission-line model.

Similar to the tabular model for the dielectric outlined in Section 10.1.2, the easiest methodology is to use a conventional 2D transmission-line calculator that assumes perfectly smooth conductors, and modify the output to establish the proper frequency-dependent relationships between values $\mathbf{R}(\omega)$ and $\mathbf{L}(\omega)$. The output of the transmission-line calculator will provide the values of the reference matrices $\mathbf{R}(\omega_{\text{ref}})$ and $\mathbf{L}(\omega_{\text{ref}})$. The reference values for each entry in the matrix are then scaled according to

$$R(f) = \begin{cases} K_{\text{SR}}(f) R_s \sqrt{\dfrac{f}{f_{\text{ref}}}} & \text{when } \delta < t \\ R_{\text{dc}} & \text{when } \delta \geq t \end{cases} \tag{10-5}$$

$$L(f) = \begin{cases} L_{\text{external}} + \dfrac{R(f)}{2\pi f} & \text{when } \delta < t \\ L_{\text{external}} + \dfrac{R(f_{\delta=t})}{2\pi f_{\delta=t}} & \text{when } \delta \geq t \end{cases} \tag{10-6}$$

where the frequency-dependent surface roughness correction factor $K_{\text{SR}}(f)$ is calculated with one of the modeling methods presented in Chapter 5 [i.e., Hammerstad equation (5-48), the hemispherical model in equation (5-58), or the Huray model in equation (5-66)], $R_s\sqrt{f}$ is the skin effect resistance assuming perfectly smooth conductors, t is the conductor thickness, δ is the skin depth, $f_{\delta=t}$ is the frequency where the skin depth equals the thickness of the conductor, and f_{ref} is the reference frequency where the seed values are calculated [Hall et al., 2007].

If the 2D transmission-line calculation does not include the internal inductance, the value of the reference frequency (f_{ref}) is immaterial because all the current is assumed to reside on the surface of the conductor and the internal inductance is zero. However, if the calculation includes the internal inductance, as described in Section 5.2.3, the reference frequency should be high enough that most of the current is confined to a small layer near the surface, so that the internal inductance is minimized. This simplifies the problem so that the reference inductance can be considered to be purely external and the internal portion can simply be added. Note that even at relatively low frequencies, the skin depth in copper is small compared to the thickness of the trace. For example, at 1 GHz, the skin depth is 2 μm, meaning that for a typical PCB trace thickness of about 1 to 2 mils

(25 to 50 μm), most of the current is confined near the surface and the inductance is mostly external.

10.2 NONIDEAL RETURN PATHS

So far, we have covered many of the advanced topics associated with high-speed digital design. Each topic has been rigorously covered analytically, beginning with fundamental theory and ending with a practical method that can be used for design. Now it is time to focus on what is often considered the most ambiguous concept of high-speed design: nonideal return paths. Many of the effects detailed in this section are very difficult or impossible to model analytically, or even with circuit simulators. Although 3D electromagnetic simulators are capable of capturing full effects for small structures, they require too much computational power to simulate the effects reliably for an entire system. Subsequently, in this section we will focus less on specific modeling techniques and more on the general impact, physical mechanisms, and trends that nonideal return current paths have on signal integrity. As a general rule, great care should be taken to ensure that nonideal current return paths are minimized.

10.2.1 Path of Least Impedance

As discussed in Chapter 3, a signal on a transmission line propagates between the signal trace and the reference plane in the form of an electromagnetic wave. Consequently, the physical characteristics of the reference plane are just as important as that of the signal trace. A very common mistake, even for experienced designers, is to focus on providing a very clean and controlled signal trace with no thought whatsoever of the reference plane. Remember that any current injected into a system must return to the source. It will do so through the path of least impedance. Figure 10-1a depicts a CMOS output buffer driving a microstrip line. The currents shown represent the instantaneous values that occur when the driver switches from a low state to a high state. Just prior to the transition (time $t = 0^-$) the line is grounded through the NMOS. Immediately after the transaction (time $t = 0^+$) the buffer switches to a high state and current flows onto the line until it is charged up to V_{DD}. As the current propagates down the line, a mirror current is induced on the reference plane, which flows in the opposite direction as described in Section 3.2.3 and shown in Figure 3-8. To complete the loop, the current must find the path of least impedance, which in the case of Figure 10-1a is the voltage supply V_{DD}.

A nonideal return path occurs when (1) a discontinuity in the reference plane causes return current to diverge from the ideal path, or (2) the return current must flow through a region of increased impedance, such as bond wires, antipads in breakout areas, or socket pins. A few examples are shown in Figure 10-1b. When a physical discontinuity exists in the return path that forces the current to diverge from the ideal path, the total area of the current loop is increased. An

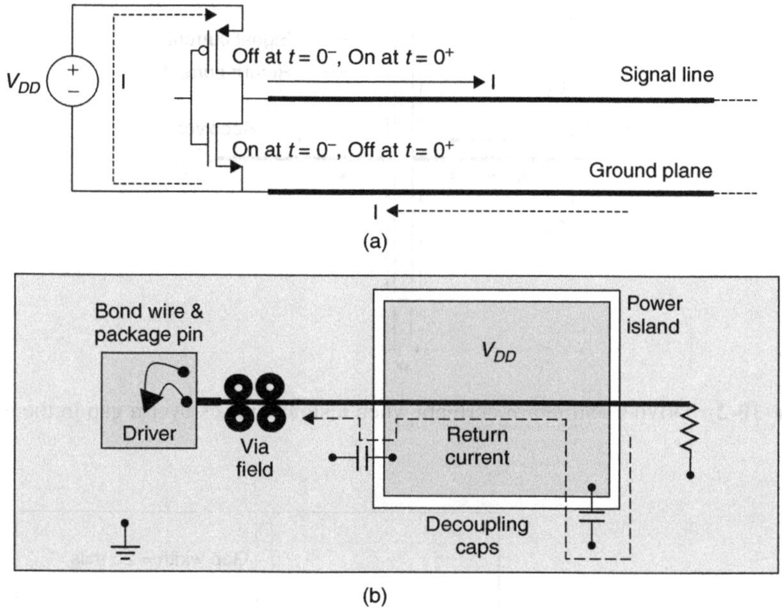

Figure 10-1 (a) Example of an ideal return current path of a ground-referenced transmission line driven by a CMOS buffer; (b) examples of common nonideal return paths.

increase in the loop area leads to an increase in inductance, which degrades the signal integrity. Subsequently, the most fundamental effect of a discontinuity in the return path is an effective increase in the series inductance with a magnitude that depends on the distance the current must diverge. When a high-impedance region such as a bond wire or inductive socket pin occurs in the return current path, it is similar to having the discontinuity in the signal path.

10.2.2 Transmission Line Routed over a Gap in the Reference Plane

To demonstrate the effect of a nonideal return current path, refer to Figure 10-2, which depicts a microstrip line traversing a gap in the reference plane. This is a convenient return path discontinuity to analyze because it is a simple structure, the return current path is well understood, and it has similar trends in the more complicated structures. As the signal current travels down the transmission line, the return current is induced on the reference plane. When the signal reaches the gap, a small portion of the return current propagates across the gap through the gap capacitance, and the other portion is forced to travel around the gap. The extra path length of the return current causes the total current loop area to increase, which increases the inductance. This means that from the driver's point of view, the gap looks inductive. To demonstrate this, consider Figure 10-3, which is a TDR measurement of a 65-Ω microstrip transmission line traversing a 25-mil

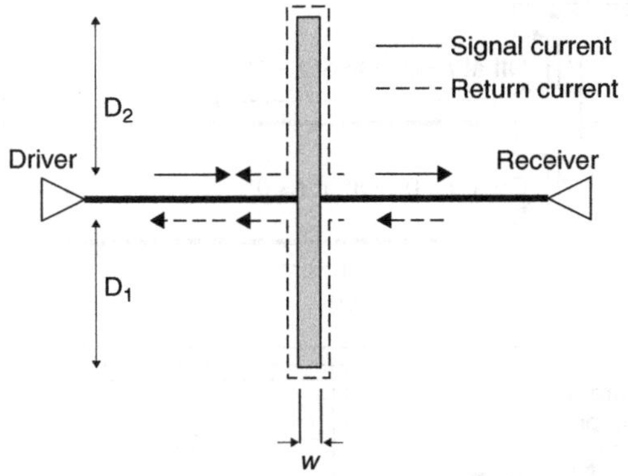

Figure 10-2 Driving and return currents when a signal passes over a gap in the ground plane.

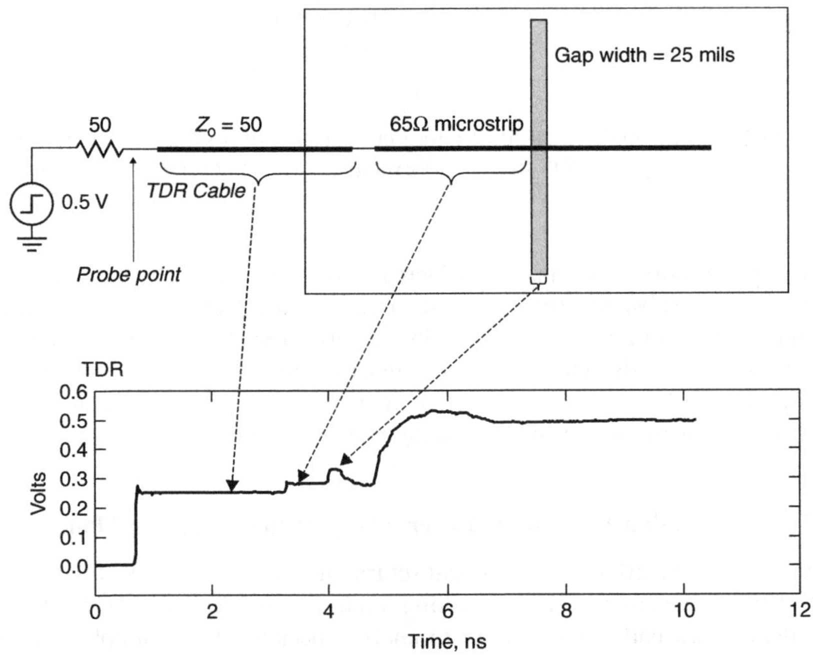

Figure 10-3 Measured TDR profile of a microstrip traversing a gap in the reference plane, showing the inductive nature of the nonideal return path.

gap in the reference plane. Notice how the gap appears inductive in the TDR profile.

The receiver waveform will be filtered by the low-pass nature of an inductor, and a portion of the signal will be delayed due to the finite time it takes the return current to propagate around the gap. A simulation of how a nonideal

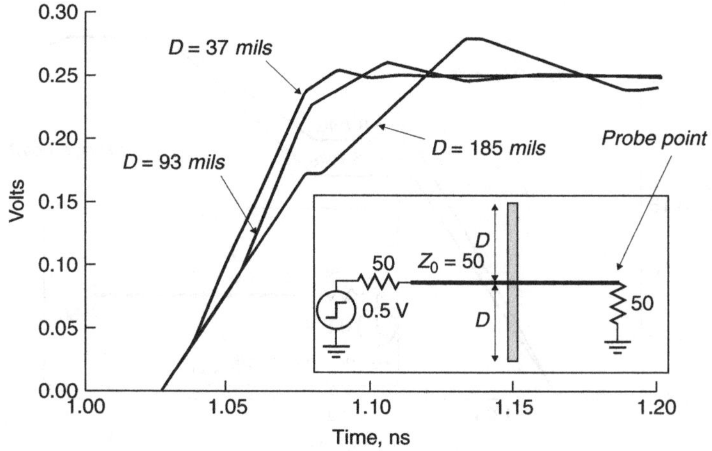

Figure 10-4 Simulated step response of the waveform at the receiver of a microstrip traversing a gap in the reference plane, showing how different return path divergent lengths D affect the waveform received.

return path can affect the receiving waveform is shown in Figure 10-4 using a loss-free transmission line. If the distance of the current return path ($2D$) is small compared to the edge rate, the gap will simply look like a series inductance in the middle of the transmission line. The extra inductance will filter out some of the high-frequency components of the signal, degrade the edge rate, and round the corners. When the electrical length of the return path becomes longer than the rise or fall times, however, a ledge will appear in the waveform. The length of the ledge (in time) will be dependent on the distance the return current must travel around the discontinuity [Hall et al., 2000].

Since the width of the gap will govern the bridging capacitance and thus the portion of current shunted across the gap, the height of the ledge will be dependent on the gap width, as shown in Figure 10-5. The larger the gap width, the less the capacitive coupling and the lower the height of the ledge.

Equivalent Circuit An equivalent circuit of a transmission line traversing a slot in the reference plane can be surmised by careful observation of the behavior described earlier. First, consider the measured TDR response shown in Figure 10-3. At the point where the transmission line crosses the slot, the TDR shows a voltage spike that represents the high-impedance, or inductive, section of the interconnect. As described earlier, the voltage spike is a function of the distance that the return current must diverge. Figure 10-6 shows how the return current flows parallel to the slot until it reaches a low-impedance pathway, where it can complete the loop (such as a short or a bridging capacitor). When current flows on one side of the slot, it induces a mirror current on the opposite side, an electric field is established, and a transmission line is formed. This type of transmission line, called a slotline, is shown in Figure 3-2. Therefore, the equivalent circuit must contain a transmission line to represent the microstrip, a

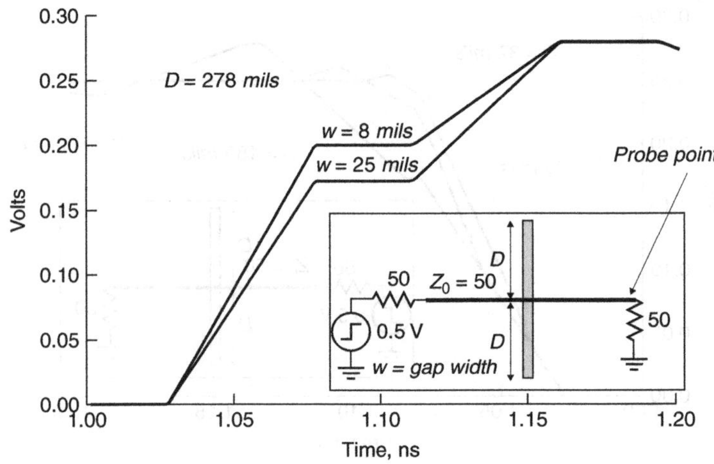

Figure 10-5 Simulated step response, showing how the height of the ledge is dependent on the gap width.

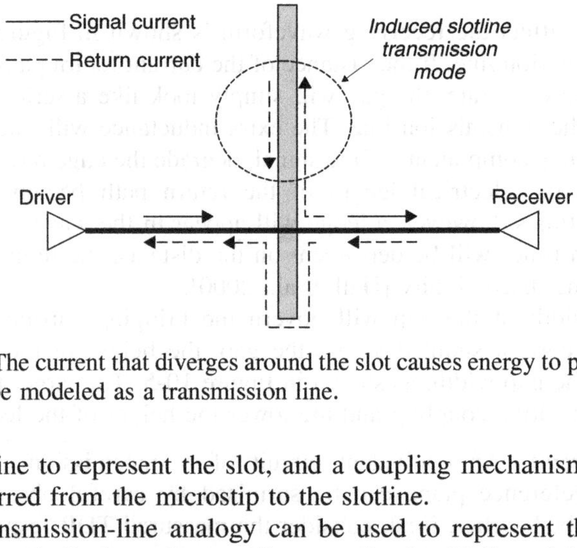

Figure 10-6 The current that diverges around the slot causes energy to propagate, allowing the slot to be modeled as a transmission line.

transmission line to represent the slot, and a coupling mechanism to model the energy transferred from the microstip to the slotline.

Since a transmission-line analogy can be used to represent the slot instead of discrete L and C, the voltage spike seen in the TDR and ledge height at the receiver can be calculated by determining how much of the total current diverges around gap using a simple current divider. Consider the transmission line traversing a slot shown in Figure 10-6, where the driving transmission line is perfectly terminated in its characteristic impedance to simplify the analyses. The current flowing around the gap is dependent on the slotline impedance and is calculated with a current divider as in

$$i_{\text{gap}} = i_{\text{drive}} \frac{2Z_0}{2Z_0 + \frac{1}{2}Z_{\text{gap}}} \tag{10-7}$$

where Z_0 is the impedance of the transmission line, Z_{gap} the impedance of the slotline, and i_{drive} the current launched onto the transmission line:

$$i_{drive} = \frac{v_i}{Z_0} = \frac{v[Z_0/(Z_0 + R_s)]}{Z_0}$$

On both sides of the slot the desired current path exists that is perfectly terminated. The current that does not jump the slot will flow around it, which represents the alternative return path. Since the transmission line is routed over the center section of the slot, current will flow in both directions, and the effective gap impedance will be $Z_{gap}/2$.

The voltage spike seen in the TDR waveform is calculated using the portion of the current that jumped the gap:

$$v_{spike} = (i_{drive} - i_{gap})Z_0 \tag{10-8}$$

The width of the voltage spike is equal to twice the delay of the distance the current must diverge, which can be calculated from the propagation delay of the slotline. Although some formulas exist to calculate the impedance and propagation delay of a simple slotline, the stackup geometry of real boards varies significantly, so an electromagnetic simulator should be used.

The ledge at the receiver is dependent on the current flowing in the gap. The ledge will stay constant for a duration that equals twice the delay of the current return path divergence:

$$v_{ledge} = i_{gap} Z_0 \tag{10-9}$$

An equivalent circuit that mimics the behavior described above can be constructed as shown in Figure 10-7. The top circuit line represents the sourcing transmission line that crosses a gap. The voltage-controlled voltage source (VCVS) represents the physical location where the transmission line crosses the slot. The bottom circuit is a transmission line that represents the slot and the current-controlled current source (CCCS) is placed at the physical location on the slot where the transmission line crosses. The current in the VCVS is mirrored into the CCCS on the slotline, which represents the current diverging around the gap. In turn, the current sourced into the gap produces a voltage, which is mirrored back into the VCVS and accurately predicts both the inductive voltage spike seen at the driver and the ledge seen at the receiver.

Although this simulation method is quite simple, it works very well and can be used to predict the signal integrity of one or more lines crossing a slot in a reference plane. An example of the model accuracy compared to a TDR measurement is shown in Figure 10-8.

Example 10-1 Calculate the step response of a 50-Ω microstrip transmission line traversing from a ground plane to a V_{DD} island as shown in Figure 10-9. Assume that the gap acts like a slotline with an impedance of 90 Ω. Since it is

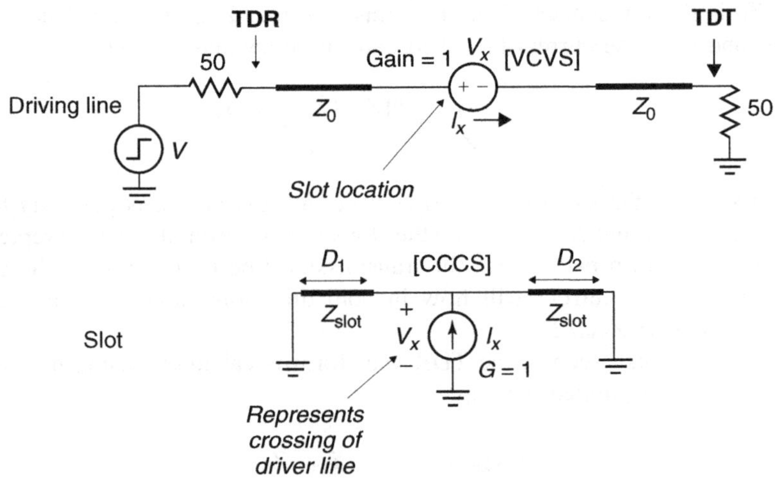

Figure 10-7 Equivalent circuit of a transmission line routed over a slot in the reference plane.

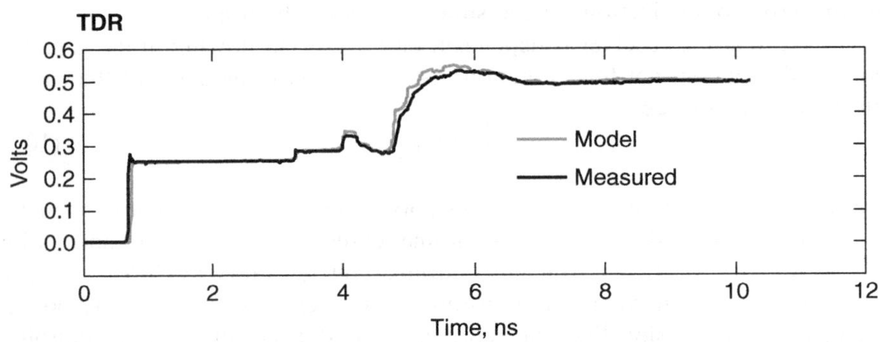

Figure 10-8 Comparison of a slot model compared to a measured response on a test board.

generally not a good idea to short V_{DD} to ground, 1-nF decoupling capacitors are used to create a high-frequency low-impedance path for the return current.

SOLUTION

Step 1: Estimate the electrical delays of the microstrip and the slotline assuming a board dielectric permittivity of 4. Since both the slotline and the microstrip will have part of the electric fields in the air and part in the dielectric, we estimate that the effective relative dielectric permittivity for both is $\varepsilon_{\text{eff}} = 3.15$. This value can be calculated rigorously using the techniques outlined in Chapter 3,

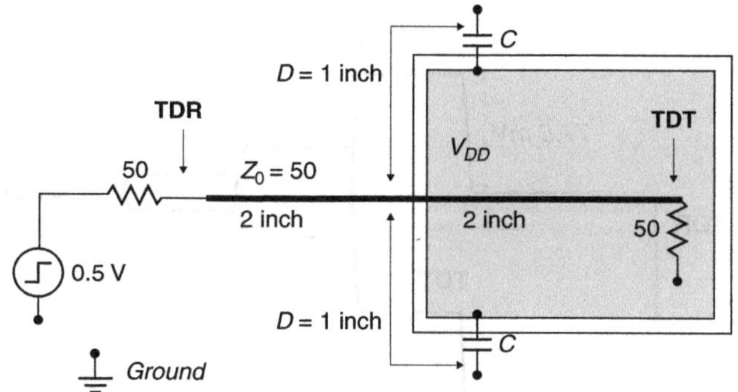

Figure 10-9 Transmission line routed onto a floating power island—a common occurrence.

or an electromagnetic simulator can be used. However, for the purposes of this example, the estimation is adequate. The propagation delay is therefore calculated for both the microstrip and slotline:

$$t_d = \frac{\sqrt{\varepsilon_{\text{eff}}}}{c} = \frac{\sqrt{3.15}}{3 \times 10^8 \text{ m/s}} \Rightarrow 150 \times 10^{-12} \text{ s/in.}$$

Step 2: Create the equivalent circuit. Using the propagation delay, the electrical length of each segment is calculated. The total length of the slot is the segment between the decoupling capacitors. Since the capacitors are large, they will look like a short for high-frequency return current. In the equivalent circuit, they are placed at the end of the slot. The equivalent circuit is shown in Figure 10-10.

Step 3: Simulate the circuit model. The step response is shown in Figure 10-11. The TDR is the waveform seen at the driver and the TDT is the waveform seen at the termination resistor at the far end of the line (as labeled in Figure 10-10).

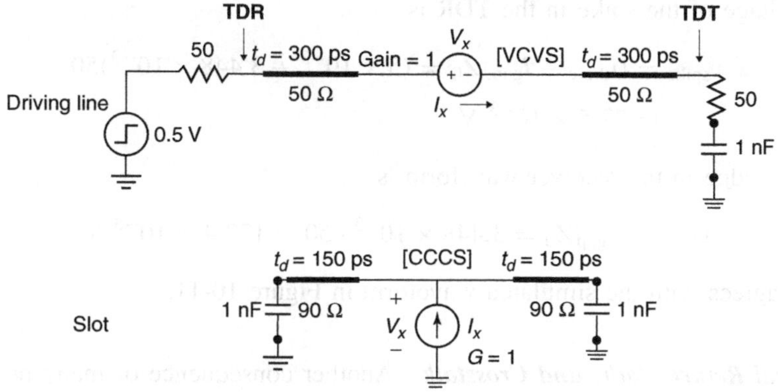

Figure 10-10 Equivalent circuit of the structure shown in Figure 10-9.

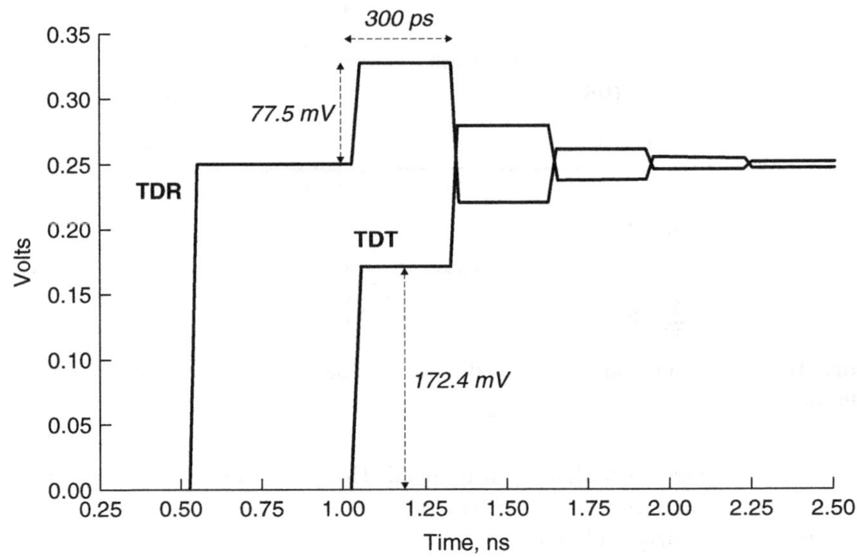

Figure 10-11 Simulated waveform of a transmission line routed onto a floating power island.

The width of the voltage spike and the ledge should be twice the return-path divergence length, which is $2 \cdot 150$ ps $= 300$ ps. The values of the initial voltage step in the TDR and the ledge in the TDT can be calculated to double-check the simulation using equations (10-7) through (10-9):

$$i_{\text{drive}} = \frac{v_i}{Z_0} = \frac{v[Z_0/(Z_0 + R_s)]}{Z_0} = \frac{0.5[50/(50+50)]}{50} = 5\times 10^{-3} \text{ A}$$

$$i_{\text{gap}} = i_{\text{drive}} \frac{2Z_0}{2Z_0 + \frac{1}{2}Z_{\text{gap}}} = 5 \times 10^{-3} \left(\frac{100}{100+45} \right) = 3.448 \times 10^{-3} \text{ A}$$

The voltage of the spike in the TDR is

$$v_{\text{spike}} = (i_{\text{drive}} - i_{\text{gap}})Z_0 = (5 \times 10^{-3} - 3.448 \times 10^{-3})50$$
$$= 77.5 \times 10^{-3} \text{ V}$$

and the ledge in the receiver waveform is

$$v_{\text{ledge}} = i_{\text{gap}} Z_0 = 3.448 \times 10^{-3} \cdot 50 = 172.4 \times 10^{-3} \text{ V}$$

which agrees with the simulated waveform in Figure 10-11.

Nonideal Return Paths and Crosstalk Another consequence of many nonideal return paths is a dramatic increase in crosstalk. Conventional crosstalk is caused by mutual inductance and capacitance. For the case of a transmission line traversing a split in the reference plane, the coupling mechanism is the slot itself. A voltage is established between each side of the slot and propagates via the slotline transmission mechanism from the sourcing line to the victim, where it couples onto the line with much more efficiency than conventional crosstalk.

Figure 10-12a shows an example of two transmission lines routed over a slot in the ground plane, and Figure 10-12b shows a measured response of a test board with a pair of 65-Ω microstrip transmission lines 1.4 in. apart traversing a 25-mil gap. Note that the coupling coefficient is approximately 20% (0.05 mV/0.25 mV), which is much higher than what would be expected between the two microstrips routed over a solid reference plane.

The structure shown in Figure 10-12a can be modeled similar to the way it was done for the single line crossing a slot, as shown in Figure 10-13. The top circuit line represents the sourcing transmission line that crosses a gap. The voltage-controlled voltage source (VCVS) represents the physical location where the transmission line crosses the slot. The middle circuit is a transmission line that represents the slot and the current-controlled current sources (CCCSs) are placed at the physical location on the slot where the transmission lines cross. The current flowing through the VCVS on the sourcing line is mirrored into the CCCS on the slotline, which represents the current diverging around the gap. The same circuit is duplicated for the victim transmission line, which effectively models the coupling by means of the slotline. Figure 10-14 shows a simulated example of loss-free slotline coupled crosstalk, where $D = 0.462$ in. and $S = 1.0$ in. with 50-Ω microstrips. Note that the coupled voltage is about 22% of the voltage

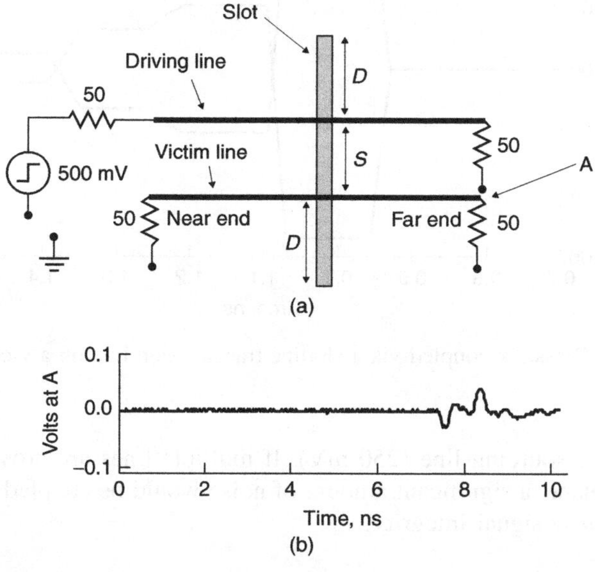

Figure 10-12 (a) Two transmission lines passing over a slot in the reference plane used to develop an equivalent circuit; (b) measurement of coupled voltage when $S = 1.4$ in., slot *width* = 25 *mils*, and 65 -Ω microstrip transmission lines on FR4.

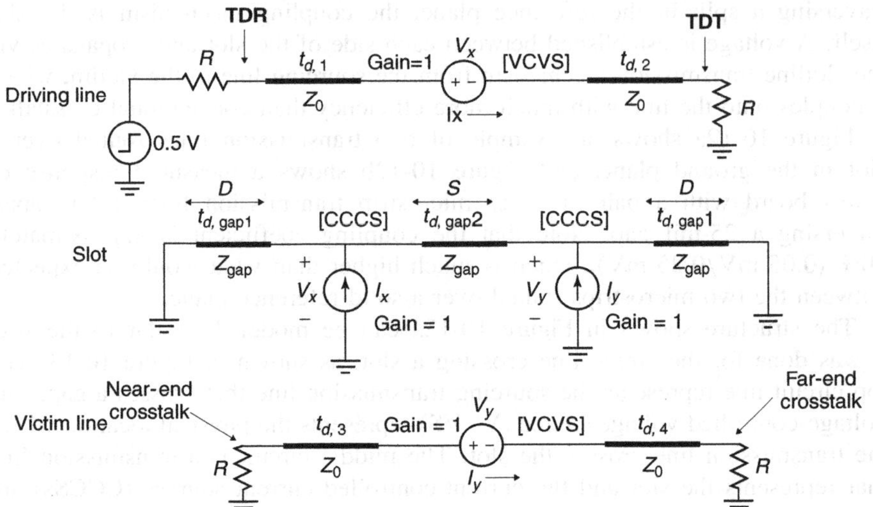

Figure 10-13 Equivalent circuit of two transmission lines crossing over a slot in the reference plane.

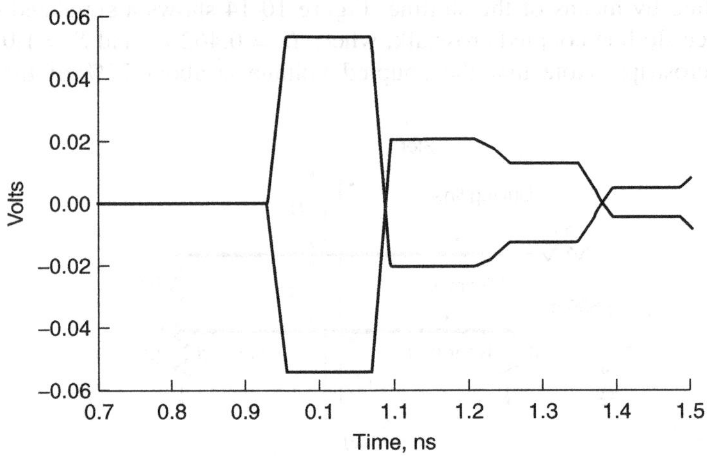

Figure 10-14 Crosstalk coupled via a slotline transmission line on a victim line 1.0 in. away.

launched on the sourcing line (250 mV). If multiple lines are crossing a gap in the reference plane, a significant amount of noise would be coupled onto the bus, which can destroy signal integrity.

Differential Differential signaling is much more immune to the harmful effects of nonideal return paths because the adjacent transmission line provides a relatively low-impedance path for the current to return. From a mathematical point of

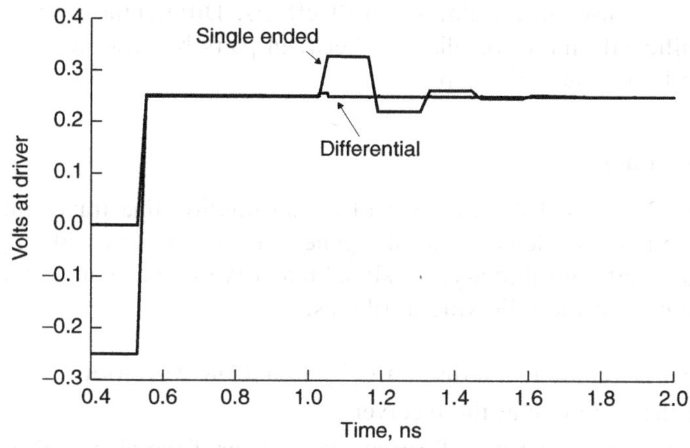

Figure 10-15 TDR of a single-ended and a differential transmission line traversing a gap in a reference plane, showing that differential routing significantly reduces the negative impact of a nonideal return path. Note the decreased inductive spike seen in the differential response.

view, the virtual reference plane that exists between the conductors in a differential pair can be thought of as the return path, as shown in Figure 7-7 and discussed in Section 7.3. The increased immunity of a differential signal to a nonideal return path is demonstrated in Figure 10-15, which shows both a single-ended and a differential TDR waveform for a signal traversing a 25-mil-wide gap in the reference plane. Note that the inductive spike is much smaller for the differential case. The step response seen at the receiver is shown in Figure 10-16, where the

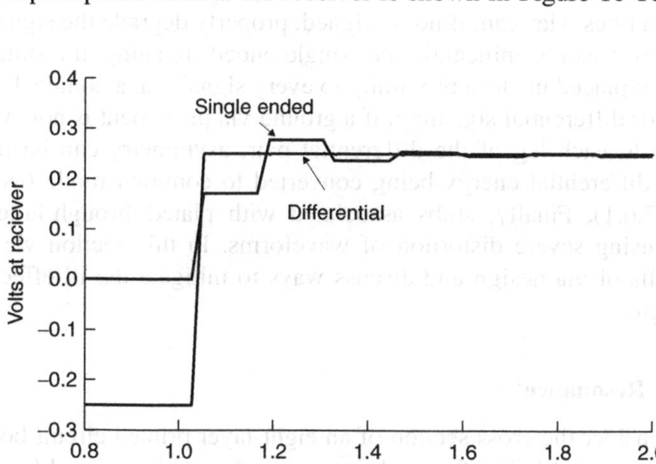

Figure 10-16 Step response at the receiver of a single-ended and a differential transmission line traversing a gap in the reference plane, showing that differential routing significantly reduces the negative impact of a nonideal return path.

differential response shows almost no ill effects. Differential routing will also reduce significantly the crosstalk seen between pairs because much less current is forced to flow around the gap.

10.2.3 Summary

Section 10.2.2 we used the example of a transmission line traversing a gap in the reference plane to demonstrate the general behavior seen with any nonideal current return path. In summary, the signal integrity of a bus with nonideal return path(s) will exhibit the following attributes:

1. Appear as an inductive discontinuity in a TDR waveform
2. Slow the edge rate at the receiver
3. Severely distort the waveform at the receiver if the electrical delay of the current divergence path is longer than the rise and fall times
4. Create unwanted transmission paths that can increase coupling or loss significantly

Differential signaling is much more immune to the effects of a nonideal return path because the adjacent signal (or the virtual reference plane) provides a continuous reference for the fields. Hall et al. [2000] discuss numerous other nonideal current return paths.

10.3 VIAS

For high data rates, vias can, if not designed, properly degrade the signal integrity of a binary bit stream significantly. For single-ended signaling, if a suitable return path via is not placed in close proximity to every signal via, a nonideal return path is created. For differential signaling, if a ground via placement is not symmetrical with respect to each leg of the differential pair, asymmetry can be introduced, resulting in differential energy being converted to common mode (as described in Section 7.6.1). Finally, stubs associated with plated-through-hole vias can resonate, causing severe distortion of waveforms. In this section we cover the largest pitfalls of via design and discuss ways to mitigate the ill effects through proper design.

10.3.1 Via Resonance†

To begin, consider the cross section of an eight-layer printed circuit board (PCB) shown in Figure 10-17. The figure shows a signal routed on signal layer 1 transitioning through a plated-through-hole (PTH) via to signal layer 4. The ground via

†The authors would like to thank Guy Barnes of Ansoft Corporation for performing the 3D simulations used in this section. His technical knowledge, modeling expertise, and patience were invaluable.

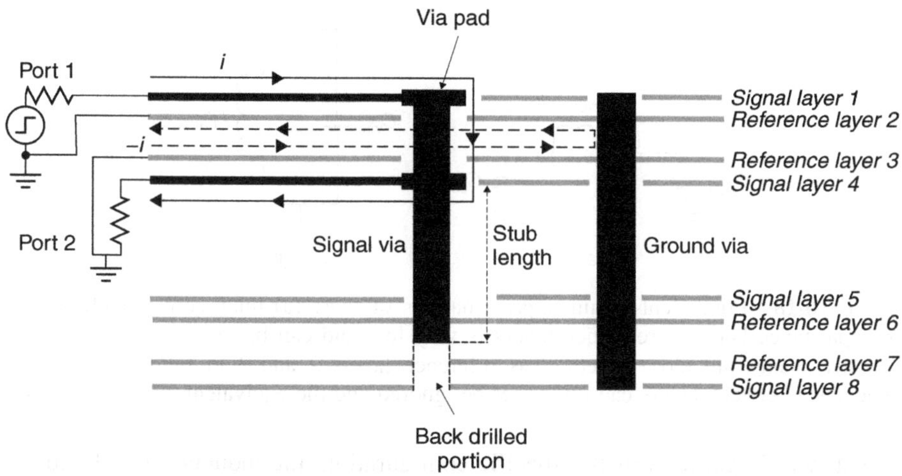

Figure 10-17 Cross section a a backdrilled PTH via with a stub showing the nonideal current return path through the ground via.

is included to provide a low-impedance path for the return current. Standard PTH vias are constructed by drilling a hole in the board and plating the walls with metal. A consequence of this manufacturing process is a "stub" that hangs off the via for all cases except when the signal transitions thorough the entire board (layers 1 to 8 in Figure 10-17). As will be shown later, the via stub has negative effects on the signal integrity. A common design practice is to shorten the length of the stub with precision depth backdrilling, as also depicted in Figure 10-17.

If a current i is driven into port 1 in Figure 10-17, it will propagate down the transmission line on signal layer 1, pass through the via down to signal layer 4, and terminate at port 2. The return current $-i$ will be mirrored on reference plane 2. Since the nearest return path from port 1 to port 2 is through the ground via, the return current must diverge away from the ideal path. The magnitude of the via inductance is therefore a function of the length of the via transition from signal layer 1 to signal layer 2 and the distance between the signal via and the ground via. The capacitance of the via will be dependent on the area of the via pads, the distance between the via pads and the closest reference layer, the capacitance of the via barrel to the adjacent planes it is passing through, and the capacitance of the stub. These observations make it possible to deduce two different forms of an equivalent circuit.

Figure 10-18a shows the equivalent circuit of a via for the case where the stub length is zero and reference layers 2 and 3 are very far apart. In this scenario, which is almost always the case with standard four-layer PCBs, the via capacitance is dominated by the pads, so a very simple pi model can be used. Figure 10-18b shows an equivalent circuit where reference layers 2 and 3 are close in proximity and a stub may or may not exist. In this case, the barrel and stub capacitances are no longer negligible and are simply combined with the pad

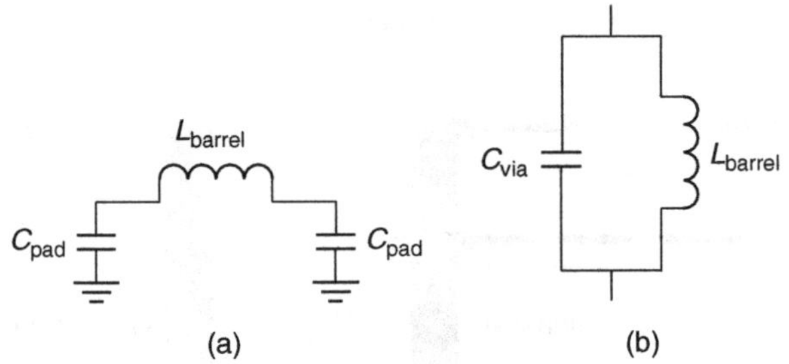

Figure 10-18 Equivalent-circuit approximations of vias: (a) thick-core model, where the capacitance between reference layers is very low and can be ignored; (b) when the reference planes are close together, as reference layers 2 and 3 in Figure 10-17, the capacitance between layers can no longer be ignored and the equivalent circuit changes.

capacitance. Regardless of the specific configuration, the inductance and capacitance of the via structures forms a tank circuit that will resonate at its natural frequency:

$$f_0 = \frac{1}{2\pi\sqrt{L_{\text{barrel}}C_{\text{via}}}} \tag{10-10}$$

For long via stubs, the resonant frequency can easily be low enough to influence significantly the signal integrity of the digital waveforms.

To demonstrate how vias behave at high frequencies, consider Figure 10-19, which shows the simulated insertion loss of a plated-through-hole via for a structure similar to Figure 10-17 with different stub lengths. The simulations were

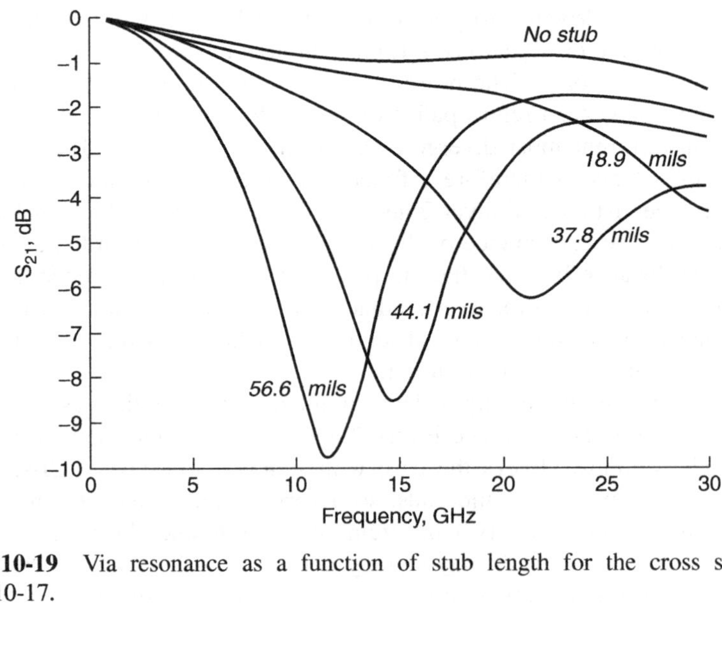

Figure 10-19 Via resonance as a function of stub length for the cross section in Figure 10-17.

TABLE 10-1. HFSS-Predicted Resonance Compared to Tank Circuit Approximation

Stub Length (mils)	HFSS-Predicted Resonance (GHz)	Q3D-Predicted L (nH)	Q3D-Predicted C (pF)	Tank Resonance (GHz)
56.6	12	0.139	0.636	16.9
44.1	15.5	0.141	0.586	17.5
37.8	22	0.141	0.419	20.7

performed using Ansoft's HFSS 3D electromagnetic simulator with loss-free dielectrics and perfect conducting metals. The longer stub lengths exhibit more capacitance and therefore resonate at lower frequencies.

The problem with simple circuit models is that it can be very difficult to calculate the via capacitance and inductance for all but the simplest configurations. Consequently, it is necessary to extract the parasitic inductance and capacitance values from measurements or simulate them with a 3D electromagnetic simulator, such as Ansoft's Q3D. Table 10-1 shows the extracted parasitic values using Q3D and the tank circuit resonance calculated with equation (10-10) for the structures simulated in Figure 10-19.

Note that the tank circuit only approximates the actual resonance within a few gigahertz. The discrepancy between the HFSS and tank circuit results is due to the electrical delay of the physical geometry of the structure. Although the extracted inductance and capacitance are true values, a simple tank circuit approximation does not account for the physical delay of the return current diverging to the ground via. As the ground via is moved farther away, the inductance will increase. However, the peaks seen in the S-parameters created from the 3D simulation are due to both the resonance between the inductance and capacitance and the phase delay of the return current, so the resonant points will change accordingly.

The approximation of the resonance using a tank circuit approach is extremely useful and accurate enough for most practical purposes. If a design has via stubs that are resonating anywhere near the frequency of operations, the primary concern is to fix the design by decreasing the length of the stubs using layout rules, microvias, or backdrilling.

10.3.2 Plane Radiation Losses

At the via stub resonance frequency, the stubs act like small antennas and "radiate" energy into the dielectric layers sandwiched between reference planes. For example, the cross section shown in Figure 10-17 will radiate a significant amount of power into the dielectric layers confined by the reference planes, where it will propagate in TEM parallel-plate waveguide mode. The energy transfer into the parallel-plate mode will be maximum at the resonant frequency. A simulation of the electric and magnetic fields between reference layers 2 and 3 in Figure 10-17 at the resonant frequency was performed in Ansoft's HFSS. The

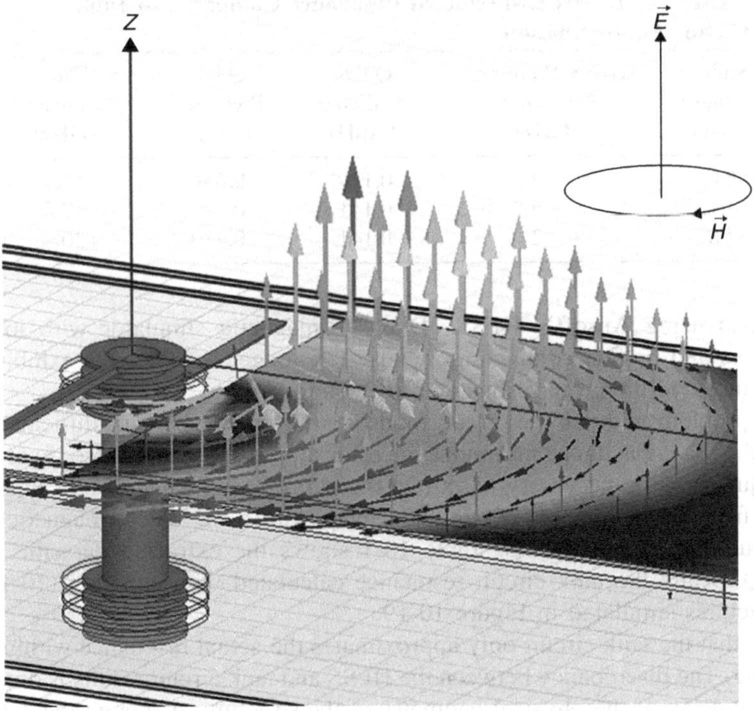

Figure 10-20 When a via does not have adequate current return paths, energy can radiate into the layers between reference planes in the TEM parallel-plate waveguide mode. Via stubs exacerbate this effect.

fields are shown in Figure 10-20. Note how the electric and magnetic fields are orthogonal to each other (and therefore TEM) and are propagating outward in a circular pattern. From the signal's point of view, the energy being sourced into the plane will look like loss. Additionally, as the energy propagates through the planes, it will be picked up by other vias similar to antennas, which can increase crosstalk dramatically. Note that this phenomenon is a direct result of a nonideal return path and is very similar to the slot example described in Section 10.2.2. The return current must diverge from the ideal path, so it can flow through the ground via, increasing the loop area, leading to an increased inductance. In Section 10.2 it was shown that energy was transferred to the slotline, where it propagated and coupled to other transmission lines crossing the gap. Similarly, in this case, the energy is transferred to the parallel-plate mode, where it propagates and could be picked up by other vias, resulting in increased crosstalk.

Figure 10-21 shows the simulated S-parameters of a signal propagating through a via with a 37.8-mil stub. Since both the dielectric and the conductors are loss-free, the percentage of the power radiated into the plane (P_{plane}) is calculated by rearranging equation (9-27) to yield

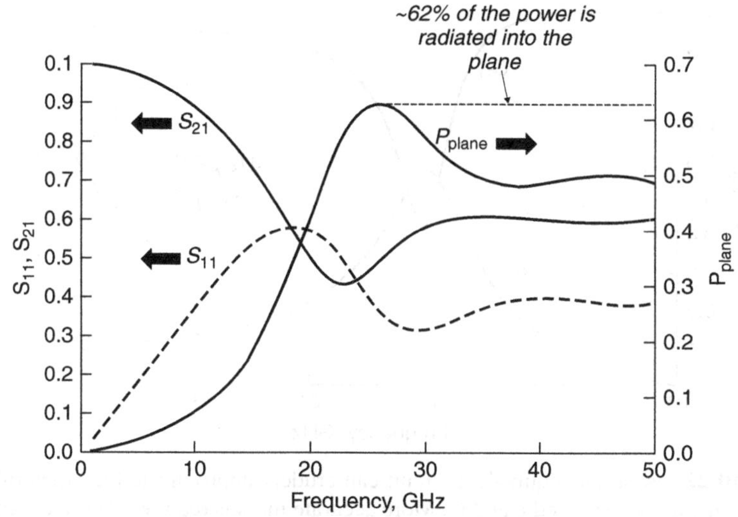

Figure 10-21 At resonance, a large percentage of the total power can be radiated into the layers between reference planes. This example shows the S-parameters and the power radiated into the plane for a 37.8-mil via stub, as shown in Figure 10-17.

$$1 - S_{11}S_{11}^* - S_{21}S_{21}^* = \frac{P_{\text{loss}}}{P_{\text{incident}}} \tag{10-11a}$$

where

$$P_{\text{plane}} = \frac{P_{\text{loss}}}{P_{\text{incident}}} \tag{10-11b}$$

Note that as the frequency nears resonance, the energy transfer into the plane is most efficient.

Figure 10-22 shows an equivalent circuit and its response. The resistor represents the losses due to energy being sourced into the parallel-plate mode between reference layers 2 and 3. The L and C values were extracted from Ansoft Q3D, and the resistor value was varied until the value of S_{21} matched the HFSS result. Note that this circuit is only a crude approximation of the via behavior. The only methodology that can model this effect accurately is to use a 3D field simulator such as Ansoft's HFSS.

10.3.3 Parallel-Plate Waveguide

To promote the understanding of how the energy propagates between reference layers after it has been sourced from a resonant via structure, it is useful to derive the equations that govern the parallel-plate waveguide. A parallel-plate waveguide is formed from two flat plates, as shown in Figure 10-23. For this analysis it is assumed that the width w is much larger than the vertical separation between the plates h, so that any fringing fields at the edges can be ignored. Since

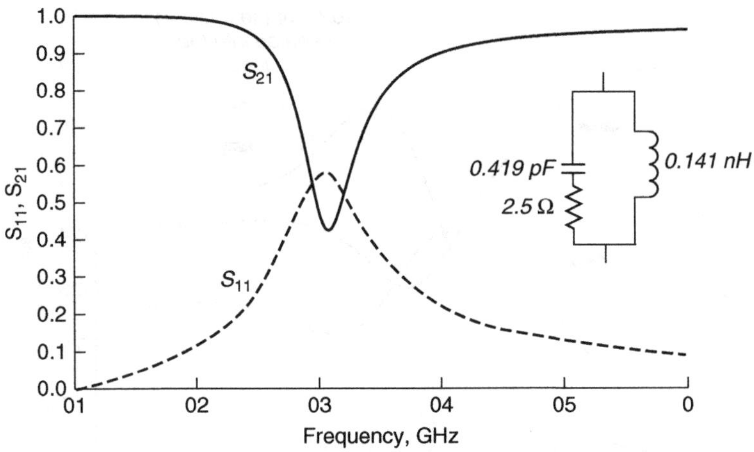

Figure 10-22 A simple equivalent circuit can crudely approximate the effect of the via resonance using the extracted L and C. More accurate models require a 3D electromagnetic simulator.

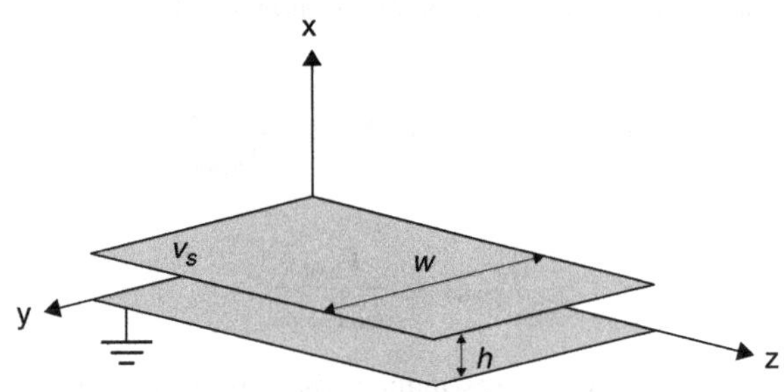

Figure 10-23 Dimension used to derive the electric and magnetic fields of a parallel-plate waveguide.

the propagation mode discussed in Section 10.3.2 is TEM, that mode is the focus of this derivation.

As was done in Section 3.4.3 for the microstrip transmission line, we begin with Laplace's equation:

$$\nabla^2\Phi = \frac{\partial^2\Phi}{\partial x^2} + \frac{\partial^2\Phi}{\partial y^2} = 0 \tag{3-48}$$

The boundary conditions for this problem are given by

$$\Phi(x = 0, y) = 0 \tag{10-12a}$$

$$\Phi(x = h, y) = v_s \tag{10-12b}$$

Since it is assumed that $w \gg h$, the variation in the y-direction is assumed constant. Therefore, the variation with y drops out of (3-48), and the general solution is given by

$$\Phi(x, y) = C_1 + C_2 x \tag{10-13}$$

The electric field is calculated using the electrostatic potential shown as discussed in Section 2.4.1.

$$\vec{E} = -\nabla\Phi(x, y) = -\vec{a}_x \frac{v_s}{h} \tag{10-14}$$

Therefore, the electric field propagating in the z-direction is

$$\vec{E}(x, y, z) = \vec{a}_x E e^{-j\beta z} = -\vec{a}_x \frac{v_s}{h} e^{-j\beta z} \tag{10-15}$$

where the propagation constant is as defined in Section 2.3.4:

$$\beta = 2\pi f\sqrt{\mu\varepsilon} = \omega\sqrt{\mu\varepsilon} \qquad \text{rad/m} \tag{10-16}$$

The magnetic field is calculated by dividing (10-15) by the intrinsic impedance of the waveguide as described in Section 2.3.4.

$$\eta \equiv \sqrt{\frac{\mu}{\varepsilon}} \tag{10-17}$$

$$\vec{H}(x, y, z) = \vec{a}_y \frac{1}{\eta}\frac{v_s}{h} e^{-j\beta z} \tag{10-18}$$

The propagation velocity is calculated from

$$v_p = \frac{1}{\sqrt{\mu_r \mu_0 \varepsilon_r \varepsilon_0}} = \frac{c}{\sqrt{\mu_r \varepsilon_r}} \qquad \text{m/s} \tag{2-52}$$

where μ_r is almost always unity for practical digital designs.

REFERENCES

Hall, Stephen, Garrett Hall, and James McCall, 2000, *High-Speed Digital System Design*, Wiley-Interscience, New York.

Hall, Stephen, Steven G. Pytel, Paul G. Huray, Daniel Hua, Anusha Moonshiram, Gary A. Brist, and Edin Sijercic, 2007, Multi-GHz causal transmission line modeling using a 3-D hemispherical surface roughness approach, *IEEE Transactions on Microwave Theory and Techniques*, vol. 55, no. 12, Dec.

Liang, Tao, Stephen Hall, Howard Heck, and Gary Brist, 2006, A practical method for modeling PCB transmission lines with conductor surface roughness and wideband dielectric properties, presented at IEEE MTT-S International, June, *Microwave Symposium Digest*, pp. 1780–1783.

PROBLEMS

10-1 For a transmission line in a bus specified to operate at 10 Gb/s with 50-ps rise and fall times, how high in frequency should the tabular model be constructed?

10-2 Derive an analytical formula to approximate the inductance and capacitance of a via. (*Hint:* Use the transmission-line techniques detained in Chapter 3.)

10-3 Derive the S-parameters directly from a tabular RLCG model. (*Hint:* Remember the telegrapher's equations?)

10-4 Derive the characteristic impedance for a parallel-plate waveguide with a width of w.

10-5 Create a model to estimate the crosstalk between two differential pairs traversing a slot in the ground plane.

10-6 For an 8-bit-wide bus crossing a slot in the ground plane, what techniques can be used to minimize the nonideal return paths? Create a model for each option and simulate it in SPICE to demonstrate.

10-7 Can the slot be used as a bus? If so, what would the architecture and topology look like? What would the advantages of this type of bus be? Create a model of the bus and simulate the pulse response.

10-8 Approximate the voltage across a via 1 in. away if E and H are launched in the TEM mode. (*Hint:* $E \cdot dl$.)

11

I/O CIRCUITS AND MODELS

Advanced Signal Integrity for High-Speed Digital Design, By Stephen H. Hall and Howard L. Heck

So far we have discussed the behavior of high-speed interconnects and have provided techniques for analyzing and modeling the key physical phenomena that affect signal quality at multi-Gb/s data rates. To fully analyze and understand the behavior of high-speed signaling links, we must include the I/O circuits that transmit and receive the digital data. Design of high-performance links demands that the circuits and interconnects be jointly optimized as a unified system. To do so successfully, the signal integrity engineer must be able to communicate with the circuit designer.

In this chapter we describe the operation and modeling of contemporary high-speed I/O circuits, including transmitters, receivers, and on-die terminations. We do not attempt to provide a complete treatment on how to design high-speed I/O; instead, we wish to give the SI engineer insight into the behavior and sensitivities of modern transceivers. This insight is fundamental to developing a sufficient understanding of the interactions between I/O circuits and physical interconnects in order to optimize a signaling system. In this chapter we identify the design parameters of the circuits for use in analyzing and optimizing designs, and describe techniques for modeling I/O circuits. In addition, we introduce the Bergeron diagram, a useful tool analyzing the time-domain behavior of a complete signaling system. Finally, we acknowledge that transceivers can be designed using either bipolar or MOSFET devices, although throughout this chapter we focus on MOSFET-based circuits.

11.1 I/O DESIGN CONSIDERATIONS

The function of a transmitter is to launch a signal representing digital data onto an interconnect for propagation to a receiver circuit. To maximize performance, engineers typically must use design techniques to provide controlled output impedance and rise and fall times. In addition, transmitters (often abbreviated as Tx) can be designed for either singled-ended or differential transmission, and to operate either as a voltage or a current source. In this section we use the term *transmitter* interchangeably with *driver* and *output buffer*, all of which are commonly used in the industry.

A range of complexity for modeling transmitters is possible. At its simplest, the model can be a simple transient voltage in series with an output resistance or a current source in parallel with an output resistance. Since they use simple circuit elements, variation of linear model parameters for the purposes of identifying a working solution and understanding sensitivity to design and process variation is extremely easy. In addition, they are easy to analyze and provide the fastest simulation times. As a result, linear models are very useful for the initial stage of the design process, when large numbers of simulations are performed in order to identify the potential solution space.

More complex nonlinear behavioral transmitter models provide improved accuracy over linear models by comprehending the nonlinear relationship between the output voltage and output current, staged switching of the output devices to control rise and fall times, and parasitic capacitances. Parametric model variation is more difficult than with linear models, but is still possible. Nonlinear behavioral models are widely used (via IBIS, the I/O Buffer Information Specification) because they allow component suppliers to provide accurate models without divulging the specifics of the circuit design and manufacturing process.

Finally, achieving maximum accuracy may require the use of full transistor models. These models are typically used only as a final check of the design, since they are more complex to construct and they require significantly more simulation time. In addition, suppliers are often reluctant to provide transistor models because they can divulge proprietary design and process information. As a result, in this chapter we focus on linear and nonlinear behavioral modeling rather than transistor-level modeling.

Options for modeling receiver circuits follow the same progression as for transmitters. Simple receiver models include only the termination resistors and input capacitance. Nonlinear behavioral models include the voltage versus current characteristics of the ESD protection circuitry and of terminations that are implemented using transistor devices. Full transistor models incorporate all device effects. Requirements and trade-offs for the various model types are summarized in Table 11-1.

In our modeling discussions in this chapter we focus heavily on linear models due to their extreme usefulness in the early design stages. We also discuss the

TABLE 11-1. Summary of Modeling Approaches and Trade-offs

Model Type	Elements	Intellectual Property	Speed	Sweepability
Linear	Voltage and/or current sources and resistors	None	Fastest	Most
Nonlinear	Current vs. voltage curves, voltage vs. time curves	Little	Fast	Some
Full transistor	All devices, including pre-driver, compensation circuits, etc.	Design and process	Slow	Very limited

limitations of such models and give guidance for when to use nonlinear behavioral models.

11.2 PUSH–PULL TRANSMITTERS

11.2.1 Operation

The simplest type of output circuit is a push–pull transmitter, which can be implemented using a simple CMOS inverter (paired with a receiver that is also an inverter in order to preserve the logic state), as shown in Figure 11-1. Push–pull transmitters were popular in the early days of CMOS ICs due to their ease of implementation and low power consumption. They can be used in interconnect systems without termination, with series termination, and/or with parallel termination. A comprehensive description of the operation of CMOS transistors and inverters is provided in a book by Rabaey et al. [2003]. We offer a brief overview here to make sure that the reader understands the fundamentals of push–pull transmitter operation.

We start by providing the expression relating the current conducted by a MOS transistor, i_D, as a function of the voltage potential applied across the terminal nodes:

$$I_D = \begin{Bmatrix} 0 & v_{GS} - v_T < 0 & \text{(subthreshold)} \\ k\left(\frac{W}{L}\right)\left[(v_{GS} - v_T)v_{DS} - \frac{v_{DS}^2}{2}\right](1 + \lambda v_{DS}) & 0 \le v_{DS} < v_{GS} - v_T & \text{(triode)} \\ \frac{k}{2}\left(\frac{W}{L}\right)(v_{GS} - v_T)^2(1 + \lambda v_{DS}) & 0 < v_{GS} - v_T \le v_{DS} & \text{(saturation)} \end{Bmatrix} \tag{11-1}$$

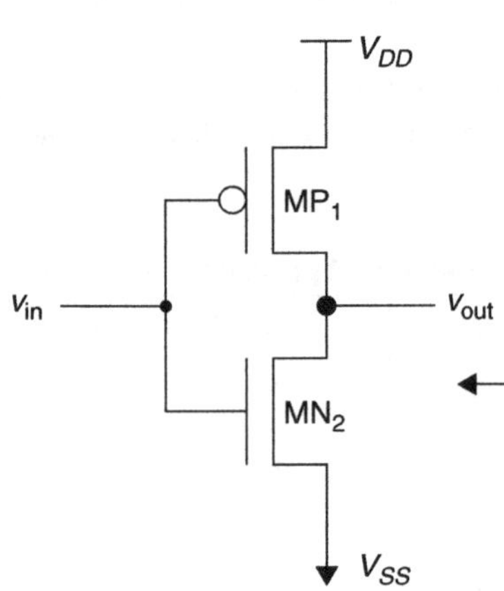

Figure 11-1 CMOS inverter transmitter circuit. (From Dabral and Maloney [1998].)

where k = process transconductance (A/V^2)
W = device width (μm)
L = gate length (μm)
v_{GS} = potential difference between the transistor gate and source nodes (V)
v_T = transistor threshold voltage (V)
v_{DS} = potential difference applied across the source and drain (V)
λ = channel length modulation parameter (V·m)

The process transconductance is

$$k = \frac{\mu \varepsilon_{\text{ox}}}{t_{\text{ox}}} \tag{11-2}$$

where μ is the device mobility (m^2/V·s), ε_{ox} the oxide permittivity (F/m), and t_{ox} the oxide thickness (m). The oxide permittivity ε_{ox} is equal to $3.97\varepsilon_0$, and the oxide thickness t_{ox} is 5.7 nm for the MOSIS 0.25-μm process. Equation (11-1) describes three distinct regions of operation. In the subthreshold region, where $v_{GS} - v_T \leq 0$, the device conducts only a very small amount of leakage current. In the triode region, the relationship between the device current and the drain–source potential is approximately linear. In the saturation region, the device enters a high-impedance state, acting like a (nearly) constant current source.

Figure 11-2 shows i_D versus v_{DS} curves for various gate–source potentials (v_{GS}) that were created from HSPICE simulation using the SPICE level 3 model for the MOSIS 0.25-μm process that is contained in Appendix F. The dimensions of the devices used to produce the figure are a width W of 222 μm and length L equal to 1 μm for the NMOS transistor and W/L = 845 μm/1 μm for the PMOS transistor. In the case of the PMOS device, v_{DS} is less than zero since v_{out} is less than V_{DD}. As a result, the current flows from the supply (source) to the drain, so that i_D is also less than zero. For the NMOS transistor, the polarities for v_{DS} and i_D are both positive. Note from the figure that positive current is defined as flowing back into the transistor device. We follow this convention throughout this chapter, although it is equally valid to define positive current flowing in the opposite direction, as long as consistency is maintained.

The voltage transfer characteristic of the inverter shown in Figure 11-3a was created from HSPICE simulation using a SPICE level 3 model for the MOSIS 0.25-μm process with device dimensions identical to those in Figure 11-2. The figure shows how the output signal varies as a function of the input signal level. When the input signal v_{in} is at ground (V_{SS}), the NMOS transistor, MN2, does not conduct current, while the PMOS device, MP1, conducts current according to equation (11-1). As a result, the output signal v_{out} is pulled up to V_{DD} (2.5 V for the 0.25-μm process) through MP1. As the input is raised to V_{DD}, the output drops to ground. At approximately one-half of V_{DD}, the inverter enters a high-gain region in which v_{out} changes rapidly as a function of v_{in}.

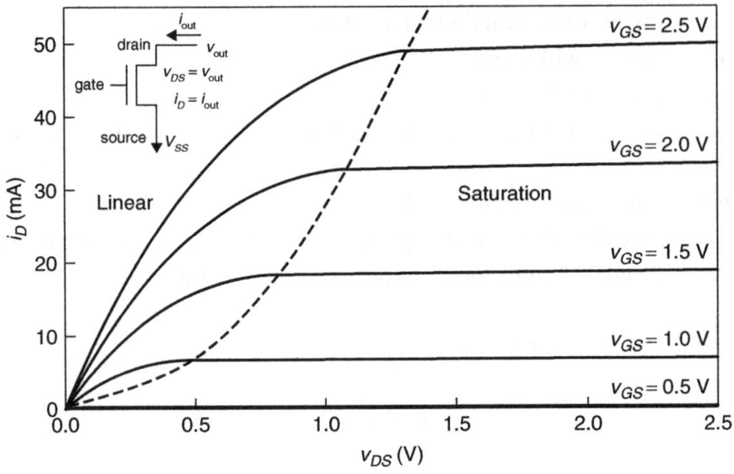

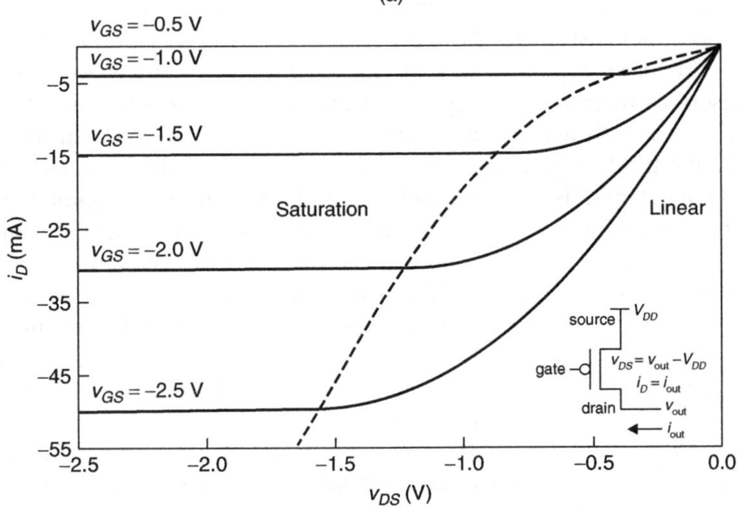

Figure 11-2 Example transmitter pull-up and pull-down i_D versus v_{DS} curves: (a) NMOS ($W = 222$ μm, $L = 1.0$ μm); (b) PMOS ($W = 845$ μm, $L = 1.0$ μm).

Figure 11-3b shows the transient response of the inverter when driving a 1-pF load.

11.2.2 Linear Models

The simplest push–pull transmitter model, shown in Figure 11-4, is the linear model. Here the transmitter behavior is modeled by a transient voltage source in series with a resistor (i.e., a Thévenin equivalent circuit). The voltage source

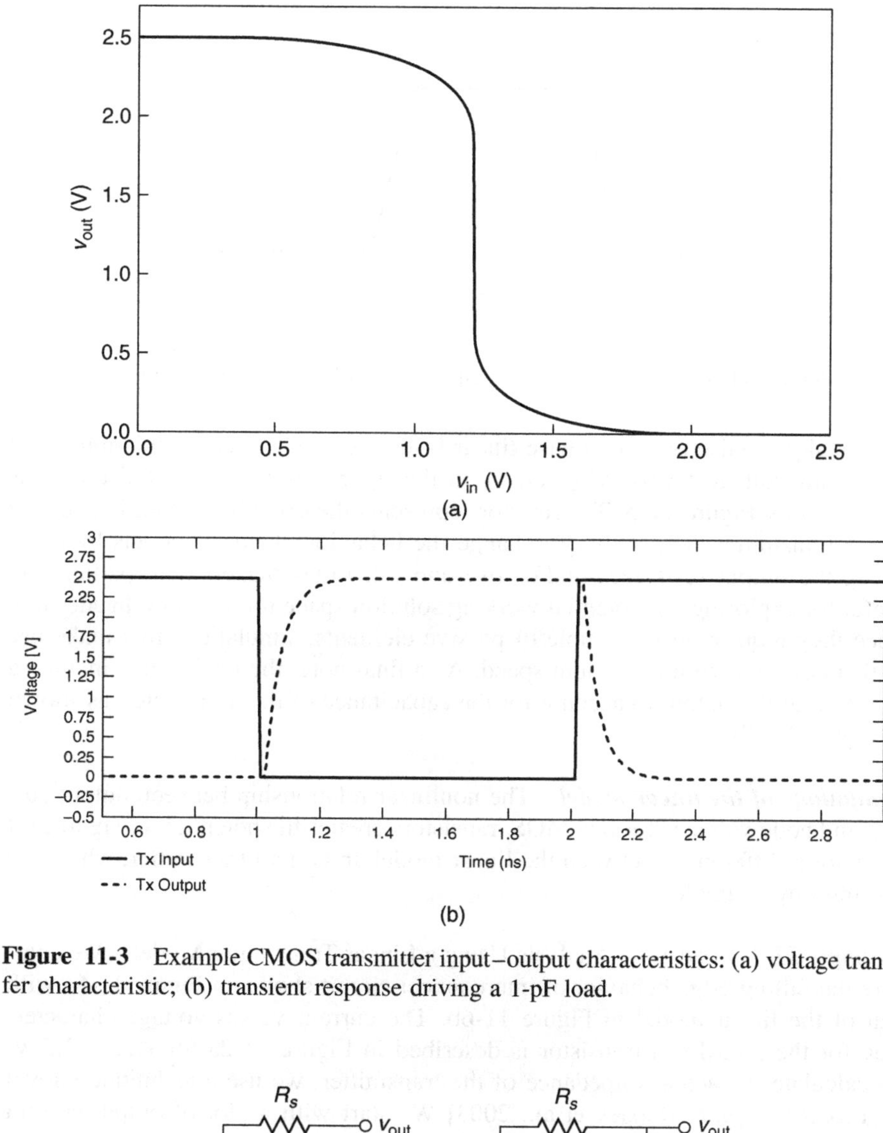

Figure 11-3 Example CMOS transmitter input–output characteristics: (a) voltage transfer characteristic; (b) transient response driving a 1-pF load.

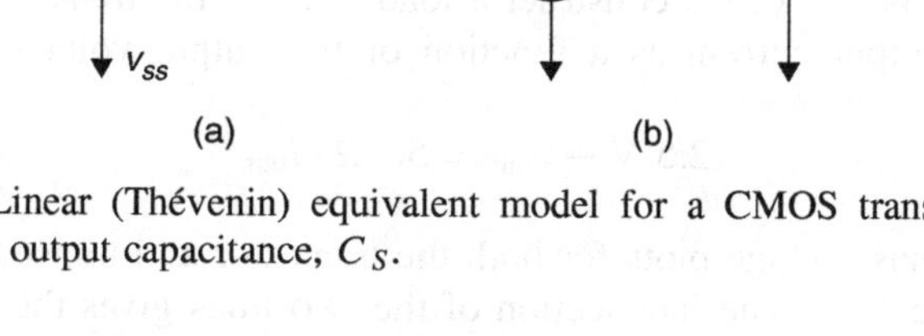

Figure 11-4 Linear (Thévenin) equivalent model for a CMOS transmitter (a) basic model; (b) with output capacitance, C_S.

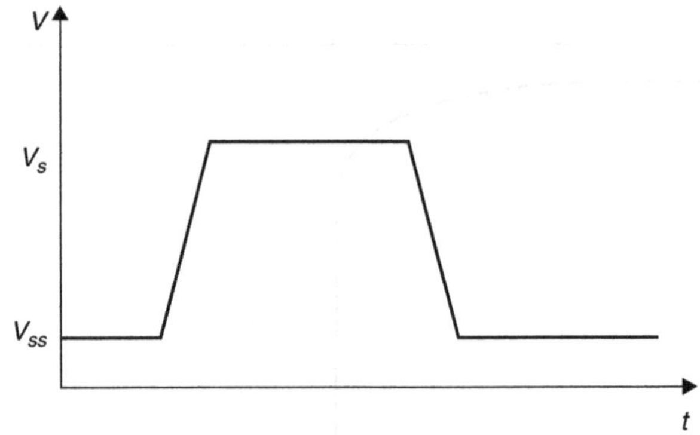

Figure 11-5 Example transient response for a linear transmitter model.

is typically a pulse or a piecewise linear source, characterized by minimum and maximum voltage levels (V_{SS} and v_S in the figure) and rise and fall times, as illustrated by Figure 11-5. The resistor represents the effective output impedance of the transmitter. The ability to change the behavior of linear models by modifying the output resistance and/or rise and fall times makes them particularly useful for exploring the potential working solution space for a design. In addition, since they require only a couple of passive elements, simulations that make use of them execute with maximum speed. As a final note, the model may include a capacitor at the output to account for the capacitance of the transmitter, as shown in Figure 11-4b.

Limitations of the linear model The nonlinear relationship between output current and node voltages for real MOS transistors creates the potential for significant behavioral differences between the linear model and a real transmitter, which we illustrate by example.

Example 11-1 Load-Line Analysis Using a Linear Transmitter Model We compare the falling-edge behavior of interconnect circuit shown in Figure 11-6a with that of the linear model in Figure 11-6b. The current versus voltage characteristic for the pull-down transistor is described in Figure 11-2a for $v_{GS} = 2.5$ V. To calculate the actual impedance of the transmitter, we use a technique known as a *load-line plot* [Rabaey et al., 2003] We start with a plot of output current versus output voltage for the NMOS transistor. Recognizing that the falling-edge transition for this circuit begins from a state in which no current flows ($i = 0$) and the voltage is 2.5 V, we construct a load line for the transmission line that expresses the output current as a function of the output voltage using Ohm's law:

$$2.5 \text{ V} - v_{\text{out}} = 50 \ \Omega \cdot i_{\text{out}}$$

The current versus voltage plots for both the transistor and transmission line are shown in Figure 11-7. The intersection of the two lines gives the values for the

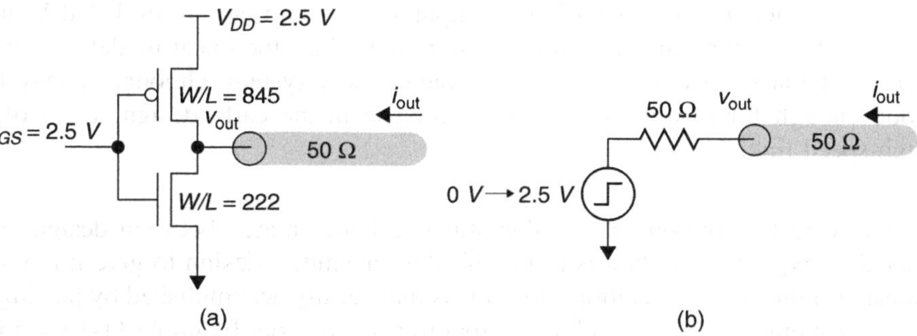

Figure 11-6 Interconnect circuit for Example 11-1: (a) circuit with push–pull transmitter; (b) linear model.

voltage and current at the transmitter output when driving the transmssion line for the falling edge (0.650 V, 37 mA). The NMOS transistor was sized to conduct 50 mA across a potential of 2.5 V, which would provide an output impedance R_S of 50 Ω. However, as the figure demonstrates, the transmitter actually conducts more than 95% of the maximum current by the time the output swing has reached one-half of the maximum value (1.25 V). As a result, the transmitter has an effective output impedance of approximately 0.65 V/37 mA = 18 Ω as seen by the transmission line.

For the linear model, we construct the voltage–current plot using Ohm's law.

$$v_{\text{out}} = 50\ \Omega \cdot i_{\text{out}}$$

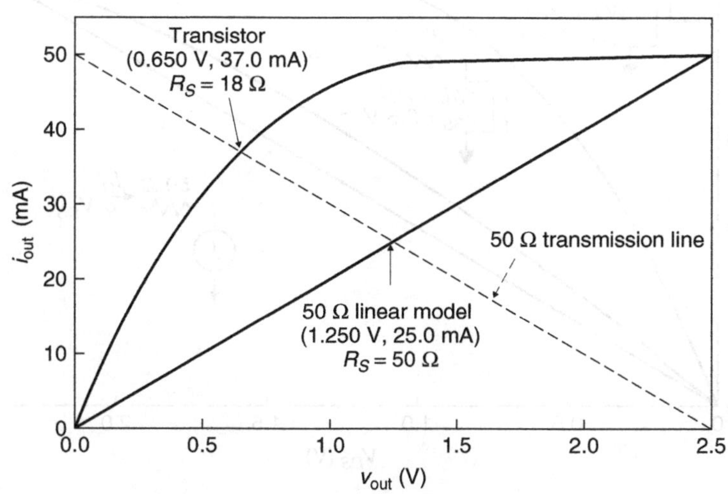

Figure 11-7 Load-line plot for Example 11-1.

The linear model provides an initial output voltage and current of 1.250 V and 25 mA when driving the 50-Ω transmission line. Thus, the linear model is a clear source of inaccuracy for modeling an interconnect system, although it may be good enough for low-speed designs or for use in the early design stages of a high-speed link.

We have two options at our disposal if a better match between design and model is required. The first is to modify the transmitter design to give it a more linear current–voltage relationship. This is most easily accomplished by placing a resistive element in series with the output transistor. From Example 11-1 we note that the transmitter actually conducts more than 95% of the maximum current by the time the output swing has reached one-half of the maximum value (1.25 V). As a result, the transmitter has an effective output impedance of approximately 25 Ω across the output voltage range 0 to 1.25 V (the linear region of operation) and a very high impedance from 1.25 to 2.5 V (the saturation region). By adding a 25-Ω resistor in series with the effective 25 Ω of the transistor, we can better approximate the desired 50-Ω impedance, as shown in Figure 11-8.

Since it requires modifying the design, this approach is typically taken only when a constant impedance over the entire voltage swing is critical to the design [Esch and Chen, 2004]. An example is the AGP4X interface, which used a series-terminated interconnect to transmit graphics data at 266 Mb/s. The interface relied on the transmitter to provide the termination, which required linearization of the voltage–current relationship [Intel, 2002].

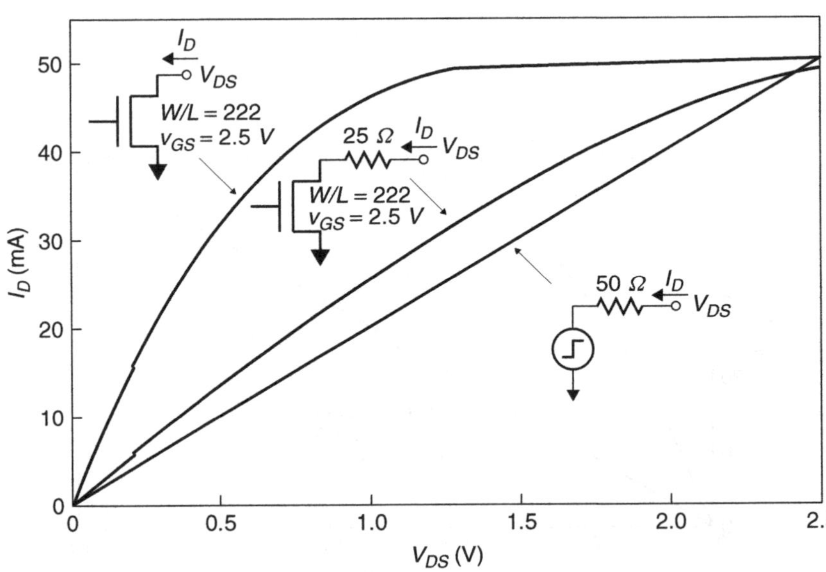

Figure 11-8 Comparison of push–pull transmitter current versus voltage characteristics to a linear model.

11.2.3 Nonlinear models

The other option for improving accuracy is to use a nonlinear model. In addition to the nonlinear current–voltage characteristic, this type of model can also comprehend the shape of the rising and falling edges of the output. The basic push–pull nonlinear model consists of an output current versus output voltage curve for both pull-up and pull-down devices and an output voltage versus time curve for both the rising and falling edges (Figure 11-9). In addition to the curves,

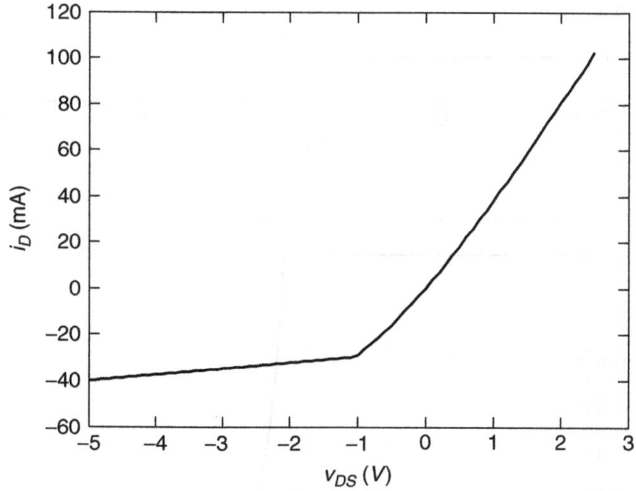

(a)

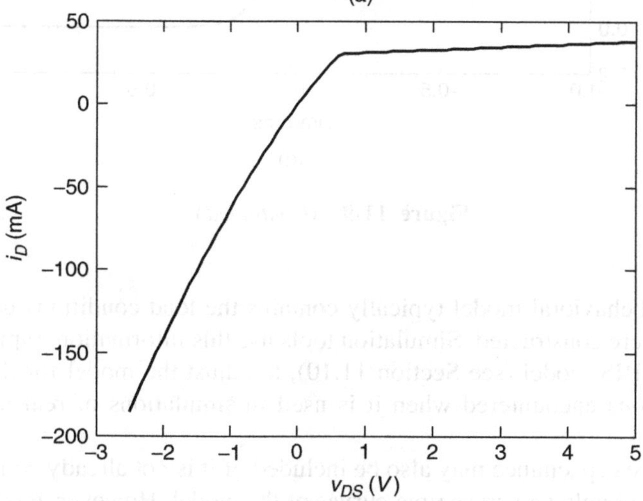

(b)

Figure 11-9 Nonlinear behavioral model components: (a) pull-up i_{out} versus v_{out}; (b) pull-down i_{out} versus v_{out}; (c) rising edge v_{out} versus time; (d) falling edge v_{out} versus time. *(Continued)*

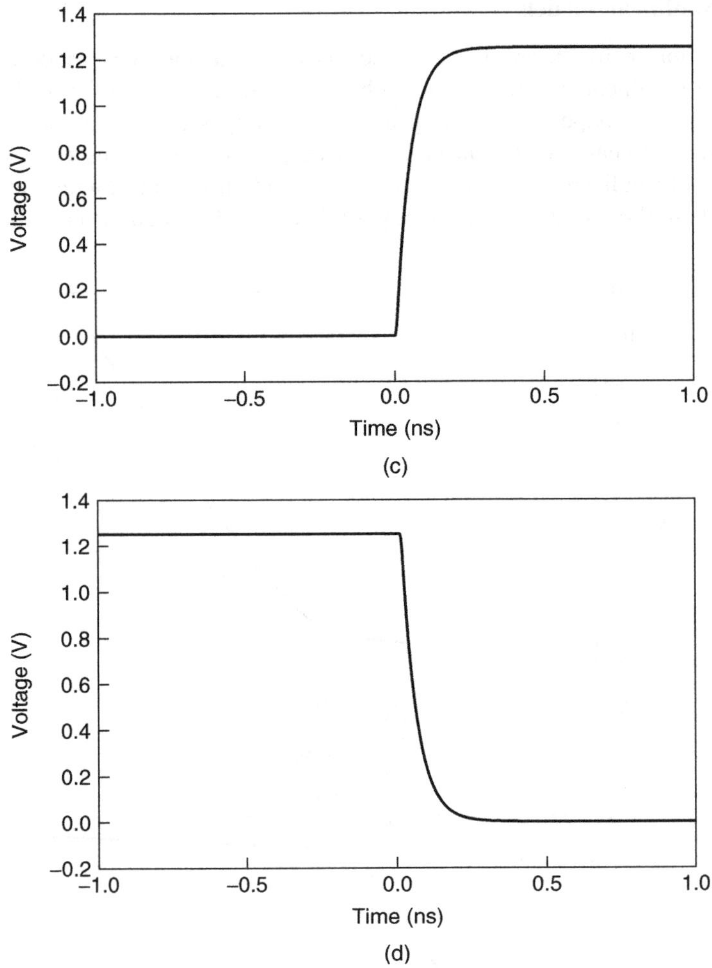

Figure 11-9 (*Continued*)

a nonlinear behavioral model typically contains the load conditions under which the curves were constructed. Simulation tools use this information, typically in the form of an IBIS model (see Section 11.10), to adjust the model for the differing load conditions encountered when it is used in simulations of real interconnect systems.

The device capacitance may also be included, if it is not already comprehended in the transient voltage versus time curves of the model. However, models created from either transistor-based simulations or from measurements typically include the effects of the capacitance, in which case it should not be explicitly called out in the model. Construction of the i_{out} versus v_{out} curves is discussed further in Section 11.10.

In general, simulation tools create multiple $i-v$ curves to account properly for the transient nature of the input signal to the model. Stated another way, a time-varying input signal, v_{in}, to the circuit in Figure 11-1 causes the device gate-to-source potential, v_{gs}, to also be transient, so that the output current–voltage relationship varies with time as well. This effect is shown in Figure 11-10 and Table 11-2 for an inverting transmitter. The rising-edge input signal starts at a value of 0.0 V, which corresponds to $v_{gs} = 0.0$ V for the NMOS device and to $v_{gs} = -2.5$ V for the PMOS device. As the input edge rises, the v_{gs} values for each device change. For example, when v_{in} is equal to 1.0 V, v_{gs} is 1.0 V for the NMOS device and -1.5 V for the PMOS device. We see that for the case of a rising-edge input signal, as the PMOS device moves from the $v_{gs} = -2.5$ V curve to the $v_{gs} = 0$ V curve, the NMOS device is transitioning from the curve for $v_{gs} = 0$ V to that for $v_{gs} = 2.5$ V.

As a final note, we point out that the $i-v$ data in the model should extend well beyond the expected operating range for the device to ensure proper operation in the event of significant signal overshoot. For example, if the signal swing expected ranges from 0 to V_{DD}, the IBIS specification expects that the $i-v$ data span a range from $-V_{DD}$ to $2V_{DD}$.

11.2.4 Advanced Design Considerations

In this section and in the counterpart sections for other transceiver types, we touch briefly on several of the issues that face circuit designers when developing transmitter circuits. For more extensive discussions of the issues and techniques, we refer the reader to books by Dabral and Maloney [1998] and Dally and Poulton [1998].

Overlap current control When the transmitter circuit of Figure 11-1 makes a rising or falling transition, the PMOS and NMOS devices will conduct current simultaneously for a brief period of time. The magnitude of this overlap current (also known as "crowbar" current) is sufficiently high that designers usually design the transmitter such that the initially conducting device turns off before the other device turns on. This is typically implemented with "pre-driver" control logic.

Tristate function Systems in which multiple devices can drive a common signal, such as a multiprocessor bus, require that the transmitter be placed in a high-impedance state when it is not actively driving the system. This is also accomplished with pre-driver logic, as shown in Figure 11-11. The circuit uses the enable signals, $en/\overline{en}$, to control whether the transmitter is connected and can actively drive the bus, or is disconnected and presents a high impedance, as described by Table 11-3.

Process and environmental compensation When fabricating large numbers of components, the physical features (e.g., gate length, oxide thickness) and

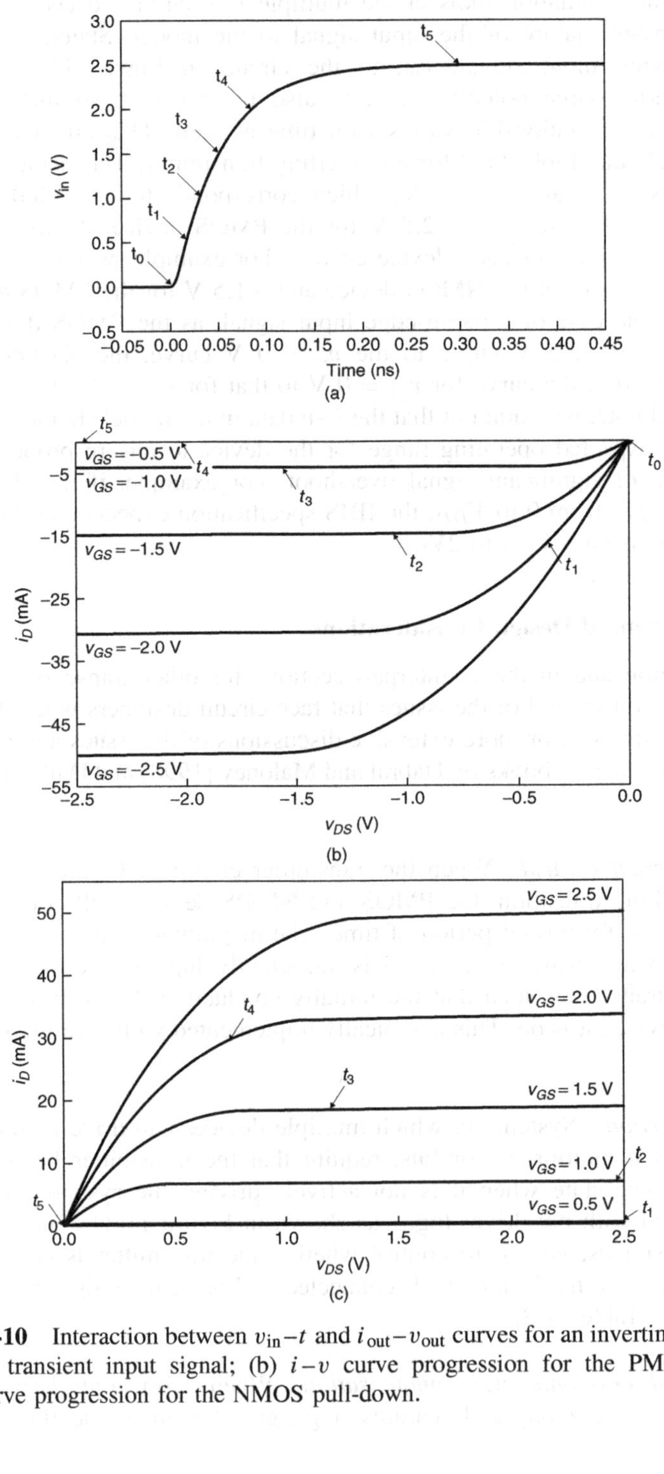

Figure 11-10 Interaction between $v_{in}-t$ and $i_{out}-v_{out}$ curves for an inverting push–pull driver: (a) transient input signal; (b) $i-v$ curve progression for the PMOS pull-up; (c) $i-v$ curve progression for the NMOS pull-down.

TABLE 11-2. Relation Between v_{in} and v_{GS} for the Pull-up and Pull-down Devices of an Inverting Transmitter

	v_{GS} (V)	
v_{in} (V)	NMOS	PMOS
0.0	0.0	−2.5
0.5	0.5	−2.0
1.0	1.0	−1.5
1.5	1.5	−1.0
2.0	2.0	−0.5
2.5	2.5	0.0

TABLE 11-3. Logic Table for a Tristate Transmitter

$en/\overline{en}$	v_{in}	v_{out}
$0/V_{DD}$	X	High Z
$V_{DD}/0$	V_{DD}	0
$V_{DD}/0$	0	V_{DD}

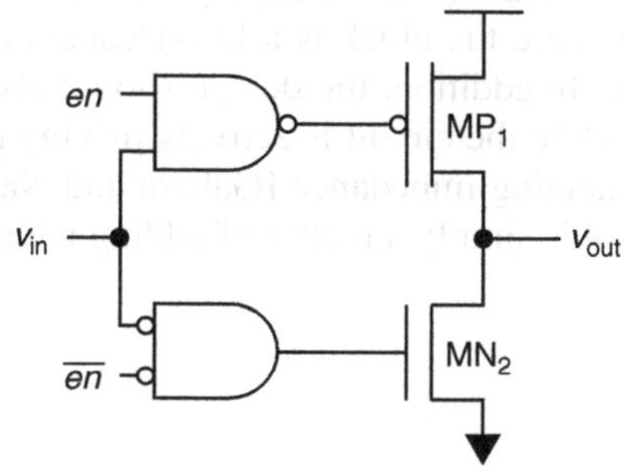

Figure 11-11 Tristate push–pull driver.

electrical characteristics will vary due to normal statistical variations in the manufacturing process. In addition, MOSFET currents are sensitive to environmental factors such as supply voltage (i_D increases as the V_{DD} increases) and device temperature (i_D decreases with increasing T). As a result, the current–voltage relationship for transistors can vary by a factor of 2 to 3 across the process and environmental extremes, leading to wide swings in the output impedance and rise–fall time. In the case of microprocessor-based systems, where yearly volumes reach into the hundreds of millions, excessive variation in the electrical characteristics can lead to failure of some of the systems to operate at full performance. *Compensation* refers to techniques that minimize the variation between different parts operating in different environments. Designers can implement compensation using both digital and analog techniques, which are used most often to provide carefully controlled impedance and/or rise and fall times. Although a detailed discussion of those techniques is beyond our scope, we present an example of digital impedance compensation to illustrate application of the concept.

Example 11-2 Digital Impedance Compensation The circuit shown in Figure 11-12 will provide controlled impedance for either the output of a transmitter or for on-die termination resistors (see Section 11.5). The circuit adjusts the impedance to closely match a desired value by comparing the strength of a binary weighted on-die replica circuit (PMOS transistors MP_0 through MP_N) with that of an external precision resistor (typically controlled to $\pm 1\%$). Both are connected to identical reference current sources (i_{ref}), forming voltage dividers whose levels are fed into the clocked comparator at the bottom of the figure. The output from the comparator is used to increment/decrement an up/down counter. The output signals (S_0 to S_2) from the counter control the turn on–turn off of devices MP_0 to MP_N, which decreases or increases the impedance of the control network to achieve the desired impedance. Signals S_0 to S_N are also connected to the transmitter circuits, each of which contains a similar structure, thus providing the controlled impedance characteristic.

Since it operates from a clock, the circuit can adjust the impedance dynamically to compensate for variation in the supply voltage and device temperature during operation. In this case the clock is a low-frequency clock that is updated about every millisecond. In addition, the design should also ensure that the control bits do not change while the circuit is actively driving in order to avoid noise on the signals due to changing impedance [Gabara and Nauer, 1992]. Achieving tighter impedance control is simply a matter of adding additional binary weighted devices.

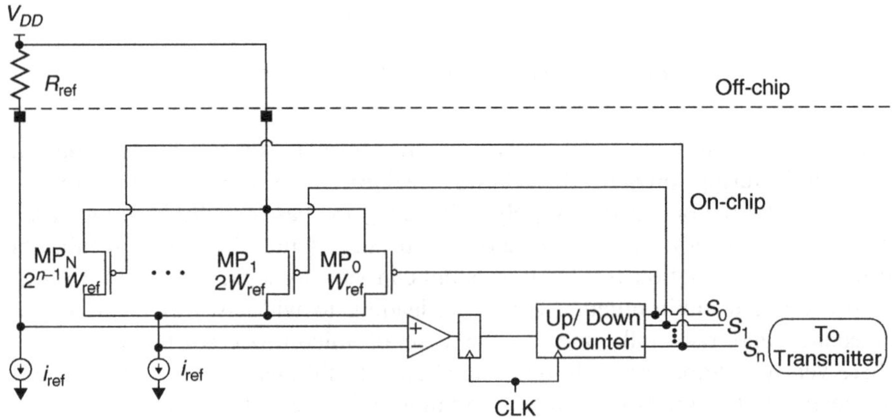

Figure 11-12 Digital impedance compensation circuit.

11.3 CMOS RECEIVERS

The most basic receiver for interchip signaling circuits is the inverter. The simplicity, low power consumption, and ease of implementation of inverters made them the receiver of choice for full-swing CMOS-based interfaces through much of the 1990s. Receivers are characterized by timing parameters (setup and hold) and by logic thresholds, which influence system noise margin and noise immunity. Our discussion of the operation of merits of receivers for high-speed signal transmission will focus on these characteristics.

11.3.1 Operation

The CMOS receiver is a low-gain inverting amplifier that provides full rail-to-rail output swings, which allows for fairly large noise margins at speeds into the hundreds of Mb/s. Figure 11-13a illustrates the voltage transfer characteristic for an example inverting receiver, which specifies the output signal as a function of the input. The input thresholds v_{il} and v_{ih} are determined by the unity gain points ($dv_{\text{out}}/dv_{\text{in}} = -1$) of the transfer characteristic. The region in between v_{il} and v_{ih} is a high-gain region in which the output signal level is extremely sensitive to variations in the input signal, making it a forbidden "keep out" area for steady-state signals.

Large noise margins provide tolerance to noise sources, and so are desirable to guarantee robust operation. As Figure 11-13b demonstrates, the high-side noise margin, v_{NMh}, is the difference between the minimum output signal when driving

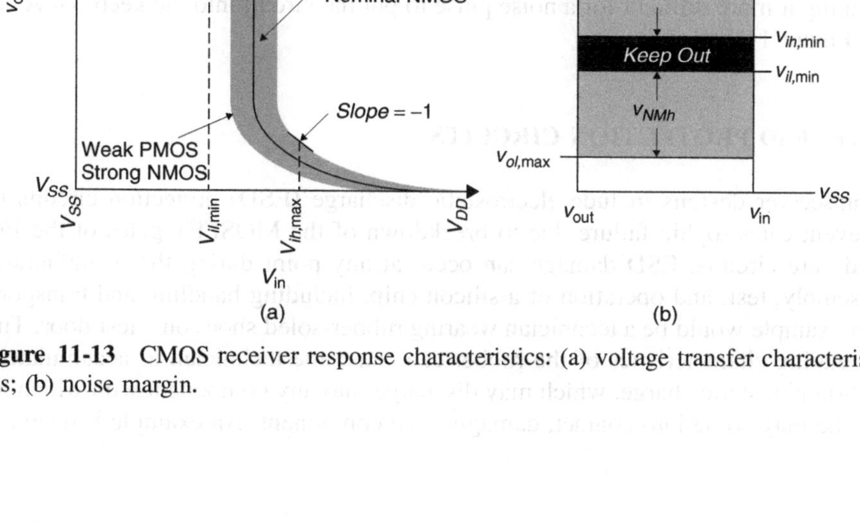

Figure 11-13 CMOS receiver response characteristics: (a) voltage transfer characteristics; (b) noise margin.

high and the minimum signal that the receiver recognizes as the high logic state. Conversely, the low-side noise margin, v_{NMl}, is the difference between the maximum output signal when driving low and the maximum signal recognized by the receiver as the logic low state. Note from the figure that the receiver input specs must comprehend variations in the signal levels caused by variability in the fabrication process. Mathematically, the noise margins are expressed as

$$v_{NMl} = v_{il,\min} - v_{ol,\max} \tag{11-3a}$$

$$v_{NMh} = v_{oh,\min} - v_{ih,\max} \tag{11-3b}$$

11.3.2 Modeling

The CMOS inverter presents a high impedance to the input signal, limited only by the input capacitance of the gate. As such, we typically model a CMOS receiver as a simple capacitance to ground.

11.3.3 Advanced design considerations

Though the relatively large swing results in high noise margins, voltage mode signaling systems possess multiple noise sources that degrade the noise immunity of the system [Dally and Poulton, 1998]. For example, process variations such as device thresholds and transconductance can cause inverter thresholds to vary by more than 10% of the signal swing (>20% if supply voltage variation is included), a phenomenon known as *receiver offset*. Other sources include power supply noise, crosstalk, reflections, and transmitter offset.

These effects may be countered by designing additional noise tolerance into the receiver through hysteresis. An example is a Schmitt trigger, which we show in Figure 11-14a [Wang, 1989]. The MP_1/MN_2 and MP_3/MN_4 transistors form a sequential pair of inverters. The hysteresis is created by feeding the output v_{out} back to the gates of MP_5 and MN_6, which shifts the voltage transfer characteristic, making it more difficult for a noise pulse to put the circuit into the keep-out zone, as Figure 11-14b shows.

11.4 ESD PROTECTION CIRCUITS

Transceiver designs include electrostatic discharge (ESD) protection circuits to prevent catastrophic failure due to breakdown of the MOSFET gates of the I/O and core circuits. ESD damage can occur at any point during the manufacture, assembly, test, and operation of a silicon chip, including handling and transport. An example would be a technician wearing rubber-soled shoes on a test floor. The insulating characteristics of the rubber can cause the technician to accumulate a substantial static charge, which may discharge into any component with which he or she may come into contact, damaging the component. An example ESD event

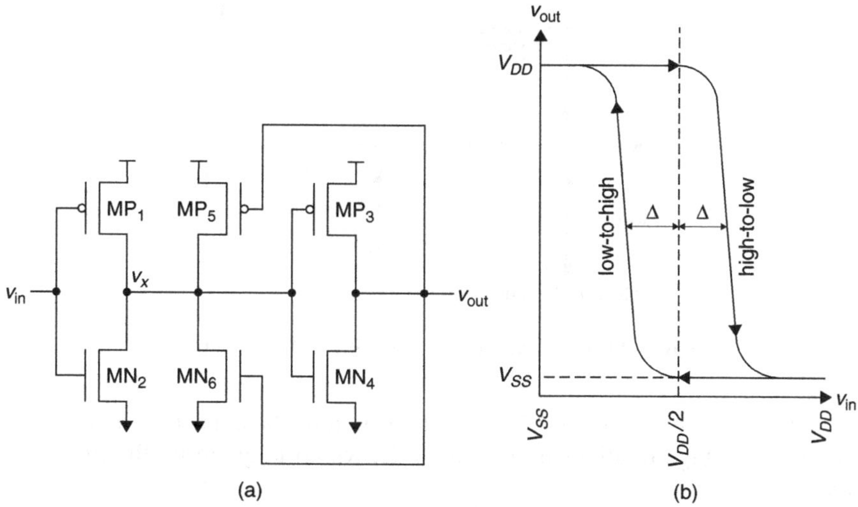

Figure 11-14 CMOS receiver with hysteresis: (a) Schmitt trigger receiver circuit; (b) voltage transfer characteristic.

based on a human body model can have a 3000 to 5000-V potential spike with a sub-10-ns rise time, 1 to 2-A peak current, and 100 to 200-ns duration [Dabral and Maloney, 1998]. The device oxide will break down at field strengths in excess of about 7×10^8 V/m, which translates to voltages in excess of approximately 4.0 V for a 25-μm silicon process. So an ESD event can exceed the process limits by several orders of magnitude, necessitating the inclusion of ESD protection in the I/O circuitry.

11.4.1 Operation

ESD protection circuits such as the one shown in Figure 11-15 protect active circuits by limiting the voltage excursions so that they do not exceed the breakdown voltage of the receiver MOSFET gates, and by limiting the amount of current that flows into the receiver device terminals. The function of the ESD diodes is to limit the voltage so that it does not exceed the gate oxide breakdown voltage and to steer the ESD current away from the internal circuits. If the voltage at the pad will be greater than a diode voltage drop ($V_{D,\mathrm{on}}$) beyond the V_{DD} supply, the upper diode will turn on, shunting the current away from the I/O device and clamping it to $V_{DD} + V_{D,\mathrm{on}}$. A negative excursion will be clamped to a value of $V_{SS} - V_{D,\mathrm{on}}$. The series resistor limits the amount of current that flows through the transceiver devices to prevent performance degradation due to device threshold shifts caused by "hot" electron tunneling [Dally and Poulton, 1998].

11.4.2 Modeling

The diodes add parasitic capacitance to the transceiver, which we include in our linear model. The series resistance is typically several hundred ohms, and

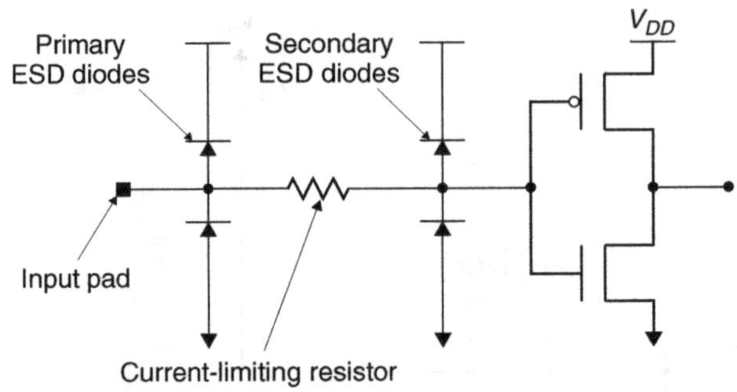

Figure 11-15 CMOS receiver with ESD protection.

does not require explicit modeling. Nonlinear models must take into account the current–voltage relationship of the diode, which is given by the *ideal diode equation*:

$$i_{\text{diode}} = i_S(e^{v_{\text{diode}}/\phi_T} - 1) \tag{11-4}$$

where i_{diode} is the current through the diode; i_S, the diode saturation current, is proportional to the diode area; v_{diode} is the bias voltage across the diode; and ϕ_T is the thermal voltage (26 mV at room temperature).

Figure 11-16 presents an example current–voltage characteristic calculated using equation (11-4) with ϕ_T equal to 26 mV and a saturation current of 10 pA. We can further approximate the diode behavior for manual analysis by setting the diode current to zero below the "turn-on" voltage (approximately 0.6 V in the figure) and allowing it to approach infinity at the turn-on voltage. This is shown

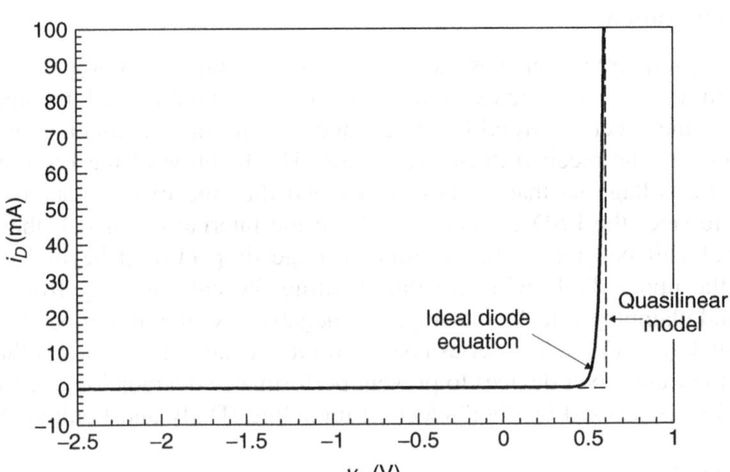

Figure 11-16 Example diode current–voltage characteristic and quasilinear approximation.

as the quasilinear model in the figure. We illustrate the use of the quasilinear model in Problem 11.6.

11.4.3 Advanced design considerations

The circuit shown in Figure 11-15 is but a simple example of ESD protection. In reality, designers have many options for implementing ESD protection, and often go to great lengths to design structures that provide sufficient protection while minimizing the parasitic capacitance for high-speed operation. We refer the reader to the books by Dally and Poulton [1998] and Dabral and Maloney [1998] for comprehensive treatment of the topic.

11.5 ON-CHIP TERMINATION

On-chip (a.k.a. on-die) termination has become the method of choice as signaling speeds continue to increase, since it eliminates a source of reflections by removing the transmission-line stub that is required to connect off-die termination.

11.5.1 Operation

Termination resistors are typically implemented using FETs. An example is shown in Figure 11-17, in which PMOS and NMOS devices are connected in parallel to create a 50-Ω termination to V_{DD} (2.5 V for the 0.25-μm process). The PMOS gate is connected to ground ($v_{gs} = -2.5$ V) and NMOS to V_{DD} ($v_{gs} = 2.5$ V) to keep the transistors in the triode region for as long as possible in order to make the current–voltage relationship as linear as possible. As the figure shows, this configuration provides a termination of approximately 46 to 58 Ω over the entire range of operation for nominal device characteristics and operating conditions. Figure 11-17b shows the waveforms obtained when using the circuit to terminate a 50-Ω transmission line when driven with a 2.5-V 12.5-Ω linear transmitter. The FET termination provides nearly equivalent performance to a perfect 50-Ω resistive termination. We also note that our choice to terminate to the positive supply rail rather than ground was also based on making the termination as linear as possible. FET termination is typically combined with the digital impedance compensation technique described in Section 11.2.4 to provide a controlled impedance termination across process, voltage, and operating temperature.

11.5.2 Modeling

The basic model for on-chip termination is a simple resistor connected to the appropriate termination supply. However, as we showed earlier, on-chip termination can exhibit significantly nonlinear response, in which case we need to model the termination as a table of current and voltage values using a nonlinear modeling format such as IBIS (refer to Section 11.10).

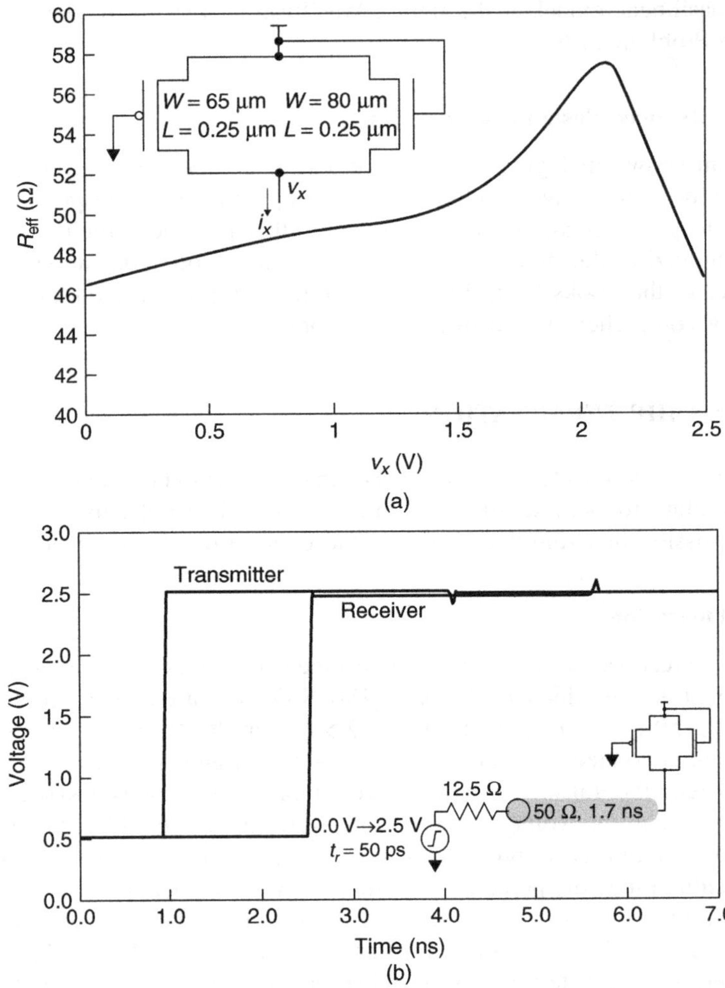

Figure 11-17 Example on-chip termination using parallel FETs: (a) FET termination circuit and resistance as a function of line voltage; (b) example waveform.

11.5.3 Advanced design considerations

The primary motivation for implementing on-chip termination is to minimize reflections by eliminating the transmission line stub that is typically required when terminating on the printed circuit board. To fully realize the potential benefit of on-chip termination, automatic impedance control similar to the impedance-matching technique described in Section 11.2.4 is commonly used. In addition, designers may implement techniques to improve the linearity of the on-chip termination [Dally and Poulton, 1997].

11.6 BERGERON DIAGRAMS

At this point we take a slight detour from our survey of transceivers to introduce the Bergeron diagram, a technique that will allow us to analyze the behavior of interconnects with nonlinear transmitter and receiver characteristics by graphically solving the simultaneous current–voltage relationships of the components of a signaling system. We introduce this technique as it is useful for furthering our understanding of transmission-line basics. We demonstrate and develop the technique through a simple example.

Example 11-3 Bergeron Diagram for a Linear Interconnect Circuit In this example we analyze the rising behavior for the circuit shown in Figure 11-18. The transmitter is a 2.5-V push–pull driver with symmetrical pull-up and pull-down impedances. The receiver has a resistive termination to a 2.5-V termination supply.

We start the analysis by plotting the current versus voltage relationships for the transmitter and receiver. The easiest way to do so is to draw the equivalent circuits and write the Ohm's law expressions. This is shown in Figure 11-19 along with the load-line plot. Note that we were careful to account correctly for the direction of the current flow in all of the equivalent circuits and that we chose to use milliamperes as the units for the y-axis. Note from the plot that the intersection of the transmitter pull-down and the receiver load lines gives the voltage and current (0.357 V and 28.6 mA) for the circuit at the steady state when driven low. Since the circuit is at steady state, this is the potential and current flow at all points on the transmission line. This gives us the starting point for our analysis, since we are studying a rising-edge transition.

Our next step is to find the initial voltage and current for the rising edge at the transmitter. We do this by drawing a load line representing the 50-Ω transmission line, starting at the intersection of the transmitter pull-down and receiver (steady-state low) and extending until it intersects with the load line for the transmitter pull-up. This load line has a slope equal to $-1/Z_0$, since the transmission line also obeys Ohm's law and the load line is a plot of current versus voltage. In effect, the load line for the transmission line is graphically depicting the Ohms' law equation:

$$v - v_0 = Z_0(i - i_0) \tag{11-5}$$

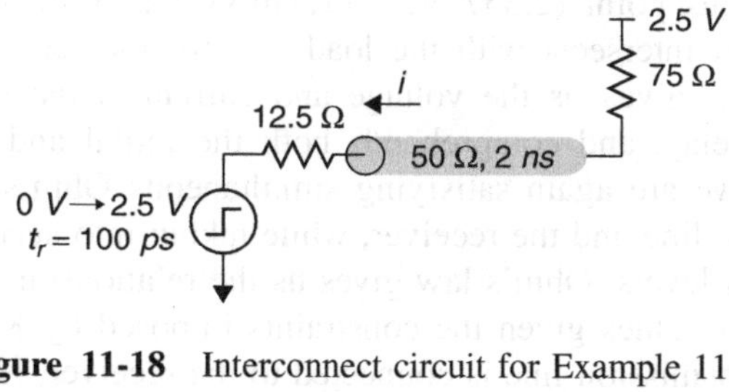

Figure 11-18 Interconnect circuit for Example 11-3.

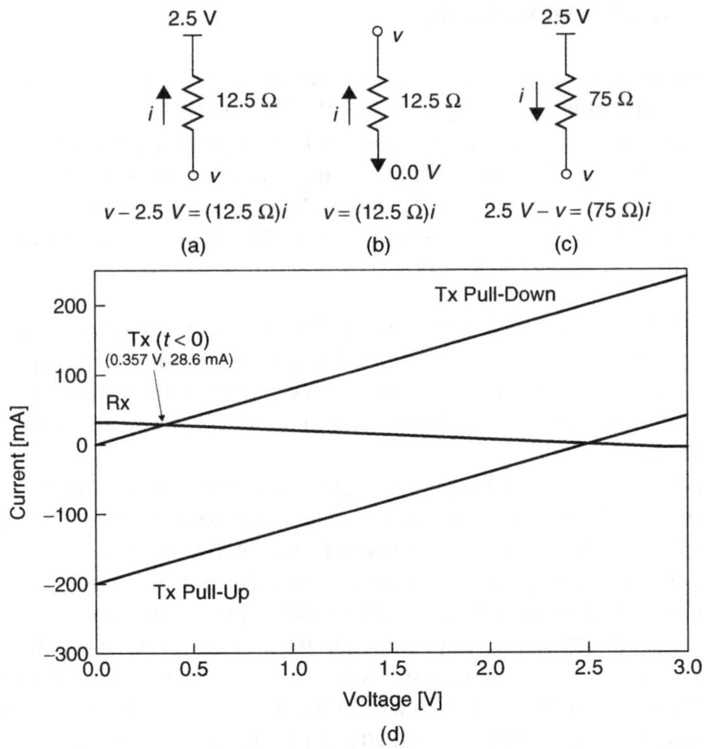

Figure 11-19 Equivalent circuits and voltage–current expressions for Example 11-3: (a) transmitter pull-up; (b) transmitter pull-down; (c) receiver; (d) load-line plot.

where v_0 and i_0 are the steady-state voltage and current for the system when driven low. The slope of the line is negative because the rising-edge transition causes current to flow from the transmitter into the transmission line, while we have defined positive current as flowing from the line back into the transmitter.

The transmitter also follows Ohm's law, so that the intersection between the transmission line and the transmitter pull-up lines gives the initial voltage and current at the transmitter for the rising edge. Figure 11-20a illustrates this, giving an initial voltage and current of 2.357 V and −11.4 mA.

The next step is to draw another load line for the transmission line. The line starts at the previous point (2.357 V, −11.4 mA), has a slope equal to $1/Z_0$, and extends until it intersects with the load line for the receiver, as shown in Figure 11-20b. This gives us the voltage and current at the receiver after the first propagation delay, and comprehends both the initial and reflected waves. It works because we are again satisfying simultaneous Ohm's law expressions for the transmission line and the receiver, while taking into account the previous voltage and current levels. Ohm's law gives us the relationships that allow us to solve for the signal values given the constraints imposed by Kirchhoff's circuit laws. Since the transmission line is connected to the receiver, Kirchhoff's circuit laws tell us that they will have equal voltages at the connection and equal currents flowing through them. Thus, our Bergeron diagram gives us the solution to two simultaneous Ohm's law equations in two unknowns (the voltage and current) via graphical means.

By continuing to draw load lines for the transmission lines with slopes that alternate between $\pm 1/Z_0$ we can find the voltage and current levels at each end of the line as the waves propagate back and forth. For example, Figure 11-20c shows the extension of the next transmission load line back to the transmitter load line, which gives the signal levels at the transmitter after a round-trip propagation delay. The transient wave components are becoming small enough

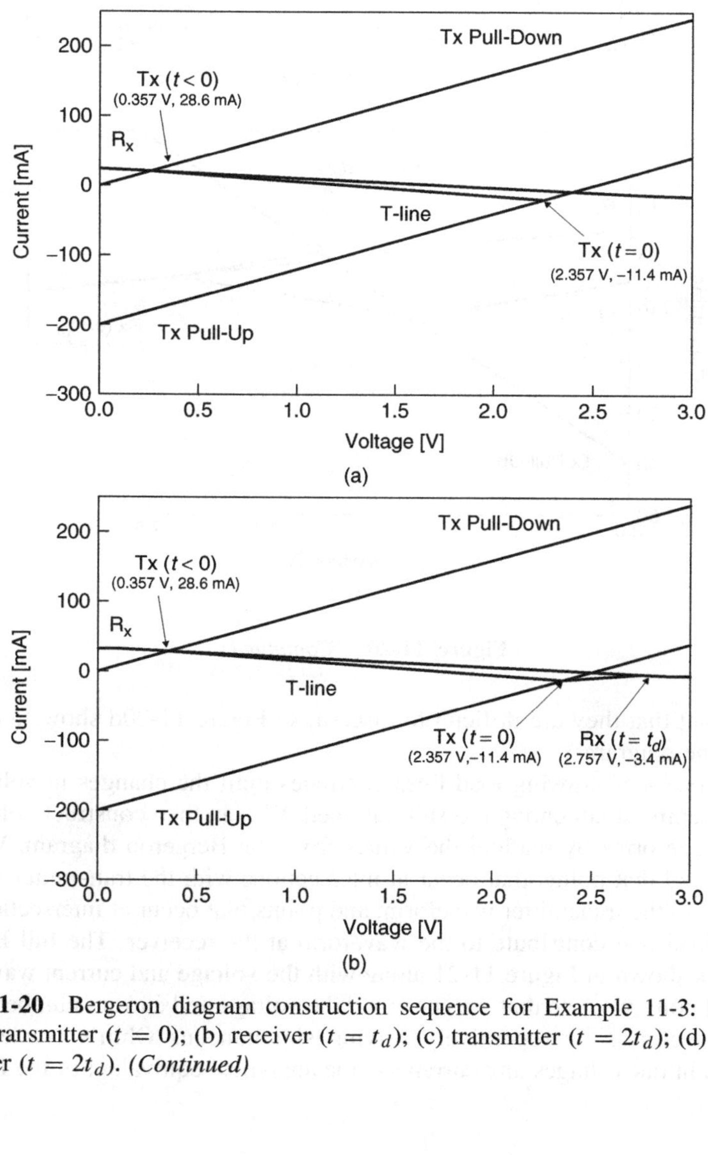

Figure 11-20 Bergeron diagram construction sequence for Example 11-3: (a) initial wave at transmitter ($t = 0$); (b) receiver ($t = t_d$); (c) transmitter ($t = 2t_d$); (d) close-up: transmitter ($t = 2t_d$). *(Continued)*

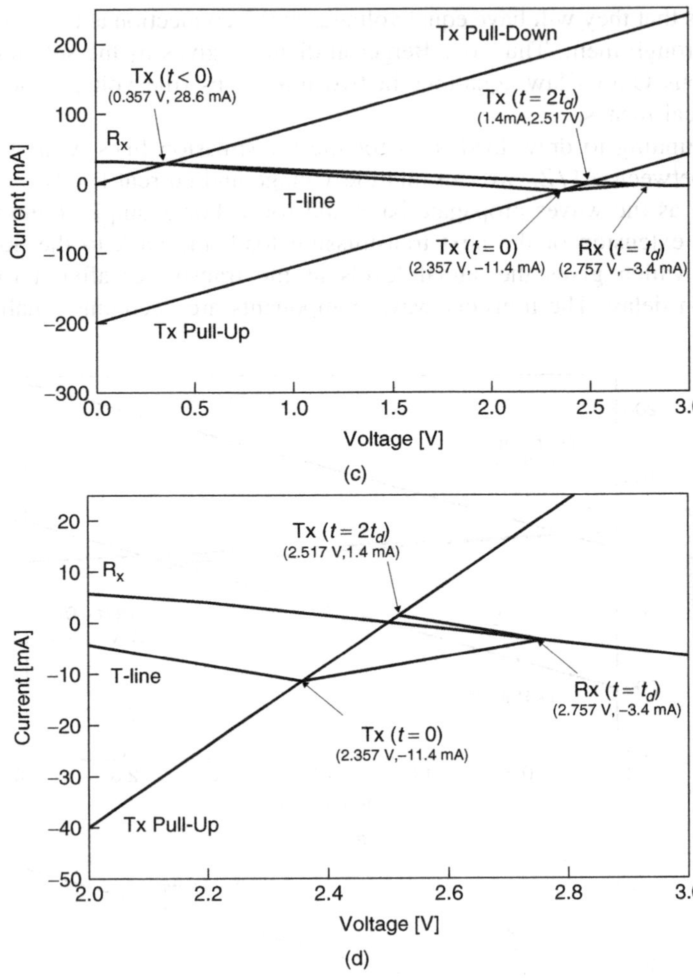

Figure 11-20 (*Continued*)

at this point that they are difficult to discern, so Figure 11-20d shows a close-up view of the diagram.

The process of drawing load lines continues until the changes in voltage and current become small enough to suit our need. We can then construct voltage and current waveforms by reading the values from the Bergeron diagram. We must keep in mind that points that occur at intersections with the transmitter load line contribute to the transmitter waveform, and points that occur at intersections with receiver load line contribute to the waveform at the receiver. The full Bergeron diagram is shown in Figure 11-21 along with the voltage and current waveforms.

At this point we note that we generated the voltage and current numbers shown in the waveforms by repetitively solving simultaneous Ohm's law equations subject to initial voltages and currents—the analytical equivalent of the Bergeron

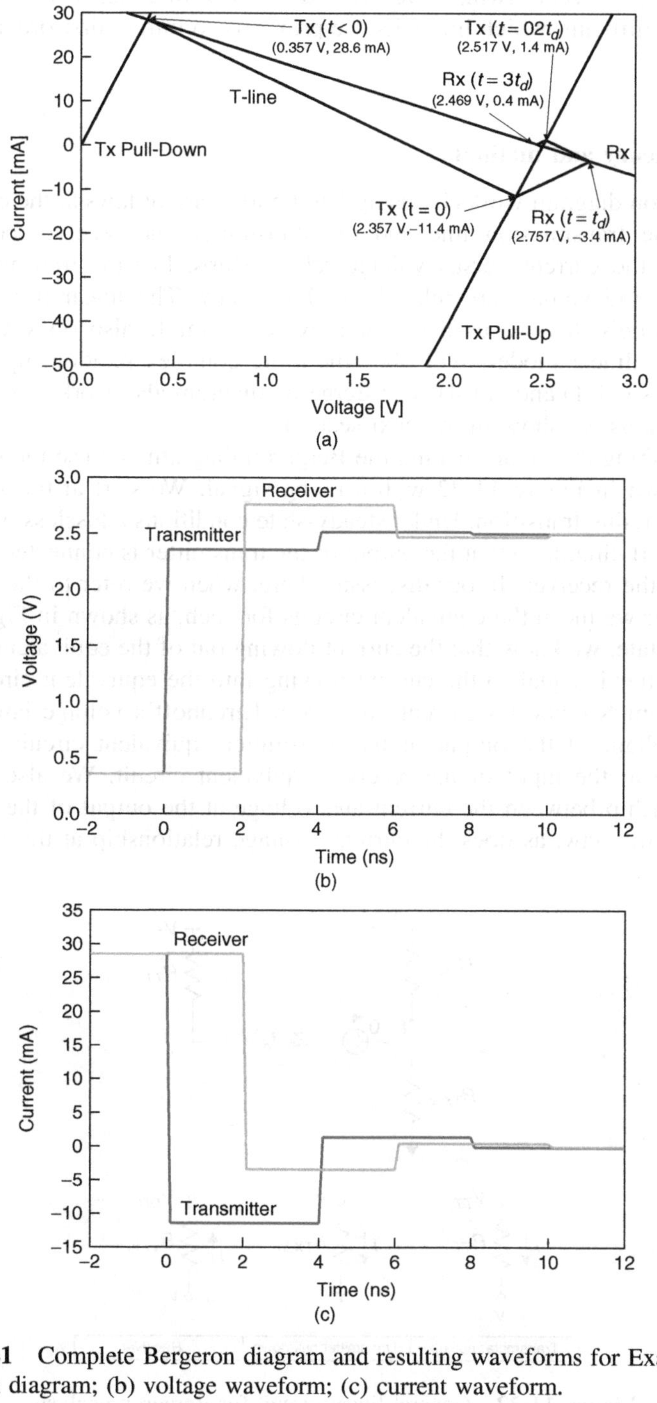

Figure 11-21 Complete Bergeron diagram and resulting waveforms for Example 11-3: (a) Bergeron diagram; (b) voltage waveform; (c) current waveform.

diagram. In practice, drawing a Bergeron diagram with better than 5 mV, 0.5 mA accuracy is difficult, which limits its use primarily to initial, first-order estimates.

11.6.1 Theory and method

The Bergeron diagram works by using Kirchhoff's circuit laws at the connections between the transmission line and the transmitter and receiver components, along with the current versus voltage relationships. For the transmission line, the current and voltage are related by Ohm's law. The linear transmitter and receiver models that we used in our previous example also obey Ohm's law, although nonlinear models will obey the more complex relationships described by equations (11-1) and (11-4). The Bergeron diagram also works with nonlinear transceivers, as we show in the next section.

In describing the theory behind the Bergeron diagram, we use the generalized circuit shown in Figure 11-22 with a rising signal. We start at the steady state prior to the rising transition. Under steady-state conditions a lossless transmission line is a short circuit, so that the output of the transmitter is connected effectively directly to the receiver. In our discussion here, when we refer to the transmitter and receiver we mean the equivalent circuits for each, as shown in Figure 11-22. At steady state, we know that the current flowing out of the equivalent circuit for the transmitter is equal to the current flowing into the equivalent circuit for the receiver from Kirchhoff's current law. From Kirchhoff's voltage law we know that the voltage at the output of the transmitter equivalent circuit is equal to the voltage at the input of the receiver equivalent circuit. We also know that the relationship between the current and voltage at the output of the transmitter follows Ohm's law, as does the current–voltage relationship at the input of the

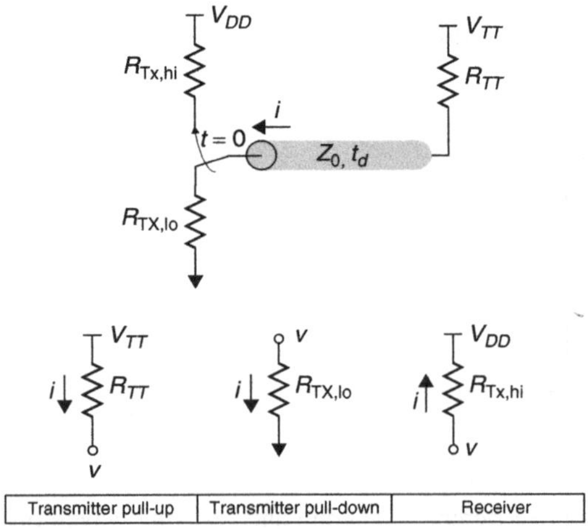

Figure 11-22 General linear circuit for Bergeron analysis.

receiver. By equating the voltages and currents and applying Ohm's law, we create a set of two simultaneous equations in two variables:

$$v(t < 0) = i(< 0)R_{\mathrm{TX,lo}} \tag{11-6}$$

$$V_{TT} - v(t < 0) = i(t < 0)R_{TT} \tag{11-7}$$

The Bergeron diagram solves the equations graphically to give the steady-state current and voltage.

To find the voltage and current values for the initial transition at the transmitter end of the circuit, we recognize that the output of the transmitter is connected to one end of the transmission line. By applying the circuit laws in the same manner as above, we calculate the initial voltage and current wave magnitudes at the transmitter. However, the Ohm's law expression for the transmission line must comprehend the initial current and voltage flowing through the line:

$$v(0) - V_{DD} = i(0)R_{\mathrm{TX,hi}} \tag{11-8}$$

$$v(t < 0) - v(t = 0) = [i(t = 0) - i(t < 0)]Z_0 \tag{11-9}$$

Note from equation (11-9) that the slope of the load line for the transmission line is equal to $-1/Z_0$, just as we constructed it in our example. So the Bergeron diagram again solves the simultaneous equations while accounting for the steady-state potential and current flow that existed on the line prior to the transition.

When the initial wave reaches the connection between transmission line and receiver, we again apply the circuit laws and account for the current and voltage of the incident wave:

$$V_{TT} - v(t = t_d) = i(t = t_d)R_{TT} \tag{11-10}$$

$$v(t = t_d) - v(t = 0) = [i(t = t_d) - i(t = 0)]Z_0 \tag{11-11}$$

The slope of this load line for the transmission line is $1/Z_0$ because it relates the current and voltage for the reflected wave, and since we must account for the current and voltage of the incident wave, it starts at the current and voltage, $i(t = 0)$ and $v(t = 0)$, that we calculated in the preceding step.

The next step in the analysis returns back to the transmitter end of the system using a load line with slope $-1/Z_0$ starting from the receiver voltage and current (i.e., the intersection between the previous transmission line load line and that of the receiver). In this case the load-line slope is negative because we are accounting for the reflected voltage and current waves at the transmitter. Since the waves flow away from the transmitter and toward the receiver, the sign of the current wave is negative.

The analysis continues in this fashion, alternating between transmitting and receiving load lines using transmission load lines with alternating slopes of

$-1/Z_0$ and $1/Z_0$, until it approaches steady state. So we see that the Bergeron is simply a graphical technique for repeatedly solving the simultaneous equations arising from circuit laws that describe the transient voltage and current signals at both ends of a transmission line. We now illustrate the application of the technique to a system with nonlinear transceiver characteristics.

Example 11-4 Bergeron Diagram with Nonlinear Transceivers In this example we analyze the falling edge for the circuit shown in Figure 11-23, which uses a CMOS push–pull transmitter with no termination at the receiver end. The system relies on the output impedance of the transmitter to provide source termination, and the transistors are sized to provide a 50-Ω output impedance to match the target impedance of the transmission line. The load-line plot shows the nonlinear transistor behavior discussed in Section 11.2.1. The load line for the receiver is simply a zero-current (infinite-impedance) line that represents the open circuit at that end. We know that the characteristic impedance of the transmission lines may vary by up to ±20% due to manufacturing tolerances. Thus, for our example we choose a 60-Ω characteristic impedance for the transmission line. With this information, we step through the analysis as follows:

1. The Bergeron diagram begins at the load-line intersection of the transmitter pull-up and the receiver. The potential and current are 2.5 V and 0 mA, respectively.
2. From that point we draw the load line for a transmission line with slope equal to $-0.0167\ \Omega^{-1}(-1/60\ \Omega)$, extending it until it intersects the nonlinear curve for the transmitter pull-down device at 0.80 V and 28.3 mA, which are the initial values at the transmitter after the falling-edge transition.
3. Draw a load line from the previous point, with slope equal to $0.0167\ \Omega^{-1}$ until it intersects with the receiver load line at $i = 0$, which gives the voltage at the receiver (−0.90 V) after the first incident wave reaches it. A simple means of checking the result at this point is to consider the magnitudes of the incident and reflected voltage waves. The incident-wave magnitude is equal to 0.80 V −2.50 V, or −1.70 V. The reflected wave is also equal to

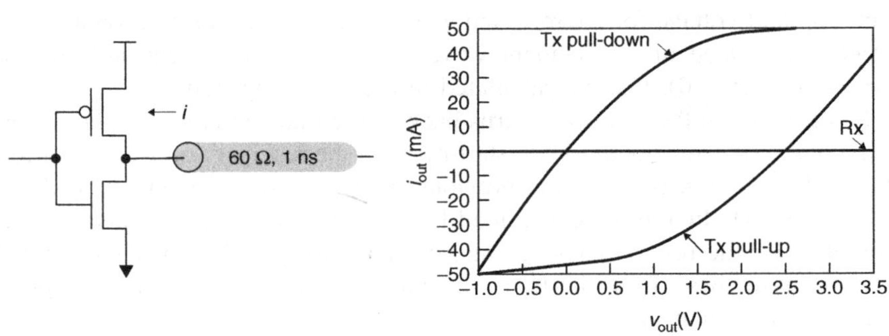

Figure 11-23 Push–pull transmitter circuit and nonlinear i–v characteristics.

−1.70 V (−0.90 V −0.80 V). Since the transmission line is open-circuited at the receiver end, this is exactly the result that we expect to find.

4. Draw the next load line, starting at −0.900 V and 0 mA with slope equal to $-0.0167\ \Omega^{-1}$ until it intersects the transmitter pull-down curve, which yields the voltage (−0.25) and current (−10.8) at the transmitter after the first reflected wave from the receiver reaches it.
5. Continue the analysis until reaching steady state (0.00 V, 0 mA).

The resulting voltage waveforms at both ends of the line are shown in Figure 11-24b.

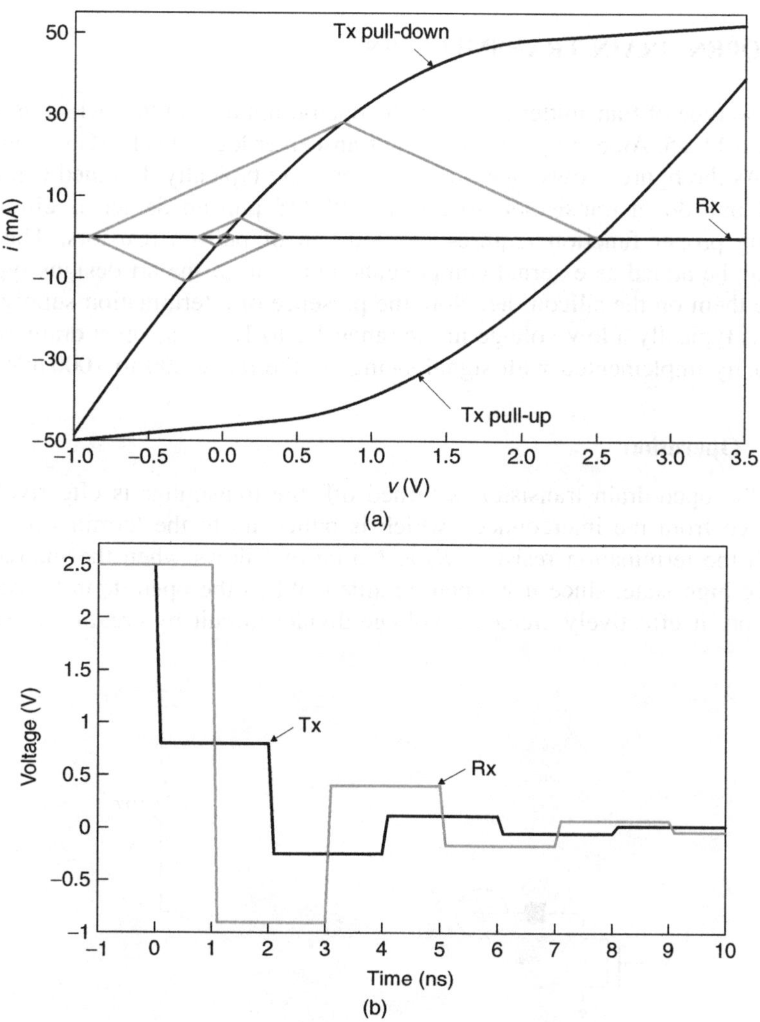

Figure 11-24 (a) Bergeron diagram and (b) transient waveforms for Example 11-4.

11.6.2 Limitations

As mentioned earlier, the accuracy of Bergeron diagrams is typically around 5 mV and 0.5 mA, limiting their use to the earliest stage of the design process. In addition, they can only handle impedance discontinuities at the ends of the lines, so they are not suitable for analysis of more realistic (and complex) topologies. Finally, they only work for lossless transmission lines, which limits their application to data rates in which losses may be ignored. Despite these limitations, Bergeron diagrams provide a means of evaluating the impact of I/O circuit nonlinearity, making them a helpful tool in the arsenal of the signal integrity engineer.

11.7 OPEN-DRAIN TRANSMITTERS

A second type of transmitter circuit is the open-drain transmitter, which is shown in Figure 11-25. An example is Gunning transceiver logic (GTL) [Gunning et al., 1992] As the figure shows, open-drain systems are typically designed using only NMOS pull-down transistors, so that the PMOS pull-up device is eliminated. Ensuring proper function requires the addition of pull-up resistors. The resistors may be added as external components, but state-of-the-art designs typically include them on the silicon die. Note the presence of a termination supply, V_{TT}, which is typically a low voltage in the range 1.2 to 1.5 V, as open-drain systems are usually implemented with signal swings in the range 800 to 1000 mV.

11.7.1 Operation

When the open-drain transistor is turned off, the transmitter is effectively disconnected from the interconnect, which is pulled up to the termination supply through the termination resistor, R_{TT}. No current flows when the interconnect is in the high state, since it is open circuited. When the open-drain transistor is turned on, it effectively creates a voltage-divider circuit by creating a path for

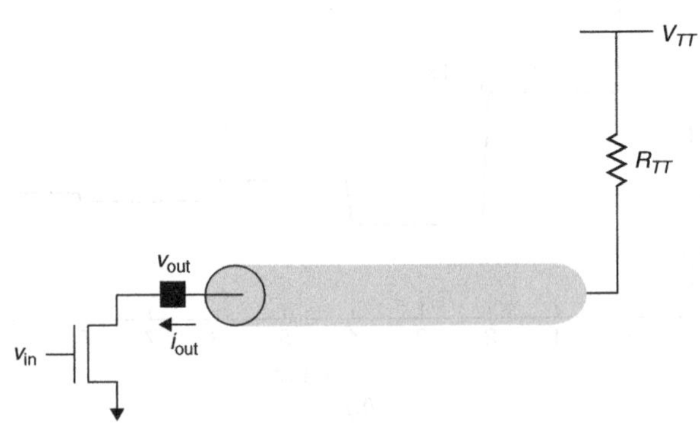

Figure 11-25 Open-drain signaling circuit.

current flow from the V_{TT} supply through the termination resistor and the NMOS transistor. As a result, open-drain systems do not use "rail-to-rail" signal swings, since the signal level does not reach ground when driving low.

Benefits Open-drain designs offer multiple potential benefits that have made them a popular design choice over the past two decades. The use of parallel termination reduces reflections for higher-speed applications, and elimination of the PMOS pull-up output transistors reduces the amount of die area consumed, which creates the potential to reduce the cost of the silicon. Since open-drain designs do not dissipate power in the high state, they are typically defined as being *active low*, in order to achieve zero power consumption when idle. In addition, in contrast to push-pull circuits, the small signal swing reduces active switching power. These features combine to make open drain a popular choice for low-power designs.

Another benefit of open-drain systems is their ability to interface components that are manufactured on different processes that have differing maximum voltage supply limits. Use of the low-voltage external termination supply allows us to decouple the signaling values from the component supplies. For example, GTL uses a 1.2-V external supply, and signals swing from 0.4 V to 1.2 V. This would allow us to connect a component that uses a 3.3-V supply rail with one that uses a 2.5-V rail.

Finally, open-drain systems provide a "wired-OR" function for multiprocessor systems, in which any one of multiple agents can assert an active signal on the bus interface (see Figure 11-26). An example is the MCERR# (machine check error) signal on the Intel Xeon processor system bus interface (a.k.a. front-side bus). MCERR# is asserted (driven low) to indicate an unrecoverable error. Since multiple agents may drive this signal at the same time, it is a wired-OR signal that must connect the corresponding pins of all processor front-side bus agents [Intel, 2005]. Wired-OR connections are susceptible to "glitches" that can occur when multiple agents drive the bus low simultaneously. In Problem 11-10 we explore this phenomenon in more detail.

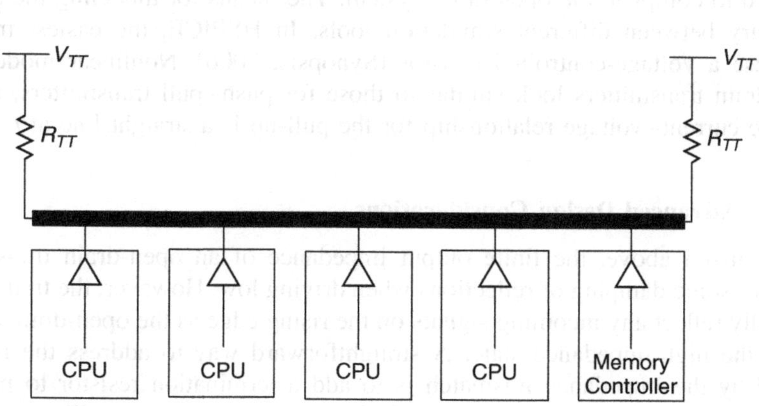

Figure 11-26 Multiprocessor system with open-drain wired-OR connection.

Limitations Despite the use of parallel termination, GTL is susceptible to ringing on rising-edge transitions. The reason for this is that when the device turns off, the transmission line is open circuited at the transmitting end. Any reflections on the interconnect that travel back to the transmitter will undergo a full reflection, as we show with an example.

Example 11-5 Rising-Edge Reflections in an Open-Drain Interconnect The interconnect circuit in Figure 11-27 has an impedance discontinuity between boards 1 and 2. To analyze the behavior of the rising edge, we follow the current-mode analysis method [Hall et al., 2000]. We model the open-drain device as an effective resistance, R_S, with a switch. The analysis starts with the circuit at steady state when driven low. Prior to $t = 1$ ns, the switch is closed, and a steady-state current flows:

$$i_{S,\text{Slo}} = \frac{V_{TT}}{R_S + R_{TT}} = \frac{1\ \text{V}}{12.5\ \Omega + 50\ \Omega} = 16\ \text{mA}$$

The switch opens at $t = 1$ ns, creating an open circuit at $z = 0$. The open circuit requires that the net current flow is zero at that point. The consequence of this is that a -16-mA current wave is launched onto the 75-Ω line, creating a 1.2-V voltage wave. From that point we can carry out the analysis using a lattice diagram, which we shown in Figure 11-27b. Figure 11-27c shows the resulting waveform at the transmitter and at the receiver, which demonstrates the ringing at both ends of the system, despite the fact that the termination matches the impedance of the second line.

11.7.2 Modeling

A linear model for an open-drain circuit requires a resistor and a switch, as shown in Figure 11-28. In addition, a termination supply and termination resistor are required to complete the open-drain system. The means for modeling the switch will vary between different simulation tools. In HSPICE, the easiest method is to use a voltage-controlled resistor [Synopsis, 2006]. Nonlinear models for open-drain transmitters look similar to those for push–pull transmitters, except that the current–voltage relationship for the pull-up is a straight line at $i = 0$.

11.7.3 Advanced Design Considerations

As discussed above, the finite output impedance of an open-drain transmitter provides some damping of reflections when driving low. However, the transmitter will fully reflect any incoming signals on the rising edge as the open-drain device enters the high-impedance state. A straightforward way to address the ringing caused by the impedance mismatch is to add a termination resistor to provide impedance matching at the transmit end of the system, as shown in Figure 11-29. Although this technique will eliminate the excessive ringing that was caused by

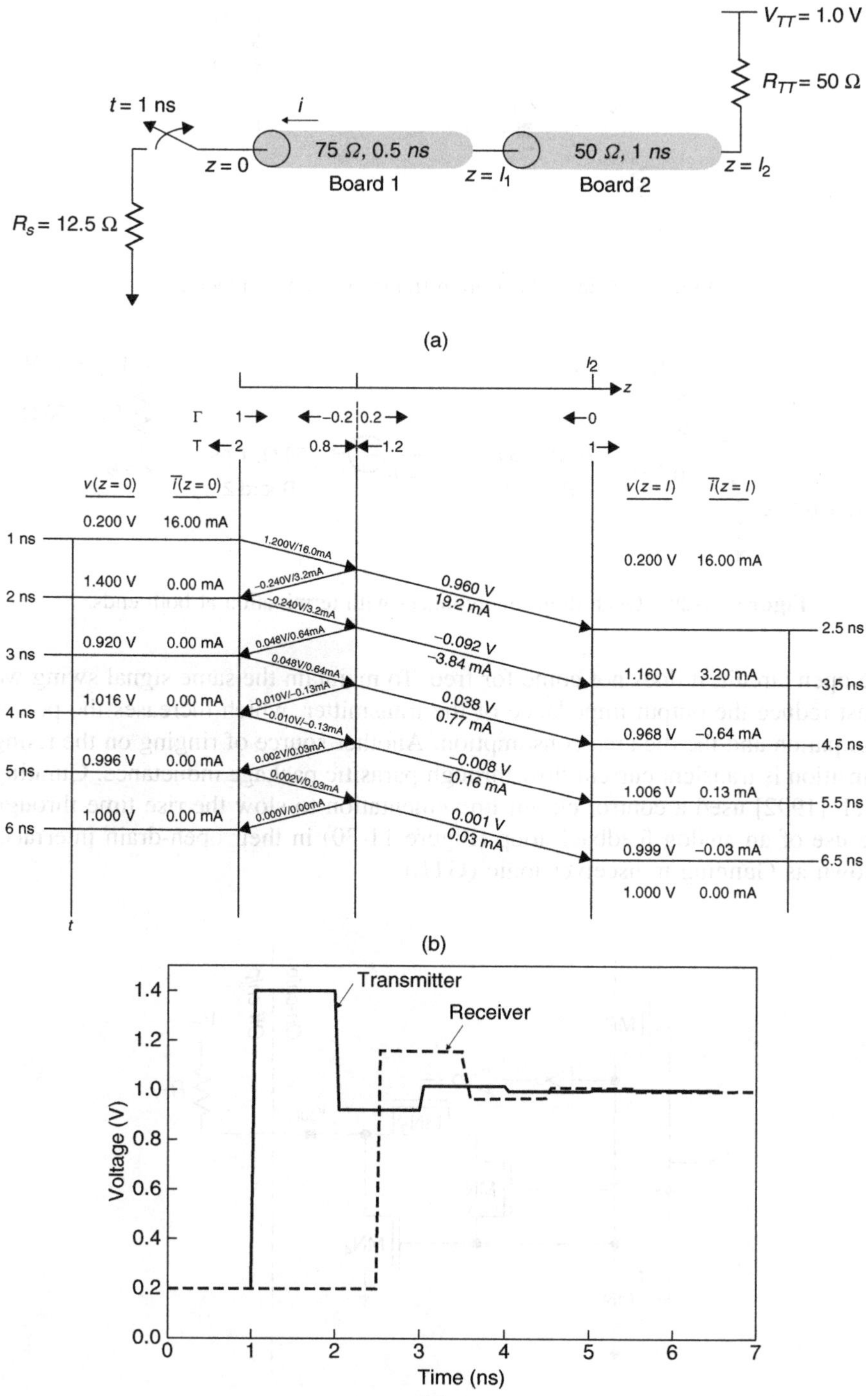

Figure 11-27 Open-drain interconnect and analysis for Example 11-5: (a) circuit; (b) lattice diagram; (c) waveform.

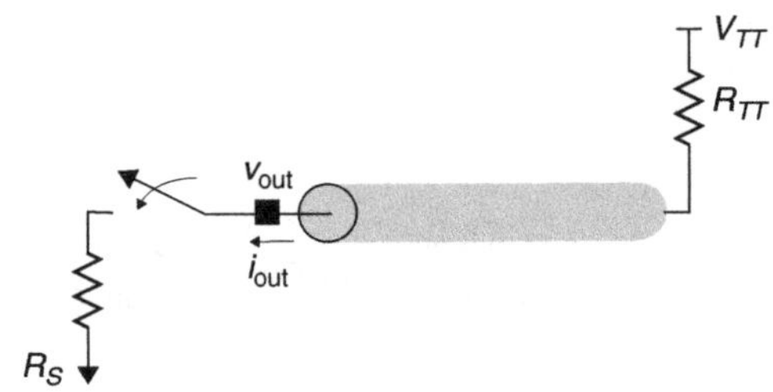

Figure 11-28 Open-drain transmitter linear model.

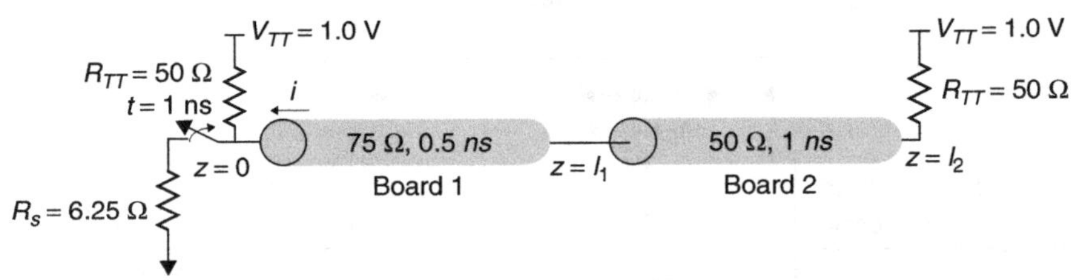

Figure 11-29 Open-drain interconnect with termination at both ends.

the open circuit, it does not come for free. To maintain the same signal swing we must reduce the output impedance of the transmitter, which increases the power dissipation and the die area consumption. Another source of ringing on the rising transition is transient current flow through parasitic package inductance. Gunning et al. [1992] used a control circuit implementation to slow the rise time through the use of an analog feedback loop (Figure 11-30) in their open-drain interface, known as Gunning transceiver logic (GTL).

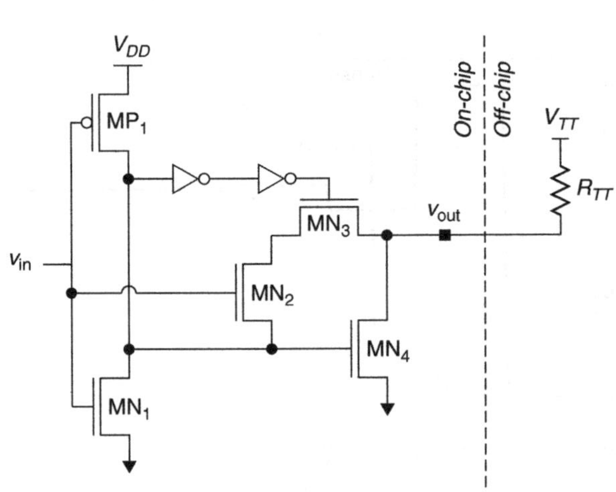

Figure 11-30 GTL transmitter circuit with analog slew rate control.

11.8 DIFFERENTIAL CURRENT-MODE TRANSMITTERS

11.8.1 Operation

Differential current-mode transmitters are often used for high-speed data transmission. In this case, the transmitter operates by injecting a current onto the transmission line. Figure 11-31a depicts a simple differential transmitter design. The transmitter uses complementary input signals, v_{in} and $\overline{v}_{in}$, which ensure that only one side of the circuit is in the conducting state at any given time. Thus, the transmitter uses the differential input signals to steer the current from the constant current source, i_S, to the desired side of the circuit. The flow of current creates a voltage drop across the source termination resistor, R_{TT}, on one side of the circuit, while the side that has no current flow is pulled up to V_{DD}, thereby creating the output signal levels v_{out} and $\overline{v}_{out}$. The relationship between input and output signals is summarized in Table 11-4. The table demonstrates that the differential signal swing has a magnitude $(2i_{T_x}R_{TT})$ that is twice that of the singled-ended swing.

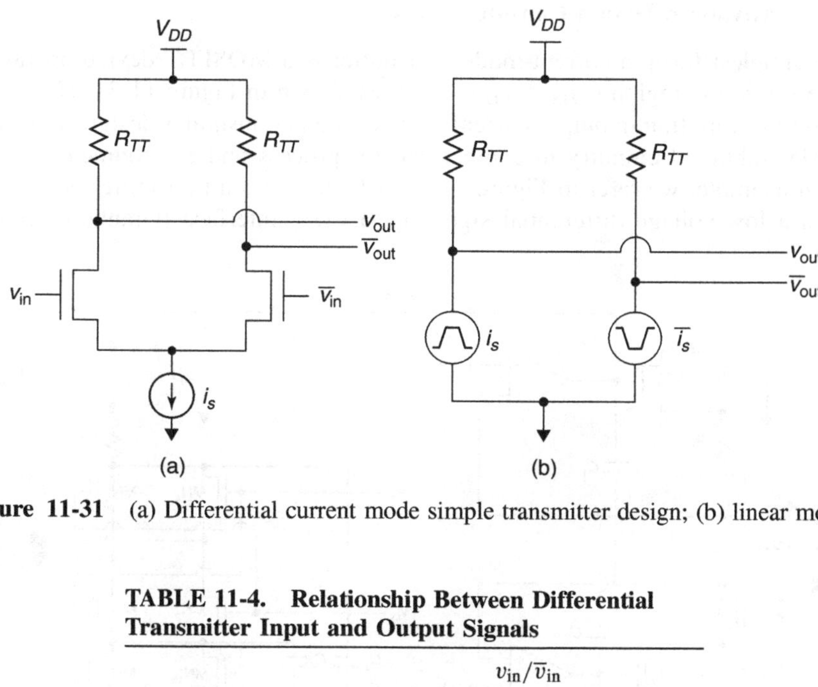

Figure 11-31 (a) Differential current mode simple transmitter design; (b) linear model.

TABLE 11-4. Relationship Between Differential Transmitter Input and Output Signals

	$v_{in}/\overline{v}_{in}$	
	Low/High	High/Low
v_{out}	V_{DD}	$V_{DD} - i_{T_x}R_{TT}$
$\overline{v}_{out}$	$V_{DD} - i_{T_x}R_{TT}$	V_{DD}
$v_{diff} = v_{out} - \overline{v}_{out}$	$i_{T_x}R_{TT}$	$-i_{T_x}R_{TT}$

Benefits As we demonstrated in Chapter 7, differential signal transmission improves signal-to-noise ratio (SNR), which offers a path to higher data rates. Part of the reason for the improved SNR is that the differential signal swing is twice that of a single-ended signal. In addition, the differential transmitter circuit draws a nearly constant current, which drastically reduces the simultaneous switching noise (SSN). In addition, differential receivers reject the vast majority of common-mode noise. All of these factors combine to provide substantial performance headroom over single-ended signaling.

11.8.2 Modeling

As shown in Figure 11-31b, a linear model of a differential current source transmitter consists simply of a pair of complementary transient current sources connected to bias/termination transistors, along with any parasitic capacitance. Nonlinear models look similar to those for push–pull transmitters, except that the current–voltage relationships follow a saturated profile rather than the more linear characteristic of the push–pull circuit.

11.8.3 Advanced Design Considerations

In its simplest form, a current-mode transmitter is a MOSFET device operating in the saturation region ($v_{DS} \geq v_{GS} - v_T$), as shown in Figure 11-31. However, to limit the variation in output current, current-mode transmitter designs typically include additional circuitry to compensate for process and environment effects. For an example, we refer to Figure 11-32, which shows a transmitter design for use in a low-voltage differential signaling (LVDS) interface [Granberg, 2004].

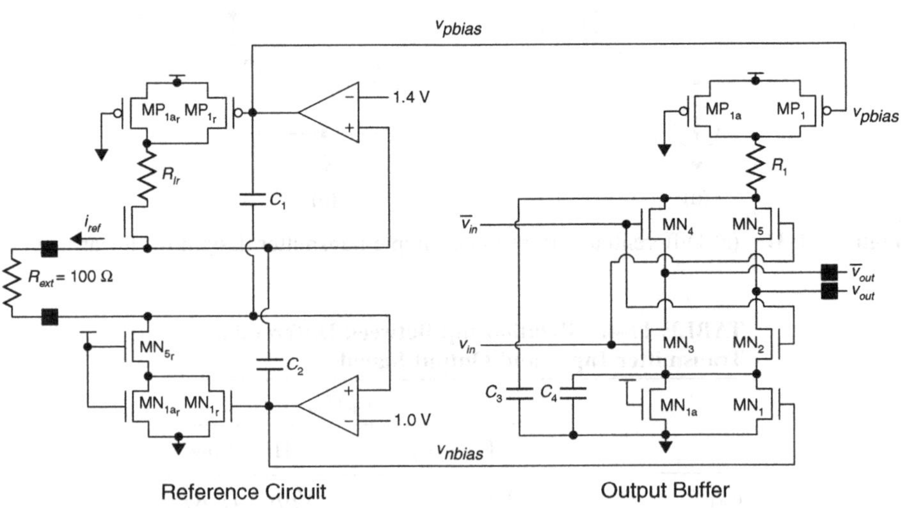

Figure 11-32 LVDS current mode transmitter circuit with controlled current reference. (From Gabara [1997].)

The circuit contains a reference current generator on the left-hand side of the figure that is used to bias the output buffer on the right-hand side of the circuit in order to produce a tightly controlled output current. The external resistor connects to the on-die reference circuit to generate a 4-mA reference current which establishes the bias levels, $v_{p,\text{bias}}$ and $v_{n,\text{bias}}$. The bias voltages feed the output buffer in order to produce the LVDS output swing (1.0 to 1.4 V) when connected by a 100-Ω load resistor, when the differential input signals are applied.

It is worth pointing out that current-mode transmission does not require differential signaling but can be applied to high-speed single-ended signals as well. Low-swing current-mode transmission systems in general provide better noise immunity and power dissipation than does voltage mode rail-to-rail signaling [Dally and Poulton, 1998]. An example of a single-ended current-mode interface is the Direct Rambus DRAM technology [Lau et al., 1998; Granberg, 2004].

11.9 LOW-SWING AND DIFFERENTIAL RECEIVERS

We treat the input receivers for low-swing and differential signaling techniques in the same section, as they both typically employ differential amplifiers, which provide fast transient response to small signal swings.

11.9.1 Operation

In principle, the differential amplifier design is the same for single-ended low-swing and differential signaling applications. In the single-ended application, one input is connected to a reference signal, v_{ref}, while the other is connected to the data signal. For differential signals, the amplifier inputs are connected to the complementary data signals. Examples of each are shown in Figure 11-33.

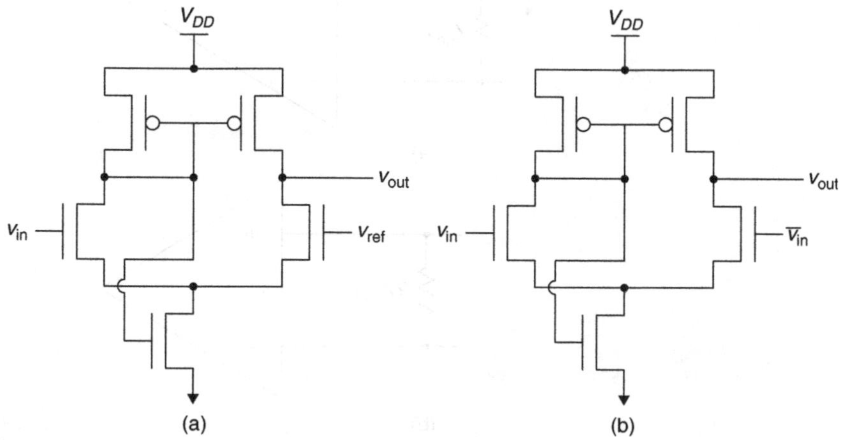

Figure 11-33 Example receiver circuits for low-swing and differential signaling: (a) single-ended; (b) differential.

The single-ended receiver is the original design for use with GTL signaling and was designed to switch within $v_{ref} \pm 50$ mV across process, voltage, and operating temperature in order to provide a high noise margin with an 800-mV swing [Gunning et al., 1992]. The differential receiver is a self-biasing Chappell amplifier that uses an internal feedback signal to adjust the bias voltage for proper operation [Chappell et al., 1998].

As mentioned above, differential amplifiers respond to small changes in input signals, as their symmetry gives low-input offset voltages and makes them relatively insensitive to power supply fluctuations. Differential receivers also reject common-mode noise, with typical common-mode rejection ratios (CMRRs) of −20 dB or more.

11.9.2 Modeling

We model differential amplifier–based receivers in the same fashion as we model single-ended receivers. The simplest model is simply a capacitance to ground in conjunction with the termination. With differential signals, we have multiple options for termination, as Figure 11-34 shows. Note that the termination for differential signals is typically implemented on the silicon die, resulting in nonlinear current versus voltage characteristics that we may need to include in a behavioral model. In addition, we need to include the parasite effects and nonlinear characteristics of the ESD protection devices.

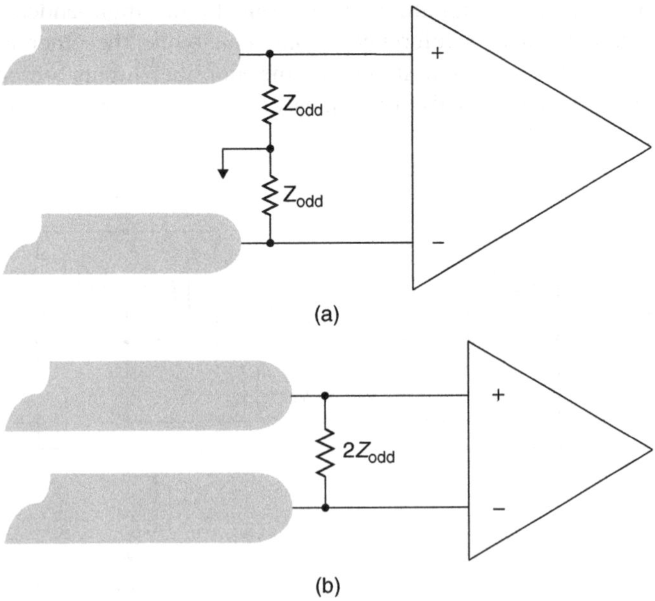

Figure 11-34 Termination options for differential signals: (a) single-ended termination; (b) differential termination.

11.9.3 Advanced Design Considerations

The examples in Section 11.9.1 are intended to illustrate the concepts, and represent only two of a wide range of choices for high-performance receiver designs. In practice, designers have multiple options for improving the various performance aspects of the receiver, such as common-mode range and input offset voltage. We refer interested readers to Dally and Poulton [1997] for more information on receiver design options and techniques.

11.10 IBIS MODELS

As we mentioned in Section 11.1, silicon suppliers do not like to provide transistor models for their I/O circuits. However, simple linear models often do not satisfy the accuracy requirements of high-speed signaling links. To meet this need, the industry has developed the I/O Buffer Information Specification (IBIS). As the name suggests, IBIS is a format for specifying I/O circuit information. Created in the early 1990s, it is now an industry standard owned by the IBIS Forum and is supported by approximately 60 companies. The diverse membership has allowed IBIS to evolve to meet the changing needs of signal integrity and I/O design engineers. In this section we give a brief overview of the IBIS standard, highlighting the major components and providing a high-level description of the model development process. The major features of IBIS include:

- Nonlinear current versus voltage curves for transmitters, ESD devices, and on-chip termination
- Separate nonlinear voltage versus time curves for transmitter pull-up and pull-down devices
- Pad capacitance for I/O circuits
- Models for minimum, typical, and maximum cases within a single model
- Description for multiple types of I/O, including differential pins, open-drain output, tristate outputs, and receivers with hysteresis
- Inclusion of signal quality specifications, including input logic thresholds, overshoots, and so on
- The Golden Parser, a tool that checks model syntax for conformance to the standard
- Backward compatibility with models created under previous revision of the standard

For further details we refer readers to the IBIS specification [IBIS, 2006] and model development "cookbook" [IBIS, 2005].

11.10.1 Model Structure and Development Process

The basic structure of an IBIS-compliant model is shown in Figure 11-35. The $i-v$ and $v-t$ curves are specified in table format, with columns for the minimum,

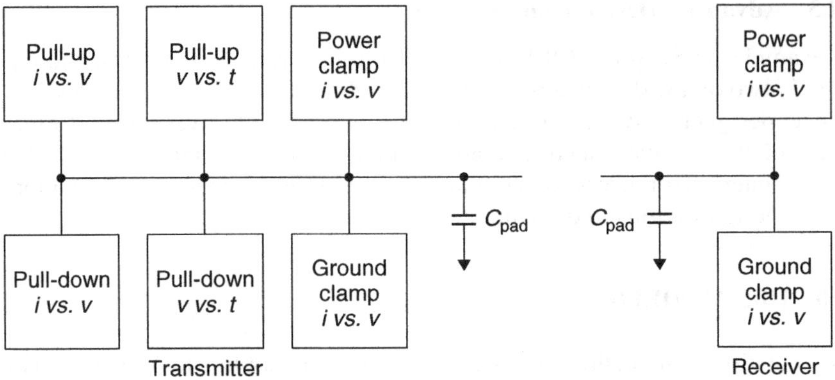

Figure 11-35 Basic structure of an IBIS model.

typical, and maximum case. The power clamp and ground clamp tables contain both ESD clamp information and the behavior of any parasitic diodes which are part of the pull-up and pull-down transistor structures. Power and ground clamp model data may be described separately, or as part of a single table since the EDS circuitry is always active. This same property allows on-die termination resistance to be incorporated into the clamp models for transmitters and receivers, as well.

The development process includes four major steps [IBIS, 2005]:

1. Determine required model features, complexity, and operating range.
2. Obtain the simulated or measured model data ($i-v$ and $v-t$ curves, parasitic capacitances). We treat these in more detail in subsequent sections.
3. Put it into IBIS format and check the file using the Golden Parser. This tool checks the model for syntax compliance with the IBIS standard and is available from the IBIS Forum website.
4. Validate the model. Initially, this is typically done by comparing the transient response of the IBIS model to the original transistor model when driving a reference load. Eventually, the model should be correlated against real silicon.

The data extraction and IBIS formatting can be done manually, though automated tools are available [Varma et al., 2003]. The model process works for both single ended and differential I/O, though it must be modified somewhat to handle the complementary outputs, and to extract both common mode and differential mode data, including capacitances.

11.10.2 Generating Model Data

Current Versus Voltage (i–v) Curves In constructing the $i-v$ curves for a transmitter circuit, the output is connected to an independent voltage source, as shown in Figure 11-36. The current flow into the pad is then measured as the source is swept across a range from $-V_{DD}$ to $2V_{DD}$. IBIS uses the convention that current flow into the transmitter is positive. Separate curves are required for pull-up and pull-down devices. For example, the input to an inverter-based push–pull transmitter would be set to ground to obtain the curve for the pull-up device.

The circuit setup for characterizing a receiver or ESD clamp is identical except that the independent voltage source is connected to the input node. As mentioned earlier, incorporating on-chip termination in an IBIS model is accomplished by including it in the $i-v$ curves for the clamp circuits. This can be accomplished by including it with the clamp circuit during the data extraction process or by extracting the curves separately and adding them together.

The data tables should use enough data points around sharply curved areas of the $i-v$ characteristics to describe the curvature accurately. IBIS does not require equally spaced points in the models, so in linear regions there is no need to include unnecessary data points. Also, as Figure 11-37 demonstrates, the $i-v$ curves for pull-ups and power clamps are V_{DD} referenced, while pull-downs and ground clamps are referenced to ground.

Voltage Versus Time Curves Transmitter output waveforms are described with tables that specify the output voltage versus time ($v-t$). These $v-t$ tables are created when the transmitter is driving into a test load, which is usually a 50-Ω resistor that is connected to the appropriate supply, as described below. The actual $v-t$ data are generated by simulating rising and falling edges, with the output of the transmitter to connected each rail through the test load, as shown in Figure 11-38.

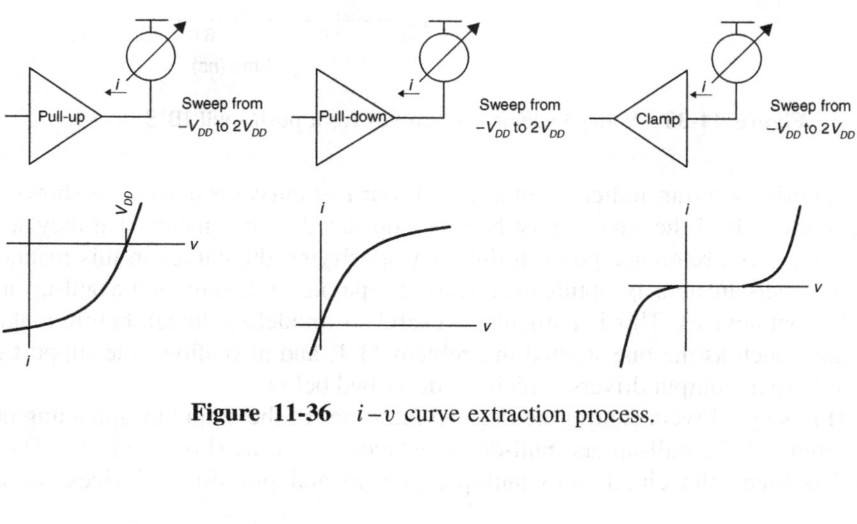

Figure 11-36 $i-v$ curve extraction process.

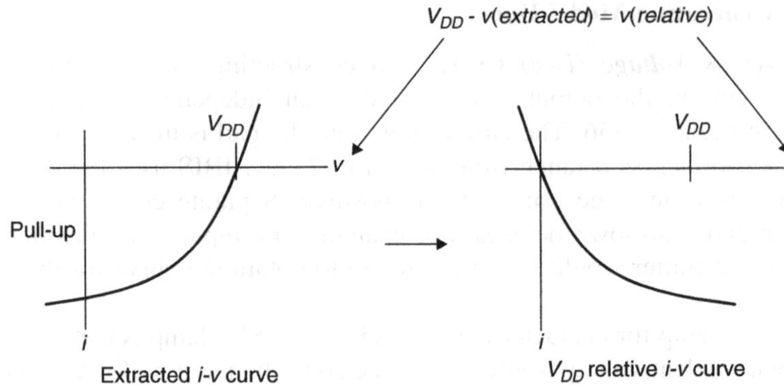

Figure 11-37 Construction of V_{DD} relative $i-v$ curves.

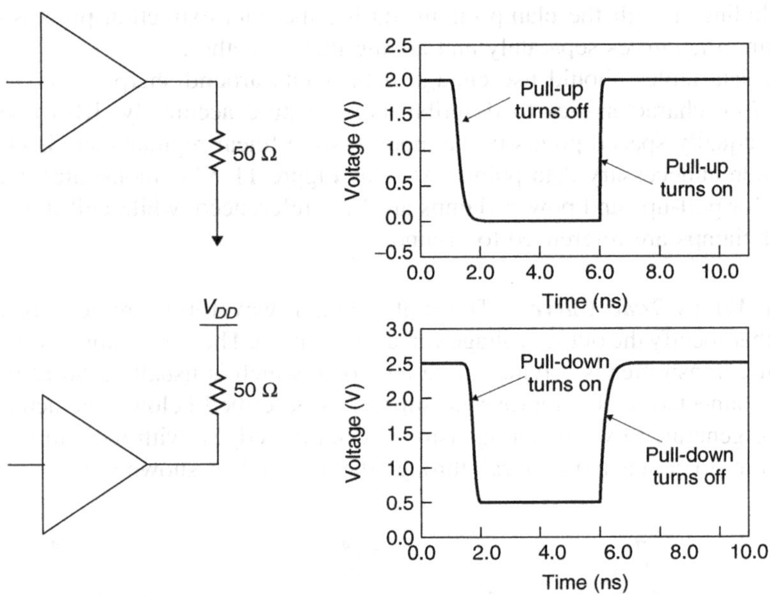

Figure 11-38 Voltage versus time curves for a push–pull IBIS model.

For a push–pull transmitter, a minimum of four $v-t$ curves is desired, as shown in the figure. All of the curves must be time correlated, which means that they start from the same reference point in time. By specifying the curves in this manner, IBIS models are able to capture the effects of separate switching for the pull-up and pull-down devices. This is particularly useful for modeling "break-before-make" circuits, such as the one studied in Problem 11.1, and also allows the support of "multistaged" output drivers, which are described below.

Multistage drivers control the rise and fall times of the output by spreading out switching of the pull-up and pull-down devices over time (Figure 11-39). These models break the circuit into multiple pull-up and pull-down devices, which

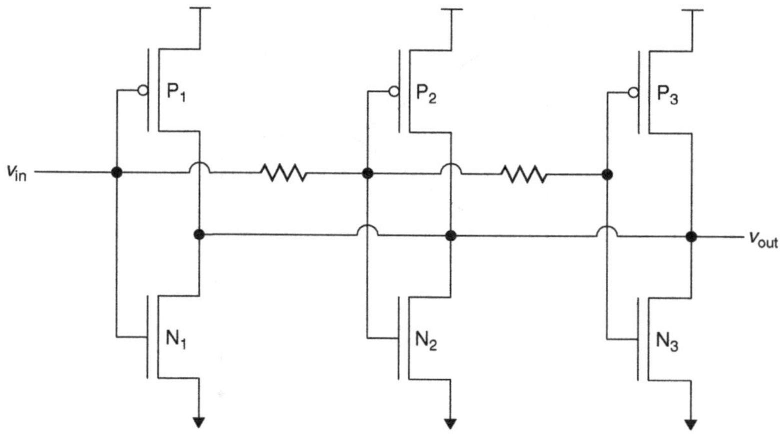

Figure 11-39 Multistage controlled-rise-time transmitter.

requires multiple sets of v–t and i–v curves to properly model the behavior of the circuit. IBIS provides a keyword to allow scheduling of the application of the multiple curves to reflect the staged switching behavior.

IBIS supports up to 1000 v–t points per table, and the time step used in the table should be minimized to provide maximum resolution. The test load that is used in generating the v–t tables must be described in the model. The signal integrity simulation tool will use this information from the model to reconcile the output waveform to reflect the behavior when driving the actual system load. As with the i–v curves, IBIS accommodates separate v–t curves for minimum, typical, and maximum cases.

I/O Capacitance The capacitance of the I/O circuit includes the transistors, on-die interconnect, and die pad, and can have separate values for minimum, typical, and maximum cases. Designers have multiple options at their disposal for extracting the I/O capacitance values. We describe one of them here. We connect the circuit to an ac voltage source at the I/O pad so that we capture all the contributors to the capacitance, as shown in Figure 11-40. By measuring the current flow into the circuit, we can calculate the capacitance using

$$C_{\text{comp}} = \frac{-\text{Im}(i_{\text{ac}})}{2\pi f v_{\text{ac}}} \tag{11-12}$$

where f is the frequency of the ac source, v_{ac} is the amplitude of the source, and $\text{Im}(i_{\text{ac}})$ is the imaginary portion of the current flowing into the ac source.

If the I/O capacitance may be affected by the bias (dc) voltage of the circuit, we must include a dc offset in the ac source. In addition, I/O capacitance typically varies with frequency and applied voltage, so performing multiple extractions while sweeping both of those parameters is recommended. For a given model, simply choose the capacitance value that corresponds to expected switching frequency and bias voltage of the application [IBIS, 2005].

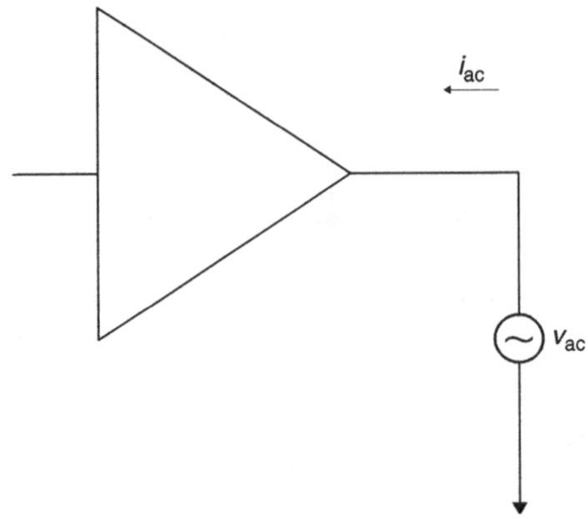

Figure 11-40 Circuit for extracting I/O capacitance.

11.10.3 Differential I/O Models

The IBIS standard includes the ability to specify models for differential transceivers, although the process is more complex than for singled-ended circuits. A differential model is specified as using a driver pair along with a series model, in the manner shown in Figure 11-41a. This structure allows the models to comprehend the differential and common-mode characteristics of a differential I/O circuit. Proper modeling requires specification of the common-mode and differential-mode $i-v$ tables. As a result, the common-mode and differential-mode currents must be must be extracted and separated. Extraction of the $v-t$ tables uses the same techniques as for single-ended transmitters, with the addition of a current source in between the output pins to cancel differential currents inside the transistor model, as shown in Figure 11-41b. The I/O capacitance must include the differential capacitance between the two signals, as Figure 11-41c shows, with the differential and total capacitance expressed as

$$C_{\text{diff}} = \frac{-\text{Im}(i_{dc})}{2\pi f v_{ac}} \tag{11-13}$$

$$C_{\text{comp}} = \frac{-\text{Im}(i_{dc}) - \text{Im}(i_{ac})}{2\pi f v_{ac}} \tag{11-14}$$

where i_{dc} is the current measured through the dc voltage source, i_{ac} is the current measured through the ac voltage source, and v_{ac} is the amplitude of the ac source. The current through the dc source will have an imaginary portion only if there is a reactive path between the two pads of the differential signal.

As a final thought, we note that multi-Gb/s differential signaling is still in the early stages of deployment and that the IBIS specification has not yet fully comprehended the needs of those high-performance links, although the standard will continue to evolve to do so in the near future.

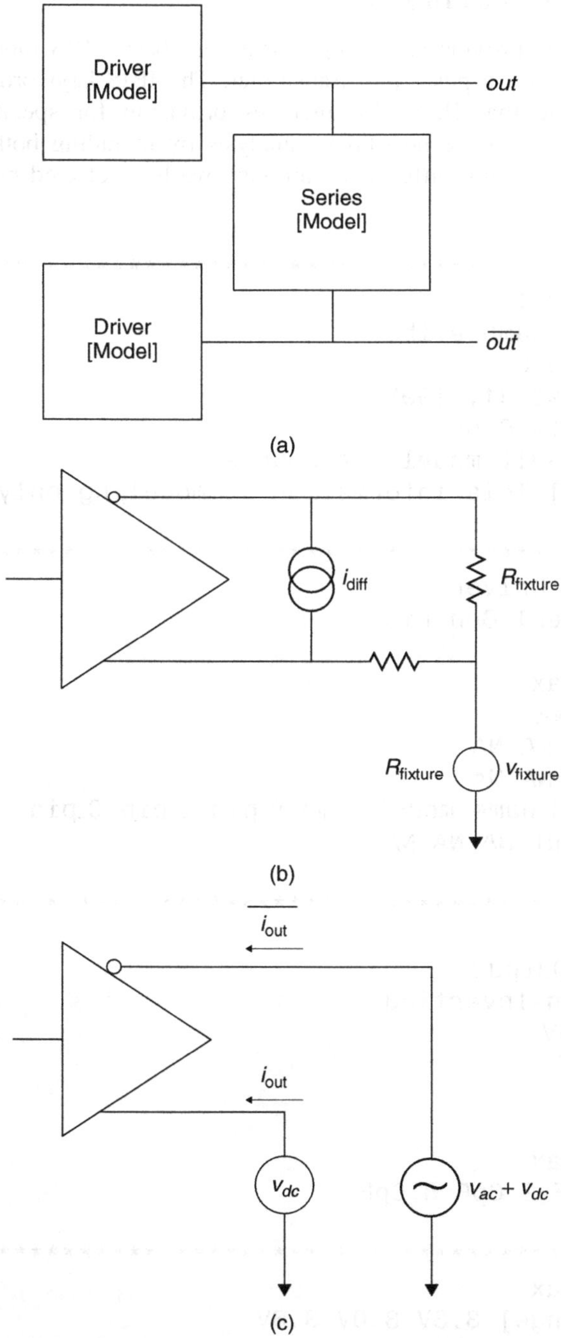

Figure 11-41 (a) IBIS differential model structure; (b) v–t extraction fixture; (c) capacitance extraction fixture.

11.10.4 Example of an IBIS File

In this section we provide an example of a very basic IBIS model file, which contains a model for a push–pull transmitter. The IBIS keywords are enclosed in brackets. Note that IBIS also includes provision for specifying package information and to enable post-layout analysis by including both package and pinlist associations with buffer data, although we have elected not to specify it in this example.

```
|*******************************************************
[IBIS Ver] 2.1
[File name] sample.ibs
[File Rev] 0.0
[Date] August 31, 1999
[Source] Data Book
[Notes] Default model for source.
[Disclaimer] This information is modeling only.

|*******************************************************
[Component] Driver
[Manufacturer] Generic
[Package]
| typ min max
R_pkg 0 NA NA
L_pkg 5.0nH NA NA
C_pkg 2.0pF NA NA
[Pin] signal_name model_name R_pin L_pin C_pin
1 UNKNOWN out NA NA NA

|*******************************************************
[Model] out
Model_type Output
Polarity Non-Inverting
Vmeas = 1.5V
Cref = 15pF
Rref = 500
Vref = 0.0
| typ min max
C_comp 5.5pF 2.0pF 8.0pF

|*******************************************************
| typ min max
[Voltage range] 3.3V 3.0V 3.6V

|*******************************************************
[Pulldown]
```

```
| Voltage I(typ) I(min) I(max)
0 0     0     0
1 0.150 0.075 0.225
2 0.230 0.115 0.345
 ...    ...    ...    ...
3 0.270 0.135 0.405
[Pullup]
| Voltage I(typ) I(min) I(max)
0  0      0      0
1 -0.240 -0.120 -0.360
2 -0.320 -0.160 -0.480
 ...    ...    ...    ...
3 -0.340 -0.170 -0.510
[GND_clamp]
| Voltage I(typ) I(min) I(max)
-1.0 -0.100 -0.050 -0.150
-0.5 -0.020 -0.010 -0.030
-0.4  0      0      0
 ...    ...    ...    ...
0     0      0      0
[POWER_clamp]
| Voltage I(typ) I(min) I(max)
-1.0 0.100 0.050 0.150
-0.5 0.020 0.010 0.030
-0.4  0     0     0
 ...    ...    ...    ...
0     0     0     0

|*******************************************************

[Ramp]
| variable typ min max
dV/dt_r 1.980/3.300n  1.800/3.750n 2.16/2.900n
dt/dt_f 1.980/3.250n  1.800/3.550n 2.16/2.860n
R_load = 50
|
[Falling Waveform]
| typ min max
R_fixture = 50
V_fixture = 3.3
V_fixture_min = 3.15
V_fixture_max = 3.45
0.000ns   3.300 3.150 3.450
0.300ns   3.250 3.110 3.390
 ...    ...    ...    ...
```

```
5.000ns    0.000 0.000 0.000
|
[Rising Waveform]
|| typ min max
R_fixture = 50
V_fixture = 3.3
V_fixture_min = 3.15
V_fixture_max = 3.45
0.000ns    0.000 0.000 0.000
0.300ns    0.050 0.040 0.060
 ...         ...     ...        ...
5.000ns    3.300 3.150 3.450
|
|****************************************************
 [End]
```

11.11 SUMMARY

In this chapter we described the operation and modeling of contemporary high-speed I/O circuits, including transmitters, receivers, and on-die terminations. Insight into the behavior of these circuits is critical to designing successful high-speed signaling solutions. The signal integrity engineer who gains sufficient understanding to interact successfully with his or her I/O circuit counterpart will have a key tool for optimizing a signaling *system* design for high-speed operation.

REFERENCES

I/O circuit design remains an area of active research, with dozens of papers published in conference proceedings and technical journals each year. We do not attempt here to provide an exhaustive survey of the published literature. For more complete treatments of I/O and ESD circuits, we refer the reader to Dabral and Maloney [1997] and Dally and Poulton [1997]. The work by Dabral and Maloney focuses more on basic techniques, whereas Dally and Poulton offer a more comprehensive approach. Granberg [2004] compiled a comprehensive reference that includes technical data on a wide variety of I/O techniques and standards, including memory and multi-Gb/s serial links.

Boni, Andrea, Andrea Pierazzi, and Davide Vecchi, 2001, LVDS I/O interface for Gb/s-per-pin operation in 0.35-μm CMOS, *IEEE Journal of Solid-State Circuits*, vol. 36, no. 4, Apr., pp. 706–711.

Chappell, Barbara, et al., 1998, Fast CMOS ECL receivers with 100-mV worst-case Sensitivity, *IEEE Journal of Solid-State Circuits*, vol. 23, No. 1, Feb., pp. 59–67.

Dabral, Sanjay, and Timothy Maloney, 1998, *Basic ESD and I/O Design*, Wiley-Interscience, New York.

Dally, William, and John Poulton, 1998, *Digital Systems Engineering*, Cambridge University Press, Cambridge, UK.

Esch, Gerald, Jr., and Tom Chen, 2004, Near-linear CMOS I/O driver with less sensitivity to process, voltage and temperature variations, *IEEE Transactions on Very Large Scale Integration Systems*, vol. 12, no. 11, Nov.

Gabara, Thaddeus, and Scott Nauer, 1992, Digitally adjustable resistors in CMOS for high performance applications, *IEEE Journal of Solid-State Circuits*, vol. 27, no. 8, Aug., pp. 1176–1185.

Gabara, Thaddeus, and David Thompson, 1998, Ground bounce control in CMOS integrated circuits, *Proceedings of the 1988 IEEE International Solid-State Circuits Conference*, pp. 88–90.

Gabara, Thaddeus, et al., 1997, *LVDS I/O Buffers with a Controlled Reference Circuit*, IEEE Publ. 1063-0988/97, IEEE Press, Piscataway, NJ, pp. 311–315.

Granberg, Tom, 2004, *Digital Techniques for High-Speed Design*, Prentice Hall, Upper Saddle River, NJ.

Gunning, Bill, et al., 1992, A CMOS low-voltage-swing transmission-line transceiver, *Proceedings of the 1992 IEEE International Solid-State Circuits Conference*, pp. 58 –59.

Hall, Stephen, Garrett Hall, and James McCall, 2000, *High-Speed Digital System Design*, Wiley-Interscience, New York.

IBIS Open Forum, 2005, *IBIS Modeling Cookbook for IBIS Version 4.0*, http://www.eigroup.org/ibis/, Sept. 15.

IBIS Open Forum, 2006, *IBIS (I/O Buffer Information Specification), Version 4.2*, http://www.eigroup.org/ibis/, June.

Intel Corporation, 2002, *AGP V3.0 Specification*, revision 1.0, Intel Press, Hellsboro, OR, Sept.

Intel Corporation, 2005, *64-Bit Intel® Xeon™ Processor with 2MB L2 Cache*, Document 306249-002, Intel Press, Hillsboro, OR, Sept.

Lau, Benedict, et al., 1998, A 2.6-Gbyte/s multipurpose chip-to-chip interface, *IEEE Journal of Solid-State Circuits*, vol. 33, no. 11, Nov., pp. 1617–1626.

Rabaey, Jan, Anantha Chandrakasan, and Borivoje Nikolić, 2003, *Digital Integrated Circuits: A Design Perspective*, 2nd ed., Prentice Hall, Upper Saddle River, NJ.

Synopsis, Inc., 2006, *HSPICE® Simulation and Analysis User Guide, Version Y-2006.3*, Synopsis, Mountain View, CA, Mar.

Texas Instruments, 1996, *The Bergeron Method: A Graphic Method for Determining Line Reflections in Transient Phenomena*, Document SDYA014, Texas Instruments, Dallars, TX, Oct.

Varma, Ambrish, et al., 2003, The development of a macro-modeling tool to develop IBIS models, *Proceedings of the IEEE 12th Topical Meeting on Electrical Performance of Electronic Packaging*, Oct. 27–29, pp. 277–280.

Wang, Niantsu, 1989, *Digital MOS Integrated Circuits*, Prentice Hall, Upper Saddle River, NJ.

PROBLEMS

Use the device parameters in Table 11-5 when solving these problems.

TABLE 11-5. Device Parameters for a 0.25 μm Process

	NMOS	PMOS
V_{T0} (V)	0.43	−0.4
γ ($V^{1/2}$)	0.4	−0.4
$V_{D,SAT}$ (V)	0.63	−1
k (A/V^2)	115×10^{-6}	-30×10^{-6}
λ (V^{-1})	0.06	−0.1

Source: Rabaey et al. [2003].

11-1 Describe operation of the break-before-make circuit shown in Figure 11-42.

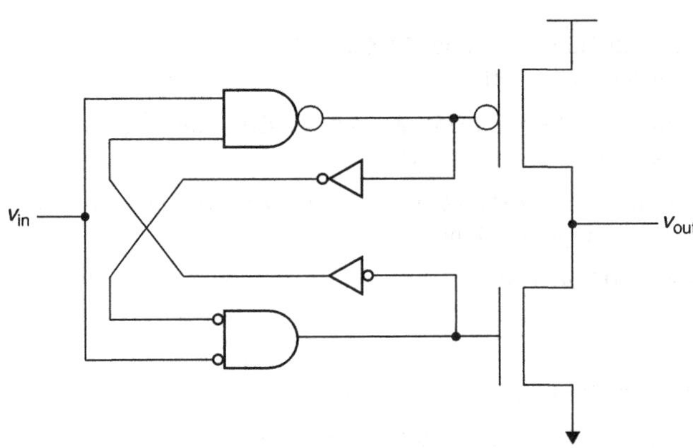

Figure 11-42 Break-before-make push–pull transmitter circuit.

11-2 Construct the i–v curve(s) and use them to calculate output impedance for the push–pull transmitter shown in Figure 11-43.

11-3 Explain how the staged circuit in Figure 11-39 provides a controlled rise time at the output of the circuit. Sketch the output waveform expected.

11-4 Describe the operation of the Schmitt trigger receiver shown in Figure 11-14.

11-5 Use a Bergeron diagram to analyze the falling-edge response of the circuit in Figure 11-18.

11-6 Use a Bergeron diagram to analyze the rising edge for the circuit shown in Figure 11-44.

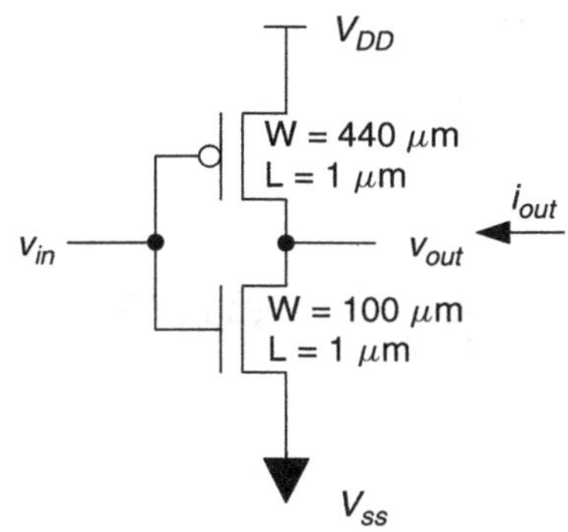

Figure 11-43 Push-pull transmitter for Problem 11-2.

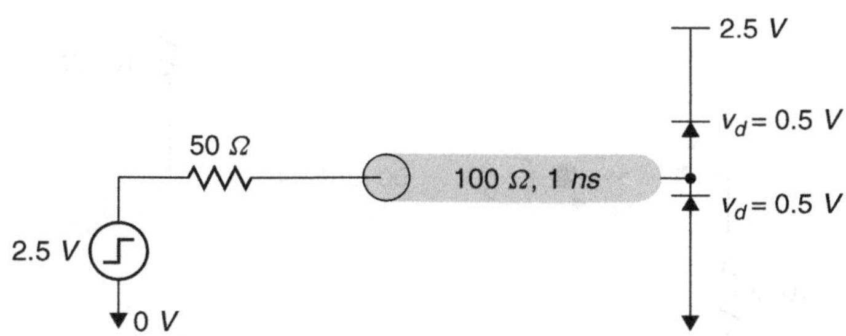

Figure 11-44 Diode-terminated circuit for Problem 11-6.

11-7 Describe how the GTL transmitter circuit shown in Figure 11-30 slows the rising-edge transition to reduce ringing due to parasitic package inductance.

11-8 Use a Bergeron diagram to analyze the far-end-terminated open-drain circuit in Figure 11-45.

11-9 Analyze the two open-drain circuits in Figure 11-45. How do the resulting waveforms differ? Which is likely to be capable of supporting a higher data transfer rate?

11-10 *Wired-OR glitch:* The circuit in Figure 11-46 contains transmitters at each end, with a receiver circuit near the middle. Initially, the right-hand transmitter pulls the circuit low while the transmitter on the left is turned off. At time $t = 0$ the left-hand transmitter turns on and the right-hand transmitter turns off. Sketch the waveform at the receiver.

11-11 Use the device parameters in Table 11.5 to calculate the current versus voltage curve for the on-die termination circuit shown in Figure 11-47 and estimate the effective termination resistance.

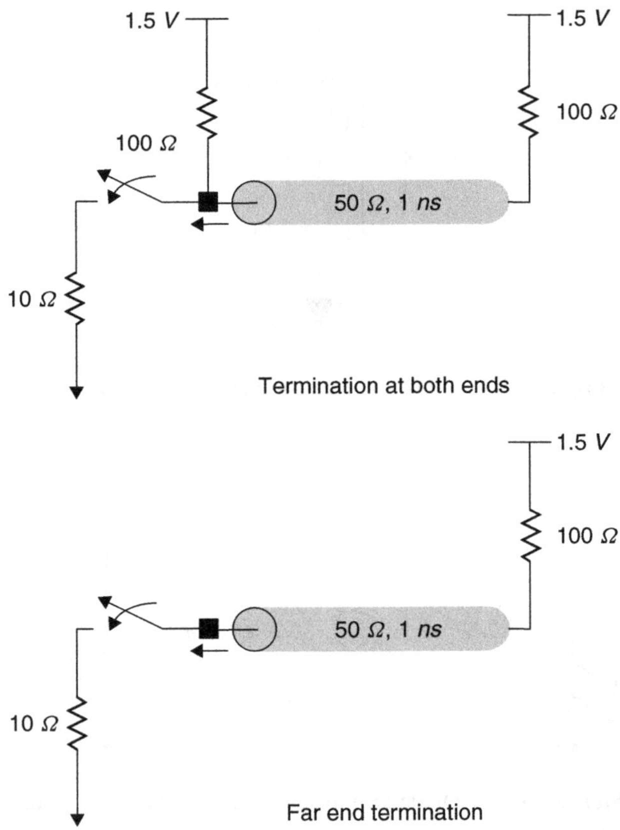

Figure 11-45 Open-drain circuits for Problem 11-8.

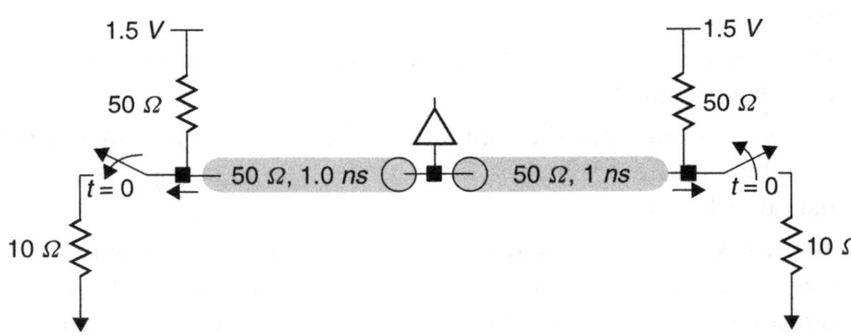

Figure 11-46 Wired-OR circuit for Problem 11-10.

11-12 Discuss the potential advantages and disadvantages of the single-ended and differential termination schemes shown in Figure 11-34.

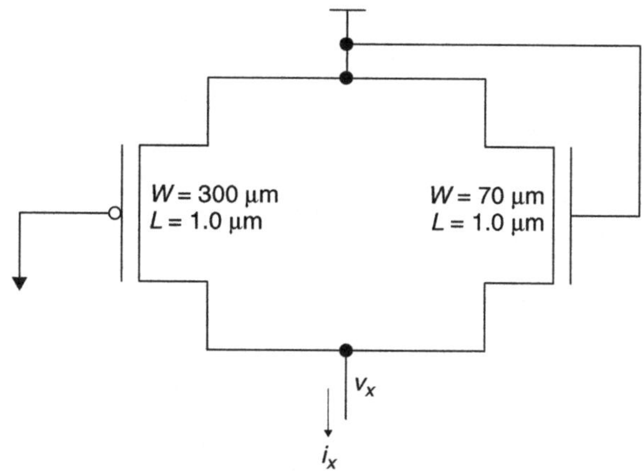

Figure 11-47 FET termination circuit for Problem 11-11.

11-13 Complete a Bergeron diagram using the transmitter and receiver load lines shown in Figure 11-48, which are used in conjunction with a 70-Ω transmission line.

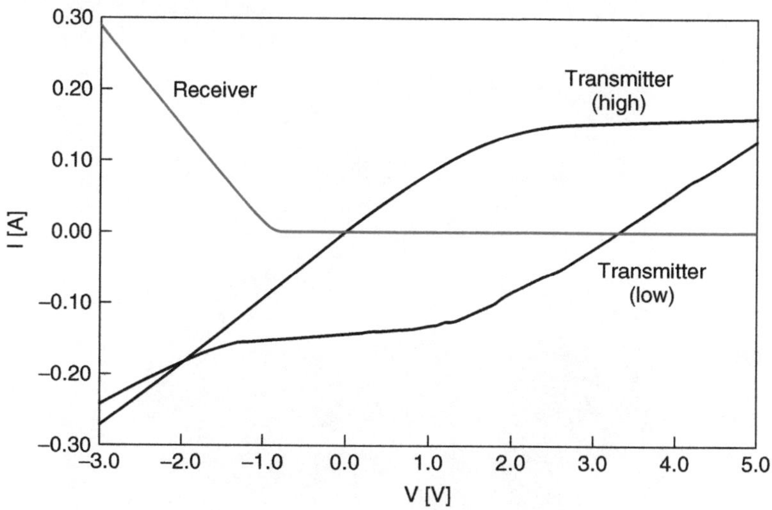

Figure 11-48 Load lines for Problem 11-13.

12

EQUALIZATION

We have already discussed the impact of Moore's law, which drives the interchip data bandwidth to continually increasing performance levels. We have also shown that nonideal aspects of transmission lines, such as crosstalk and losses, can have a significant impact on signal integrity and timing. These impacts dominate at multi-Gb/s speeds, causing "smearing" of signals so that their energy is spread over multiple bit positions, a phenomenon known as *intersymbol interference* (ISI). The impact of ISI is an increase in the jitter that degrades the timing margin and a distortion in the signal levels that degrades the voltage margin of the interchip signaling link. Equalization is a circuit technique that reduces the

Advanced Signal Integrity for High-Speed Digital Design, By Stephen H. Hall and Howard L. Heck

ISI-induced timing jitter and voltage margin loss by compensating for nonideal aspects, in particular the loss of interconnects at high speed.

In this chapter we adopt a communications channel–based approach to analyzing our signaling interfaces. Communications engineers view the I/O circuits and the interconnect (also known as the *channel*) as filters, as shown in Figure 12-1. In previous chapters we noted the low-pass filtering effect of the I/O and interconnect. By viewing the system as a series of filters, including the equalizer, we can determine the desired characteristics for a given equalizer.

12.1 ANALYSIS AND DESIGN BACKGROUND

Before considering equalization analysis and design, we must cover some necessary background material. We first offer some motivation for employing equalizers in high-speed signaling systems by examining the requirements for maximizing data transfer rates. We then present the notion of linear time-invariant (LTI) systems and show how to use their characteristics to analyze the behavior of high-speed signals. Finally, we contrast the characteristics of an ideal interconnect with that of a physically realizable interconnect.

12.1.1 Maximum Data Transfer Capacity

Shannon's capacity theorem [Shannon, 1949] describes the upper limit on the information rate that can be transmitted over a communications channel. The theorem is widely accepted in the scientific community and has never been exceeded in practice. We provide a heuristic derivation of the capacity equation with the motivation of developing an understanding of how closely we can approach the theoretical maximum with conventional interconnects, and we then demonstrate how equalization techniques allow us to come closer to achieving the maximum rates.

We start by defining the data transfer rate in bits per second as the product of the number of symbols transmitted per second (S) and the number of bits per symbol (B):

$$D = SB \tag{12-1}$$

The symbol transfer rate is related directly to the channel bandwidth by the Nyquist rate [Nyquist, 1928]:

$$S = 2\text{BW} \tag{12-2}$$

where BW is the bandwidth in hertz. To develop an intuitive feeling for (12-2), consider a simple binary nonreturn-to-zero (NRZ) signaling scheme. In Figure 12-2 we show a periodic pulse train with a repetition frequency f and a random data sequence with the same fundamental frequency. A single symbol is contained within successive edge positions, each positioned a width of T_{symbol} apart, and the full cycle is twice the symbol width. For the periodic signal, the

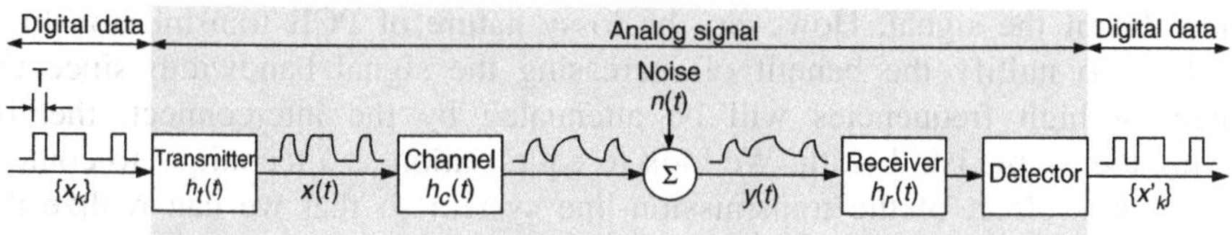

Figure 12-1 Communications channel view of a high-speed signaling interface.

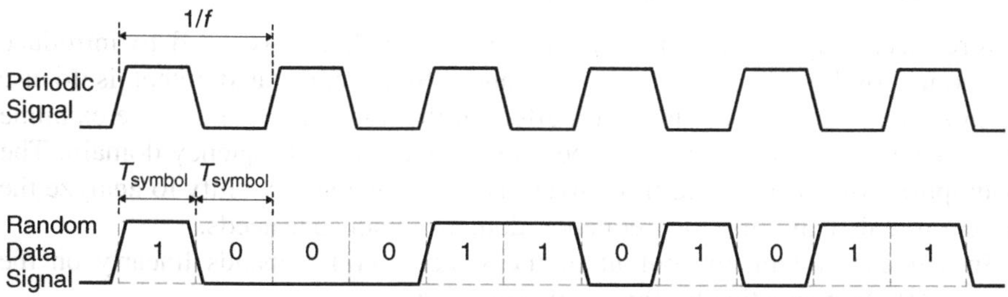

Figure 12-2 Symbol rate illustration for a binary NRZ signal.

bandwidth of the signal is equal to the repetition frequency.† Thus, we have two symbols per cycle.

One of the outcomes of Shannon's work states that the maximum number of bits per symbol, B, that can be transmitted without error is given by

$$B = \frac{1}{2}\log_2\left(1 + \frac{P_s}{P_n}\right) \tag{12-3}$$

where P_s is the average signal power and P_n is the noise power. The quantity P_s/P_n is also known as the *signal-to-noise ratio* (SNR). Equation (12-3) assumes that the noise is Gaussian, meaning that it is constant at all frequencies within the channel bandwidth, which is a reasonable approximation for digital systems [Sklar, 2001].

Combining the preceding equations gives the *Shannon–Hartley theorem*, which expresses the maximum data transfer rate in bits per second (b/s) as a function of the interconnect channel bandwidth and the SNR:

$$D = \mathrm{BW}\log_2(1 + \mathrm{SNR}) \tag{12-4}$$

Equation (12-4) shows that we can increase throughput across an interchip interconnect either by increasing the signal-to-noise ratio or by increasing the

†A real digital signal contains energy at harmonic frequencies above the fundamental, which we can estimate from the rise time ($\mathrm{BW} \cong 0.35/t_r$) as derived in Section 8.1.3. In this analysis, however, we are considering only the fundamental frequency.

bandwidth of the signal. However, the lossy nature of PCB transmission lines will tend to nullify the benefit of increasing the signal bandwidth since the content at high frequencies will be attenuated by the interconnect, thereby limiting the usable bandwidth. We look to equalization as a technique to counter the low-pass effect of the transmission-line system so that we can realize the performance benefit of increasing signal bandwidth.

12.1.2 Linear Time-Invariant Systems

Before discussing the design and analysis of equalizers we need to introduce the notion of linearity and time invariance. Since our stated intent is to use equalization to counter the low-pass effect of the transmission lines, we examine the behavior of the equalizer and the interconnect in the frequency domain. The assumptions of linearity and time invariance give us the flexibility to analyze the system in either the time or frequency domain to suit our needs.

In a linear system, the output of the system, $y(t)$, depends linearly on the input, $x(t)$. Mathematically, this is expressed as

$$y(t) = f[cx(t)] = cf[x(t)] \tag{12-5}$$

In addition, linearity also means that the relationship between the input and the output of the system satisfies the superposition property. That is, if the system input can be expressed as the sum of multiple input components, $x(t) = \sum_i x_i(t)$, the output is equal to the sum of the output values that are obtained for each of the input components:

$$y(t) = \sum_i f[x_i(t)] = \sum_i y_i(t) \tag{12-6}$$

These are interesting results that we will use many times in analyzing multi-Gb/s systems, but we have only begun to scratch the surface. Additional aspects of LTI systems that we will exploit are the notions of the impulse response, transfer function, and equivalence between the time and frequency domains. In the time domain, the impulse response of the system, $h(t)$, relates the input to the output via convolution:

$$y(t) = \int_{t=-\infty}^{\infty} h(t-\tau)x(\tau)\,d\tau = h(t) * x(t) \tag{12-7}$$

where $*$ signifies the convolution operation. The impulse response is the response of the system to an impulse function. The impulse function $\delta(t)$ has the properties that it is zero everywhere except at $t = 0$ and that the area under it is equal to 1 (see Figure 12-3):

$$\delta(t) = \begin{cases} 0, t \neq 0 \\ \infty, t = 0 \end{cases} \quad \text{and} \quad \int_{t=-\infty}^{\infty} \delta(t)\,dt = 1 \tag{12-8}$$

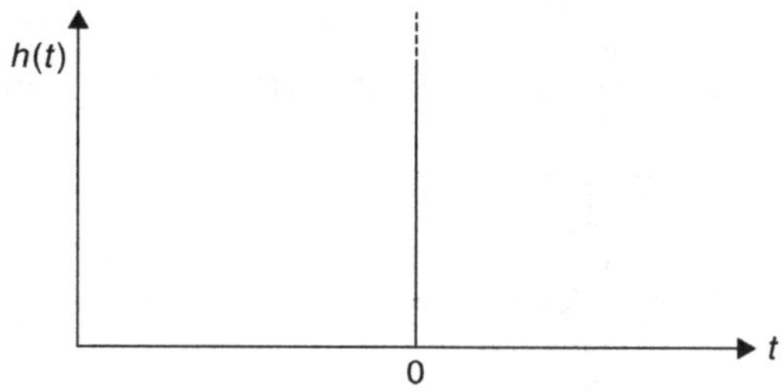

Figure 12-3 Impulse function.

Examination of equation (12-7) reveals the convolution operation to be an integral equation that may be difficult to solve. A simpler alternative is to perform the operation in the frequency domain, where it becomes a simple multiplication:

$$Y(f) = H(f)X(f) \tag{12-9}$$

where $Y(f), H(f)$, and $X(f)$ are the frequency-domain representations of $y(t), h(t)$, and $x(t)$.

We obtain the frequency-domain forms by applying the Fourier transform:

$$H(f) = \int_{t=-\infty}^{\infty} h(t)e^{2\pi jft}\, dt \tag{12-10}$$

where f is the frequency in hertz. Notice that the units of $H(f)$ will be the units of $h(t)$ multiplied by seconds. For example, if $h(t)$ is dimensionless, $H(f)$ has units of seconds. Restoring results back to the time domain requires application of the inverse Fourier transform:

$$h(t) = \int_{f=-\infty}^{\infty} H(f)e^{-2\pi jft}\, df \tag{12-11}$$

The time–frequency equivalence and the relationships between input and output are summarized in Figure 12-4. Note that the frequency-domain representation of the impulse response $H(f)$ is known as the *transfer function*. In general, the frequency-domain representations will be complex quantities, whereas time-domain quantities are real.

Example 12-1 Transfer Function for a PCB Differential Transmission-Line Pair Figure 12-5 shows the transfer function for a 0.381-m-long differential pair on a printed circuit that has the following characteristics at a reference frequency, f_0, of 1 GHz:

$$L = \begin{bmatrix} 3.299 & 0.407 \\ 0.407 & 3.299 \end{bmatrix} \text{nH/cm} \qquad C = \begin{bmatrix} 1.098 & 0.085 \\ 0.085 & 1.098 \end{bmatrix} \text{pF/cm}$$

$$R = \begin{bmatrix} 509.1 & 60.63 \\ 60.64 & 509.1 \end{bmatrix} \text{m}\Omega\text{/cm} \qquad G = \begin{bmatrix} 0.131 & 0.012 \\ 0.012 & 0.131 \end{bmatrix} \text{mS/cm}$$

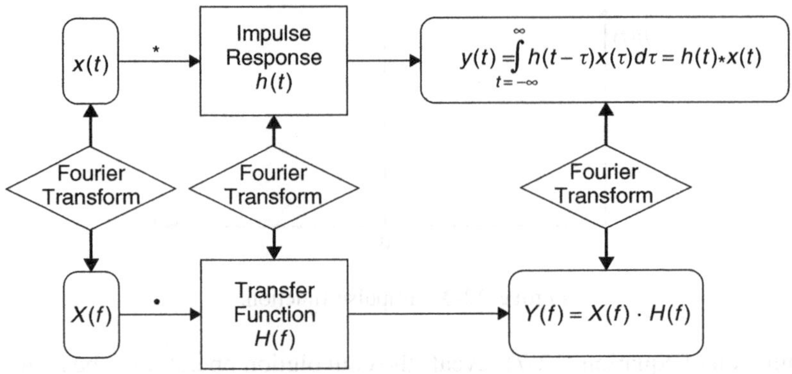

Figure 12-4 Time/frequency-domain relationship for LTI systems.

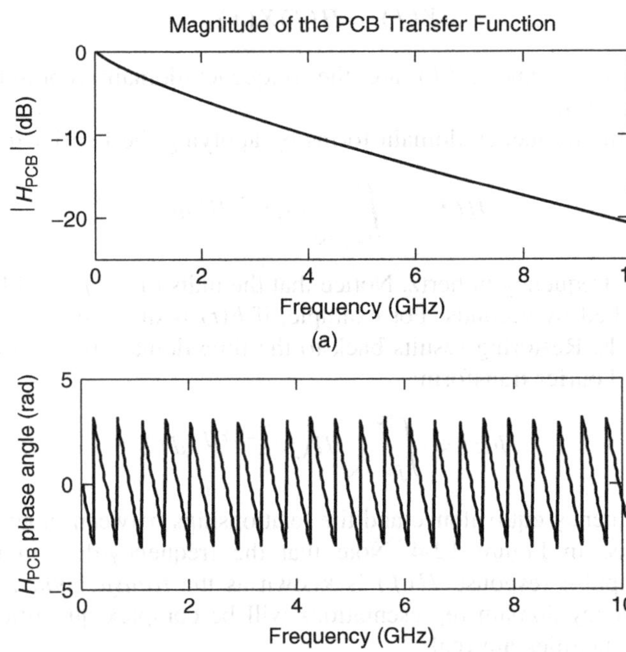

Figure 12-5 Frequency response for a PCB-based lossy differential transmission-line pair: (a) magnitude; (b) phase.

The differential transfer function is calculated from equation (6-49) assuming no reflections and the odd-mode values for the propagation constant using the method described by Johnson and Graham [2003], which gives

$$H_{\text{PCB}}(f) = \frac{v_{\text{out}}}{v_{\text{in}}} = e^{-\gamma(f)l} \tag{12-12}$$

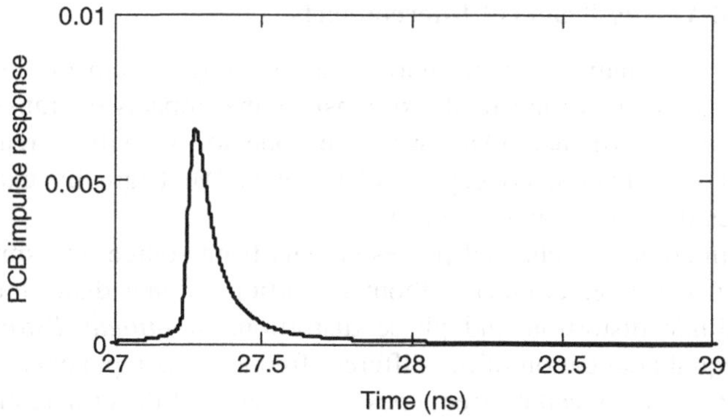

Figure 12-6 Impulse response for a PCB-based lossy differential transmission-line pair.

where

$$\gamma(f) = \sqrt{[R_{\text{odd}}(f) + j2\pi f L_{\text{odd}}(f)][G_{\text{odd}}(f) + j2\pi f C_{\text{odd}}(f)]} \qquad (12\text{-}13)$$

The frequency-dependent transmission-line parameters in equation (12-13) are calculated using the causal modeling methodology presented in Chapter 10. Equation (12-12) assumes that the line is perfectly terminated so that there is no reflected wave.† The plot in Figure 12-5a shows the magnitude of the transfer function plotted in decibels:

$$|H_{\text{PCB}}| = 20\log[\sqrt{\text{Re}(H_{\text{PCB}})^2 + \text{Im}(H_{\text{PCB}})^2}] \qquad \text{dB}$$

Figure 12-5b shows the phase angle formed by the imaginary and real components:

$$\angle H_{\text{PCB}} = \tan^{-1}\frac{\text{Im}(H_{\text{PCB}})}{\text{Re}(H_{\text{PCB}})} \qquad \text{rad}$$

Figure 12-6 shows the impulse response for the differential pair that was calculated from equation (12-11).

†The transfer function for a terminated transmission line is equal to

$$\frac{1}{\left(1 + \frac{R_{\text{Tx}}}{R_{\text{Rx}}}\right)\frac{e^{\gamma l}+e^{-\gamma l}}{2} + \left(\frac{Z_0}{R_{\text{Rx}}} + \frac{R_{\text{Tx}}}{Z_0}\right)\frac{e^{\gamma l}-e^{-\gamma l}}{2}}$$

where R_{Tx} and R_{Rx} are the termination values at the transmitter and receiver, respectively, and Z_0 is the frequency-dependent characteristic impedance. If we assume that $R_{\text{Tx}} = R_{\text{Rx}} = Z_0$, the transfer function becomes $\frac{1}{2}e^{-\gamma l}$. The differential transfer function will have twice the swing, so that $H(f) = e^{-\gamma l}$.

12.1.3 Ideal Versus Practical Interconnects

Now that we can analyze the behavior of a signaling system in either the time or frequency domain, we are ready to consider the impacts of transmission-line losses on their performance. Once we have done so, we will be ready to begin our study of equalization concepts and designs. We begin by examining the characteristics of an ideal interconnect.

An ideal interconnect channel passes signals from source (transmitter circuit) to destination (receiver circuit) without distortion. Signal distortion takes two forms, amplitude distortion and phase distortion. *Amplitude distortion* results from the unequal attenuation of the different frequency components of the signal. *Phase distortion* is induced by frequency dependence of the propagation velocity, which causes the phase relationship of the frequency components of the signal to change as it propagates on the transmission line. Amplitude distortion shows up clearly in a plot of the transfer function magnitude, while phase distortion is evident from a plot of the phase of the transfer function versus frequency.

Distortion-free transmission is demonstrated in Figure 12-7, which shows the transfer function and impulse response for the differential transmission lines of the example from the preceding section, except that conductor and dielectric losses have been eliminated and the dielectric permittivity is constant across all frequencies. Since losses have been eliminated, the magnitude of the transfer function is equal to 1 (or to 0 dB) at all frequencies. In addition, the impulse response shows a very sharp peak with almost no tail.

Example 12-2 Maximum Data Rate Capacity of an Ideal Interconnect In this example we use the Shannon–Hartley theorem to calculate the theoretical maximum transfer capacity of a channel possessing ideal transmission characteristics up to a frequency of 10 GHz, with the assumption that the signal transmitted contains no energy above 10 GHz. Recalling the Shannon–Hartley theorem,

$$D = \mathrm{BW} \log_2 \left(1 + \frac{P_s}{P_n}\right)$$

we see that we have the bandwidth information, but we also need the power spectral densities for the signal and noise.

We assume that the noise on the channel is *white noise*, so that it is spread uniformly across the entire frequency band [Sklar, 2001]. A reasonable figure for the noise spectrum in terms of the voltage is $V_{\text{noise}} = 10^{-7}\ \mathrm{V} \cdot \mathrm{s}^{0.5}$. The power spectrum can then be approximated as $P_{\text{noise}} = V_{\text{noise}}^2 = 10^{-14}\ \ \mathrm{V}^2 \cdot \mathrm{s}$ (since power is proportional to V^2, $P = V^2/R$). Since our goal is to estimate the theoretical maximum data rate, we assume that the signal is spread uniformly across the entire band with a 1-V amplitude. The signal spectrum is equal to

$$V_s = \frac{1\ \mathrm{V}}{\sqrt{10\ \mathrm{GHz}}} = 10^{-5}\mathrm{V} \cdot \mathrm{s}^{0.5}$$

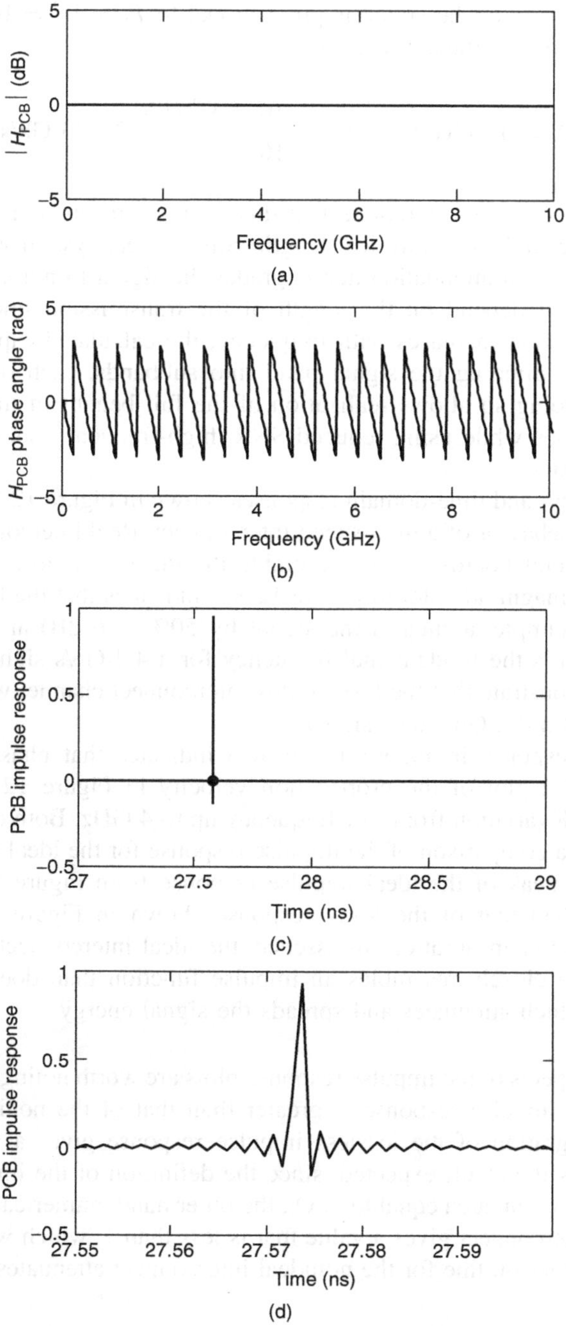

Figure 12-7 Transfer function and impulse response for a lossless, dispersionless PCB-based differential transmission line pair: (a) transfer function magnitude in decibels; (b) transfer function phase in radians; (c) impulse response; (d) impulse response (close-up).

The power spectrum of the signal is proportional to $P_s = V_s^2 = 10^{-10}\ \text{V}^2 \cdot \text{s}$.

The maximum data transfer capacity is then

$$D = (10\ \text{GHz}) \log_2 \left(1 + \frac{10^{-10}\text{V}^2 \cdot \text{s}}{10^{-14}\text{V}^2 \cdot \text{s}}\right) \cong 133\ \text{Gb/s}$$

The assumption of ideal interconnect in this example means that the maximum data rate is independent of physical length. Since a real system will experience per unit length signal attenuation that degrades the signal-to-noise ratio, the theoretical limit will depend on the length of the transmission line. In addition, since the attenuation increases with frequency, the calculated capacity can also be increased by dividing the signal band into subbands, each of which has a unique SNR. Doing so allows realization of the full benefit of the higher SNR at low frequencies while using reduced-SNR, high-frequency subbands to carry some of the data.

The frequency- and time-domain responses shown in Figure 12-7 represent the desired (ideal) behavior of a high-speed interconnect. Real interconnects, such as FR4 printed circuit boards, do not resemble the ideal behavior, as Figure 12-8 illustrates. The magnitude plot in Figure 12-8a[†] indicates that the lossy line from our previous example attenuates the signal by 50% (−6 dB) at approximately 2.1 GHz, which is the fundamental frequency for a 4.2-Gb/s signal. In the next section we demonstrate that the loss on this interconnect channel will completely close the eye of a 4.2-Gb/s data signal.

The phase response in Figure 12-8b also indicates that phase distortion is significant, and a plot of the propagation velocity in Figure 12-8c reveals an approximate 9% variation from low frequency up to 4 GHz. Both of these effects are apparent in a comparison of the impulse response for the ideal and real cases. Notice that the peak of the ideal impulse response from Figure 12-8d is many times greater than that of the lossy response shown in Figure 12-8e and the tail is much shorter in duration. In essence, the ideal interconnect gives a result that much more closely resembles an impulse function than does the nonideal interconnect, which attenuates and spreads the signal energy.

Two other aspects of the impulse response plots are worth noting. First, the area under the ideal impulse response is greater than that of the nonideal response. Numerical integration of the lossless impulse response gives a value of 1 for the area. This is the result expected, since the definition of the impulse function specifies that it has an area equal to 1. On the other hand, numerical integration of the nonideal interconnect gives a value that is less than 1, which we expect since the lossy transmission line for the nonideal interconnect attenuates the impulsive input signal.

[†]The figure plots the magnitude of the transfer function as a power ratio in decibels as $20 \log[H(f)/H_0]$, where $H(f)$ is the voltage transfer function at frequency f and H_0 is equal to 1.

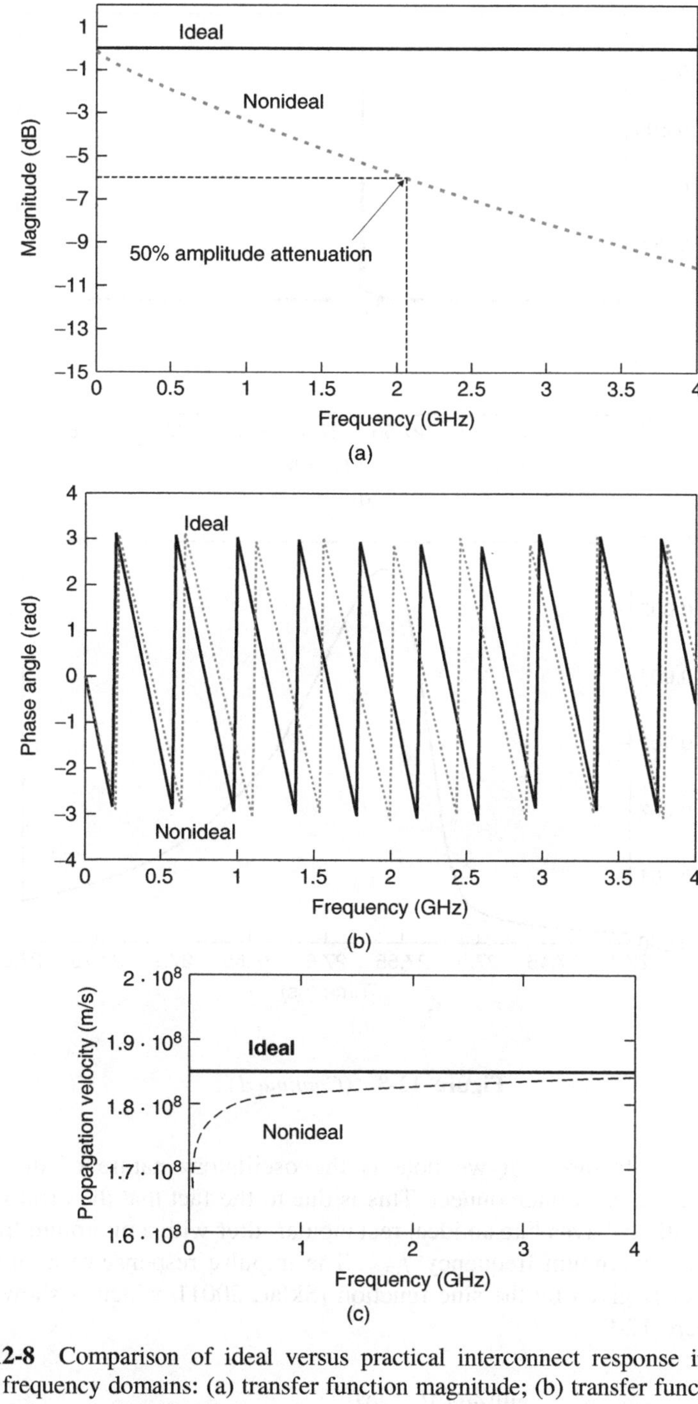

Figure 12-8 Comparison of ideal versus practical interconnect response in both the time and frequency domains: (a) transfer function magnitude; (b) transfer function phase; (c) propagation velocity; (d) ideal interconnect impulse response; (e) nonideal interconnect impulse response. *(Continued)*

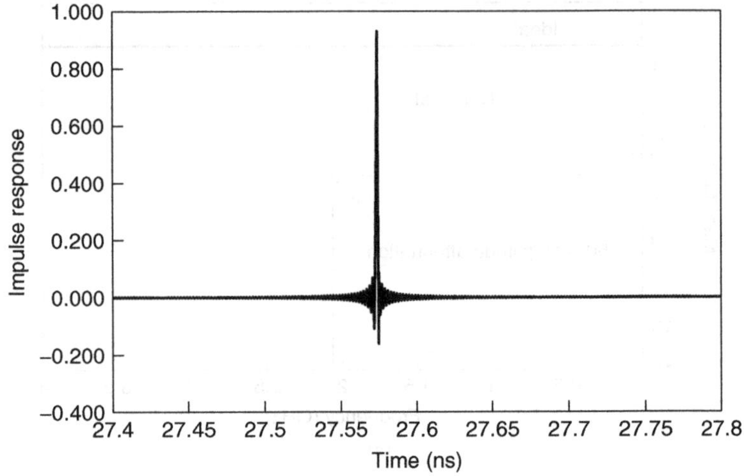

(d)

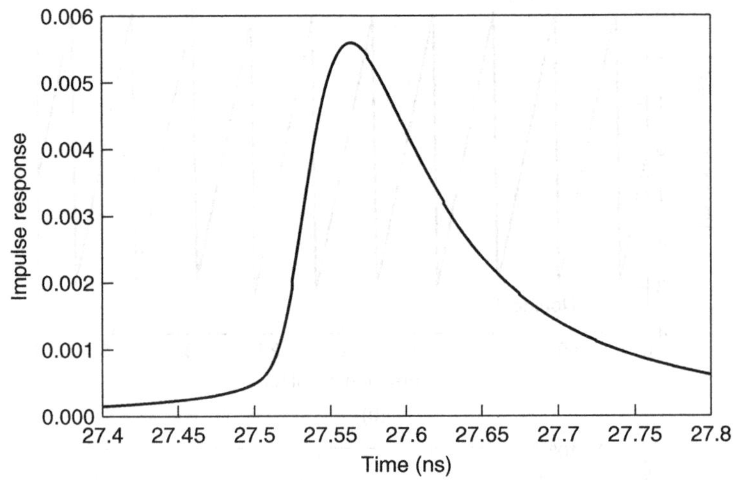

(e)

Figure 12-8 *(Continued)*

The second feature that we note is the oscillatory nature of the impulse response for the ideal interconnect. This is due to the fact that the ideal interconnect essentially behaves like an ideal rectangular filter with a minimum frequency at zero and a maximum frequency f_{max}. The impulse response of an ideal rectangular filter is given by the sinc function [Sklar, 2001], which is shown below and in Figure 12-9.

$$h(t) = 2f_{max}\frac{\sin[2\pi f(t - t_0)]}{2\pi f(t - t_0)} = 2f_{max}\,\text{sinc}\,2\pi f(t - t_0)$$

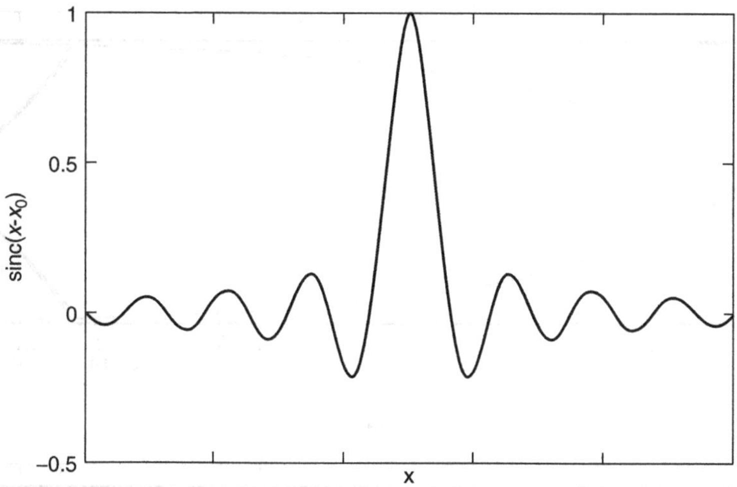

Figure 12-9 Sinc function.

12.1.4 Equalization Overview

Our study of high-frequency interconnects has established that they typically impart a low-pass effect on high-speed signals, which causes amplitude and phase distortion. The typical result is that the interconnect channel closes the data "eye." An *eye diagram* is constructed by superimposing numerous consecutive bits in a data stream. The opening of the eye is a metric often used to judge the quality of the signal integrity; an open eye generally indicates a recoverable bit stream, and a closed eye indicates a distorted bit stream that is not recoverable. The eye diagram is discussed fully in Chapter 13. Figure 12-10a shows an open eye at the output of a transmitter. Figure 12-10b shows the eye at the receiver after the data stream has propagated across a lossy interconnect. Note that the interconnect has reduced the area of eye opening dramatically. To develop our conceptual understanding further, we consider the waveforms in Figure 12-11. Figure 12-11a shows the transmitter output for a bit pattern that contains sections with alternating logic levels for each consecutive bit, along with sections that maintain a given logic level for multiple bit positions. As a result, the signal power spectrum spans a wide range of frequencies, with significant content from dc to beyond the 5-GHz fundamental frequency. The high-frequency components get attenuated much more severely than do the low-frequency and dc components, as Figure 12-11b demonstrates. A bit pattern of alternating logic states (101010) will have a fundamental frequency that is higher than that of bit patterns that have fewer transitions (such as 110011001100).

Since losses increase with frequency, bit patterns that have a higher frequency content will be attenuated more. In other words, for fast bit patterns, the time it takes for the signal to "charge up the interconnect" and transition to its maximum value is greater than the switching rate of the transmitter. This is the intersymbol

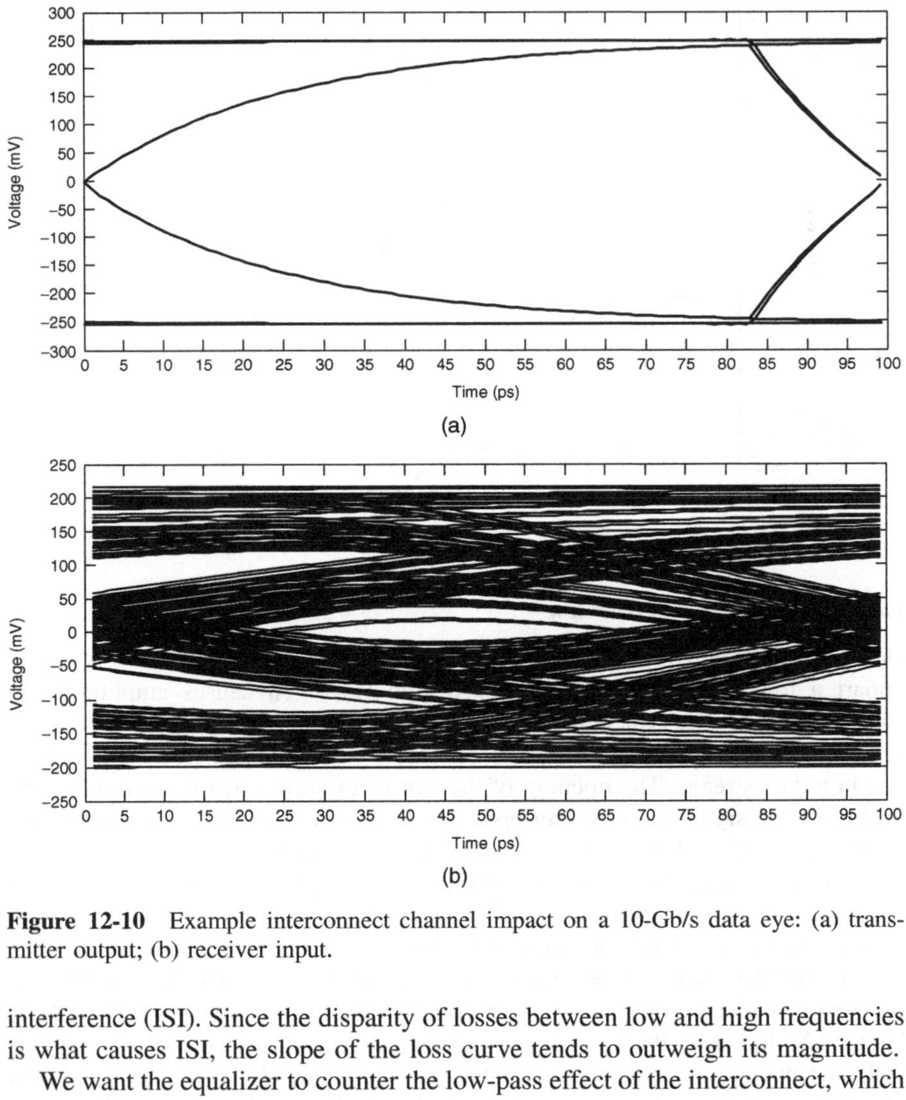

Figure 12-10 Example interconnect channel impact on a 10-Gb/s data eye: (a) transmitter output; (b) receiver input.

interference (ISI). Since the disparity of losses between low and high frequencies is what causes ISI, the slope of the loss curve tends to outweigh its magnitude.

We want the equalizer to counter the low-pass effect of the interconnect, which causes the high-frequency signal content to be attenuated much more severely than the low-frequency content. Thus, the desired operation of the equalizer is to amplify the signal in such a way as to perfectly counter the attenuation at each frequency. This is expressed mathematically as

$$H_{\text{eq}}(f) = H_{\text{channel}}^{-1}(f) \tag{12-14}$$

Simply stated, the ideal equalizer has a transfer function that is the inverse of the channel transfer function, making it a high-pass filter. Figure 12-12 illustrates this concept, showing that the ideal equalizer counters not only the amplitude distortion caused by the interconnect, but the phase distortion as well. As

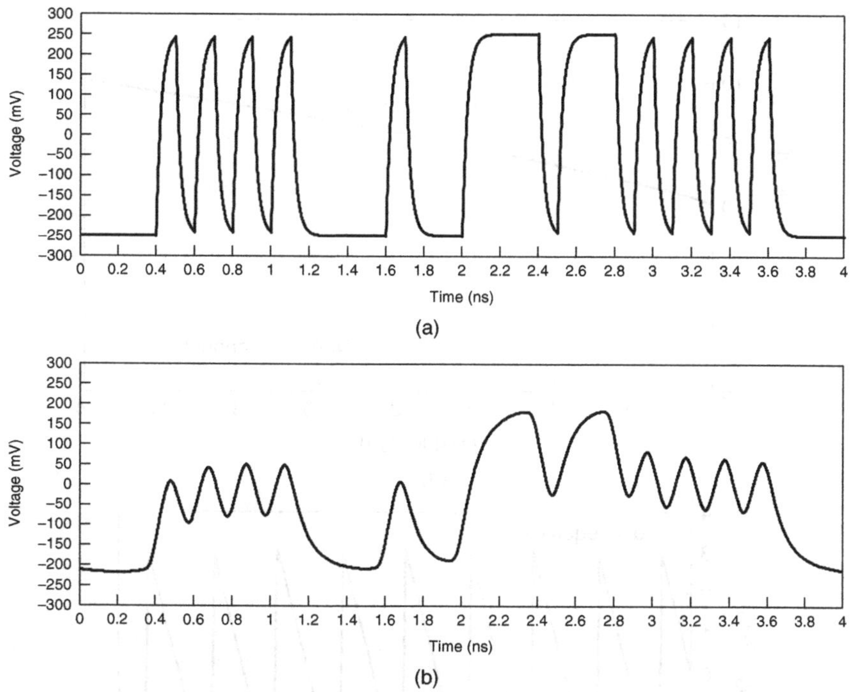

Figure 12-11 Example interconnect channel impact on a 10-Gb/s signal waveform: (a) transmitter output; (b) receiver input.

Figures 12-13 and 12-14 demonstrate, an idealized equalizer would completely reverse the effects of the interconnect, restoring the signal to its original form, exactly matching the waveform and eye at the transmitter output.

In practice, power and device count limitations make the design of a perfect equalizer impractical. However, we do not require ideal implementation in order to realize the benefits of equalization, as we shall see in the remaining sections of this chapter. In subsequent sections we explore the design, operation, and limitations of the different types of equalizers that find use in multi-Gb/s signaling systems now and in the future. Equalization is an area in which there is considerable ongoing research and one that has many possible implementations. Although we present some representative examples, the nuances of equalizer design implementations are beyond our scope. Instead, we focus on developing a solid understanding of their behavior in terms of fundamental building blocks so that we may use them effectively to maximize the performance of our signaling systems.

12.2 CONTINUOUS-TIME LINEAR EQUALIZERS

Continuous-time linear equalizers (CTLEs) are analog in nature, operating continuously as the name implies. This is in contrast to discrete-time equalizers,

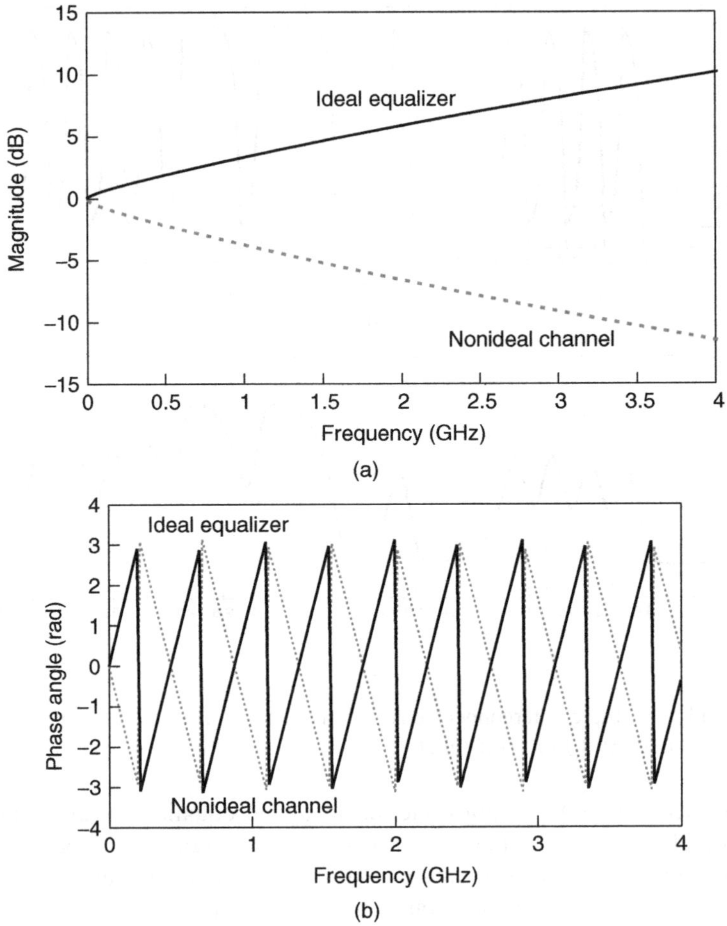

Figure 12-12 Transfer function of an ideal equalizer: (a) magnitude; (b) phase.

which we discuss in subsequent sections. CTLEs do not require digital devices for implementation but are constructed from simple analog components, both passive (resistors, capacitors, inductors) and active (amplifiers). In this section we use the CTLE to expand on our previous discussion of equalizer behaviors, and we develop an understanding of their operation through the presentation of some representative designs.

12.2.1 Passive CTLEs

An example of a passive CTLE is shown in Figure 12-15. As the name suggests, this type of equalizer does not amplify any components of the signal that pass through it. Instead, to achieve the desired high-pass effect the equalizer attenuates the low-frequency signal components. The main sections of the equalizer are the

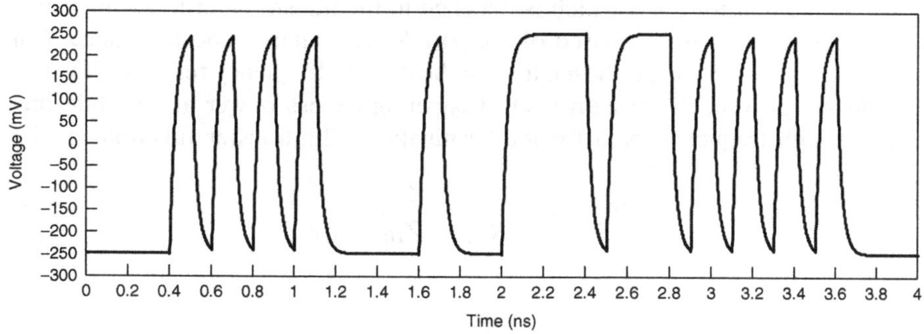

Figure 12-13 Waveform showing the restoration of the lossy signal from Figure 12-11 after processing with an ideal equalizer.

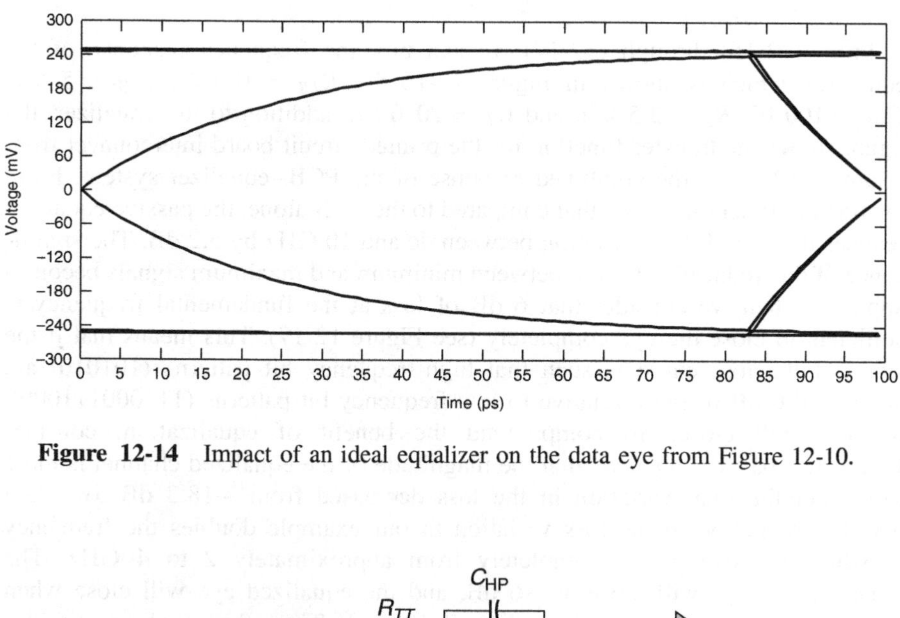

Figure 12-14 Impact of an ideal equalizer on the data eye from Figure 12-10.

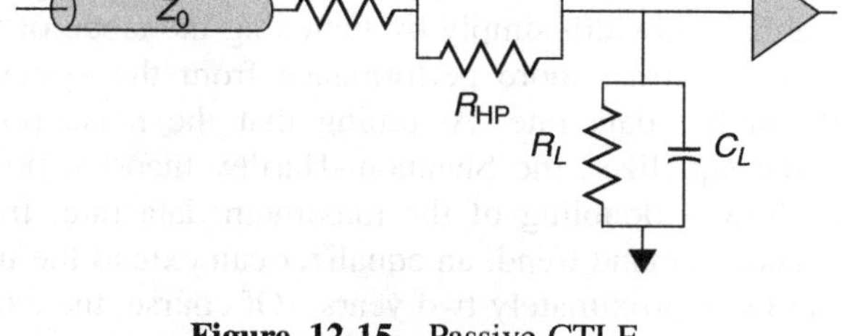

Figure 12-15 Passive CTLE.

termination, high-pass filter, and dc power-limiting filter. Termination is provided by the single resistor R_{TT}, which is matched to the transmission-line impedance. The high-pass filter is composed of the parallel capacitance and resistance, C_{HP} and R_{HP}, whose values are chosen to give the desired frequency response. Finally, C_L and R_L prevent the system from dissipating excess power at dc while providing a high-frequency path for the termination. The transfer function for this circuit is

$$H_{\mathrm{eq}}(f) = \frac{Z_L}{R_{TT} + Z_{\mathrm{HP}} + Z_L} \tag{12-15}$$

where

$$Z_L(f) = \frac{R_L}{1 + j2\pi f R_L C_L} \tag{12-16}$$

$$Z_{\mathrm{HP}}(f) = \frac{R_{\mathrm{HP}}}{1 + j2\pi f R_{\mathrm{HP}} C_{\mathrm{HP}}} \tag{12-17}$$

From (12-15) through (12-17) we can plot the frequency response of the equalizer, which is shown in Figure 12-16 for $R_{TT} = 100\ \Omega$, $R_{\mathrm{HP}} = 5\ \mathrm{k}\Omega$, $C_{\mathrm{HP}} = 100$ fF, $R_L = 2.5\ \mathrm{k}\Omega$, and $C_L = 20$ fF. In addition to the equalizer, the figure shows the transfer function for the printed circuit board interconnect from Example 12-1 and the combined response of the PCB–equalizer system. From the transfer function we see that compared to the PCB alone, the passive equalizer reduces the overall loss variation between dc and 10 GHz by 5.2 dB. The significance of this reduced variation between minimum and maximum signals becomes apparent when we consider that 6 dB of loss at the fundamental frequency is sufficient to close the eye completely (see Figure 12-17). This means that if the slope of the loss curve is such that high-frequency bit patterns (101010) are attenuated 6 dB or more relative to low-frequency bit patterns (111000111000), the eye will close. To comprehend the benefit of equalization, consider Figure 12-16a, which shows that the magnitude of the equalized channel is more lossy, but the total variation in the loss decreased from −18.3 dB to −23.5 dB. The reduction in the loss variation in our example doubles the frequency at which the eye closes completely from approximately 2 to 4 GHz. The nonequalized eye will close at −6 dB, and the equalized eye will close when there is −6 dB of variation from the −7-dB peak. Thus, the equalizer provides an increase in usable bandwidth simply by flattening the slope of the loss curve that will allow us to extract more performance from the system by running at a substantially higher data rate. Assuming that the noise power spectrum is unaffected by the equalizer, the Shannon–Hartley theorem predicts that the equalizer would allow a doubling of the maximum data rate. In terms of the expected performance demand trend, an equalizer can extend the useful life of a given interconnect by approximately two years. (Of course, the exact benefit for a given system will depend on the specifics of the channel loss characteristics, the equalizer design, and actual performance demand.)

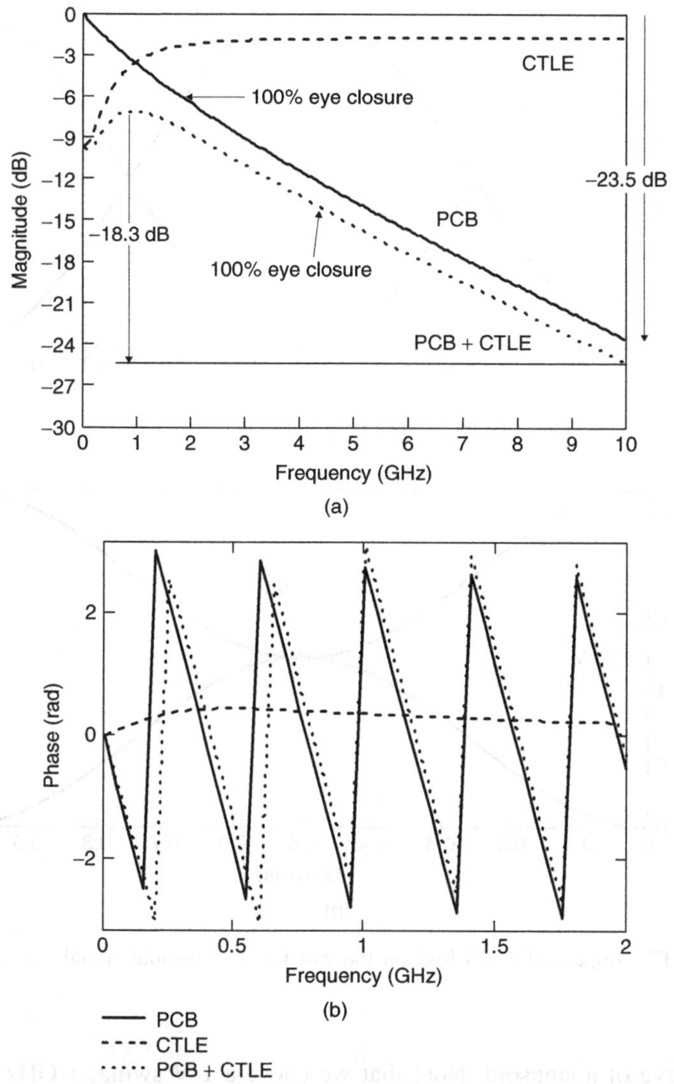

Figure 12-16 Passive CTLE transfer function: (a) magnitude; (b) phase.

The other impact of the equalizer is that the overall swing is reduced in order to achieve the reduction in minimum-to-maximum loss for the system. This "flattening" of the magnitude of the frequency response is a good qualitative way to view the benefit of equalization. Even though we reduce the maximum signal swing of the system, we realize a net gain because we have reduced the variation between the maximum signal swing (low frequency) and the minimum signal swing (high frequency). Figures 12-17 through 12-19 each illustrate this concept in slightly different ways. In Figure 12-17 we see that a 6-dB loss is sufficient to completely

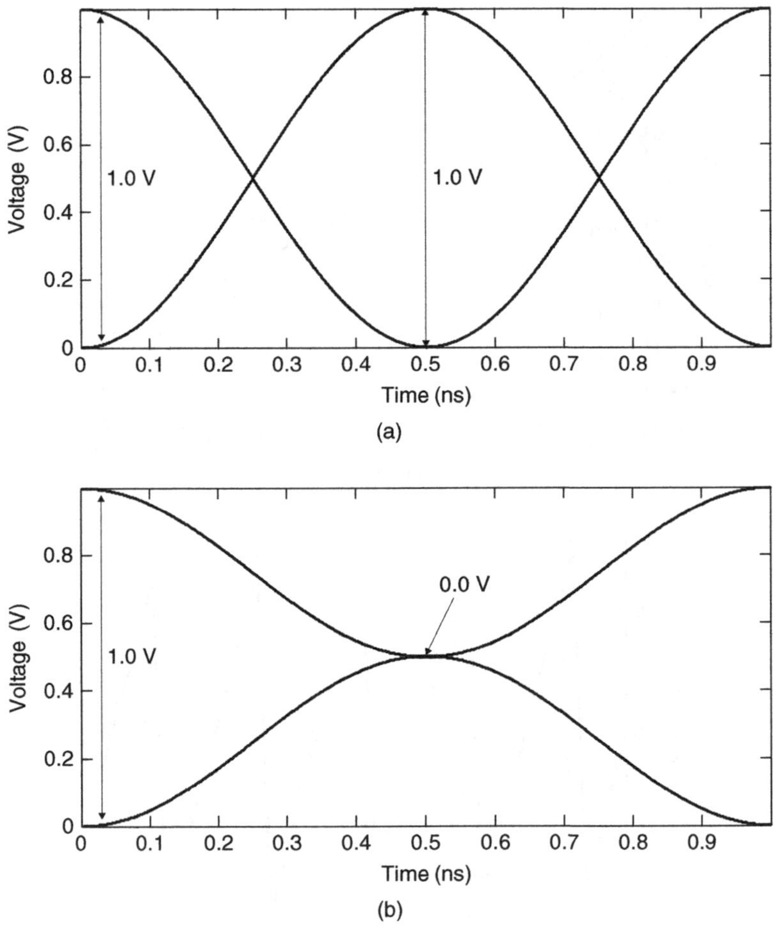

Figure 12-17 Impact of 6-dB loss on the eye for a sinusoidal signal: (a) lossless; (b) 6-dB loss.

close the eye of a sinusoid. Note that we chose a 1-V swing, 1-GHz signal for illustration, but that the 6-dB rule is independent of swing and frequency. What does depend on frequency is where the 6-dB loss point actually occurs.

Figure 12-18 portrays the eye diagram for a 10-Gb/s signal traveling across the 0.381-m-long differential PCB transmission-line pair from Example 12-1. Given a 500-mV signal swing and a 300-bit random sequence at the transmitter output, we find that the eye at the receiver in Figure 12-18a is completely closed. Figure 12-18b shows that the equalizer improves the situation considerably, providing an eye opening of approximately 55 ps and 40 mV. Comparison of the eyes shows that the equalizer reduces the maximum signal swing from 430 mV at the receiver (Figure 12-18a) input to 250 mV at the output of the equalizer (Figure 12-18b).

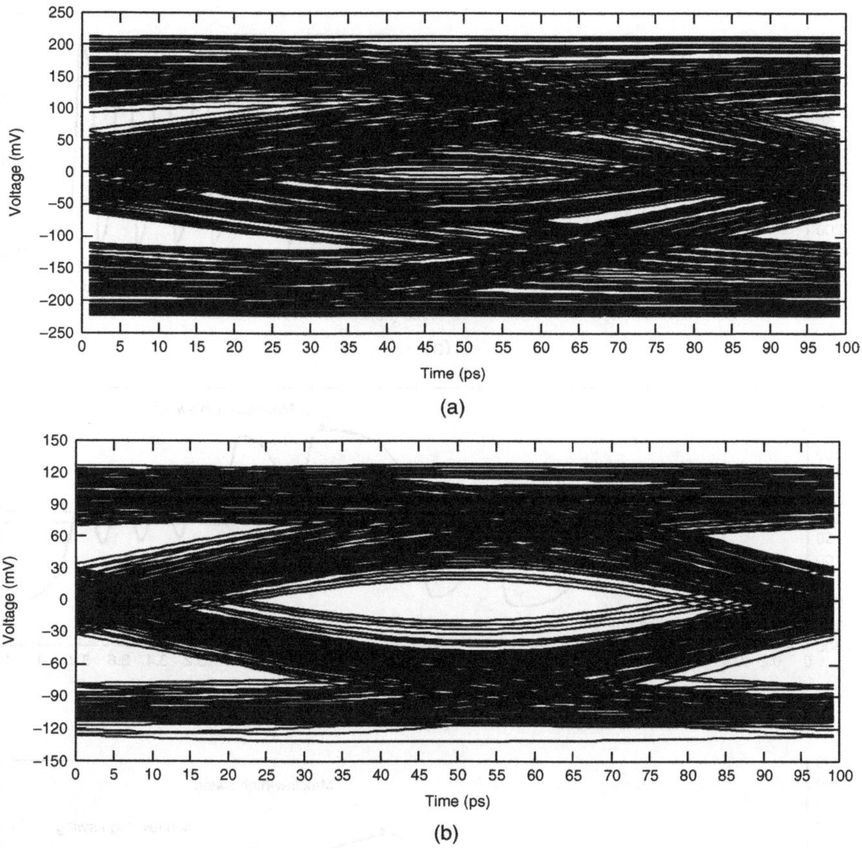

Figure 12-18 Received eye diagram with passive CTLE (a) before and (b) after equalization.

Finally, Figure 12-19a shows the signal waveform for the same bit pattern as in Figure 12-11a. The waveform at the receiver in Figure 12-19b shows the nature of the behavior that leads to the closed data eye. The signal shows a maximum range of approximately −210 mV to approximately 200 mV. However, the minimum swings for the low-to-high and high-to-low transitions reach only 5 mV and −5 mV, respectively. So for this abbreviated bit pattern, there is only a 10-mV eye opening. A longer bit sequence will further degrade the eye to the point where it is completely closed. Moving to the passively equalized signal in Figure 12-19c reveals a maximum range of approximately −105 mV to approximately 125 mV. The minimum swings for the low-to-high and high-to-low transitions reach 30 mV and −20 mV, respectively. So while the equalizer reduces the maximum signal swing from 410 mV to 230 mV, it increases the minimum eye from 10 mV to 50 mV. Clearly this is a favorable trade-off.

Before moving to discussion of additional CTLE design options, we pause to note that the transfer function for our passive equalizer does not come close

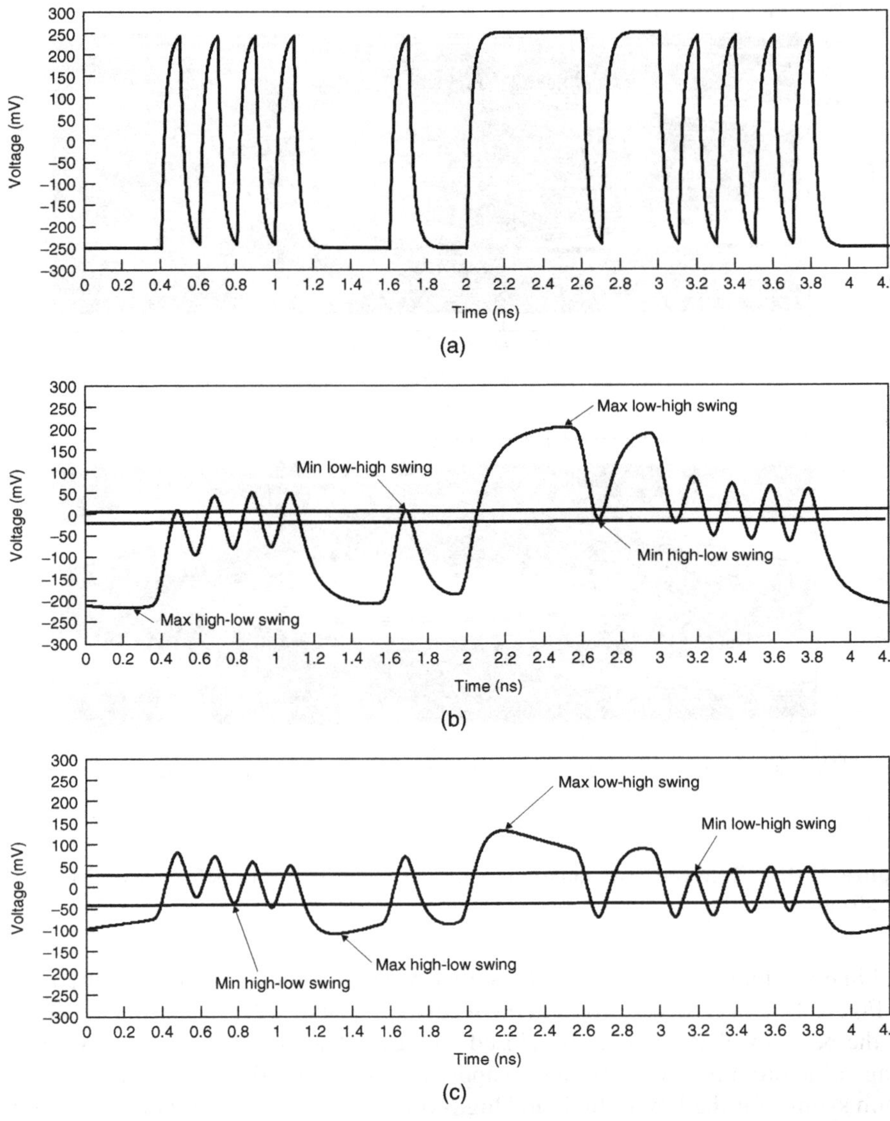

Figure 12-19 Example signal waveforms with passive CTLE: (a) transmitter output; (b) received input; (c) equalized output.

to providing perfect equalization (compare Figures 12-12 and 12-16). However, the equalizer compensates sufficiently for the loss of the 0.381-m PCB trace to increase the eye height by more than 400% for a 10-Gb/s data signal (from 10 to 50 mV).

Another example of a passive equalizer from Sun et al. [2005] based on a high-pass *RLC* filter is shown in Figure 12-20. In this case we show versions

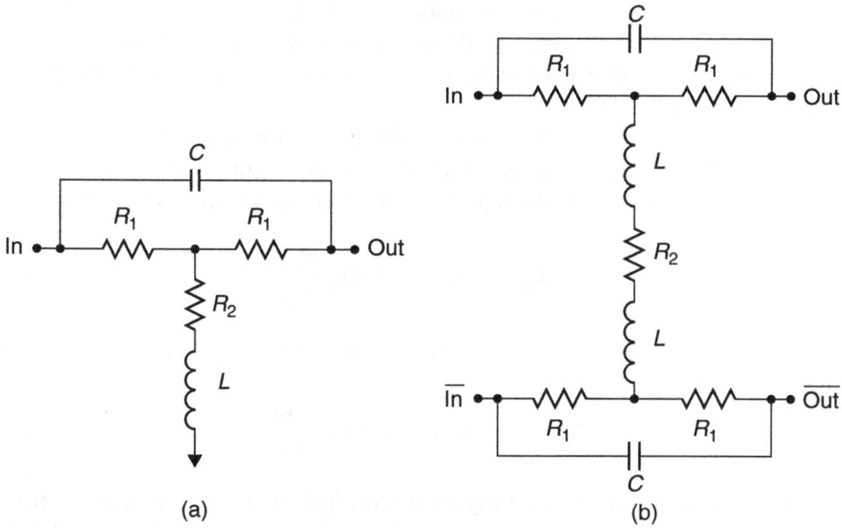

Figure 12-20 Alternate passive CTLE implementation: (a) single-ended; (b) differential.

for both singled-ended and differential signals. Analysis of this equalizer is left as a problem at the end of the chapter. Passive equalizers offer the advantage of improved performance with no additional power consumption. In addition, multi-Gb/s data rates drive smaller, higher-frequency passive devices, making integration into the silicon an attractive design option. However, they demand tighter control of the components values than for typical digital applications. In addition, the frequency response of passive equalizer circuits is not directly tunable without additional active control circuitry, which will tend to degrade the power benefit.

12.2.2 Active CTLEs

Equalizers can also be constructed with active components (amplifiers) to provide some signal gain. This type of equalizer is often done using a split-path approach, as shown in Figure 12-21 [Liu and Ling, 2004]. The incoming signal is fed into a unity-gain path and a high-frequency boost path, which are then summed to create the output. The transfer function for this equalizer is

$$H(f) = \frac{1 + R_2/R_3}{1 + 1/2\pi f R_1 C_1} + 1 \tag{12-18}$$

The high-pass filter has a voltage gain of $1 + R_2/R_3$, with a corner frequency equal to $1/R_1C_1$. The equalizer transfer function for a high-pass filter voltage gain of -3.5 dB and a 5-GHz corner frequency is plotted along with that of the passive equalizer in Figure 12-22. The magnitude plots for the two equalizers show a very similar shape, with an offset of approximately 5 dB. The resulting data eye for the equalized 200-bit random pattern is shown in Figure 12-23. As expected, the equalizer gives an open data eye with approximately twice the signal amplitude and eye height than those of the data eye from the passive equalizer in Figure 12-18.

A potential source of confusion that merits discussion is the use of power gain versus that of voltage gain. In our transfer function plots, we use power gain. Since power is a function of the square of the voltage (or current) signal,

$$\text{gain}_{\text{power}}(\text{dB}) = 10 \log \frac{P_{\text{out}}}{P_{\text{in}}} \tag{12-19}$$

Conversely, amplifier manufacturer datasheets often specify gain in terms of the voltage:

$$\text{gain}_{\text{voltage}}(\text{dB}) = 20 \log \frac{V_{\text{out}}}{V_{\text{in}}} \tag{12-20}$$

The moral here is that when working with gain figures, be sure to know whether you are dealing with voltage or power gains, as mixing the two can lead to erroneous results.

Active CTLEs provide gain, so they must dissipate some power, although careful design can keep the power to less than 10 mW. In addition, active CTLEs can be designed using higher-order high-pass filters, but the performance gain typically may not justify the additional power consumed depending on the channel response of the specific application. The active CTLE has some fundamental limitations, notably the limited bandwidth of the amplifier and phase mismatch between the two amplifier paths. For example, Kudoh et al., [2003] show −3-dB bandwidths of 3 GHz for a conventional feedback amplifier and 10 GHz for an improved design. Finally, tailoring the frequency response and the gain of the filter requires the inclusion of additional control circuitry, just as it did with the passive equalizer.

12.3 DISCRETE LINEAR EQUALIZERS

In Section 12.2 we dealt with equalizers that were completely analog in nature. Contemporary high-speed components such as microprocessors, graphics processors, or memory controllers are manufactured using processes that are highly optimized for digital circuit applications. Whereas the equalization problem is very analog in nature, the discrete linear equalizer makes use of a combination of digital and analog techniques. This allows us to better utilize the economies of scale provided by Moore's law, thus satisfying the performance demand at minimum cost. As such, they find wider use in computing equipment than do their analog counterparts.

The linear time-invariant nature of our signaling system has another aspect that we have not yet covered that will give us flexibility in implementing equalizer solutions. In particular, the response of an LTI system does not depend on the

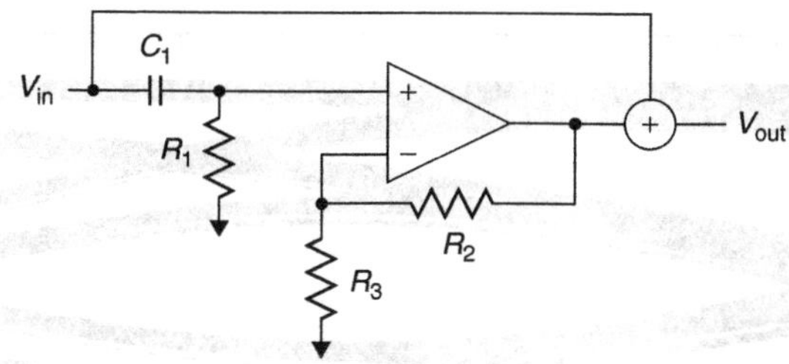

Figure 12-21 Example active equalizer using a first-order high-pass filter.

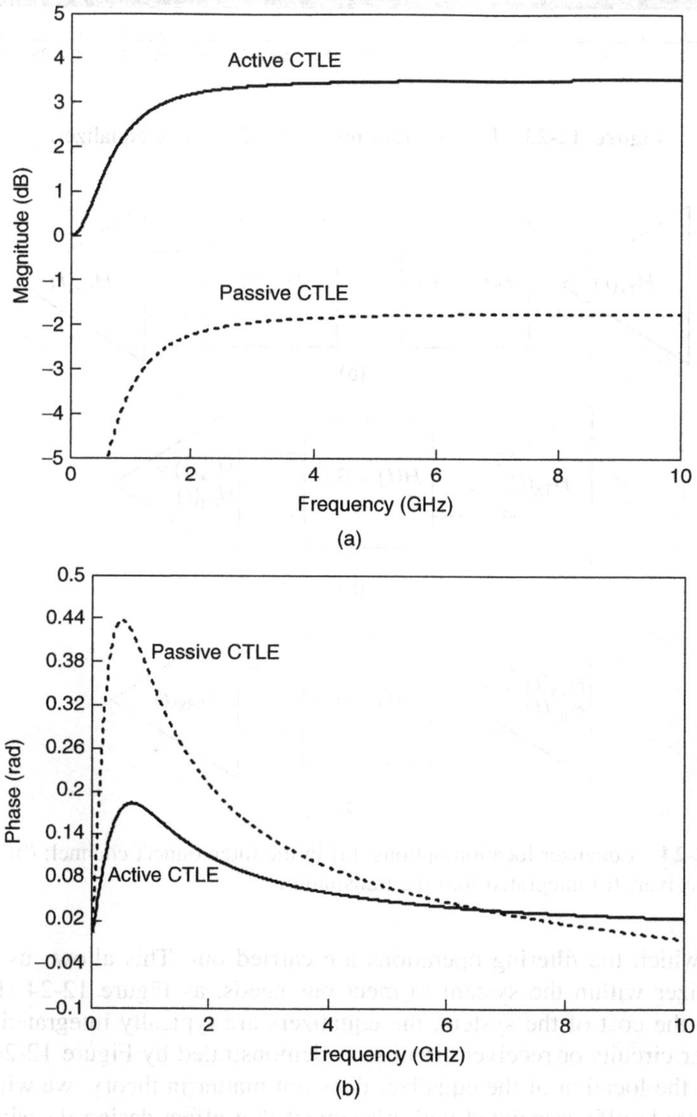

Figure 12-22 Transfer function for the active equalizer: (a) magnitude; (b) phase.

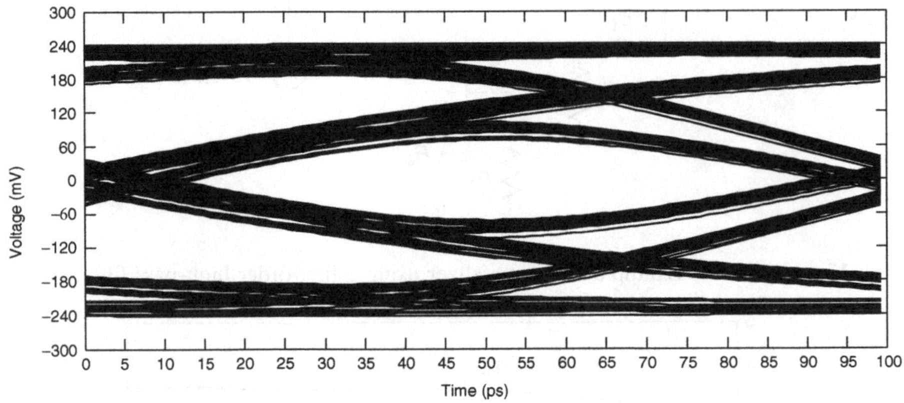

Figure 12-23 Eye diagram results for the active equalizer.

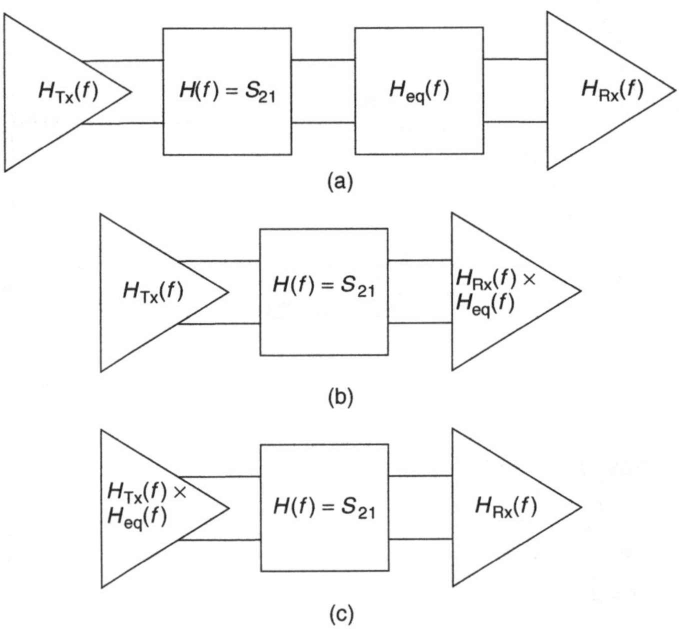

Figure 12-24 Equalizer location options: (a) in the interconnect channel; (b) integrated into the receiver; (c) integrated into the transmitter.

order in which the filtering operations are carried out. This allows us to locate the equalizer within the system to meet our needs, as Figure 12-24 shows. To minimize the cost of the system, the equalizers are typically integrated with the transmitter circuits or receiver circuits, as demonstrated by Figure 12-24b and c. Although the location of the equalizer does not matter in theory, we will see that there are trade-offs associated with placement that affect design decisions.

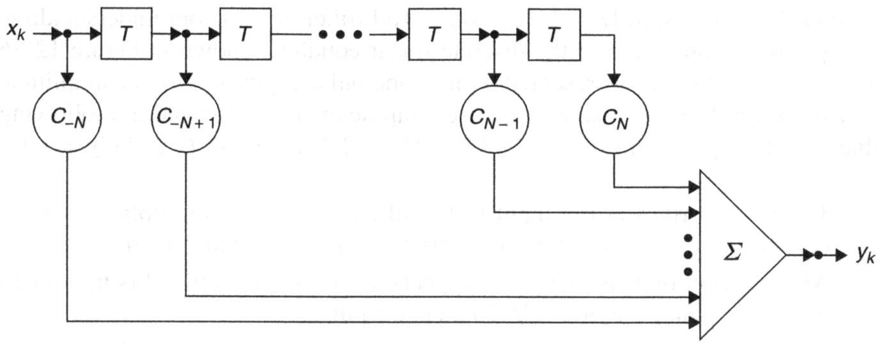

Figure 12-25 Finite impulse response filter.

12.3.1 Transmitter Equalization

The basic architecture of a discrete linear equalizer is the *transversal filter*, also known as a *finite impulse response* (FIR) *filter*, which is shown in Figure 12-25. In the figure the rectangles represent delay elements, such as the stages of a shift register. The circles represent the filter *taps*. In this filter, input samples (typically, voltage samples), x_k, propagate through the delay elements, each of which has a delay value of T, which is also known as the *tap spacing*. At each stage, the input samples are multiplied by the filter tap coefficient, C_i, where i is simply the index into the tap subscripts. With each cycle the outputs from the taps are then summed to provide the filter output, y_k. In effect, the current and past values of the signal are linearly weighted with the equalizer coefficients (also known as *tap weights*) and then summed to produce the output.

The figure shows a total of $2N$ taps in the filter, numbered from $-N$ to N. The main contribution comes from the *cursor tap*, C_0. This tap is intended to amplify the main portion of the signal. Filter taps with negative coefficient indices are known as *precursor taps*; those with positive coefficient indices are called *postcursor taps*. Figure 12-25 shows symmetry in the number of precursor and postcursor taps, but equalizers are typically designed with unequal numbers of precursor and postcursor taps. Precursor taps compensate for dispersion-induced phase distortion, which typically requires only a single tap. Postcursor taps compensate for the ISI caused by amplitude distortion and may require multiple taps, depending on the length of the channel with respect to the width of a data bit.

The output of the equalizer, $y(k)$, is expressed as the discrete convolution of the input signal, $x(k)$, with the equalizer filter coefficients:

$$y(k) = \sum_{k=-N}^{N} x(k-n)c_n \tag{12-21}$$

where k is the sample number of the discretely sampled signal (i.e., the time position of a given sample is $t_k = kT$, where T is the tap spacing of the equalizer).

Example 12-3 Transmit DLE Operation To further develop our understanding, we step through operation of the discrete linear equalizer shown in Figure 12-26 for the case in which it is presented with a lone pulse input of 600 mV amplitude and 1 ns width. For this example we use a tap delay T of 1 ns and the following values for the tap weights: $C_{-1} = -1/12$, $C_0 = 2/3$, $C_1 = -1/6$, and $C_2 = 1/12$.

1. The pulse arrives at the input to the filter, where it is multiplied by C_{-1}, generating a precursor pulse of -50 mV amplitude and 1 ns duration.
2. After a delay of 1 ns, the pulse appears at the second tap and is multiplied by C_0, generating a 400-mV 1-ns cursor pulse.
3. One nanosecond later, the pulse is at the third tap, where it is weighted with C_1, generating the first postcursor pulse of -100 mV amplitude and 1 ns duration.
4. After an additional 1-ns delay, the input comes to the final tap, where it undergoes multiplication with C_2, creating a $+50$-mV, 1-ns postcursor pulse.

The summing element operates during the entire sequence. Since our input is a lone pulse, the signal inputs to the filter taps on either side of our pulse do not generate any "echoes," so that the output from the equalizer is a linearly weighted time-delayed version of the original input, as Figure 12-26b shows.

Equalization at the transmitter is often called transmitter *preemphasis* to reflect the effect of the filter operation. As we have discussed, the function of the equalizer is to provide a high-pass filtering effect to the signal. In a digital signal, the highest-frequency content is contained in rapid transitions between logic states, while low-frequency content is contained in portions of the signal that do not make transitions. The manifestation in a discrete linear equalizer is increased amplitude for the first bit after a logic transition relative to successive bits. This effect is shown in Figure 12-27 for a 500-mV transmitter with 20% equalization:

$$\frac{v_{\text{max.swing}} - v_{\text{min.swing}}}{v_{\text{total swing}}} \times 100 = \frac{400\text{ mV} - 300\text{ mV}}{500\text{ mV}} \times 100$$
$$= 20\% = -13.97\text{ dB}$$

The full signal swing occurs for the case when the signal transitions between logic states for at least two successive bit positions. It ranges from -100 mV on the low side to 400 mV on the high side, for a total of 500 mV. To find the minimum swing we look at the waveform regions that contain multiple consecutive zeros (0 mV) and 1's (300 mV). The variation in swing from 500 mV maximum to 300 mV minimum corresponds to 20% equalization.

High-speed signaling systems typically use as much drive current as possible in order to maximize speed, the limitation being the maximum voltage swing that

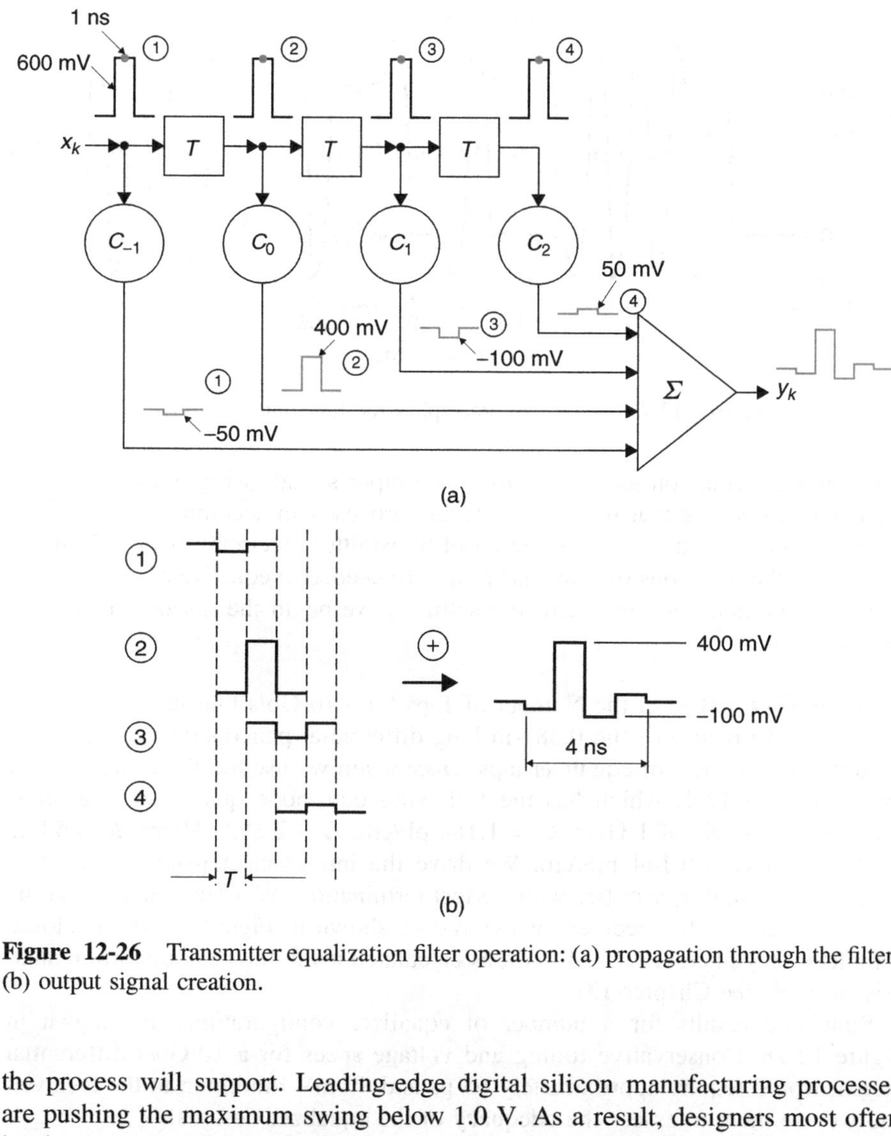

Figure 12-26 Transmitter equalization filter operation: (a) propagation through the filter; (b) output signal creation.

the process will support. Leading-edge digital silicon manufacturing processes are pushing the maximum swing below 1.0 V. As a result, designers most often implement transmit equalization by attenuating the low-frequency bits rather than by amplifying the high-frequency bits. This is often called *deemphasis* rather than preemphasis, to reflect the fact that the equalization is done by reducing the amplitude of the repetitive bits.

The maximum signal swing limit (for both pre- and deemphasis) places a constraint on the coefficient settings for an equalizer according to

$$\sum_i |c_i| = 1 \tag{12-22}$$

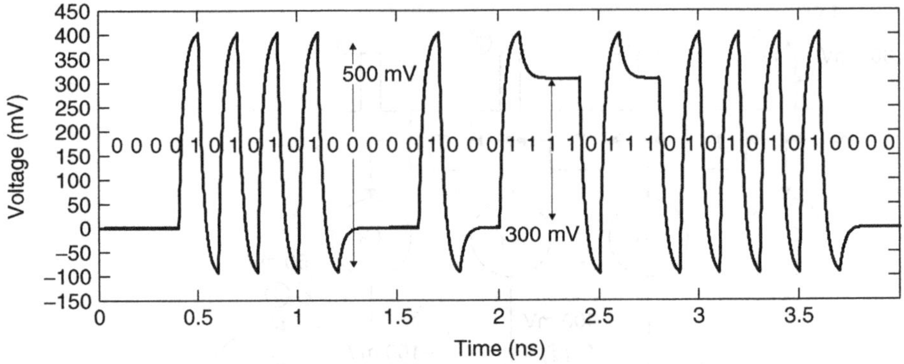

Figure 12-27 Example preemphasized transmitter output.

Satisfying the equation guarantees that the output signal swing does not exceed the maximum swing that the manufacturing process can accommodate.

Having developed an understanding of transmitter preemphasis operation, we can turn to the questions of how many taps are needed to equalize a given system and how to determine the coefficient settings. We begin the discussion with an example.

Example 12-4 Effect of the Number of Taps for a 10-Gb/s Interface We compare the performance of the 0.381-m-long differential pair operating at 10 Gb/s for different numbers of equalizer taps. Once again we use the PCB interconnect from Example 12-1, which has the following odd-mode quantities at a reference frequency, f_0, of 1 GHz: $C = 1.184$ pF/cm, $L = 2.892$ nH/cm, $R = 448.2$ mΩ/cm, and $G = 0.144$ mS/cm. We drive the interconnect using a ± 2.5-mA 0.5-pF differential transmitter with perfect termination. Without equalization the worst-case eye at the receiver for the system shown in Figure 12-18b is closed completely (eye height $= -34$ mV), as calculated using the peak distortion analysis method (see Chapter 13).

Simulated results for a number of equalizer configurations are shown in Figure 12-28. Conservative timing and voltage specs for a 10-Gb/s differential system would require approximately 65 ps and 80 mV minimum at the receiver, and we use them to assess the adequacy of the equalizer designs.

In calculating the response of the equalizer, we use the expression for the transfer function of the equalizer (the derivation of which is left as a problem at the end of the chapter):

$$H(f) = \sum_{k=-N_{\text{pre}}}^{N_{\text{post}}} c_k e^{-j2\pi f(k-N_{\text{pre}})T} \tag{12-23}$$

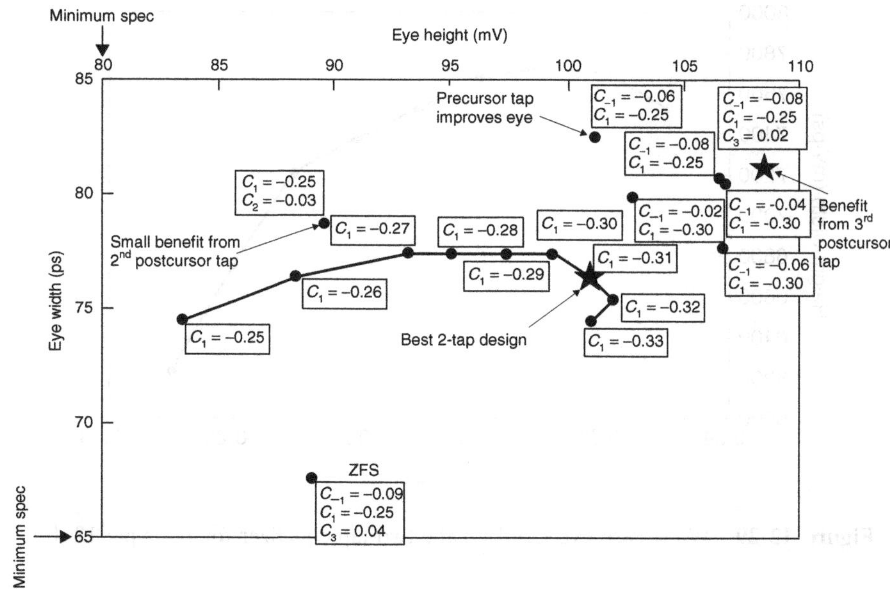

Figure 12-28 Worst-case received eyes as a function of equalizer design for Example 12-4.

The results for two tap designs (cursor tap, c_0, plus one postcursor tap, c_1) are connected via lines on the figure and show that a two-tap design will meet the specs proposed when the postcursor tap coefficient (c_1) is -0.25. In fact, the effectiveness of the equalizer increases up to a value of approximately -0.30 to -0.32. Beyond that, the eye margin begins to degrade. The question of which coefficient value works best will depend on whether voltage margin or timing margin is more important to the design. If we were to assume that they are equally important, we could evaluate the results by calculating the product of the eye height and width. This trend is plotted in Figure 12-29 for the two-tap equalizer, which shows that -0.31 gives the best results for this criterion, providing a minimum eye height of 101 mV and width of 76 ps.

Figure 12-28 also shows that adding taps can improve the voltage and timing margin, although it will not necessarily do so. In particular, the addition of a second postcursor (c_2) tap offers very little benefit, as shown in the figure. Adding a precursor tap (c_{-1}) provides incremental improvement in the eye size (101 mV, 83 ps). With a third postcursor tap the eye increases to 109 mV and 81 ps. All of these results assume that we can optimize the equalizer coefficients with a step size (granularity) of 0.01. Note that all equalizer configurations meet the criteria specified by equation (12-22).

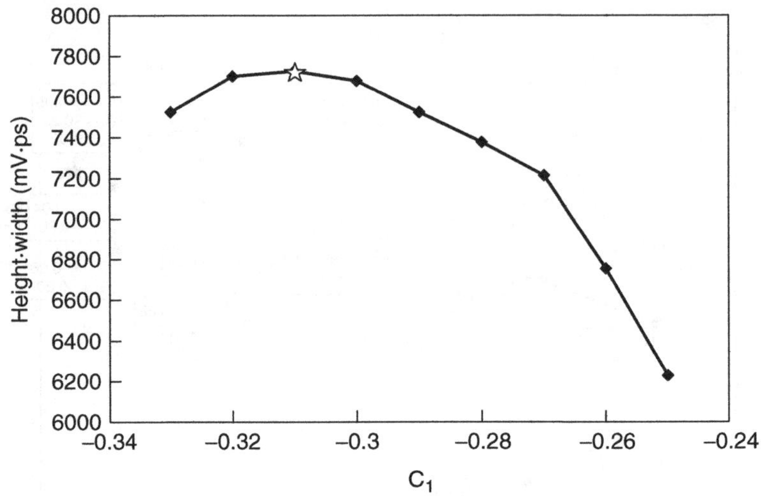

Figure 12-29 Worst-case eye trend for the two-tap equalizer for Example 12-4.

In summary, the two-tap equalizer is capable of meeting the voltage and timing specs proposed, with some room for adjustment. Addition of a precursor improves the margins by 7.5% for eye height and 3.8% for eye width. Addition of the additional postcursor taps provides an additional 2.4% increase in eye width but no improvement in eye height.

The final step in this example is to use the results to decide how many taps to include in the equalizer design. The two-tap design provides a sufficient eye to meet the design specs, and it can be implemented with a very simple design. In general, designers will choose the simplest equalizer design that meets the requirements for a given application. In our case, the two-tap equalizer will work just fine. Longer channels will typically have more attenuation, causing the ISI to intrude on additional postcursor bit positions and may therefore require additional equalizer taps. Higher data rates may also require more taps.

12.3.2 Coefficient Selection

At this point the question of how to determine the tap coefficients seems appropriate to consider. In this section we present a method that can be used to establish coefficient values based on specific performance criteria. The method, called a *zero forcing solution* (ZFS), sets the tap coefficients to force the equalizer output to match the values desired at all sample points. [Qureshi, 1985; Sklar, 2001]. Development of the algorithm follows.

Given a pulse response input to the equalizer, we start by extracting samples from the input stream x_i, where i runs from $-n_{\text{pre}}$ to n_{post}, with n_{pre} being the number of precursor taps and n_{post} the number of postcursor taps in the equalizer. This gives a total of $n_{\text{pre}} + n_{\text{post}} + 1$ samples. We can express the input samples in matrix form:

$$\mathbf{x} = \begin{bmatrix} x(0) & x(-1) & \cdots & x(-n_{\text{pre}}) & 0 & \cdots & 0 \\ x(1) & x(0) & \cdots & x(-n_{\text{pre}}+1) & x(-n_{\text{pre}}) & 0 & \vdots \\ \vdots & \vdots & & \vdots & & \vdots & 0 \\ x(n_{\text{post}}) & x(n_{\text{post}}-1) & & \vdots & & x(-n_{\text{pre}}+1) & x(-n_{\text{pre}}) \\ 0 & x(n_{\text{post}}) & & \vdots & & x(-n_{\text{pre}}+2) & x(-n_{\text{pre}}+1) \\ \vdots & \vdots & & \vdots & \ddots & \vdots & \vdots \\ 0 & \cdots & x(n_{\text{post}}) & x(n_{\text{post}}-1) & \cdots & x(0) & x(-1) \\ 0 & \cdots & 0 & x(n_{\text{post}}) & \cdots & x(1) & x(0) \end{bmatrix} \tag{12-24}$$

The columns of $\mathbf{x}$ represent the taps of the equalizer and the rows represent consecutive time steps with an interval between steps that is equal to the tap spacing of the equalizer. The matrix is square with $n_{\text{pre}} + n_{\text{post}} + 1$ rows and columns. In this form, $\mathbf{x}$ shows the propagation of the input samples through the equalizing filter. For example, $x(0)$ appears in the first row and the first column, which indicates that it is at the first tap at the first time sample. It also appears in the second row and second column, which corresponds to the second tap and second time sample. This is exactly what we expect for a discrete linear equalizer.

In matrix form, equation (12-21) is written

$$\mathbf{y} = \mathbf{xc} \tag{12-25}$$

where $\mathbf{y}$ is the vector containing the output from the equalizer and $\mathbf{c}$ is the vector of equalizer tap coefficients. The number of elements in both $\mathbf{y}$ and $\mathbf{c}$ is $n_{\text{pre}} + n_{\text{post}} + 1$. We have essentially expressed the discrete convolution of the data stream and equalizer as a matrix multiplication, which we now apply to our problem.

By transmitting a lone pulse, we define the expected output results from the equalizer according to Nyquist's first method for the elimination of ISI [Nyquist, 1928; Couch, 1987]:

$$\mathbf{y}_{\text{target}} = \begin{cases} 0 & \text{for } k \neq 0 \\ 1 & \text{for } k = 0 \end{cases} \tag{12-26}$$

In (12-26) the expected value is 1 for the cursor sample and zero for all others. The equalizer coefficients that give the zero forcing solution are then

$$\mathbf{c}_{\text{ZFS}} = \mathbf{x}^{-1}\mathbf{y}_{\text{target}} \tag{12-27}$$

The coefficients calculated are not constrained by power or by the maximum signal swing that can be achieved. Application of such constraints requires that the coefficients be adjusted using equation (12-22), as we illustrate next.

Example 12-5 Zero Forcing Solution We wish to find the coefficients for an equalizer with a single precursor tap and three postcursor taps that give the zero forcing solution for the PCB differential pair from Example 12-1 when operating at 10 Gb/s. In Example 12-3 we saw that the worst-case eye height without equalization is −34 mV for this interconnect. Using the 5-mA transmitter from our earlier example gives a differential swing of ±250 mV. The driving pulse is 100 ps wide, corresponding to a single bit of data in a 10-Gb/s data stream.

From the simulated differential pulse response in Figure 12-30, we extract the input sample vector for a tap spacing of 100 ps:

$$\mathbf{x_{in}} = \begin{bmatrix} -214 \\ -20 \\ -152 \\ -213 \\ -232 \end{bmatrix} \qquad \text{mV}$$

The expected output from the equalizer with $n_{pre} = 1$ and $n_{post} = 3$ is

$$\mathbf{y_{target}} = \begin{bmatrix} 0 \\ 1 \\ 0 \\ 0 \\ 0 \end{bmatrix}$$

In order to apply equations (12-26) and (12-27), we must apply a level shift to the differential signal so that the minimum output voltage at the transmitter is zero. We do this by adding one-half of the differential swing (250 mV) to the input vector to produce an adjusted input sample vector:

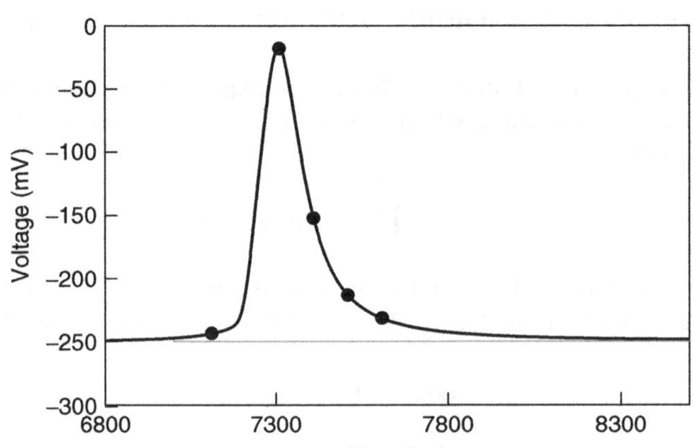

Figure 12-30 Differential pulse response before equalization for Example 12-5.

$$\mathbf{x_{in,adj}} = \begin{bmatrix} 36 \\ 230 \\ 97 \\ 37 \\ 18 \end{bmatrix} \quad \text{mV}$$

We then construct the input sample matrix and calculate the equalizer coefficients:

$$\mathbf{x} = \begin{bmatrix} 230 & 36 & 0 & 0 & 0 \\ 97 & 230 & 36 & 0 & 0 \\ 37 & 97 & 230 & 36 & 0 \\ 18 & 37 & 97 & 230 & 36 \\ 0 & 18 & 37 & 97 & 230 \end{bmatrix} \quad \text{mV}$$

$$\mathbf{c} = \mathbf{x}^{-1}\mathbf{y_{target}} = \begin{bmatrix} -0.77 \\ 4.93 \\ -1.98 \\ 0.14 \\ -0.13 \end{bmatrix}$$

We apply the maximum signal swing constraint by using equation (12-22) to adjust the equalizer coefficients:

$$\mathbf{c_{ZFS_{\mathit{i}}}} = \frac{\mathbf{c}_i}{\sum |\mathbf{c}_i|} = \begin{bmatrix} -0.097 \\ 0.624 \\ -0.250 \\ 0.017 \\ -0.016 \end{bmatrix}$$

Figure 12-31 shows the resulting equalized pulse response, which demonstrates the effect of the zero forcing equalizer. Since the width of the nonequalized pulse in Figure 12-30 is much wider than a single bit (which is 100 ps wide), significant ISI is expected. The equalized pulse response is much more narrow and the ISI has been removed (i.e., forced to be zero) at the equalizer sample points. However, the figure also reveals the fact that the ZFS approach eliminates the ISI only at the sampling points that correspond to the equalizer taps. The equalized pulse shows ISI in the intervals between the sample points and at sample points outside the equalizer.

The simulated output from the ZFS equalizer in Figure 12-32 shows an eye opening of approximately 110 mV and 85 ps for a 300-bit pseudorandom bit sequence. In addition, the worst-case data eye calculated from peak distortion analysis on the equalized pulse response is 107 mV high and 80 ps wide. As we can see, the zero forcing equalizer provides significant improvement in this case over that of the nonequalized channel.

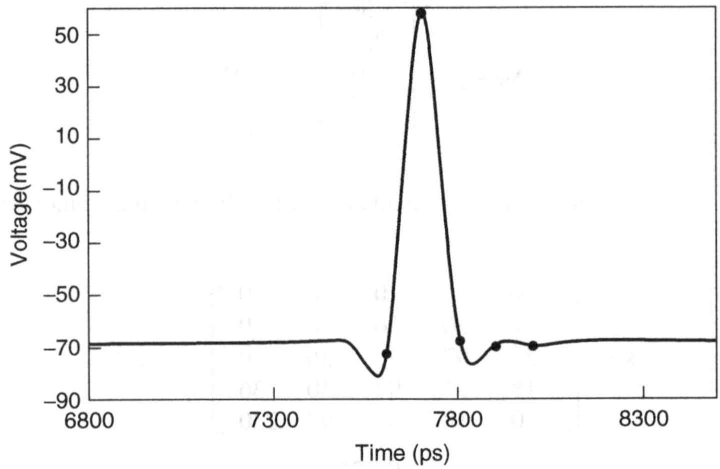

Figure 12-31 Differential pulse response with equalization for Example 12-5.

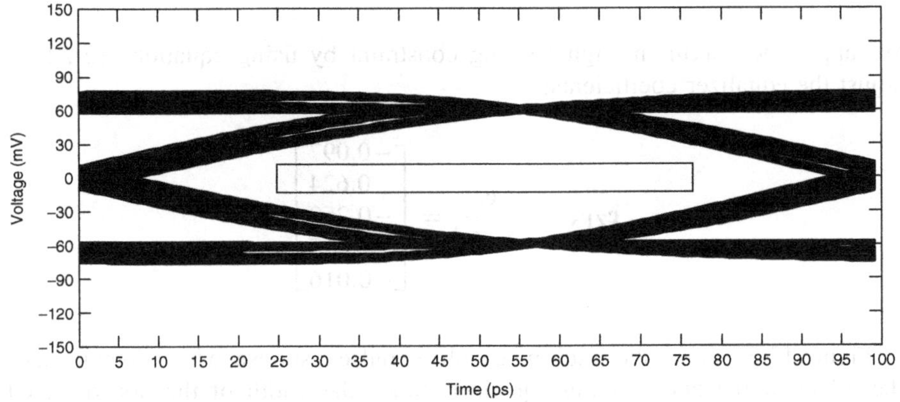

Figure 12-32 Equalized differential data eye for Example 12-5.

The chief limitation of the zero forcing approach is that for a finite-length equalizer, minimization of the ISI is guaranteed only if the eye is initially open, a condition that is not necessarily true for long, lossy channels such as backplanes. An approach that avoids the limitation of the ZFS algorithm is the minimum mean-square error (MMSE) algorithm. This approach sets the equalizer coefficients such that the mean-square error of the ISI is minimized at the output of the equalizer. Examination of the MMSE equalizer is left for Problem 12-7, and a thorough treatment of both the ZFS and MMSE methods is given by Proakis [2001].

12.3.3 Receiver Equalization

As we showed in Section 12.1.2, the equalizer may be located anywhere in the channel under the LTI assumption. Although equalization is often implemented at the transmitter, it may also be done at the receiver. The main reason that discrete linear equalizers are most often located at the transmitter is ease of implementation. Figure 12-33 provides a block-level diagram of a discrete linear equalizer implementation at the receiver which shows that the equalizer looks like the transmit equalizer with the addition of a sample-and-hold circuit on the front of the equalizer (represented by the switch in the figure). The input to the receiver equalizer is an analog voltage waveform rather than a binary data pattern. In addition, transmitter equalization can be accomplished in a straightforward manner by using a multiplexer to multiply the tap coefficient by the bit value.

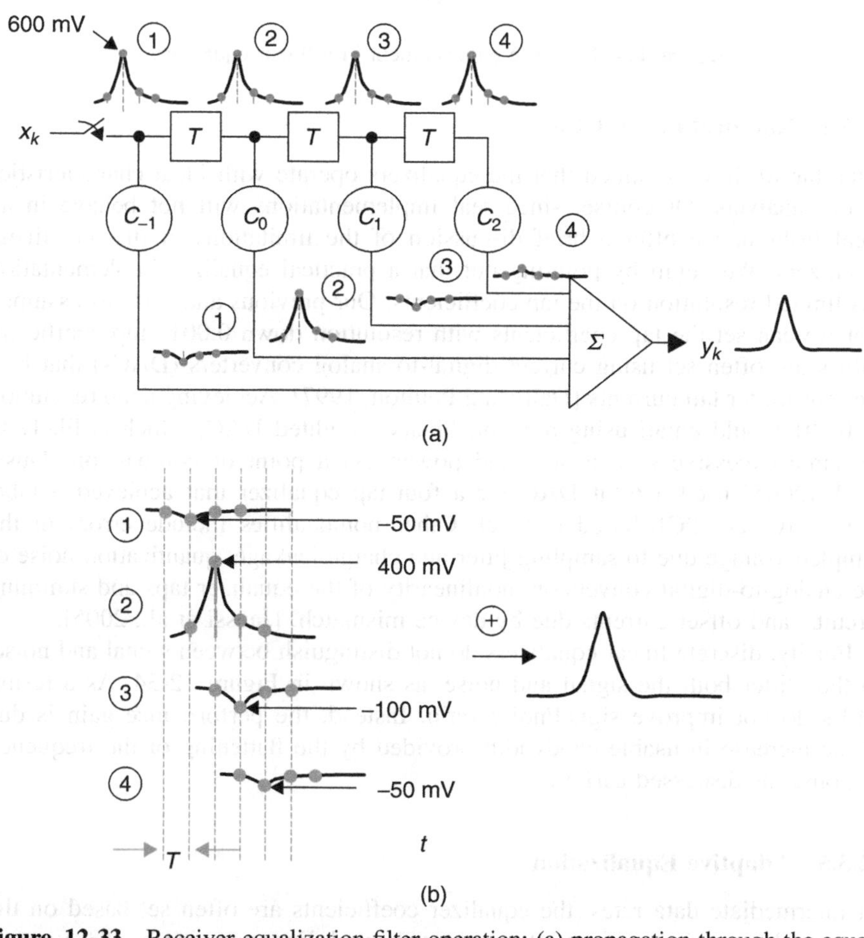

Figure 12-33 Receiver equalization filter operation: (a) propagation through the equalizer; (b) output signal creation.

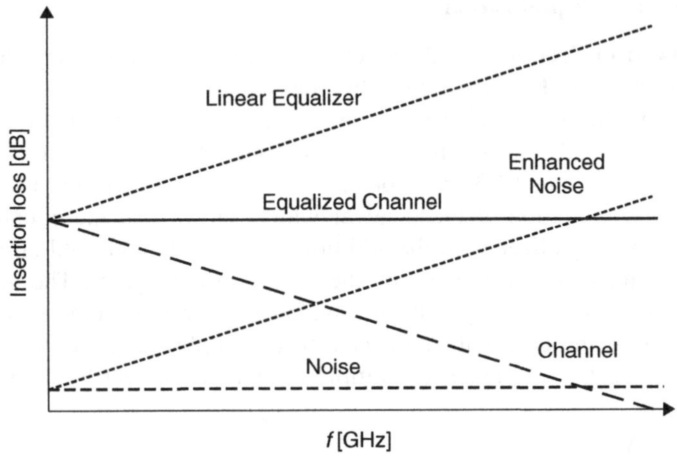

Figure 12-34 Noise enhancement in a linear equalizer.

12.3.4 Nonidealities in DLEs

Thus far we have assumed that the equalizers operate with ideal characteristics in our analysis. Of course, since real implementations will not behave in an ideal fashion, we offer a brief discussion of the limitations of discrete linear equalizers. We begin by pointing out that a practical equalizer implementation has limited resolution on the tap coefficients. Our previous analysis has assumed that we can set the tap coefficients with resolution down 0.001. Tap coefficient values are often set using current digital-to-analog converters (DACs) that bias the transmitter tail currents [Dally and Poulton, 1997]. Achieving a tap resolution of 0.001 would entail using a 10-bit binary weighted DAC, which is likely to consume excessive silicon area and power. As a point of comparison, Jaussi et al. [2005] used a 6-bit DAC for a four-tap equalizer that achieved 8 Gb/s over a 102-cm PCB-based channel. Other nonidealities include errors in the sampled voltage due to sampling jitter and charge leakage, quantization noise of the analog-to-digital conversion, nonlinearity of the equalizer taps and summing circuits, and offset currents due by device mismatch. [Jaussi et al., 2005].

Finally, discrete linear equalizers do not distinguish between signal and noise, so they filter both the signal and noise, as shown in Figure 12-34. As a result, DLEs do not improve signal/noise ratio. Instead, the performance gain is due to the increase in usable bandwidth provided by the flattening of the frequency response, as discussed earlier.

12.3.5 Adaptive Equalization

At intermediate data rates, the equalizer coefficients are often set based on the average characteristics of the interconnect channel. For example, the PCI Express interface calls for −3.5 dB of equalization for PCB-based interconnects of up to 15 in. in length [Coleman et al., 2004; PCI-Sig 2005]. However, effectively equalizing channels that may exhibit wide variation in frequency-dependent loss (e.g., a range of PCB lengths) requires that the equalizer design contain the flexibility to set the equalizer coefficients adaptively to minimize the ISI. Such an equalizer, called an *adaptive equalizer*, was invented by Lucky in 1964 to improve data transmission rates over telephone lines from 1200 b/s (using non-adaptive equalizers based on average channel characteristics) to 9600 b/s [Lucky, 2006].

The benefit of adaptive equalization is the flexibility to accommodate a range of interconnect lengths and/or data rates. It does, however, add significant complexity to the design and consumes more power and chip real estate. A high-level schematic of the adaptive equalizer structure is shown in Figure 12-35. The intent of the adaptation is illustrated in Figure 12-36. The figure depicts the deviation of the equalized signal from desired value as a function of equalizer coefficients. The deviation is plotted as a set of error contours in which the error is the difference in the equalized output, $y(t)$, from the training data, $\hat{y}(t)$. The error is a convex function of the equalizer coefficients, so that it has a global minimum. The goal of the adaptive algorithm is to converge on a set of coefficient values that minimize the error in a small number of iterations. The general approach of an adaptive equalizer to updating the equalizer tap coefficient is

$$c_{\text{new}} = c_{\text{old}} + (\text{step size})(\text{error function})(\text{input function}) \tag{12-28}$$

The error function is typically based on the difference between the actual equalized signal y and the desired equalizer output $\hat{y}$. The input function is based on the signal at the input to the equalizer, and step size is a design parameter. Designers have many options for implementing adaptive equalizers, the range of which extends beyond our scope. However, we present a pair of examples to provide some insight into the operation of adaptive equalizers.

The first example is the adaptive implementation of the zero forcing equalizer. In this approach a known *data pattern* (a.k.a. *training sequence*) of equal or larger length than that of the equalizer is transmitted and equalized. The coefficients are updated from the equalized results using

$$c_k(n+1) = c_k(n) + \Delta_k[\hat{y}(n) - y(n)]x(n-k) \tag{12-29}$$

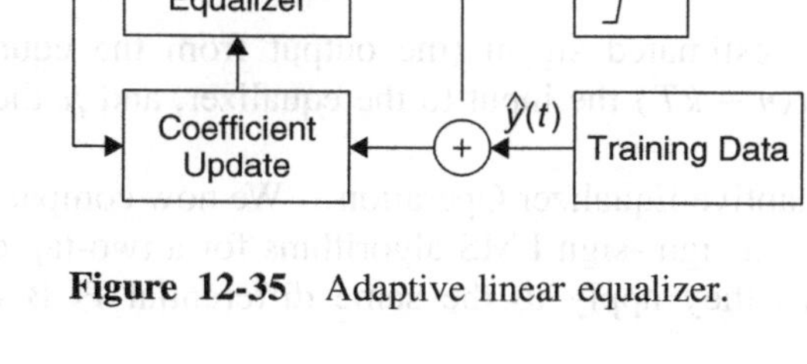

Figure 12-35 Adaptive linear equalizer.

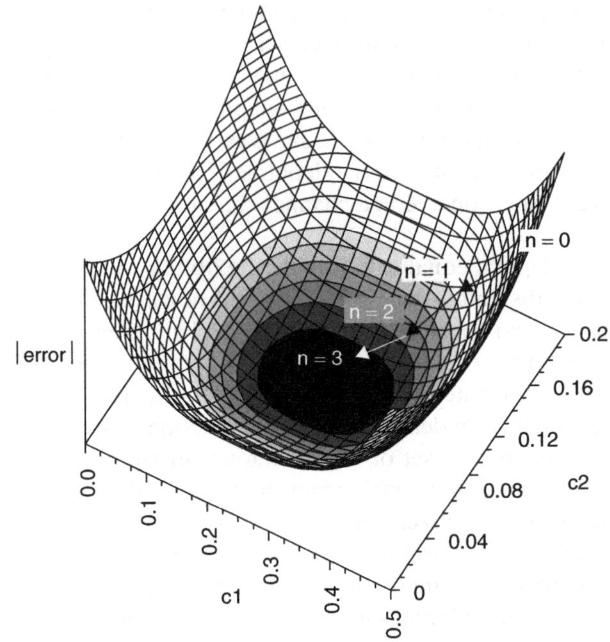

Figure 12-36 Adaptive equalization error "contour" and coefficient convergence.

where $c_k(n)$ is the value of the kth coefficient at time $t = nT$, $y(n)$ the equalized signal, $\hat{y}(n)$ the training signal, and Δ_k a scaling factor that controls the rate of coefficient adjustment. The adaptive ZFS approach has the advantage that it is very easy to implement but has the drawback that it does not comprehend ISI that occurs outside the length of the equalizer.

Another class of adaptive algorithms is the least mean square (LMS) set of algorithms. They attempt to minimize the mean square error of the equalizer output at all times and are commonly used in adaptive equalizers because they typically achieve better performance than ZFS algorithms and are relatively easy to implement. An example is the sign–sign least mean square approach [Kim et al., 2005]. Sign–sign LMS updates the equalizer coefficients based on the sign of the error of the equalized signal and the sign of the input signal, as described by

$$c_k(n+1) = c_k(n) + \mu \ \text{sign}[y(n) - \hat{y}(n)] \ \text{sign}[x(n-kT)] \tag{12-30}$$

where $y(n)$ is the estimated signal (the output from the equalizer), $\hat{y}(n)$ the reference signal, $x(n-kT)$ the input to the equalizer, and μ the scaling factor.

Example 12-6 Adaptive Equalizer Operation We now compare the behavior of the adaptive ZFS and sign–sign LMS algorithms for a two-tap equalizer (cursor plus postcursor) as they apply to the same differential PCB channel that we have studied in previous examples in this chapter. The input samples and desired output values are

$$x = [6, 36, 233, 99, 37, 18, 11, 7, 5, 4, 3, 2, 2, 2, 1] \text{ mV}$$

$$y_{\text{target}} = [0, 0, 150, 0, 0, 0, 0, 0, 0, 0, 0, 0, 0, 0, 0] \text{ mV}$$

For the ZFS equalizer we choose $\Delta = 12$, and for the sign–sign LMS equalizer we set μ equal to -0.025. Applying equations (12-29) and (12-30) iteratively gives the results shown in Figure 12-37, which shows the trajectories of each algorithm in terms of the mean square error of the equalizer output. The mean

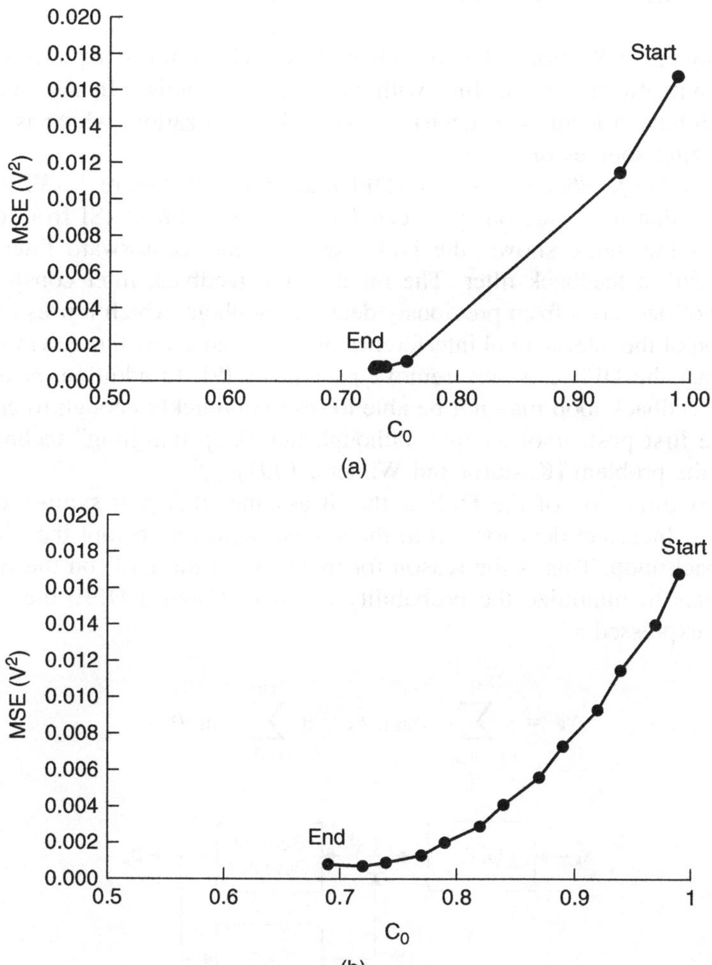

Figure 12-37 Adaptive algorithm convergence for Example 12-6: (a) adaptive ZFS; (b) adaptive sign–sign LMS.

square error is calculated as

$$\mathrm{MSE} = \sum_{i=0}^{n} [y(i) - \hat{y}(i)]^2 \tag{12-31}$$

Note that by applying the maximum voltage swing constraint of equation (12-22), which makes C_0 and C_1 interdependent for the two-tap equalizer, we make the optimization problem a function of a single variable.

12.4 DECISION FEEDBACK EQUALIZATION

As mentioned in Section 12.3, the chief drawback of linear equalizers is that they are not effective at dealing with most types of noise. In this section we describe briefly a technique, decision feedback equalization, which is effective at minimizing sources of noise.

The *decision feedback equalizer* (DFE), as shown in Figure 12-38, is a nonlinear filter that uses previously detected symbols to subtract ISI from the input stream. As the figure shows, the DFE uses a linear feedforward filter in conjunction with a feedback filter. The input to the feedback filter consists of the sequence of decisions from previously detected symbols, which it uses to remove the portion of the intersymbol interference that was caused by those symbols. This implies that the DFE can only remove postcursor ISI. In addition, at high data rates the feedback loop may not be able to respond quickly enough to cancel the ISI of the first postcursor sample, although the "loop unrolling" technique can alleviate the problem [Kasturia and Winters, 1991].

Another limitation of the DFE is that it assumes that past symbol decisions are correct. Incorrect decisions from the symbol detector corrupt the filtering of the feedback loop. This is the reason for inclusion of the DLE on the front end, which helps to minimize the probability of error. Using a DFE, the equalized output is expressed as

$$y_k = \sum_{i=-n_{\mathrm{DLE}}}^{0} c_{\mathrm{DLE}_i} x_{k-i} + \sum_{i=1}^{n_{\mathrm{FBF}}} c_{\mathrm{FBF}_i} b_{k-i} \tag{12-32}$$

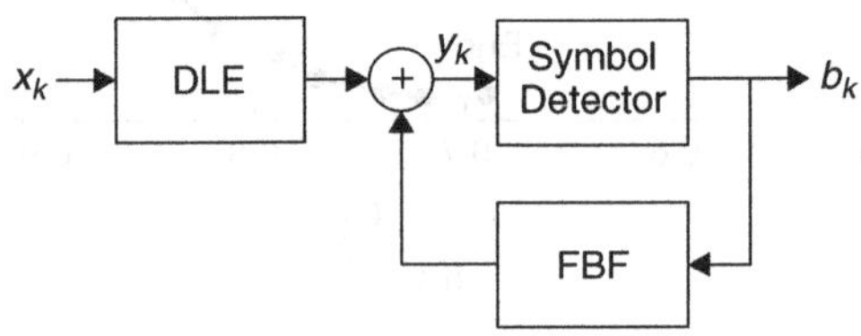

Figure 12-38 Decision feedback equalization.

where y_k is the input to the symbol detector, c_{DLE_i} the coefficients of the feed-forward filter, c_{FBF_i} the coefficients of the feedback filter, and x_k the input to the DFE. The previously detected symbols, b_k, are expressed as

$$b_k = \begin{cases} 0 & \text{if } y_k < y_{\text{threshold}} \\ 1 & \text{if } y_k > y_{\text{threshold}} \end{cases} \tag{12-33}$$

where $y_{\text{threshold}}$ is the decision threshold.

Decision feedback equalization is often used to remove ISI that is caused by reflections due to impedance mismatch and may also use zero forcing and LMS adaptive algorithms [Proakis, 2001]. Note that because of the nonlinearity of the DFE response, it must be modeled in the time domain.

Example 12-7 DFE Operation for a 10-Gb/s Signaling Link   We now apply a DFE to the differential transmission lines that we have used throughout this chapter. To guarantee successful operation at 10 Gb/s, the system must meet minimum differential received specs of 80 mV and 80 ps for the height and width, respectively. Our system, shown in Figure 12-39, uses 5.057-mA current-mode transmitters with 5 MΩ output resistance and 0.5 pF output capacitance. The lines are terminated to ground in 50 Ω, and the receiver has an input capacitance of 0.5 pF. The model for the differential pair is generated using the methodology presented in Chapter 10 from the distributed transmission-line parameters listed below that were calculated at a reference frequency of 1 GHz:

$$L = \begin{bmatrix} 3.299 & 0.407 \\ 0.407 & 3.299 \end{bmatrix} \text{nH/cm} \qquad C = \begin{bmatrix} 1.098 & 0.085 \\ 0.085 & 1.098 \end{bmatrix} \text{pF/cm}$$

$$R = \begin{bmatrix} 509.1 & 60.63 \\ 60.64 & 509.1 \end{bmatrix} \text{m}\Omega/\text{cm} \quad G = \begin{bmatrix} 0.131 & 0.012 \\ 0.012 & 0.131 \end{bmatrix} \text{mS/cm}$$

As the figure shows, the system employs the linear equalizer at the transmitter. The transmit equalization does not include adaptive capability, and is set based on an expected channel loss characteristic. However, while the transmit equalizer coefficients open the data eye, they are too strong for this particular system,

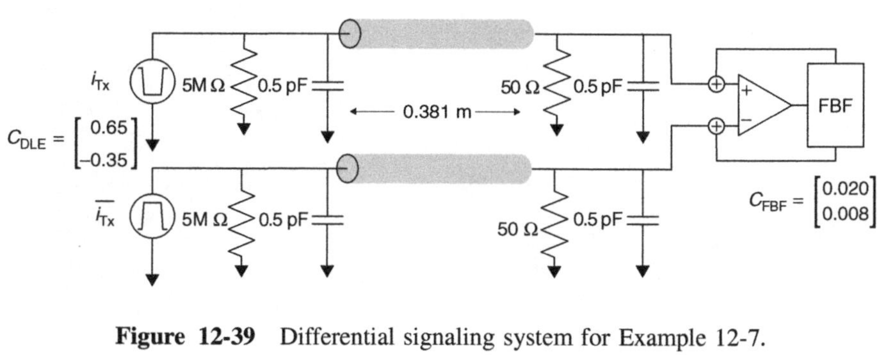

Figure 12-39 Differential signaling system for Example 12-7.

leading to overequalization. The DFE taps, on the other hand, are adaptive, based on the zero forcing criteria for the first two postcursor samples.

Figure 12-40 compares the pulse responses and the worst-case data eyes for the system without equalization, with the transmitter preemphasis only, and with DFE. As before, the worst-case data eyes are calculated using the peak distortion analysis method. The figure shows that without equalization, the data eye is closed completely. Transmit equalization opens the eye, providing a 96-mV minimum height and 76-ps minimum width, which meets the specs for eye height but not for width. The DFE allows the system to meet the specification requirements by trading off eye height margin by approximately 6% in order to obtain an 11.5% improvement in eye width. This results in minimum eye height and width of 90 mV and 88 ps.

As in the case of the linear equalizer, further improvements in both eye height and width are possible with the addition of taps to the DFE. For example, a 10-tap zero forcing DFE provides a minimum eye of 95 mV and 91 ps. However, since the system meets the specifications, the additional taps are not likely to justify the additional design complexity and power that they would add to the design.

12.5 SUMMARY

In this chapter we described the application of equalization to signaling systems whose performance is limited by intersymbol interference caused by frequency-dependent interconnect losses. Passive and active continuous linear equalizers and discrete linear equalizers all reduce ISI by flattening the frequency response of the interconnect system. By flattening the frequency response, equalization increases the usable bandwidth of the interconnect, making it possible to increase data transfer rates. Decision feedback equalizers can further cancel ISI by using information from symbol decisions as input to a feedback filter, and adaptive equalization offers a way to improve performance further by optimizing filter performance based on the data rate and the characteristics of the interconnect system. Equalization may be incorporated into the transmitter or the receiver, or is split between both.

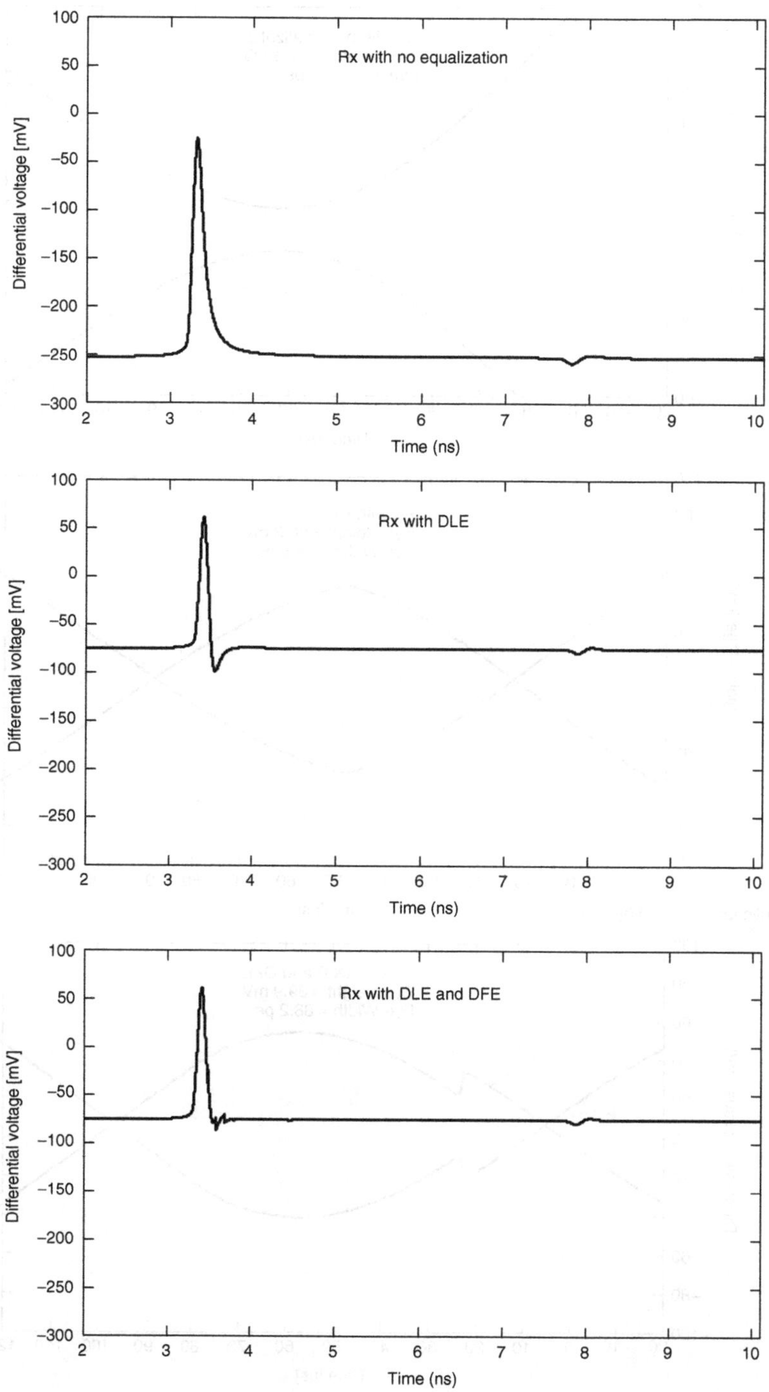

(a)

Figure 12-40 (a) Pulse responses and (b) worst-case data eyes for Example 12-7. *(Continued)*

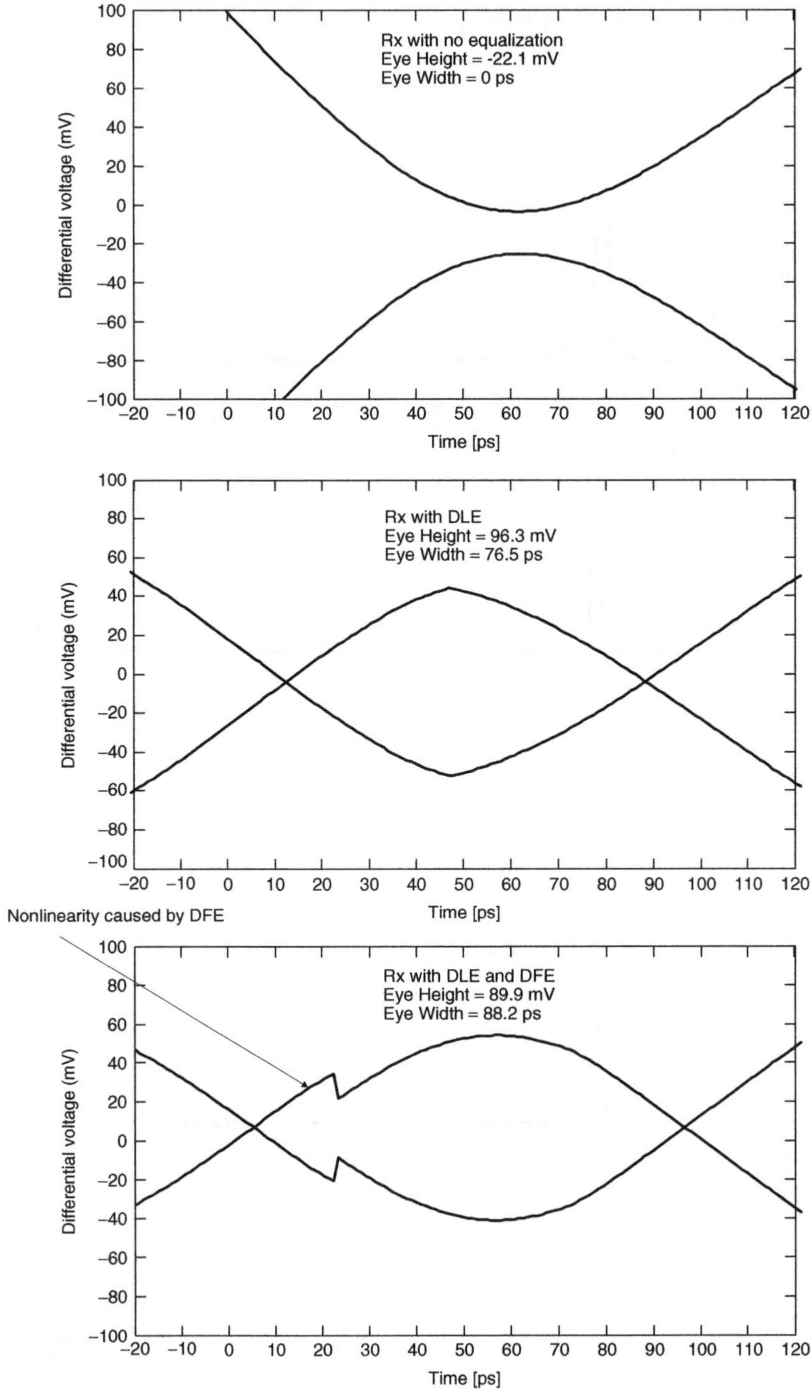

(b)

Figure 12-40 *(Continued)*

REFERENCES

With data rates moving into the multi-Gb/s range, dozens of research papers are being published in conference proceedings and technical journals each year. We make no attempt to provide an exhaustive survey of the publish literature here. For tutorial treatments of equalization, we refer the reader to books by Couch [1987], Dally and Poulton [1997], and Liu and Ling [2004]. Books by Qureshi [1985] and Proakis [2001] give adaptive equalization a thorough and rigorous treatment. Finally, Lucky [2006] gives a historical account of the invention of the adaptive equalizer that provides insight into the nature of the discovery and development of a groundbreaking advance in communications technology.

Coleman, Dave, Scott Gardiner, Mohammad Kolbehdari, and Stephen Peters, 2004, *PCI Express Electrical Interconnect Design*, Intel Press, Hillsboro, OR.

Couch, Leon, 1987, *Digital and Analog Communication Systems*, 2nd ed., Macmillan, New York.

Dally, William, and John Poulton, 1997, Transmitter equalization for 4-Gbps signaling, *IEEE Micro*, Jan.–Feb., pp. 48–56.

Jaussi, James, et al., 2005, 8-Gb/s source-synchronous I/O links with adaptive receiver equalization, offset cancellation, and clock de-skew, *IEEE Journal of Solid-State Circuits*, vol. 40, no. 1, Jan., pp. 80–88.

Johnson, Howard, and Martin Graham, 2003, *High-Speed Signal Propagation: Advanced Black Magic*, Prentice Hall, Upper Saddle River, NJ.

Kasturia, Sanjay, and Jack Winters, 1991, Techniques for high-speed implementation of nonlinear cancellation, *IEEE Journal on Selected Areas in Communications*, vol. 9, no. 5, June, pp. 711–717.

Kim, Jinwook, et al., 2005, A four-channel 3.125-Gb/s/ch CMOS serial-line transceiver with a mixed-mode adaptive equalizer, *IEEE Journal of Solid-State Circuits*, vol. 40, no. 2, Feb., pp. 462–471.

Kudoh, Yoshiharu, Muneo Fukaishi, and Masayuki Mizuno, 2003, A 0.13 μm CMOS 5-Gb/s 10-m 28AWG cable transceiver with no-feedback-loop continuous-time post-equalizer, *IEEE Journal of Solid-State Circuits*, vol. 38, no. 5, May, pp. 741–746.

Liu, Jin, and Xiaofen Ling, 2004, Equalization in high-speed communication systems, *IEEE Circuits and Systems Magazine*, vol. 4, no. 2, pp. 4–17.

Lucky, Robert, 2006, The adaptive equalizer, *IEEE Signal Processing Magazine*, May, pp. 104–107.

Nyquist, H., 1928, Certain topics in telegraph transmission theory, *Proceedings of the AIEE*, vol. 47, Apr., pp. 617–644. Reprinted as a classic paper in *Proceedings of the IEEE*, vol. 90, no. 2, Feb. 2002.

PCI-SIG, 2005, *PCI Express™ Base Specification*, revision 1.1, PCI, Wakefield, MA, Mar. 28.

Proakis, John, 2001, *Digital Communications*, 4th ed., McGraw-Hill, New York.

Qureshi, Shahid, 1985, Adaptive equalization, *Proceedings of the IEEE*, vol. 73, no. 9, Sept., pp. 1349–1387.

Shannon, Claude, 1949, Communication in the presence of noise, *Proceedings of the Institute of Radio Engineers*, vol. 37, Jan., pp. 10–21.

Sklar, Bernard, 2001, *Digital Communications: Fundamentals and Applications*, Prentice Hall, Upper Saddle River, NJ.

Sun, Ruifeng, Jaejin Park, Frank O'Mahony, and C. Patrick Yue, 2005, A low-power, 20-Gb/s continuous-time adaptive passive equalizer, *IEEE Symposium on Circuits and Systems*, May 23–26, pp. 920–923.

PROBLEMS

12-1 Given the channel transfer function shown in Figure 12-41, calculate the maximum achievable data rate, assuming a signaling scheme that uses a 10-GHz bandwidth. Use the same signal and noise power spectra figures as in Example 12-2.

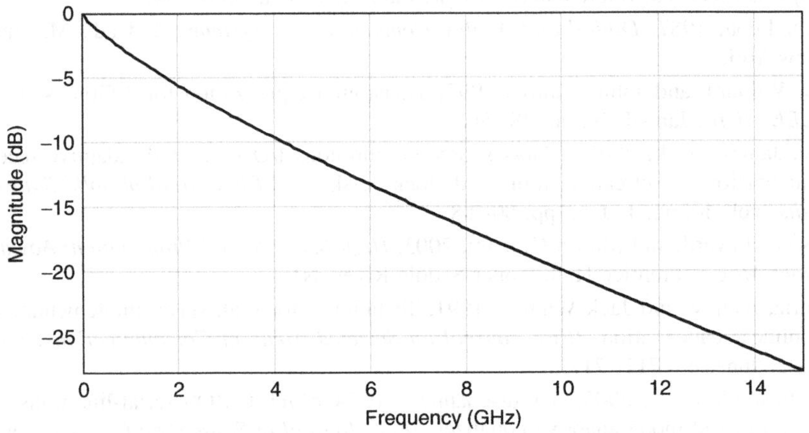

Figure 12-41 Channel transfer function for Problem 12-1.

12-2 Given the sampled differential voltages at the receiver input and the coefficients for the discrete linear equalizer shown below, calculate the values of the output samples:

$$\mathbf{x} = \begin{bmatrix} -246 \\ -190 \\ -31 \\ -165 \\ -218 \\ -235 \\ -242 \\ -245 \\ -247 \\ -248 \end{bmatrix} \text{mV} \quad \text{and} \quad \mathbf{c} = \begin{bmatrix} -0.05 \\ -0.20 \\ 0.70 \\ -0.05 \end{bmatrix}$$

12-3 Given the received pulse response below, use the ZFS algorithm to calculate the coefficients for a five-tap discrete linear equalizer that has one precursor and three postcursor taps. Calculate the sample values of the resulting equalized response.

$$\mathbf{y} = \begin{bmatrix} -75 \\ -74 \\ -74 \\ -68 \\ 18 \\ -41 \\ -62 \\ -68 \\ -71 \\ -72 \\ -72 \\ -73 \\ -73 \\ -74 \\ -74 \end{bmatrix} \text{mV}$$

12-4 Using the received pulse response from Problem 12-3 and the coefficients for the DLE and DFE below, calculate the sample values of the equalized output response.

$$\mathbf{C}_{\mathbf{DLE}} = \begin{bmatrix} 0.65 \\ -0.35 \end{bmatrix} \qquad \mathbf{C}_{\mathbf{DFE}} = \begin{bmatrix} 0.010 \\ 0.005 \end{bmatrix}$$

12-5 Derive the transfer function for a discrete linear equalizer, starting from equation (12-21). The time-shift property of the Fourier transform will prove useful: If $y(t)$ has transform $F[y(t)](f) = Y(f)$, then

$$\int_{-\infty}^{\infty} y(t - t_0)e^{-2\pi jft}\, dt = e^{-2\pi jft_0}Y(f)$$

12-6 Derive the transfer function of the alternate passive equalizer shown in Figure 12-20.

12-7 An alternative to the ZFS approach is the minimum mean square error (MMSE) equalizer. Finding the coefficients using the MMSE requires first calculating the autocorrelation matrix, $\mathbf{R}_{\mathbf{xx}} = \mathbf{x}^{\mathrm{T}}\mathbf{x}$, and the cross-correlation matrix, $\mathbf{R}_{\mathbf{xz}} = \mathbf{x}^{\mathrm{T}}\mathbf{y}$. The equalizer coefficients are then calculated as $\mathbf{c} = \mathbf{R}_{\mathbf{xx}}^{-1}\mathbf{R}_{\mathbf{xz}}$. Use the pulse response input samples from the Problem 12-3 to calculate the coefficients for a linear equalizer with one precursor and three postcursor taps using the MMSE algorithm. Calculate the sample values of the resulting equalized response.

12-8 Given the pulse response in Problem 12-3, calculate the equalizer coefficient progression obtained using the adaptive ZFS algorithm.

12-9 Given the pulse response in Problem 12-3, calculate the equalizer coefficient progression obtained using the adaptive sign–sign LMS algorithm.

12-10 Determine the output from a DFE applied to a 1-V signaling system whose sampled voltages at the receiver are

$$\mathbf{x} = \begin{bmatrix} -0.12 \\ 0.9 \\ 0.1 \\ -0.05 \\ 0.02 \end{bmatrix} \text{V} \quad \text{and} \quad \mathbf{c}_{\mathbf{DFE}} = \begin{bmatrix} -0.090 \\ 0.030 \\ -0.001 \end{bmatrix}$$

13

MODELING AND BUDGETING OF TIMING JITTER AND NOISE

This chapter ties together the concepts presented in previous chapters with the intent of providing the reader with a method for managing timing noise and voltage noise successfully in order to create a successful multi-Gb/s design. We start by introducing the eye diagram as a tool for evaluating the performance of a signaling interface, introducing eye width and eye height as key metrics. The eye diagram is also a fundamental prerequisite to understanding the bit error rate (BER) of a link. Modern high-speed interfaces are designed to a specified BER and data rate, which often determines the amount of voltage and timing noise allowed on a link. Accordingly, the chapter is centered on the concept of bit error rate, and we relate each of the analysis techniques described in the chapter to the BER to provide a useful design methodology.

Following the introduction to bit error rate, we discuss sources of timing variation, known as *jitter*, that degrade the eye width, and describe the method for creating a jitter budget. We follow that with a description of the sources of voltage noise that degrade the eye height, and describe a method for analyzing and creating a system noise budget. We then present the peak distortion analysis (PDA) technique for comprehending voltage and timing noise in determining the worst-case received eye from the pulse response of a high-speed interconnect channel. In providing a method for determining the maximum jitter and voltage noise, PDA accounts for intersymbol interference (ISI) caused by losses and reflections, and for degradation of the data eye due to crosstalk.

13.1 EYE DIAGRAM

Most high-speed designs use the eye diagram to evaluate system performance. We show an example eye diagram for a 10-Gb/s 100-bit data sequence in Figure 13-1. An eye diagram is constructed by slicing the time-domain signal waveform into sections that are a small number of symbols in length, and overlaying them. The horizontal axis of the eye diagram represents time and is typically one or two symbols wide, while the vertical axis represents the amplitude of the signal. Figure 13-2 illustrates the eye diagram construction process for both a "perfect" eye and one that is distorted by losses and/or reflections.

As the figures show, distortion of the signal causes the data eye to close. Conceptually, we want the eye to be as "open" as possible, as a larger eye opening implies that we have more margin to the voltage and timing requirements. From a

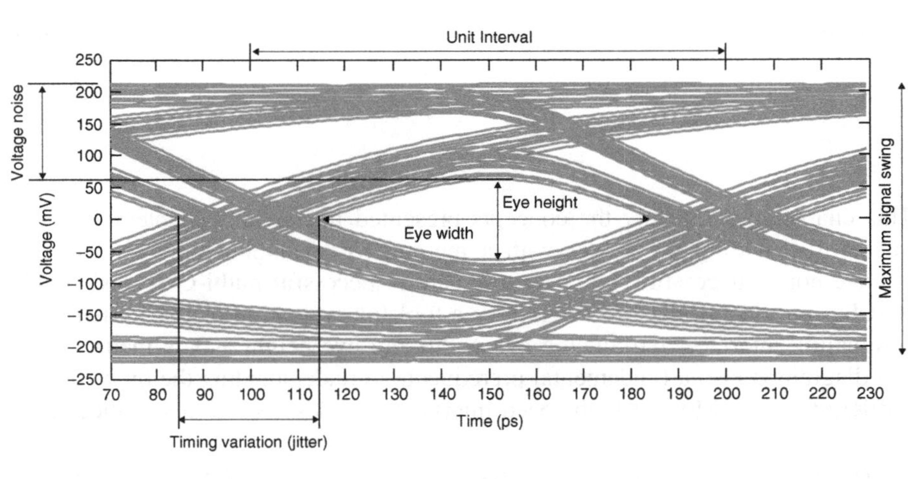

Figure 13-1 Example received eye diagram for a 10-Gb/s 100-bit data sequence.

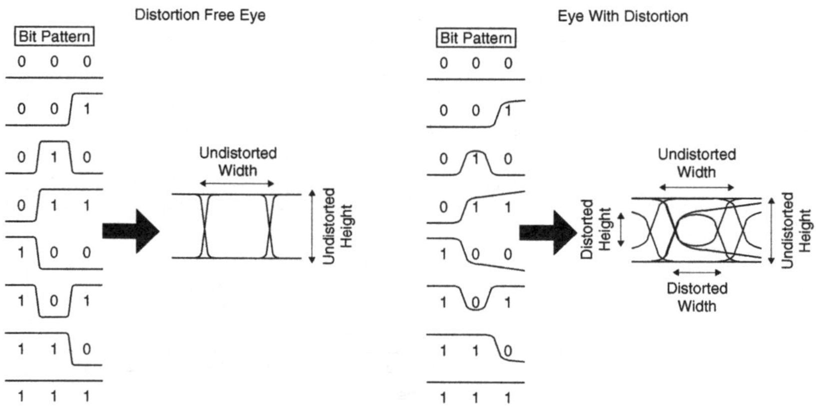

Figure 13-2 Eye diagram construction process and the impact of signal distortion.

quantitative standpoint, the minimum height and width of the data at the receiver are key metrics for evaluating link performance. The eye must be wide enough to provide adequate time to satisfy the setup and hold requirement of the receiver, and have sufficient height to ensure that the voltage levels meet v_{ih} and v_{il} requirements in a system that may possess multiple sources of noise. This allows the receiver to resolve the input signals successfully into digital values.

Using the eye diagram properly is critical, as we cannot assess performance simply by comparing the eye diagram to the receiver's setup and hold window and v_{ih} and v_{il} requirements. In addition to the signal distortion induced by the interconnect channel, we must also comprehend the variations in the voltage and timing induced by the clock used to capture data at the receiver and by the transmitter and transmit clock in order to ensure proper operation. We discuss the sources of the timing uncertainties (jitter) and voltage uncertainties (noise), and provide methods for creating jitter and noise budgets in Sections 13.3 and 13.4. A widely used method for evaluating whether or not the eye meets the system timing and noise requirements is to apply an *eye mask*, as shown in Figure 13-3. The mask represents a forbidden region that the actual eye must not cross, and it includes the receiver setup and hold window and voltage specs, and all jitter and noise terms. With it we can evaluate the performance of a given design by comparing it to worst-case eye obtained using the peak distortion analysis (PDA) method described in Section 13.5. PDA is a deterministic method for finding the minimum received eye height and width in a signaling system that has significant sources of distortion, such as loss, reflections, and crosstalk.

We can also use the eye to estimate the probability of receiving erroneous bits, which is known as the *bit error rate* (BER) or *bit error ratio*. The BER is expressed as the ratio of the number of erroneous received bits to the total number of transmitted bits over a sufficiently long interval:

$$\mathrm{BER}(t_s, v_s) = \lim_{N\to\infty} \frac{N_{\mathrm{err}}(t_s, v_s)}{N} \tag{13-1}$$

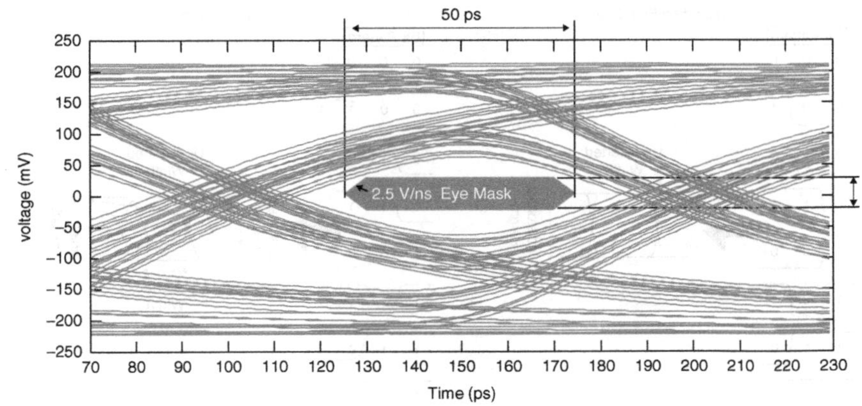

Figure 13-3 Example eye diagram and spec eye mask.

where (t_s, v_s) represents the relative voltages and times at which the signal is sampled, N_{err} is the number of erroneous bits received, and N is the number of bits transmitted over the same time interval.

Equation (13-1) suggests that the BER depends on when we sample the data and at what voltage level (i.e., the sampling point position). By varying the sample point, we can generate a diagram that consists of a set of contours at a set of BER values, as shown in Figure 13-4. Ideally, the receiver will sample the data signal in the center of the eye. However, jitter and noise can cause nonideal data sampling, which can result in transmission errors. For example, if the receiver samples in the middle of the data eye, at the point corresponding to 50 mV and 50 ps in the figure, the error rate is less than 1 bit in 100 trillion (10^{-14}). On the other hand, sampling at the edge of the eye, such as 10 mV and 10 ps, gives a BER in excess of 1 bit per million (10^{-6}). At a transmission rate of 10 Gb/s, a BER of 10^{-14} will produce one error in approximately 2 hours and 45 minutes, while the 10^{-6} BER results in 10,000 errors per second! Communicating at high bit error ratios requires that error detection and correction capabilities be designed into the I/O circuits, which adds complexity and consumes power. Errors also degrade performance, as they must be detected and then the data re-transmitted. We discuss BER estimation in Section 13.2.

As a final thought, we note that the eye diagram applies to both single-ended and differential interfaces. In the case of a differential interface, the eye diagram should plot the differential voltage ($v_{\text{diff}} = v - \overline{v}$) at the receiver.

13.2 BIT ERROR RATE

13.2.1 Worst-Case Analysis

Historically, timing analysis for signaling interfaces in PC systems has employed a worst-case methodology. In this approach, all sources of timing uncertainty that

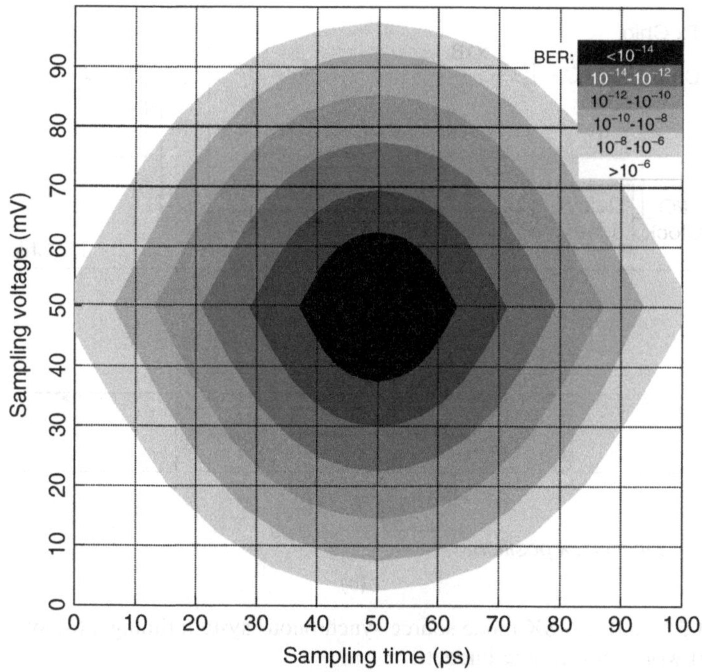

Figure 13-4 Example BER contours.

would degrade the eye opening are treated as bounded sources that do not exceed a specified amount. An example of an interface design based on the worst-case approach is the AGP 8X mode interface, which we describe here [Intel, 2002].

The AGP 8X mode is a 533-Mb/s source synchronous interface, which is shown in Figure 13-5a. The AGP source synchronous transmitter sends a clock signal along with a group of 16 data signals. The delay line in the transmitting chip offsets the clock signal by 90° from the data signal in order to center it in the data eye. Source synchronous designs achieve high performance by preserving the clock-to-data phase relationship. Doing so requires that the delays of the data signals be matched to those of the clock signal, which is accomplished by using identical transmitters and by matching the interconnect lengths. In theory, the maximum transmission rate of a source synchronous system is limited only by the setup-and-hold window of the receiver, which is described by Dally and Poulton [1998]. In practice, however, variation in the delay of the transmitters, interconnects, and receivers will reduce maximum achievable rate to a substantially lower value. For example, sources of variation in the transmitter delay include differences in the clock distribution path to the various circuits, process variation within the chip, and noise. Timing variation terms for source synchronous interfaces are typically specified as relative delays between the data and clock signals, so they include variations of each.

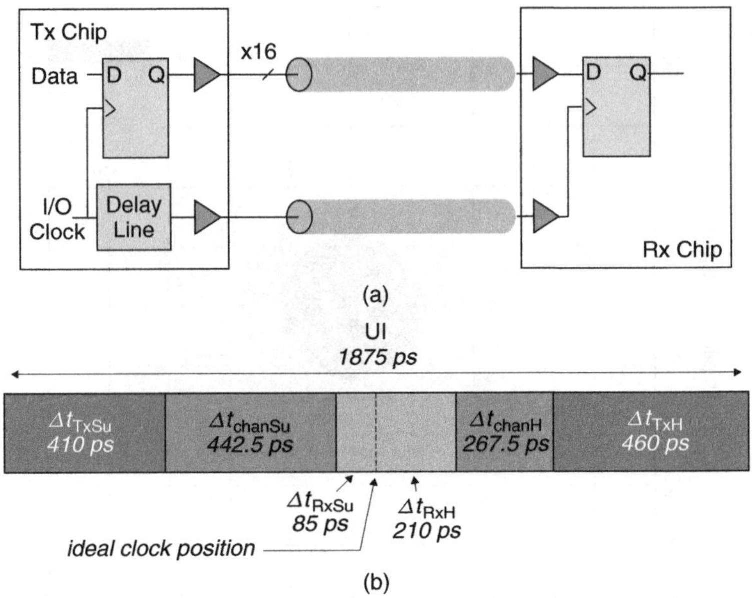

Figure 13-5 AGP 2.0 8X mode source synchronous system timings: (a) system configuration; (b) worst-case timing budget.

The worst-case timing equations for the AGP source synchronous link are

$$t_{\text{marSu}} = \frac{\text{UI}}{2} - \Delta t_{\text{TxSu}} - \Delta t_{\text{chanSu}} - \Delta t_{\text{RxSu}} \tag{13-2}$$

$$t_{\text{marH}} = \frac{\text{UI}}{2} - \Delta t_{\text{TxH}} - \Delta t_{\text{chanH}} - \Delta t_{\text{RxH}} \tag{13-3}$$

where t_{marSu} and t_{marH} = timing margins for the setup and hold conditions (ps)

UI = unit interval, which is the width of a single bit (ps)

$\Delta\ t_{TxSu}$ and $\Delta\ t_{TxH}$ = variation in transmitter delay relative to the clock path for the setup and hold cases (ps)

$\Delta\ t_{IntSu}$ and t_{IntH} = variation in interconnect delay relative to the clock path for the setup and hold cases (ps)

$\Delta\ t_{RxSu}$ and $\Delta\ t_{RxH}$ = variation in receiver delay relative to the clock path for the setup and cases (ps)

At 533 Mb/s, the length of a single bit, known as the *unit interval* (UI), is 1875 ps. The AGP 8X specification at the receiver is 85 ps for the setup case and 210 ps for the hold case. The spec budgets 410 ps for the worst-case variation of the transmitter delay for the data edge that precedes the clock (the setup case), and 460 ps of worst-case transmitter delay variation for the data edge that follows the clock (hold case). The spec allows for 442.5 ps and 267.5 ps of interconnect delay variation in the setup and hold cases, respectively. The timings are depicted graphically in Figure 13-5b. The total window for setup and hold time at the receiver is 295 ps. This defines a minimum UI that would give a maximum transfer rate of approximately 3.4 Gb/s if there were no variation in the transmitter and interconnect delays. However, the transmitter and interconnect delay variations add a total of 1560 ps to the unit interval, degrading the maximum transfer rate to the 533 Mb/s final value.

Although the worst case approach treats the sources of timing variation as though they are bounded, in reality this is a faulty assumption. Some sources, for example channel induced jitter due to intersymbol interference, are in fact bounded. Others, however, such as phase-locked loop (PLL) jitter induced by power supply noise, are random in nature. These sources are not bounded, but instead typically fit a Gaussian distribution, in which the timing uncertainty is described by

$$\mathrm{RJ}(t) = \frac{1}{\sqrt{2\pi}\,\sigma_{\mathrm{RJ}}} e^{-t^2/2\sigma_{\mathrm{RJ}}^2} \tag{13-4}$$

where RJ(t) is the probability of having a timing jitter of t ps due to a random source and σ_{RJ} is the root-mean-square *timing uncertainty* (a.k.a. *jitter*) (ps).

With a Gaussian distribution, even very large uncertainties have a nonzero (although extremely small) probability of occurrence, as Figure 13-6 shows. This has the consequence of rendering the notion of worst-case timings meaningless. Instead, we must interpret the timings in terms of the bit error rate, which is really just a probability that the timing uncertainties exceed the width of the unit interval. The significance of this discussion is that it implies that prior system designs based on worst-case timings were not really designed to a worst case, since it has no meaning. Why, then, did these designs work? The truth is that worst-case timing-based systems were in reality designed to achieve immeasurably low BER.

To illustrate, we examine the mean time between errors for the AGP 8X interface, assuming a BER of 10^{-18}. The AGP data bus is 32 bits wide, running at 533 Mb/s. At the specified bit error rate, the mean time between errors would be approximately two and one-half years for the entire bus (assuming that errors on different data lines are uncorrelated). The 2.5-year estimate also assumes continuous operation and 100% bus utilization. If we assume that the system is a personal computer, and is only in use half of the time, the mean time between errors increases to five years. If we assume further that the average traffic on the bus is unlikely to exceed 50%, the mean time between errors increases to 10 years, which far exceeds the expected lifetime of the computer.

13.2.2 Bit Error Rate Analysis

As signaling speeds continue to increase, maintaining sufficient margins to guarantee immeasurable bit error rates becomes prohibitive. As a result, high-speed

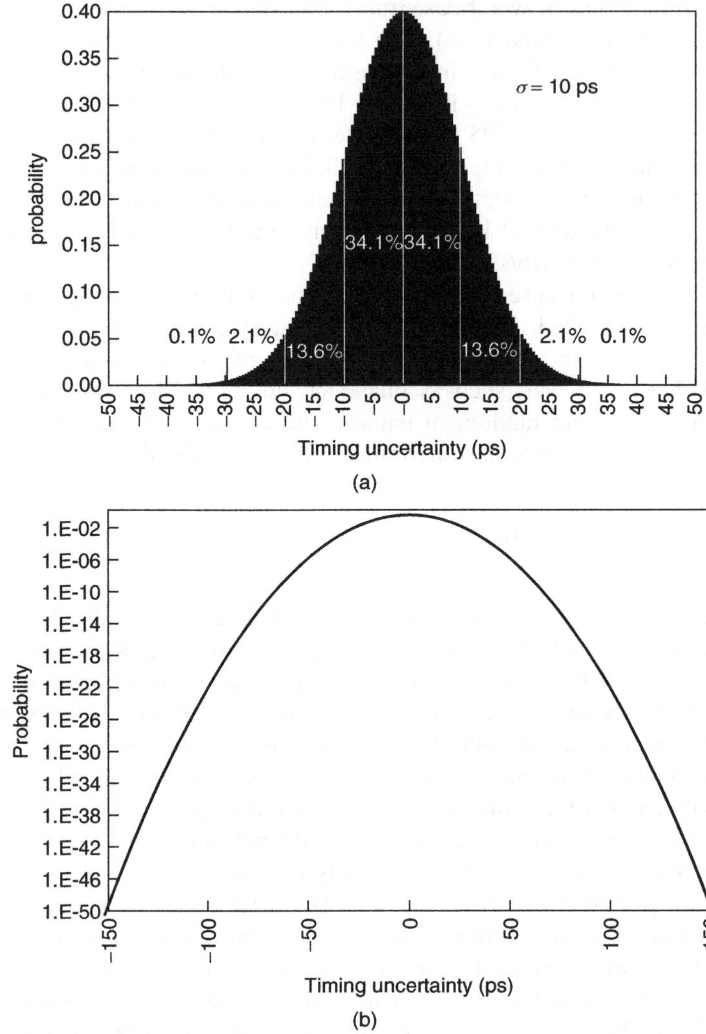

Figure 13-6 Gaussian distribution of timing uncertainties with $\sigma = 10$ ps: (a) linear plot; (b) semilog plot.

links are moving to a BER-based budgeting approach. Our discussion proceeds with the premise that future high-speed signaling links will be designed to achieve a finite bit error rate. As such, we need to develop a method for calculating the BER from the timing distributions. We start by formulating an expression for the bit error rate as a function of the distribution of timing jitter:

$$\mathrm{BER}(t) = \rho_T \int_{-\infty}^{\infty} J(t)\, dt \tag{13-5}$$

where ρ_T is the transition density, which is the ratio of the number of logic transitions to the total number of bits transmitted (typically, ρ_T is equal to 0.5), and $J(t)$ is the jitter distribution. The equation is the cumulative distribution function of the timing jitter, which takes into account the probability of actually making a transition on any given bit (since the jitter will be zero if the signal remains at a given level).

To calculate the BER we need a model for the jitter distribution. We can use equation (13-4) for the random jitter sources, but it will not be accurate for systems that also include deterministic (bounded) jitter sources such as ISI, as Figure 13-7 demonstrates. In particular, the jitter histograms extracted from the zero crossing points of the eye show bimodal distributions, which suggests that the jitter is comprised of a combination of deterministic and Gaussian jitter sources. So we need a model for deterministic jitter that we can combine with the Gaussian random jitter model.

Deterministic jitter sources can fit a variety of distributions, examples of which are shown in Figure 13-11 and described in the next section. In creating system-level jitter budgets, we employ the dual Dirac model in equation (13-6) to express the probability density function (PDF) of the deterministic jitter, DJ(t):

$$\mathrm{DJ}(t) = \frac{\delta\left(t - \mathrm{DJ}_{\delta\delta}/2\right)}{2} + \frac{\delta\left(t + \mathrm{DJ}_{\delta\delta}/2\right)}{2} \tag{13-6}$$

where $\mathrm{DJ}_{\delta\delta}$ is the dual Dirac deterministic jitter (ps) and $\delta(t)$ is the Dirac delta function,

$$\delta(t) = \begin{cases} 0 & t \neq 0 \\ 1 & t = 0 \end{cases}$$

As the equation shows, the dual Dirac model, which is widely used in industry, treats the deterministic jitter as though it is equally distributed at extreme values. The deterministic jitter in real systems does not fit a dual delta distribution function. The usefulness of the model is that it allows us easily to combine the deterministic and random jitter distributions. The reason that the dual Dirac model works is that we really only need to be accurate in estimating the jitter at low BER, which is governed by the random jitter. As a result, we can use the dual Dirac distribution model to shift the "tails" of the jitter distributions (the RJ) to their proper locations. Conceptually, the dual Dirac model gives us a Gaussian approximation to the outer edges of the jitter distribution when displaced by $\mathrm{DJ}_{\delta\delta}$. For a more in-depth treatment of jitter distributions and the dual Dirac model, we refer the reader to a report by Stephens [2004] and a book by Li [2008]. We provide an example that demonstrates the application of the dual Dirac model in Section 13.3.

The total jitter PDF, JT(t), is created by convolving the DJ and RJ models:

$$\mathrm{JT}(t) = \mathrm{RJ}(t) * \mathrm{DJ}(t) = \int_{-\infty}^{\infty} \left[\frac{1}{\sqrt{2\pi}\,\sigma_{\mathrm{RJ}}} e^{-t^2/2\sigma_{\mathrm{RJ}}^2}\right] \left(\frac{t - \mathrm{DJ}_{\delta\delta}}{2} + \frac{t + \mathrm{DJ}_{\delta\delta}}{2}\right) dt$$

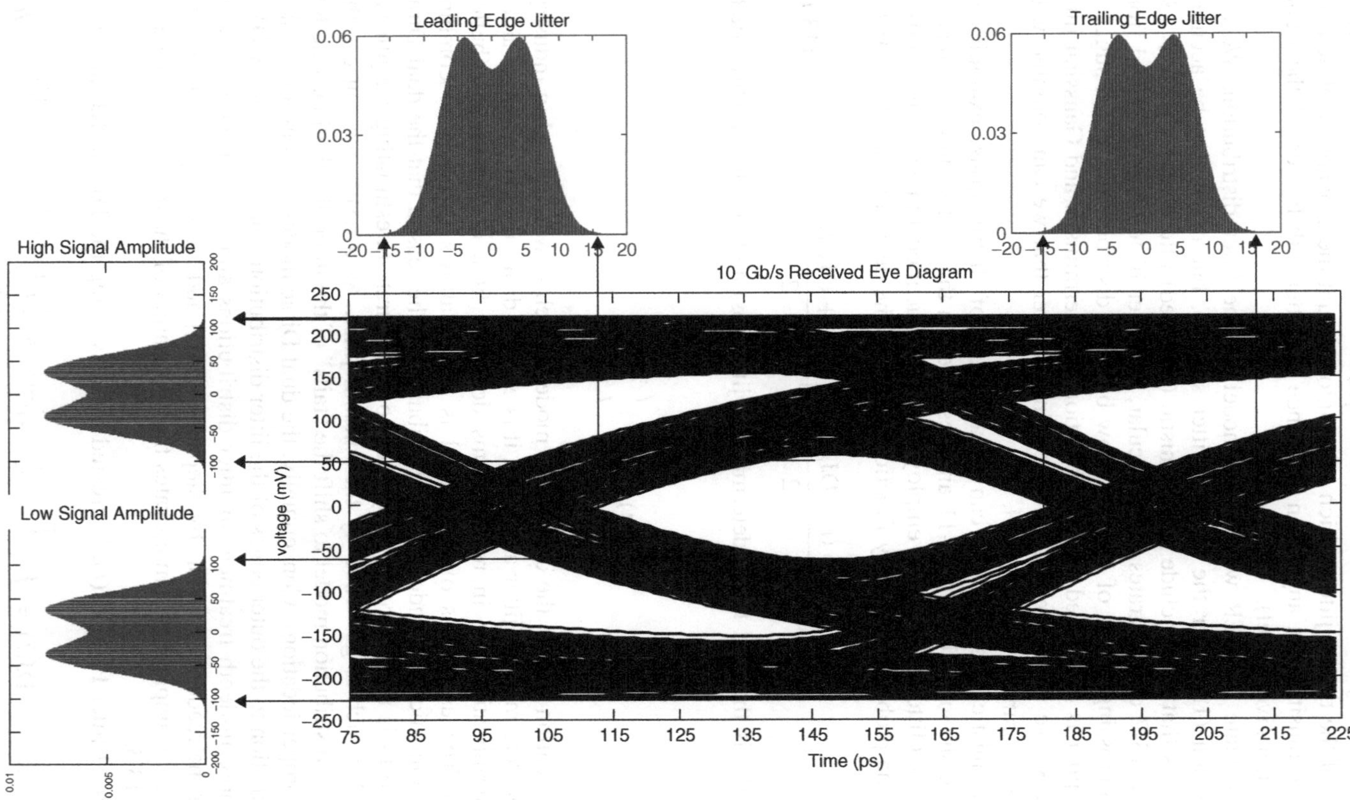

Figure 13-7 Example eye diagram with jitter and noise histograms.

We can use the convolution property of the delta function, $\int_{-\infty}^{\infty} f(t)\delta(t-a)\,dt = f(a)$, to simplify the total jitter model:

$$\mathrm{JT}(t) = \frac{1}{2\sqrt{2\pi}\,\sigma_{\mathrm{RJ}}}\left[e^{-(t-\mathrm{DJ}_{\delta\delta}/2)/2\sigma_{\mathrm{RJ}}^2} + e^{-(t+\mathrm{DJ}_{\delta\delta}/2)/2\sigma_{\mathrm{RJ}}^2}\right] \qquad (13\text{-}7)$$

The resulting jitter distribution is bimodal and is composed of a pair of Gaussian distributions with equal variance (σ_{RJ}^2) whose mean values are separated by $\mathrm{DJ}_{\delta\delta}$.

As an example, we return to the timings for the AGP 8X interface. To illustrate the distributions, we assume that $\mathrm{DJ}_{\delta\delta} = 1000$ ps and $\sigma_{\mathrm{RJ}} = 50$ ps, which includes the jitter contributions of the transmitter, receiver, and interconnect. We allocated the jitter distributions so that the sum of the DJ plus $17.5 \times \sigma_{\mathrm{RJ}}$ is equal to the unit interval, which will result in a BER of approximately 10^{-18}. The dual Dirac model jitter distributions for the leading edge of the data eye are shown in Figure 13-8.

Using the probability density function for the total system jitter, we calculate the bit error rate as the area under the PDF. The expression for the bit error rates

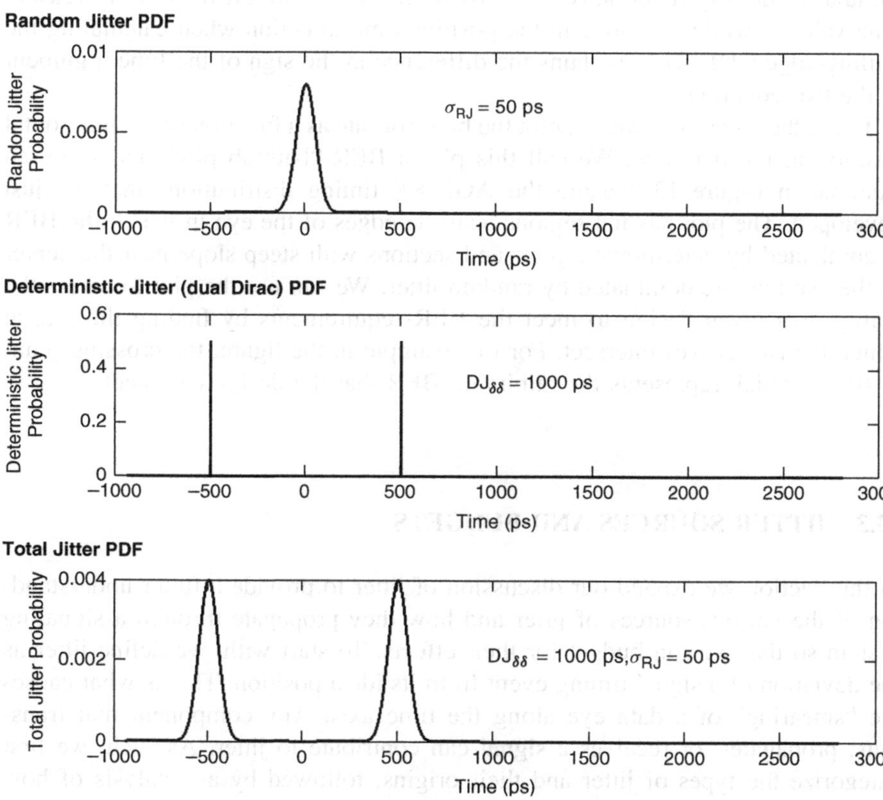

Figure 13-8 Hypothetical jitter PDF models for AGP 8X interface (UI = 1.875 ns).

of the leading and trailing edges of the eye are given by

$$\text{BER}_{\text{lead}}(t) = 0.5\left[\text{erfc}\left(\frac{t - \text{DJ}_{\delta\delta}/2}{\sqrt{2}\,\sigma_{\text{RJ}}}\right) + \text{erfc}\left(\frac{t + \text{DJ}_{\delta\delta}/2}{\sqrt{2}\,\sigma_{\text{RJ}}}\right)\right] \tag{13-8}$$

$$\text{BER}_{\text{trail}}(t) = 0.5\left[\text{erfc}\left(\frac{\text{UI} - t - \text{DJ}_{\delta\delta}/2}{\sqrt{2}\,\sigma_{\text{RJ}}}\right) + \text{erfc}\left(\frac{\text{UI} - t + \text{DJ}_{\delta\delta}/2}{\sqrt{2}\,\sigma_{\text{RJ}}}\right)\right] \tag{13-9}$$

where erfc(t) is the complementary error function, which is defined as

$$\text{erfc}(t) = \frac{2}{\sqrt{\pi}}\int_t^{\infty} e^{-x^2}\,dx$$

Note that in developing the equation for the leading-edge BER, we calculate the cumulative density function (CDF) moving from right to left (toward decreasing time values), while we move in the positive time direction when calculating the trailing-edge CDF. This explains the difference in the sign of the time argument in the two equations.

Using the equations, we can plot the bit error rate as a function of the horizontal position in the data eye. We call this plot a BER "bathtub plot" and show an example in Figure 13-9 using the AGP 8X timing distributions that we just developed. The plot has flat regions near the edges of the eye in which the BER is dominated by deterministic jitter and sections with steep slope near the center of the eye that are dominated by random jitter. We can use the plot to assess the ability of a given design to meet the BER requirements by finding the rate at which the two curves intersect. For the example in the figure, the crossing point is 10^{-18}, which represents the minimum BER that the design can meet.

13.3 JITTER SOURCES AND BUDGETS

In this section we expand our discussion of jitter to provide a fuller understanding of the various sources of jitter and how they propagate through a signaling system so that we can budget for their effects. To start with, we define jitter as the deviation of a signal timing event from its ideal position. This is what causes the "smearing" of a data eye along the time axis. Any component that transmits, propagates, or receives a signal can contribute to jitter. As such, we first categorize the types of jitter and their origins, followed by an analysis of how the different components combine to produce the jitter signature for the entire system.

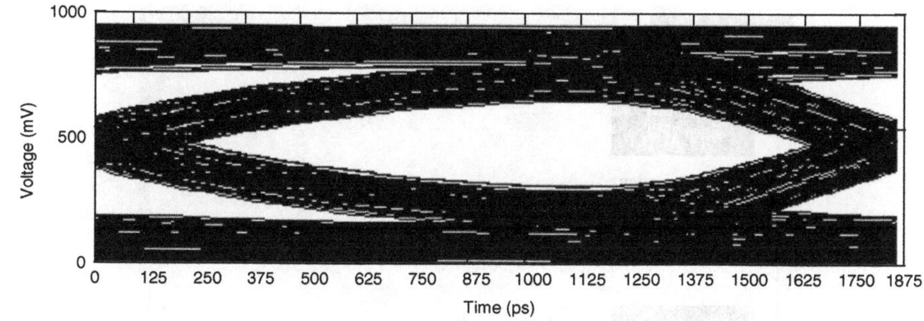

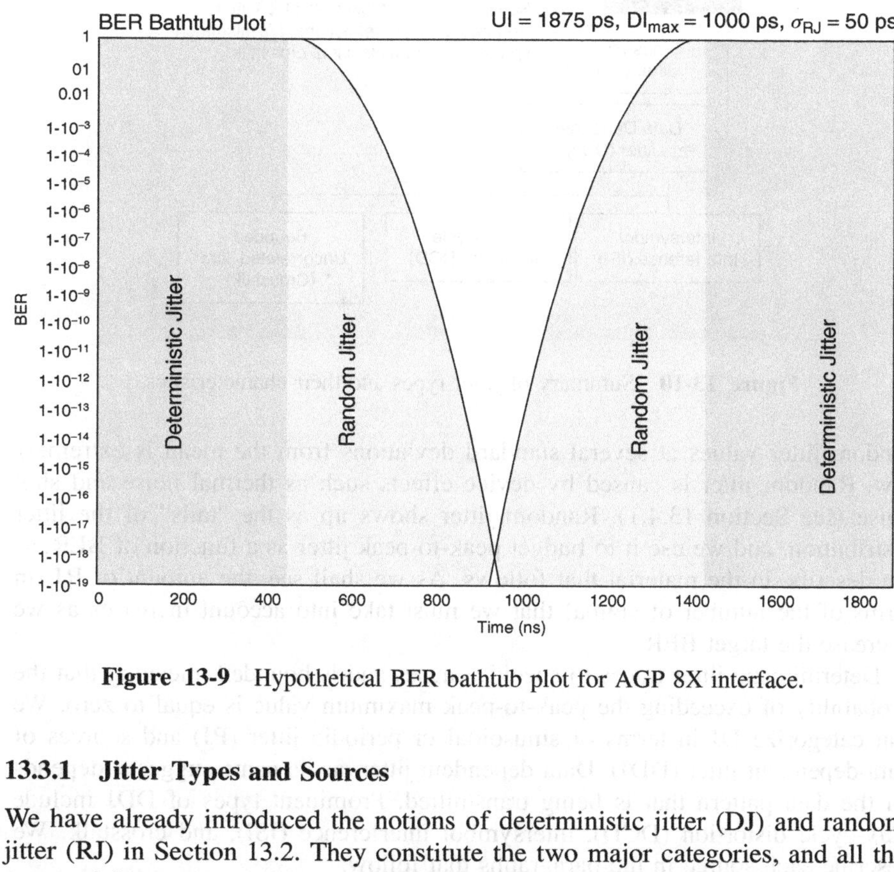

Figure 13-9 Hypothetical BER bathtub plot for AGP 8X interface.

13.3.1 Jitter Types and Sources

We have already introduced the notions of deterministic jitter (DJ) and random jitter (RJ) in Section 13.2. They constitute the two major categories, and all the types of jitter that relate to specific causes that we present in this section are subcategories of either DJ or RJ. The key characteristics of deterministic and random jitter sources are described in Figure 13-10 and are discussed in detail below.

As we have shown, RJ is expressed as a Gaussian distribution which is described by a mean value, μ (typically equal to zero), and a standard deviation, σ_{RMS}. As a result, RJ is not bounded, although the probability of having

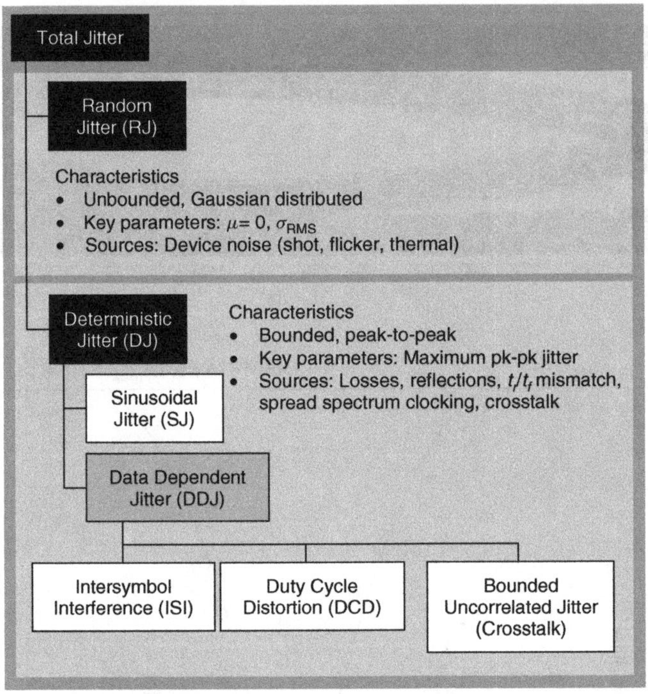

Figure 13-10 Summary of jitter types and their characteristics.

random jitter values at several standard deviations from the mean is extremely low. Random jitter is caused by device effects such as thermal noise and shot noise (see Section 13.4.1). Random jitter shows up as the "tails" of the jitter distribution, and we use it to budget peak-to-peak jitter as a function of BER, as we describe in the material that follows. As we shall see, the amount of RJ (in terms of the number of sigma) that we must take into account increases as we decrease the target BER.

Deterministic jitter traces to specific causes and is bounded, meaning that the probability of exceeding the peak-to-peak maximum value is equal to zero. We can categorize DJ in terms of sinusoidal or periodic jitter (PJ) and sources of data-dependent jitter (DDJ). Data-dependent jitter, as its name suggests, depends on the data pattern that is being transmitted. Prominent types of DDJ include duty-cycle distortion (DCD), intersymbol interference (ISI), and crosstalk. We describe each source in the paragraphs that follow.

Periodic Jitter PJ repeats at a fixed frequency and is caused by modulating effects, such as spread-spectrum clocking. For a system with multiple periodic sources, we model the total periodic jitter as

$$PJ(t) = \sum_i A_i \cos(\omega_i t + \theta_i) \tag{13-10}$$

where A_i is the amplitude of source i, ω_i the frequency of source i, and θ_i the phase of source i. The probability density function for an individual periodic jitter source is given by (see Figure 13-11a):

$$\mathrm{PDF_{PJ}}(t) = \begin{cases} \dfrac{1}{\pi\sqrt{A^2 - t^2}} & A > |t| \\ 0 & A \leq |t| \end{cases} \tag{13-11}$$

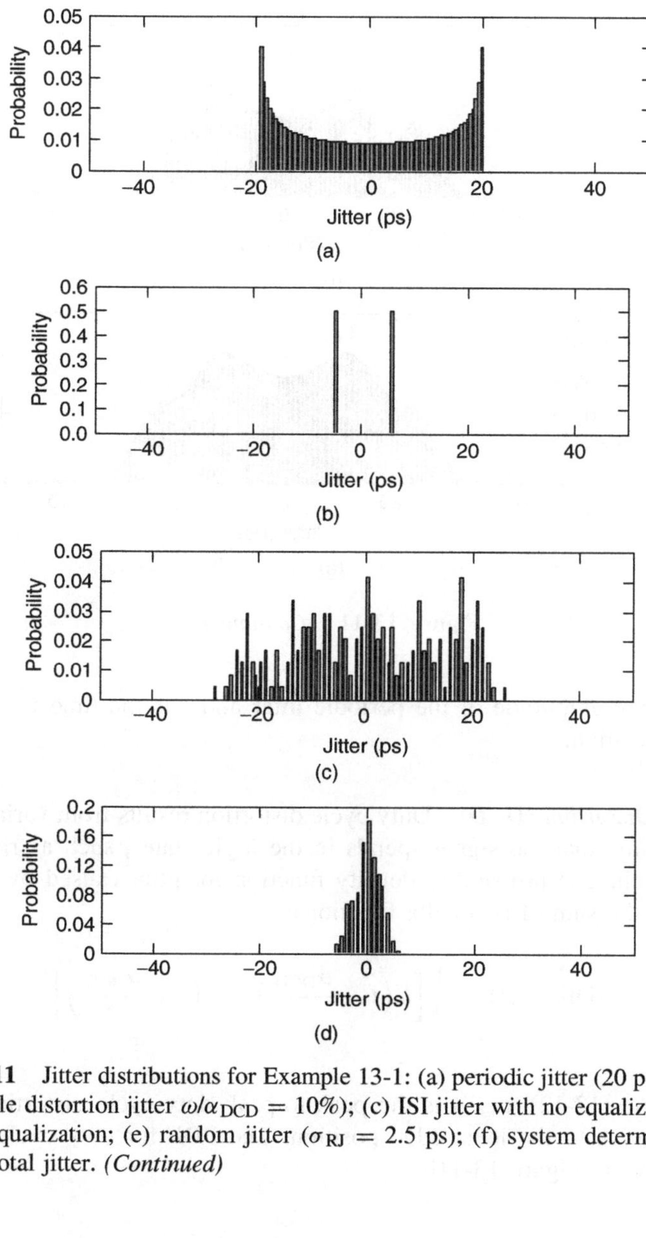

Figure 13-11 Jitter distributions for Example 13-1: (a) periodic jitter (20 ps amplitude); (b) duty cycle distortion jitter $\omega/\alpha_{\mathrm{DCD}} = 10\%$); (c) ISI jitter with no equalization; (d) ISI jitter with equalization; (e) random jitter ($\sigma_{\mathrm{RJ}} = 2.5$ ps); (f) system deterministic jitter; (g) system total jitter. *(Continued)*

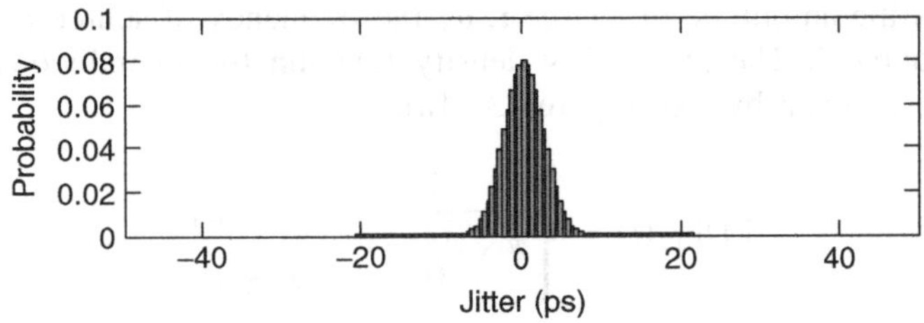

(e)

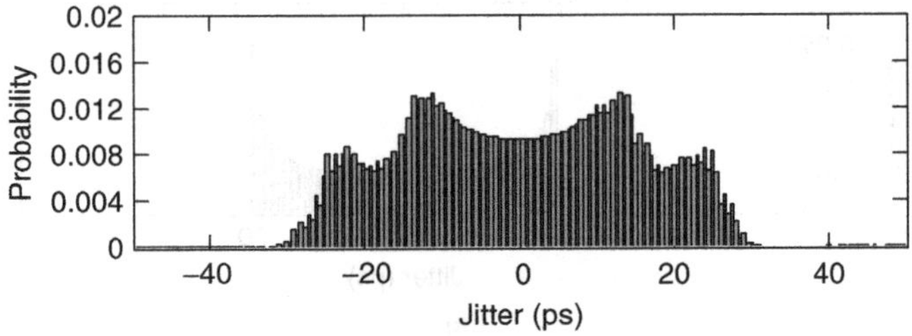

(f)

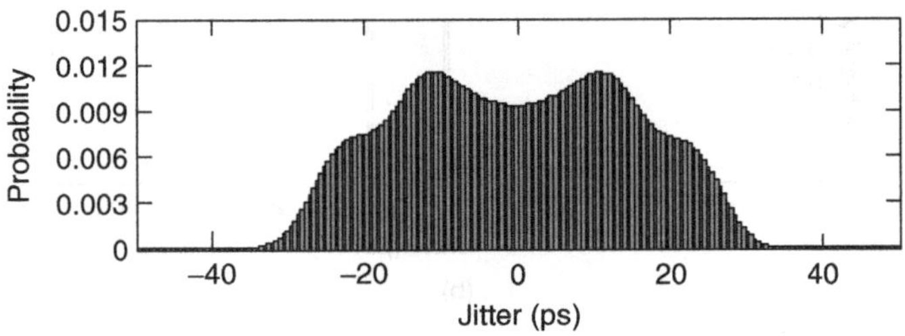

(g)

Figure 13-11 (*Continued*)

where A is the amplitude of the periodic jitter and t is the time relative to the ideal edge position.

Duty Cycle Distortion (DCD) Duty cycle distortion results from variation in the amount of time that the signal spends in the logic states, such as rise and fall time mismatch. The probability density function for jitter caused by duty cycle distortion is the sum of two delta functions:

$$\text{PDF}_{\text{DCD}}(t) = \frac{1}{2}\left[\delta\left(t - \frac{\alpha_{\text{DCD}}}{2}\right) + \delta\left(t + \frac{\alpha_{\text{DCD}}}{2}\right)\right] \tag{13-12}$$

In equation (13-12), α_{DCD} is the peak-to-peak duty cycle distortion and t is the time relative to the ideal edge position. The PDF for duty cycle distortion jitter is shown in Figure 13-11b.

Intersymbol Interference As we have seen, ISI is caused by losses, dispersion, and reflections on the transmission lines that make up the interconnect channel. We show examples of probability density functions for channels with and without equalization in Figure 13-11c and d. In general, the ISI PDF for a nonequalized channel has multiple peaks, whereas that for an equalized channel looks like a truncated Gaussian distribution.

Bounded Uncorrelated Jitter BUJ is deterministic jitter that is not aligned in time to the data stream. The most common source of BUJ is crosstalk. Recall that we have shown that crosstalk affects signal delay and amplitude through coupling of signals between neighboring transmission lines. As a result, crosstalk can be a significant source of jitter in high-speed systems. Crosstalk-induced jitter is correlated to the data on neighboring signal, but is not correlated to the data pattern on the signal itself. Hence, it is uncorrelated. The key characteristics of deterministic and random jitter types are described in Figure 13-10.

Example 13-1 Jitter Distributions for a 10-Gb/s Interface We now present an example running at 10 Gb/s to demonstrate multiple aspects of system jitter and bit error rate calculations. We start by showing how the various jitter sources combine to give the total system jitter. We then develop the dual Dirac model parameters and show how the model provides a suitable approximation of the system jitter at low error rates.

Our system contains sources for periodic jitter, duty cycle distortion (DCD) jitter, ISI jitter, and random jitter. The amplitude of the periodic jitter is 20 ps, and it has the PDF in Figure 13-11a. The DCD jitter is caused by a 10% duty cycle variation ($\alpha_{\text{DCD}} = 0.1$), and the resulting PDF is shown in Figure 13-11b. The ISI jitter in our example is obtained from a signaling system that is based on the design that is presented in Section 14.2. This particular case is an 84-Ω differential pair that is driven by a 5-mA current source with 40-Ω termination at each end. The system uses a single bit of equalization with a coefficient value of −0.27. The resulting ISI PDFs for the nonequalized and equalized cases are shown in Figure 13-11c and d, respectively. For this analysis we focus on the equalized case, which has a distribution that lies between ±6 ps.

The random jitter has an RMS value of 2.5 ps, resulting in the Gaussian jitter PDF shown in Figure 13-11e. We calculate the PDFs for the system-level DJ and total jitter via convolution using equations (13-13) and (13-14). The resulting distributions are shown in Figure 13-11f and g.

$$\text{PDF}_{\text{DJ}}(t) = \text{PDF}_{\text{PJ}}(t) * \text{PDF}_{\text{DCD}}(t) * \text{PDF}_{\text{ISI}}(t) \tag{13-13}$$

$$\text{PDF}_{\text{TJ}}(t) = \text{PDF}_{\text{DJ}}(t) * \text{PDF}_{\text{RJ}}(t) \tag{13-14}$$

Our next step is to use the distribution in Figure 13-11f to find the maximum (peak-to-peak) deterministic jitter. By plotting the PDF on a log scale, as shown in Figure 13-12, we see that distribution shows a steep slope at the lower and upper jitter extremes. This indicates that we have reached the bounds of the

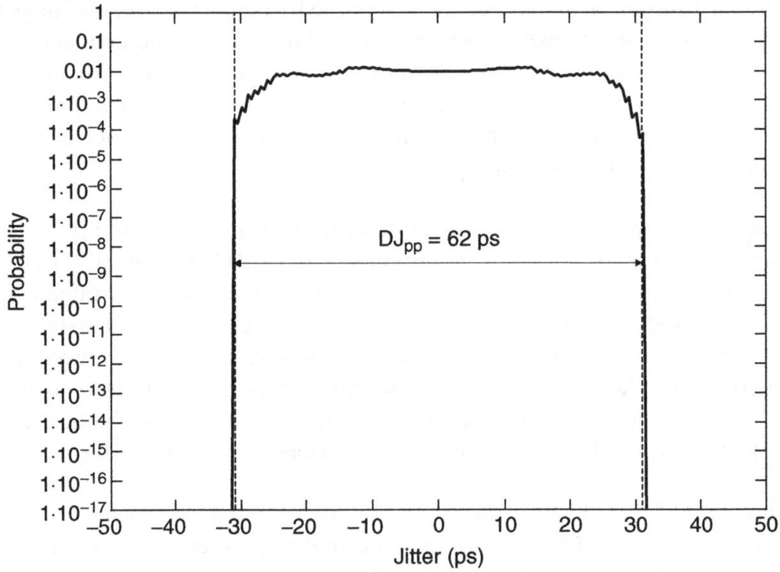

Figure 13-12 Extraction of system-level deterministic jitter for Example 13-1.

DJ distribution, which allows us to estimate that the peak-to-peak deterministic jitter is 62 ps. We can check the accuracy of the system DJ distribution by comparing the peak-to-peak DJ against the sum of the peak-to-peak jitter for the individual DJ components. The sum of PJ (±20 ps), ISI (±6 ps), and DCD (±5 ps) components also equals 62 ps, indicating that our system DJ distribution is accurate.

To estimate the $DJ_{\delta\delta}$ terms we must look at the PDF for the total system jitter. Recall that the dual Dirac model treats the DJ as being distributed as a pair of delta functions. The model works by setting $DJ_{\delta\delta}$ such that the tails of the total jitter distribution model accurately reflect the tails of the actual distribution. We start by converting the probability values from the total jitter PDF to Q-scale values. The Q scale, denoted as Q_{BER}, specifies the amount of eye closure due to random jitter that we must account for at a given BER and is described by

$$Q_{\mathrm{BER}}(\mathrm{BER}) = \sqrt{2}\,\mathrm{erf}^{-1}\left(1 - \frac{\mathrm{BER}}{\rho_T}\right) \tag{13-15}$$

Values of Q_{BER} over a wide range of error rates are listed in Table 13-1. Figure 13-13 shows the Q_{BER} versus jitter plot. The utility of the plot is that it is approximately linear at low error rates, with a slope equal to $\pm\sigma_{\mathrm{RJ}}^{-1}$. We obtain the deterministic jitter term for the dual Dirac model by extrapolating the linear slope starting at very small BER. The $DJ_{\delta\delta}$ is simply the jitter value obtained via the linear extrapolation to a BER value of 1. From the figure we calculate that the values are −28.5 ps and 27.7 ps for the left- and right-hand side of the plots, so that $DJ_{\delta\delta} = 56.2$ ps.

TABLE 13-1. Q_{BER} as a Function of the Bit Error Rate

BER	Q_{BER}	BER	Q_{BER}	BER	Q_{BER}
1×10^{-3}	6.180	1×10^{-10}	12.723	1×10^{-17}	16.987
1×10^{-4}	7.438	1×10^{-11}	13.412	1×10^{-18}	17.514
1×10^{-5}	8.530	1×10^{-12}	14.069	1×10^{-19}	18.026
1×10^{-6}	9.507	1×10^{-13}	14.698	1×10^{-20}	18.524
1×10^{-7}	10.399	1×10^{-14}	15.301	1×10^{-21}	19.010
1×10^{-8}	11.224	1×10^{-15}	15.882	1×10^{-22}	19.484
1×10^{-9}	11.996	1×10^{-16}	16.444	7.7×10^{-24}	20.000

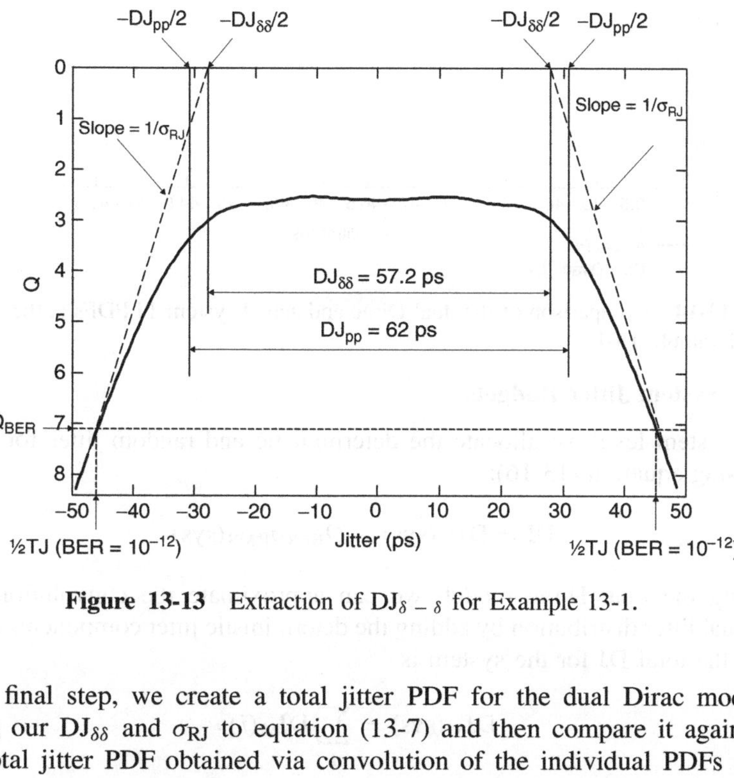

Figure 13-13 Extraction of $DJ_{\delta-\delta}$ for Example 13-1.

As a final step, we create a total jitter PDF for the dual Dirac model by applying our $DJ_{\delta\delta}$ and σ_{RJ} to equation (13-7) and then compare it against the actual total jitter PDF obtained via convolution of the individual PDFs at low error rates. If we are truly operating in the region of the plot that is dominated by RJ, we expect exact agreement. Figure 13-14 shows good, though not perfect agreement between the model and the actual system distribution. The error in the comparison is attributed to numerical errors in our frequency-domain convolution algorithm. We have thus illustrated the application of the dual Dirac model for estimating the jitter of high-speed signaling systems. Finally, we can use the model to estimate the total jitter as a function of bit error rate. For example, in Figure 13-13 we show that the total jitter is approximately 91 ps for a BER of $10^{-12}(\frac{1}{2}Q_{BER} = 7.034)$, leaving us with 9 ps of margin in our design.

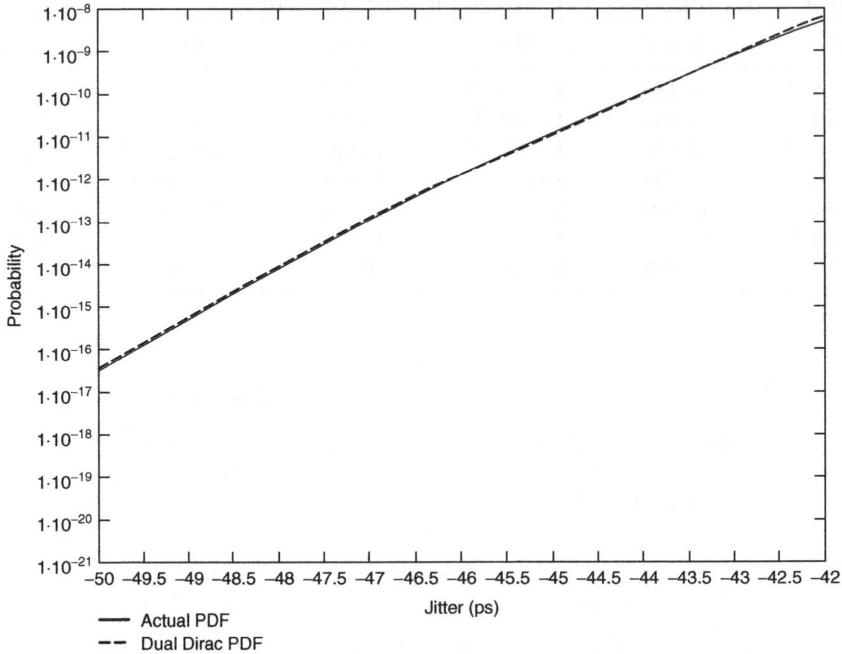

Figure 13-14 Comparison of the dual Dirac and actual system TJ PDFs at the low BER tail for Example 13-1.

13.3.2 System Jitter Budgets

At the system level we allocate the deterministic and random jitter for a given BER using equation (13-16):

$$\mathrm{UI} = \mathrm{DJ}_{\delta\delta}(\mathrm{sys}) + Q_{\mathrm{BER}}\sigma_{\mathrm{RMS}}(\mathrm{sys}) \tag{13-16}$$

Applying the dual Dirac model, we can approximate the convolution of the individual jitter distribution by adding the deterministic jitter components linearly, so that the total DJ for the system is

$$\mathrm{DJ}_{\delta\delta}(\mathrm{sys}) = \sum_i \mathrm{DJ}_{\delta\delta}(i) \tag{13-17}$$

The convolution of random jitter components results in a root-sum-of-squares relationship, leading to the following equation for the system RJ:

$$\sigma_{\mathrm{RMS}}(\mathrm{sys}) = \sqrt{\sum_i \sigma^2_{\mathrm{RMS}}(i)} \tag{13-18}$$

As mentioned, any component that transmits, propagates, or receives signals will add jitter to the system. Although specific implementations will vary, all signaling systems include data transmitters and receivers, interconnect channels, and clock sources. In designing a system, we must account for the DJ and RJ from each.

Example 13-2 2.5-Gb/s PCI Express System Jitter Budget To illustrate, we use the system jitter budget for the first-generation PCI Express (PCIe) interface. Figure 13-15a depicts the main components of a PCIe link. PCI Express systems do not distribute a global high-frequency I/O clock, but instead, use I/O clocks that are locally generated from a low-frequency reference clock, often using phase-locked loops (PLLs). In addition, a clock-and-data recovery (CDR) circuit at the receiver handles synchronization of the data signals [Martwick, 2005].

Figure 13-15b contains a block diagram that describes the system model for the generation and propagation of jitter in a PCI Express link. The 100-MHz reference clock generates jitter, TJ_{refclk}, which enters the phase-locked loops for both the transmitter and receiver. The PLLs act as high-pass filters that allow high-frequency jitter to pass through them. In addition, each will also act as a source of additional jitter ($TJ_{Tx,gen}$ and $TJ_{Rx,gen}$) caused by noise introduced both locally (e.g., thermal noise) and by the system (e.g., supply noise). The channel adds jitter (TJ_{chan}) due to ISI and crosstalk, but is assumed not to filter the incoming jitter noticeably. The comparator function reflects the fact that both PLLs track the jitter, so that the jitter propagation due to the transmitter and receiver is a function of the difference in their transfer functions. Intuitively, if both PLLs have the same impulse response, they will pass identical jitter, given the same input, TJ_{refclk}. As a result, the receiver will tend to track the jitter passed by the transmitter, and the total jitter from the reference clock that gets passed by the PLLs is determined by the difference in their transfer functions. Finally, the CDR circuit will also filter the jitter.

Having identified the main sources of jitter, we can write expressions for the system DJ and RJ:

$$\mathrm{DJ}_{\delta\delta}(\mathrm{sys}) = \mathrm{DJ}_{\delta\delta}(\mathrm{Tx}) + \mathrm{DJ}_{\delta\delta}(\mathrm{channel}) + \mathrm{DJ}_{\delta\delta}(\mathrm{Rx}) + \mathrm{DJ}_{\delta\delta}(\mathrm{clock}) \quad (13\text{-}19)$$

$$\sigma_{\mathrm{RMS}}(\mathrm{sys}) = \sqrt{\sigma^2_{\mathrm{RMS}}(\mathrm{Tx}) + \sigma^2_{\mathrm{RMS}}(\mathrm{channel}) + \sigma^2_{\mathrm{RMS}}(\mathrm{Rx}) + \sigma^2_{\mathrm{RMS}}(\mathrm{clock})} \quad (13\text{-}20)$$

Equations (13-10), (13-13), and (13-14) provide the fundamental jitter relations that we use to budget the jitter for a PCI Express system, which is shown in Table 13-2. What they do not do, however, is to tell us how to use them to develop a jitter budget that results in a successful system design. Doing so successfully requires that we use the equations within the context of a design methodology. We offer an example of a contemporary design methodology in Figure 13-16 and we describe the individual steps below.

Step 1: Negotiate initial targets. The targets are most often based on engineering judgment and experience with prior designs. In addition to providing the I/O and signal integrity engineers with targets for their designs, they also provide the team with the means for making trade-offs in order to keep the budget balanced as the design proceeds.

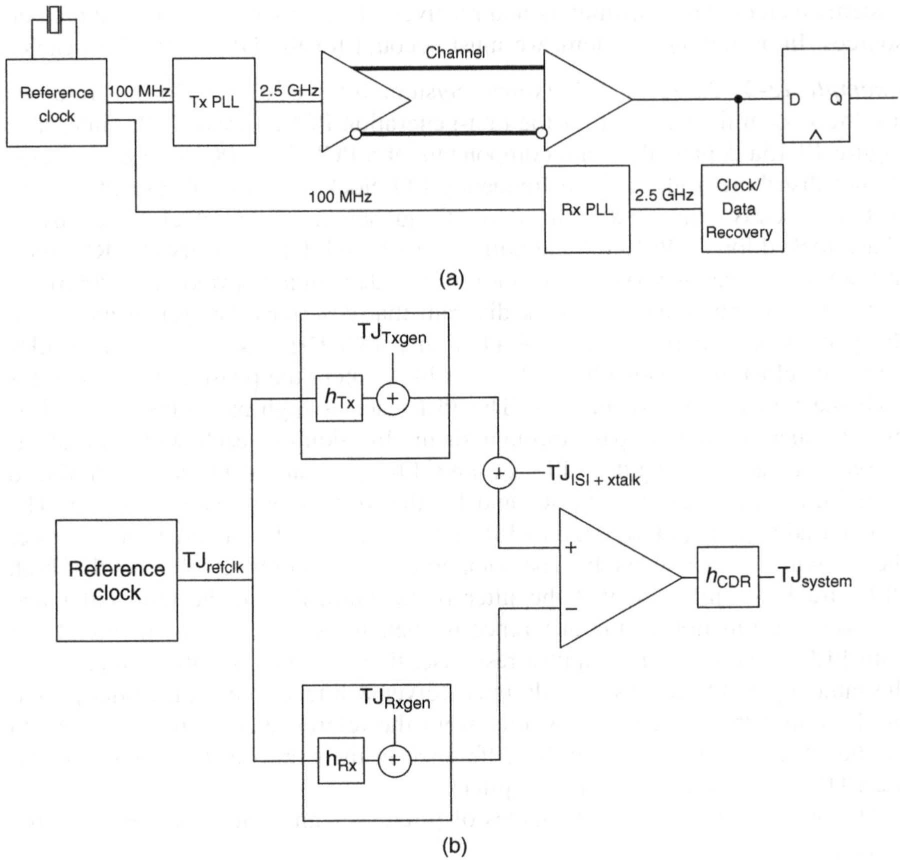

Figure 13-15 PCI Express system and jitter model: (a) physical architecture; (b) jitter generation and propagation.

Step 2: Evaluate design options for each of the individual components in order to satisfy the targets for jitter and other design metrics, such as received voltage (see Section 13.4). This typically involves performing a number of simulations that yield insight into sensitivity of the jitter to variation in the design parameters. Current design practice typically involves using statistical analysis methods, which we discuss in Chapter 14.

Step 3: Once the exploratory work is done, the predicted jitter is compared against the targets. Some trade-offs between individual components may be required, based on the analysis. At this point, it may not be possible to show that the jitter budget can meet the BER target at the data rate planned. That is not necessarily cause for alarm, as long as the owners of the components have plans for improving the timings as the design proceeds. As the figure shows, this stage may require multiple iterations between steps 2 and 3, although a solid

TABLE 13-2. PCI Express 2.5-Gb/s Jitter Budget at 10^{-12} BER

Component	Term	σ_{RJ} (ps)	$DJ_{\delta\delta}$ (ps)	TJ (ps)
Reference clock	TJ_{clock}	4.7	41.9	108
Transmitter	TJ_{TX}	2.8	60.6	100
Channel	$TJ_{channel}$	0	90	90
Receiver	TJ_{Rx}	2.8	120.6	147
Linear TJ				458
RSS TJ		86.5	313.1	399.6

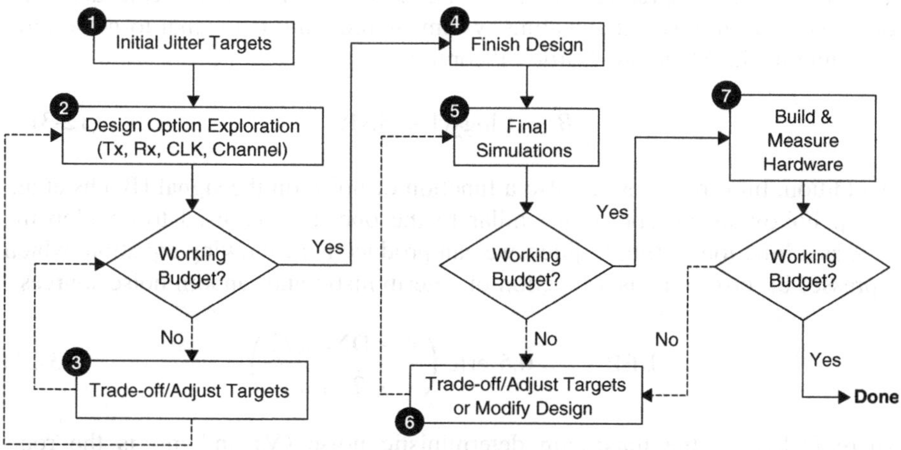

Figure 13-16 System design flow.

understanding of the design behavior provided by statistical analysis methods is especially helpful in minimizing the number of iterations.

Step 4: The component and board designs are completed after closing the budget. This includes layout, design rule check, and extraction of final models from the board, package, and I/O circuit designs.

Step 5: A final set of simulations with the extracted models for verification purposes is recommended prior to building hardware. This step provides the design team with a final chance to identify and correct design issues. Finding issues in simulation is faster and less expensive than waiting until hardware is in the lab.

Step 6: As was the case with the exploratory analysis, it may be necessary to adjust the timing targets or to modify some parts of the system design at this stage, although the use of statistical analysis methods reduces the likelihood that changes will be required at this stage.

Step 7: Finally, we reach the last major step in the process: hardware building and measurement. The design process is structured to minimize the risk of finding issues at this stage, but the ultimate measure of success is functioning systems.

Although it is beyond our scope here, we note that the hardware measurement step can be quite extensive. For example, statistical techniques may be employed to extrapolate the measured behavior for the purpose of validating that the design will meet the timing requirements at the extremes of the manufacturing process [Norman, 2003].

13.4 NOISE SOURCES AND BUDGETS

As discussed in Chapter 12, voltage noise limits the maximum achievable performance of a high-speed signaling system by reducing the signal-to-noise ratio according to the Shannon–Hartley theorem:

$$B = \tfrac{1}{2}\log_2(1 + \mathrm{SNR}) \tag{12-3}$$

In addition, bit error rates are also a function of noise on the signal [Buchs et al., 2004]. Following an approach similar to the one that we used to develop the BER as a function of timing jitter, we can produce the following equation, which expresses bit error rate as a function of deterministic and random noise sources:

$$\mathrm{BER}(v) = 0.5\ \mathrm{erfc}\left(\frac{v - \mathrm{DN}_{\max}/2}{\sqrt{2}\,\sigma_{\mathrm{RN}}}\right) \tag{13-21}$$

where $\mathrm{DN}_{\max}$ is the maximum deterministic noise (V) and σ_{RJ} is the root-mean-square Gaussian noise (V). Finally, voltage noise is a chief source of timing noise (jitter) and therefore limits the maximum data rates that we can achieve in practice. We describe the main sources of voltage noise in the next section.

13.4.1 Noise Sources

The primary sources of noise in digital signaling systems include crosstalk, intersymbol interference, supply noise, circuit input offsets and resolution, thermal noise, and shot noise. We discuss crosstalk and ISI extensively elsewhere in the book, so we focus on the remaining sources in this section.

Supply Noise As discussed in Chapter 10, power supply noise as it applies to high-speed signals arises from two sources. The first is externally generated supply noise, which occurs in the distribution system between the voltage regulator and the local I/O circuits. This external supply noise is typically on the order of 5 to 10% of the nominal supply. Current-mode transmitters typically have a large-impedance connection to the power supply, effectively isolating them from the supply noise.

The second source is local supply noise generated by transient current in the I/O circuits. Because it depends on the transient current demand according to

$v_{\mathrm{PS,noise}} = -L(di/dt)$, the characteristics of single-ended and differential signaling systems will be dramatically different. In particular, the current transients can be very large for single-ended systems, causing local supply noise of approximately 10% of the supply voltage for the signals. By contrast, since current-mode differential transmitters draw constant current locally, they generate essentially zero local supply noise. In addition, supply noise is largely a common-mode source, which means that we rely on the differential I/O circuits to reject most of the noise.

Receiver Offset and Sensitivity Variation in the threshold voltage of a receiver circuit is caused by process variation that results in device mismatch, which we treat as a static noise source. For a single-ended inverter, the source of the variation is the dependence on the ratio of NMOS and PMOS device characteristics (gate length and width, mobility, etc.), which can be largely independent of one another. We calculate the offset by finding the variation in the inverter threshold voltage, $v_{t,\mathrm{inv}}$, which is given by [Dally and Poulton, 1998]

$$v_{t,\mathrm{inv}} = \frac{v_{TN} + \sqrt{\beta_P/\beta_N}(V_{DD} + v_{TP})}{1 + \sqrt{\beta_P/\beta_N}} \quad \text{volts} \qquad (13\text{-}22)$$

where v_{TN} is the NMOS device threshold voltage (V), β_P is the PMOS device transconductance parameter (μA/V^2), β_N is the NMOS device transconductance parameter (μA/V^2), V_{DD} is the supply voltage (V), and v_{TP} is the PMOS device threshold voltage (V).

The device transconductance is a function of the process transconductance and the device dimensions:

$$\beta = k\frac{W}{L} \qquad (13\text{-}23)$$

where k is the process transconductance (μA/V^2), W the device width (μm), and L the device length (μm).

Equation (13-22) demonstrates that the source of the threshold variation is the dependence on the ratio of NMOS and PMOS device transconductance parameters, which can be largely independent of one another. The maximum offset will occur when either the NMOS or PMOS device is at a process extreme (usually called a "fast" or "slow" corner), while the other is at the opposite extreme. We approximate the device variations for the fast and slow corner cases by

$$\beta_{\substack{\text{fast}\\\text{slow}}} = (k \pm \Delta k)\frac{W \pm \Delta W}{L \mp \Delta L} \qquad (13\text{-}24)$$

where Δk is the worst-case variation in the process transconductance (μA/V^2), ΔW the worst-case variation in the device width (μm), and ΔL the worst-case variation in the device length (μm).

We can then estimate the offset for the inverter as the average offset from the typical case:

$$v_{\text{offset}} = \frac{v_{t,\text{inv}}(\text{max}) - v_{t,\text{inv}}(\text{min})}{2} \quad \text{(volts)} \qquad (13\text{-}25)$$

With differential amplifiers the input offsets depend on matching of devices of the same type. The transconductance parameters for devices of the same type will be much more closely matched, leading to much smaller offsets in the differential receiver. The input offset for a differential receiver that is based on an NMOS source coupled pair is given by

$$v_{\text{offset}} = \Delta v_{TN} + 2(v_{GS} - v_{TN})\left(\frac{\Delta k_N}{k_N}\frac{\Delta W}{W}\frac{L}{\Delta L}\right) \quad \text{volts} \qquad (13\text{-}26)$$

Receiver sensitivity is the amount of input voltage swing that is required to get a specified output swing [Dally and Poulton, 1998]. Reasonable sensitivity values for designs on a 0.25- μm process are 200 to 250 mV for a CMOS inverter-based receiver and approximately 10 mV for a differential receiver.

Example 13-3 Input Offsets for Single-Ended and Differential Receivers In this example we calculate the input offsets for an inverter and a differential receiver that are designed on a 0.25- μm process, whose parameters and variation are summarized in Table 13-3.

The worst-case device-length variation, ΔL, does not scale with the length. As a result, the inverter design can reduce the impact of the length variation on the input offset by using 0.5- μm gate lengths. The NMOS and PMOS design widths are 1.12 and 4μm, respectively. Table 13-4 summarizes the parameter conditions for the minimum and maximum inverter threshold cases. Note that the table includes variation in the supply voltage. This is because equation (13-22) shows that the inverter threshold depends directly on the supply voltage. Since the supply voltage may vary by up to ± 10%, we choose to include the variation in our offset calculation. To calculate the input offsets for the inverter, we use

TABLE 13-3. Example Device Parameters and Variation for a 0.25-μm Process

Parameter	Typical Value	Variation
V_{TN} (V)	0.43	± 0.01
V_{TP} (V)	−0.4	± 0.01
k_N (A/V^2)	115×10^{-6}	$\pm 12 \times 10^{-6}$
k_P(A/V^2)	30×10^{-6}	$\pm 3 \times 10^{-6}$
W (μm)		±0.025
L (μm)		±0.025

the values in the table in conjunction with equations (13-22) through (13-25):

$$\beta_{P,\text{typ}} = (30\ \mu\text{A/V}^2)\left(\frac{4\ \mu\text{m}}{0.5\ \mu\text{m}}\right)$$

$$= 240.000\ \mu\text{A/V}^2$$

$$\beta_{P,\min} = [(30-3)\ \mu\text{A/V}^2]\left[\frac{(4-0.025)\ \mu\text{m}}{(0.5+0.025)\ \mu\text{m}}\right]$$

$$= 204.429\ \mu\text{A/V}^2$$

$$\beta_{P,\max} = 279.632\ \mu\text{A/V}^2$$

$$\beta_{P,\text{typ}} = (115\ \mu\text{A/V}^2)\left(\frac{1.12\ \mu\text{m}}{0.5\ \mu\text{m}}\right)$$

$$= 257.600\ \mu\text{A/V}^2$$

$$\beta_{N,\min} = 215.871\ \mu\text{A/V}^2$$

$$\beta_{N,\max} = 304.932\ \mu\text{A/V}^2$$

$$v_{t,\text{typ}} = \frac{v_{TN,\text{typ}} + \sqrt{\beta_{P,\text{typ}}/\beta_{N,\text{typ}}}\ (V_{DD} + V_{TP,\text{typ}})}{1 + \sqrt{\beta_{P,\text{typ}}/\beta_{N,\text{typ}}}}$$

$$= \frac{0.43\ \text{V} + \sqrt{257.600\ \mu\text{A/V}^2/240.000\ \mu\text{A/V}^2}(2.5\ \text{V} + 0.4\ \text{V})}{1 + \sqrt{257.600\ \mu\text{A/V}^2/240.000\ \mu\text{A/V}^2}}$$

$$= 1.250\ \text{V}$$

$$v_{t,\min} = \frac{v_{TN,\min} + \sqrt{\beta_{P,\min}/\beta_{N,\max}}\ (V_{DD,\min} + V_{TP,\min})}{1 + \sqrt{\beta_{P,\min}/\beta_{N,\max}}}$$

$$= \frac{[(0.43-0.01)\ \text{V}] + \sqrt{204.429\ \mu\text{A/V}^2/304.932\ \mu\text{A/V}^2}\ [(2.5-0.25+0.4-0.01)\ \text{V}]}{1 + \sqrt{204.429\ \mu\text{A/V}^2/304.932\ \mu\text{A/V}^2}}$$

$$= 1.059\ \text{V}$$

$$v_{t,\max} = \frac{v_{TN,\max} + \sqrt{\beta_{P,\max}/\beta_{N,\min}}(V_{DD} + V_{TP,\max})}{1 + \sqrt{\beta_{P,\max}/\beta_{N,\min}}}$$

$$= \frac{[(0.43+0.01)\ \text{V}] + \sqrt{279.632\ \mu\text{A/V}^2/215.871\ \mu\text{A/V}^2}\ [(2.5+0.25+0.4+0.01)\ \text{V}]}{1 + \sqrt{279.632\ \mu\text{A/V}^2/215.871\ \mu\text{A/V}^2}}$$

$$= 1.462\ \text{V}$$

From the calculations, we see that the inverter has an input offset of approximately ± 201 mV.

TABLE 13-4. 0.25- μm Process Parameter Variation for Inverter Offset Estimation

	$v_{t,\min}$	$v_{t,\max}$
v_{TN} (V)	0.042	0.044
v_{TP} (V)	0.039	0.041
k_N (μA/V^2)	127	103
k_P (μA/V^2)	27	33
W_N (μm)	0.87	1.37
W_P (μm)	3.75	4.25
L_N (μm)	0.25	0.75
L_P (μm)	0.75	0.25
V_{DD} (V)	2.25	2.75

Equation (13-26) shows that the offset of the differential receiver depends on the ratio of variation in device width to device length. As a result, feature size variations will cancel each other. This allows us to reduce the offset by designing the receiver with the minimum gate length, which has the added benefit of minimizing the device area and input capacitance. We can also reduce the receiver offset by increasing the device width. Choosing a device width of 1 μm with a minimum length gives us an estimated offset of ± 30 mV for the differential receiver, for an input gate bias, $v_{GS} - v_{TN}$, of 400 mV:

$$
\begin{aligned}
v_{\text{offset}} &= \Delta v_{TN} + 2(v_{GS} - v_{TN})\left(\frac{\Delta k_N}{k_N}\frac{\Delta W}{W}\frac{L}{\Delta L}\right) \\
&= 0.01\ \text{V} + 2(0.4\ \text{V} - 0.01\ \text{V})\left[\frac{(12\ \mu\text{A/V}^2)(0.025\ \mu\text{m})(0.25\ \mu\text{m})}{(115\ \mu\text{A/V}^2)(1\ \mu\text{m})(0.025\ \mu\text{m})}\right] \\
&= 30\ \text{mV}
\end{aligned}
$$

V_{REF} Noise As we described in Chapter 12, we can use differential amplifiers to receive signal-ended signals by connecting one of the inputs to a reference voltage (see Figure 11-33). Examples of high-speed single-ended interfaces that employ this approach include the GTL + technology used with Intel microprocessors and the stub series terminated logic (SSTL) interface used with dual-data-rate (DDR) memory devices [Intel, 1997; JEDEC, 2002]. The reference voltage is often generated as part of a low current voltage regulator and distributed to the on-chip circuits through the printed circuit board and packages, making it susceptible to noise from nearby signals and reference planes. Typical ac noise specifications are $\pm 2\%$ of the reference voltage. In addition, some standards allow for dc offset in the reference voltage. For example, the high-speed transceiver logic (HSTL) specification allows the dc value of the reference to vary between 0.68 and 0.90 V from a nominal value of 0.75 V while restricting the ac noise to a maximum of $\pm 2\%$ of the actual reference (e.g., 18 mV for $V_{\text{ref}} = 0.9$ V) [EIA, 1995].

Equalizer Quantization Error Discrete linear equalizers typically use D/A converters (DACs) to generate the equalizer coefficients. A DAC has a finite resolution that is a function of the number of bits, which leads to a minimum step size (granularity) in setting the tap coefficients. For a current steering DAC used in transmit equalization of differential signals, we estimate the minimum resolution of the output current to be limited to the least significant bit (LSB) of the DAC:

$$i_{\text{res}} = \frac{i_{\text{DAC}}}{2^{n_{\text{DAC}}}} \qquad \text{(amperes)} \tag{13-27}$$

where i_{DAC} is the maximum output current and n_{DAC} is the number of bits in the DAC.

We treat the limited resolution as a voltage "noise" on an output voltage signal, whose value is equal to one-half of the resolution:

$$v_{\text{eq,noise}} = \frac{i_{\text{DAC}}}{2^{n_{\text{DAC}}+1}} Z_0 \qquad \text{volts} \tag{13-28}$$

Equation (13-28) is an approximation that does not account for other nonidealities in DACs. An example is the differential nonlinearity, which is the difference between the ideal and the measured output responses for successive DAC codes. A thorough treatment of the sources of nonidealities in DACs is provided by Razavi [1995].

Example 13-4 Equalizer DAC Noise Estimation A current-mode differential transmitter with a single postcursor equalization tap is designed to have a maximum tap coefficient of −0.2. The transmitter is a 5-mA transmitter that drives a differential transmission pair whose differential impedance is 100 Ω. The equalizer uses a 4-bit DAC to set the coefficient value. The differential voltage swing is $v_{\text{swing}} = (5\text{ mA})(100\ \Omega) = 0.500$ V. The noise due to the DAC resolution is $v_{\text{eq,noise}} = [0.2(5\text{ mA})/2^5](100\Omega) = 3.1$ mV, which is 0.6% of the total signal swing.

Thermal Noise and Shot Noise Thermal and shot noise are random in nature and so are modeled as Gaussian sources [Gray et al., 2001]. *Thermal noise*, also known as *Johnson noise*, is a result of power dissipation in devices, and has a root-mean-square (RMS) power spectral density of

$$\text{PSD}_{\text{therm}} = 4k_{\text{B}}TR \qquad \text{V}^2/\sqrt{\text{Hz}} \tag{13-29}$$

where k_{B} is Boltzmann's constant (1.38×10^{-23} J/K), T the temperature (K), and R the device resistance (Ω). Over a given bandwidth, we calculate the RMS thermal voltage noise with equation (13-3):

$$\sigma_{\text{therm}} = \sqrt{4k_{\text{B}}TR\Delta f} \qquad \text{volts} \tag{13-30}$$

where Δf is the bandwidth over which the noise is measured (Hz).

Shot noise arises from the quantization of device current to individual device carriers and has a current spectral density $\overline{i^2}$ of

$$\overline{i^2} = 2qi \qquad \text{A}^2/\sqrt{\text{Hz}} \tag{13-31}$$

where q is the charge of an electron (1.6×10^{-19} C) and i is the current through the device.

Shot noise arises from current fluctuations caused by quantization of current to individual charges. The RMS voltage due to shot noise generated by a device with current i flowing through an effective resistance R over a bandwidth Δf is

$$\sigma_{\text{shot}} = \sqrt{2qi\Delta f}R \qquad \text{volts} \tag{13-32}$$

Since thermal noise and shot noise are Gaussian in nature, we must treat them in a probabilistic fashion when creating a noise budget. Gaussian noise sources are described by the normal probability density function (PDF), which has the form

$$P(v) = \frac{1}{\sqrt{2\pi}\,\sigma_{\text{noise}}} e^{-v^2/2\sigma_{\text{noise}}^2} \tag{13-33}$$

Equation (13-33) yields the probability of having noise of a given value. The probability that the noise does not exceed a given amount is

$$P(v_{\text{noise}} < v) < 1 - e^{-v^2/2\sigma_{\text{noise}}^2} \tag{13-34}$$

Example 13-5 Estimation of Thermal and Shot Noise We have a 2.5-V 50-Ω transmitter driving a 50-Ω transmission line at 10 Gb/s (5 GHz fundamental frequency). The circuit operates at room temperature (300 K). From load-line analysis we can show that the output current is

$$i_{\text{Tx}} = \frac{2.5\text{ V}}{50\ \Omega + 50\ \Omega} = 25\text{ mA}$$

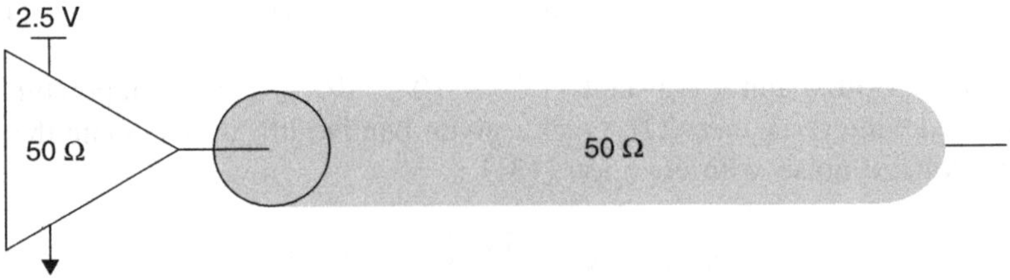

Figure 13-17 Circuit for Example 13-5.

The RMS thermal noise is

$$\sigma_{\text{therm}} = \sqrt{4\left(1.38\times10^{-23}\ \text{J/K}\right)(300\ \text{K})(50\ \Omega)(5\times10^{9}\ \text{Hz})}$$
$$= 6.43\times10^{-5}\ \text{V}$$

The RMS shot noise is

$$\sigma_{\text{shot}} = \sqrt{2(1.6\times10^{-19}\ \text{C})(0.025\ \text{A})(5\times10^{9}\ \text{Hz})}(50\ \Omega) = 3.16\times10^{-4}\ \text{V}$$

The two noise sources combine in an RMS relationship:

$$\sigma_{\text{total}} = \sqrt{(6.43\times10^{-5}\ \text{V})^2 + (3.16\times10^{-4}\ \text{V})^2} = 3.23\times10^{-4}\ \text{V}$$

From equation (13-34) we find that the probability that the noise from thermal and shot sources at any instant is less than 1 mV:

$$P(v_{\text{noise}} < v) < 1 - e^{-(1\ \text{mV})^2/2(0.323\ \text{mV})^2} = 0.992 = 99.2\%$$

The small value suggests that thermal noise and shot noise are generally not significant sources of noise for single-ended signaling systems, which have relatively large signal swings.

13.4.2 Noise Budgets

The existence of so many noise sources in a high-speed signaling system demands that they be managed to ensure proper operation. The method for managing noise is to construct a noise budget and use it to design sufficient noise margin into the system. Most of the sources that we have discussed are bounded, and in constructing a budget we assume that each of them is at the worst-case value. In practice, this is a conservative approach, since the probability of all noise sources being at their worst-case extremes is remote. The benefit of the worst-case approach is that by satisfying such a conservative budget, we minimize the probability of having a noise problem in our system, and noise issues can be exceeding difficult to diagnose.

Thermal noise and shot noise, which are Gaussian sources, do not directly fit the worst-case noise method. However, we can calculate an approximate worst-case value for Gaussian sources by choosing a maximum probability of exceeding the worst-case value that we are willing to tolerate. We then use the probability curve shown in Figure 13-18 to determine how many standard deviations we must take into account. This is similar in concept to the BER calculation from the preceding section. (Of course, BER applies to signal amplitudes as well as timing. Noise-based BER calculation is left as a problem at the end of the chapter.) For example, if we choose a maximum probability of exceeding the worst-case noise of 1 in 1 trillion (10^{-12}), we set the noise to a minimum of

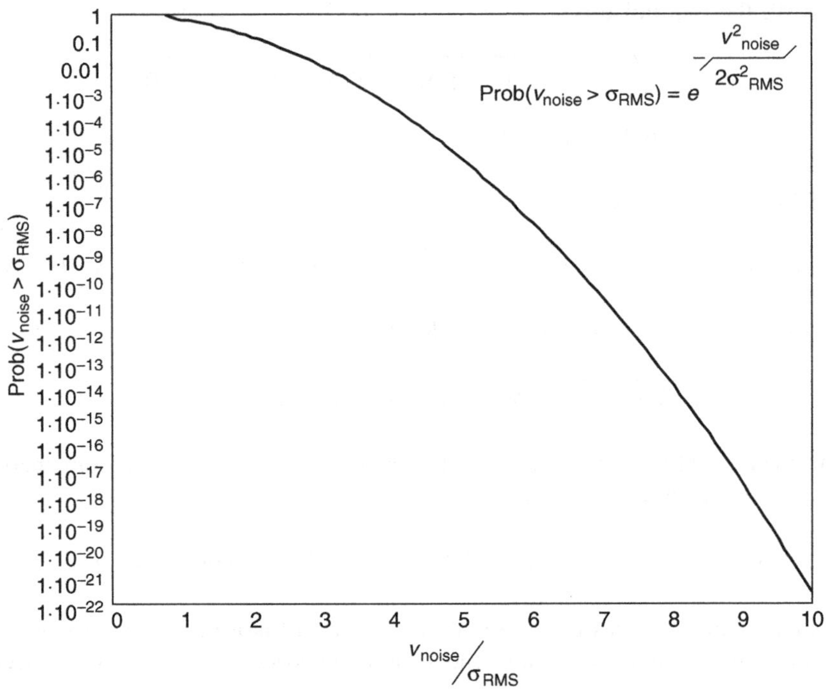

Figure 13-18 Probability curve for Gaussian noise sources.

7.4 times the RMS noise voltage. Using the RMS noise that we calculated earlier ($\sigma_{total} = 0.233$ mV), gives an estimated worst-case Gaussian noise of 1.7 mV.

We usually do not include crosstalk and ISI as explicit terms in the noise budget, since we comprehend them in our signal integrity simulations. As a general comment, we recommend simulation as the preferred means for comprehending noise sources wherever we have accurate models and sufficient computing power. The ability to simulate (and correlate to measurement) gives us a better understanding of the behavior of the system, which allows us to reduce margins, resulting in higher performance designs at lower cost.

Example 13-6 Noise Budgets We want to construct noise budgets and compare the noise margins and noise immunity of a pair of hypothetical signaling systems, which we show in Figure 13-19. The single-ended system in Figure 13-19a is designed to operate at 1 Gb/s with a 2.0-V single swing created by a voltage-mode transmitter. The swing is centered around a reference voltage of 1.0 V that has an allowed dc shift of up to ±5% along with ±2% ac noise. The reference voltage is connected to one input of the differential receiver. Since we are demonstrating the budget process without performing simulations, we explicitly budget for crosstalk and ISI at 10% of the signal swing for each. We power the circuit with a 2.5-V supply and budget the supply noise budget at 10% of the supply value.

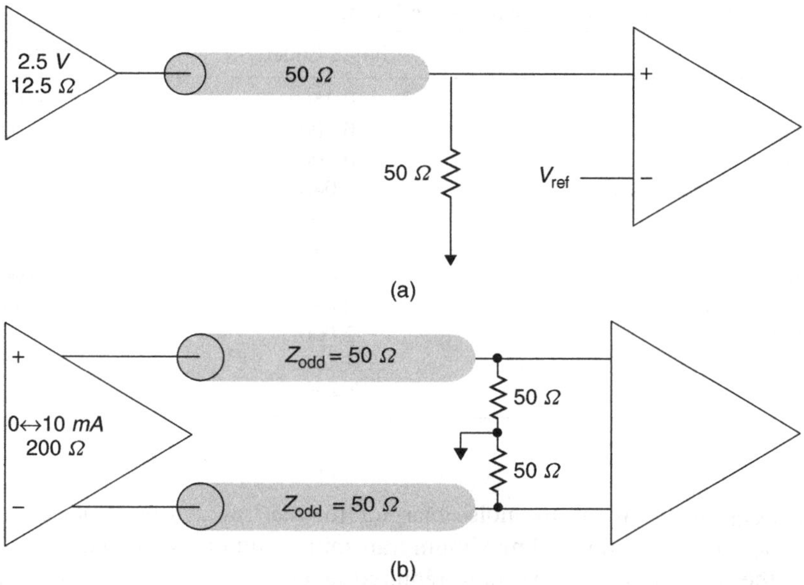

Figure 13-19 Circuits for Example 13-6: (a) single-ended system; (b) differential system.

The differential signaling system shown in Figure 13-19b is designed to operate at 5 Gb/s over a lossy interconnect that requires equalization. It employs a 10-mA current-mode transmitter to provide a singled swing of 400 mV when driving the parallel combination of transmitter impedance and transmission line. The differential circuit also uses a 2.5 V supply with a 10% budget for supply noise. However, the combination of high transmitter impedance and the differential receiver is expected to reject more than 99% of the supply noise. The crosstalk budget is reduced to 5% of the signal swing due to reduced coupling of the differential transmission lines and the common-mode rejection of the receiver. The ISI budget remains at 10% of the signal swing. The equalizer uses a 4-bit DAC and is designed to provide a maximum of 20% equalization. The receivers for both designs are differential and have similar offsets and sensitivities which we estimate at 35 and 10 mV, respectively. Finally, we include thermal and shot noise in both interfaces, specifying that the probability of exceeding the budgeted noise cannot exceed 10^{-12}.

Table 13-5 summarizes the contributions of the individual noise sources that we discussed above. From the table we see that the projected worst-case noise is 835 mV for the single-ended system and 114 mV for the differential system. A widely used metric for judging the noise characteristics of a digital system is the noise margin, which we introduced in Chapter 12. We can use equations (12-3a) and (12-3b) to develop a system noise margin:

$$v_{\mathrm{NM}} = v_{\mathrm{NMl}} + v_{\mathrm{NMh}} = (v_{oh,\mathrm{min}} - v_{ol,\mathrm{max}}) - (v_{ih,\mathrm{max}} - v_{il,\mathrm{min}})$$

TABLE 13-5. Noise Budget Analysis for Example 13-6

Source	Single-Ended	Differential
Supply noise (V)	0.250	0.003
Crosstalk (V)	0.200	0.020
ISI (V)	0.200	0.040
Rx offset + sensitivity (V)	0.045	0.045
Gaussian noise (V)	0.000	0.001
V_{ref} noise (V)	0.140	—
Equalizer resolution (V)	—	0.005
Total noise (V)	0.835	0.114
Signal swing (V)	2.000	0.400
Noise margin (V)	1.165	0.286
Noise immunity	2.4 : 1	3.5 : 1

where v_{NMl} and v_{NMh} are the noise margins for the low and high sides, $v_{ol,\max}$ and $v_{oh,\min}$ the minimum and maximum transmitter output levels, and $v_{ih,\min}$ and $v_{ih,\max}$ the minimum and maximum required levels at the receiver. Recognizing that the difference between minimum and maximum output levels is equal to the signal swing and that the difference between minimum and maximum input levels at the receiver is equivalent to the worst-case noise, we can express the system noise margin as

$$v_{\text{NM}} = v_{\text{swing}} - v_{\text{noise}} \tag{13-35}$$

Since the single-ended system swing is five times greater than for the differential system, it has a much higher noise margin (1165 versus 286 mV). From these figures we might be tempted to conclude that the single-ended system offers superior noise performance to the differential system. However, the noise margin metric has a serious limitation; it does not account for the relative difference between the signal swing and the noise. Several of the biggest noise sources (e.g., crosstalk, supply noise, ISI) increase in proportion to the signal swing, while the low swing differential system typically reduces the nonproportional sources (e.g., offset, sensitivity, common-mode rejection). This implies that despite potentially higher noise margins, increasing signal swing does not guarantee improved noise performance. An alternate metric, the *noise immunity*, addresses this limitation by taking the ratio of the signal swing to the worst-case noise:

$$r_{\text{NI}} = \frac{v_{\text{swing}}}{v_{\text{noise}}} \tag{13-36}$$

If we take the worst case noise to be a measure of the minimum signal swing required for successful operation, the noise immunity expresses the ratio of the actual signal to the required swing. The differential system has a noise immunity

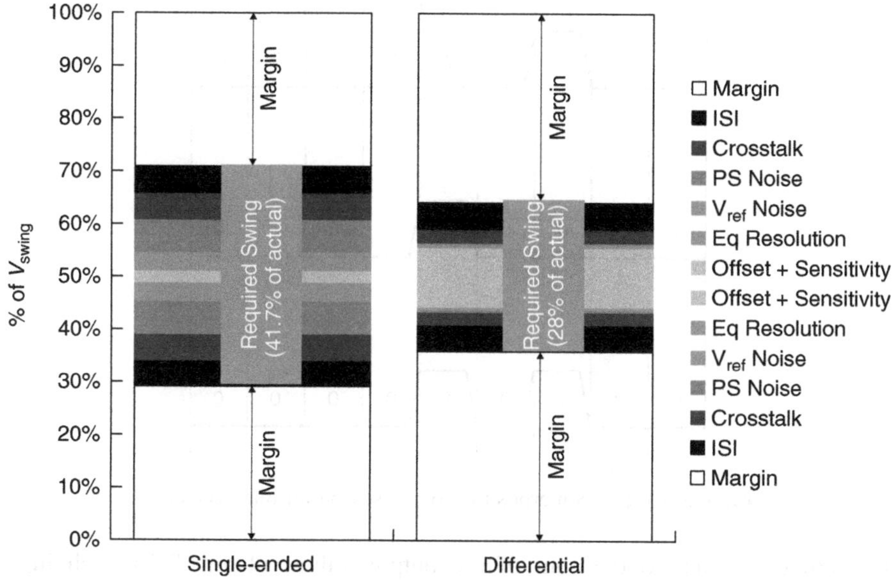

Figure 13-20 Noise immunity comparison for Example 13-6.

ratio of 3.5:1, as opposed to 2.4:1 for the single-ended system. Figure 13-20 shows that the singled-ended system has a minimum required swing that is 41.7% of the actual swing, while the differential system requirements a minimum swing that is only 28% of the actual. Viewed in this light, the differential system offers improved noise performance over single-ended signaling [Dally and Poulton, 1998].

13.5 PEAK DISTORTION ANALYSIS METHODS

Since our primary focus in this book is signal integrity, we now describe a method for assessing the peak jitter and noise caused by the system interconnect. Recall that the jitter and noise caused by crosstalk and ISI are deterministic, so that we can budget them using worst-case values. The method that we use to obtain the worst-case values is called *peak distortion analysis* (PDA) [Casper et al., 2002].

13.5.1 Superposition and the Pulse Response

We start by elaborating on the linear time invariance (LTI) concept that we introduced in Chapter 12. In particular, we make use of the superposition property, which says that if the system input can be expressed as the sum of multiple input components,

$$x(t) = \sum_i x_i(t)$$

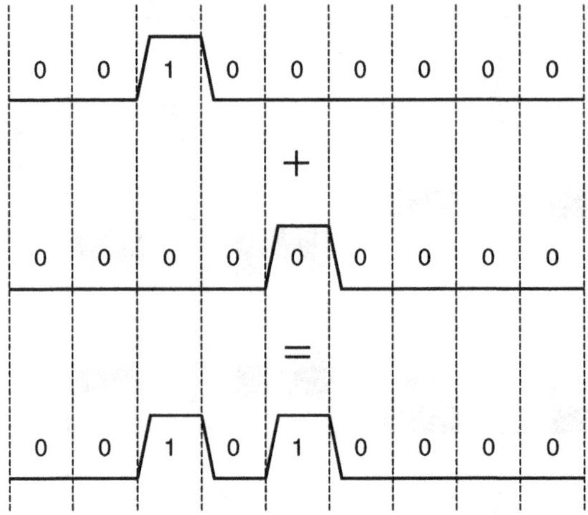

Figure 13-21 Superposition of pulses on an interconnect.

the output is equal to the sum of the output values obtained for each input component according to

$$y(t) = \sum_i f[x_i(t)] = \sum_i y_i(t) \tag{13-37}$$

The equation allows us to use superposition to calculate the received signal waveform for an arbitrary bit pattern using only the response of the system to a single pulse. Figure 13-21 illustrates the application of superposition to construct the waveform for a 0 101 000 pattern from a pair of pulse responses simply by summing them at each point.

Using superposition, building a mathematical model for constructing the received waveforms for arbitrary bit patterns from a received pulse response is a straightforward matter. First, we express the transmitted pulse signal for the ith bit position as

$$x_i(t) = x(t - i \cdot \text{UI}) \tag{13-38}$$

where UI, the unit interval, is equal to the width of a data bit. We can then write an equation for the transmitted waveform for an n-bit data sequence as

$$x'(t) = \sum_{i=0}^{n-1} b_i x(t - i \cdot \text{UI}) \tag{13-39}$$

where b_i is the logical value of the ith bit.

We express the received signal waveform of the same n -bit data sequence as

$$y'(t) = \sum_{i=0}^{n} b_i y(t - i \cdot \text{UI}) \tag{13-40}$$

where $y(t)$ is the received pulse response. Equation (13-40) represents the superposition of pulse responses, in which the response for an individual bit is shifted in time by an amount equal to the bit position multiplied by the unit interval. Let's illustrate with an example.

Example 13-7 Superposition of Pulse Response Waveforms Figure 13-22a shows a simple interconnect circuit operating at 1 Gb/s. The transmission line is lossless with a characteristic impedance of 50 Ω and a 1-ns electrical length. Impedance mismatches at the ends of the line cause reflections on the waveform shown in Figure 13-22b. We want to find the received response at the far end of the line when stimulated with the following pattern:

Bit	Value
0	0
1	0
2	1
3	0
4	1
5	0
6	0
7	0
8	0
9	0
10	0
11	0
12	0
13	0

Figure 13-23a provides a graphical representation of the superposition of pulses in order to construct the desired waveform. In this case we copy the pulse response and shift it two bit positions to the right. We then add the two pulse responses together using the voltage values extracted from the pulse response at each bit position, which are listed in the second column of Table 13-6. The third column of the table contains the shifted pulse response values, and the final column contains the resulting values for the desired waveform. Figure 13-23b shows the waveform generated using the superposition calculation, while Figure 13-23c contains the waveform obtained by explicitly simulating the bit pattern desired. The two waveforms give an exact match, and we have thus demonstrated the applicability of superposition for analyzing high-speed signaling interfaces.

13.5.2 Worst-Case Bit Patterns and Data Eyes

Having demonstrated the applicability of superposition, we now show how we can use it to find the worst-case bit pattern and resulting data eye for a high-speed

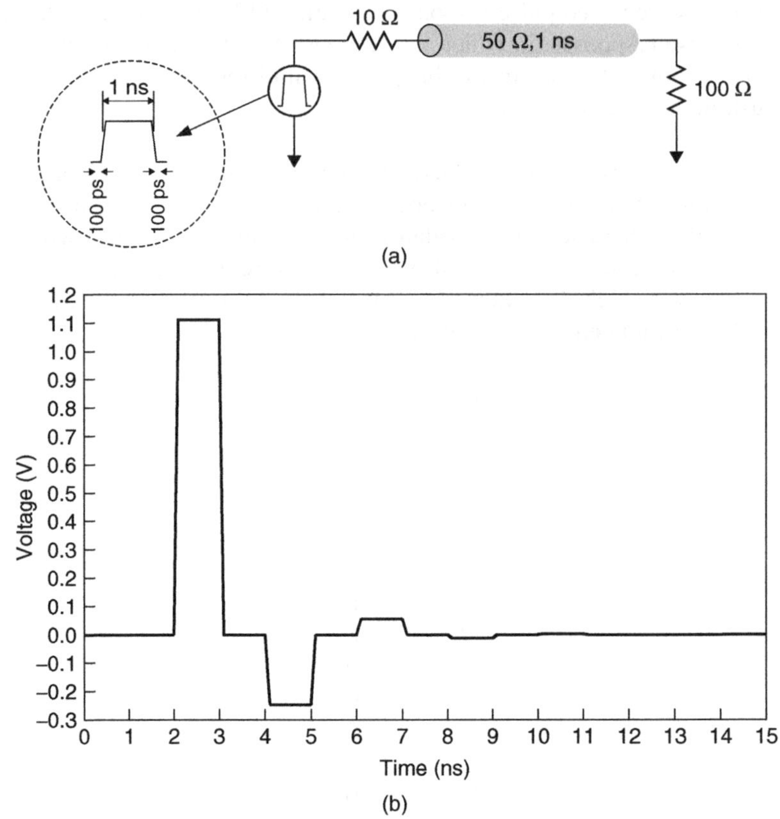

Figure 13-22 Example circuit and pulse response for superposition application: (a) circuit and signal characteristics; (b) 1-Gb/s pulse response.

signaling system. The obvious approach would be to create each of the possible waveforms for the system based on our knowledge of the pulse response. To illustrate, we reexamine the pulse response from the previous example in Figure 13-24. We have annotated the figure to show deviations in the pulse response from the ideal case. For example, the second bit position after the pulse has a value of −0.247 V. Ideally, the pulse response would be zero at all bit positions except for the one at the interval from 2 to 3 ns. We call that position the *cursor*, which identifies it as the bit position of interest. The positions that follow it are called *postcursor* positions, and those that precede the cursor are *precursor* positions. From the figure we see that the pulse shows deviations from ideal that cause intersymbol interference out to the twelfth postcursor bit. The bits after that one are all equal to zero and will not affect any signal waveform that we generate from the pulse response.

Covering all possible combinations would require that we create 2^{13} different waveforms in order to comprehend the cursor bit plus the 12 postcursors. Once we generate them, we then have to analyze each to find the pattern that gives

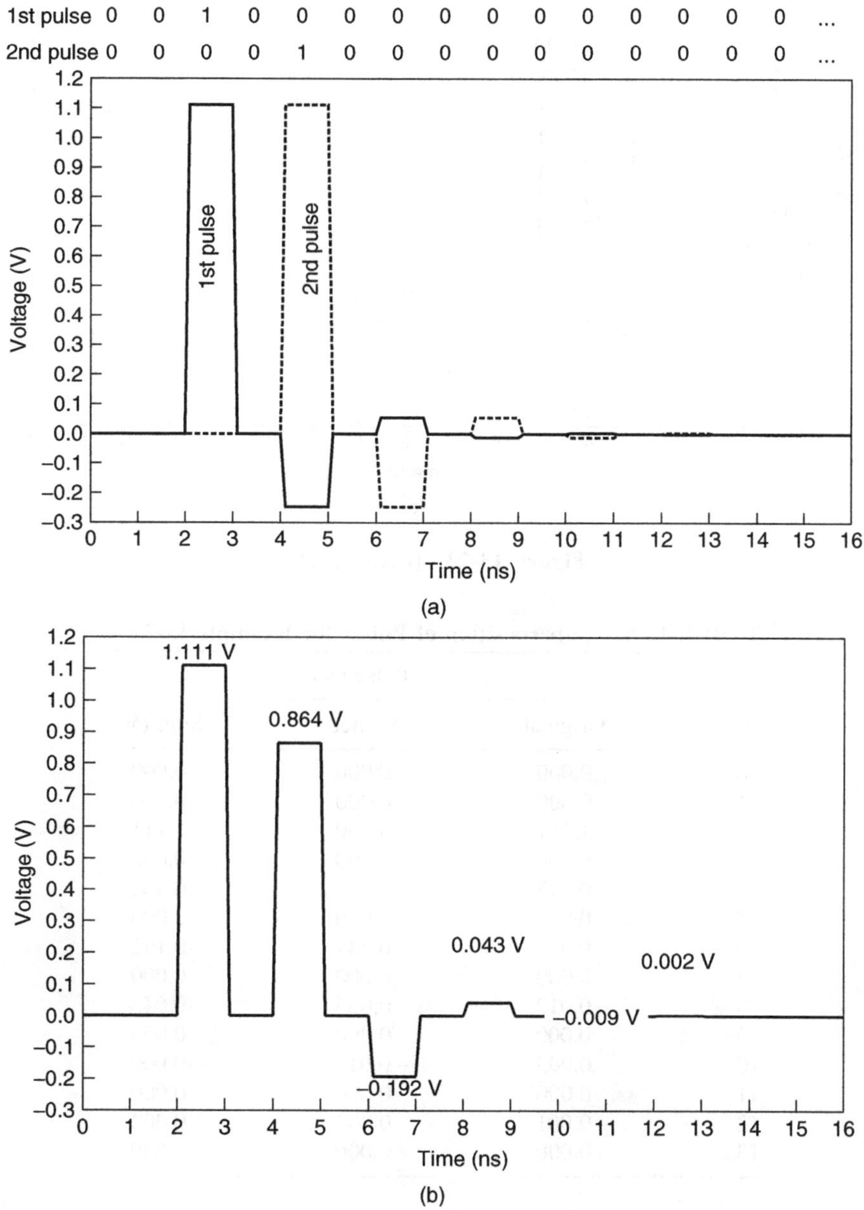

Figure 13-23 Example use of superposition to calculate the waveform for a bit pattern from the pulse response in Figure 13-22: (a) superposition of pulses to emulate the bit pattern for Example 13-7; (b) waveform calculated from superposition; (c) waveform from simulated bit pattern.

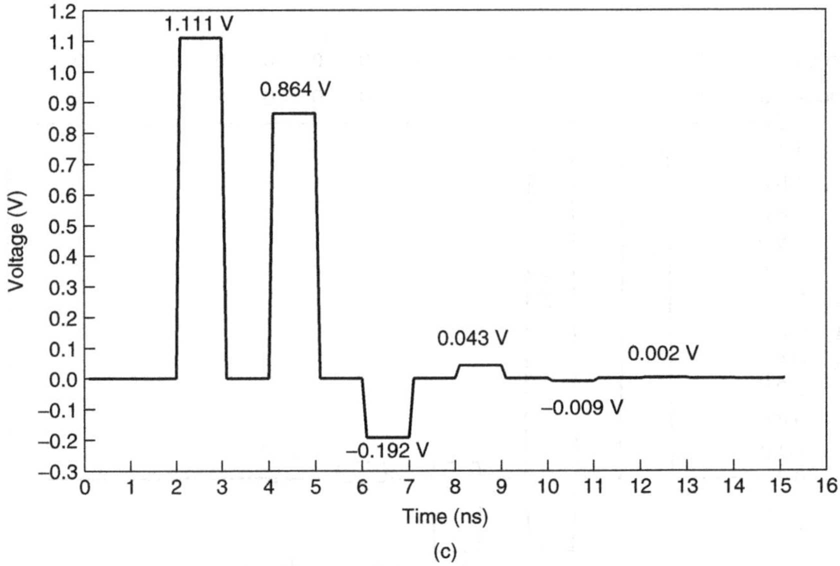

Figure 13-23 (*Continued*)

TABLE 13-6. Superposition of Pulses for Example 13-7

	Pulse (V)		
Bit	Original	Shifted	Sum (V)
0	0.000	0.000	0.000
1	0.000	0.000	0.000
2	1.111	0.000	1.111
3	0.000	0.000	0.000
4	−0.247	1.111	0.864
5	0.000	0.000	0.000
6	0.055	−0.247	−0.192
7	0.000	0.000	0.000
8	−0.012	0.055	0.043
9	0.000	0.000	0.000
10	0.003	−0.012	−0.009
11	0.000	0.000	0.000
12	−0.001	0.003	0.002
13	0.000	0.000	0.000

us the minimum eye opening. Further, for a complex, lossy signaling channel, the worst-case bit pattern may be different for different locations within the eye. So while this approach is technically feasible, it does not seem terribly efficient. Fortunately, by using superposition in a different way, we can find both the worst-case bit pattern and minimum eye opening without having to create the

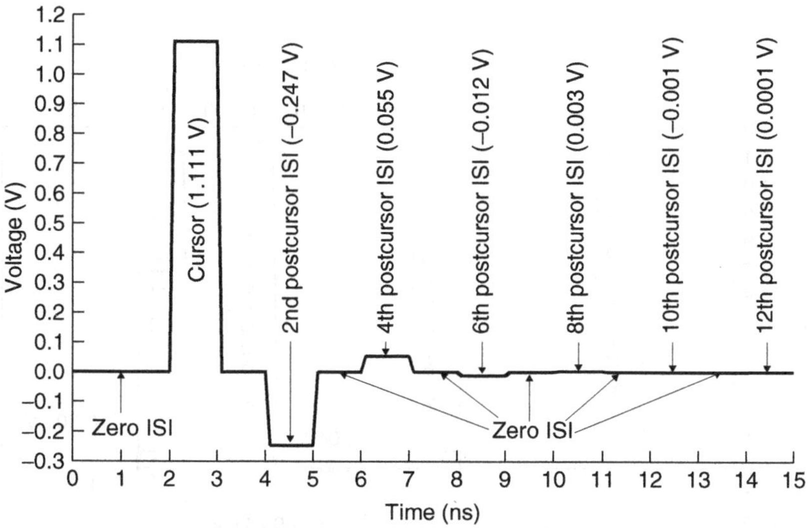

Figure 13-24 Pulse response with ISI terms.

8000+ waveforms. To begin to understand how we can use superposition, we present another example.

Example 13-8 Using Superposition to Find the Worst-Case Bit Pattern Referring to Figure 13-24, we observe that if we were to transmit a second pulse at $t = 4$ ns (the second postcursor position), superposition will cause the signal level of the second pulse to be reduced by the amount of negative ISI at that position. As a result, the value of the second pulse is 1.111 V − 0.247 V = 0.864 V. This tells us that the worst-case pattern for a logical 1 should include a 101 subpattern.

If we launch a third pulse starting at $t = 8$ ns (the sixth postcursor position), it will be affected negatively by the − 0.247-V ISI from the second pulse and by the − 0.012-V ISI from the first pulse, degrading it to a value of 0.852 V. Thus, we expand the worst-case bit pattern so that it now includes 1000101.

Launching a fourth pulse at the tenth postcursor position adds an additional 1 mV of degradation to a logical 1 signal. The pulse shows no negative ISI beyond the tenth postcursor, so we don't need to carry the analysis any further. The waveform for a worst-case pattern will contain 10001000101. We demonstrate the analysis graphically in Figure 13-25a. The voltage will be

$$1.111 \text{ V} - 0.247 \text{ V} - 0.012 \text{ V} - 0.001 \text{ V} = 0.851 \text{ V}$$

which corresponds to the value of a worst-case 1.

We find the pattern that results in the worst-case zero using the same approach. However, in this case, we focus on the bit positions that exhibit positive ISI. The fourth, eighth, and twelfth postcursors have positive ISI of 0.055, 0.003, and 0.0001 V, respectively. Figure 13-25b shows the analysis and the worst-case bit

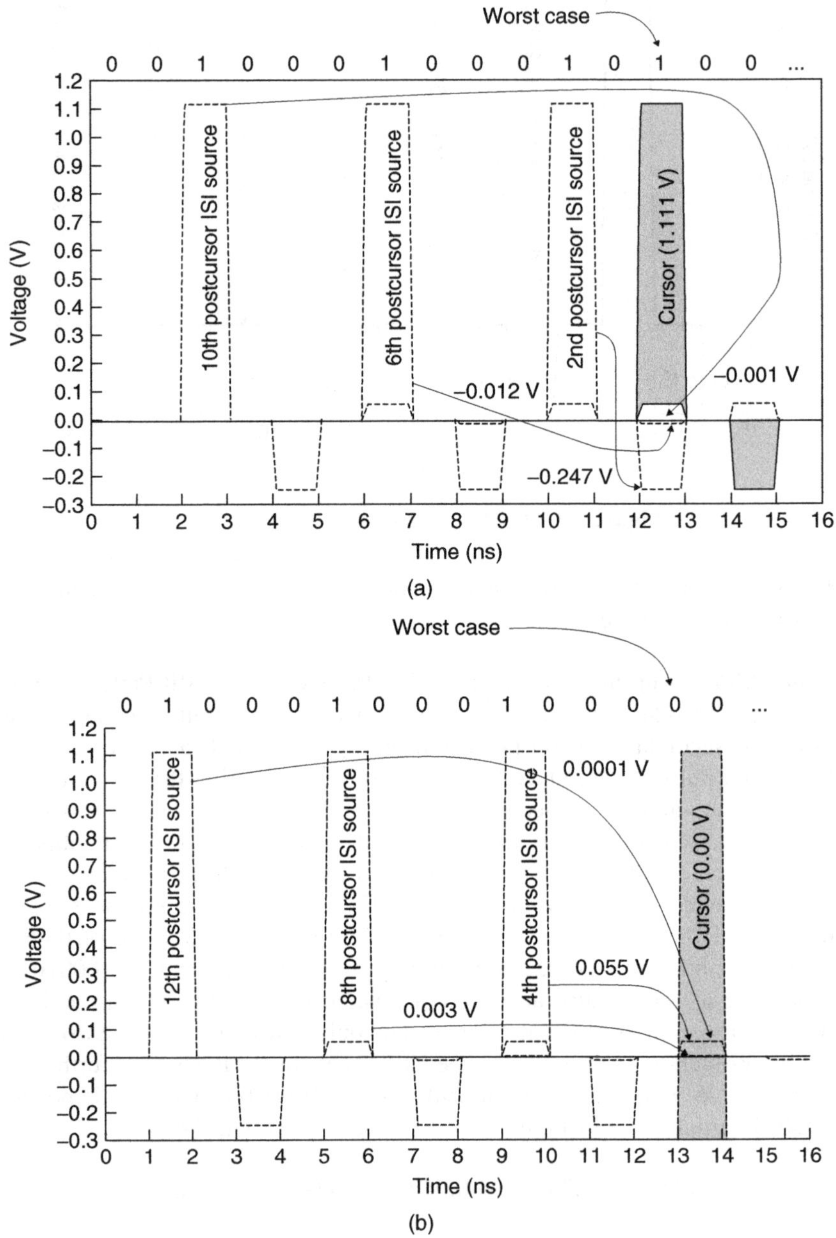

Figure 13-25 Construction of the worst-case eye waveforms: (a) superposition of pulse to create a worst-case 1; (b) superposition of pulses to create a worst-case 0; (c) worst-case waveforms; (d) worst-case eye diagram; (e) close-up view of eye at the trailing edge.

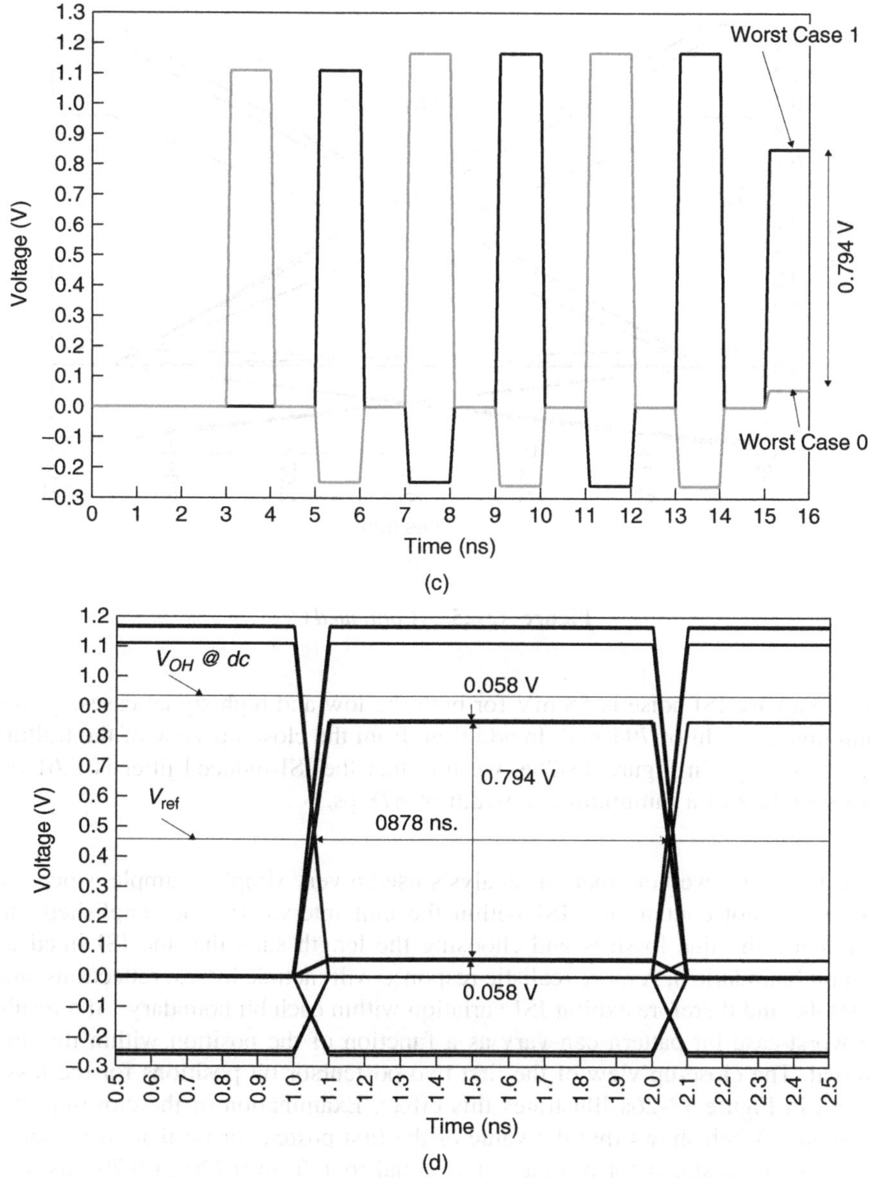

Figure 13-25 (*Continued*)

pattern for a zero (1000100010000). The voltage value of the worst-case zero is 0.055 V + 0.003 V + 0.0001 V = 0.058 V.

We show the waveforms for both worst-case patterns in Figure 13-25c, and the resulting eye diagram in Figure 13-25d. From the eye diagram we see that

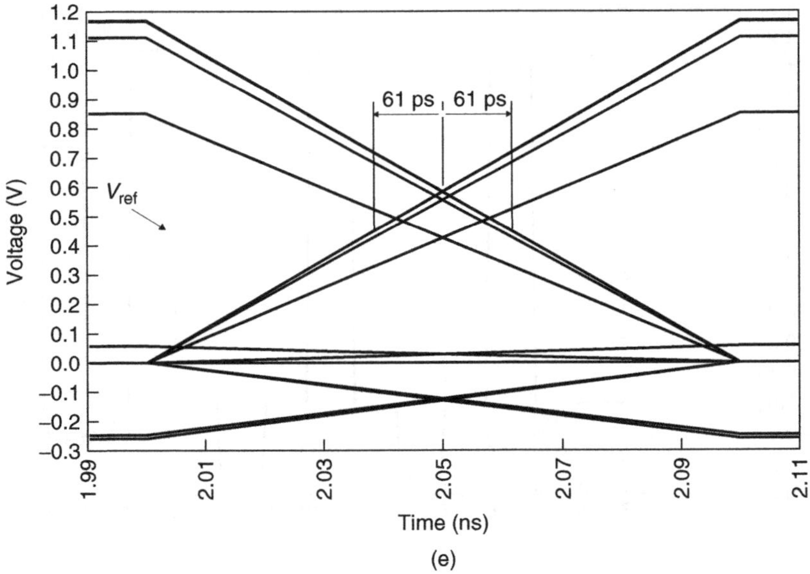

Figure 13-25 (*Continued*)

the worst-case ISI noise is 58 mV for both the low and high signal cases, giving a minimum height of 794 mV. In addition, from the close-up view of the trailing edge of the eye in Figure 13-25e, we find that the ISI-induced jitter is ±61 ps, corresponding to a minimum eye width of 878 ps.

At this point we note that our analysis used a very simple example, since our system did not contain any ISI within the unit interval (we accomplished this by making the line lossless and choosing the length such that the ISI lined up with bit boundaries). A more realistic response will include losses, reflections, and crosstalk, and therefore exhibit ISI variation within each bit boundary. At a result, the worst-case bit pattern can vary as a function of the position within the unit interval. The close-up view of the first two postcursor bit positions for the lossy system in Figure 13-26a illustrates this effect. Examination of the close-up plot in Figure 13-26b shows that the value of the first postcursor bit that corresponds to a worst-case signal for a logical 1 is equal to 1 from 0.720 to 0.795 ns, and is equal to zero from 0.795 to 0.820 ns. However, the corresponding value of the second postcursor bit is zero from 0.820 to 0.872 ns, and 1 from 0.872 to 0.920 ns. This pulse response implies, therefore, that we have multiple potential worst-case patterns to comprehend.

Clearly, the situation has become much more complicated. As Figure 13-26 illustrates, the positional dependence of the worst-case bit pattern makes the analysis for a realistic channel too complex to carry out in the manual fashion that we have described. Fortunately, we can translate the method into a set of

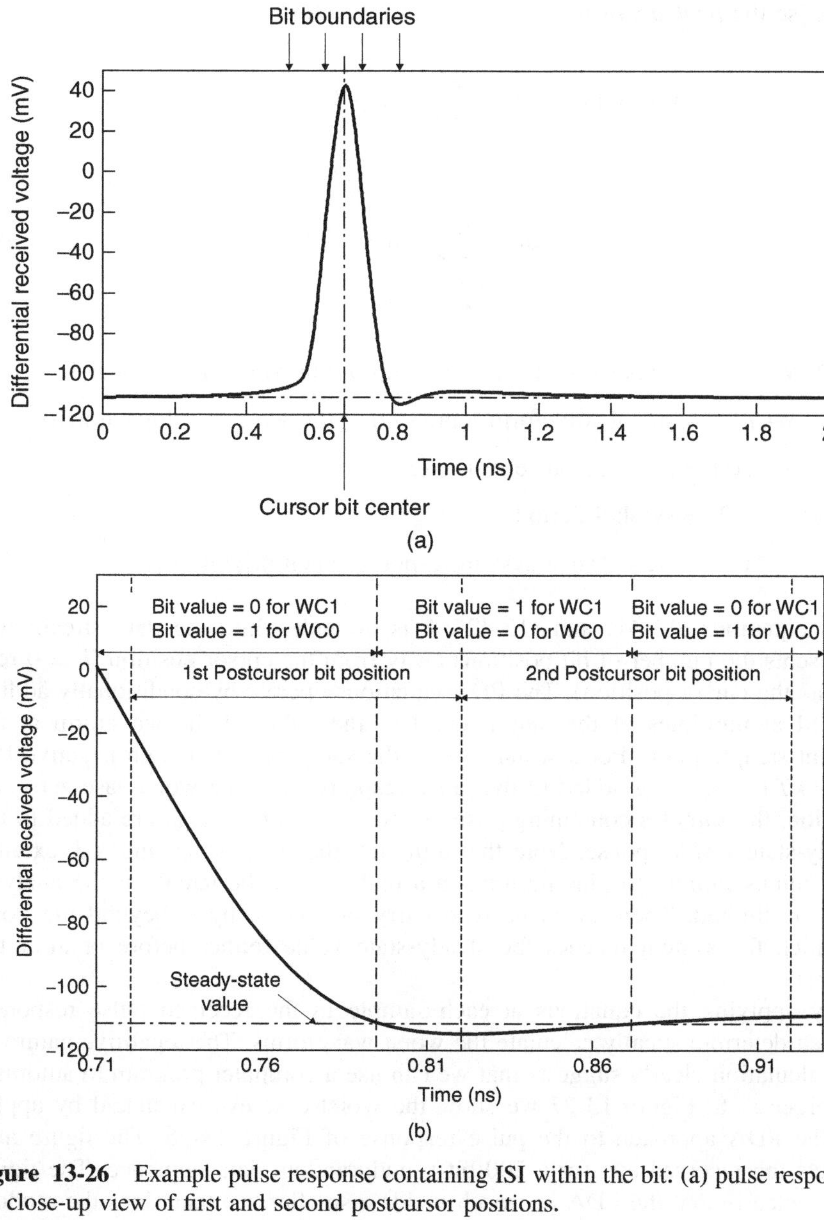

Figure 13-26 Example pulse response containing ISI within the bit: (a) pulse response; (b) close-up view of first and second postcursor positions.

equations that implement the peak distortion calculation of the worst-case waveforms given the pulse response at the receiver [Casper et al., 2002]. Equations (13-41) and (13-42) describe the waveforms for the worst-case 1 and zero, respectively, including the ISI due to losses and reflections. The equations, along with determination of the worst-case bit patterns (left as a problem for the reader), comprise the *peak distortion analysis* (PDA) method.

$$WC_1(t) = y(t) + \sum_{\substack{k=-\infty \\ k\neq 0}}^{\infty} y(t-kT)|_{y(t-kT)<v_{ss0}} \tag{13-41}$$

$$WC_0(t) = v_{ss0} + \sum_{\substack{k=-\infty \\ k\neq 0}}^{\infty} y(t-kT)|_{y(t-kT)>v_{ss0}} \tag{13-42}$$

where $WC_1(t)$ = signal waveform representing the worst-case logical 1

$WC_0(t)$ = signal waveform representing the worst-caselogical zero

$y(t)$ = received pulse response

T = symbol period

v_{ss0} = steady-state system response when driving low

In equations (13-41) and (13-42), k is an index into the data stream that represents the number of bit positions away from the cursor position ($k = 0$ represents the cursor position). The PDA equations operate by conditionally adding the ISI at multiples of the unit interval to the value of the waveform at the current sample point. For a worst-case 1, the samples that contain negative ISI, $y(t - kT) < v_{ss0}$, are added to the pulse response. For the worst-case zero calculation, the samples containing positive ISI, $y(t - kT) > v_{ss0}$, are added to the steady-state low response. Note that although the bit position index k extends from minus infinity to plus infinity, in a real system the length of the analysis must be limited. There is no need to carry out the analysis beyond the point at which the system reaches the steady-state value (either before or after the cursor).

By applying the equations at each sample in the received pulse response, we can deterministically calculate the worst waveforms. The repetitive nature of the calculation clearly suggests that we can use a computer program to automate the process. In Figure 13-27 we show the worst-case eye calculated by applying the PDA approach to the pulse response of Figure 13-26. The figure also depicts the received eye for a 200-bit pseudorandom data sequence. The figure shows clearly that the PDA approach yields a smaller eye than does the random sequence, thus providing a better estimate of the worst-case noise and jitter from a single simulation!

13.5.3 Peak Distortion Analysis Including Crosstalk

In Chapter 4 we showed that crosstalk also affects the behavior of signals in a coupled channel. As a consequence, our methodology must include potential

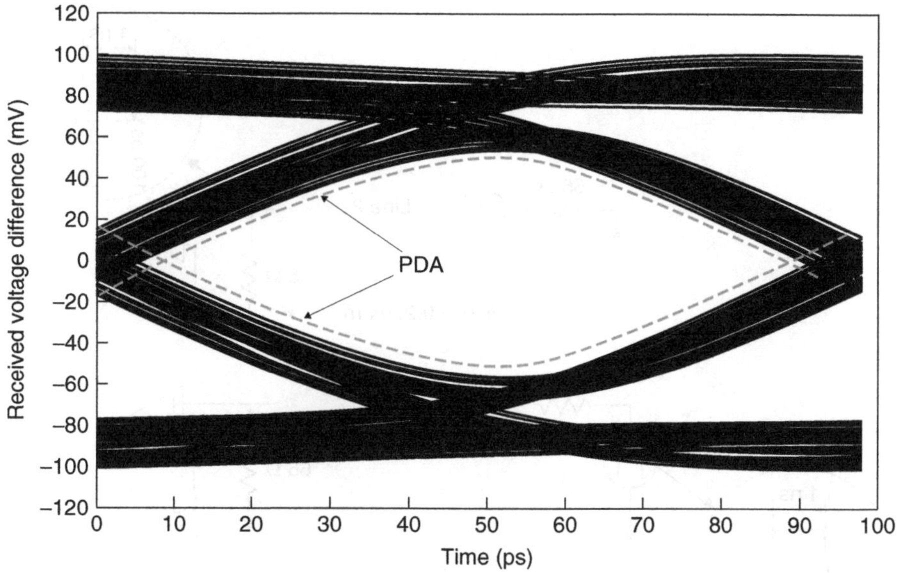

Figure 13-27 Comparison of eyes for a 200-bit random sequence and the worst case from peak distortion analysis.

impacts of crosstalk on the data eye to make sure that it is comprehended in the system jitter and noise margin estimates. Fortunately, the superposition principle makes it possible to extend the PDA method to include crosstalk using the same approach that we employed for a single signal.

To illustrate the concept, we return to the coupled system from Example 4-2. Shown in Figure 13-28, the system contains a 0.2794-m-long coupled pair that is terminated with 65 Ω at each end. The lossless, single-ended signals are driven by a 1-V source with a 100-ps rise time and 1-Gb/s switching rate. The signals are detected by a high-speed differential receiver that has one end connected to a reference voltage of 0.25 V. To perform peak distortion analysis, we again simulate the system pulse response, which is shown in Figure 13-29a. When including crosstalk, we apply the active pulse to the line of interest (the "aggressor") and capture the pulse response at the receiver for all signals in the coupled system.

We start by using PDA to calculate the worst-case signal waveforms that account for ISI from the pulse response on the aggressor line. To incorporate the effect of crosstalk, we use the response on the victim line in the same manner as we did for the ISI calculation. That is, we conditionally add the crosstalk at multiples of the unit interval to the worst-case signals that include ISI at the current sample point. For a worst-case 1, the samples that contain negative crosstalk are added to the pulse response. For the worst-case zero calculation, the samples containing positive crosstalk are added to the steady-state low response.

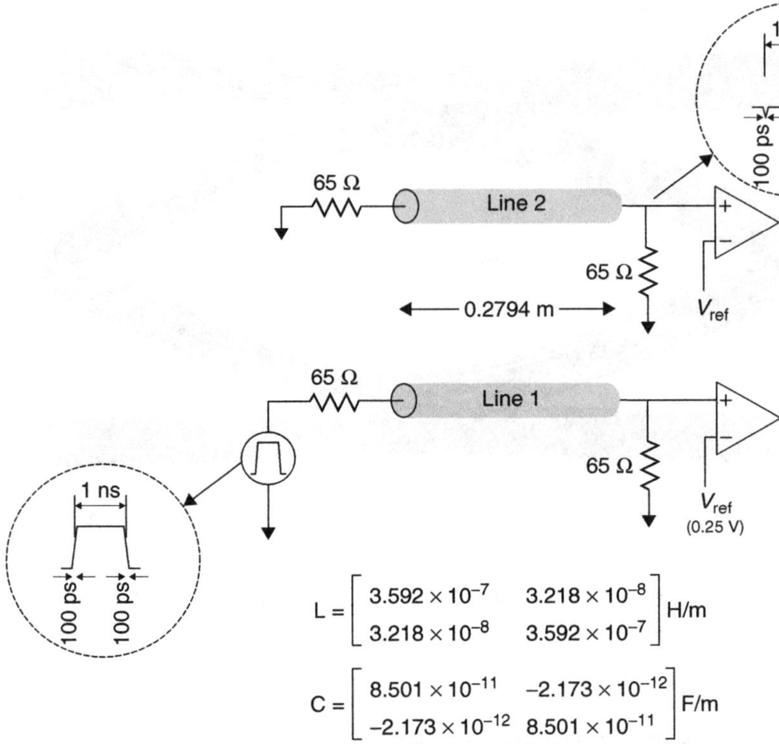

Figure 13-28 Coupled system from Example 4-2.

This method is illustrated in Figure 13-29b and shown as follows:

$$WC_1(t) = y(t) + \sum_{\substack{k=-\infty \\ k\neq 0}}^{\infty} y(t-kT)|_{y(t-kT)<v_{ss0}} + \sum_{i=1}^{n}\sum_{k=-\infty}^{\infty} y^i(t-kT-t_i)|_{y^i(t-kT-t_i)<v_{ss0}} \tag{13-43}$$

$$WC_0(t) = v_{ss0} + \sum_{\substack{k=-\infty \\ k\neq 0}}^{\infty} y(t-kT)|_{y(t-kT)>0}v_{ss0} + \sum_{i=1}^{n}\sum_{k=-\infty}^{\infty} y^i(t-kT-t_i)|_{y^i(t-kT-t_i)>v_{ss0}} \tag{13-44}$$

where $y^i(t - kT - t_i)$ is the pulse response received on line i (i.e., crosstalk source), also known as co-channel interference (CCI), and t_i is the relative sample point of each crosstalk source.

Note that in incorporating crosstalk into our analysis, we are taking advantage of the symmetry of the **L** and **C** matrices for the transmission-line model. Because the matrices are symmetric with respect to the mutual (off-diagonal) terms, the coupling between traces is also symmetric. Therefore, the crosstalk pulse that

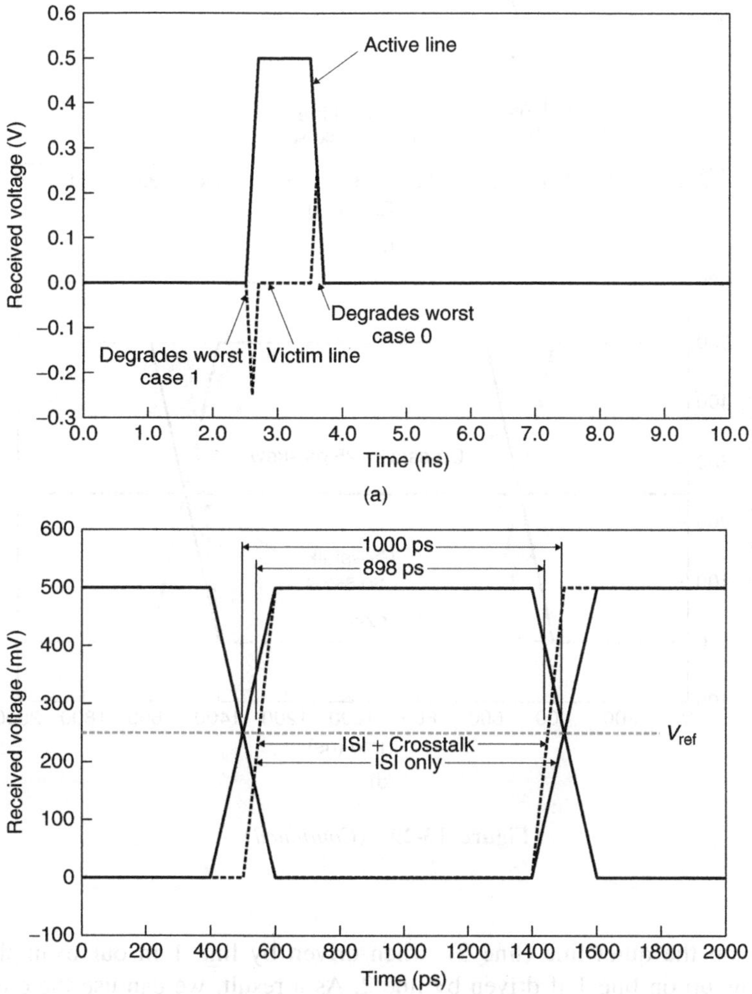

Figure 13-29 Pulse response and PDA for the coupled system of Example 13-3: (a) pulse response for aggressor and victim traces; (b) worst-case eyes calculated via PDA; (c) pulse response and PDA eye with +25 ps of transmit skew; (d) PDA eye with +25 ps of transmit skew. *(Continued)*

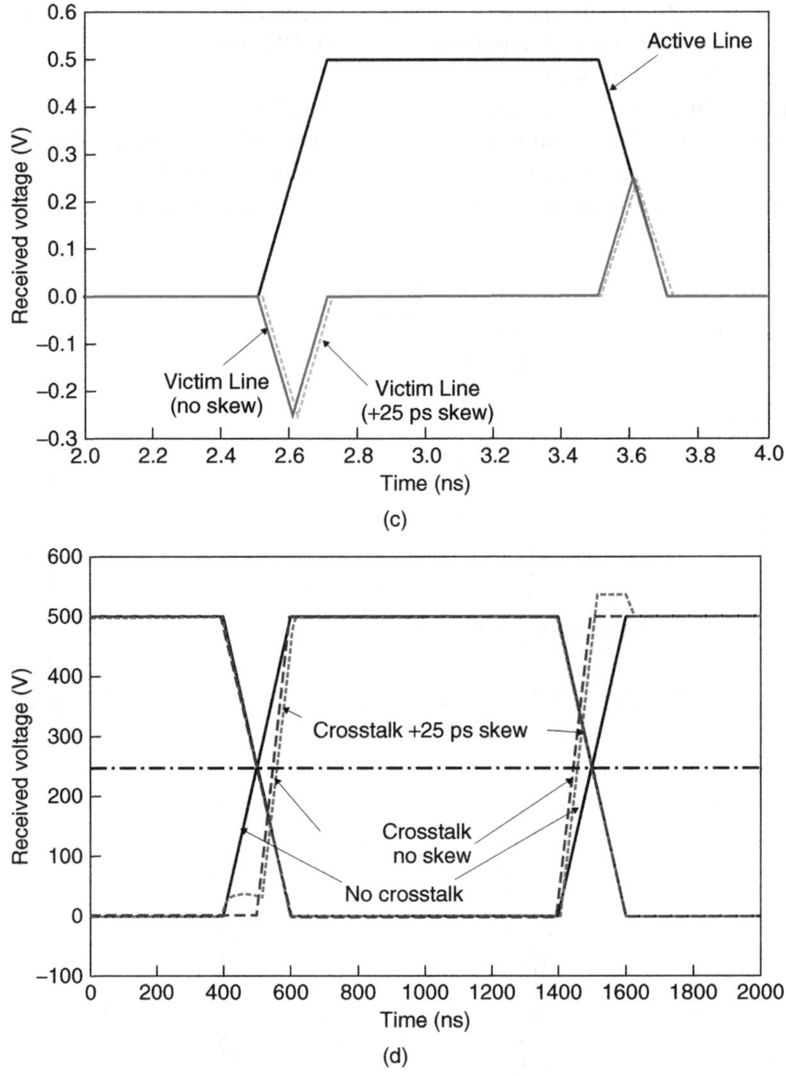

Figure 13-29 (*Continued*)

we saw on the quiet line (line 2), when driven by line 1 in our example will also show up on line 1 if driven by line 2. As a result, we can use the crosstalk response on line 2 when driven by line 1 to determine the impact of crosstalk on line 1.

The relative sample point t_i allows us to find the worst-case data eye, including crosstalk in the presence of skew on the data signal transmitter outputs. For example, Figure 13-29c shows the pulse response and resulting data eye under the condition that the transmitter for the second data line has a +25-ps skew relative to line 1.

13.5.4 Limitations

PDA is a powerful tool for analyzing the jitter and noise characteristics of a high-speed signaling link. However, it has limitations that we must comprehend to ensure robust products while minimizing overdesign. We summarize them briefly below.

Linearity Assumption As discussed, peak distortion analysis explicitly assumes that the signaling system is linear. However, the transceiver circuits may show significant nonlinearity. This suggests that PDA results should be checked for accuracy. The easiest way to do so would be to run some check simulations using full transistor or IBIS transceiver models using the worst-case bit patterns obtained via the PDA algorithm.

Dual Dirac Model The treatment of the channel jitter distribution as though it were a delta function located at the maximum value is an approximation. For a random data stream, the channel jitter will actually be distributed across a range that extends from zero to the maximum value calculated by PDA. As a result, treating the channel jitter as though it is distributed only at the extreme values obtained from PDA is inherently conservative. Although it is certainly possible to calculate the actual channel jitter distribution and to incorporate the distribution into the total system jitter model (e.g., Casper et al. [2002] and Sanders et al. [2004]), we leave it as an advanced design project for the student.

Differential Signaling Although not a limitation, we note that the PDA method works equally well for both differential and single-ended signals. Use with differential signals requires that the differential voltage signal be calculated from the individual signal components on the pair prior to applying PDA.

Simulation Noise Some simulators may leave small voltage fluctuations on the line due to numerical error or convergence effects that last for many unit intervals. These fluctuations can artificially close the data eye, leading to highly pessimistic results. A simple means of eliminating the problem is to apply a "noise floor" to the PDA calculation that effectively ignores any value that falls closer to the steady-state value than the noise floor.

Jitter Amplification Due to the lossy nature of the interconnect channel, it will tend to amplify any jitter that the transmitter circuit injects onto the channel. Since the PDA method that we presented uses only a single pulse in calculating the worst-case jitter, it does not comprehend the amplification of the transmit jitter. Techniques to comprehend the amplification of transmitter jitter by the channel are an area of active research [Casper et al., 2007].

13.6 SUMMARY

Contemporary high-speed signaling links specify performance in terms of both the maximum data rate and the bit error rate. In this chapter we introduced the concept of bit error rate and described the necessary mathematics for relating the BER to timing jitter and voltage noise. We described the various types of jitter and noise that are present in systems, along with their sources, and developed methods for creating system jitter and noise budgets. Finally, we described the peak distortion analysis technique for evaluating the worst-case jitter and noise caused by the interconnect channel over which high-speed signals propagate.

REFERENCES

Jitter characterization has been an area of active research over the past decade. Refer to Li [2008] and Derickson [2008] for comprehensive treatment. In addition, Buchs et al. [2004] and Stephens [2004] offer primers on BER estimation from noise and jitter, while Martwick [2005] contains a jitter budgeting case study. The topic of noise budgeting is covered by Dally and Poulton [1998]. The peak distortion analysis method is spelled out by Casper et al. [2002], while Sanders et al. [2004] and Casper et al. [2007] deal with probabilistic methods for analyzing channel jitter and noise. Finally, we refer the reader who wishes to develop further understanding of worst-case timing analysis for synchronous and source synchronous timings to a book by Hall et al. [2000].

Buchs, Kevin, Pat Zabinski, and Jon Coker, 2004, *Basic Bit Error Rate Analysis for Serial Data Links*, Document Mayo-R-04-07-R0, Mayo Clinic, Rochester, MN, June 3.

Casper, Bryan K., Matt Haycock, and Randy Mooney, 2002, An accurate and efficient analysis method for multi-Gb/S chip-to-chip signaling schemes, Symposium on VLSl Circuits, June 13–15, *Digest of Technical Papers*, pp. 54–57.

Casper, Bryan K., et. al., 2007, Future microprocessor interfaces: analysis, design and optimization, *Proceedings of the IEEE 2007 Custom Integrated Circuits Conference*, Sept. 16–19, pp. 479–486.

Dally, William, and John Poulton, 1998, *Digital Systems Engineering*, Cambridge University Press, Cambridge, UK.

Derickson, Dennis, and Marcus Müller, 2008, *Digital Communications Test and Measurement*, Prentice Hall.

EIA 1995, *High-Speed Transceiver Logic (HSTL): A 1.5V Output Buffer Supply Voltage Based Interface Standard for Digital Integrated Circuits*, Document EIA/JESD8-8, Electronic Industries Association, Aug. city

Gray, Paul R., Paul J. Hurst, Stephen H. Lewis, and Robert G. Meyer, 2001, *Analysis and Design of Analog, Integrated Circuits*, 4th ed., Wiley, New York.

Hall, Stephen, Garrett Hall, and James McCall, 2000, *High-Speed Digital System Design*, Wiley-Interscience, New York.

Intel Corporation, 1997, *Pentium® II Processor Developer's Manual*, Document 243502-001, Intel Press, Hillsbors, OR, Oct.

Intel Corporation, 2002, *AGP V3.0 Interface Specification*, revision 1.0, Intel Press, Hillsbors, OR, Sept.

JEDEC Solid State Technology Association, 2002, *Stub Series Terminated Logic for 1.8V (SSTL_18)*, Document JESD8-15, JEDEC, Arlington, VA, Oct.

Li, Mike Peng, 2008, *Jitter, Noise and Signal Integrity at High Speed*, Prentice Hall, Upper Saddle River, NJ.

Martwick, Andy, ed., 2005, *PCI Express™ Jitter and Bit Error Rates*, revision 1.0, PCI-SIG, Beaverton, OR, Feb. 11.

Norman, Adam, et al., 2003, Application of design of experiments (DOE) methods to high-speed interconnect validation, *Proceedings of the 12th IEEE Topical Meeting on Electrical Performance of Electronic Packaging*, October 25, pp. 223–226.

Razavi, Behzad, 1995, *Principles of Data Conversion System Design*, IEEE Press, Piscataway, NJ.

Sanders, Anthony, Mike Resso, and John D'Ambrosia, 2004, Channel compliance testing using novel statistical eye methodology, presented at *DesignCon 2004*, Feb. 2–5.

Stephens, Ransom, 2004, *Jitter Analysis: The Dual-Dirac Model, RJ/DJ, and Q-Scale*, Document 5989-3206EN, Agilent Technologies, Santa Clara, CA, Dec. 31.

PROBLEMS

13-1 Explain how the BER contour in Figure 13-4 works. Why is the bit error rate lowest at the center of the eye, with increasing BER as we move away from the center in either the horizontal or vertical direction?

13-2 Estimate the BER that we can achieve for a 10-Gb/s link, given DJ = 40 ps and $\sigma_{RJ} = 3.5$ ps. Assuming that the DJ and RJ contributions remain constant, what is the maximum data rate at which we can operate if the maximum tolerable BER is 10^{-12}?

13-3 Estimate the PDFs for deterministic and total jitter for a 5-Gb/s system containing the following sources:

- Periodic jitter with 30 ps of amplitude
- 10% duty cycle jitter
- Random jitter with $\sigma_{RJ} = 6$ ps
- ISI jitter in the form of a truncated Gaussian with $\sigma = 20$ ps and $-50 \text{ ps} \le DJ_{ISI} \le 50$ ps

13-4 Create a dual Dirac model for the total jitter distribution of Problem 13-3, and use it to estimate the maximum bit error rate for the system.

13-5 Explain why the DAC-induced equalizer noise is equal to one-half of the DAC resolution.

13-6 Develop a noise budget for the 10-Gb/s differential system shown in Figure 13-30. Include all deterministic and random noise sources following the guidelines presented in Section 13.4.

13-7 Use the noise budget developed in Problem 13-6 to estimate the bit error rate for the system in Figure 13-30.

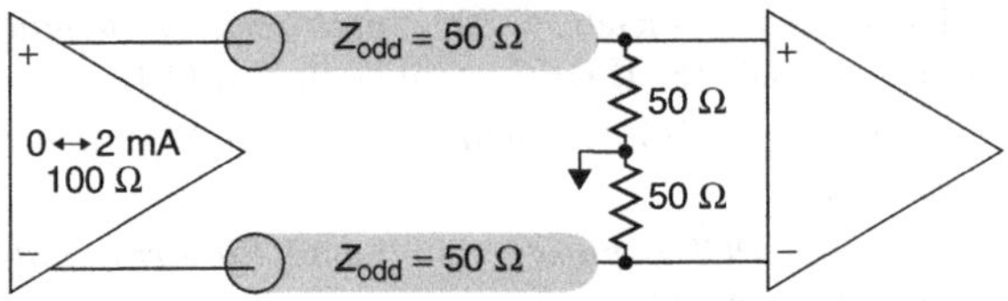

Figure 13-30 Open-drain signaling system for Problem 13-6.

13-8 Use the pulse response data from the second column of Table 13-6 to calculate the waveform for a 1101000 pattern.

13-9 An interesting aspect of the PDA equations is that they account automatically for the positional dependence of the worst-case bit pattern without explicitly determining the actual patterns. However, sometimes we may wish to know the worst-case pattern(s). Develop an algorithm for finding the bit patterns that result in the worst-case signal waveform for both a logical zero and a logical 1.

13-10 Calculate the actual total jitter distribution for the PCI Express system in Example 13-2. Use the DJ and RJ values from Table 13-2, assume that the individual DJ components fit a dual Dirac distribution, and convolve the DJ distributions to find the total TJ.

13-11 Explain why PDA is not applicable for the open-drain system shown in Figure 13-31.

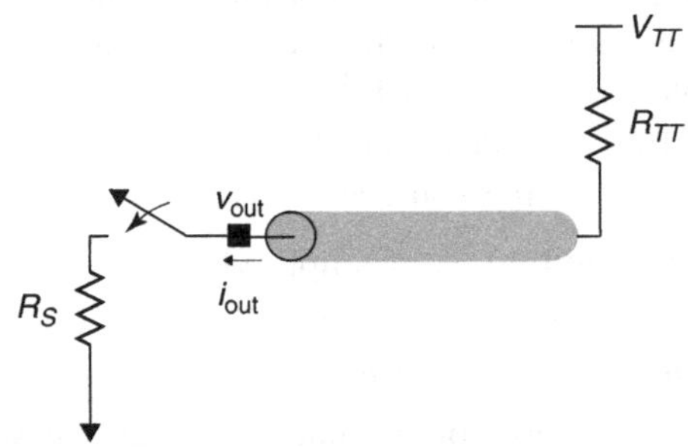

Figure 13-31 Open-drain signaling system for Problem 13-11.

13-12 Use peak distortion analysis to determine the worst-case waveform from the pulse response shown in Figure 13-32.

13-13 Outline the method that you would use to perform peak distortion analysis on a differential system, including the effects of crosstalk.

13-14 Propose a method for performing peak distortion analysis using the step response rather than the pulse response. What would be the advantage of using the step response?

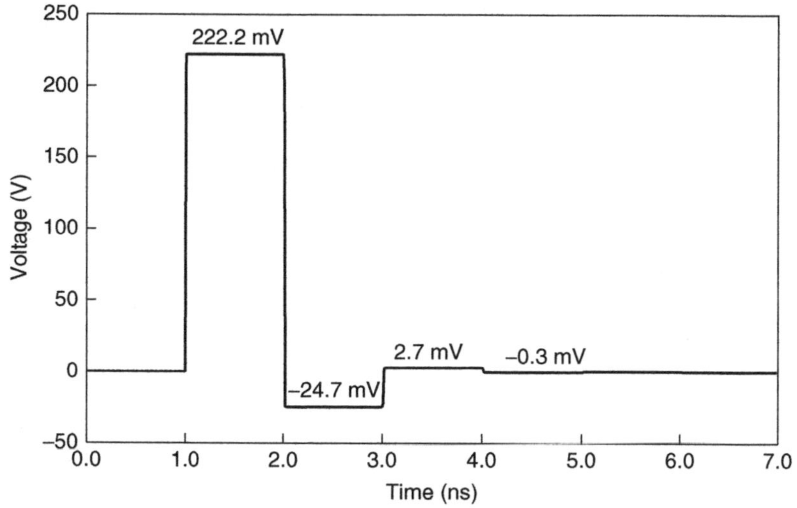

Figure 13-32 Pulse response for Problem 13-12.

13-15 *Project:* Develop a software tool to perform peak distortion analysis from a received pulse response.

13-16 *Project:* Casper et al. [2002] outline a probabilistic approach to determining the ISI and crosstalk jitter and noise distributions from the received pulse response. Develop a software tool to perform the probabilistic channel analysis.

14

SYSTEM ANALYSIS USING RESPONSE SURFACE MODELING

By this point, we have covered a wide range of important topics for the study and design of high-speed signaling systems. We now understand the fundamentals of transmission line behavior, including losses and crosstalk, modeling of I/O circuits, equalization, and modeling of jitter and noise. In this chapter we tie those concepts together with a method for analyzing and predicting the behavior of high-speed signaling interfaces using response surface modeling (RSM). Once we have added RSM to our repertoire, we will possess the necessary ingredients for the successful design of signaling systems operating at multi-Gb/s speeds.

Advanced Signal Integrity for High-Speed Digital Design, By Stephen H. Hall and Howard L. Heck

14.1 MODEL DESIGN CONSIDERATIONS

Our understanding of channel noise and jitter modeling, combined with the peak distortion analysis technique, gives us nearly all the tools we need in order to design a multi-Gb/s signaling interface. However, we still need a method that allows us to understand how our design behaves as the physical and electrical characteristics of the system components change. Without the means to evaluate the signal behavior in a systematic fashion as the system characteristics vary, finding a combination of transmitter, package, PCB, termination, and receiver characteristics that result in a functional and reliable system becomes an intractable problem at multi-Gb/s rates.

In addition, we must also account for the impacts of manufacturing large volumes of high-speed systems. The variation that is an inherent part of all manufacturing processes means that different systems will behave differently, even if they were built from the same design. For example, the differential impedance of signal traces on our printed circuit boards may show as much as ±20% variation around the nominal value. This manufacturing variability may manifest itself as a variation in electrical performance. In particular, we are interested in the impact of manufacturing variation on eye height and width. If we do not comprehend these variations in our design, a significant percentage of our systems may fail to operate properly. The impact of such a poor design is typically increased cost, due either to returns from customers or to yield loss during manufacture.

Fortunately, the *response surface modeling* (RSM) technique provides a tool that gives us the ability to model the behavior of our signaling system as the circuit and interconnect characteristics vary. RSM works by fitting a statistical model of the output response as a function of changes in the input variables. For example, a useful RSM would provide predictions of the eye height and width as a function of board impedance, termination resistance, line length, and equalization. Response surface modeling is a broad topic for which entire textbooks exist [Myers and Montgomery, 1995]. As such, we cannot provide a comprehensive treatment here. Instead, we endeavor to provide a sufficient overview to allow the reader to begin applying RSM techniques in the design of signaling links.

The basic idea behind RSM is to apply linear regression techniques to create a statistical model that predicts the system response (output) as the inputs vary. We do this by first creating a set of observed outputs in response to specific sets of input conditions that we determine via a designed experiment. The system model is a linear equation that is constructed by fitting the observed responses and inputs via the least squares fitting technique. Once the model is created, we can use it to predict the output of the system in response to arbitrary combinations of inputs.

Since the model represents a statistical fit, we expect that there may be some error in the values predicted. As a result, understanding how well the model fits the observed data and determining how much uncertainty exists in the predicted responses become important topics that we must also comprehend in our analysis. Before describing the creation and use of the response surface model, we introduce an example signaling link that we use throughout the chapter to demonstrate the application and benefit of the RSM approach.

14.2 CASE STUDY: 10-GB/S DIFFERENTIAL PCB INTERFACE

We want to develop a pair of response surface models that estimate the eye height and eye width for a 10-Gb/s differential link. Our ultimate goal is to use the models to identify a working solution that accommodates the desired range of trace lengths, comprehends the expected manufacturing variation, and meets a defect rate of less than 1000 parts per million (ppm). We have complete system budgets for timing and noise that require our interconnect channel to meet minimum specs of 60-mV eye height and 70-ps eye width as determined from peak distortion analysis in order to achieve a 10^{-12} bit error rate. We use this example system throughout the remainder of the chapter to illustrate the model-fitting process, analysis of the model fit, and application to prediction of design limits and defect rates.

The system characteristics are summarized in Figure 14-1 and Table 14-1. We will fit our model to the five input variables listed in Table 14-1. The variability in the models for termination and characteristic impedance represent the expected variation due to the manufacturing process. The range of the transmission-line lengths is based on experience with prior systems, which indicates the need to accommodate a range of trace lengths from 10 to 20 in. Finally, the link uses a two-tap equalizer with a coefficient range from −0.30 to −0.10, which was determined from results of initial simulations. The equalizer is not adaptive, so the tap weights must be set prior to operation. The weights are controlled by a 4-bit digital-to-analog converter, so that the granularity between adjacent taps settings is equal to 0.0133. The valid equalizer settings are summarized in Table 14-2.

The differential transmission-line characteristics are calculated from the physical cross-sectional dimensions using a 2D field solver. To generate the eye data, we first simulate the pulse response in the frequency domain using the causal transmission-line modeling method (Chapter 8) and frequency-domain equalizer (Chapter 12) model. After transformation to the time domain, we determine the worst-case eye width and height for each observation using peak distortion analysis (Chapter 13), which we show in columns 7 and 8 in Table 14-3.

14.3 RSM CONSTRUCTION BY LEAST SQUARES FITTING

The general form of the response surface model is

$$y = \beta_0 + \beta_1 x_1 + \beta_2 x_2 + \cdots + \beta_k x_k + \varepsilon \tag{14-1}$$

where y is the system response (output), β_i the model fit coefficients, x_i the system inputs, k the number of terms in the model, and ε the error in the prediction from the model. The response surface model is a linear function of the fit

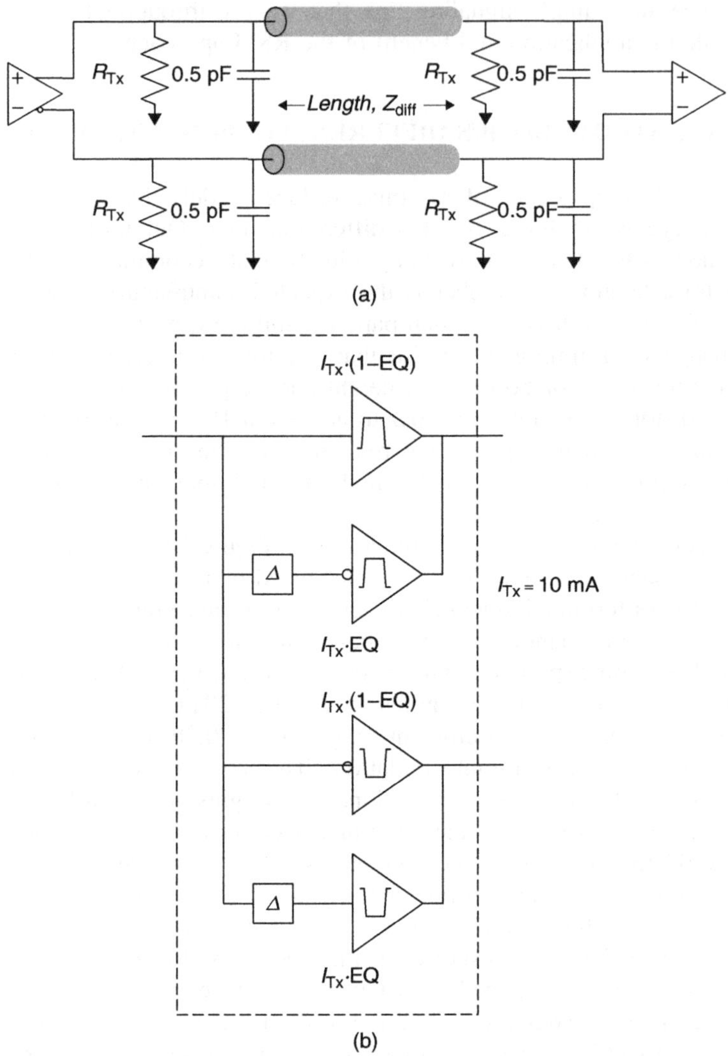

Figure 14-1 Signaling system for construction of a response surface model: (a) system simulation model; (b) transmitter model.

coefficients. As such, it is highly flexible, allowing us to fit curved response surfaces by using higher-order combinations of input variables (e.g., x_1^2). In general, second-order models are sufficient for high-speed signaling links. They have the form

$$y = \beta_0 + \sum_{i=1}^{n_{\text{var}}} \beta_i x_i + \sum_{i=1}^{n_{\text{var}}} \beta_{ii} x_i^2 + \sum_{i=1}^{n_{\text{var}}} \sum_{j \neq i}^{n_{\text{var}}} \beta_{ij} x_i x_j \tag{14-2}$$

where n_{var} is the number of independent input variables.

TABLE 14-1. System Input Variables for the Signaling Example of Chapter 14

Model Term	Minimum	Typical	Maximum
Transmitter termination, $R_{\text{Tx}}(\Omega)$	40	50	60
Receiver termination, $R_{TT}(\Omega)$	40	50	60
Differential impedance, $Z_{\text{diff}}(\Omega)$	84.0	100.3	117.9
Trace length, L (m)	0.254	0.381	0.508
Equalization coefficient, EQ	−0.1	−0.2	−0.3

TABLE 14-2. Valid Equalization Settings

Setting	bit_3	bit_2	bit_1	bit_0	EQ
0	0	0	0	0	−0.3000
1	0	0	0	1	−0.2867
2	0	0	1	0	*−0.2733*
3	0	0	1	1	**−0.2600**
4	0	1	0	0	*−0.2467*
5	0	1	0	1	−0.2333
6	0	1	1	0	−0.2200
7	0	1	1	1	−0.2067
8	1	0	0	0	−0.1933
9	1	0	0	1	−0.1800
10	1	0	1	0	−0.1667
11	1	0	1	1	−0.1533
12	1	1	0	0	−0.1400
13	1	1	0	1	−0.1267
14	1	1	1	0	−0.1133
15	1	1	1	1	−0.1000

The number of terms in the second-order model, k, is

$$k = 1 + 2n_{\text{var}} + \frac{n_{\text{var}}(n_{\text{var}} - 1)}{2} \tag{14-3}$$

In creating the model, we use response data from multiple observations that are generated from different combinations of model inputs. The number of observations and choice of input combinations is typically determined via the *design of experiments* (DOE) approach. As its name implies, DOE is a discipline that allows us to design experiments that will result in the creation of accurate models that require a minimal number of observations to obtain a good fit. The latter characteristic is particularly important for analyzing and designing signaling links. Simulations for fully coupled interconnects with nonlinear transceiver models can take several minutes to complete. With that in mind, a technique that minimizes the number of simulations needed to create a reliable response surface model will reduce development time and CAD infrastructure costs. As was the case with response surface modeling, DOE is a topic for which entire textbooks

TABLE 14-3. Model Fitting Observations

(1)	(2)	(3)	(4)	(5)	(6)	(7)	(8)	(9)	(10)	(11)	(12)
								Predicted		Residual	
Run	R'_{Tx}	Z'_{diff}	R'_{TT}	L'	EQ'	Eye Height (mV)	Eye Width (ps)	Eye Height (mV)	Eye Width (ps)	Eye Height (mV)	Eye Width (ps)
0	0	0	0	0	1	79.43	77	80.61	73.87	−1.18	3.13
1	−1	−1	1	1	1	30.93	52	29.93	52.65	1.00	−0.65
2	1	1	−1	−1	1	118.70	91	119.35	91.15	−0.65	−0.15
3	1	−1	1	−1	1	148.40	89	149.25	89.15	−0.86	−0.15
4	−1	1	1	−1	1	133.38	90	133.72	89.66	−0.34	0.34
5	1	−1	−1	−1	−1	76.89	88	77.63	87.37	−0.74	0.63
6	0	0	−1	0	0	83.83	90	85.44	89.86	−1.61	0.14
7	−1	−1	−1	1	−1	56.80	84	55.69	83.87	1.11	0.13
8	1	1	−1	1	−1	58.40	79	58.23	79.33	0.17	−0.33
9	−1	1	1	1	−1	72.24	83	71.75	82.84	0.48	0.16
10	0	0	0	−1	0	127.60	92	123.36	95.83	4.24	−3.83
11	0	0	1	0	0	105.96	89	104.71	89.08	1.25	−0.08
12	0	0	0	1	0	57.13	78	61.73	74.12	−4.60	3.88
13	1	−1	−1	1	1	17.43	40	16.77	41.18	0.66	−1.18
14	1	1	1	1	1	21.64	45	21.57	46.05	0.07	−1.05
15	−1	0	0	0	0	90.41	89	93.30	90.90	−2.90	−1.90
16	−1	1	−1	−1	−1	68.65	88	68.88	86.94	−0.24	1.06
17	0	−1	0	0	0	93.19	89	94.20	88.92	−1.01	0.08
18	−1	−1	−1	−1	1	107.62	91	107.40	90.68	0.22	0.32
19	−1	1	−1	1	1	32.63	62	31.56	62.61	1.07	−0.61
20	1	−1	1	1	−1	79.85	84	79.82	84.34	0.03	−0.34
21	−1	−1	1	−1	−1	93.97	87	94.38	85.84	−0.41	1.16
22	0	0	0	0	0	96.24	89	95.52	89.11	0.72	−0.11
23	0	0	0	0	0	96.24	89	95.52	89.11	0.72	−0.11
24	0	0	0	0	−1	82.02	85	81.20	88.08	0.82	−3.08
25	1	1	1	−1	−1	106.52	86	107.75	85.38	−1.23	0.62
26	1	0	0	0	0	100.64	89	98.10	87.05	2.54	1.95
27	0	1	0	0	0	95.09	90	94.44	90.03	0.65	−0.03

are available, and the details of DOE theory are beyond our scope. However, those readers who adopt the RSM methodology will undoubtedly wish to acquire a more in-depth understanding of DOE fundamentals. We refer them to books by Myers and Montgomery [1995], Steppan et al. [1998], and Montgomery [2005] for further study.

The response surface model is a linear function of the model coefficients. For a given model-fitting experiment, the input variables have predetermined values. As a result, we can replace each of the higher-order terms with single variables that have the same values without affecting the fit. For example, if our response model is

$$\hat{y} = \beta_0 + \beta_1 x_1 + \beta_2 x_2 + \beta_{11} x_1^2 + \beta_{22} x_2^2 + \beta_{12} x_1 x_2 + \varepsilon$$

we can create an equivalent representation

$$\hat{y} = \beta_0 + \beta_1 x_1 + \beta_2 x_2 + \beta_3 x_3 + \beta_4 x_4 + \beta_5 x_5 + \varepsilon$$

by defining

$$\beta_{11} \equiv \beta_3, \beta_{22} \equiv \beta_4, \beta_{12} \equiv \beta_5, x_1^2 \equiv x_3, x_2^2 \equiv x_4, \text{and} x_1 x_2 \equiv x_5$$

By expressing the model this way, we can recast it in matrix form:

$$\mathbf{y} = \mathbf{X}\boldsymbol{\beta} + \boldsymbol{\varepsilon} \tag{14-4}$$

where

$$\mathbf{y} = \begin{bmatrix} y_1 \\ y_2 \\ \vdots \\ y_n \end{bmatrix} \text{ is the } n \times 1 \text{ vector of observed responses}$$

$$\mathbf{X} = \begin{bmatrix} 1 & x_{11} & x_{12} & \cdots & x_{1k} \\ 1 & x_{21} & x_{22} & \cdots & x_{2k} \\ \vdots & \vdots & \vdots & & \vdots \\ 1 & x_{n1} & x_{n2} & \cdots & x_{nk} \end{bmatrix} \text{ is the } n \times k \text{ matrix of inputs}$$

$$\boldsymbol{\beta} = \begin{bmatrix} \beta_0 \\ \beta_1 \\ \vdots \\ \beta_k \end{bmatrix} \text{ is the } k \times 1 \text{ vector of model coefficients}$$

$$\boldsymbol{\varepsilon} = \begin{bmatrix} \varepsilon_1 \\ \varepsilon_2 \\ \vdots \\ \varepsilon_n \end{bmatrix} \text{ is the } n \times 1 \text{ vector of random errors}$$

and n is the number of observations for fitting the model and k is the number of model terms.

Each column of the input matrix corresponds to a model term. The first column represents the intercept for the model, from which the β_0 coefficient is estimated. Each row of the input matrix corresponds to an experimental observation. The response vector contains a row for each observation, as does the residual vector. The model coefficient vector contains a row for each term in the model.

With the form of the model established, we can fit it to the experimental data. The most widely used fitting technique is known as the *method of least squares*, which calculates the coefficients from the system response and inputs such that the sum of the squares of the errors (ε_i) is minimized. Omitting the derivation, the least squares fit algorithm yields the following equation, which provides an estimate of β that satisfies the least squares criterion:

$$\mathbf{b} = (\mathbf{X}^{\mathrm{T}}\mathbf{X})^{-1}\mathbf{X}^{\mathrm{T}}\mathbf{y} \tag{14-5}$$

where $\mathbf{b}$ is the $k \times 1$ vector containing an estimate of $\boldsymbol{\beta}$, the vector of true model fit coefficients. The fitted regression model is then expressed as

$$\hat{\mathbf{y}} = \mathbf{X}\mathbf{b} \tag{14-6}$$

where $\hat{\mathbf{y}}$ is the vector of estimated system responses for the given input matrix, $\mathbf{X}$, and vector of estimated fit coefficients, $\mathbf{b}$.

When fitting the model, we have multiple choices for the form of the model factors and responses. Rather than fitting to the raw data, regression tools often fit the model to a transformed version of the data. Throughout this chapter we fit the model to input variables that are coded according to

$$x'_{ik} = \text{round}\left[\frac{2(x_{ik} - \overline{x}_k)}{x_{k,\max} - x_{k,\min}}\right] \tag{14-7}$$

where x_{ik} is the value of the ith observation of the input for the kth model term, $\overline{x}_k$ the mean value of the observations for the kth model term, $x_{k,\max}$ the maximum value of the observations for the kth model term, $x_{k,\min}$ the minimum value of the observations for the kth model term, and round(x) rounds x to the nearest integer.

The variable coding maps the input variables such that the minimum, nominal, and maximum values for each variable correspond to coded values of -1, 0, and 1, respectively. By fitting the model to the coded variables, we reduce the variation in magnitude of the individual coefficients to avoid instabilities in the model.

Case Study Application We now apply the least squares fitting procedure to our case study. A frequently used experiment design for second-order response surface models is a *central composite design* [Montgomery, 2005]. For a set of five input variables, the central composite experiment requires a total of $n = 28$ observations, which are summarized in the first eight columns of Table 14-3. Since we have five independent variables, equation (14-3) specifies that the input matrix will have 21 columns. The form of the input matrix is shown in equation (14-8), and the expressions for the individual columns are defined in Table 14-4. For example, the term $x_{1,1}$ would be equal to the R'_{Tx} value of zero from run 0 in Table 14-3, $x_{1,2}$ would be equal to the R'_{Tx} value of -1 from run 1, $x_{1,3}$ would be equal to the R'_{TX} value of 1 from run 2, and so on. To carry out the least squares fit, we put the observations from the table into the coded input matrix ($\mathbf{X}$) and response vectors ($\mathbf{y}_{\mathbf{eyeH}}$ and $\mathbf{y}_{\mathbf{eyeW}}$). Note that the eye height and width are expressed in units of mV and ps, respectively.

$$\mathbf{X} = \begin{vmatrix} 1 & x_{1,1} & x_{2,1} & \cdots & x_{20,1} \\ 1 & x_{1,2} & x_{2,2} & \cdots & x_{20,2} \\ \vdots & \vdots & \vdots & \ddots & \vdots \\ 1 & x_{1,28} & x_{2,28} & \cdots & x_{20,28} \end{vmatrix} \tag{14-8}$$

With the observations in matrix form, application of the model fitting equations

$$\mathbf{b}_{\mathbf{eyeH}} = (\mathbf{X}^{\mathrm{T}}\mathbf{X})^{-1}\mathbf{X}^{\mathrm{T}}\mathbf{y}_{\mathbf{eyeH}} \tag{14-9}$$

$$\mathbf{b}_{\mathbf{eyeW}} = (\mathbf{X}^{\mathrm{T}}\mathbf{X})^{-1}\mathbf{X}^{\mathrm{T}}\mathbf{y}_{\mathbf{eyeW}} \tag{14-10}$$

gives the estimated fit coefficient vectors:

$$\mathbf{b}_{\mathbf{eyeH}} = \begin{vmatrix} 95.51799 \\ 2.32444 \\ 0.12056 \\ 9.55222 \\ -30.81556 \\ -0.28778 \\ 0.18751 \\ -2.18063 \\ -1.19749 \\ 1.26063 \\ -2.43813 \\ -0.44249 \\ -3.88563 \\ -0.02937 \\ -4.43812 \\ -2.97249 \\ -1.77437 \\ 0.22937 \\ -2.11688 \\ -20.42063 \\ -14.61249 \end{vmatrix} \qquad \mathbf{b}_{\mathbf{eyeW}} = \begin{vmatrix} 89.10651 \\ -1.94444 \\ 0.55556 \\ -0.44444 \\ -10.83333 \\ -7.05556 \\ -0.13314 \\ -0.56250 \\ 0.36686 \\ 1.18750 \\ -1.56250 \\ 0.36686 \\ -1.93750 \\ 0.56250 \\ 0.31250 \\ -4.13314 \\ -1.56250 \\ 1.43750 \\ -0.56250 \\ -8.93750 \\ -8.13314 \end{vmatrix}$$

Before proceeding further, it makes sense that we should apply the estimated fit coefficients, $\mathbf{b}_{\mathbf{eyeH}}$ and $\mathbf{b}_{\mathbf{eyeW}}$, to the input observed using equation (14-6) to obtain estimates of the response. These estimates are shown in columns (9) and (10) of Table 14-3. Comparing the predicted and observed responses, we see that our models agree to within approximately ±4 mV and ±4 ps.

TABLE 14-4. Model Terms for the RSM Example[a]

Model Term	Expression	Model Term	Expression	Model Term	Expression	Model Term	Expression
x_1	R'_{Tx}	x_6	$(R'_{\text{Tx}})^2$	x_{11}	$(R'_{TT})^2$	x_{16}	$R'_{\text{Tx}}\text{EQ}'$
x_2	Z'_{diff}	x_7	$R'_{\text{Tx}}Z'_{\text{diff}}$	x_{12}	$R'_{\text{Tx}}L'$	x_{17}	$Z'_{\text{diff}}\text{EQ}'$
x_3	R'_{TT}	x_8	$(Z'_{\text{diff}})^2$	x_{13}	$Z'_{\text{diff}}L'$	x_{18}	$R'_{TT}\text{EQ}'$
x_4	L'	x_9	$R'_{\text{Tx}}R'_{TT}$	x_{14}	$R'_{TT}L'$	x_{19}	$L'\text{EQ}'$
x_5	EQ'	x_{10}	$Z'_{\text{diff}}R'_{TT}$	x_{15}	$(L')^2$	x_{20}	$(\text{EQ}')^2$

[a] R'_{Tx}, Z'_{diff}, R'_{TT}, L', and EQ' are calculated using equation (14-7).

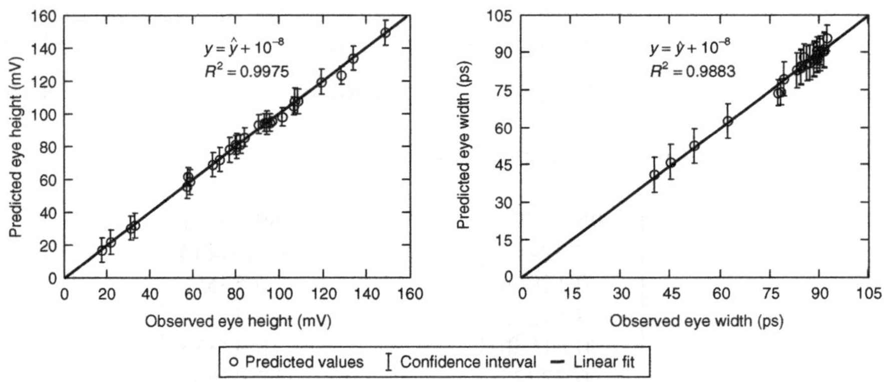

Figure 14-2 Comparison of actual versus predicted values for eye height and width.

The plots in Figure 14-2 demonstrate an alternative way to look at the agreement between prediction and observation. For the predictions to match the model perfectly, we expect the plot to show a linear relationship with a slope equal to 1 and intercept at zero (i.e., $y = \hat{y}$). In addition, we would expect the value of R-squared (R^2), which is a measure of the proportion of variability explained by the linear relationship of the sample data, to be nearly equal to 1. In subsequent sections we develop metrics, including a revised definition of R^2 for a multiple variable regression, to assess whether the agreement between model and observation is good enough to allow us to predict the response with enough confidence use it in the design of a high-speed signaling interface.

The ability to predict the system response for an arbitrary set of input variables is a very powerful tool. Using our model fit equations, we can explore a potential design space in order to find the working solution limits, or we can use it to project the rate at which systems will fail to meet our performance specifications. However, before doing so, we need to develop the metrics to assess the accuracy and limitations of models generated using the RSM technique so that we use the models properly when making design decisions.

14.4 MEASURES OF FIT

The metrics that allow us to evaluate the accuracy of the response surface model, known as *measures of fit*, are explained in this section.

14.4.1 Residuals

We start exploring model accuracy by examining the errors, known as *residuals*, which we obtain when we apply the model to the same set of inputs that we used in generating the model coefficients. The residual vector is simply the difference between the vector of estimated responses and the vector of responses observed:

$$\mathbf{e} = \hat{\mathbf{y}} - \mathbf{y} \tag{14-11}$$

If the model contains no systematic errors, we expect a plot of the residuals as a function of the responses observed to be randomly scattered with no apparent correlation. In addition, we can glean some idea as to how well the model fits the data by using the standardized residuals, **d**, to look for outliers. The standardized residual is a scaled version of the raw residual which has zero mean and unit variance. As a result, we expect that most of the standardized results will fall within the interval $-3 \leq d_i \leq 3$. The standardized residual vector is defined as

$$\mathbf{d} = \frac{\mathbf{e}}{\hat{\sigma}} \tag{14-12}$$

The estimate of the root-mean-square error, $\hat{\sigma}$, is

$$\hat{\sigma} = \sqrt{\frac{\sum_{i=1}^{n} (\hat{y}_i - \overline{y})^2}{n - (k+1)}} \tag{14-13}$$

where $\overline{y}$ is the mean of the observed responses.

The numerator of equation (14-13) is the error sum of squares, while the denominator is the associated degrees of freedom:

$$\mathrm{SS}_{\mathrm{error}} = \sum_{i=0}^{n} (\hat{y}_i - \overline{y})^2 = \mathbf{y}^{\mathrm{T}}\mathbf{y} - \mathbf{b}^{\mathrm{T}}\mathbf{x}^{\mathrm{T}}\mathbf{y} \tag{14-14}$$

$$\mathrm{dF}_{\mathrm{error}} = n - (k+1) \tag{14-15}$$

For the model to fit the observed data, we expect the residuals to be distributed normally with zero mean and to fall within ± 3 standard deviations.

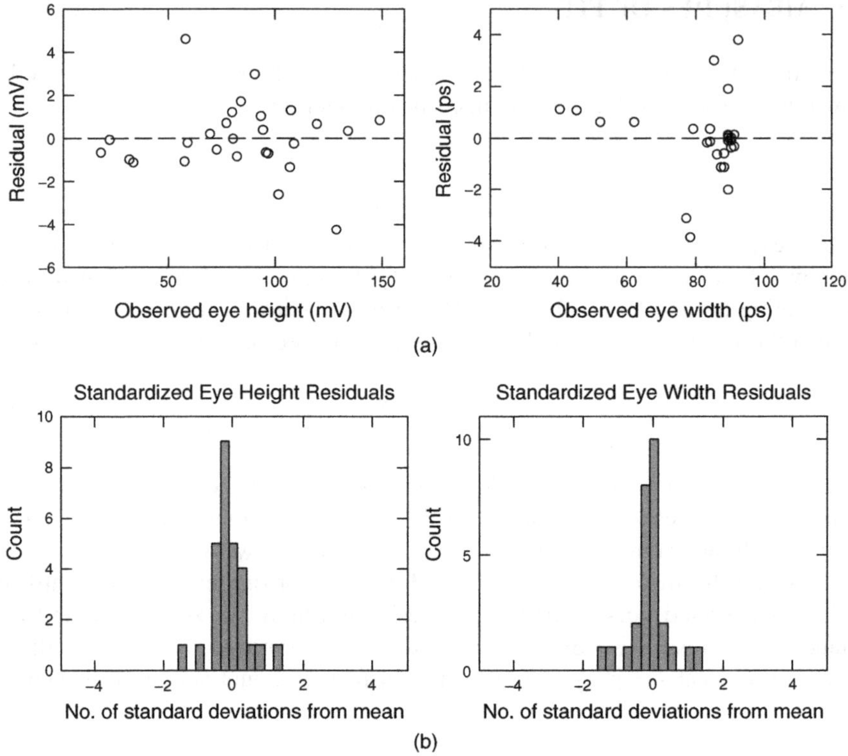

Figure 14-3 Residuals for the response surface model: (a) scatterplots of residuals (e_H, e_W) versus actuals (y_H, y_W); (b) standardized residual histograms (d_H, d_W).

Case Study Application The residuals for our model are summarized in columns 11 and 12 of Table 14-3 and are plotted against the observed responses in Figure 14-3a, while the histograms of standardized residuals are shown in Figure 14-3b. The residual scatterplot shows no apparent correlation to the responses observed, and the histogram meets our expectation that it be normally distributed with zero mean and residuals that lie within 3 standard deviations of the mean. So the residuals give us our first indication that we have a good model.

14.4.2 Fit Coefficients

A common metric for the quality of model fit is the coefficient of multiple determination, R^2. It describes the amount of variability of the response that is explained by the model, and is equal to the ratio of the model sum of squares to the total sum of squares:

$$R^2 = \frac{SS_{\text{model}}}{SS_{\text{total}}} = 1 - \frac{SS_{\text{error}}}{SS_{\text{total}}} \qquad (14\text{-}16)$$

where SS_{model} is the sum of squares of the model, SS_{error} the sum of squares of the residual, and SS_{total} the total sum of squares.

Since the response of the system is determined by the combined effect of the model and the error in estimation, we know that the sum of squares for the system (SS_{total}) is made up of the model sum of squares and the error sum of squares. These quantities are expressed as

$$SS_{total} = \mathbf{y}^T\mathbf{y} - \frac{\left(\sum_{i=1}^{n} y_i\right)^2}{n} \tag{14-17}$$

$$SS_{model} = SS_{total} - SS_{error} = \mathbf{b}^T\mathbf{x}^T\mathbf{y} - \frac{\left(\sum_{i=1}^{n} y_i\right)^2}{n} \tag{14-18}$$

Equation (14-16) suggests that a perfect model fit will have $SS_{error} = 0$, resulting in $R^2 = 1$. Although a near-unity value of R^2 is a good sign, it does not guarantee that the model is good. This is because adding a term to the model will always increase R^2, even if the term is not statistically significant (we discuss significance testing in Section 14.5). An additional correlation metric that is frequently employed is the adjusted coefficient

$$R^2_{adj} = \frac{SS_{error}/dF_{error}}{SS_{total}/dF_{total}} = 1 - \frac{n-1}{n-k-1}(1 - R^2) \tag{14-19}$$

where

$$dF_{total} = n - 1 \tag{14-20}$$

We can use (14-15) and (14-20) to calculate the number of degrees of freedom of the model as

$$dF_{model} = k - 1 \tag{14-21}$$

The inclusion of the degrees of freedom of the system and the error means that the adjusted R^2 statistic can decrease if an insignificant term is added to the model. Thus, dramatically different values of R^2 and R^2_{adj} indicate that the model may contain unnecessary terms.

For signaling systems, we recommend using $R^2 \geq 0.95$ and $R^2_{adj} \geq 0.90$ as fit criteria. Systems that contain significant nonlinearity in response due to physical effects such as reflections may have difficulty in meeting these criteria, requiring either adjustment to the criteria or adjustments to the model.

A final check that we often use is to check that the root-mean-square error (RMSE):

$$RMSE = \sqrt{\frac{SS_{error}}{dF_{error}}} \tag{14-22}$$

is less than one-tenth of the difference between the minimum and maximum responses observed.

TABLE 14-5. Summary of Model Fit Results

	Eye Height	Eye Width
SS_{error}	71.823	62.62
SS_{model}	28,435.558	5344.06
SS_{total}	28,507.381	5406.68
R^2	0.9975	0.9884
R^2_{adj}	0.9899	0.9537
RMSE	3.203	2.991
Range	130.97	52.00

Case Study Application For our example the degrees of freedom are $dF_{total} = 27$ and $dF_{error} = 7$. The coefficient calculations are shown in Table 14-5, including the sum-of-squares terms. We can calculate the RMSE metrics from the sum-of-squares values in Table 14-5 and extract the ranges of observed responses from Table 14-3. The RMSE values are 3.203 mV and 2.991 ps for eye height and width, respectively, while the corresponding ranges from Table 14-3 are 130.97 mV and 52.00 ps. The calculated values of R^2, R^2_{adj}, and RMSE for our example system all meet the fit criteria for both eye height and width.

14.5 SIGNIFICANCE TESTING

Thus far we have shown that our response surface model fits the data. A good fit does not, however, guarantee that the model has a significant impact on the response of the system, nor does it indicate that all of the model terms are statistically significant. Drawing conclusions about the significance of the model is a matter of determining whether or not the model equation is meaningful compared with the error. Stated in simple terms, the significance of a model term determines whether or not the term can be removed from the model without degrading the results. Answering these questions requires the use of tests for the significance of the model (F-test) and of the individual model coefficients (t-test).

14.5.1 Model Significance: The F-Test

To test the significance of the regression, we first calculate the F-ratio:

$$F_0 = \frac{SS_{model}/dF_{model}}{SS_{error}/dF_{error}} \tag{14-23}$$

The F-ratio is used to determine whether or not any of the model inputs contribute significantly to the model. In other words, the test for significance of regression evaluates whether or not there is a linear relationship between the response and some subset of the model terms. We do this by using F_0 in a hypothesis test:

$$\begin{aligned} &H_0\colon\ \beta_i = 0 \qquad \text{for all } i \\ &H_1\colon\ \beta_i \neq 0 \qquad \text{for at least one } i \end{aligned} \tag{14-24}$$

In equation (14-24), H_0 is the *null hypothesis*, which states that none of the model terms are significant since their fit coefficients are all statistically indistinguishable from zero. The null hypothesis is essentially a statement that the model is not useful because the response of the model cannot be distinguished from the noise. The *alternative hypothesis*, H_1, is that at least one of the coefficients is nonzero. Rejection of the null hypothesis means that at least one of the terms in the model has a significant contribution to the model response.

We reject the null hypothesis if the F_0 that we calculate for our model meets the requirement in equation (14-25), which indicates that the component of variance from the model is significantly greater than the component of variance from the error:

$$F_0 > F_{\alpha,k,n-k-1} \tag{14-25}$$

where $F_{\alpha,k,n-k-1}$ is the critical value for the F-distribution, $\alpha = 1 - \text{conf} = 1$ minus the desired confidence level, $k = \text{df}_{\text{model}} =$ the model degrees of freedom, and $n - k - 1 = \text{df}_{\text{error}} =$ the error degrees of freedom. For high-speed signaling systems we often want a 95% confidence level, so we typically set $\alpha = 0.05$.

An alternative, perhaps more intuitive way to look at the test for significance of regression is that rejection of the null hypothesis tells us that the response from the model is determined largely by the model, rather than by the error, at a given level of confidence. This is exactly what we want for the model to be useful.

Case Study Application The values for SS_{model}, SS_{error}, F_0, and $F_{0.5,20,7}$ are listed in Table 14-6. The value of $F_{0.05,20,7}$ is obtained from Appendix C. The F-ratio values far exceed the critical value for both eye height and eye width. As a result, we reject H_0 for both eye width and eye height and conclude that at least one of the model terms contributes significantly to the response with 95% confidence. Before moving on to discuss parameter significance, we summarize the model fit and significance metrics and criteria in Table 14-7.

14.5.2 Parameter Significance: Individual *t*-Tests

The methods described in previous sections provide measures of how well the model fits the data, which is important if we are to have high confidence in the predictions that we make using the model. However, to optimize a system through use of the model, we also need a method for determining which terms

TABLE 14-6. F-Test Results for the Model Fit Example

		Eye Height			Eye Width		
Source	dF	Sum of Squares	Mean Square	F-Ratio	Sum of Squares	Mean Square	F-Ratio
Model	20	28,435.56	1,421.78	138.568	5,344.06	267.203	29.869
Error	7	71.823	10.26		62.62	8.946	
Total	27	28,507.381			5,406.679		
		$F_{0.05,20,7} = 3.445$					

TABLE 14-7. Summary of Model Fit and Significance Criteria

Metric	Criteria
R^2	≥ 0.95
R^2_{adj}	≥ 0.90
RMSE	$<$ Range(y)/10
Residuals	Normally distributed Mean $\cong 0$ Residuals within mean ± 3 standard deviations
F_0	$\geq F_{\alpha,k,n-k-1}$

in the model have a significant effect on the response. To so do, we again use a hypothesis test. In this case, the hypotheses that we test are

$$H_0\colon\ \beta_i = 0$$
$$H_1\colon\ \beta_i \neq 0 \tag{14-26}$$

We can perform the test on any or all of the coefficients in the model. In essence, failure to reject the null hypothesis for a given model term means that the term can be deleted without significantly degrading the predictive capability of the model. The t-statistic used to test the hypothesis is

$$t_{0i} = \frac{b_i}{\sqrt{\hat{\sigma}^2 C_{ii}}} = \frac{b_i}{\text{SE}_i} \tag{14-27}$$

where t_{0i} is the t-statistic for the ith term in the model; b_i is the estimated fit coefficient for the ith term; C_{ii} is the diagonal element of the covariance matrix $(\mathbf{X}^{\text{T}}\mathbf{X})^{-1}$, which corresponds to b_i; and $\hat{\sigma}^2$ is the estimated variance of the model error, which is equal to the ratio of the sum of squares of the error and its associate degrees of freedom:

$$\hat{\sigma}^2 = \frac{\text{SS}_{\text{error}}}{\text{dF}_{\text{error}}} \tag{14-28}$$

The denominator of equation (14-27) is also known as the *standard error of regression* (SE_i) for coefficient b_i. So the t-statistic is the ratio of the fit coefficient to the standard error for a given model term.

For this test we reject the null hypothesis for coefficient b_i if

$$|t_{0i}| > t_{\alpha/2,n-(k+1)} \tag{14-29}$$

where $t_{\alpha/2,n-k-1}$ is the critical value for the t-distribution at a confidence level $1-\alpha$ with $n-k-1$ degrees of freedom.

In essence, we are testing whether or not the fit coefficient, the numerator of (14-27), has a greater effect on the model response than does the error (the denominator) introduced by a given model term. If it does, we reject the null hypothesis and conclude that the term has a significant effect. If not, we conclude that the term does not have a significant effect and can exclude it from the model. Appendix D provides critical values for the t-distribution at various confidence levels and degrees of freedom.

Case Study Application Hypothesis testing involves calculating the individual t-statistic for 21 fit coefficients each for the eye height and eye width, the details of which are omitted. (Clearly, this suggests the use of a computer tool to automate the process!) We have actually already calculated the values for $\hat{\sigma}^2$ in Section 14.5.1. They are the mean-squared-error terms in Table 14-6, which have values of 10.260 and 8.946 for the eye height and width, respectively. Table 14-8 summarizes the calculations for the individual t-statistics. The diagonal elements of the covariance matrix $\mathbf{C}_{i,i}$, listed in the second column of the table, are used along with the error variance terms to calculate the standard error for each term. The t-ratios are calculated from the estimated fit coefficients and standard errors according to equation (14-27).

At a 95% confidence level, the critical value for the t-statistic is 2.365 obtained from Appendix D. Results of the hypothesis tests are listed in the rightmost columns of the table, which show that 11 of the 20 model terms are significant for the eye height model, while seven terms are significant for the eye width model. All other terms can be removed from the model without significant degradation in the model fit (see Problems 14-1 and 14-2).

14.6 CONFIDENCE INTERVALS

Our model allows us to estimate the predicted response to a given set of input conditions. In creating the model, we have assumed that the errors are random, uncorrelated, and fit a normal distribution. It follows that any prediction $\hat{y}$ will also be a random variable that has an associated probability distribution. In fact, the value predicted is the mean of a probability distribution that is determined by the experiment design, the mean-squared error, and the set of input values for which we are predicting the response.

TABLE 14-8. Summary of Significance Test Results for Individual Model Terms

Model Term	Covariance, $C_{i,I}$	Fit Coefficient		Standard Error		t-Ratio		Reject H_0?	
		Eye Height	Eye Width	Eye Height	Eye Width	Eye Height	Eye Width	Eye Height	Eye Width
Intercept	0.12130	95.51799	89.10651	1.116	1.042	**85.58959**	**85.51488**	✓	✓
R_{Tx}	0.05556	2.32444	−1.94444	0.755	0.705	**3.07873**	**−2.75808**	✓	✓
Z_{diff}	0.05556	0.12056	0.55556	0.755	0.705	0.15968	0.78802		
R_{TT}	0.05556	9.55222	−0.44444	0.755	0.705	**12.65195**	−0.63042	✓	
L	0.05556	−30.81556	−10.83333	0.755	0.705	**−40.81531**	**−15.36643**	✓	✓
EQ	0.05556	−0.28778	−7.05556	0.755	0.705	−0.38116	**−10.00788**		✓
$R_{Tx} \cdot R_{Tx}$	0.40828	0.18752	−0.13314	2.047	1.911	0.09160	−0.06967		
$R_{Tx} \cdot Z_{diff}$	0.06250	−2.18063	−0.56250	0.801	0.748	**−2.72238**	−0.75201	✓	
$Z_{diff} \cdot Z_{diff}$	0.40828	−1.19749	0.36686	2.047	1.911	−0.58500	0.19197		
$R_{Tx} \cdot R_{TT}$	0.06250	1.26063	1.18750	0.801	0.748	1.57381	1.58757		
$Z_{diff} \cdot R_{TT}$	0.06250	−2.43813	−1.56250	0.801	0.748	**−3.04385**	−2.08890	✓	
$R_{TT} \cdot R_{TT}$	0.40828	−0.44249	0.36686	2.047	1.911	−0.21616	0.19197		
$R_{Tx} \cdot L$	0.06250	−3.88563	−1.93750	0.801	0.748	**−4.85097**	**−2.59024**	✓	✓
$Z_{diff} \cdot L$	0.06250	−0.02938	0.56250	0.801	0.748	−0.03667	0.75201		
$R_{TT} \cdot L$	0.06250	−4.43813	0.31250	0.801	0.748	**−5.54073**	0.41778	✓	
$L \cdot L$	0.40828	−2.97249	−4.13314	2.047	1.911	−1.45212	−2.16281		
$R_{Tx} \cdot$ EQ	0.06250	−1.77438	−1.56250	0.801	0.748	−2.21520	−2.08890		
$Z_{diff} \cdot$ EQ	0.06250	0.22938	1.43750	0.801	0.748	0.28636	1.92179		
$R_{TT} \cdot$ EQ	0.06250	−2.11688	−0.56250	0.801	0.748	**−2.64279**	−0.75201	✓	
$L \cdot$ EQ	0.06250	−20.42063	−8.93750	0.801	0.748	**−25.49391**	**−11.94853**	✓	✓
EQ · EQ	0.40828	−14.61249	−8.13314	2.047	1.911	**−7.13849**	**−4.25596**	✓	✓

As a result, we must account for the error sources when using the model to make decisions based on the responses predicted. The way in which we do so is to determine a confidence interval around the response, which is estimated from the t- distribution at a specified confidence level $(1 - \alpha)$ and the number of degrees of freedom associated with the error. In simple terms, a 95% confidence interval accounts for 95% of the statistical variation in the prediction. Intuitively, requiring increased confidence that a prediction falls within the calculated confidence interval means that the width of the interval must also increase, so that it accounts for more of the statistical variability.

Since the confidence interval depends on the input values, we start by constructing the input vector from the independent variables using the method outlined in Section 14.3:

$$\mathbf{x_{in}} = |1, x_1', x_2', \ldots, x_k'| \tag{14-30}$$

The predicted value $\hat{y}$ is

$$\hat{y} = \mathbf{x_{in}} \cdot \mathbf{b} \tag{14-31}$$

We calculate the confident interval $\mathrm{CI}_{\hat{y}}$ from the t-distribution, estimated error variance, input vector, and the covariance matrix:

$$\mathrm{CI}_{\hat{y}} = t_{\alpha/2,n-k-1}\sqrt{\hat{\sigma}^2[1 + \mathbf{x}_{\mathbf{in}}^{\mathrm{T}}(\mathbf{X}^{\mathrm{T}}\mathbf{X})^{-1}\mathbf{x_{in}}]} \tag{14-32}$$

In equation (14-32), the term under the square root is the standard error of prediction. At a given confidence level, the predicted response has a confidence interval of

$$\hat{y} - \mathrm{CI}_{\hat{y}_i} \le y \le \hat{y} + \mathrm{CI}_{\hat{y}_i} \tag{14-33}$$

We can also calculate confidence intervals on the mean response for the experimental observations using

$$\mathrm{CI}_{\overline{y}} = t_{\alpha/2,n-k-1}\sqrt{\hat{\sigma}^2[\mathbf{x}_{\mathbf{in}}^{\mathrm{T}}(\mathbf{X}^{\mathrm{T}}\mathbf{X})^{-1}\mathbf{x_{in}}]} \tag{14-34}$$

The confidence interval around the response observed is smaller than those around the responses predicted. Figure 14-2 includes confidence intervals calculated with (14-34) for our example signaling system.

14.7 SENSITIVITY ANALYSIS AND DESIGN OPTIMIZATION

As stated in Section 14.1, the power of the response surface model is that it provides a tool for understanding the behavior of a complicated system so that we can adjust the design to create a working solution. This involves identifying the factors that have the greatest influence on system performance and adjusting them, within the limits of the manufacturing capability, to maximize the robustness of the design. Developing an understanding of the sensitivities suggest that we make use of graphical techniques wherever possible. We develop several of them in this section and use them to help us to identify a working solution for our example system.

Prediction Profiles The JMP software package includes a prediction profiler that shows how the responses predicted change as each factor is varied while holding the others constant. Figure 14-4 shows an implementation of the prediction profile capability. The figure depicts the response trends when the input parameters are set at nominal values ($R_{TT} = 50\ \Omega$, $Z_{\text{diff}} = 100\ \Omega$, $R_{TT} = 50\ \Omega$, $L = 0.381$ m, and EQ $= -0.20$. The figure indicates that the eye height increases in a roughly linear fashion as a function of R_{Tx} and R_{TT} and decreases linearly as the length increases. It shows nonlinear dependence on Z_{diff} and the equalization setting (justifying our choice of a second-order model). The eye width plots also show linear dependence on R_{Tx} and R_{TT} (although with opposite slopes), and nonlinear variation with Z_{diff}. Another feature to notice from the figure is that the confidence intervals tend to get wider as we move toward the edges of the model space.

Now that we have an initial feeling for the behavior of the system, our next step is to look for our "worst case." Since we have two responses, the height and width of the eye, we can very easily end up with different conditions that cause the worst-case response for each. Since both eye width and height show strong dependence on the differential trace length, and since we know that some designs will require the maximum length, we start by setting the length to 0.508 m in the model. At this condition the model predicts 61.7 mV and 74.1 ps for height and width.

Next, we see that R_{TT} has a strong influence on eye height but affects eye width only weakly. Setting $R_{TT} = 40\ \Omega$ further degrades the eye height to 56 mV. Continuing in this way, we end up with a worst-case condition that has $R_{TT} = 40\ \Omega$, $R_{\text{Tx}} = 40\ \Omega$, and $Z_{\text{diff}} = 84\ \Omega$ for a 0.508-m-long system, resulting in eye height and width of 53 mV and 76 ps, respectively. At this corner, the system exhibits a 7-mV violation of the 60-mV minimum eye height spec.

However, we can improve the eye height by adjusting the equalization coefficient. So far, we have set our model conditions to their worst-case values either because of design requirement (trace length) or because they are variables that contain natural sources of variation in manufacturing (R_{TT}, R_{Tx}, Z_{diff}). Equalization, however, is a design feature that we control and can thus use to optimize the system response. After some trial and error, we can find that we maximize the eye height by setting the equalization coefficient to a value of -0.26 (see Figure 14-5). At this condition, the worst-case eye height that we predict is equal to 58.3 mV, which still does not meet the spec.

From Table 14-2 we see that -0.26 exactly matches a valid equalization setting. In addition, we allocate ± 1 bit of error to accommodate inaccuracies cause by nonideal effects in the equalizer. As a result, we must make sure that we can meet our defect rate requirements over an equalization range of -0.2467

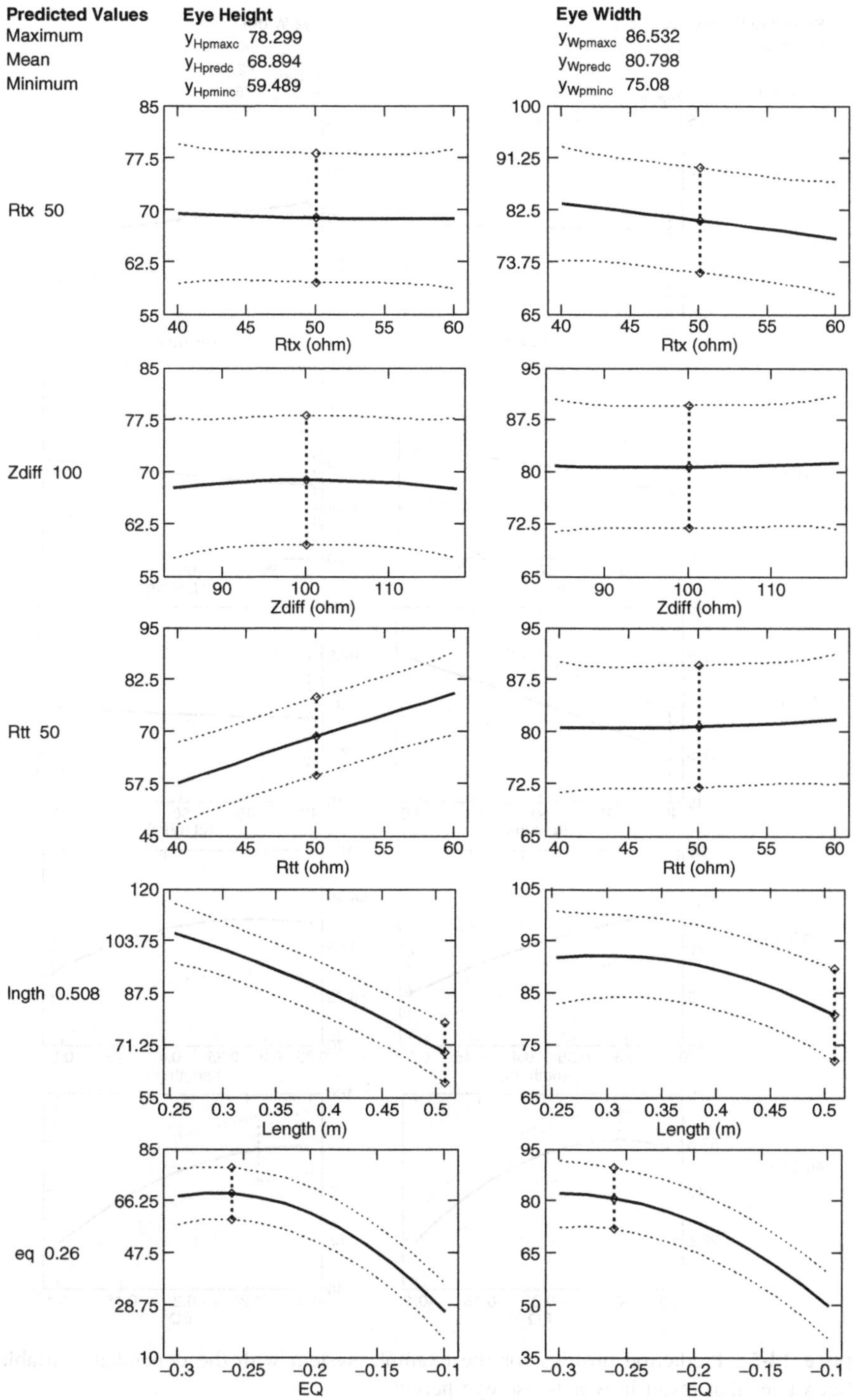

Figure 14-4 Prediction profiles for the example system with all dependent variables set at their nominal values.

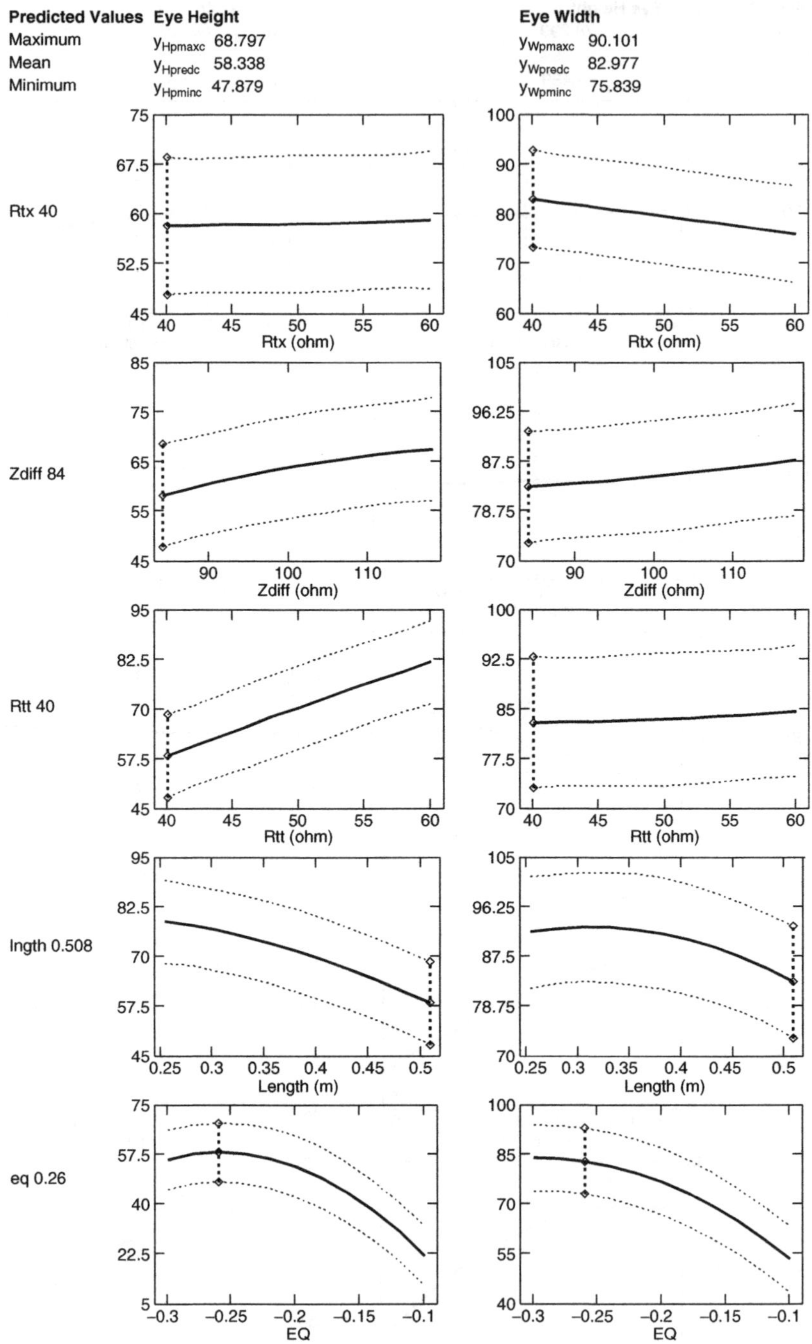

Figure 14-5 Prediction profiles for the example system with the dependent variables set at values that result in worst-case eye height.

to −0.2733. Repeating the process to find the worst-case corner for eye width, we find that the worst-case corner is $R_{TT} = 60\ \Omega$, $R_{\text{Tx}} = 40\ \Omega$, and $Z_{\text{diff}} = 84\ \Omega$ for the 0.508-m length. At an equalization setting of −0.2467, the worst-case eye width is greater than 74 ps, which meets the spec.

Before moving on, we note that an alternative approach to finding the worst-case corner would be to look at the worst-case observations from the original experiment. They will often provide insight into the system trends and response sensitivities.

Contour Plots Another technique for visualizing the behavior of the system that provides insight into the interactions between variables is the contour plot. We show an example in Figure 14-6, in which the impedances are set at the worst-case values that we determined previously. As the figure shows, there is a clear interaction between the equalization setting and the length of the system in determining the eye height. The longer the trace length, the more equalization we need in order to maximize the eye height. (This is, of course, what we expect for a lossy transmission-line pair.) The contour plot confirms our conclusion that we want the equalization value to be set at −0.26 for the maximum-length case. It also shows that even though a setting of −0.26 is not optimum for shorter lengths, the eye widths still improve as the length decreases.

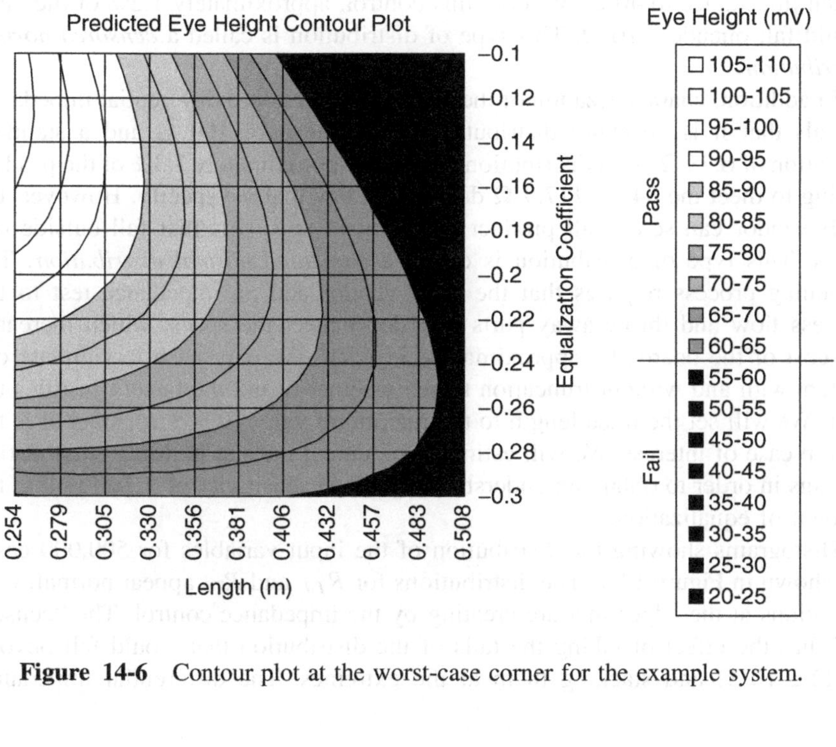

Figure 14-6 Contour plot at the worst-case corner for the example system.

14.8 DEFECT RATE PREDICTION USING MONTE CARLO SIMULATION

At this point we have identified the worst-case corners and the optimum equalizer settings. We have also seen that our worst-case corner does not meet the eye height spec, which means that we do not yet have a working design. However, we can use our response surface model to evaluate whether or not our design can meet the 1000-ppm defect rate target through the *Monte Carlo simulation* method. In Monte Carlo simulation, we generate random samples that adhere to the input sample distributions for a large number of cases of each input variable. To do so, we require knowledge of the probability distributions of the input variables. We then use the fitted RSM equations to calculate the predicted output responses. By doing so for a large number of samples (we use 500,000 for our examples), we generate distributions for the responses from which we can then estimate the defect rates.

For our example system, the circuits that provide the on-die termination for both transmitter and receiver are identical designs with built-in impedance control features that limit the variation to $\pm 10\ \Omega$. Characterization of the impedance variability due to process shows that it is normally distributed with a 50-Ω mean and a 4-Ω standard deviation. The way that the impedance control circuit works is to "clamp" the impedance of any circuit that would exceed the limits to the corresponding limiting value. For example, if the termination impedance that occurs for a given part due to natural variation were 38 Ω, the circuit will set the actual impedance to 40 Ω. Without this control, approximately 1.2% of the parts would fall outside $\pm 10\ \Omega$. This type of distribution is called a *censored normal distribution*.

In addition, characterization of the printed circuit board differential impedance reveals that it is normally distributed with a mean of 100 Ω and a standard deviation of 6.78 Ω. This distribution results in approximately 1.3% of the product failing to meet the 84- to 117.9-Ω design window that we specify. However, the PCB vendor can screen all product and remove any parts that fall outside our specs. This type of distribution is called a *truncated normal distribution*. The screening process requires that the PCB vendor add an impedance test to the process flow and throw away parts that don't meet the specs, which increases the cost of the boards by approximately $1 each. We may wish to simulate our system with and without truncation to see whether or not the benefit justifies the cost. We will set the trace length to the maximum value (0.508 m), since it is the design case of interest. We will estimate the defect rates at multiple equalization settings in order to obtain an understanding of the sensitivity of defect rate to the amount of equalization.

Histograms showing the distribution of the input variables for 500,000 cases are shown in Figure 14-7. The distributions for R_{TT} and R_{Tx} appear normal, with deviations at the edges that are creating by the impedance control. The "censoring" has the effect of taking the tails of the distribution that would fall beyond $50\ \Omega \pm 10\ \Omega$ and stacking them at the extremes. The differential impedance distribution also appears normal, except that it is truncated so that no values fall below 84 Ω or above 117.9 Ω. This is because our PCB vendor has implemented a "screen" to ensure that their boards comply with our specifications. The screen, which involves the use of time-domain reflectometry measurements, is a labor-intensive process. The extra processing step adds $1 to the cost of each board. To reduce our cost we would like to eliminate the screen, if possible. Later in our analysis we will use our RSM model to determine whether or not we can eliminate the screen and still meet our performance and defect rate targets.

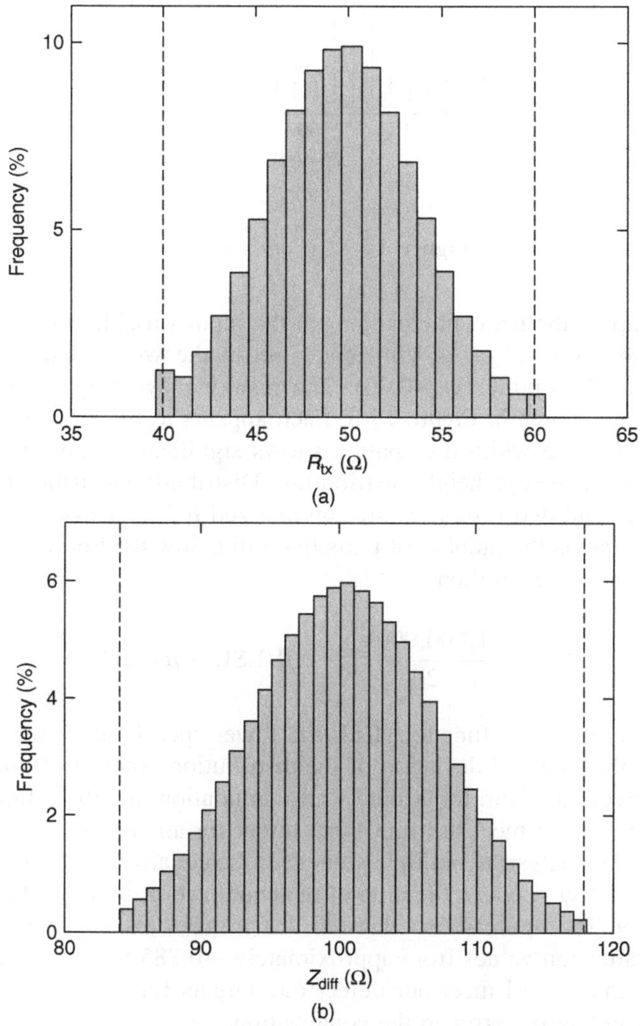

Figure 14-7 Input variable distributions from 500,000-case Monte Carlo simulation: (a) R_{Tx} (normal censored); (b) Z_{diff} (normal truncated); (c) R_{TT} (normal censored).(*Continued*)

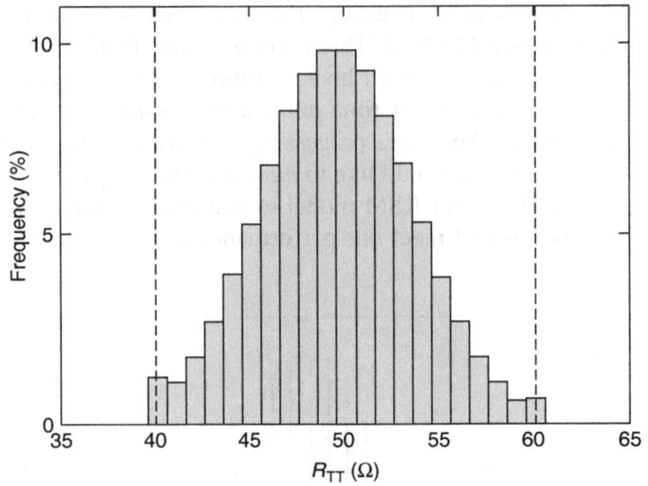

(c)

Figure 14-7 *(Continued)*

Next, we apply the RSM fit equations to the input variables that we generated for each of the 500,000 cases with length set at the worst case (0.508 m) and equalization at the optimum (−0.26). The resulting eye height and eye width distributions are shown in Figure 14-8. Each appears to be normally distributed. As expected, the eye width distribution shows significantly more margin in the spec than that for the eye height distribution. Distribution statistics for the three input variables and two responses are summarized in Table 14-9.

Our defect rate is the number of parts that fall below the lower spec limit. We estimate it in parts per million:

$$D_{\text{ppm}} = \frac{1{,}000{,}000}{2}(1 + \text{erf}[(\text{LSL} - \mu)/\sqrt{2}\sigma]) \tag{14-35}$$

where erf(x) is the error function, LSL the lower spec limit, μ the distribution mean, and σ the standard deviation of the distribution. Table 14-9 includes both the actual defect rate from the Monte Carlo simulation and the estimated defect rate, showing that we meet the 1000-ppm target for an equalizer.

Repeating the analysis at multiple values for the equalization coefficient allows us to plot the defect rate as a function of the equalization setting, which we show in Figure 14-9. The figure shows that our design has sub-1000 ppm fallout for a range of equalization values from approximately −0.285 to −0.245. As a result, we conclude that it will meet our defect rate targets for an equalization setting of −0.26, including the error in the equalization.

As a final step, we repeat the analysis without the impedance screening to assess our ability to eliminate the PCB impedance screen. Figure 14-10 shows that although the defect rates increase slightly, we can still meet the target without

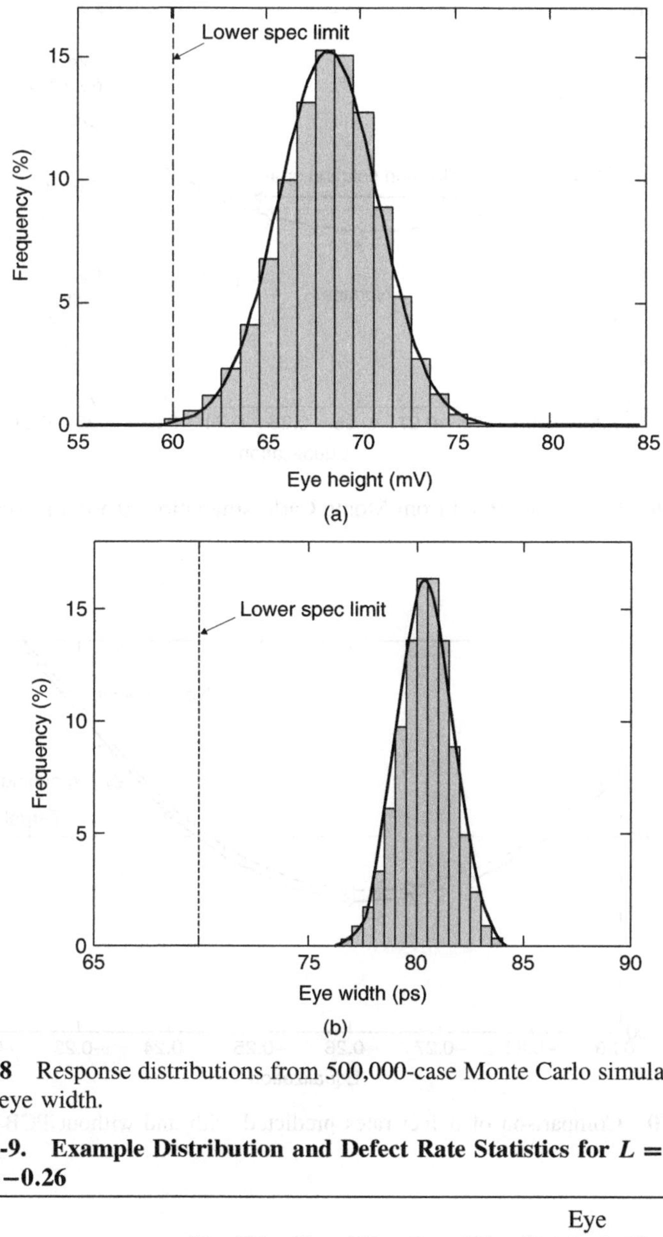

Figure 14-8 Response distributions from 500,000-case Monte Carlo simulation: (a) eye height; (b) eye width.

TABLE 14-9. Example Distribution and Defect Rate Statistics for $L = 0.508$ m and EQ = −0.26

	R_{Tx} (Ω)	Z_{diff} (Ω)	R_{TT} (Ω)	Eye Height (mV)	Eye Width (ps)
Mean	50.004	100.36	49.996	68.682	80.905
Standard deviation	3.959	6.46	3.961	2.361	1.212
Actual defect rate (ppm)				544	0
Estimated defect rate (ppm)				484	$< 10^{-10}$

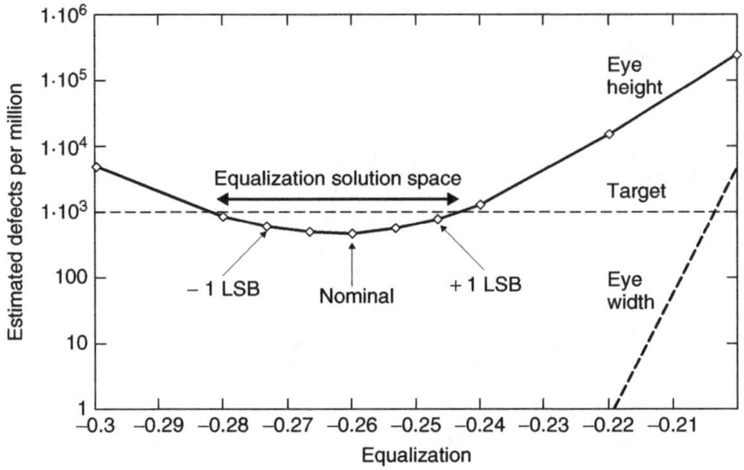

Figure 14-9 Defect-rate trend from Monte Carlo simulation (0.508-m trace length).

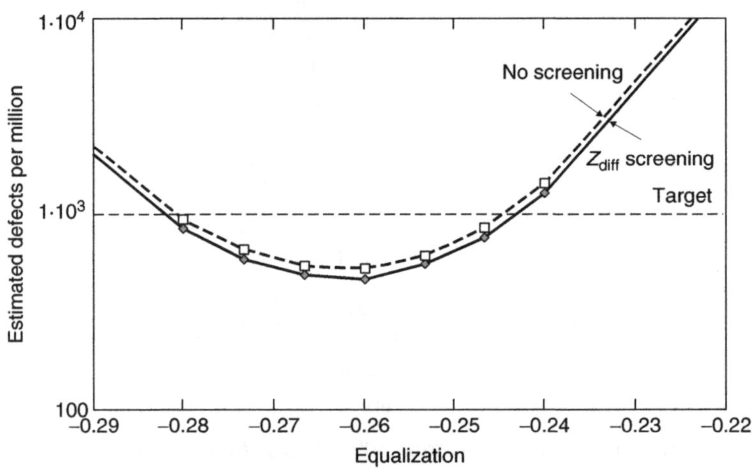

Figure 14-10 Comparison of defect rates predicted with and without PCB impedance screening.

screening, thus saving us $1 per system. If this seems like a small savings, consider that many systems sell hundreds of thousands or million of units (e.g., personal computers and laptops). A signal integrity engineer can therefore save his or her company millions of dollars per year by findings ways to reduce the cost of each system that he or she designs by $1 per system shipped.

14.9 ADDITIONAL RSM CONSIDERATIONS

As this chapter is intended to provide an introduction to the application of response surface modeling to high-speed signaling systems, our treatment does not attempt to cover all aspects of the technique. Before closing the chapter, we note some additional aspects of the modeling process that interested readers may find useful. These aspects are covered in detail by Myers and Montgomery [2005].

In practical applications the RSM construction process is often carried out in a number of steps. The process often starts with a screening experiment, in which a first-order experimental design is used to determine which input variables have a significant impact on the response. Subsequent experiments fit higher-order models using only the important variables. Finally, a more accurate model may be created, if necessary, by fitting over a smaller region of interest.

Response surface models also have additional statistics to describe:

Goodness of Fit Along with the metrics we presented, there are other ways of looking at the residuals, such as the prediction error sum of squares (PRESS).

Prediction Variance The confidence interval for a predicted value is a function of both the location of the input variables within the design space and of the model itself. The scaled prediction variance quantifies the variability across the design:

$$\overline{\text{Var}}[\hat{y}(\mathbf{x})] = N\mathbf{x}_0^{\mathrm{T}}(\mathbf{X}^{\mathrm{T}}\mathbf{X})^{-1}\mathbf{x}_0 \tag{14-36}$$

Experimental Error versus True Lack-of-Fit The error of any statistical model can be dissected into a component from *random noise* (pure error) and residuals from functional dependencies not accounted for by the model (lack of fit). The functional relationship between the error components is

$$\text{SS}_{\text{error}} = \text{SS}_{\text{pure error}} + \text{SS}_{\text{lack of fit}} \tag{14-37}$$

Our model is based on simulated results, which are repeatable, so that there is no experimental error and all of the error is attributed to lack of fit.

14.10 SUMMARY

In this chapter we presented the response surface model, a powerful tool that allows us to model the behavior of a high-speed signaling system and to use the model to optimize system cost and performance. Next we summarize the process for creating and using the response surface model in the design of a high-speed signaling link:

1. *Design the experiment.* Although we did not cover this step in any depth, it is where the process starts. A good design ensures the construction of a model that fits the data without requiring excessive numbers of observations. We recommend using a software package, such as JMP [SAS, 2007] or Essential Experimental Design [Steppan, et al., 1998] when designing experiments.
2. *Perform the regression.* This step creates the model, fit metrics, and significance parameters. It is also typically done via statistical software package, since high-speed systems may have 10 or more significant variables, which means more than 60 model terms.
3. *Check the model.* Before using the model, we need to make sure that it fits the data (residuals, R^2, R^2_{adj}), has a significant impact on the predicted results (F-test for significance of regression), and contains the appropriate variables (t-test for parameter significance).
4. *Use the model.* Use the prediction profile to develop a working solution by identifying the worst-case conditions for uncontrolled variables and the optimum settings for design parameters. If needed, apply Monte Carlo simulation to the model to estimate defect rates.

Limitations The RSM should only be used over the ranges of input variables to which the model was fit. Response surface models are essentially Taylor series approximations, and as such they are not guaranteed to fit outside the experimental range.

REFERENCES

Despite over 50 years of applications, statistical modeling and optimization of complex systems remains an area of active research. The list of references below includes several texts that present fundamental theory and concepts: Myers and Montgomery [1996], Steppan et al. [1998], Montgomery [2005]. The remaining references deal more with the application of RSM and alternatives (e.g., artificial neural networks, genetic algorithms). Papers by Beyene (2007), Hsu (2008), Matoglu (2004), and Mutnuray (2006) provide examples of statistical model fitting to signal integrity design and modeling. Norman (2003) gives an example of the application of the RSM methodology to the validation of high-speed signality systems.

Beyene, Wendem, 2007, Application of artificial neural networks to statistical analysis and nonlinear modeling of high-speed interconnect systems, *IEEE Transactions on Computer-Aided Design of Integrated Circuits and Systems*, vol. 26, no. 1, Jan.

Hsu, Ku-Ten, et al., 2008, Design of reflectionless vias using neural network-based approach, *IEEE Transactions on Advanced Packaging*, vol. 31, no. 1, Feb., pp. 211–218.

Matoglu, Erdem, Nam Pham, Daniel N. de Araujo, Moises Cases, and Madhavan Swaminathan, 2004, Statistical signal integrity analysis and diagnosis methodology for high-speed systems, *IEEE Transactions on Advanced Packaging*, vol. 27, no. 4, Nov., pp. 611–629.

Montgomery, Douglas C., 2005, *Design and Analysis of Experiments*, 6th ed., Wiley, Hoboken, NJ.

Mutnuray, Bhyrav, et al., 2006, Genetic algorithms with scalable I/O macromodels to find the worst case corner in high-speed server electrical analysis, *Proceedings of the 2006 IEEE Symposium on Electrical Performance of Electronic Packaging*, Oct., pp. 73–76.

Myers, Raymond H., and Douglas C. Montomery, 2002, *Response Surface Methodology*, 2nd ed., Wiley, Hoboken, NJ.

Norman, Adam, et al., 2003, Application of design of experiments (DOE) methods to high-speed interconnect validation, *Proceedings of the 12th IEEE Topical Meeting on Electrical Performance of Electronic Packaging*, Princeton, NJ, Oct., pp. 223–226.

SAS Institute, 2007, *JMP™ Release 7 Statistics and Graphics Guide*, SAS, Cary, NC.

Steppan, Dave, Joachim Werner, and Bob Yeater, 1998, *Essential Regression and Experimental Design*, http://www.jowerner.homepage.t-online.de/download.htm.

PROBLEMS

14-1 Re-fit the eye height response surface model presented in this chapter with nonsignificant parameters removed, and compare the results against the original model presented in the text.

14-2 Re-fit the eye width response surface model presented in this chapter with nonsignificant parameters removed, and compare the results against the original model presented in the text.

14-3 Create a response surface model for the differential impedance of a differential pair on a printed circuit using the cross section shown in Figure 14-11 and the data in Table 14-10.

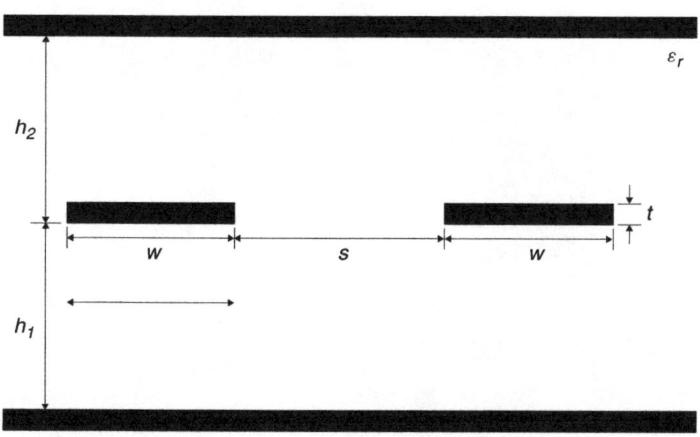

Figure 14-11 PCB cross section for Problem 14-3.

TABLE 14-10. PCB Geometrical Data for Problem 14-3

Parameter	Minimum	Typical	Maximum
h_1 (in.)	0.004	0.005	0.006
h_2 (in.)	0.004	0.005	0.006
W (in.)	0.004	0.005	0.006
S (in.)	0.006	0.008	0.010
T (in.)	0.0005	0.001	0.0015
ε_r	3.5	4.0	4.5

14-4 *Project:* Using suitable analysis software, carry out your own response surface model fitting project for a signaling system with more than five input variables.

14-5 *Project:* Develop a response surface modeling software tool using your choice of language (e.g., Matlab, Mathematic, Mathcad). Include the calculation of fit coefficients, measures of fit, confidence intervals, parameter significance testing.

14-6 *Project:* Develop a tool for modeling and optimizing signaling systems using the artificial neural network (ANN) approach, using the article by Beyene [2007] as a starting point.

APPENDIX A

USEFUL FORMULAS, IDENTITIES, UNITS, AND CONSTANTS

Constants

$$\text{Speed of light in a vacuum}: \quad c = 2.99792 \times 10^8 \text{ m/s}$$

$$\text{Permittivity of free space}: \quad \varepsilon_0 = 8.854 \times 10^{-12} \text{ F/m}$$

$$\text{Permeability of free space}: \quad \mu_0 = 4\pi \times 10^{-7} \text{ H/m}$$

Decibels

$$\text{Voltage:} \quad \text{dB} = 20 \log_{10} \frac{v_2}{v_1}$$

$$\text{Current:} \quad \text{dB} = 20 \log_{10} \frac{i_2}{i_1}$$

$$\text{Power:} \quad \text{dB} = 10 \log_{10} \frac{P_2}{P_1}$$

Units

Time

$$\text{picosecond (ps)} = 1 \times 10^{-12} \text{ s}$$

$$\text{nanosecond (ns)} = 1 \times 10^{-9} \text{ s}$$

Frequency

$$\text{megahertz (MHz)} = 1 \times 10^6\,\text{Hz}$$
$$\text{gigahertz (GHz)} = 1 \times 10^9\,\text{Hz}$$

Length

$$1\ \text{micron}\ (\mu\text{m}) = 1 \times 10^{-6}\,\text{m}$$
$$1\ \text{mil} = 1 \times 10^{-3}\,\text{in.}$$
$$1\ \text{mil} = 25.4\,\mu\text{m}$$
$$1\ \text{inch} = 0.0254\,\text{m}$$
$$1\ \text{millimeter (mm)} = 1 \times 10^{-3}\,\text{m}$$
$$1\ \text{millimeter} = 39.37\,\text{mils}$$

Circuits

$$\text{picohenry (pH)} = 1 \times 10^{-12}\,\text{H}$$
$$\text{nanohenry (nH)} = 1 \times 10^{-9}\,\text{H}$$
$$\text{picofarad (pf)} = 1 \times 10^{-12}\,\text{F}$$

Vector Formulas

$$\vec{A} \cdot (\vec{B} \times \vec{C}) = \vec{A} \cdot (\vec{C} \times \vec{A}) = \vec{C} \cdot (\vec{A} \times \vec{B})$$
$$\vec{A} \times (\vec{B} \times \vec{C}) = (\vec{A} \cdot \vec{C})\vec{B} - (\vec{A} \cdot \vec{B})\vec{C}$$
$$(\vec{A} \times \vec{B}) \cdot (\vec{C} \times \vec{D}) = (\vec{A} \cdot \vec{C})(\vec{B} \cdot \vec{D}) - (\vec{A} \cdot \vec{D})(\vec{B} \cdot \vec{C})$$
$$\nabla \times \nabla \psi = 0$$
$$\nabla \cdot (\nabla \times \vec{A}) = 0$$
$$\nabla \times (\nabla \times \vec{A}) = \nabla(\nabla \cdot \vec{A}) - \nabla^2 \vec{A}$$
$$\nabla \times (\psi \vec{A}) = \nabla \psi \times \vec{A} + \psi \nabla \times \vec{A}$$
$$\nabla(\vec{A} \cdot \vec{B}) = (\vec{A} \cdot \nabla)\vec{B} + (\vec{B} \cdot \nabla)\vec{A} + \vec{A} \times (\nabla \times \vec{B}) + \vec{B} \times (\nabla \times \vec{A})$$
$$\nabla \cdot (\vec{A} \times \vec{B}) = \vec{B} \cdot (\nabla \times \vec{A}) - \vec{A} \cdot (\nabla \times \vec{B})$$
$$\nabla \times (\vec{A} \times \vec{B}) = A(\nabla \cdot B) - B(\nabla \cdot A) + (B \cdot \nabla)A - (A \cdot \nabla)B$$
$$\nabla \cdot (\psi \vec{A}) = \vec{A} \cdot \nabla \psi + \psi \nabla \cdot \vec{A}$$

Theorems from Vector Calculus

$$\int_V (\nabla \cdot \vec{F})\, dV = \oint_S \vec{F} \cdot d\vec{s} \quad \text{(divergence theorem)}$$
$$\int_S (\nabla \times \vec{F}) \cdot d\vec{s} = \oint_l \vec{F} \cdot d\vec{l} \quad \text{(Stokes's theorem)}$$

Explicit Forms of Vector Operations

Cartesian (x, y, z)

$$\nabla f = \vec{a}_x \frac{\partial f}{\partial x} + \vec{a}_y \frac{\partial f}{\partial y} + \vec{a}_z \frac{\partial f}{\partial z}$$

$$\nabla \cdot \vec{F} = \frac{\partial F_x}{\partial x} + \frac{\partial F_y}{\partial y} + \frac{\partial F_z}{\partial z}$$

$$\nabla \times \vec{F} = \vec{a}_x \left(\frac{\partial F_z}{\partial y} - \frac{\partial F_y}{\partial z} \right) + \vec{a}_y \left(\frac{\partial F_x}{\partial z} - \frac{\partial F_z}{\partial x} \right) + \vec{a}_z \left(\frac{\partial F_y}{\partial x} - \frac{\partial F_x}{\partial y} \right)$$

$$\nabla^2 f = \frac{\partial^2 f}{\partial x^2} + \frac{\partial^2 f}{\partial y^2} + \frac{\partial^2 f}{\partial z^2}$$

Cylindrical (r, φ, z)

$$\nabla f = \vec{a}_r \frac{\partial f}{\partial r} + \vec{a}_\phi \frac{1}{r} \frac{\partial f}{\partial \phi} + \vec{a}_z \frac{\partial f}{\partial z}$$

$$\nabla \cdot \vec{F} = \frac{1}{r} \frac{\partial (r F_r)}{\partial r} + \frac{1}{r} \frac{\partial F_\phi}{\partial \phi} + \frac{\partial F_z}{\partial z}$$

$$\nabla \times \vec{F} = \vec{a}_r \left[\frac{1}{r} \frac{\partial F_z}{\partial \phi} - \frac{\partial F_\phi}{\partial z} \right] + \vec{a}_\phi \left[\frac{\partial F_r}{\partial z} - \frac{\partial F_z}{\partial r} \right] + \vec{a}_z \left[\frac{1}{r} \frac{\partial (r F_\phi)}{\partial r} - \frac{1}{r} \frac{\partial F_r}{\partial \phi} \right]$$

$$\nabla^2 f = \frac{1}{r} \frac{\partial}{\partial r} \left(r \frac{\partial f}{\partial r} \right) + \frac{1}{r^2} \frac{\partial^2 f}{\partial \phi^2} + \frac{\partial^2 f}{\partial z^2}$$

Spherical (r, θ, φ)

$$\nabla f = \vec{a}_r \frac{\partial f}{\partial r} + \vec{a}_\theta \frac{1}{r} \frac{\partial f}{\partial \theta} + \vec{a}_\phi \frac{1}{r \sin\theta} \frac{\partial f}{d\phi}$$

$$\nabla \cdot \vec{F} = \frac{1}{r^2} \frac{\partial (r^2 F_r)}{\partial r} + \frac{1}{r \sin\theta} \frac{\partial (\sin\theta F_\theta)}{\partial \theta} + \frac{1}{r \sin\theta} \frac{\partial F_\phi}{\partial \phi}$$

$$\nabla \times \vec{F} = \vec{a}_r \left[\frac{1}{r \sin\theta} \left(\frac{\partial \sin\theta F_\phi}{\partial \theta} - \frac{\partial F_\theta}{\partial \phi} \right) \right] + \vec{a}_\theta \left[\frac{1}{r \sin\theta} \frac{\partial F_r}{\partial \phi} - \frac{1}{r} \frac{\partial (r F_\phi)}{\partial r} \right]$$
$$+ \vec{a}_\phi \frac{1}{r} \left[\frac{\partial (r F_\theta)}{\partial r} - \frac{\partial F_r}{\partial \theta} \right]$$

$$\nabla^2 f = \frac{1}{r^2} \frac{\partial}{\partial r} \left(r^2 \frac{\partial f}{\partial r} \right) + \frac{1}{r^2 \sin\theta} \frac{\partial \left(\sin\theta (\partial f / \partial \theta) \right)}{\partial \theta} + \frac{1}{r^2 \sin^2\theta} \frac{\partial^2 f}{\partial \phi^2}$$

$$\frac{1}{r^2} \frac{\partial}{\partial r} \left(r^2 \frac{\partial f}{\partial r} \right) \equiv \frac{1}{r} \frac{\partial^2 (r f)}{\partial r^2}$$

Coordinate Transformations

Rectangular to Cylindrical

	$\vec{a}_x$	$\vec{a}_y$	$\vec{a}_z$
$\vec{a}_r$	$\cos\phi$	$\sin\phi$	0
$\vec{a}_\phi$	$-\sin\phi$	$\cos\phi$	0
$\vec{a}_z$	0	0	1

Rectangular to Spherical

	$\vec{a}_x$	$\vec{a}_y$	$\vec{a}_z$
$\vec{a}_r$	$\sin\theta\cos\phi$	$\sin\theta\sin\phi$	$\cos\theta$
$\vec{a}_\theta$	$\cos\theta\cos\phi$	$\cos\theta\sin\phi$	$-\sin\theta$
$\vec{a}_\phi$	$-\sin\phi$	$\cos\phi$	1

Cylindrical to Spherical

	$\vec{a}_r$	$\vec{a}_\phi$	$\vec{a}_z$
$\vec{a}_r$	$\sin\theta$	0	$\cos\theta$
$\vec{a}_\theta$	$\cos\theta$	0	$-\sin\theta$
$\vec{a}_\phi$	0	1	0

APPENDIX B

FOUR-PORT CONVERSIONS BETWEEN *T*-AND *S*-PARAMETERS

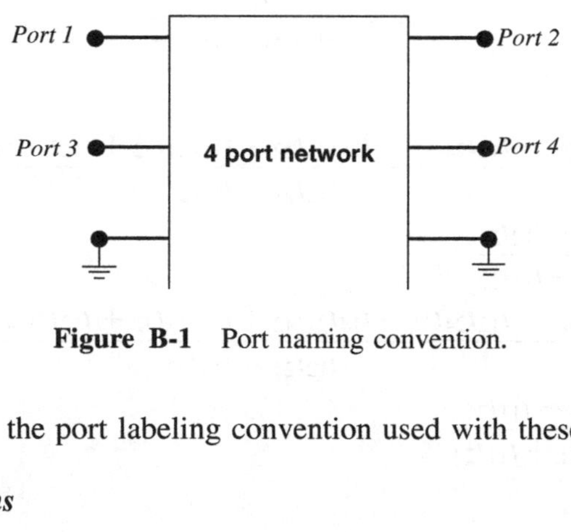

Figure B-1 Port naming convention.

Figure B-1 shows the port labeling convention used with these conversions.

T-to-S Conversions

$$s_{11} = \frac{t_{22}t_{31} - t_{21}t_{32}}{-t_{12}t_{21} + t_{11}t_{22}}$$

$$s_{21} = \frac{t_{22}}{-t_{12}t_{21} + t_{11}t_{22}}$$

$$s_{31} = \frac{t_{22}t_{41} - t_{21}t_{42}}{-t_{12}t_{21} + t_{11}t_{22}}$$

$$s_{41} = \frac{t_{21}}{t_{12}t_{21} - t_{11}t_{22}}$$

$$s_{12} = \frac{t_{13}t_{22}t_{31} - t_{12}t_{23}t_{31} - t_{13}t_{21}t_{32} + t_{11}t_{23}t_{32} + t_{12}t_{21}t_{33} - t_{11}t_{22}t_{33}}{t_{12}t_{21} - t_{11}t_{22}}$$

$$s_{22} = \frac{t_{13}t_{22} - t_{12}t_{23}}{t_{12}t_{21} - t_{11}t_{22}}$$

$$s_{32} = \frac{t_{13}t_{22}t_{41} - t_{12}t_{23}t_{41} - t_{13}t_{21}t_{42} + t_{11}t_{23}t_{42} + t_{12}t_{21}t_{43} - t_{11}t_{22}t_{43}}{t_{12}t_{21} - t_{11}t_{22}}$$

$$s_{42} = \frac{t_{13}t_{21} - t_{11}t_{23}}{-t_{12}t_{21} + t_{11}t_{22}}$$

$$s_{13} = \frac{t_{12}t_{31} - t_{11}t_{32}}{t_{12}t_{21} - t_{11}t_{22}}$$

$$s_{23} = \frac{t_{12}}{t_{12}t_{21} - t_{11}t_{22}}$$

$$s_{33} = \frac{t_{12}t_{41} - t_{11}t_{42}}{t_{12}t_{21} - t_{11}t_{22}}$$

$$s_{43} = \frac{t_{11}}{-t_{12}t_{21} + t_{11}t_{22}}$$

$$s_{14} = \frac{t_{14}t_{22}t_{31} - t_{12}t_{24}t_{31} - t_{14}t_{21}t_{32} + t_{11}t_{24}t_{32} + t_{12}t_{21}t_{34} - t_{11}t_{22}t_{34}}{t_{12}t_{21} - t_{11}t_{22}}$$

$$s_{24} = \frac{t_{14}t_{22} - t_{12}t_{24}}{t_{12}t_{21} - t_{11}t_{22}}$$

$$s_{34} = \frac{t_{14}t_{22}t_{41} - t_{12}t_{24}t_{41} - t_{14}t_{21}t_{42} + t_{11}t_{24}t_{42} + t_{12}t_{21}t_{44} - t_{11}t_{22}t_{44}}{t_{12}t_{21} - t_{11}t_{22}}$$

$$s_{44} = \frac{t_{14}t_{21} - t_{11}t_{24}}{-t_{12}t_{21} + t_{11}t_{22}}$$

S-to-T Conversions

$$t_{11} = \frac{s_{43}}{-s_{23}s_{41} + s_{21}s_{43}}$$

$$t_{21} = \frac{s_{41}}{s_{23}s_{41} - s_{21}s_{43}}$$

$$t_{31} = \frac{s_{13}s_{41} - s_{11}s_{43}}{s_{23}s_{41} - s_{21}s_{43}}$$

$$t_{41} = \frac{s_{33}s_{41} - s_{31}s_{43}}{s_{23}s_{41} - s_{21}s_{43}}$$

$$t_{12} = \frac{s_{23}}{s_{23}s_{41} - s_{21}s_{43}}$$
$$t_{22} = \frac{s_{21}}{-s_{23}s_{41} + s_{21}s_{43}}$$
$$t_{32} = \frac{s_{13}s_{21} - s_{11}s_{23}}{-s_{23}s_{41} + s_{21}s_{43}}$$
$$t_{42} = \frac{s_{23}s_{31} - s_{21}s_{33}}{s_{23}s_{41} - s_{21}s_{43}}$$

$$t_{13} = \frac{s_{23}s_{42} - s_{22}s_{43}}{-s_{23}s_{41} + s_{21}s_{43}}$$
$$t_{23} = \frac{s_{22}s_{41} - s_{21}s_{42}}{-s_{23}s_{41} + s_{21}s_{43}}$$
$$t_{33} = \frac{s_{13}s_{22}s_{41} - s_{12}s_{23}s_{41} - s_{13}s_{21}s_{42} + s_{11}s_{23}s_{42} + s_{12}s_{21}s_{43} - s_{11}s_{22}s_{43}}{-s_{23}s_{41} + s_{21}s_{43}}$$
$$t_{43} = \frac{s_{23}s_{32}s_{41} - s_{22}s_{33}s_{41} - s_{23}s_{31}s_{42} + s_{21}s_{33}s_{42} + s_{22}s_{31}s_{43} - s_{21}s_{32}s_{43}}{s_{23}s_{41} - s_{21}s_{43}}$$

$$t_{14} = \frac{s_{24}s_{43} - s_{23}s_{44}}{s_{23}s_{41} - s_{21}s_{43}}$$
$$t_{24} = \frac{s_{24}s_{41} - s_{21}s_{44}}{-s_{23}s_{41} + s_{21}s_{43}}$$
$$t_{34} = \frac{s_{14}s_{23}s_{41} - s_{13}s_{24}s_{41} - s_{14}s_{21}s_{43} + s_{11}s_{24}s_{43} + s_{13}s_{21}s_{44} - s_{11}s_{23}s_{44}}{s_{23}s_{41} - s_{21}s_{43}}$$
$$t_{44} = \frac{s_{24}s_{33}s_{41} - s_{23}s_{34}s_{41} - s_{24}s_{31}s_{43} + s_{21}s_{34}s_{43} + s_{23}s_{31}s_{44} - s_{21}s_{33}s_{44}}{-s_{23}s_{41} + s_{21}s_{43}}$$

APPENDIX C

CRITICAL VALUES OF THE *F*-STATISTIC

	dF1																		
$dF2$	1	2	3	4	5	6	7	8	9	10	12	15	20	24	30	40	60	120	∞
1	161.448	199.500	215.707	224.583	230.162	233.986	236.768	238.883	240.543	241.882	243.906	245.950	248.013	249.052	250.095	251.143	252.196	253.253	254.314
2	18.513	19.000	19.164	19.247	19.296	19.330	19.353	19.371	19.385	19.396	19.413	19.429	19.446	19.454	19.462	19.471	19.479	19.487	19.496
3	10.128	9.552	9.277	9.117	9.014	8.941	8.887	8.845	8.812	8.786	8.745	8.703	8.660	8.639	8.617	8.594	8.572	8.549	8.526
4	7.709	6.944	6.591	6.388	6.256	6.163	6.094	6.041	5.999	5.964	5.912	5.858	5.803	5.774	5.746	5.717	5.688	5.658	5.628
5	6.608	5.786	5.410	5.192	5.050	4.950	4.876	4.818	4.773	4.735	4.678	4.619	4.558	4.527	4.496	4.464	4.431	4.399	4.365
6	5.987	5.143	4.757	4.534	4.387	4.284	4.207	4.147	4.099	4.060	4.000	3.938	3.874	3.842	3.808	3.774	3.740	3.705	3.669
7	5.591	4.737	4.347	4.120	3.972	3.866	3.787	3.726	3.677	3.637	3.575	3.511	3.445	3.411	3.376	3.340	3.304	3.267	3.230
8	5.318	4.459	4.066	3.838	3.688	3.581	3.501	3.438	3.388	3.347	3.284	3.218	3.150	3.115	3.079	3.043	3.005	2.967	2.928
9	5.117	4.257	3.863	3.633	3.482	3.374	3.293	3.230	3.179	3.137	3.073	3.006	2.937	2.901	2.864	2.826	2.787	2.748	2.707
10	4.965	4.103	3.708	3.478	3.326	3.217	3.136	3.072	3.020	2.978	2.913	2.845	2.774	2.737	2.700	2.661	2.621	2.580	2.538
11	4.844	3.982	3.587	3.357	3.204	3.095	3.012	2.948	2.896	2.854	2.788	2.719	2.646	2.609	2.571	2.531	2.490	2.448	2.405
12	4.747	3.885	3.490	3.259	3.106	2.996	2.913	2.849	2.796	2.753	2.687	2.617	2.544	2.506	2.466	2.426	2.384	2.341	2.296
13	4.667	3.806	3.411	3.179	3.025	2.915	2.832	2.767	2.714	2.671	2.604	2.533	2.459	2.420	2.380	2.339	2.297	2.252	2.206
14	4.600	3.739	3.344	3.112	2.958	2.848	2.764	2.699	2.646	2.602	2.534	2.463	2.388	2.349	2.308	2.266	2.223	2.178	2.131
15	4.543	3.682	3.287	3.056	2.901	2.791	2.707	2.641	2.588	2.544	2.475	2.403	2.328	2.288	2.247	2.204	2.160	2.114	2.066
16	4.494	3.634	3.239	3.007	2.852	2.741	2.657	2.591	2.538	2.494	2.425	2.352	2.276	2.235	2.194	2.151	2.106	2.059	2.010
17	4.451	3.592	3.197	2.965	2.810	2.699	2.614	2.548	2.494	2.450	2.381	2.308	2.230	2.190	2.148	2.104	2.058	2.011	1.960
18	4.414	3.555	3.160	2.928	2.773	2.661	2.577	2.510	2.456	2.412	2.342	2.269	2.191	2.150	2.107	2.063	2.017	1.968	1.917
19	4.381	3.522	3.127	2.895	2.740	2.628	2.544	2.477	2.423	2.378	2.308	2.234	2.156	2.114	2.071	2.026	1.980	1.930	1.878
20	4.351	3.493	3.098	2.866	2.711	2.599	2.514	2.447	2.393	2.348	2.278	2.203	2.124	2.083	2.039	1.994	1.946	1.896	1.843
21	4.325	3.467	3.073	2.840	2.685	2.573	2.488	2.421	2.366	2.321	2.250	2.176	2.096	2.054	2.010	1.965	1.917	1.866	1.812
22	4.301	3.443	3.049	2.817	2.661	2.549	2.464	2.397	2.342	2.297	2.226	2.151	2.071	2.028	1.984	1.938	1.889	1.838	1.783
23	4.279	3.422	3.028	2.796	2.640	2.528	2.442	2.375	2.320	2.275	2.204	2.128	2.048	2.005	1.961	1.914	1.865	1.813	1.757
24	4.260	3.403	3.009	2.776	2.621	2.508	2.423	2.355	2.300	2.255	2.183	2.108	2.027	1.984	1.939	1.892	1.842	1.790	1.733
25	4.242	3.385	2.991	2.759	2.603	2.490	2.405	2.337	2.282	2.237	2.165	2.089	2.008	1.964	1.919	1.872	1.822	1.768	1.711
26	4.225	3.369	2.975	2.743	2.587	2.474	2.388	2.321	2.266	2.220	2.148	2.072	1.990	1.946	1.901	1.853	1.803	1.749	1.691
27	4.210	3.354	2.960	2.728	2.572	2.459	2.373	2.305	2.250	2.204	2.132	2.056	1.974	1.930	1.884	1.836	1.785	1.731	1.672
28	4.196	3.340	2.947	2.714	2.558	2.445	2.359	2.291	2.236	2.190	2.118	2.041	1.959	1.915	1.869	1.820	1.769	1.714	1.654
29	4.183	3.328	2.934	2.701	2.545	2.432	2.346	2.278	2.223	2.177	2.105	2.028	1.945	1.901	1.854	1.806	1.754	1.698	1.638
30	4.171	3.316	2.922	2.690	2.534	2.421	2.334	2.266	2.211	2.165	2.092	2.015	1.932	1.887	1.841	1.792	1.740	1.684	1.622
40	4.085	3.232	2.839	2.606	2.450	2.336	2.249	2.180	2.124	2.077	2.004	1.925	1.839	1.793	1.744	1.693	1.637	1.577	1.509
60	4.001	3.150	2.758	2.525	2.368	2.254	2.167	2.097	2.040	1.993	1.917	1.836	1.748	1.700	1.649	1.594	1.534	1.467	1.389
120	3.920	3.072	2.680	2.447	2.290	2.175	2.087	2.016	1.959	1.911	1.834	1.751	1.659	1.608	1.554	1.495	1.429	1.352	1.254
∞	3.842	2.996	2.605	2.372	2.214	2.099	2.010	1.938	1.880	1.831	1.752	1.666	1.571	1.517	1.459	1.394	1.318	1.221	1.000

Note: dF1, model degrees of freedom $= k$; dF2, error degrees of freedom $= n - k - 1$.

APPENDIX D

CRITICAL VALUES OF THE *T*-STATISTIC

df	0.4	0.25	0.1	0.05	0.025	0.01	0.005	0.0005
1	0.32492	1	3.077684	6.313752	12.7062	31.82052	63.65674	636.6192
2	0.288675	0.816497	1.885618	2.919986	4.30265	6.96456	9.92484	31.5991
3	0.276671	0.764892	1.637744	2.353363	3.18245	4.5407	5.84091	12.924
4	0.270722	0.740697	1.533206	2.131847	2.77645	3.74695	4.60409	8.6103
5	0.267181	0.726687	1.475884	2.015048	2.57058	3.36493	4.03214	6.8688
6	0.264835	0.717558	1.439756	1.94318	2.44691	3.14267	3.70743	5.9588
7	0.263167	0.711142	1.414924	1.894579	2.36462	2.99795	3.49948	5.4079
8	0.261921	0.706387	1.396815	1.859548	2.306	2.89646	3.35539	5.0413
9	0.260955	0.702722	1.383029	1.833113	2.26216	2.82144	3.24984	4.7809
10	0.260185	0.699812	1.372184	1.812461	2.22814	2.76377	3.16927	4.5869
11	0.259556	0.697445	1.36343	1.795885	2.20099	2.71808	3.10581	4.437
12	0.259033	0.695483	1.356217	1.782288	2.17881	2.681	3.05454	4.3178
13	0.258591	0.693829	1.350171	1.770933	2.16037	2.65031	3.01228	4.2208
14	0.258213	0.692417	1.34503	1.76131	2.14479	2.62449	2.97684	4.1405
15	0.257885	0.691197	1.340606	1.75305	2.13145	2.60248	2.94671	4.0728
16	0.257599	0.690132	1.336757	1.745884	2.11991	2.58349	2.92078	4.015

(*continued overleaf*)

(*Continued*)

df	0.4	0.25	0.1	0.05	0.025	0.01	0.005	0.0005
17	0.257347	0.689195	1.333379	1.739607	2.10982	2.56693	2.89823	3.9651
18	0.257123	0.688364	1.330391	1.734064	2.10092	2.55238	2.87844	3.9216
19	0.256923	0.687621	1.327728	1.729133	2.09302	2.53948	2.86093	3.8834
20	0.256743	0.686954	1.325341	1.724718	2.08596	2.52798	2.84534	3.8495
21	0.25658	0.686352	1.323188	1.720743	2.07961	2.51765	2.83136	3.8193
22	0.256432	0.685805	1.321237	1.717144	2.07387	2.50832	2.81876	3.7921
23	0.256297	0.685306	1.31946	1.713872	2.06866	2.49987	2.80734	3.7676
24	0.256173	0.68485	1.317836	1.710882	2.0639	2.49216	2.79694	3.7454
25	0.25606	0.68443	1.316345	1.708141	2.05954	2.48511	2.78744	3.7251
26	0.255955	0.684043	1.314972	1.705618	2.05553	2.47863	2.77871	3.7066
27	0.255858	0.683685	1.313703	1.703288	2.05183	2.47266	2.77068	3.6896
28	0.255768	0.683353	1.312527	1.701131	2.04841	2.46714	2.76326	3.6739
29	0.255684	0.683044	1.311434	1.699127	2.04523	2.46202	2.75639	3.6594
30	0.255605	0.682756	1.310415	1.697261	2.04227	2.45726	2.75	3.646
∞	0.253347	0.67449	1.281552	1.644854	1.95996	2.32635	2.57583	3.2905

APPENDIX E

CAUSAL RELATIONSHIP BETWEEN SKIN EFFECT RESISTANCE AND INTERNAL INDUCTANCE FOR ROUGH CONDUCTORS

To derive a causal relationship between the skin effect resistance calculated with rough copper and the internal inductance, we start with the equation for the surface impedance, which was derived in equation (5-29):

$$Z_s(\omega) = R_{\text{ac}}(\omega) + j\omega L_{\text{internal}}(\omega) \tag{5-29}$$

For a function to be causal, it must be zero for negative time, as described in Section 8.2.1.[†]. To be causal, the function must obey the relationship.

$$h(t) = h_e(t) + \text{sgn}(t) h_e(t) \tag{8-17}$$

From Section 8.2.1, equation (8-20) shows that the function $H(\omega)$ [which is the Fourier transform of $h(t)$] can be written in terms of only the even part:

$$H(\omega) = H_e(\omega) - j\hat{H}_e(\omega) \tag{8-20}$$

where$\hat{H}_e(\omega)$ is the Hilbert transform of the even part of the function.

[†]For practical purposes, it can also be said that a system is not causal if a portion of energy is traveling faster than the propagation velocity in the board allows $\left(v_p = c/\sqrt{\varepsilon_r}\right)$. This means that time t = 0 is defined by the length of the interconnect and the speed of light in the material.

Advanced Signal Integrity for High-Speed Digital Design, By Stephen H. Hall and Howard L. Heck

The similarity of (8-20) and (5-29) allows us to conclude that$H_e = R_{ac}$ and $-\hat{H}_e = \omega L_{\text{internal}}$. So, quite simply, the internal inductance for a causal relationship is

$$L_{\text{internal}} = -\frac{\hat{R}_{ac}}{\omega} \tag{E-1}$$

where the hat indicates the Hilbert transform of the skin resistance.

The Hilbert transform can be implemented in many ways. Some tools, such as Matlab, have a convenient Hilbert transform function. Otherwise, two possible ways of implementing the Hilbert transform are:

1. Solve the integral directly:

$$\hat{R}_{ac}(\omega) = \frac{1}{\pi}\int_{-\infty}^{\infty} \frac{R_{ac}(\omega')}{\omega - \omega'} d\omega'$$

2. Use the fast Fourier transform (FFT) to convolve $R_{ac}(\omega)$ and $1/\pi\omega$:

$$\hat{R}_{ac}(\omega) = \frac{1}{\pi\omega} * R_{ac}(\omega) = \text{FFT}^{-1}\left[\text{FFT}\left\{\frac{1}{\pi\omega}\right\} \cdot \text{FFT}\{R_{ac}(\omega)\}\right]$$

Care must be taken when using an FFT to solve the Hilbert transform. At $\omega = 0$ the term $1/\pi\omega$ is infinite. This means that the Fourier transform is an improper integral, so the Cauchy principal value must be considered when implementing the FFT. Since singular functions are not often considered during signaling analysis, care must be taken to ensure that the singularity at $\omega = 0$ is handled properly by the software being used. Furthermore, the analysis above assumes that the signal processing convention ($a = 0$, $b = -2\pi$) of the Fourier integral (8-1a) is used (as explained in Section 8.2.1). Since different tools may assume different conventions for the Fourier transform, consistency must be ensured.

When calculating the internal inductance of a rough conductor, it is important to test the implementation of the Hilbert transform to ensure that it is behaving correctly. This can be done by calculating the internal inductance with (E-1), assuming a smooth conductor (where R_{ac} is proportional to $\sqrt{f}$) and comparing the result to the internal inductance calculated directly from the diffusion equation in Section 5.2.3 [equation (5-30)]. Therefore, for a smooth conductor, $L_{\text{internal}} = -\hat{R}_{ac}/\omega = R_{ac}/\omega$. For a rough conductor, this relationship is not true (although it is often very close).

When using this method to calculate the frequency-dependent properties of a transmission line, the conductor thickness must be accounted for, as shown in Chapter 5. When the skin depth is larger than the conductor thickness, then the dc value of the resistance and low-frequency inductance where the skin depth is equal to the conductor thickness should be used:

$$R(f) = \begin{cases} K R_s \sqrt{f} & \text{when } \delta < t \\ R_{\text{dc}} & \text{when } \delta \geq t \end{cases} \tag{E-2}$$

$$L_H(f) = \begin{cases} L_{\text{external}} + \dfrac{\hat{R}_{\text{ac}}(f)}{2\pi f} & \text{when } \delta < t \\ L_{\text{external}} + \dfrac{\hat{R}_{\text{ac}}(f_{\delta=t})}{2\pi f_{\delta=t}} & \text{when } \delta \geq t \end{cases} \tag{E-3}$$

where K is the frequency-dependent surface roughness correction factor calculated in Sections 5.3.1, 5.3.2, or 5.3.3, t is the conductor thickness, δ is the skin depth, $f_{\delta=t}$ is the frequency where the skin depth equals the thickness of the conductor, $R_s\sqrt{f}$ is the classic skin resistance for a smooth conductor as calculated in (5-17) and (5-18), $R_{\text{ac}}(f) = K R_s \sqrt{f}$, and $\hat{R}_{\text{ac}}(f)$ is the Hilbert transform of$R_{\text{ac}}(f)$. Note that for a smooth conductor, $K = 1$.

APPENDIX F

SPICE LEVEL 3 MODEL FOR 0.25 μM MOSIS PROCESS

F-1 DEVICE MODELS

The device models used in the text were supplied by MOSIS, a supplier of prototyping and small-volume production services for VLSI circuit development. Since 1981, MOSIS has fabricated more than 50,000 circuit designs for commercial firms, government agencies, and research and educational institutions around the world. MOSIS provides access to a wide variety of semiconductor processes offered by multiple silicon foundries.

The NMOS and PMOS level 3 device models shown below were used for all transistor level simulation and analysis examples and problems in this text.

```
*
* DATE: Jun 11/01
* LOT: T14Y                  WAF: 03
* DIE: N_Area_Fring          DEV: N3740/10
* Temp= 27
.MODEL CMOSN NMOS (                               LEVEL = 3
+ TOX = 5.7E-9             NSUB = 1E17            GAMMA = 0.4317311
+ PHI = 0.7                VTO = 0.4238252        DELTA = 0
+ UO = 425.6466519         ETA = 0                THETA = 0.1754054
+ KP = 2.501048E-4         VMAX = 8.287851E4      KAPPA = 0.1686779
+ RSH = 4.062439E-3        NFS = 1E12             TPG = 1
+ XJ = 3E-7                LD = 3.162278E-11      WD = 1.232881E-
```

Advanced Signal Integrity for High-Speed Digital Design, By Stephen H. Hall and Howard L. Heck

```
8
+ CGDO = 6.2E-10           CGSO = 6.2E-10          CGBO = 1E-10
+ CJ = 1.81211E-3          PB = 0.5                MJ = 0.3282553
+ CJSW = 5.341337E-10      MJSW = 0.5              )

.MODEL CMOSP PMOS (                                LEVEL = 3
+ TOX = 5.7E-9             NSUB = 1E17             GAMMA = 0.6348369
+ PHI = 0.7                VTO = -0.5536085        DELTA = 0
+ UO = 250                 ETA = 0                 THETA = 0.1573195
+ KP = 5.194153E-5         VMAX = 2.295325E5       KAPPA = 0.7448494
+ RSH = 30.0776952         NFS = 1E12              TPG = -1
+ XJ = 2E-7                LD = 9.968346E-13       WD = 5.475113E-
9
+ CGDO = 6.66E-10          CGSO = 6.66E-10         CGBO = 1E-10
+ CJ = 1.893569E-3         PB = 0.9906013          MJ = 0.4664287
+ CJSW = 3.625544E-10      MJSW = 0.5              )
```

REFERENCE

The MOSIS Service, SPICE Level 3 Model Parameters for Classroom Instructional Purposes TSMC (0.25 micron), available at http://www.mosis.com/Technical/Testdata/t14y_tsmc_025_level3.txt.

INDEX

Advanced Signal Integrity for High-Speed Digital Design, By Stephen H. Hall and Howard L. Heck

Printed in the USA/Agawam, MA
November 25, 2025

896663.004